Advanced Microwave and Millimeter Wave Technologies: Semiconductor Devices, Circuits and Systems

Advanced Microwave and Millimeter Wave Technologies: Semiconductor Devices, Circuits and Systems

Edited by
Moumita Mukherjee

Advanced Microwave and Millimeter Wave Technologies: Semiconductor Devices, Circuits and Systems,

Edited by Moumita Mukherjee

CBS, Edition 2016

Published by InTech

Janeza Trdine 9, 51000 Rijeka, Croatia

Additional hard copies can be obtained from orders@intechopen.com

Advanced Microwave and Millimeter Wave Technologies:
Semiconductor Devices, Circuits and Systems,
Edited by Moumita Mukherjee
p. cm.
ISBN 978-953-307-031-5

Preface

Today, the development and use of microwave (3-30 GHz) and millimeter wave (30-300 GHz) band is being actively promoted. Microwave has been used extensively since the Second World War when the sources were based on vacuum devices. Microwaves are presently playing a vital role in RADAR, land and satellite based communication and also have wide civilian and defence applications. Two typical areas of application of millimeter-wave are information communication and remote sensing. This wide spectrum of application is making the microwave and millimeter wave system development one of the most advanced technologies of radio science, especially in view of the ever increasing demand of communication. Studies on Microwave and Millimeter waves go back a long way. Advanced studies on MM-wave were first conducted about 100 years ago by Acharya J. C. Bose of the Presidency College, University of Calcutta in India. He measured the refractive index of natural crystal in the 60 GHz band, developing a variety of MM-wave components in the process.

Now-a-days researchers all over the world are focusing their attention in the terahertz frequency region of the electromagnetic spectrum, which is typically defined in the frequency range 100 GHz to 10 THz, corresponding to a wavelength range of 3 mm to 30 microns. The Millimeter-Wave region overlaps a portion of the Terahertz region. Following the development of coherent sources and detectors, there has been growing interest in the role of terahertz technology for security and defence. The terahertz region offers a huge expanse of unused bandwidth, which currently presents a significant advantage for both security and defense initiatives. The ability of terahertz radiation to probe intermolecular interactions, large amplitude vibrations and rotational modes, in addition to showing polarization sensitivity makes terahertz radiation a unique and diverse region of the electromagnetic spectrum. The additional ability of both Terahertz and MM-Wave radiation to see through common materials, such as thick smoke, fog and dust, which are often considered as opaque in other regions of the electromagnetic spectrum offers further advantages over other optical techniques.

This book is planned to publish with an objective to provide a state-of-the-art reference book in the areas of advanced microwave, MM-Wave and THz devices, antennas and systemtechnologies for microwave communication engineers, Scientists and post-graduate students of electrical and electronics engineering, applied physicists. This reference book is a collection of 30 Chapters characterized in 3parts: Advanced Microwave and MM-wave devices, integrated microwave and MM-wave circuits and Antennas and advanced microwave computer techniques, focusing on simulation, theories and applications. This book provides a comprehensive overview of the components and devices used in microwave and MM-Wave circuits, including microwave transmission lines, resonators, filters, ferrite devices, solid state devices, transistor oscillators and amplifiers, directional couplers, microstripeline components, microwave detectors, mixers, converters and harmonic generators, and microwave solid-state switches, phase shifters and attenuators. Several applications area also discusses here, like consumer, industrial, biomedical, and chemical applications of microwave technology. It also covers microwave instrumentation and measurement, thermodynamics, and applications in navigation and radio communication.

Editor:
Moumita Mukherjee

Contents

The present and future trends in High Power Microwave and Millimeter Wave Technologies

S. Azam[1, 3] and Q. Wahab[1, 2, 4]
[1) Department of Physics (IFM), Linköping University, SE-581 83, Linköping, Sweden]
[2) Swedish Defense Research Agency (FOI), SE-581 11, Linköping, Sweden]
[3) Department of Electrical Engineering, Linköping University, SE-581 83, Linköping, Sweden]
[4) Department of Electronic Engineering, ECE Faculty, NED University of Engineering and Technology, 75270 Karachi, Pakistan]

1. Introduction

Microwave and millimeter wave high-power vacuum electron devices (VEDs) are essential elements in specialized military, scientific, medical and space applications. They can produce mega watts of power which would be equal to the power of thousands of solid state power devices (SSPDs). Similarly, in most of today's T/R-Modules of active phased array antennas for radars and electronic warfare applications GaAs based hybrid and MMIC amplifiers are used. The early applications of millimeter-wave MMICs were in military, space and astronomy systems. They are now also utilized for civil applications, such as communications and automotive radars. As transmission speeds in next-generation wireless communications have become faster, wireless base stations that operate in the microwave frequency range consume an ever-increasing amount of power. The mm waves (above 30 GHz) deliver high speed and good directionality and have a large amount of available bandwidth that is currently not being used. They have the potential for use in high-speed transmissions. Point-to-point wireless is a key market for growth since it can replace fiber-optic cable in areas where fiber is too difficult or costly to install. But the real high volume action at mm-wave will likely be in the MMICs for automobile radar systems devices for short-range radar (24 GHz) and long-range radar (77 GHz). Such radars will not only be used for collision avoidance and warning, but also for side- and rear-looking sensors for lane changing, backup warning and parking assistance. While only available in high-end automobiles at present, cost reductions in MMIC chip manufacturing could lead to significant deployment in all cars in the future.

SiC MESFETs and GaN HEMTs have wide bandgap features of high electric breakdown field strength, high electron saturation velocity and high operating temperature. The high power density combined with the comparably high impedance attainable by these devices also offers new possibilities for wideband power microwave systems. The SiC MESFETs has high cost and frequency limitation of X band. On the other hand the GaN transistors have the potential to disrupt at least part of the very large VEDs market and could replace at least

some microwave and millimeter wave VEDs.The hybrid and MMIC amplifiers based on AlGaN/GaN technology has demonstrated higher output power levels, broader bandwidth, increased power added efficiency and higher operating voltages compare to GaAs for performance improvement to meet future requirements. Very promising results up to 35 GHz are demonstrated by GaN HEMT technology [1]-[8]. Resulting power density is about ten times higher than that demonstrated in GaAs.
To make GaN cost competitive with other technologies, Nitronex Corp has developed GaN transistors on low-cost 100 mm silicon substrates (GaN-on-silicon growth technology). These transistors are commercially available which cover cellular phones, wireless LANs and other applications at the lower end of the microwave frequency spectrum (1-5 GHz). The devices for high frequencies and powers are in progress. This is believed to have a major impact in the future development of millimeter-wave systems. Since low-cost mass-production potential pushes forward the technology, a very high integration of circuit functions on a single chip is possible.
Si-based other solid-state transistor amplifiers are typically fabricated using a combination of silicon bipolar and laterally diffused metal oxide semiconductor (LDMOS) technologies. LDMOS technology works well in UHF and VHF frequencies up to around 3.5 GHz. Typical power levels for these devices are usually in the <200 W range; however multi-die modules can offer power levels up to 1000 W [9]. Although LDMOS transistors are also low cost but they have the power handling and frequency limitations.

2. Classification of Power Devices

RF power devices can be broadly classified into three families:

2.1 Electron beam devices (EBD)

Travelling-wave tubes (TWTs), klystrons and the inductive output tubes ("IOTs") all belong to the EBD family. They all require multiple operating voltages, one of which is a high DC voltage (tens of kV) that accelerates the electron beam.
The TWTs are presently produced for all common microwave communication and radar bands. It has been recently shown that it is feasible to build an active 2-D phased array at X-band using TWTs that fit within the array lattice, one TWT per element [10]. The DC-to-RF conversion efficiency is poor, 25-35 %, implying severely increased operating costs compared to other devices.
The Klystrons has very high output power per device. At 30 - 2000 kW per device, the output power is 30...1000 times greater than that needed to drive an individual array element, thus requiring a very complicated system of power dividers and high-power phase-shifters to distribute and control the power flow to as many as 1000 elements per klystron. In a feed system of this kind, variable power tapering is almost impossible to realize. Also a single failed device will result in a large fraction of the array losing power at once. Also, the instantaneous power bandwidth of a large klystrons is only marginally sufficient, or even insufficient, to meet the range resolution requirement.
The *IOTs* also has output power levels in the 30 - 70 kilowatt range and are subject to the same complications as the klystrons with regard to the RF power distribution / feed / beam-steering system.

2.2 Power grid tubes (PGTs)

These tubes come in many shapes and sizes. There should be no problem finding a tube in the power range of one-kilowatt. A kilowatt is in the right power range for feeding an individual phased-array element, so tubes of this class could be used as the active elements of element-level power amplifiers. However, power grid tubes need multiple operating voltages, one of which is always a medium high DC voltage (> 2 kV), thus necessitating a fairly complicated power supply system, relatively short lifetimes and the more long-lived directly heated filament cathode types instead consume substantial amounts of filament-heating power, which reduces the overall DC-to-RF conversion efficiency significantly.

2.3 Solid-state Semiconductor Power Devices (SSPDs)

The maximum output power that can be obtained from an RF power transistor is limited by the physical properties of the semiconductor material, in particular the safe junction/channel power density. Increasing the junction/channel area and reducing the device thickness in an attempt to increase power also increases the junction/gate capacitance, consequently reducing frequency and power gain. The heat resistance between the semi-conductor die and the heat sink determines how much dissipated power can be transported away from the die at the maximum allowed device temperature and is often the factor that the ultimately limits the output power. Until recently, these factors combined to limit the practical output power of CW-rated semiconductor devices to about 150 watts at all frequencies from VHF upwards. But during the last decade, demands from industry for better devices for the base stations for 3rd generation mobile telephone systems have generated much R&D to push the upper frequency power limit to 100 Watt and even higher. When operated within their ratings, RF power semiconductors show excellent lifetimes, upwards of many tens of thousands of hours, primarily limited by slow electro-migration of the metal used in contact pads and bonds. Semiconductor devices often operated by a single power supply in the 28 - 50 volt range, thus simplifying the power supply problem dramatically as compared to all electron devices. An additional advantage of FETs is that, being majority carrier devices, they do not suffer from thermal runaway effects. Biasing is also very simple, requiring only a source of adjustable positive voltage; the bias voltage can be derived from the main power supply through a voltage divider or a small regulator IC.
A comparison of these devices on the basis of device characteristics is given in Table 1.

Device Type	P_{max} (kW)	Drain Eff. %	Gain (dB)	Operating voltage (kV)	Life time (hours)
SSPDs	0.5	50-65	10-17	0.025-0.1	50×10^3
PGTs	0.5-10	50-60	10-13	0.5-10	$(3-10) \times 10^3$
EBDs	0.1-2000	25-60	20-40	25-100	$(10-20) \times 10^3$

Table 1. Comparison of power devices

3. VEDs vs SSPDs

Following are the main fundamental physical differences in SSPDs and VEDs;

1: In vacuum microwave electronic devices the electron stream moves without collision through an evacuated region between anode and cathode. As electrons pass without any collision, there is no loss in their energy (hence less efficiency loss) and thus no heat is generated during electron current flow through the device. The only heat is produced in the collector of VEDs, due to that energy of electrons which is not converted into microwaves.

In SSPDs, the electron current drifts between Emitter/Source and collector/Drain through a solid material and experience collisions. The electrons current waste some of its KE inside the device. Thus these devices have lower electron mobility compare to vacuum devices, which is an advantage for VEDs in terms of high power at high frequencies.

2: At long term high operating temperatures the performance of the device is degraded specially mobility is reduced which reduces performance at high frequencies. To keep the active region temperature of a microwave power transistor at acceptable low levels, the solid state devices need larger heat sink compare to VEDs, because the interaction region in VEDs is surrounded by Vacuum. For this purpose the base plate for solid state devices must be kept at or below 30 C, while VEDs can operate with base plate temperatures of 250 C.

3: In solid state devices the long term ionizing radiations must be avoided to prevent device degradation, while VEDs are virtually immune to ionizing radiation fluxes which make them suitable choice for the applications in space.

4: The VEDs have high electric field and power densities compare to solid state devices.

5: The SSPDs are smaller in size and low cost compare to VEDs.

6: The SSPDs are easy to fabricate compare to VEDs.

4. Why GaN transistors but not GaAs?

GaAs-based amplifiers are well-known devices currently used as pre-driver, driver, and even final-stage amplifiers for radar applications. GaN transistors and MMICs challenge GaAs technology mostly in high-bandwidth, high power applications, because, due to the smaller required device periphery for a given specified output power, good impedance matching can be achieved for GaN FETs over a broader frequency range than for GaAs pHEMTs. Also, Practical manufacturing of much higher power GaAs FETs than those currently available is facing significant technical difficulties.

The wide bandgap of GaN increases the breakdown field by five times and the power density by a factor of 10 to 20, compared with GaAs-based devices. The GaN components are therefore smaller and have a lower capacitance for the same operating power, which means that amplifiers can operate over a wider bandwidth while exhibiting good input and output matching.

GaN devices are also highly efficient because they can operate at higher voltages (24–35 V, compared with 5–8 V for GaAs-based devices at millimeter-wave frequencies), as well as having a lower on resistance. The high voltage also improves the power supply efficiency,

while the two dimensional electron gas (2DEG) produces a high electron velocity, ensuring good signal gain at K, Q and even W band frequencies.
The unique attribute of the AlGaN/GaN structure is the possibility of building high channel charge, which increases the device's current handling capability. Because GaN is a strongly polar material, the strain resulting from growing lattice-mismatched AlGaN on GaN induces a piezoelectric charge. This supplies additional electrons to the HEMT channel. This total channel charge is roughly four to five times higher than for AlGaAs/GaAs HEMTs. This piezoelectric property is a unique power-boosting bonus factor for AlGaN/GaN HEMTs.
GaN devices built on SiC substrates have a thermal conductivity 10 times higher than those fabricated using GaAs, which means that these wide bandgap devices can operate at higher power densities. GaN HEMTs can also work at higher temperatures, which reduce the need for cooling and allows for a more compact module design. The comparison in GaAs and GaN on the basis of parameters required for high power performance is summarized in Table 2.
The introduction of GaN on Silicon (most highly refined semiconductor substrates in the world are silicon wafers) is another great advantage in terms of cost. High volume production is possible because of growth on large silicon substrate. This GaN-on-silicon approach yields a low-cost, high-performance platform for high-frequency, high-power products, which is a potentially exciting combination.
The most important is the process similarities of HEMT in both technologies; hence GaN HEMT can share production process with GaAs HEMT.

Parameter	GaAs	GaN
Maximum Operating Voltage (Volts)	20	48
Maximum Current (mA)	500	~1000
Maximum Breakdown Voltage (Volts)	40	>100
Maximum Power Density (W/mm)	1.5	>8

Table 2. Comparison of GaN and GaAs

5. New Developments in GaN Technology

In only 16 years (since 1993), GaN-based transistors have evolved tremendously from a poor initial performance [11] to worldwide commercialization as power amplifiers in the *S* and *X* bands [12]. To increase their frequency of operation to millimeter and sub millimeter wave frequencies, improved growth in combination with the introduction of new device structures [13]-[15] have been reported. These new structures have allowed devices with a current gain cutoff frequency f_T in excess of 150 GHz and a maximum oscillation frequency

f_{max} of 230 GHz in AlGaN/GaN HEMTs with a gate length of 100 nm [13]. GaN MMICs up to Ka-Band have been presented [16-19], showing power densities up to 5 W/mm at 50 Ω load impedance.

AlGaN/GaN HEMTs grown on silicon (111) high-resistivity substrates with cutoff frequencies f_T = 90 GHz and f_{MAX} = 105 GHz have been demonstrated [20]. The results indicate that GaN-on-Si technology is a viable low-cost alternative to mm-wave transistors and that it suffers no significant raw speed disadvantages in terms of channel electron transport in comparison to devices fabricated on sapphire or SiC substrates. Further device scaling and improvements in epitaxial layer design are expected to lead to f_T values well in excess of 100 GHz for AlGaN/GaN on Si technology.

Fujitsu Develops World's First Gallium-Nitride HEMT able to cut power in standby mode and achieve high output of over 100 W, that features a new structure ideal for use in amplifiers for microwave and millimeter-wave transmissions, frequency ranges for which usage is expected to grow. This technological advance will contribute to higher output and lower power consumption in microwave and millimeter-wave transmission amplifiers for high-speed wireless communications [21]. A record power density of 10.5 W/mm with 34% power added efficiency (PAE) has been measured at 40 GHz in MOCVD-grown HEMTs biased at DS = 30 V [22]. A commercial company Aethercomm believes that if the trends in GaN advancement are maintained at their current rate, the predicted performance of GaN HEMTs in the year 2010 will be as depicted in ***Figure 1***. As shown, GaN will soon overtake all of its competitors in every category [23].

The low parasitic capacitance and high breakdown voltage of GaN HEMTs makes them ideal for class-E and class-F high efficiency amplifier modes. Recently, several GaN transistor vendors have implemented class-E & F amplifiers in hybrid form. Typical results are ten watts output power with efficiencies above 80 percent [24], [25].

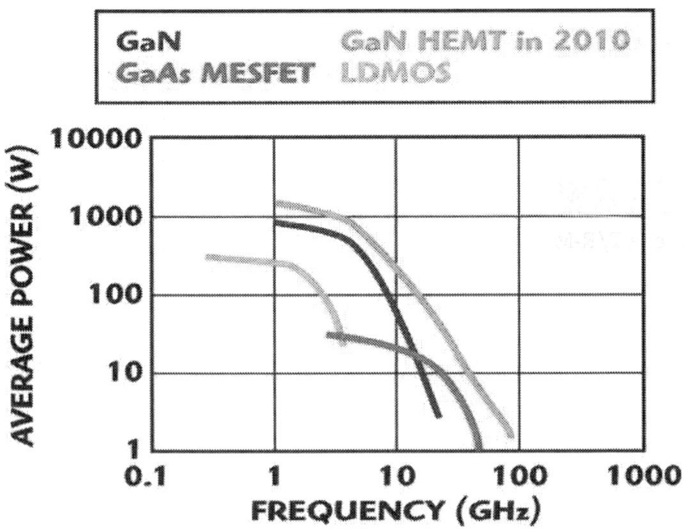

Fig. 1. Evolution of GaN FET performance [23]

A Comtech PST company has released a new high power 500 W broadband amplifier based on latest Gallium Nitride (GaN) device technology biased in class-AB mode at an input power of 0 dBm, covering the frequency range of 1-3 GHz. The amplifier offers excellent efficiency, high gain (minimum 57 dB), and linear dynamic range [26].
An S-band, 800 W GaN HEMT is released from Eudyna Device Co. Ltd. An output power of 851 W and a drain efficiency of 57.4 percent were reported at 2.9 GHz, with a 200 μs pulse width, a 10 percent duty cycle and 65 V drain-source voltage supply (Vds) [27].
GaN devices are now becoming available for pulse operated applications. A high power amplifier developed for X-band weather radar [28]. It delivers over 250 W of output power in the range of 9.1 to 9.6 GHz with at least 38 dB gains and a PAE of 21 percent. ***Figure* 2** shows a photograph of a GaN SSPA transmitter for radar that uses GaN HEMT amplifiers and a photo of the weather radar using that amplifier [29]. SSPAs, have successfully reduced the equipment size to one sixth of that of the existing equipment, using electronic tubes. It is the first practical weather radar using SSPA.
Power amplifiers for a next generation of T/R modules in future active array antennas are realized as monolithically integrated circuits on the bases of novel AlGaN/GaN HEMT structures. Both, driver and high power amplifiers were designed for X-band frequencies. Amplifier chains integrated on multi-layer LTCC substrates demonstrated an output power levels up to 30W [30]. A photo of another X-band 20 W T/R module is shown in Fig. 3 [31].

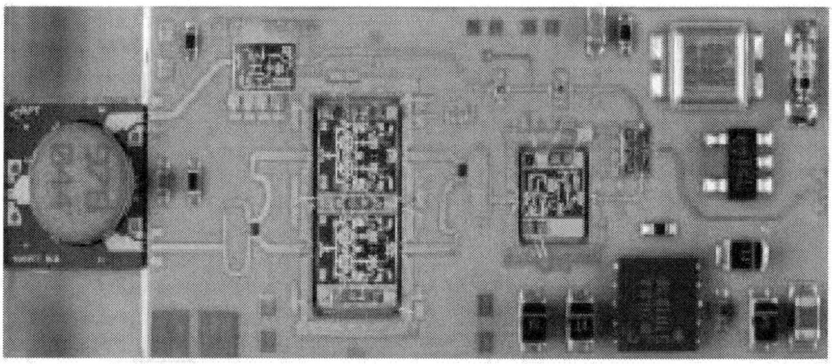

Fig. 2. Photo of a T/R-Module front-end with GaN MMIC chips [31]

SPECIFICATIONS of GaN TRANSMITTER

FREQUENCY	X-BAND (500 MHz BW)
OUTPUT POWER	≥166 W
NF	≤5 dB
SUPPLY VOLTAGE	DC 48 V
SIZE	214 × 450 × 153 mm

Fig. 3. Weather radar with GaN SSPA transmitter [29].

6. Emerging applications

Millimeter (mm) wavelengths reside at 30-300 GHz. The current and emerging applications are in the early stages of creating a demand for MMICs based on gallium arsenide (GaAs) and GaN technologies. Digital radio transceivers for cellular communications backhaul and ground terminal transceivers for very small aperture terminals (VSATs) already employ mm-wave band MMICs. Most VSATs now operate in the Ku band (12 GHz to 18 GHz) but in the future will be moving higher in frequency to Ka band (26 GHz to 40 GHz). Most of the excitement, however, for the future growth of mm-wave technology lies in E-band (60 GHz to 90 GHz).

These bands are intended to encourage a range of new products and services including point-to-point wireless local-area networks and broadband Internet access. Point-to-point wireless is a key market for growth since it can replace fiber-optic cable in areas where fiber is too difficult or costly to install. But the real high volume action at mm-wave will likely be in the automotive radar market at 77 GHz. While only available in high-end automobiles at present, cost reductions in MMIC chip manufacturing could lead to significant deployment in all cars in the future. Such radars will not only be used for collision avoidance and warning, but also for side- and rear-looking sensors for lane changing, backup warning and parking assistance.

Similarly active antenna arrays and radar transmitters operating at W-band, especially 94 GHz, offer superior performance through clouds, fog, and smoke. W band spans roughly 70 to 110 GHz and can be used for communications, radar and non-lethal weapons systems.

Novel wide bandgap RF circuit technology is sought for radar operation at W-band in brownout and degraded visibility conditions. This need has led to interest in the development of W-band high power, high efficiency amplifiers, which are currently realized almost exclusively in gallium arsenide (GaAs) and indium phosphide (InP) material systems due to their high transition frequency (Ft) performance [32], [33]. However, use of these devices has resulted in larger device peripheries for a given specified output power, more combining structures, higher combining losses, and lower power densities. These device technologies are not capable of meeting future peak power requirements. On the other hand, wide bandgap device technologies such as gallium nitride (GaN) can overcome these limitations as they can operate at higher voltages and have demonstrated power handling capabilities on the order 10 xs greater than that of GaAs or InP technologies. **A** three stage GaN MMIC power amplifiers for E-band radio applications is demonstrated that produce 500 mW of saturated output power in CW mode and have > 12 dB of associated power gain. The output power density from 300 µm output gate width GaN MMICs is seven times higher than the power density of commercially available GaAs pHEMT MMICs in this frequency range [34].

7. Millimeter band is not yet widely used. Why?

Due to faster transmission speeds in next-generation wireless communications, wireless base stations consume an ever-increasing amount of power. The millimeter wave frequency range above 30 GHz has a large amount of available bandwidth, because it delivers high speed and good directionality, its potential for use in high-speed transmissions is significant. However, due to millimeter-wave frequencies being higher than frequencies for conventional wireless transmissions, it has been difficult to develop amplifiers for practical use that are both compact and economical, and thus the millimeter band is not yet widely used.

8. CONCLUSIONS

Future Communication, EW and radar systems such as Base station, auto radars, the active phased-array radar (APAR) etc. will require increasingly smaller, more highly efficient SSPAs. In case of APAR, the desire for extremely fast scanning rates, much higher range, the ability to track and engage a tremendous number of targets, low probability of intercept and the ability to function as EW system, will require an innovative and cost-effective SSPD technology. The EBDs and PGTs are seen to be poor alternatives for the power amplifier of radars and other communication electronics in respect of power supply requirements, output power, bandwidth, fabrication and potential for graceful degradation compare to SSPDs especially PAs and MMICs based on wideband gap GaN technology transistors. Recent developments in the GaN HEMT have made it possible to realize highly efficient amplifiers at microwave frequencies. The results of GaN technology in terms of f_T, f_{max}, power density, efficiency, band width etc. both at microwave and mm waves indicate that it will be the possible first choice for applications in future microwave and mm wave technologies.

9. Acknowledgement

The authors wish to acknowledge efforts of the Government of Oman for the financial support of this work and creating and financing the Sultan Qabos IT Chair at NED University of Engineering and Technology, Karachi, Pakistan.

10. References

[1] A. M. Darwish, K. Boutros, B. Luo, B. D. Huebschman, E. Viveiros, and H. A. Hung. Algan/gan *ka*-band 5-w mmic amplifier. *IEEE Transactions on Microwave Theory and Techniques*, 54(12):4456- 4463, 2006.

[2] K.S. Boutros, W.B. Luo, Y. Ma, G. Nagy, and J. Hacker. 5 W GaN mmic for millimeter-wave applications. IEEE *Compound Semiconductor Integrated Circuit Symposium, 2006*, pages 93–95, 2006.

[3] M. van Heijningen, F.E. van Vliet, R. Quay, F. van Raay, R. Kiefer, S. Muller, D. Krausse, M. Seelmann-Eggebert, M. Mikulla, and M. Schlechtweg. Ka-band algan/gan hemt high power and driver amplifier mmics. *Gallium Arsenide and Other Semiconductor Application Symposium, 2005. EGAAS 2005. European*, pages 237–240, 2005.

[4] Y.-F.Wu, A. Saxler, M. Moore, T.Wisleder, U.K. Mishra, and P. Parikh. "Field-plated gan hemts and amplifiers." *IEEE Compound Semiconductor Integrated Circuit Symposium (IEEE Cat. No.05CH37701)*, page 4, 2005.

[5] M. Nishijima, et al.; "A k-band algan/gan hfet mmic amplifier on sapphire using novel superlattice cap layer." *Microwave Symposium Digest, 2005 IEEE MTT-S International*, 2005.

[6] M. Micovic, et al.; "Ka-band MMIC power amplifier in GaN HFET technology." *Microwave Symposium Digest, 2004 IEEE MTT-S International*, pages 3:1653–1656, 2004.

[7] Y.-F.Wu, M. Moore, A. Saxler, P. Smith, P.M. Chavarkar, and P. Parikh. 3.5-watt algan/gan hemts and amplifiers at 35 ghz. *Electron Devices Meeting, 2003. IEDM '03 Technical Digest. IEEE International*, page 23.5.1, 2003.

[8] W.L. Pribble, J.W. Palmour, S.T. Sheppard, R.P. Smith, S.T. Allen, T.J. Smith, Z. Ring, J.J. Sumakeris, A.W. Saxler, and J.W. Milligan. Applications of sic mesfets and gan hemts in power amplifier design. *Microwave Symposium Digest, 2002 IEEE MTT-S International*, 3:1819–1822 vol.3, 2002.

[9] www.freescale.com/files/rf_if/doc/data_sheet/MRF6VP11KH.pdf

[10] B. Levush and E.J. Dutkowski, "Vacuum Electronics: Status and Trends," 2007 IEEE Radar Conference, April 17–20, 2007, Boston, MA.

[11] M. A. Khan, A. Bhattarai, J. N. Kuznia, and D. T. Olson, Appl. Phys. Lett. **63**, 1214, 1993.

[12] Cree, Inc., www.cree.com.

[13] T. Palacios, A. Chakraborty, S. Heikman, S. Keller, S. P. DenBaars, and U. K. Mishra, "AlGaN/GaN high electron mobility transistors with InGaN back-barrier," *IEEE Electron Device Lett.*, vol. 27, no. 1, pp. 13–15, Jan. 2006.

[14] M. Micovic, A. Kurdoghlian, P. Hashimoto, M. Hu, M. Antcliffe, P. J. Willadsen, W. S. Wong, R. Bowen, I. Milosavljevic, A. Schmitz, M. Wetzel, and D. H. Chow, BGaN HFET for W-band power applications,[in IEEE International Electron Devices Meeting, 2006.
[15] M. Higashiwaki, T. Matsui, and T. Mimura, IEEE Electron Device Lett. **27**, 16, 2006.
[16] J.W. Palmour, J.W. Milligan, J. Henning, S.T. Allen, A. Ward, P. Parikh, R.P. Smith, A. Saxler, M. Moore and Y. Wu, "SiC and GaN Based Transistor and Circuit Advances", Proc. GAAS 2004, Amsterdam, pp. 555–558.
[17] T. Inoue, Y. Ando, H. Miyamoto, Ta Nakayama, Y. Okamoto, K. Hataya and M. Kuzuhara, "30GHz-band 5.8 W High-Power AlGaN/GaN Heterojunction-FET", MTTS 2004, Fort Worth, pp. 1649-1651.
[18] M. Micovic, Ara Kurdoghlian, H.P. Moyer, P. Hashimoto, A. Schmitz, I. Milosavljevic, P. J. Willadsen, W.-S. Wong, J. Duvall, M. Hu, M. J. Delaney, D. H. Chow, "Ka-band MMIC Power Amplifier in GaN HFET Technology", MTT-S 2004, Fort Worth, pp. 1653-1656.
[19] Y.-F. Wu, M. Moore, A. Saxler, P. Smith, P.M. Chavarkar, P. Parikh, "3.5-Watt AlGaN/GaN HEMTs and Amplifiers at 35 GHz", 2003 IEEE Int. Electron Device Meeting.Dig., pp. 579-581, December 2003.
[20] H.F. Sun, A.R. Alt, H. Benedickter and C.R. Bolognesi, "100 nm gate AlGaN/GaN HEMTs on Silicon with f_T = 90 GHz", ELECTRONICS LETTERS 26th March 2009 Vol. 45 No. 7
[21] Fujitsu Limited and Fujitsu Laboratories Ltd. International Symposium on Compound Semiconductors (ISCS), held in Rust, Germany from September 21 - 24, 2008.
[22] T. Palacios, A. Chakraborty, S. Rajan, C. Poblenz, S. Keller, S. P. DenBaars, J. S. Speck, and U. K. Mishra, "High-Power AlGaN/GaN HEMTs for Ka-Band Applications", IEEE ELECTRON DEVICE LETTERS, VOL. 26, NO. 11, pp. 781-783, NOVEMBER 2005
[23] http://www.mwjournal.com/search/article.asp?HH_ID=AR_5370,"Gallium Nitride Microwave Transistor Technology for Radar Applications", Technical feature, Microwave Journal, Vol. 51 | No. 1 | January 2008 | Page 106
[24] Yong-Sub Lee *, Mun-Woo Lee, Yoon-Ha Jeong, "A 1-GHz GaN HEMT based class-E power amplifier with 80% efficiency" DOI 10.1002/mop.23803, 2008.
[25] David Schmelzer and Stephen I. Long, "A GaN HEMT Class F Amplifier at 2 GHz with > 80 % PAE" Compound Semiconductor Integrated Circuit Symposium, CSIC 2006. IEEE, pages: 96-99, 2006.
[26] http://www.comtechpst.com/products/specs/114.pdf
[27] E. Mitani, et al, "An 800 W AlGaN/GaN HEMT for S-band High-power Application," *2009 CS Mantech Conference Digest*, p. 213.
[28] K. Kanto, et al, "An X-band 250 W Solid-state Power Amplifier Using GaN Power HEMTs," *2008 IEEE RWS Conference Digest*, p. 77.
[29] Toshiba Press Release, http://www.toshiba.co.jp/about/press/2009_11/pr_j2801.htm.
[30] Schuh, P. et al "Advanced High Power Amplifier Chain for X-Band T/R-Modules based on GaN MMICs," The 1st European Microwave Integrated Circuits Conference, 2006. Page(s):241 - 244.

[31] Schuh, P. et al "GaN MMIC based T/R-Module Front-End for X-Band Applications," The 3rd European Microwave Integrated Circuits Conference, 2008. Page(s):274 - 277.

[32] L. Marosi, M. Sholley, et al "94 GHz Power Amplifier using PHEMT Technology," Microwave Symposium Digest, 1995, IEEE MTT-S International, 16-20 May 1995 Page(s):1597 - 1600 vol.3.

[33] Pin-Pin Huang; Tian-Wei Huang; et al.; Elliott, J.H, "A 94-GHz 0.35-W power amplifier module", Microwave Theory and Techniques, IEEE Transactions on Volume 45, Issue 12, Part 2, Dec. 1997 Page(s):2418 – 2423.

[34] M. Micovic, et al.; "GaN MMIC PAs for E-Band (71 GHz - 95 GHz) Radio", Compound Semiconductor Integrated Circuits Symposium, 2008. CSICS '08. IEEE, pp. 1-4.

2

Explosive pulsed plasma antennas for information protection

Igor V. Minin and Oleg V. Minin
Novosibirsk State Technical University
Russia

1. Introduction

Since the discovery of radio frequency ("RF") transmission, antenna design has been an integral part of virtually every communication and radar application. In its most common form, an antenna represents a conducting metal surface that is sized to emit radiation at one or more selected frequencies. Antennas must be efficient so the maximum amount of signal strength is expended in the propogated wave and not wasted in antenna reflection. The modern requirements to antenna include compactness and conformality, rapid reconfigurability for directionality and frequency agility and should also allow low absolute or out-of-band radar cross-section and facilitate low probability of intercept communications. The need for an antenna that is "invisible" (thus not detectable while not in operation) has, already in the 1980's, sparked work on the feasibility of using an atmospheric discharge plasma[1] as an RF antenna. Moreover, data communications can be made more secure if the antenna only "exists" during the transmission of each data packet. Such antennas use plasma formations as the receiver or transmitter elements. The characteristics of the plasma formations are determined by purpose of the specific antenna. Plasmas have two important properties that are relevant for interaction with electromagnetic waves:

- For frequencies above the plasma frequency, a semi-infinite plasma transmits EM waves with a wavelength, l / e_r where l is the free space wavelength. Thus plasmas can in principle be used for electronic tuning or control of a radiation pattern by varying the plasma density. For the densities typical of discharge tubes, this phenomenon appears especially useful at microwave frequencies.
- For frequencies below the plasma frequency, however, the dielectric constant is $e_r < 0$, the plasma behaves as a metal, free space EM waves cannot penetrate, and are reflected. Radio communications via the ionosphere rely on this effect.

Plasma technology can be utilized to create secure WiFi data transmission capability for use in different applications up to 100 GHz [33]. WiFi has enabled a wide array of inexpensive communication devices that are utilized in desk-top computing, networking, PDA's etc. Its

---

[1] Sir William Crookes, an English physicist identified a fourth state of matter, now called plasma, in 1879.

biggest drawback is data transmission security. When plasma is not energized, it is difficult to detect by radar. Even when it is energized, it is transparent to the transmissions above the plasma frequency, which falls in the microwave region. This is a fundamental change from traditional antenna design that generally employs solid metal as the conducting element. Additionally, a transient antenna does not interfere with any other antenna based communication system. The term transient antenna, used here, refers to an antenna that changes radiation characteristics over time. This may be accomplished by varying the dimensions, impedance, or conductivity of the antenna or by changing its position with respect to other radiating elements. Transient antenna technology is relatively new. Such a thin plasma channel can be produced using high explosives.

It has long been known that plasma, ionized gas, can be used as an antenna, however, further investigations have not come until recently, because it was believed to be impossible to produce long plasma objects with high electron densities ($>10^{20}$ cm^{-3}) [1]. Several experiments and investigations have shown that high electron density levels are achievable. Numerous investigations around the world have been conducted in order to characterize the operation of this class of antennas.

Plasma antennas offer several advantages for different applications, having:

1. compactness and conformability;
2. rapid configurability for directionality and frequency agility;
3. low allowance of absolute or out-of-band radar cross section;
4. low probability of communications interceptions;
5. higher digital performance in commercial applications;
6. very large frequency capabilities, ranging from a fraction of a Hertz to several Giga Hertz or more;
7. practically zero mass;
8. capacity for instant creation and rapid disappearance [1].

Also a problem with metal antennas is their tendency to "ring". That is once you turn off the drive frequency, they continue to radiate as the oscillations die down. This can pose a serious problem for the short range ground penetrating radars used in petrochemical and mineral exploration. However because of their rapid switchability, plasma antennas don't ring. So a fundamental distinguishing feature is that after sending a pulse the plasma antenna can be deionized (or time of life of plasma antenna is about a pulse), eliminating the ringing associated with traditional metal elements. Ringing and the associated noise of a metal antenna can severely limit capabilities in high frequency short pulse transmissions. In these applications, metal antennas are often accompanied by sophisticated computer signal processing. By reducing ringing and noise, we believe plasma antenna provides increased accuracy and reduces computer signal processing requirements. These advantages are important in cutting edge applications for impulse radar and high-speed digital communications.

The design allows for extremely short pulses, important to many forms of digital communication and radars. The design further provides the opportunity to construct an antenna that can be compact and dynamically reconfigured for frequency, direction, bandwidth, gain and beamwidth. Plasma antenna technology will enable antennas to be designed that are efficient, low in weight and smaller in size than traditional solid wire antennas.

Today there are well known the following main types of plasma antennas:

- Laser Induced antenna. As it was known by the authors the laser induced antenna was offered by Askar'yan G. A. [2]. Possibilities of the plasma application for antenna parameters control have been proposed in the sixties of 20 century. The transmission was realized along a plasma channel that was created by the atmosphere breakdown. The atmosphere breakdown was created by the focused laser emission. Later, for example, Dwyer et al. [3] discussed the use of a laser to assist in the ionization of paths up to several meters. The laser is used to designate the path of the antenna while an electrical discharge is employed to create and sustain the plasma. As a rule plasma antenna produced by the discharge of a Marx generator.
- Plasma Antennas Using Tube Structures. At Australian National University, a sealed-glass tube design, fed by a capacitive coupler was employed [4]. When the plasma creating voltage is turned off, the antenna effectively disappears. An efficiency of 50% was observed, the radiation patterns were predictable, and low-base-band noise for HF and VHF transmissions was recorded. The glass tube design can be very effective in providing the desires previously mentioned. When the glass tube is not energized, no plasma exists; therefore the antenna is non-conducting and incapable of coupling an EMP. If the antenna is energized with a low plasma frequency, an EMP will simply pass through the plasma without coupling into the device.
- Explosively Formed Plasma Dielectric Antennas. Another approach to creating plasma is with the use of explosives. Altgilbers et al. [5] discussed the possibilities of using an explosive driven plasma jet as an antenna. A simple explosive charge design, called a plasma cartridge (Figure 1), can be used to generate a column of ionized gas. In this design, 1-3 grams of seeded explosive charge, which contained Fe, Pb, C, N, K, CI, and O was used to create plasma. The jet obtained a distance of 4 m in 1 mks. Due to the high temperatures generated by the explosive material, the surrounding gases became ionized, forming a plasma column. Altgilbers et al. [5] stated that the plasma generated by the plasma cartridge had a temperature of 3650 K with an estimated plasma density of 5×10^{19} cm^{-3}. The temperatures required to produce these plasmas are a direct result of the specific explosive constituents employed. Most likely, though it is not stated, this plasma cartridge design used potassium perchlorate or some other high temperature producing oxidizer. For the fuel in the explosive to bum, oxygen is required and the amount provided in the atmosphere is minimal compared with oxidizing agents. Typical high explosives contain an oxidizer in order to provide an ample amount of oxygen for the fuel, which ensures that all of the fuel contributes to the explosive process [6]. The maximum attainable temperature that can be achieved is dependent upon the available oxygen for fuel recombination. It has been proven that a plasma jet antenna is feasible, but the details of such a design are not yet fully understood [5].

It is shown [5] that the pulsed current in the solenoid couples its energy through excite transiently varied electric and magnetic fields in the plasma jets, which are produced by explosives. The electric and magnetic fields accelerate/decelerate electrons and cause the plasma jets radiate the energy out. The possible radiation mechanisms, which have been discussed above include the line emission, the continuum radiation, and the plasma radiation. After studying their radiation powers and frequencies, it is seen that the line emission and continuum radiation including Bremsstrahlung and recombination are not influenced by the pulse current in the solenoid. the frequencies of these types of radiation are in a wide range up to 105 GHz. The energy of the plasma radiation such as electron cyclotron radiation, pulse oscillation radiation, sequence dipole oscillation radiation, and surface wave radiation are coupled from the pulse current in the solenoid. The frequencies of these radiations are in a rather narrow range with an order about GHz. The electron cyclotron radiation and pulse oscillation radiation are shown to be much weaker than the dipole oscillation radiation and surface wave radiation. The same investigations of a plasma dielectric antenna with other design was also made in [7]. It were shown that the main properties of such an antenna are: 1) plasma antenna action has many general with the dielectric antenna action; 2) plasma antenna is designed for pulse operation.

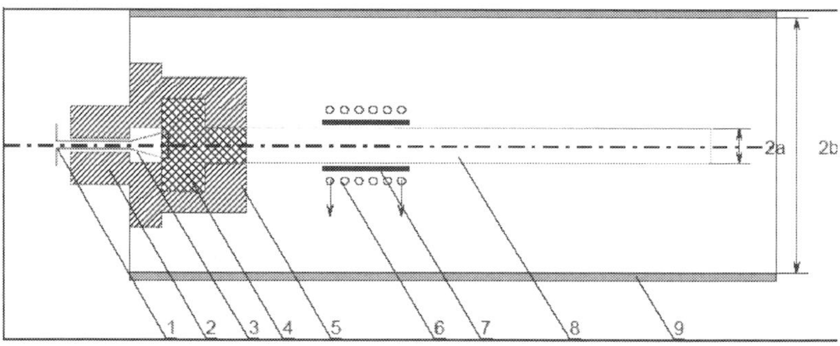

Fig. 1. Diagram of a plasma cartridge [5]. I- input wires from power source, 2- connecters, 3- bridgewire, 4- explosives, 5-cartridge housing, 6- lew frequency solenoid, 7- high frequency solenoid, 8- ferrite inserts, 9- plasma jet, and 10- cyfindrical metal casing.

Experimental investigations of multi-jet plasma dielectric antennas are described at [8]. A plasma antenna with four jets has been designed, built, and tested. The design of multi-jet plasma antenna with the H-waveguide excitation system is shown at the Fig. 2.

Comparison of the simulation results for the antenna, operating at the frequency of 700 MHz, with four and six HE-shaped jets, shown that the basic characteristics of both antennas are virtually same. Analyzing the results of computer simulation, shown at the Fig. 3, it is possible to see that at the frequencies of 700 MHz the influence of plasma jets is substantial enough for their length of 200 mm.

It was demonstrated that this antenna can achieve more than 6-7 dB of gain at a frequency of 700 MHz.

So according to [5] today there are well known three main physical types of plasma antennas: 1) plasma dielectric antennas, 2) plasma horn antennas and 3) plasma mirror antennas. Additional a plasma jet produced by an explosive cartridge, if properly designed, is the one of the most suitable form of this antenna. The disadvantage is the chemical composition of the explosive must be carefully chosen to ensure that the antenna operates in a nonlinear mode so as to enrich its high-frequency spectral output. But the use of explosives in order to generate transient antennas is net limited to plasma designs alone. It has been shown that the use of a high velocity metal jet resulting from a shaped charge explosive design is capable of generating such an antenna [9-11].

Fig. 2. Plasma dielectric antenna with four jet [8].

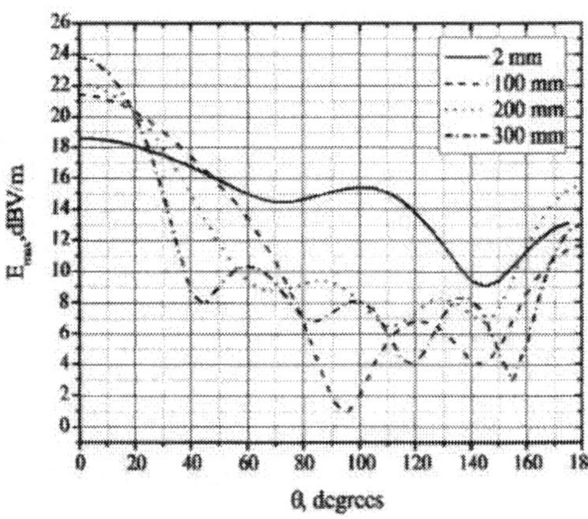

Fig. 3. E-field strength of a four-jet plasma antenna in E-plane for different length of a conductive plasma jet. Frequency is 700 MHz [8].

2. Shaped Charges as Transient Antennas

The shaped charge antenna design, frozen in time, resembles a monopole antenna. Even though the jet produced is not a perfect cylinder, as discussed later, the model for an ideal monopole can be used to predict the results. In the same way, two shaped charges placed back to back resemble a dipole antenna. In fact, the metal jets can be arranged in several different configurations in order to produce several different antenna designs. One example of shaped charge 9 jet configurations is shown in the figure 4 [12].

Fig. 4. The shaped charge 9 jet experimental configuration developed by D.R. Kennedy [12].

2.1. BRIEF SHAPED CHARGE HISTORY

A cylinder of explosive with a hollow cavity in one end and a detonator at the opposite end is known as a hollow charge. The hollow cavity, which may assume almost any axis-symmetric geometric shape such as a hemisphere, cone, ellipse, tulip, trumpet, dual angle cone, pyramid, or the like, causes the gaseous products formed from the initiation of the explosive at the end of the cylinder opposite the hollow cavity to focus the energy of the detonation products. The focusing of the detonation products creates an intense localized force. This concentrated force, when directed against a metal plate, is capable of creating a deeper cavity than a cylinder of explosive without a hollow cavity, even though more explosive is available in the latter case.

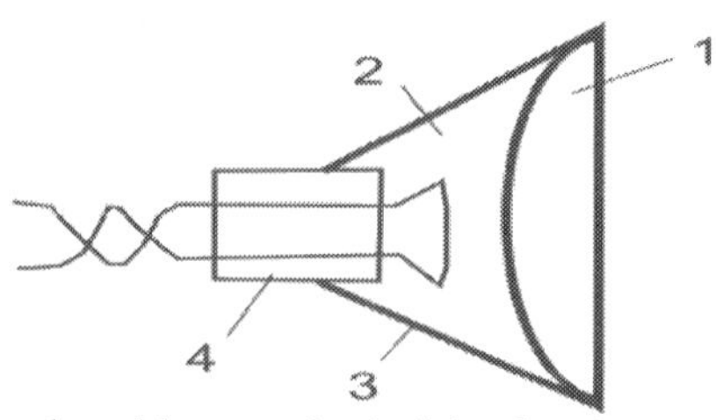

Fig. 5. Shaped charge developed by captain Andrievsky: 1– sawdust, 2 - gunpowder, 3 - body, 4 - electroigniter.

This phenomenon is known in the U. S. and Britain as the Munroe effect (1885) and in Europe as the von Foerster (1883) or Neumann (1910, but no practical applications were developed) effect [12]. This phenomenon also was discovered by Russian general Boreskov in 1864 [13]. Later Russian captain Andrievsky in 1865 developed a special charge for detonation of explosive (see figure 5).

If the hollow cavity is lined with a thin layer of metal, plastic, ceramic, or similar materials, the liner forms a jet when the explosive charge is detonated. Upon initiation, a spherical wave propagates outward from the point of initiation for the basic case of a single point initiated charge, initiated along the axis of symmetry. This high pressure wave moves at a very high velocity, typically around 8 km/s. As the detonation wave engulfs the lined cavity, the liner material is accelerated under the high detonation pressure, collapsing the liner. During this process, for a typical conical liner, the liner material is driven to very violent distortions over very short time intervals (microseconds) at strain rates of 10^4 to 10^7 /s. Maximum strains greater than 10 can be readily achieved since superimposed on the deformation are very large hydrostatic pressures (peak pressures of approximately 200 GPa, decaying to an average of approximately 20 GPa). The collapse of the liner material on the centerline forces a portion of the liner to flow in the form of a jet where the jet tip velocity can travel in excess of 10 km/s. The conical liner collapses progressively from apex to base under point initiation of the high explosive. A portion of the liner flows into a compact slug (sometimes called a carrot), which is the large massive portion at the rear of the jet. Slugs constitute 80 to 85% of the jet mass and typically travel at about 1 km/s. The pressures generated during the liner collapse far exceed the yield strength of the liner material and thus the liner behaves approximately as an in viscid, incompressible fluid [14].

Many efforts have been made over the years to understand the process of jet formation regarding the various modes of jet formation, i.e., formation of jets from conical, hemispherical, and etc. liners. Early work on shaped charges showed that a range of alternative constructions, including modifying the angle of the liner or varying its thickness, would result in a faster and longer metal jet. These research and development efforts to maximize penetration capabilities were based largely on trial and error. While the concept of a metal surface being squeezed forward may seem relatively straightforward, the physics of shaped charges is very complex and even today is not completely understood.

Now in the theory of functioning of cumulative charges there was a quite certain sight on a nature of a without the jet collapse of facing of a cumulative liner and formation of a dispersion of cumulative jets. In formation of this sight the basic role the works [15, 16, 17] have played.

The hydrodynamical theory of a cumulative jet offered by G.Taylor and G.Birkhoff [18] and M.Lavrent'ev [19] , allows to receive the analytical decision of a task as a first approximation at the assumption, that the process of formation of a jet follows the laws of the stationary expiration.

In [17] on the basis of the analysis of various experimental data and results of numerical accounts the criterion of formation of compact cumulative jets is offered which says: the speeds impacting of jets in system of coordinates moving with phase speed, should be subsonic.

Problems concerning the stationary impacts of flat contracted jets are also resolved by employing a model that is based on the behaviors of ideal fluids [15]. The geometrical and kinematic parameters of jets can be defined when there no derivated inverse (cumulative) jets at their impact point. In a rotationally symmetric case the criterion of Wolsh is inapplicable, as is indicated by the possible lack of inverse (back) jet formation due to the development of dissipative processes. The embodiment of such a condition in practice is possible at small jet impact velocities. In a rotationally symmetric case the criterion for formation of compact cumulative jets is offered on the basis of the analysis of different data resources (both experimental and calculation) [17]. The fundamental criterion states: the velocities of impacting jets in a frame moving with a phase velocity should be subsonic.

In mid-50 years of the 20th century Wolsh and his co-authors published papers dedicated to the study of steady-state symmetrical impacts of flat supersonic jets. This work was an essential addition to the hydrodynamic theory of a cumulative jet formation. It established criteria regarding the formation of a cumulative jet, with these being related to observations of the behavior a flat shaped-charge impact. Subsequently, a ranking of criteria were formulated for jet formation that incorporate all the current nomenclature related to shaped charges, including both non-stationary and rotationally symmetric cases. The expressed criticism of this approach's practical applicability in a rotationally symmetric case has not had any effect on its general "authority". Indeed, Wolsh et al.'s criteria are justified by practical examinations of the conditions needed for the formation and maintenance of compact cumulative jets, and are, as a rule, of great interest. Examining these conditions demands the observance of rigorous requirements, and even more so than was the case in Wolsh et al's pioneering work. However, the essential observable and replicable condition is that the formation of a cumulative jet requires that the impact of jets happen at subsonic speeds

The application of Wolsh's jet formation criterion [15] to a typical non-stationary case and to cases wherein currents exist with rotational symmetry is not substantiated by the relevant and precise theoretical results.

In the works [20-21] the criterion of Wolsh is distributed on a non-stationary case. It was shown, that in a rotationally symmetric case the criterion of Wolsh is inapplicable. Also it was described, that in a rotationally symmetric case that is produced by a current associated with a symmetrical axis produced by a shock wave can never be realized. That is, that cumulative jets will always be formed. This depends on the velocity function and angle of impact, wherein the formation of continuous, partially dispergated and completely dispergated jets (stream of macroscopic particles) is possible. So the quality of a jet will depend on distance of a shock wave up to an axis of symmetry. At rather small distances such jet will be intensively dispersive and to represent a flow of high-speed particles.

2.2. Shaped charge for plasma antenna design consideration

The goal of shaped charge design [22] was to produce a metal jet capable of radiating electromagnetic energy effectively into free space. It was determined that the frequency range for operation would be around 1 GHz, thus a jet length of 75 mm (quarter wavelength) was required. The jet had to be capable of carrying current the entire distance required. Therefore, the shaped charge had to be designed so that the jet did net particulate or become non-coherent before it reached 75 mm in length. Composition-C4 was the explosive chosen for use at shaped charge. The decision for the liner material was copper; the casing was made of aluminum. Diameter of shaped charge liner was 29.8mm, the mass of the jet was 3.84 g, the mass of the slug -18.64 g, total mass of the liner was 22.48 g, and explosive mass was 40.37g. It was determined that the velocity of the jet tip was approximately 3.9 km/s. The length of the jet was approximately 103 mm (figure 6).

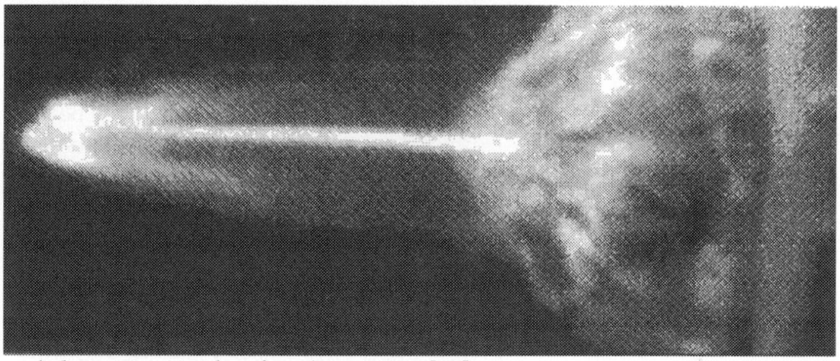

Fig. 6. Shaped charge jet 40 mks after detonation [22].

It was concluded that the metal jet created by this shaped charge design was capable of achieving a conductive length of at least the required 1 GHz quarter wavelength. Calculations shown that in the ideal case, the receiving antenna should collect 33 mW of power if 1 W of power is successfully radiated into free space by the shaped charge antenna. Previously, it was assumed that the resonant length of the shaped charge would be 75 mm based on known antenna characteristics. Given that the shaped charge jet did not represent an ideal monopole exactly, it was expected that the actual resonant length would be slightly different.

Due to the corresponding changes in reflected and received power with relation to time, it was concluded that transmission of energy from the transient antenna to the dipole was successful [22] - see figure 7. The maximum value received by the dipole in each test (approximately 4 mW) was much less than the measured value of 14 mW for the mockup. A couple of possibilities exist that may explain these apparent losses.

The tests revealed an apparent power dissipation of up to 90% over duration of 5 mks is for the 1 GHz frequency regime. It is not known exactly how much of that dissipated power is divided between heat and radiation; however preliminary calculations suggest that ohmic losses are minimal and radiation efficiency is close to 100%. Similar results were discovered in the 915 MHz thin wall and thick wall tests. Each of the tests revealed comparable values for the maximum power received by the dipole antennas.

The effective use of a shaped charge as a transient antenna greatly depends en the ability of the design to limit particulates and off-axis breakup. These two factors dictate the maximum length that can be achieved by the design and the path taken by the current. The coherency of the jet determines the current path, which is a major contributing factor to the resulting polarization and the subsequent field pattern of the antenna. Given that the apex angle value affects all of the critical velocities, the particulates event, and coherency status of the jet, future research should focus primarily en the dependence of success on this parameter [22].

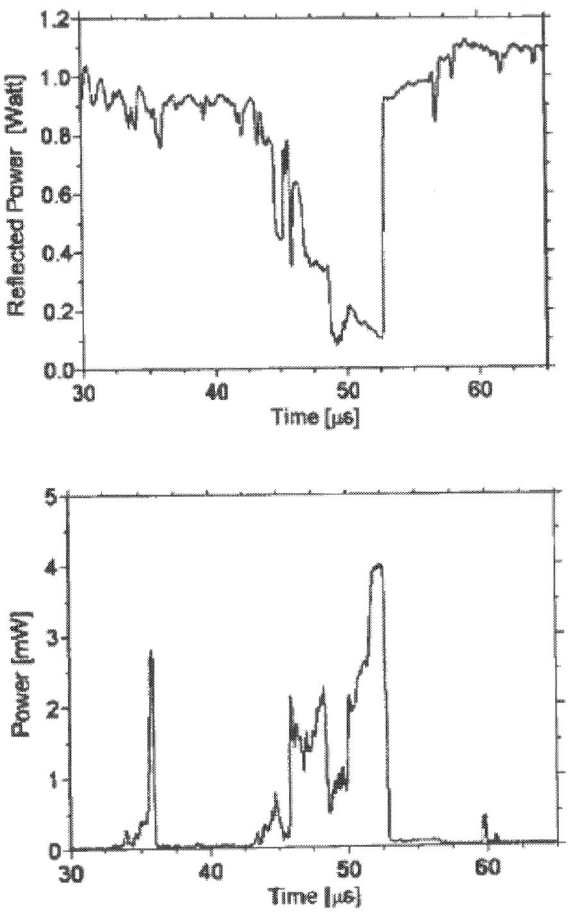

Fig. 7. Reflected (top) and received (bottom) power for 1 GHz, 1.95 mm liner thickness test [22].

2.3. The main disadvantages

Over time, the jet begins to follow a phenomenon referred to as particulation. As the jet propagates, the liner material particulates into several smaller pieces along the axis of the jet. It is assumed that particulation first occurs between the neck and the head of the jet. At the interface between these two parts, a considerable velocity gradient (several km/s differences) exists and eventually, the head pulls away from the neck. After this initial particulation, the rest of the jet begins to breakup and if uninterrupted, will particulate completely. Characterization of the time before particulation, or breakup time, is of great importance to shaped charge designers.

The breakup time of the jet can be increased by:

1. increasing the jet radius,
2. decreasing the jet stretching rate,
3. decreasing the jet strength,
4. increasing the jet density,
5. increasing the dynamic ductility [14].

Another concern that can have a dramatic effect on the length of a shaped charge is jet coherency. This refers to the coherency of the jet in the radial direction, which occurs symmetrically about the jet axis. A so-called coherent jet is one that has very little deviation away from that axis. Similar to particulation, a noncoherent jet is one that results in breakup into smaller particles. The particles are off-axis, therefore, spreading the force of the jet to a larger area.

3. Pulsed Explosive Dielectric Plasma Antenna

During the flaying in air the metal jet of shaped charge produced plasma. The flight of hypersonic bodies in the air is accompanied by intense processes of gas and body surface heating, its erosion and ablation. At the same time, temperature near the body surface and the intensity of these processes are higher, for greater density and pressure of the environmental gas [10].

The detailed results of experimental investigations of properties of low-temperature air plasma that is initiated by hypersonic bodies moving at velocities close to the first cosmic velocity (M~17) was described at [10]. It was discovered [23] a new physical regularity in that the plasma trace parameters could be characterized by the grain size of the surface microstructure of the hypersonic body (at a fixed velocity, shape and surface material of the hypersonic body). The experiments show the following properties of non-ideal air plasma (non-ideal parameter γ =0.15) with a condensed dispersed phase. To intensify the processes and for studying the character features of the flight in extreme conditions, the model experiments were conducted in air in the normal initial state. The technique of X-ray and mm-wave diagnostics of plasma properties and microstructural analysis of hypersonic body material were utilized in the experiments. The investigations were made in the air with normal initial conditions. Through the experiment, the body, formed in the process of compression by products of detonation of cumulative shell-hole (CSH), moved between central transmitting and receiving high-frequency QDE's antennas [11, 24] (offset zone plates Reley-Wood with the relative hole ≈1, diameter 40λ, were λ=0.004 m of wavelength).

A photograph of the equipment is shown in Fig. 8. The radar chamber consists of a 3.5 meter-long tank, 1.5 meter in diameter, located at the middle of the ballistic range. The uprange end of the tank is lined with microwave absorbing material to minimize the effect of range wall reflections on the radar. The radar reflector, which is made of thin sheet aluminum, is suspended across the flight line at 45° to the flight axis. The supporting framework is designed to enable the sheet to be mounted easily and accurately in position. The sheet is replaced after each firing. The time of body flight to the high-frequency section was several times longer than the time of relaxation of the local temperature of the body, taken from the equation of heat conduction:

$$t \approx \rho c_p L^2 \chi^{-1},$$

where ρ- density, c_{p}- specific heat, χ- the coefficient of specific heat of body material, L- the scale of irregular warning up of the material in the given condition, probably having the size similar to the size of the grain on the surface of the flying body.

For the full-scale experimental investigations the explosive chambers developed under the leadership of professor V.F.Minin as described at [25] were used (see figure 2).

We have used orientation imaging microscopy (OIM) to study the evolution of the grain boundary character distribution in copper as a function of annealing time. The copper shaped-charge liner material was mechanically processed to obtain the usual conical shape. A metallographic sample was cut from the wall of the cone and polished on a through-thickness plane parallel to the axis of the cone. The grain boundary character distribution was characterized using OIM. The average size of the grain of microstructure of CSH was observed also by it`s length in three cross-sections on the microedges. The size of the grain was regulated by finding regimes of warmer CSH material. A metallographic sample was polished on a through-thickness plane parallel to the axis of symmetry. The grain boundary character distribution was characterized using orientation imaging microscopy. The observed trend was that the factor of "random" grain boundaries generally decreased with annealing while the grain size increased by less then a factor of 2.5.

At the first stage exploratory laboratory experiments were carried out using own plasma jets expelled from 17-gram conical shaped charges (without the metal liner). Measured conductivities were in the range of 4.5 kS/m for unseeded Octol charges and 23 kS/m for seeded Octol charges containing 2.2% potassium carbonate by mass.

Fig. 8. The experimental plant.

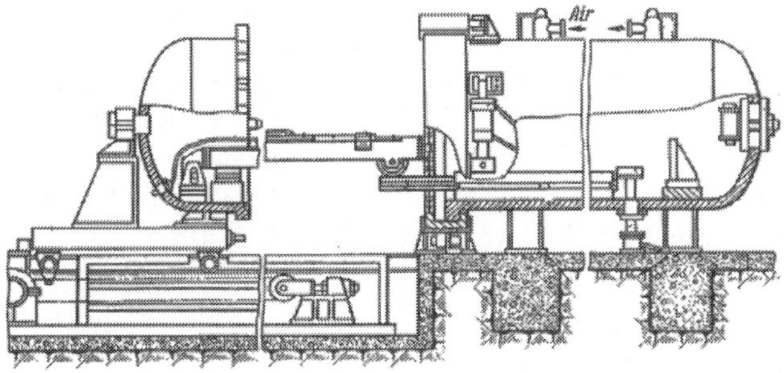

Fig. 9. The explosive chamber [25].

In Fig. 10, typical signals from the Cu and Al body are shown. Fig. 10 shows the dependence of the time of existence of the plasma material with CDP on the size of microstructure grain of CSH material, which was obtained through processing of experimental data for the Cu body. While making the given dependence the data of three series of experiments were used, in which there were 10-15 experiments. The sizes of the CSH material were δ=7-10, 15-25, 80-100 micron. Taking account that the size of the grain on the inner surface of CSH is equal to the size of the surface on the body, fig. 11. From this data, we can see that the size of grain structure on the surface of a high-velocity body greatly influences the electrophysical properties of plasma with CDP. Introducing the particles of metal into the CDP in the plasma track influence it`s electromagnetic properties through thermoelectronic emission from their surface. Thus the increasing of the grain size of metal microstructure on the body surface probably leads to the increasing of the charged particles in plasma. And the surface of a shaped charge jet is in a solid state. It could be mentioned that in a classic experiment von Holle and Trimble, see [14], measured the surface temperature of a copper liner near the tip and concluded the jet surface temperature was around 500 C. They developed their temperature measurement technique for shocked materials based on two-color infrared radiometry. However, the use of only two colors (wavelengths) indicates that the measured results are not as accurate as they would be if more colors were obtained.

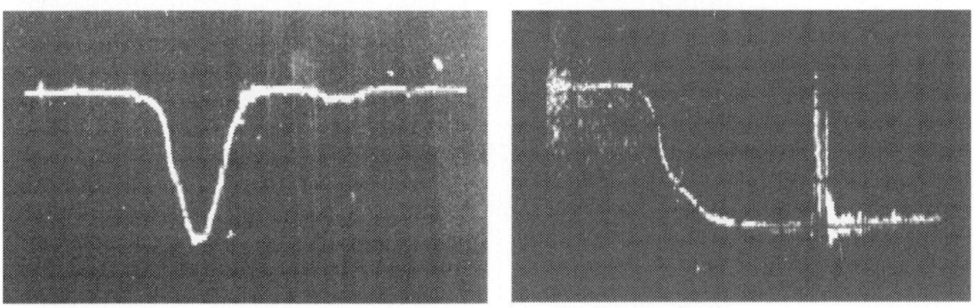

Fig. 10. The typical signals from Cu-body (left) and Al-body (right).

Thus, the above mentioned results of the experiments show that while obtaining of plasma with CDP, appearing in the air though movement of high-velocity bodies and studying their electrophysical properties, it is necessary to take into account the influence of the microstructure of the surface of the body. This phenomenon may be used, for example, to create remote quasistationary, impulse or moving reflection of microwaves. This impulse antenna is shown in Fig. 12-14. The length of a plasma body is about several tens meters.

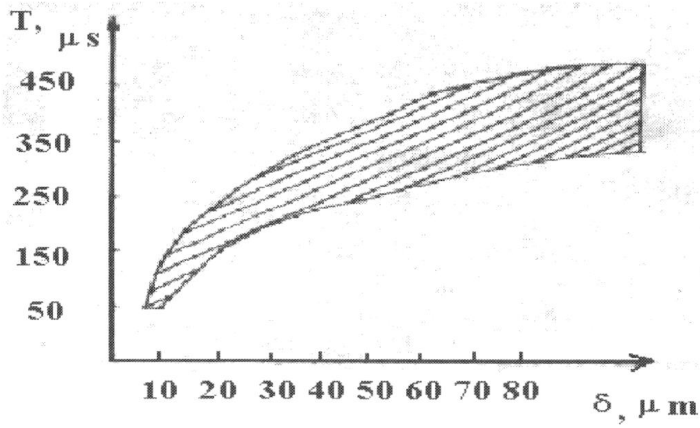

Fig. 11. The dependence of the time of existance of plasma with the CDP on the size of microstructure grain of CSH material for Cu body.

Fig. 12. The plasma impulse antennas created in the air by a hypervelocity Al body.

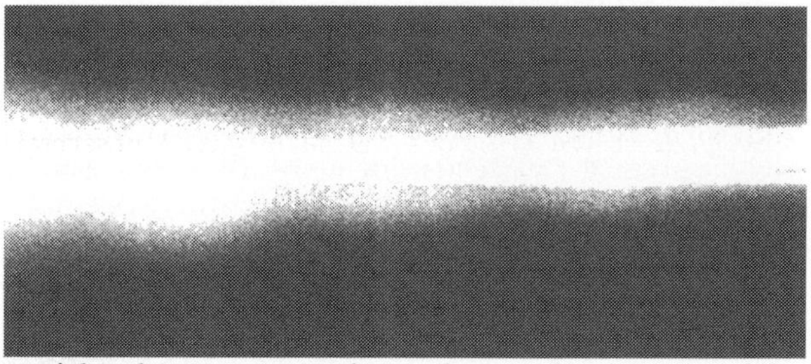

Fig. 13. Fragment of the plasma antenna shown the impulse character of the surface particle erosion.

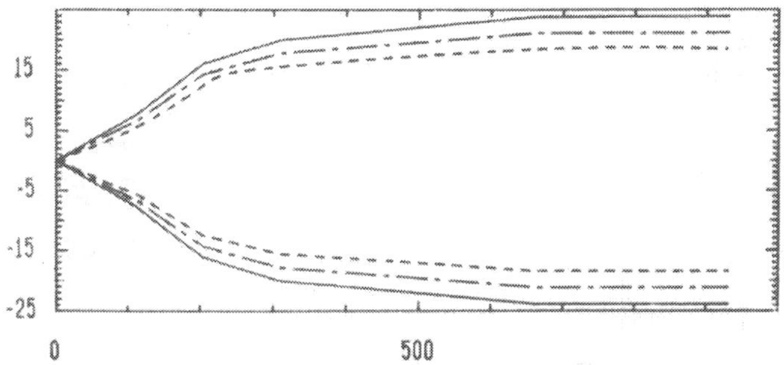

Fig. 14. The typical example of the plasma body renovated from the data of the transmitted coefficient of the mm-waves. The initial stage is shown.

For the Al body with speed about 6 km/sec it was shown [10] that the length of a plasma antenna can reach several tens meters. The principal scheme of such type of antenna is shown in the figure 15.

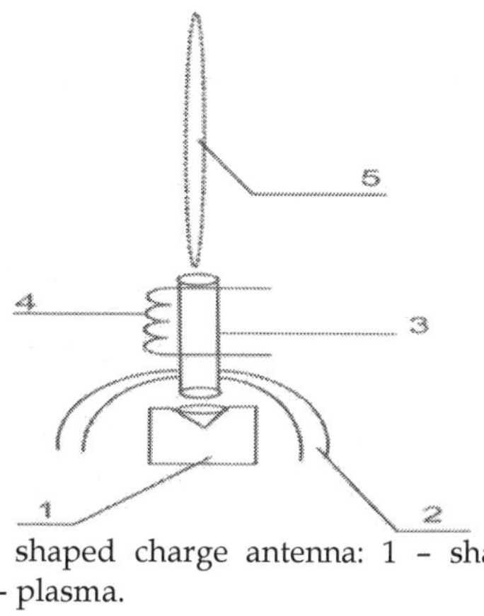

Fig.15. Principal scheme of shaped charge antenna: 1 - shaped charge, 2 - explosive chamber, 3 - tube, 4 - coil, 5 - plasma.

4. Future investigations

It could be noted [26] detailed discussions of the shaped charge concept and an extensive list of sources (too numerous to list here) are available elsewhere. This concept is not well understood by people outside the warhead community. For example, the jet is not "cutting plasma", it is not a liquefied or molten metal jet, the jet temperature is not 20,000 C, and the density of the jet is not several times that of steel, etc. [26].

The main physical problems today in applications to plasma antenna design are as follows. Why do alloys not perform as well as pure metals (with a few exceptions)? Along these same lines, it is known that fine grain liner materials perform better than coarse grain materials. What is the optimum grain size (for materials other than copper, where fine grain liners perform better than coarse grain liners [27])? What is the preferred grain orientation/texture? What about metal purity? (Note that determining the optimal microstructure and mechanical properties for the liner would determine the preferred fabrication method). Should the grain size of the explosive components be fine or coarse to match the grain size present in the liner? Another area of interest is jet particulation or breakup. Can one predict the necking and fracture process of a hypervelocity jet stretching due to its velocity gradient?

Other areas of interest are scaling. Typically, shaped charges can be scaled (homologous scaling) over a charge diameter range of about 1.5 inches to 7 inches. For small charges, the precision/tolerance requirements are hard to achieve and the detonation physics involved does not scale. For large charges, the liner metallurgy is harder to control and the explosive loading is more difficult (may have to be done in stages). Are other, not homologous, scaling laws possible?

Finally, the jet temperature is in general unknown and its prediction by hydrocodes (large computer codes) is difficult and driven by the constitutive equation and equation of state used in the calculation. What is the gradient of temperature, tip to tail, for a shaped charge jet? Recovered jet particles and slugs indicate regions of at least localized melting. What is the interior temperature of the jet? [14]

As it was mentioned below the length of metal shaped charge jet is limited by the value of $L < 10D$, where D is a liner diameter. Today as it was shown in [28] the largest shaped charges are available. For example, in [28] a parametric study was conducted using the Shaped Charge Analysis Program code to design 28 inch (711 mm) outside diameter of shaped charge. The total charge weight was about 900 pounds. The total weight of Octol explosive was about 600 pounds. So the length of shaped charge metal jet (antenna) about 7-10m is available.

The length of plasma dielectric shaped charge antenna is also depends on the dimensions of shaped charge. As it was shown in [10] the length of such type of antenna about several hundreds meters are available (figure 16).

Also new physical principles of shaped charge design in applications to plasma antenna must be developed [29-32]. For example, academic prof. V.F.Minin [34] at the beginning of 1960th discovered a new shaped charge with cylindrical liner [13, 29]. Such type of shaped charge could be also applied to plasma antenna design (figure 17).

Fig. 16. Dielectric shaped charge antenna.

Fig. 17. Plasma dielectric antenna based on shaped charge with cylindrical liner.

As it was mentioned below the jet properties depend on the charge case and liner shape, released energy, and the liner mass and composition. Also the plasma properties depend on the type of material on the surface of the jet and the square of a jet surface. To increase the erosion process from the surface of a jet the unconventional shaped charge can be investigated.

As previously described, conventional shaped charges initiate an explosive material to collapse a liner material about a cavity defined by the liner. The collapsing liner material moves axially inwardly toward the longitudinal axis and simultaneously moves outwardly in the direction of the detonation wave to generate a high velocity jet. Energy from the detonation wave is transferred to the individual particles of the collapsing liner material. For example to increase the jet diameter the *forced jet formation* principle [29, 31] could be apply -

by resisting collapse of the liner toward the longitudinal axis, and by maintaining a jet diameter greater than conventional jets.

Also it is not a direct subject of a present chapter it could be mentioned that some new perspective shaped charge liners for the plasma antenna development were discovered and investigated under the scientific leadership of academic Prof. Dr. Vladilen F. Minin [34], they are:

- Star-shaped liners: in the cross section from the charge butt end, the liner has a four-ray shape. The distance from the charge axis to the liner decreases with the charge length in an unproportional manner. Therefore, the two-dimensional cross sections of the liner are different in different planes. That is why the liner has four planes of symmetry. Some of the experimental geometric dimensions were as follows: the liner thickness 3.5 mm, the largest radius of the liner 27.5 mm, the shell thickness 5 mm, the external radius of the shell 32.5 mm, the length of the HE charge 70 mm (see figure 18).
- W-liner: some results of preliminary studies of cumulative charges; with the linings that we name W-shaped linings. Tests with the W-shaped charges have shown that massive jets (without the formation of a slug) with velocities of 1.5-5 km/s can be formed at large focusing distances (see figure 19). It could be noted that the W-shaped lining can be bimetallic, i.e., its central and peripheral parts are made of different metals. So the metal in inner surface of a jet can be made from the metal with the specific properties to plasma formation [13, 29, 30].

Fig. 18. Star-like liner for the shaped charge and the target.

Fig. 19. X-ray of initial stage of a jet formation from W-type liner.

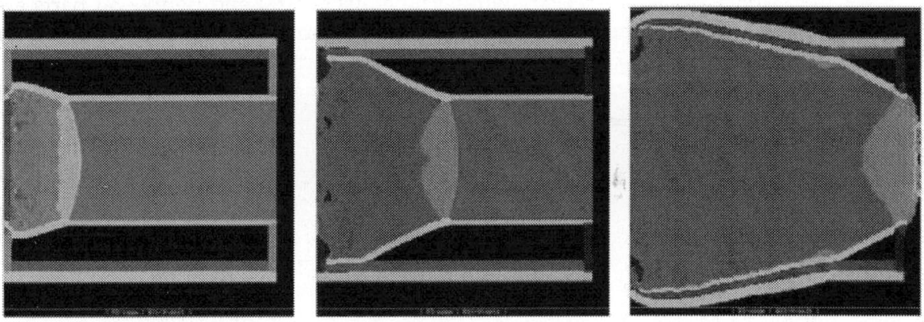

Fig. 20. Simulation of the FCG operation developed under the scientific leadership of Prof. V.F.Minin.

Other aspect of the future explosive plasma antenna development is as follows. Pulse power levels obtained now have the units of GW values. That power is determined by high values of the antenna input voltage (units of megavolts). It can leads to breakdowns in the feeder and antenna. Such a problem also arise in the making of mobile sources that have power of tens megawatt value. When such sources are made on the magnetic cumulative generator (MCG) or flux compression generator (FCG) bases the task of decreasing of dimensions and weight of them keeps being relevant. FCGs use the chemical energy from high explosives to accelerate a metallic conductor that traps and compresses a magnetic field initially created by a seed energy source such as a capacitor bank, battery, or another pulsed generator (figure 20) [35, 36].
The output voltage in such generators as a rule exceeds 30KV value under there volume in 0.5l and mass in 300g values.

6. Acknowledgement

The author would like to thanks Vedran Kordic for invitation me as an editor of the present book. The preparation of this chapter would not have been possible without the support of our father and mother.

7. References

1. Anishchenko, Y. V. (1997). Radiation Initiated by a Surface Wave Propagating along a Long Plasma Column with a Varying Impedance. *Plasma Physics Reports,* Vol. 23 No. 12, pp. 1001-1006.
2. Askar'yan G. A. (1982). *Letters to journal of technical physics (JTF),* Vol. 8, pp. 1131.
3. Dwyer, T.J., Greig, J.R., Murphy, D.P., Perin, J.M., Pechacek, R.E., and Raleigh, M. (1984). On the Feasibility of Using an Atmospheric Discharge Plasma as an RF Antenna. *IEEE Transactions on Antennas and Propagation,* Vol. AP-32. No.2, pp.78-83.
4. Alexeff, I., Kang, W. L., Rader, M., Douglass, C, Kintner, D., Ogot, R., and Norris, E. (2000). A Plasma Stealth Antenna for the U. S. Navy-Recent Results. *Plasma Sources and Applications of Plasmas II,* November 18.
5. Larry L. Altgilbers et al. (1998). Plasma antennas: theoretical and experimental conciderations. *Plasmadynamics and Lasers Conference,* 29th, Albuquerque, NM, June 15-18. AIAA-1998-2567.
6. Zhang T. X., Wu S. T., Altgilbers L. L., Tracy P., and Brown M. Radiation Mechanisms of Pulsed Plasma Dielectric Antennas, 2002, AIAA-2002-2104.
7. Novikov V.E., Puzanov A.O., Sin'kov V.V., Soshenko V.A. (2003). Plasma antenna for magneto cumulative generator. *Int. Conf. On antenna theory and techniques,* Sept. 9-12. Ukraine, pp. 692-695.
8. Shkilyov A.L., Khristenko V.M., Somov V.A., Tkach. Yu.V. (2003). Experimental Investigation of Explosive Plasma Antennas. *Electromagnetic phenomenon's,* Vol. 3, N 4(12), pp.521-528.
9. Schoeneberg N.J. (2003). Generation of transient antennas using cylindrical shaped charges, *A THESIS IN ELECTRICAL ENGINEERING,* Submitted to die Graduate Faculty of Texas Tech University in Partial Fulfillment of the Requirements for the Degree of MASTER OF SCIENCE IN ELECTRICAL ENGINEERING.
10. Minin I., Minin O. (2002). The possibility of impulse plasma antenna creation, *Proceeding of the 6th Russian-Korean Int. Symp. On Science and Technology,* June 24-30, Novosibirsk, Russia. v.2, pp. 289 - 292.
11. Minin I.V., Minin O.V. (1998). *Diffractional quasioptics.* 180 p. Moskow: ImformTei.
12. Kennedy, D. R. (1983). *History of the Shaped Charge Effect, the First 100 Years,* 75p. U. S. Department of Commerce, AD-A220 095.
13. Minin I.V. and Minin O.V. (2003). *World's history of shaped charge.* Proceeding of the Russian conference "Science, Industry and defense", Novosibirsk, April 23-25, pp. 51-53.
14. Walters, W.P. and Zukas J.A. (1989). *Fundamentals of Shaped Charges.* 130 p. CMCPress. Baltimore, MD.
15. Wolsh J., Shreffler, Willing F. (1954). *The limiting conditions for jet formation at high speed.* Moskoy.: Mechanics, 1(23), (in Russian).

16. Godunov S., Deribas A., Mali V. (1975). *About the influences of viscous of metall to the jet formation process.* Fisika gorenia i vzriva (in Russian), Vol. 11, № 1.
17. Pei Chi Chon, J.Carleone, R.Karpp. (1976). *Criteria for jet formation from impinging shell and plates.* J. Appl. Phys., Vol. 47.
18. Birkhoff G., McDougall D., Pugh E., Taylor G. (1948). *Explosives with lined cavities.* J. Of Appl. Phys. Vol. 19, pp. 563-582.
19. Lavrent'ev M. (1957). *The shaped charge and principles of it operations.* Uspehi matem. Nauk (in Russian). Vol. 12, № 4, pp.41-56.
20. Minin I.V., Minin O.V. (2003). *New criterion of cumulative jet formation.* 7th Korea-Russia International Symposium on Science and Technology "KORUS 2003",June 29-July 2, 2003. University of Ulsan, Ulsan, Korea, vol.3, Pages: 93 - 94.
21. V.F.Minin, I.V.Minin, O.V.Minin. Criterium of jet formation for the axisymmetrical shaped charge//Izvestia Vuzov, Povoljskii region, 2006, № 6 (27), pp. 380-389 (in Russian).
22. Neuber, A.; Schoeneberg, N.; Dickens, J.; Kristiansen, M. (2002). *Feasibility study of an explosively formed transient antenna.* Power Modulator Symposium, 2002 and 2002 High-Voltage Workshop. Conference Record of the Twenty-Fifth International Volume , Issue , 30 June-3 July 2002, pp. 374 - 377.
23. Minin O.V. and Minin I.V. (2000). *The influence of the grain size of microstructure of the surface layer material of a hypersonic body on the properties of air plasma.*- The 10th Electromagnetic Launch Technology Symposium, Institute for Advanced Technology, San Francisco, California, USA, April 25-28, 2000. The book of abstracts, pp. 160. See also: Minin O.V. and Minin I.V. (2000). *The influence of the grain size of microstructure of the surface layer material of a hypersonic body on the properties of air plasma. //* Computer optics, N20, pp.93-96. http://www.computeroptics.smr.ru/KO/PDF/KO20/ko20221.pdf
24. Minin I.V., Minin O.V. (2003). *Diffraction optics of millimeter waves.* - IOP Publisher, Boston-London.
25. Patent of the USA № 4100783. Minin V.F. et al. *Installation for explosion machining of articles.*, Jul.18, 1978.
26. Walters. W.P. *An Overview of the Shaped Charge Concept* http://www.scribd.com/doc/6193899/An-Overview-of-the-Shaped-Charge-Concept
27. Dante, J. G. and Golaski, S. K. (1985). *Micrograin and Amorphous Shaped Charge Liners.* Proceedings of ADPA Bomb and Warhead Section, White Oak, MD, May 1985.
28. Manuel G. Vigil. (2003). *Design of Largest Shaped Charge: Generation of Very Large Diameter, Deep Holes in Rock and Concrete Structures.* SANDIA REPORT SAND2003-1160, Unlimited Release, Printed April 2003.
29. Minin I.V., Minin O.V. (2002). *Physical aspects of shaped charge and fragmentational warheads.* 84 p. Novosibirsk, NSTU.
30. Minin I.V., Minin O.V. (1999). *Some new principles of cumulative jet formation.* Collection of works NVI (in Russian), Vol. 7, pp. 19-26. Patent SU № 1508938 (1987). Minin V.F., Minin I.V., Minin O.V. and et. *Devise for plasma jet forming.*
31. Minin I.V., Minin O.V. (1992). Analytical and computation experiments on forced plasma jet formation. *Proc. of the 2nd Int. Symp. on Intense Dynamic Loading and Its Effects.* Chengdu, China, June 9-12, 1992, pp. 588-591.

32. Minin I.V., Minin O.V. (2005). Cumulative plasna jet formation for acceleration of macroparticles, *9th Korea-Russia International Symposium on Science and Technology /* KORUS 2005, June 26-July 2, 2005, NSTU, Russia.
33. Minin I.V., Minin O.V. (2006). Experimental research on reactive type plasma antenna for secure WiFi networks, *8th Int. Conf. On actual problems on electronics instrument engineering, Proceeding*, APIEE-2006, v.2, Novosibirks, Sep.26-28, 2006.
34. Prof. Dr. V.F.Minin http://www.famous-scientists.ru/2677/
35. Minin F.V., Minin I.V., Minin O.V. (1992) Technology of calculation experiments // Mathematical modeling, v.4, N 12, pp. 78-86 (in Russian).
36. Minin F.V., Minin I.V., Minin O.V. (1992) The calculation experiment technology, *Proceedings of the 2nd Int. Symp. on Intense Dynamics loading and its effects*, Chengdu, China, July 9-12, pp.581-587.

Exploiting the semiconductor-metal phase transition of VO2 materials: a novel direction towards tuneable devices and systems for RF-microwave applications

Crunteanu Aurelian[1], Givernaud Julien[1], Blondy Pierre[1], Orlianges Jean-Christophe[2], Champeaux Corinne[2] and Catherinot Alain[2]
[1]*XLIM, CNRS/ Université de Limoges*
[2]*SPCTS, CNRS/ Université de Limoges*
France

1. Introduction

Increasing demands for reconfigurable microwave and millimeter-wave circuits are driven for their high-potential integration in advanced communication systems for civil, defense or space applications (multi-standard frequency communication systems, reconfigurable / switchable antennas, etc.). A wide range of tunable and switchable technologies have been developed over the past years to address the problems related to the overlapping of the frequency bands allocated to an ever-increasing number of communication applications (cellular, wireless, radar etc.). Usually, the reconfiguration of such complex systems is realized by using active electronics components (semiconductor-based diodes or transistors) (Pozar, 2005) or, at an incipient stage, RF MEMS (Micro-electro-mechanical systems)-based solutions (Rebeiz, 2003). However, the performances of these systems are sometimes limited by the power consumption and non-linear behaviour of the semiconductor components or by the yet-to-be-proved reliability of the MEMS devices (switches or variable capacitors).
Current research towards the development of smart multifunctional materials with novel, improved properties may be a viable solution for realizing electronic devices and/ or optical modules with greater functionality, faster operating speed, and reduced size. Smart materials are those materials whose optical and electrical properties (transmittance, reflectance, emittance, refractive index, electrical resistivity etc.) can be controlled and tuned by external stimuli (applied field or voltage, incident light, temperature variation, mechanical stress, pressure etc.). In the RF-microwave fields, materials that are relevant towards the fabrication of tuneable components (resistors, capacitors, inductors), can be classified according to their tuneable properties as: *tuneable resistivity materials* (semiconductors, phase change materials), *tuneable permittivity materials* (ferroelectrics,

liquid crystals, pyrochlores, multiferroics) or *tuneable permeability materials* (ferromagnetics, multiferroics etc.) (Gevorkian, 2008). They can be used to build intelligent components for a broad range of applications: phase shifters/ modulators, delay lines, switches, filters and matching networks, tuneable loads, agile antennas, sensors, detectors etc.

Among the most attractive class of smart materials are those exhibiting a phase transition or a metal- insulator transition. The metal-insulator transition is a large area of research that covers a multitude of systems and materials (chalcogenides, colossal magnetoresistance manganites, superconducting cuprates, nickelates, ferroelectrics, etc.) (Mott, 1968; Edwards et al., 1998). In particular, certain transition metal oxides exhibit such phase transition (Rice &McWhan, 1970), and among these, the vanadium oxide family (V_2O_5, V_2O_3, VO_2) shows the best performance, in particular, presenting a noticeable resistivity change between the two phases. Among these, vanadium dioxide, VO_2, has been studied intensely in the last decade because of his large, reversible change in its electrical, optical and magnetical properties at a temperature close to room temperature, of ~68°C (Morin, 1959) which makes it a potential candidate for introducing advanced functionalities in RF-microwave devices.

Within the present chapter, we want to offer an insight on the *amazing properties of the VO_2 materials* (focusing on the electrical ones) and to *give practical examples of their integration in advanced adaptive devices* in the RF-microwave domain, as developed in the last years at the XLIM Institute in collaboration with the SPCTS laboratory, both from CNRS/ University of Limoges, France (Crunteanu et al., 2007; F. Dumas-Bouchiat et al., 2007, 2009, Givernaud et al., 2008).

We will focus in a first step, on the fabrication using the laser ablation (or the pulsed laser deposition -PLD) method of the VO_2 thin films, on its structural, optical and electrical characterization (speed and magnitude of phase transition induced by temperature or an external electrical field). In a second step we will show the practical integration of the obtained VO_2 films in RF- microwave devices (design, simulation and realisation of VO_2-based switches and tuneable filters in the microwave domain etc.) and we will conclude by presenting the latest developments we are pursuing, namely the demonstration of VO_2-based, current-controlled broadband power limiting devices in the RF- microwave frequency domains.

2. VO_2 material properties and applications

As mentioned before, vanadium dioxide is one of the most interesting and studied members of the vanadates family performing a metal-insulator (or, more correctly, a semiconductor to metal phase transition- SMT) (Morin, 1959; Mott, 1968). At room temperature (low temperature state) VO_2 is a semiconductor, with a band gap of ~1 eV. At temperatures higher than 68°C (341 K) VO_2 undergoes an abrupt transformation to a metallic state, which is reversible when lowering the temperature below 65°C (VO_2 becomes again semiconductor). This remarkable transition is accompanied by a large modification of its electrical and optical properties: the electrical resistivity decreases by several orders of magnitude between the semiconductor and the metallic states while the reflectivity in the near-infrared optical domain increases (Zylbersztejn & Mott, 1975; Verleur et al., 1968). The reversible SMT transition can be triggered by different external excitations: temperature, optically (Cavalleri et al., 2001, 2004, 2005; Ben-Messaoud et al., 2008; Lee et al., 2007), electrically- by charge injection (Stefanovich et al., 2000; Chen et al., 2008, Kim et al., 2004,

Guzman et al., 1996, Dumas-Bouchiat et al., 2007) and even pressure (Sakai & Kurisu, 2008). Recent studies showed that the electrically- and optically- induced transitions can occur very fast (Stefanovich et al., 2000; Cavalleri et al., 2001-2005) (down to 100 fs for the optically- triggered ones (Cavalleri et al., 2005)) and that the transition is more typical of a rearrangement of the electrons in the solid (electron- electron correlations) than it is a an atomic rearrangement (crystalline phase transition from semiconductor monoclinic to a metallic rutile structure).

Although a large number of studies have been devoted to the understanding of the SMT in VO_2, there is still no consensus concerning the driving mechanisms of this phase transition (Pergament at el., 2003; Laad et al., 2006, Qazilbash et al., 2007, Cavalleri et al., 2001). The two mechanisms believed to be responsible for the phase transition (the Peierls mechanisms-electron-phonon interactions and the Mott-Hubard transition - strong electron-electron interactions) are still elements under debate (Morin, 1959; Mott, 1968; Cavalleri et al., 2001, Stefanovich et al., 200, Pergament et al. 2003, Kim, 2004; Kim, 2008).

The transition temperature of the VO_2 layers can be shifted to lower temperatures e.g. by applying an electric field or an incident light beam to a planar two-terminal device (Kim et al., 2004; Lee et al., 2007, Qazilbash et al., 2008, Chen et al., 2008). It is believed that an electric field application to VO2 or an incident beam influences the electron or holes concentrations resulting in a shift of the transition temperature. According to the Mott-Hubard mechanism (Laad et al., 2006), the SMT transition should be driven by the increase in electron concentration (once the electrons reach a critical concentration, the VO_2 pass from semiconductor to metallic). Also, the transition temperature of the VO_2's SMT can be increased or decreased by doping with metals like W, Cr, Ta or Al (Kitahiro & Watanabe, 1967; Kim et al., 2007). VO_2 has a high voltage breakdown, which can be exploited for transmission of high power levels in microwave devices.

In the last years, en ever increasing number of papers have been published and discussed VO_2-based applications, most of which are on microbolometers applications (Yi et al., 2002; Li et al., 2008), smart thermochromic windows (Manning et al., 2002), spatial light modulators (e.g. Richardson and Coath, 1998; Jiang and Carr, 2004; Wang et al., 2006) or electrical switches development (thin films and single-crystal structures) (e.g. Guzman et al., 1996; Stefanovich et al., 2000; Qazilbash et al., 2007; Kim et al., 2004), but the functioning of the proposed devices is based mainly on the thermal activation of the MIT transition which is far more slow than the purely electric or optical- activated ones (massive charge injection or optical activation). The very few reports concerning the possible integration of VO_2 thin films in devices and systems for RF and millimetre wave applications concerns their dielectric properties in this domains (Hood & DeNatale, 1991), the fabrication of submillimeter -wave modulators and polarizers (Fan et al., 1977), of thermally controlled coplanar microwave switches (Stotz et al., 1999) and numerical simulations of VO_2-based material switching operation in the RF-microwave domain (Dragoman et al., 2006). The operating frequency for VO_2-based switches was estimated to be beyond 1 THz (Stefanovich et al., 2000), which makes them very attractive for realizing broadband devices in the millimetr-wave domain.

In the last few years we successfully integrated PLD-deposited VO_2 thin films in several types of components and more complex devices such as thermally and electrically-activated microwave switches (Crunteanu et al., 2007; Dumas-Bouchiat et al., 2007 and 2009), tunable band stop filters including VO_2-based switches (Givernaud et al., 2008) and recently, we proposed an original approach for the design and fabrication of self-resetting power limiting devices based on microwave power induced SMT in vanadium dioxide (Givernaud et al., 2009). As an illustration of our current activities towards the integration of VO_2 layers in RF-microwave (RF- MW) devices, we will present the design, fabrication and caracterization of thermally activated MW switches and their integration in a new type of thermally triggered reconfigurable 4-bit band stop filter designed to operate in the 9- 11 GHz frequency range.

3. PLD deposition and structural, optical and electrical characterization of the VO_2 thin films

Several deposition methods have been proposed for fabrication of VO_2 thin films: sputtering, evaporation pyrolysis or chemical reaction techniques (Hood & DeNatale, 1991; Stotz et al., 1999; Manning et al., 2002; Li et al., 2008 etc.). According to the multivalency of vanadium ion and its complex oxide structure (Griffiths & Eastwood, 1974), numerous phases with stoechiometries close to VO_2 can exist (from V_4O to V_2O_5) and the synthesis of phase pure VO_2 thin films is an important challenge. Reactive pulsed laser deposition (PLD) is a suitable technique for obtaining high-purity oxide thin films (Chrisey & Hubler, 1994; Eason, 2007), very well adapted for obtaining the stoichiometric VO_2 layers. However, careful optimisation of the working parameters is necessary to obtain thin films of the pure VO_2 stabilized phase without any post-treatment.

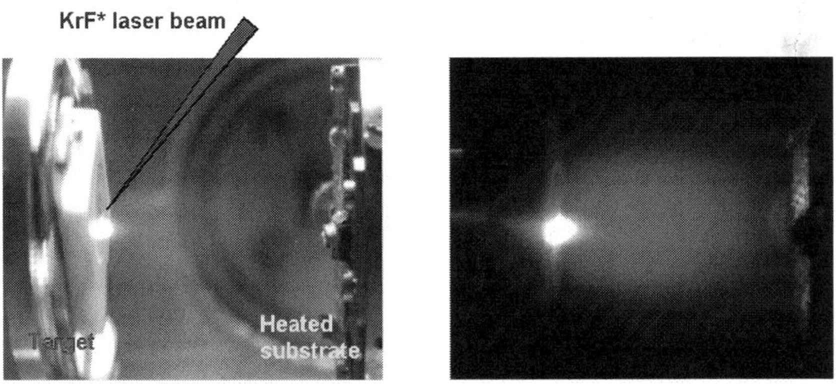

Fig. 1. Photography of the PLD set-up showing schematically the inside of the deposition chamber (left-hand side) and the expansion of the plasma plume towards the substrate after the laser pulse (right-hand side).

In our case, VO_2 thin films were deposited using reactive pulsed laser deposition from a high purity grade (99.95%) vanadium metal target under an oxygen atmosphere. The experimental set-up (picture shown in Fig.1) was described elsewhere (Dumas-Bouchiat et al., 2006) and is based on an excimer KrF laser (with a wavelength of 248 nm and a pulse duration of 25 ns), operating at a repetition rate of 10 Hz. The laser beam is focused on a rotating target in order to obtain fluences (i.e. energies per irradiated surface unit) in the order of 5 to 9 J/cm^2. The plasma plume expands in the ambient oxygen atmosphere (total pressure in the chamber maintained at 2×10^{-2} mbar). Since it has a relatively low lattice parameter mismatch (4.5%) as compared to VO_2 monoclinic phase, monocristalline Al_2O_3(C) is a good candidate to deposit mono-oriented VO_2 films (Garry et al., 2004). The substrate is heated by an halogen lamp at about 500°C and the deposition duration is changing from 10 to 45 minutes leading to thickness in the range 100 - 600 nm. VO_2 thin films have been also deposited on sapphire R-type substrates (Al_2O_3(R)), quartz or 100 Si substrates (bare or oxidized with a 1-μm thick layer of SiO_2).

Irrespective on the substrate we used, the obtained films show a smooth surface with very low-density or no particulates at all, as indicated by scanning electron microscopy analysis, see Fig. 2a. Their morphology (as revealed by atomic force microscopy, AFM, Fig. 2b) consists of compact quasispherical crystallites with typical dimensions (root mean square roughness) between 5 and 15 nm. The non-dependence of film morphology on the substrate nature may be an indication that the growth mechanism is governed mainly by the laser beam/ target interaction.

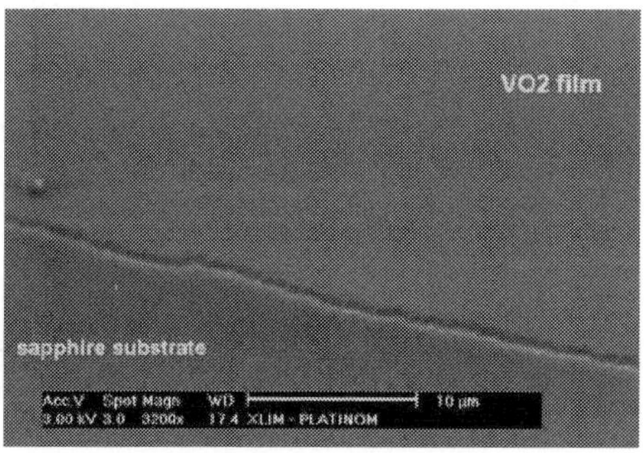

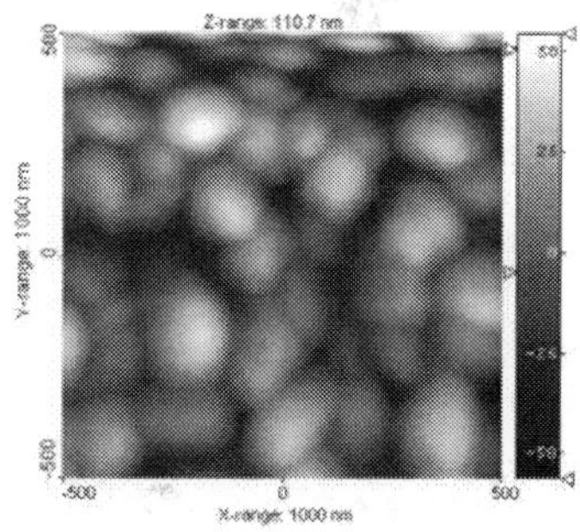

a. b.

Fig. 2. a) SEM image of a VO_2 thin film growth on a sapphire substrate showing a smooth surface and b) AFM image obtained on a VO_2 film (75-nm thickness) onto a sapphire R substrate showing compact crystallites.

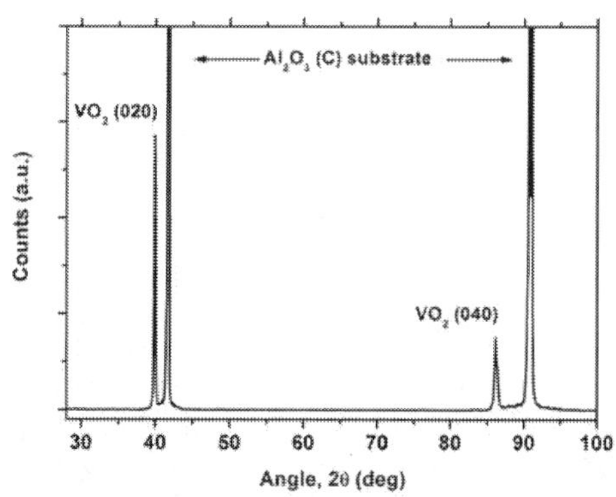

Fig. 3. Typical XRD scan for a 200-nm thick VO_2 thin film deposited on an Al_2O_3 (C) substrate showing characteristic peaks ((020) and (040) of the monoclinic phase of VO_2.

X-Ray diffraction -XRD investigations (in θ, 2θ configuration) performed on VO_2/Al_2O_3(C) thin films reveal two peaks located near 40.2° and 86.8° corresponding respectively to the (020) and (040) planes of the monoclinic VO_2 phase. In certain cases, and especially for amorphous substrates (SiO_2/ Si substrates), depending on the deposition parameters, a peak appears near 28° corresponding to the (011) planes of VO_2 with an orthorhombic structure (Youn et al., 2004).

3.1 Temperature-induced SMT of VO_2 thin films

For the obtained VO_2 films we recorded the variation of their electrical optical and properties (resistivity and optical transmission variation) with the applied temperature in order to rapidly assess the amplitude of their temperature-activated SMT transition.

The electrical resistance/ resistivity of the VO_2 thin films was recorded in the 20-100°C temperature range using a two-terminal device (two metallic contacts deposited nearby on a rectangular VO_2 pattern). A typical resistance hysteresis cycle (heating- cooling loop) of a 200-nm thick VO_2 thin films deposited on a C-type sapphire substrate can be observed in Fig. 4 (the VO2 pattern between the two measurements electrodes was, in this case, 70 μm long x 45 μm wide and 200 nm thick). One may observe a huge change in its resistance as the temperature is cycled through the phase transition (R~ 450 kΩ at 20°C down to R~75 Ω· at 100°C). The width of the hysteresys curve (heating- cooling cycle) is very small: the transition occurs in the 72-74°C range when heating the sample (transformation from semiconductor to metal) and in the 65-68°C range when cooling down at room temperature, and is witnessing on the high quality of the obtained material.

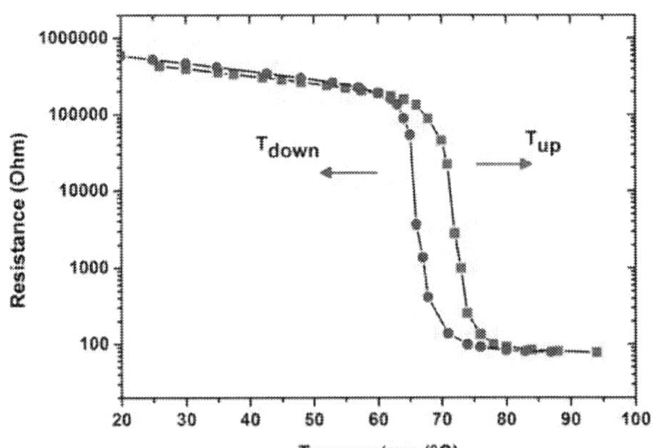

Fig. 4. Resistance variation with temperature for a VO_2 film (two terminal device of 70 mm long, 45 mm wide and 200 nm thick) fabricated by PLD on a C-type sapphire substrate

The optical transmission measurements of VO_2 layers on different substrates as a function of the temperature were done in the UV-visible- mid-IR regions of the spectrum using a Varian Carry 5000 spectrophotometer equipped with a sample heater. They were recorded for different temperatures in the 20-100° C domain. As observed on Fig. 5, the VO_2 films deposited on Al2O3 (R) and on SiO_2/ Si substrates showed a very sharp phase transition witnessing of abrupt change (transmission change factors between 4 and 8) of its optical properties (drastic modification of its refractive index and absorption coefficient). One may notice on the graph on Fig. 5a that the temperature- dependent transmission curves intersect in a particular point, the isosbestic point (at ~850 nm) where the transmittance is constant for all temperatures (Qazilbash et al., 2007).

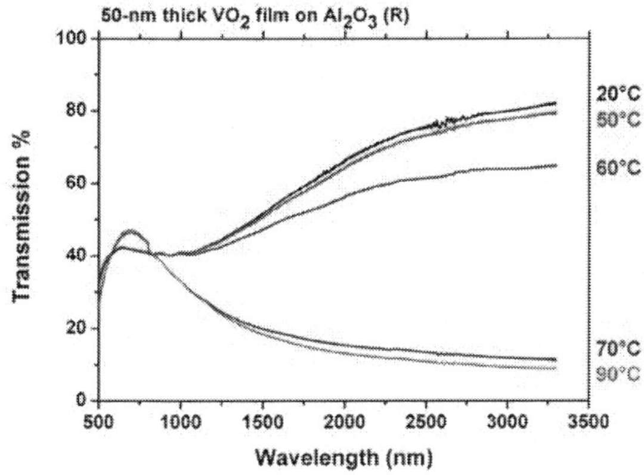

a.

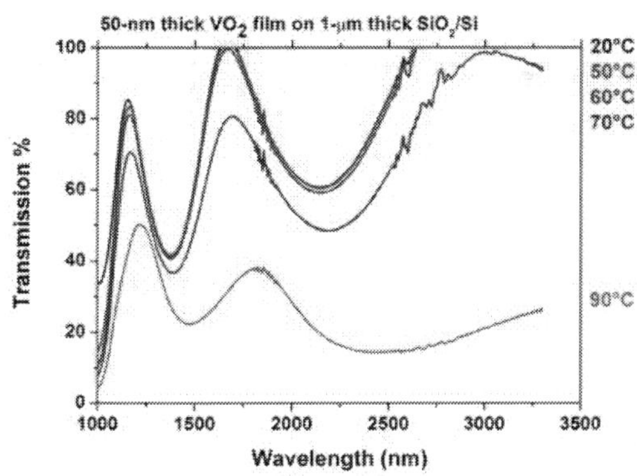

b.

Fig. 5. Optical transmission spectra vs. temperature for 50-nm thick VO_2 films made by PLD on R-type sapphire substrates (a) and 1-µm thick SiO_2/ Si substrate (the oscillations visible on these spectra are interference patterns due to the SiO_2/ Si stack layers)(b).

We also investigated the reflectivity variation of the VO_2 films versus the temperature. Typically, a substrate covered with a VO_2 thin layer was placed on a heating stage and the optical power of a reflected fiber laser beam (at 1550 nm) directed at almost normal incidence onto the film surface was recorded during temperature variation in the 20-100°C domain. On Fig. 6 is presented a typical hysteresys cycle of film's reflectivity (heating-cooling cycle). The VO_2 films showed a very sharp, abrupt phase transition that occurs irrespective of the used substrate or of their thickness. As in the case of the electrical resistivity measurements, the width of the hysteresys curve is very small.

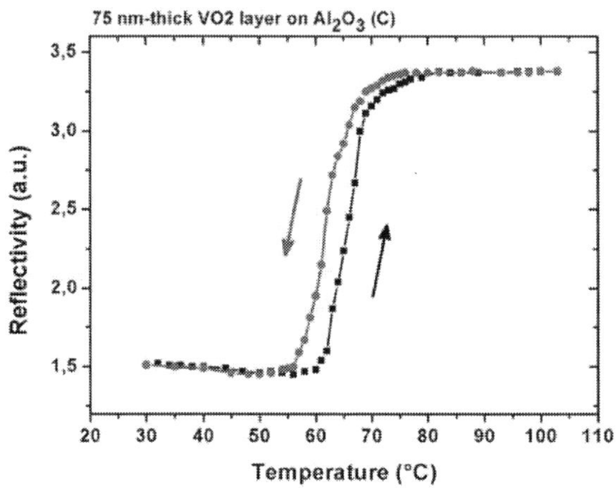

Fig. 6. Hysteresis cycle of reflectivity (at 1550 nm) vs. temperature for a 75-nm thick VO_2 film made by PLD on C-type sapphire substrate showing the sharp phase transition of the VO_2 material.

3.2 Electrically- induced SMT of VO_2 thin films

The proof of concept of thermally induced SMT of VO_2 thin films for realising microwave (and optical) switching devices shown above represents already an innovative, interesting field of research both from theoretically and practical points of view. However, the electrically driven SMT of the VO_2 material will results in more practical devices (without the need of a additional temperature source for the phase transition activation) that, theoretically, can be activated several orders of magnitude faster (Mott, 1968; Cavalleri et al., 2001; Stefanovich et al., 2000; Kim et al., 2004).

We therefore initiated investigations for evaluating the electrically induced phase transition of VO_2 thin films integrated in two-terminal switching devices. The VO_2 pattern is included in an electrical circuit (Fig. 7a) with a c.c. voltage source (applied voltage, V_{ap}), an amperemeter (measuring the current in the circuit, I) and a resistor (Rs, with typical values between 100 and 1500 Ω) for limiting the overall current in the circuit since high values of the current may damage the VO_2 switch. The first results (I-V_{ap} and I-V_{VO2} characteristics) of the electrically actuated VO_2- based two-terminal device (rectangular pattern, 40- μm long, 95-μm wide and 200 nm thick) are presented on Figs. 7 b, c. It may be seen that at a given threshold voltage (V_{ap} between 11 and 14 V for the c.c. voltage source, and V_{VO2}~ 10.5 to 13 V for the voltage on the VO_2 circuit, depending on the Rs value) the current increase abruptly, indicating that the resistivity of the VO_2 layer decreased. This phenomenon is indicative on the onset of the phase transition, VO_2 passes from a high resistive state (semiconductor) in a low-resistive one (it becomes metallic).

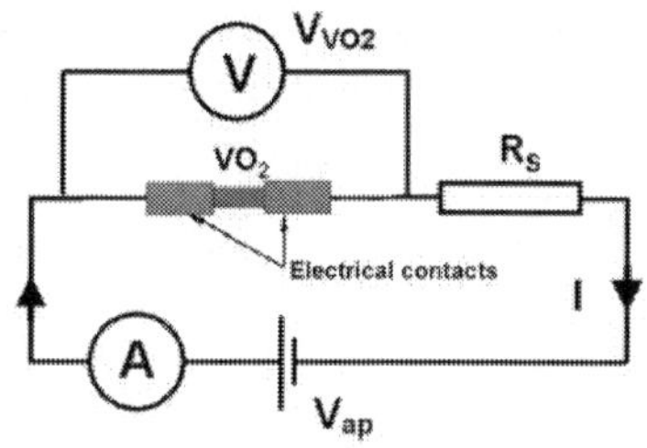

a.

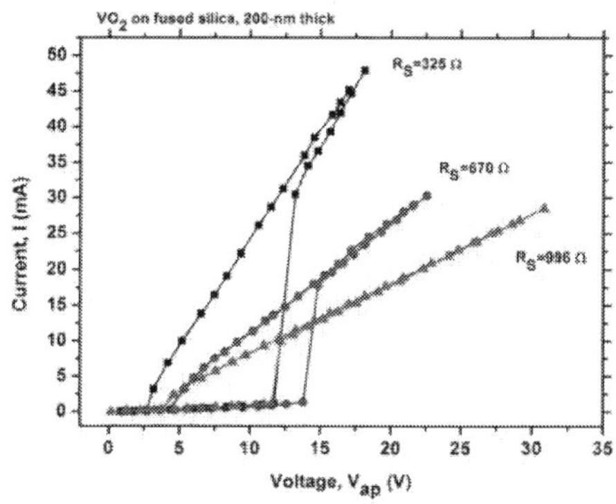

b.

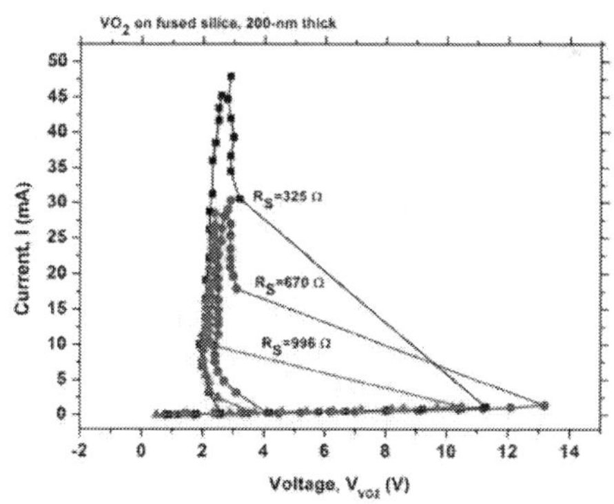

c.

Fig. 7. a) Electrical circuit set-up for investigating the Electrically- induced SMT transition of a two terminal switching device based on a VO_2 thin film (200-nm thick on C-type sapphire); b) I-V_{ap} hysteresis characteristic as the V_{ap} is swept between 0 and maximum of 30V and backwards and c) the typical S-shape of the I-V_{VO2} characteristic of the device.

The nonlinear, S-shaped, negative differential resistance (NDR) I-V_{VO2} characteristic, typical for the VO_2 material (and whose shape can be tuned with external applied temperature) is of high interest from the viewpoint of fundamental physics as well as of a broad range of applications (NDR based oscillators, transistors, filters etc.).

The device remains in the activated state as long as the voltage or the current is maintained in the circuit. For evaluating the switching time of the electrically induced transition we used a similar activation schema (Fig. 8a) but including an a.c. voltage actuation with a square-type waveform (voltage pulses for which the temporal width were varied from 1 to 20 μm).

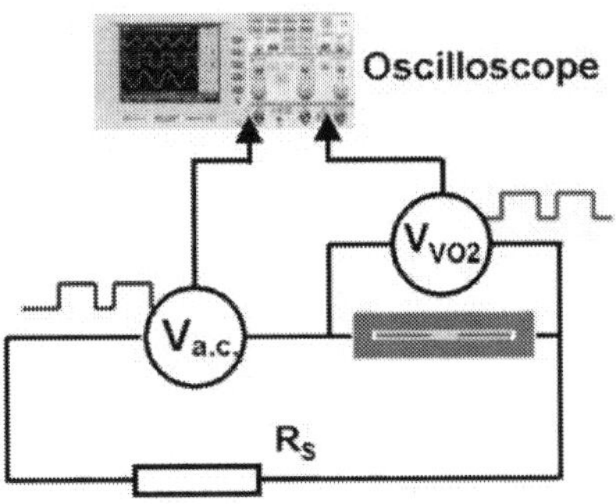

a.

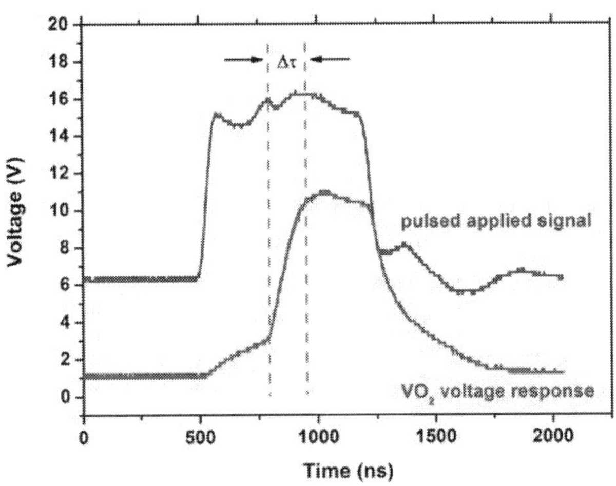

b.

Fig. 8. a) Set-up for electrical activation of the SMT transition and evaluation of the switching time of a VO_2-based two-terminal switch b) applied squared waveform (16 V amplitude, 1.5 μs in width) and the voltage variation through the VO2 switch (in series with a resistor having R_S= 278 Ω) showing installation of the VO2's SMT with activation times Δτ which varies between 100 and 250 ns.

As indicated in Fig. 8 b, c, preliminary results indicate switching times values as low as several hundreds of nano-seconds, which are, however, situated well above the electronically induced VO_2 transition (supposed to occur in the ps domain).

Although the theoretical calculations for a current-induced temperature initiation of the SMT transition (by the Joule heating effect) on the tested device lies in the order of the micro-second scale time (higher than the switching times we recorded), it is prematurely to asses on a purely electrical-induced phase transition (by charge injection). More likely we recorded a switching time describing a mixture of the two potentially present mechanisms (Joule effect heating and charge injection). Nevertheless, the key point of these experiments is that the switching time values are better than those of devices employing fast MEMS-based solutions (Lacroix et al., 2007) and not far from the switching times values of the semiconductors currently used in millimeter domain-switching devices.

We should point out that the electrical activation of VO_2 thin films is also accompanied by changes in their optical properties, easily perceived using optical microscopy and recorded using a CCD camera, as reflectivity change periodically with the applied a.c. signal. These findings are currently exploited in our group for fabrication of variable reflectivity micro mirrors and attenuators in the optical domain for high-speed modulators in novel laser systems (results not reported here).

To resume the preliminary results presented above we may say that the VO_2 is a very interesting and exciting phase transition material. Its electrical and optical properties may be tuned in a static or dynamical way by external factors such as the temperature or an applied electrical field or voltage. These results were further exploited for the realization of rapid electrically switching of microwave coplanar waveguide (CPW) lines or the fabrication of band-stop-type MW filters.

4. Integration of VO_2 thin films in microwave switches and filters

The enormous resistivity change (3 to 4 order of magnitude) of the VO_2 material undergoing the SMT induced by the temperature or by an applied voltage was exploited to fabricate and characterize simple microwave switches based on a coplanar microwave waveguide integrating VO_2 thin films. We obtained temperature activated switching functions (in both shunt and series configurations) with relatively low losses and more than 25 dB transmission variations between the ON/OFF states, on a very large bandwidth (50 MHz-35 GHz) (Crunteanu et al., 2007; Dumas-Bouchiat et al., 2009). The concept was successfully implemented for more complex devices, such as tuneable band stop filters operating around 10 GHz in the microwave frequency domain (Givernaud et al., 2008).

4.1 Microwave switching based on VO_2 films two terminal devices

In the followings we will present a novel concept of VO_2-based electrical switch by using the discrete (and local) thermal activation of a VO2 two-terminal device using a miniature heating element. The micro-heater is based on a thin-film resistor fabricated from a Ni-doped tetrahedral carbon layer (Ni:ta-C). Nickel-doped ta-C layers are currently used in our laboratory and efficiently integrated in radio frequency micro electro mechanical systems (RF MEMS) and in other tunable components (Orlianges et al., 2005). These thin films allows the realization of localized, high value, planar, easily patterned resistances, leading to significant improvement of insertion losses of MEMS switches integrated in electronic devices. Such thin-film resistors are often used under high value of electrical current, which generate important heating of these devices. Our previous investigations on ta-C layers doped with 5%- 30% wt. Ni showed that the layers preserve their integrity for current

densities as high as 1.5.10^{5}A/cm^2 (Orlianges et al., 2004). This caracteristic of the Ni:ta-C layers can be exploited for fabrication of localized, micrometer-range heating elements which may be used to discretely activate VO_2-based two terminal switches (the important ammount of heat generated into the Ni:ta-C layers will be transmitted to the VO2 patterns placed underneath). The amount of the heat generated by the micro heater element can be adjusted by changing the dimensions and the doping level of the Ni:ta-C pattern.
The design of a fabricated VO_2-based switch which can be activated by the heat generated in a Ni:ta-C thin film is presented in the optical micrscope image on Fig. 9 a.

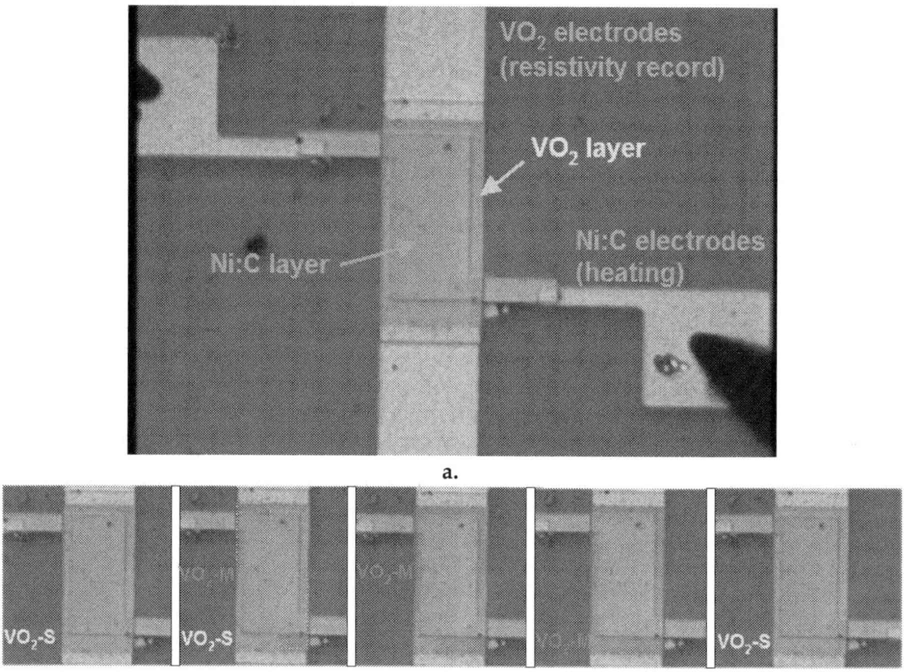

a.

b.

Fig. 9. a) Optical microscopy image of a VO_2-based two terminal switch (400-μm long, 200-μm wide, 200-nm thick pattern between two gold electrodes) which is activated by the current induced heating in a 10% wt. Ni:ta-C pattern situated above it (340-μm long, 150-μm wide and 100-nm thick) and b) optical images showing the sequential activation (phase transition) of the underneath VO_2 layer when applying periodical squared voltage pulses (80V amplitude, 1Hz) on the Ni:ta-C pattern (VO_2-S-semiconducting phase and VO_2-M - the mettalic state of the VO_2 layer.

The device was fabricated in a clean room environment using classical micro fabrication technology. The 200-nm thick VO_2 films were deposited using PLD from a vanadium target in oxygen atmosphere on C-cut sapphire substrates (500-μm thickness) in the conditions described above. The VO2 layers were further patterned using optical lithography and wet

etching for defining the rectangular patterns. It follows the partial masking of the substrate with a photoresist layer for deposition of the Ni:ta-C layers (~100-nm thick) precisely above the VO_2 patterns (the lift-off technique). The nickel doped ta-C films have been deposited under high vacuum by KrF laser ablation of alternating C and Ni targets at ambient temperature (Orlianges et al., 2004). At the end, we fabricated the metallic electrodes: a Ti/ Au layer (6-nm/ 1-μm thick Ti is used as adhesion layer) is deposited using thermal evaporation; the shape of the electrodes are defined by photoresist masking using optical lithography followed by the partial wet etching of the Ti/ Au layer. We tested different pattern dimensions for the VO2 switch (from 200 to 400-μm long and 100 to 200-μm wide) and for the heating Ni:ta-C thin film resistors (100 to 350-μm long and 50 to 150-μm wide).
For the device shown in Fig. 9 a (VO_2 pattern of 400-μm long, 200-μm wide, 200-nm thick pattern between the two metallic electrodes), when applying a current (up to 10 mA) to the Ni:ta-C heating element (340-μm long, 150-μm wide, with an overall resistance of ~11 kΩ) the heat generated in the micro-heater will dissipate to the underneath VO_2 layer and will raise its temperature above the SMT's transition temperature (around 68°C). The VO_2 will therefore pass from a semiconductor to a metal state. As in the case of an optical switch, the transition is easily observed using the optical microscopy as clear changes of the VO_2 layer's reflectivity. These sequential reflectivity changes were recorded using a CCD camera (Fig. 9 b) as we applied to the micro heating layer (Ni:ta-C) a pulsed periodical squared signal (80V amplitude, 1Hz). The onset of the VO_2's phase transition was also recorded electrically by monitoring the resistance of the two-terminal device as a c.c. voltage was progressively applied on the Ni:ta-C heater (Fig. 10).

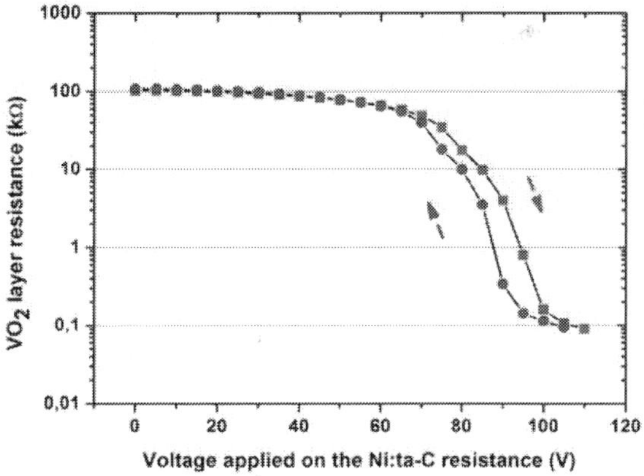

Fig. 10. VO_2's two-terminal device transversal resistance versus the voltage applied on the Ni:ta-C heating resistance: heating phase (red), cooling phase (blue)

One may easily noticed the great variation of the VO_2's resistivity (onset of the SMT) as the Ni:ta-C element dissipate the resistive heating. Work is in progress in order to simulate the heating transfer processes in the overall device, which will allow for optimum design in term of lowering the power consumption. The obtained thermal switching device allows for discrete, localized activation of micrometer-sized VO_2 patterns and may be easily integrated in more complex functions (filtering module), as it will be demonstrated in the next sub-chapter.

4.2 Design and performances of tuneable band-stop filters including VO_2-based switches.

We used the large resistivity change of the device presented above for realising a tuneable 4-pole band stop filter designed to operate in the 9- 11 GHz frequency range with a large signal attenuation in the attenuated band (> 20 dB) (Givernaud et al., 2008). The filter (realised in the micro strip geometry) consists in a 50 Ω transmission line coupled with four U-shaped resonators (Fig. 11). Each resonator is "closed" by a VO_2-based pattern which can be independently activated from the semiconductor to the metallic phase by the Ni:ta-C thin film micro heater. At room temperature, the VO_2 patterns are insulating (VO_2 pattern resistance of 98 kΩ), the resonators are "opened" and each of them will introduce a specific absorption band in the transmission spectrum of the filter (Fig. 13 a). The design of the filter was done using the ADS Momentum simulator and the dimensions and position of each of the resonators (position and distance from the transmission line) was optimised in such a way that the sum of each absorption band result in an broad absorption band between 10 and 11 GHz while maintaining a high signal attenuation (> 20 dB), as visualized in Fig. 13 a.

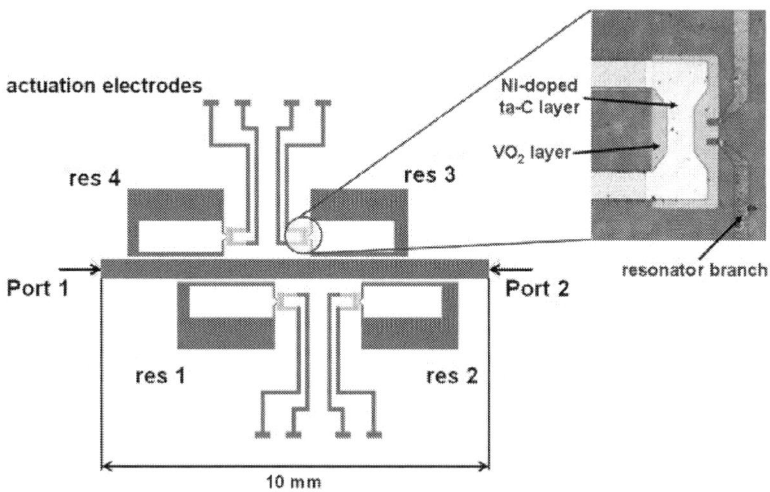

Fig. 11. Design of the four-pole band stop filter that can be discretely tuned by thermally activated each of the VO_2-charged resonators; the insert shows the design of the VO_2-based switch (activated by a Ni:ta-C pattern) which was adapted to the filter's design.

When individually activated, the metallic VO_2 pattern (resistance of 78 Ω) will electrically closed its corresponding U-shaped resonator. The design of the filter (dimensions, resonators dimensions etc.) was done in such a way that the absorption band of the activated resonator would be then shifted far away from the operation frequency band of the filter. The response of the filter will change: shift of the absorption band (tuneability), bandwidth decrease and even disappearance of the attenuation band (Fig. 13 b when all the resonators are activated). This concept was already applied (Givernaud et al., 2008; Dumas-Bouchiat et al., 2009) and results in innovative, discretely tuned filtering functions in the microwave domain.

The filter was fabricated in a clean room environment using classical micro fabrication technology in the conditions described elsewhere (Givernaud et al., 2008). The obtained device was placed using a conductive epoxy paste (for defining the ground plane of the micro strip geometry) in a metallic package and the transmission line ends are electrically connected to SMA-type connectors for measurements (Fig. 12).

Fig. 12. Photography of the realized VO_2-based four-pole filter inserted in a metallic housing and connected to SMA connectors for measuring its response.

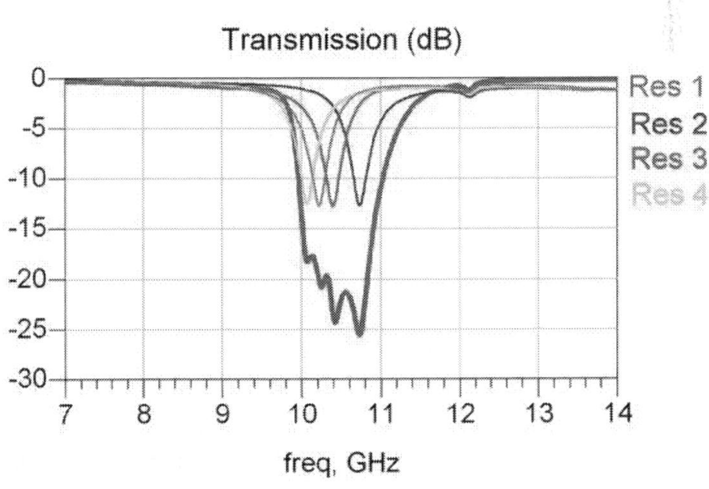

a.

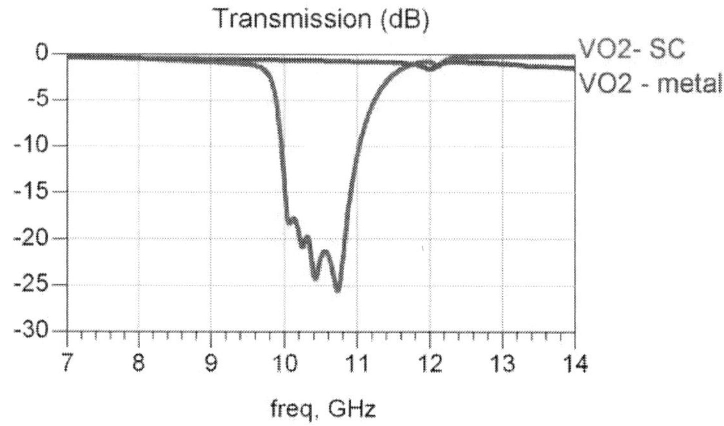

b.

Fig. 13. a) ADS Momentum simulation of the S21 transmission parameter for the overall filter (red curve), showing the absorption band contributions of each resonators and b) the simulated S_{21} transmission parameter of the four-pole band stop filter when the VO_2-based resonators are "opened" (red curve, VO_2-SC) and "closed" (blue curve, VO_2-metal).

The response of the packaged filter was measured using a calibrated four-ports vectorial network analyser (VNA, HP 8722 ES) in the 7 to 14 GHz frequencies range. The measured response of the filter is presented on the graph in Fig. 14 in the two extreme cases: when all the VO_2 patterns are insulating, red curve, and when all the VO_2 patterns are activated by the Ni:ta-C micro-heating elements.

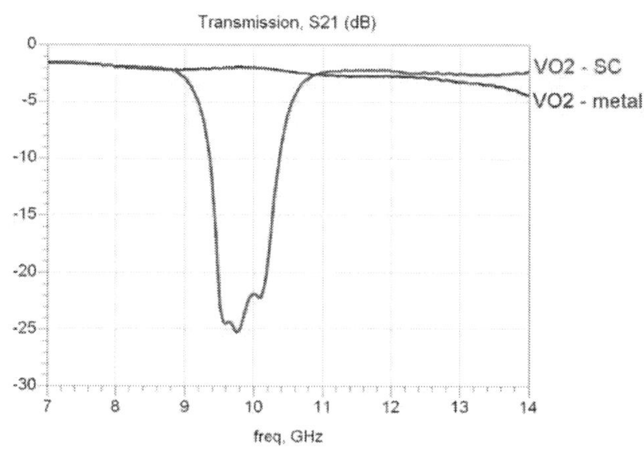

Fig. 14. Measured responses (transmission S21 parameter) of the four pole band stop filter, at room temperature (red curve VO2-SC) and activated (blue curve, VO2-metal, all of the resonators are activated)

One may notice a good agreement of the measured filter responses with the simulations (Fig. 13b). Although the operation band is shifted towards the low frequencies this can be easily corrected for future design by taking into account the deviation from the theoretical values of the materials constants used for the simulation response and by taking care to the micro fabrication tolerances).The tunability of the filter can be demonstrated by individual activation (using the micro-heaters) of specific resonators. When the VO_2-switch of two resonators, for example resonators 1 and 4 (as marked on the Fig. 11) becomes low resistive (VO_2 in the metallic state), the rejection band of the filter will change: its central frequency will shift towards higher frequencies, at 10.6 GHz and its full width at half maximum (FWHM) will lower from ~ 1 GHz to about 0.4 GHz, as shown for the simulated response on Fig. 15 a. A similar behaviour was recorded for the measured response (Fig. 15 b) although the decreasing of the rejection bandwidth was less marked.

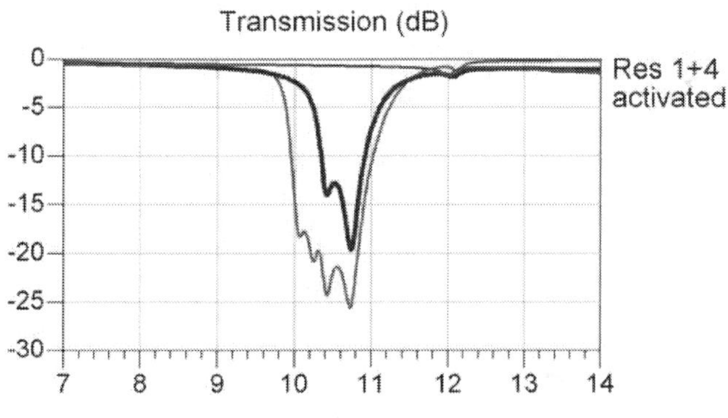

a.

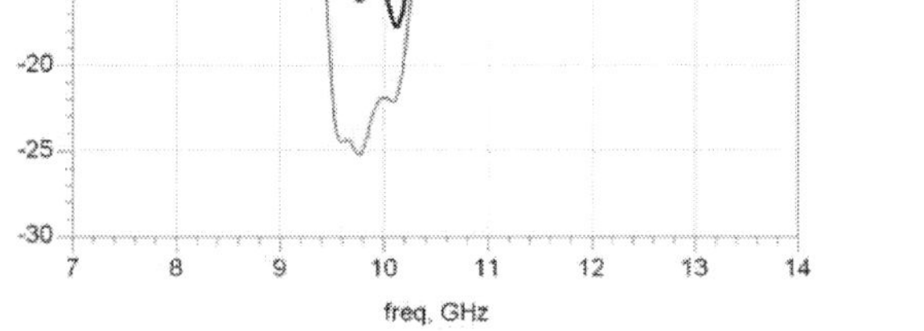

b.

Fig. 15. ADS Momentum simulation (a) and measurement results (b) of the four-pole band stop filter when resonators 1 and 4 (as indicated on the Fig. 11) are simultaneously activated (blue curve, compared with the initial response of the non-activated filter, the red curve).

The simultaneously activation of two others resonators (those numbered 2 and 3 on Fig. 11), leads to a displacement of the central frequency of the rejection band of the filter towards lower frequencies (measurements shown on Fig. 16) at 9.6 GHz (FWHM = 0.4 GHz).

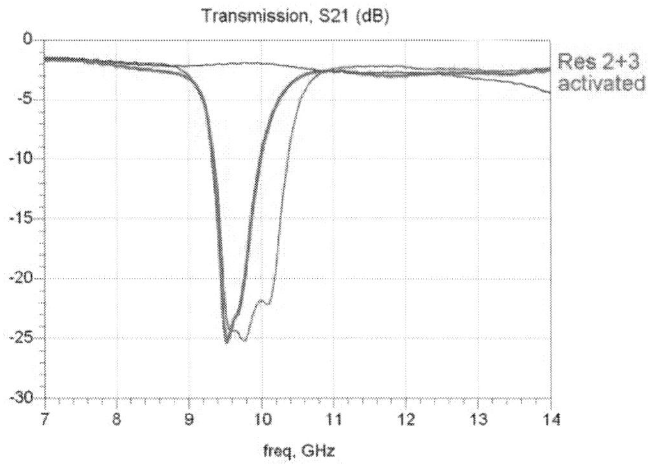

Fig. 16. S21 transmission measurement of the four-pole band stop filter when resonators 2 and 3 (as indicated on the Fig. 11) are simultaneously activated (curve in magenta, compared with the response of the non-activated filter, the red curve and the response of the totally activated filter, blue curve).

We presented above the concept of an VO_2- based electrical switch which can be discretely activated using a Ni:ta-C micro-heating element. These resistivity-switching functions were introduced in a more complex design for fabrication of a band-stop filter with tunable absorption bands and bandwidth operating in the 9-11 GHz frequency domain. Although the device can be further optimized in for obtaining better performancesz we wanted to demonstrate that VO_2 material-based components are serious candidates for RF-microwave switching and microwave reconfigurable devices.

5. Conclusion

The VO_2 material fabricated using the PLD technique results in crystalline thin films performing sharp, high amplitude SMT phase transition and with very good electrical properties. The electrical and temperature-activated VO_2-based switches are promising devices for fabrication of tunable filters and other complex functions in the RF/ microwave domains. The results presented so far are of the state-of-the-art international level concerning the elaboration and characterization of thin films VO_2 of and their integration in practical microwave (and optical) devices. Further applications we are currently developing concerns complex broadband devices employed in the telecommunication networks in the millimeter-wave domain (3-terminal type fast switches, phase shifters, broadband power limiting devices based on microwave power induced SMT in vanadium dioxide, tunable bandpass filter that combines split ring resonators (SRRs) and vanadium dioxide (VO_2)-

based microwave switches etc.) as well as the design of optical devices for applications using miniature laser systems or optoelectronics (optical switches, optical filters, variables attenuators and modulators etc.). The realization of these new type of devices widens a new and extremely rich activity in the field of device fabrication for millimeter-wave reconfigurable systems or for integrated optics and optical communication systems.

6. References

Ben-Messaoud, T.; Landry, G.; Gariépy, J.P.; Ramamoorthy, B.; Ashrit, P.V. & Haché, A. (2008). High contrast optical switching in vanadium dioxide thin films. *Opt. Commun.* 281 (24), pp. 6024-6027

Cavalleri, A.; Tóth, Cs.; Siders, C.W.; Squier, J.A.; Ráksi, F.; Forget P. & Kieffer, J. C. (2001). Femtosecond Structural Dynamics in VO2 during an Ultrafast Solid-Solid Phase Transition. *Physical Review Letters,* 87(23) 237401-1 - 237401-4

Cavalleri, A.; Dekorsky, Th.; Chong, H.H.W.; Kieffer, J.C. & Schoenlein, R.W. (2004). Evidence for a structurally-driven insulator-to-metal transition in VO2: A view from the ultrafast timescale. *Phys. Rev. B. rapid Commun.* 70, 161102(R)

Cavalleri, A.; Rini, M.; Chong, H.H.W.; Fourmaux, S.; Glover, T.E.; Heimann, P.A.; Kieffer, J.C. & Schoenlein, R.W. (2005). Band-selective measurement of Electronic Dynamics in VO2 using Femtosecond Near Edge X-ray Absorption. *Phys. Rev. Lett.* 95, 067405,

Chen, C.; Wang, R.; Shang, L. & Guo, C. (2008). Gate-field-induced phase transitions in VO2: Monoclinic metal phase separation and switchable infrared reflections. *Appl. Phys. Lett.* 93, art. No.171101

Chrisey, D.B. & Hubler, G.K. (1994). *Pulsed Laser Deposition of Thin Films,* Wiley/ Interscience New York, 1994, ISBN 0-471-59218-8

Crunteanu, A.; Dumas-Bouchiat, F.; Champeaux, C.; Catherinot, A.; Pothier, A. & P. Blondy. (2007). Microwave Switching Functions Using Reversible Metal-Insulator Transition (MIT) in VO2 Thin Films. *37th European Microwave Conference (EuMC),* paper EuMC01-4, 8-12 October 2007, Munich, Germany

Dumas-Bouchiat, F.; Nagaraja, H. S.; Rossignol, F.; Champeaux, C.; Trolliard, G.; Catherinot, A. & Givord, D. (2006). Cobalt cluster-assembled thin films deposited by low energy cluster beam deposition: Structural and magnetic investigations of deposited layers. *J. Appl. Phys.* 100 (6), 064304

Dumas-Bouchiat, F.; Champeaux, C.; Catherinot, A.; Crunteanu, A. & Blondy, P. (2007). RF-Microwave switches based on reversible semiconductor-metal transition of V02 thin films synthetized by pulsed laser deposition. *Applied Physics Letters,* 91, pp.223505-1-3

Dumas-Bouchiat, F.; Champeaux, C.; Catherinot, A.; Givernaud, J.; Crunteanu, A. & P. Blondy (2009). RF Microwave Switches Based On Reversible Metal-Semiconductor Transition Properties Of VO2 Thin Films: An Attractive Way To Realise Simple RF Microelectronic Devices. *Materials and Devices for Smart Systems III,* edited by J. Su, L-P. Wang, Y. Furuya, S. Trolier-McKinstry, J. Leng, MRS Symp. Proc. Vol. 1129, Warrendale, PA, 2009, paper 1129-V14-01

Dragoman, M.; Cismaru, A.; Hartnagel, H. & Plana, R. (2006). Reversible metal-semiconductor transitions for microwave switching applications. *Appl. Phys. Lett.* 88, 073503-1 - 073503-3

Eason, R. (2007). *Pulsed laser deposition of thin films: applications-led growth of functional materials*, Wiley/ Interscience, ISBN 978-0-471-44709-2

Fan, J.C.C.; Feterman, H.R.; Bachner, F.J.; Zavracky, P.M. & Parker, C.D. (1977). Thin-film VO_2 submillimiter-wave modulators and polarizers. *Appl. Phys. Lett.* 31 (1), 11-13

Garry, G.; Durand, O. & Lordereau, A. (2004). Structural, electrical and optical properties of pulsed laser deposited VO_2 thin films on R- and C-sapphire planes. *Thin Solid Films* 453, 427

Gevorgian, S. (2008). Tuneable Materials for Agile Microwave devices, an overview. *38th European Microwave Conference* Workshop WWE-6 (EuMC/EuMIC/EuWiT) "Reconfigurable RF Systems", paper WWE-6-6, 27-31 October 2008, Amsterdam, The Netherlands

Givernaud J.; Champeaux, C.; Catherinot, A.; Pothier, A.; Blondy, P. & Crunteanu A. (2008). Tunable band stop filters based on metal insulator transition in vanadium dioxide thin films. *IEEE MTT-S International Microwave Symposium Digest, IMS 2008*, paper WEP1D-02, 15-20 June 2008, Atlanta, GA, USA

Givernaud, J.; Crunteanu, A.; Pothier, A.; Champeaux, C.; Catherinot, A. & Blondy, P. (2009). CPW Self-resetting Power Limiting Devices Based on Microwave Power Induced Semiconductor-Metal Transition in Vanadium Dioxide. *IEEE MTT-S International Microwave Symposium, IMS 2009*, paper TU2E-5, 7-12 June 2009, Boston, MA, USA.

Griffiths, C. H. & Eastwood, H. K. (1974). Influence of stoichiometry on the metal-semiconductor transition in vanadium dioxide. *J. Appl.Phys.* 45, 2201–2206

Guzman, G.; Beteille, F.; Morineau R. & Livage, J. (1996). Electrical switching in VO2 sol-gel films. *J. Mater. Chem.* 6(3), 505-506

Hood, P. J. & DeNatale, J. F. (1991). Millimeter-wave dielectric properties of epitaxial vanadium dioxide thin films. *Appl. Phys. Lett.,* Vol. 70, No. 1, pp. 376-381

Jiang L. & Carr, W.N. (2004). Design, fabrication and testing of a micromachined thermo-optical light modulator based on a vanadium dioxide array. *J. Micromech. Microeng.* 14 833–840

Kim, C.; Shin J.S. & Ozaki, H. (2007). Effect of W doping in metal-insulator transition material VO_2 by tunnelling spectroscopy. *J. Phys. Condens. Matter* 19, 096007-1-7

Kim, H.-T.; Chae, B.-G.; Youn, D.-H.; Maeng, S.-L.; Kim, G.; Kang, K.-Y. & Lim, Y.-S. (2004). Mechanism and observation of Mott transition in VO2-based two- and three-terminal devices. *New J. Phys.* 6, 52-70

Kim, H.-T.; Kim, B.-J.; Lee, Y. W.; Chae, B.-G. & Yun, S.J. (2008). Switching of the Mott transition based on hole-driven MIT theory. *Physica B* 403, 1434–1436

Kitahiro I. & Watanabe, A. (1967). Shift of transition temperature of vanadium dioxide crystals. *Jpn. J. Appl. Phys.* 6, 1023–1024

Laad, M. S.; Craco, L. & Muller-Hartmann, E. (2006). Metal-insulator transition in rutile-based VO_2. *Phys. Rev. B* 73, 195120

Lacroix, B.; Pothier, A.; Crunteanu, A.; Cibert, C.; Dumas-Bouchiat, F.; Champeaux, C.; Catherinot, A. & Blondy, P. (2007). Sub-microsecond RF MEMS switched capacitors. *IEEE Trans. Microwave Theory Tech.* 55, 1314

Lee, Y.W.; Kim, B.J.; Choi, S.; Kim, H.-T.; & Kim, G. (2007). Photo-assited electrical gating in a two-terminal device based on vanadium dioxide film. *Optics Express* 15 (19), 12108-12113

Li, G.; Wang, X.; Liang, J.; Ji, A.; Hu, M.; Yang, F.; Liu, J.; Wu, N. & Chen, H. (2008). Low Temperature Deposited Nano-structured Vanadium Oxide Thin Films for Uncooled Infrared Detectors. *2nd IEEE International Nanoelectronics Conference (INEC 2008)*, pp. 921- 923, March 24-27, 2008, Pudong- Shanghai, China

Manning, T.D.; Parkin, I.P.; Clark, R.J.H.; Sheel, D.; Pemble, M.E. & Vernadou, D. (2002). Intelligent window coatings: atmospheric pressure chemical vapour deposition of vanadium oxides. *J. Mater. Chem.* 12, 2936–2939

Morin, F. (1959). Oxide which shows a metal-to-insulator transition at the high temperature. *Phys. Rev. Lett.* 3, 34

Mott, N.F. (1968). Metal-Insulator Transition. *Review of Modern Physics* 40(4), pp. 677-683

Orlianges, J.C.; Champeaux, C.; Catherinot, A.; Pothier, A.; Blondy, P.; Abelard, P. & Angleraud, B. (2004). Electrical properties of pure and metal-doped pulsed laser deposited carbon films. *Thin Solid Films* 453-454, pp. 291-295

Orlianges, J.C.; Pothier, A.; Mercier, D.; Blondy, P.; Champeaux, C.; Catherinot, A.; De Barros, M.I. & Pavant, S. (2005). Application of aluminum oxide and ta-C thin films deposited at room temperature by PLD in RF-MEMS fabrication. *Thin Solid Films* 482 (1-2), pp. 237-241

Pergament, A. (2003). Metal–insulator transition: the Mott criterion and coherence length. *J. Phys.: Condens. Matter* 15, 3217–3223

Pozar, D. M. (2005). *Microwave Engineering – 3rd ed.*, J. Wiley & Sons

Qazilbash, M. M.; Brehm, M.; Chae, B.-G.; Ho, P.-C.; Andreev, G. O.; Kim, B.-J.; Yun, S.J.; Balatsky, A.V.; Maple, M. B.; Keilmann, F.; Kim, H.-T. & Basov, D. N. (2007). Mott transition in VO_2 revealed by infrared spectroscopy and nano-imaging. *Science* 318, 1750

Qazilbash, M.M.; Li, Z.Q.; Podzorov, V.; Brehm, M.; Keilmann, F.; Chae, B.G.; Kim, H.T. & Basov, D.N. (2008). Electrostatic modification of infrared response in gated structures based on VO2. *Appl. Phys. Lett.* 92 (24), art. no. 241906

Rebeiz, G. M. (2003). *RF MEMS Theory, Design, and Technology*, New Jersey: J. Wiley & Sons

Richardson, M.A. & Coath, J.A. (1998). Infrared optical modulators for missile testing. *Optics & Laser Technology* 30, 137-140

Sakai, J. & Kurisu, M. (2008). Effect of pressure on the electric-field-induced resistance switching of VO2 planar-type junctions. *Phys. Rev. B* 78(3), art. No. 033106

Stefanovich, G.; Pergament A. & Stefanovich, D. (2000). Electrical switching and Mott transition in VO_2. *J. Phys.: Condens. Matter*, Vol. 12, pp. 8837-8845

Stotz, M.; Fritze, S.-D.; Downar, H. & Wenger, J. (1999). Thermally Controlled Coplanar Microwave Switches. *29th European Microwave Conference Proceedings*, pp.415-418, 5-7 October 1999 Munich, Germany

Verleur, H.W.; Barker Jr., A. S. & Berglund, C.N. (1968). Optical properties of VO_2 between 0.25 and 5 eV. *Phys. Rev.* 172 (3), 788-798

Wang, W.; Luo, Y.; Zhang, D. & Luo, F. (2006). Dynamic optical limiting experiments on vanadium dioxide and vanadium pentoxide thin films irradiated by a laser beam. *Applied Optics* Vol. 45, No. 14, pp.3378-3381

Yi, X.; Chen, S.; Wang, Y.; Xiong, B. & Wang, H. (2002). VO2-based infrared microbolometer array. *Intl. J. of Infrared and Millimeter Waves* 23(12), 1699- 1704

Youn, D.; Lee, J.; Chae, B.; Kim, H.; Maeng, S. & Kang, K. (2004). Growth optimization and electrical characteristics of VO_2 films on amorphous SiO_2/Si substrates. *J. Appl. Phys.* 95, 1407

Zylbersztejn, A. & Mott, N.F. (1975). Metal-insulator transition in vanadium dioxide. *Phys. Rev. B* 11(11), 4383-4395

4

Analysis of Parasitic Effects in AlGaN/GaN HEMTs

Kazushige Horio
Shibaura institute of Technology
Japan

1. Introduction

AlGaN/GaN high electron mobility transistors (HEMTs) are now receiving great attention because of their potential applications to high-power and high-frequency devices (Mishra et al., 2008). An output power of more than 32 W/mm is reported at 4 GHz for the 0.55 μm gate-length device (Wu et al., 2004), and a current-gain cutoff frequency (f_T) of 163 GHz is obtained for 0.06 μm gate length (Higashiwaki et al., 2006). However, slow current transients are often observed even if the gate voltage or the drain voltage is changed abruptly (Binari et al., 2002). This is called gate lag or drain lag, and is problematic in circuit applications. The slow transients indicate that the dc current-voltage (*I-V*) curves and the RF *I-V* curves become quite different, resulting in lower RF power available than that expected from the dc operation (Binari et al., 2002; Mishra et al., 2008). This is called power slump or current collapse. This current reduction in RF *I-V* curves or pulsed *I-V* curves is also referred to as current slump, RF dispersion and knee-walkout behavior. These parasitic effects are serious problems, and there are many experimental works reported on these phenomena (Khan et al., 1994; Daumiller et al., 2001; Ventury et al., 2001; Koley et al., 2003; Mizutani et al., 2003; Koudymov et al., 2003; Meneghesso et al., 2004; Desmaris et al., 2006), but, only a few theoretical works are reported recently (Braga et al., 2004; Meneghesso et al., 2004; Tirado et al., 2007). The literature suggests that the surface properties (surface states) play an important role in these phenomena, but traps in a buffer layer could also affect the characteristics (Binari et al., 2002; Desmaris et al., 2006). It is also shown that the gate lag and current collapse can be reduced by introducing a field plate (Koudymov et al., 2005). This is considered due to a decrease in surface-state effects. It is well recognized that the field plate can improve the breakdown voltage and the power performance, because the electric field at the drain edge of the gate is reduced (Karmalkar & Mishra, 2001; Ando et al., 2003; Xing et al., 2004; Saito et al., 2005; Pala et al., 2008). However, it is not well understood whether the field plate affects buffer-related lag phenomena and current collapse.
In the previous theoretical works by device simulation, effects of a donor-type surface state (near the valence band) on gate lag and pulsed *I-V* curves of AlGaN/GaN HEMTs were studied (Meneghesso et al., 2004; Tirado et al., 2007), and a bulk deep-acceptor effect (~ 1 eV) above the midgap of GaN was studied for the gate lag (Braga et al., 2004). However, the types of traps and their energy levels seem to be artificial. Therefore, in this article, we have

made two-dimensional transient simulations of AlGaN/GaN HEMTs with a buffer layer in which trap levels based on experiments are considered, as in our previous work on GaN MESFETs (Horio et al, 2005), and showed that the lag phenomena and current collapse could be reproduced (Horio & Nakajima, 2008). Also, we have studied dependence of current collapse on the impurity densities in the buffer layer and on an off-state drain voltage. Additionally, we have analyzed effects of introducing a field plate on buffer-related lag phenomena and current collapse (Nakajima et al., 2009).

In Section 2, we describe physical models used here, such as analyzed device structures, traps in the buffer layer, and basic equations for the device analysis. Calculated drain-current responses are described in section 3 in terms of drain lag, gate lag and pulsed *I-V* curves (current collapse). The dependence of current collapse on the impurity densities in the buffer layer and on an off-state drain voltage is also described. In section 4, effects of introducing a field plate on buffer-related lag phenomena and current collpase are described.

2. Physical Model

Fig.1 shows modeled AlGaN/GaN HEMT structures analyzed in this study. Fig.1(a) is a normal structure without a field plate, and Fig.1(b) is a structure with a field plate. The gate length L_G is 0.3 μm, and the source-to-gate distance L_{SG} is 0.5 μm. The gate-to-drain distance L_{GD} is typically set to 1 μm in Fig.1(a) and 1.5 μm in Fig.1(b). Note that in Fig.1(b), the gate electrode extends on to SiN passivation layer. This is called field plate. The field-plate length L_{FP} is typically set to 1 μm. The thickness of SiN layer d is varied as a parameter between 0 and 0.1 μm. Polarization charges of 10^{13} cm^{-2} are set at the heterojunction interface, and the surface polarization charges are assumed to be compensated by surface-state charges, as in (Karmalkar & Mishra, 2001). As a model for the buffer layer, we use a three level compensation model which includes a shallow donor, a deep donor and a deep acceptor (Horio et al., 2005). Some representative experiments show that two levels (E_C - 1.8 eV, E_C - 2.85 eV) are associated with current collapse in GaN-based FETs with a semi-insulating buffer layer (Klein et al., 1999; Binari et al., 2002). Therefore, we use an energy level of E_C - 2.85 eV (E_V + 0.6 eV) for the deep acceptor, and for convergence problem, we use E_C - 1.7 eV for the deep donor. (Note that the origin of these deep levels is not well known and our treatment is only an assumption.) Other experiments show shallower energy levels for deep

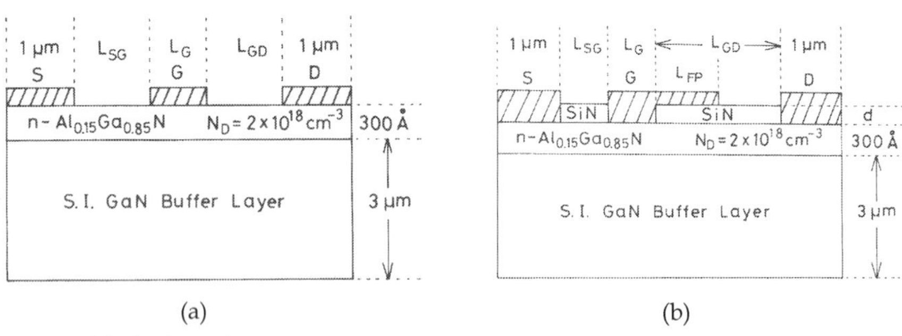

Fig. 1. Modeled AlGaN/GaN HEMT structures analyzed here. (a) normal structure without a field plate. (b) structure with a field plate

donors in AlGaN/GaN system or in GaN (Kruppa et al., 1995; Morkoc, 1999), so that we vary the deep donor's energy level (E_{DD}) as a parameter. Here, the deep-donor density (N_{DD}) and the deep-acceptor density (N_{DA}) are typically set to 5x10^{16} cm^{-3} and 2x10^{16} cm^{-3}, respectively. The electron and hole capture cross sections for the deep donor are set to 10^{-13} cm^2 and 10^{-15} cm^2, respectively, and the electron and hole capture cross sections for the deep acceptor are both set to 10^{-15} cm^2. The shallow-donor density in the buffer layer N_{Di} is set to 10^{15} cm^{-3}. Here, it should be noted that when $N_{DD} > N_{DA}$, the deep acceptors are usually fully occupied by electrons that are supplied from the deep donors, and the ionized (empty) deep donors act as electron traps (the ionized deep-donor density N_{DD}^+ is nearly equal to N_{DA} under equilibrium).

Basic equations to be solved are Poisson's equation including ionized deep-level terms, continuity equations for electrons and holes which include carrier loss rates via the deep levels, and rate equations for the deep levels (Horio et al., 2000; Horio et al., 2005). They are expressed as follows.
1) Poisson's equation

$$\nabla \bullet (\varepsilon \nabla \psi) = -q(p - n + N_{\mathrm{D}} + N_{\mathrm{Di}} + N_{\mathrm{DD}}^{+} - N_{\mathrm{DA}}^{-}) \quad (1)$$

2) Continuity equations for electrons and holes

$$\frac{\partial n}{\partial t} = \frac{1}{q}\nabla \bullet J_n - (R_{n,\mathrm{DD}} + R_{n,\mathrm{DA}}) \quad (2)$$

$$\frac{\partial p}{\partial t} = -\frac{1}{q}\nabla \bullet J_p - (R_{p,\mathrm{DD}} + R_{p,\mathrm{DA}}) \quad (3)$$

where

$$R_{n,\mathrm{DD}} = C_{n,\mathrm{DD}} N_{\mathrm{DD}}^{+} n - e_{n,\mathrm{DD}} (N_{\mathrm{DD}} - N_{\mathrm{DD}}^{+}) \quad (4)$$

$$R_{n,\mathrm{DA}} = C_{n,\mathrm{DA}} (N_{\mathrm{DA}} - N_{\mathrm{DA}}^{-}) n - e_{n,\mathrm{DA}} N_{\mathrm{DA}}^{-} \quad (5)$$

$$R_{p,\mathrm{DD}} = C_{p,\mathrm{DD}} (N_{\mathrm{DD}} - N_{\mathrm{DD}}^{+}) p - e_{p,\mathrm{DD}} N_{\mathrm{DD}}^{+} \quad (6)$$

$$R_{p,\mathrm{DA}} = C_{p,\mathrm{DA}} N_{\mathrm{DA}}^{-} p - e_{p,\mathrm{DA}} (N_{\mathrm{DA}} - N_{\mathrm{DA}}^{-}) \quad (7)$$

3) Rate equations for the deep levels

$$\frac{\partial}{\partial t}(N_{\mathrm{DD}} - N_{\mathrm{DD}}^{+}) = R_{n,\mathrm{DD}} - R_{p,\mathrm{DD}} \quad (8)$$

$$\frac{\partial}{\partial t} N_{\mathrm{DA}}^{-} = R_{n,\mathrm{DA}} - R_{p,\mathrm{DA}} \quad (9)$$

where N_{DD}^+ and N_{DA}^- represent ionized densities of the deep donors and the deep acceptors, respectively. C_n and C_p are the electron and hole capture coefficients of the deep levels, respectively, e_n and e_p are the electron and hole emission rates of the deep levels, respectively, and the subscript (DD, DA) represents the corresponding deep level.
The above basic equations are put into discrete forms, and are solved numerically. We have calculated drain-current responses of the AlGaN/GaN HEMTs when the drain voltage and/or the gate voltage are changed abruptly.

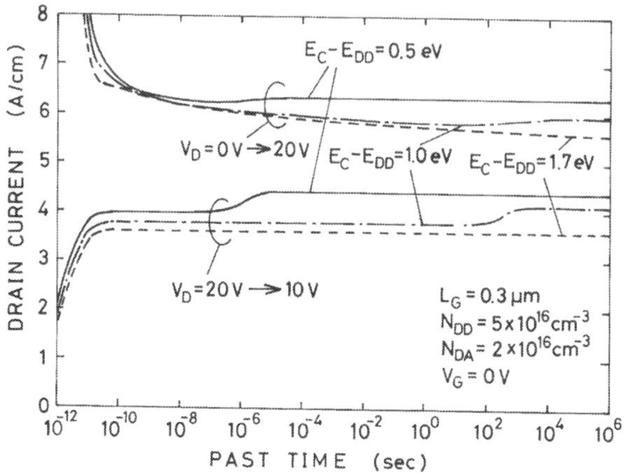

Fig. 2. Calculated drain-current responses of normal AlGaN/GaN HEMTs as a parameter of deep donor's energy level E_{DD} when V_D is raised abruptly from 0 V to 20 V (upper) or when V_D is lowered from 20 V to 10 V (lower). V_G = 0 V. L_{GD} = 1 μm, N_{DD} = 5x10^{16} cm^{-3} and N_{DA} = 2x10^{16} cm^{-3}.

3. Parasitic Effects in AlGaN/GaN HEMTs

3.1 Drain lag

First, we describe a case when only the drain voltage V_D is changed. Fig.2 shows calculated drain-current responses of normal AlGaN/GaN HEMTs when V_D is raised abruptly from 0 V to 20 V (upper) or when it is lowered abruptly from 20 V to 10 V (lower), where the gate voltage (V_G) is kept constant at 0 V. Here N_{DD} is 5x10^{16} cm^{-3} and N_{DA} is 2x10^{16} cm^{-3}, and three cases with different E_C - E_{DD} are shown. In the cases when V_D is raised, initially an extremely large transient overshoot is observed (Horio and Fuseya, 1994). This is because the drain voltage is initially applied along the gate-to-drain bulk region. This initial current decays as the space-charge region at the gate edge extends toward the drain. After this extreme overshoot, the drain currents decrease gradually, reaching steady-state values, although slight undershoot is observed just before reaching the steady state for E_C - E_{DD} = 0.5 eV and 1.0 eV. On the other hand, when V_D is lowered, initially a large transient undershoot is observed. This is also due to the abrupt change of drain voltage, as described above (Horio and Fuseya, 1994). Then, the drain currents remain at low values for some periods ("quasi-steady state") and begin to increase slowly, showing drain-lag behavior. The response is faster for shallower E_{DD}, and for E_C - E_{DD} = 1.7 eV, the drain current I_D remains at a low value even at 10^6 s. The change of drain current (ΔI_D) between the quasi-steady state and the steady state is not so dependent on E_C - E_{DD}.

Fig.3 and Fig.4 show electron density profiles and ionized deep-donor density N_{DD}^+ profiles, respectively, as a function of time t when V_D is raised from 0 V to 20 V, where E_C - E_{DD} = 1.0 eV. From Fig.3(b), we can see that electrons are injected into the buffer layer when V_D is raised, and at this time (t = 10^{-10} s), the deep donors are not almost responding (see Figs.4(a) and (b)). It is understood that I_D decreases due to electron capturing by the deep donors. In

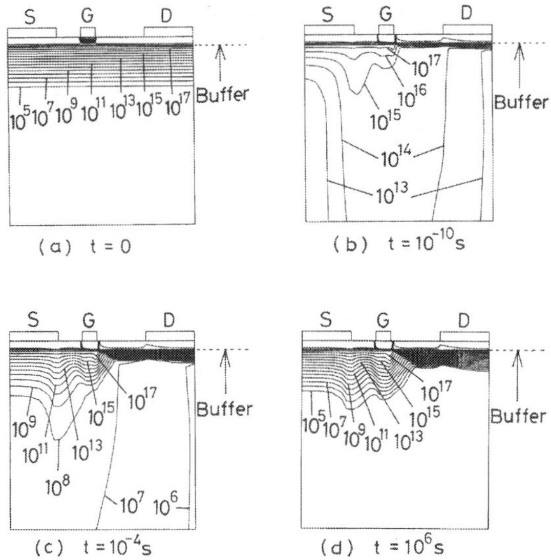

Fig. 3. Change of electron density profiles with time t when V_D is raised from 0 V to 20 V for the case of $E_C - E_{DD} = 1.0$ eV, corresponding to Fig.2. (a) initial state ($t = 0$), (b) $t = 10^{-10}$ s, (c) $t = 10^{-4}$ s, (d) $t = 10^{6}$ s.

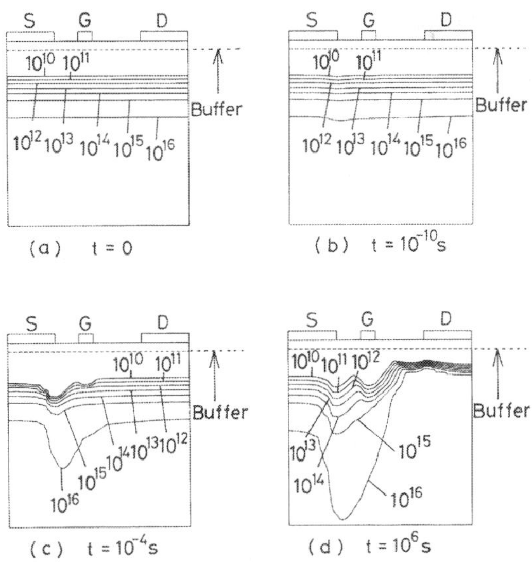

Fig. 4. Change of ionized deep-donor density N_{DD}^{+} profiles with time t when V_D is raised from 0 V to 20 V for the case of $E_C - E_{DD} = 1.0$ eV, corresponding to Fig.2. (a) initial state ($t = 0$), (b) $t = 10^{-10}$ s, (c) $t = 10^{-4}$ s, (d) $t = 10^{6}$ s.

fact, from Figs.3(c) and 4(c), it is seen that electron densities in the buffer layer are decreasing and N_{DD}^+ is decreasing particularly under the source-to-gate region. The undershoot or slight increase in I_D before reaching the steady state, which was observed in Fig.2, is regarded as a result of over capturing or due to electron emission by deep donors around the drain side of the gate region (see Figs.4(a) and (d)).
On the other hand, in the case when V_D is lowered, the response is simply understood by (captured) electron emission process of the deep donors. In fact, the current rise time is roughly consistent with the deep donor's electron-emission time constant given by $1/e_{n,DD}$, which becomes 3.9x10^{-5} s and 9.8x10^3 s for E_C - E_{DD} = 0.5 eV and 1.0 eV, respectively. The time constant for E_C - E_{DD} = 1.7 eV becomes quite long (> 10^{10} s), so that I_D remains at a low value even at t = 10^6 s, as was seen in Fig.2.
Here, it should be mentioned that above drain lag phenomena (overshoot and undershoot behavior) are also reported experimentally in AlGaN/GaN HEMTs (Binari et al., 2002; Meneghesso et al., 2004).

3.2 Gate lag and pulsed *I-V* curves

Next, we describe a case when the gate voltage V_G is also changed (from an off point). Fig.5 shows calculated turn-on characteristics of a normal AlGaN/GaN HEMT when V_G is changed from the threshold voltage V_{th} (= - 9.24 V) to 0 V. Off-state drain voltage V_{Doff} is 20 V and the parameter is an on-state drain voltage V_{Don}. V_{th} is defined here as a gate voltage when I_D becomes 5x10^{-3} A/cm. Here, N_{DD} = 5x10^{16} cm^{-3}, N_{DA} = 2x10^{16} cm^{-3} and E_C - E_{DD} = 1.0 eV. V_{th} becomes rather deep because the current component via the buffer layer exists. The characteristics are similar to those shown in Fig.2 where V_D is lowered. However, as seen in the uppermost curve of Fig.5 (V_{Doff} = V_{Don} =20 V), some transients are observed when only V_G is changed. This indicates that gate lag as well as drain lag could occur due to deep levels in the buffer layer. We will describe below why the gate lag arises.

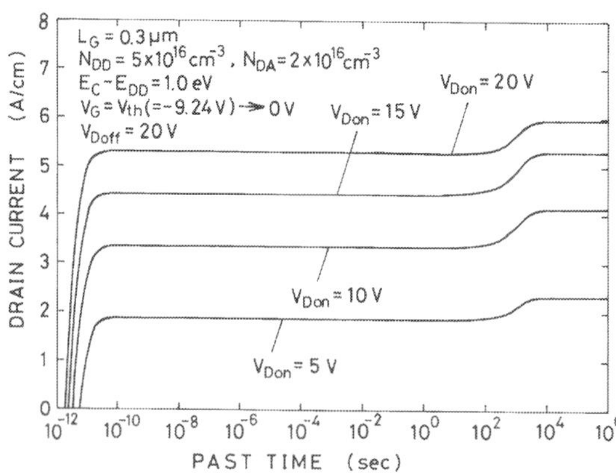

Fig. 5. Calculated turn-on characteristics of the normal AlGaN/GaN HEMT when V_G is changed from threshold voltage V_{th} (= - 9.24 V) to 0 V, with on-state drain voltage V_{Don} as a parameter. Off-state drain voltage V_{Doff} is 20 V. L_{GD} = 1 μm and E_C - E_{DD} = 1.0 eV.

Fig.6 shows a comparison of (a) conduction-band-edge energy profiles, (b) electron density profiles, and (c) N_{DD}^{+} profiles between the on state (left: V_D = 20 V, V_G = 0 V) and the off state (right: V_D = 20 V, V_G = V_{th} = - 9.24 V). Note that only V_G is different here. From Fig.6(a), in the on state, some potential drops are observed between source and gate (and between gate and drain). The potential drop is given by I (current) x R (resistance), and hence the large current and relatively large resistance become a cause of the visible potential drop. This indicates that source access resistance (and drain access resistance) is not negligible. It is understood that due to this potential drop at the source side, when V_G becomes negative and the channel is depleted, electrons under the gate are not all pushed into the source and drain electrodes, but can be injected into the buffer layer as seen in Fig.6(b). (Note that the energy barrier at the channel-buffer interface can be weakened under the gate when V_G becomes strongly negative.) These electrons are captured by deep donors, and hence N_{DD}^{+} decreases in the off state, as seen in Fig.6(c). Because of this increase in negative space

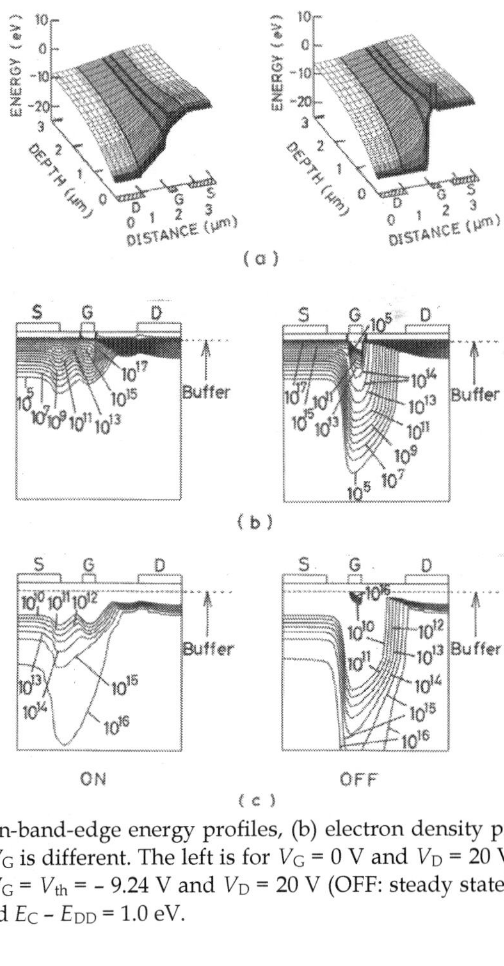

Fig. 6. (a) Conduction-band-edge energy profiles, (b) electron density profiles, and (c) N_{DD}^{+} profiles when only V_G is different. The left is for V_G = 0 V and V_D = 20 V (ON: steady state), and the right is for V_G = V_{th} = - 9.24 V and V_D = 20 V (OFF: steady state). N_{DD} = 5x10^{16} cm^{-3}, N_{DA} = 2x10^{16} cm^{-3} and E_C - E_{DD} = 1.0 eV.

charges in the buffer layer, even if V_G is switched on, I_D remains at a low value until the deep donors begin to emit electrons, showing gate-lag behavior. It should be mentioned that this type of gate lag is not observed in the similar simulation for GaAs MESFETs with deep donors "EL2" and shallow acceptors in the substrate (Horio et al., 2000), where visible potential drops are not observed in the on state between source and gate. This is because the current density is relatively low and the parasitic resistance is low due to the higher electron mobility. Therefore, in the case of AlGaN/GaN HEMTs, we can say that relatively high source access resistance is correlated to the gate lag. The high access resistance in AlGaN/GaN HEMTs is considered problematic because it degrades the high-frequency performance (Palacios et al., 2005).

Next, we describe *I-V* characteristics. Fig.7 shows calculated I_D-V_D curves of the normal AlGaN/GaN HEMT with a semi-insulating buffer layer, where N_{DD} = 5x10^{16} cm^{-3}, N_{DA} = 2x10^{16} cm^{-3} and E_C - E_{DD} = 1.0 eV. The solid line is the steady-state *I-V* curve. In this figure, we plot by point (x) the drain current at t = 10^{-8} s (after V_G is switched on) as a parameter of V_{Don} (V_D). This is obtained from calculated turn-on characteristics like Fig.5, and hence this curve corresponds to a quasi-pulsed *I-V* curve with pulse width of 10^{-8} s. (In this figure, for reference, we are also plotting other quasi-pulsed *I-V* curves (○, Δ) when only V_D is changed, which reflect overshoot and undershoot (cf. Fig.2).) It is seen that the drain currents in the pulsed *I-V* curves are rather lower than those in the steady state, and the current reduction due to drain lag is regarded as predominant in this case (V_{Doff} = 40 V). This clearly indicates that the current collapse could occur due to the slow response of deep levels in the buffer layer. This type of current reduction (current collapse) is commonly observed experimentally in AlGaN/GaN HEMTs.

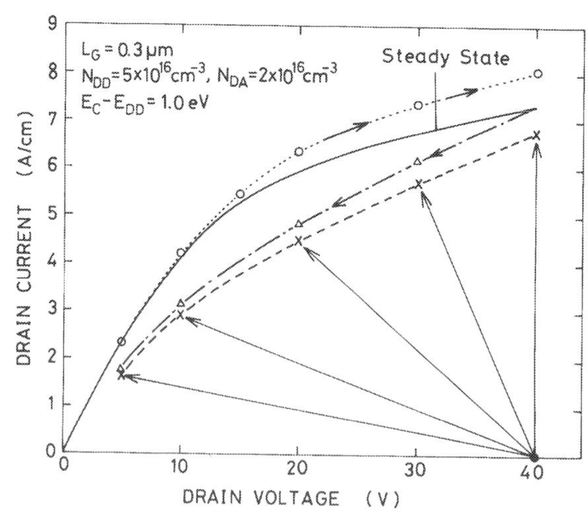

Fig. 7. Steady-state *I-V* curve (solid line) and quasi-pulsed *I-V* curves of normal AlGaN/GaN HEMT, with L_{GD} = 1 μm, N_{DD} = 5x10^{16} cm^{-3}, N_{DA} = 2x10^{16} cm^{-3} and E_C - E_{DD} = 1.0 eV. (x): V_{Doff} = 40 V and V_{Goff} = V_{th} (t = 10^{-8} s), (○): V_D is raised from 0 V (t = 10^{-9} s), (Δ): V_D is lowered from 40 V (t = 10^{-8} s).

3.3 Current collapse

In this section, we describe dependence of current collapse on the impurity densities in the buffer layer and on the off-state drain voltage.

3.3.1 Dependence on deep-acceptor density

We have calculated dependence of drain-current responses and *I-V* curves on impurity densities in the buffer layer, and found that these are essentially determined by the deep-acceptor density N_{DA} when $N_{DD} > N_{DA}$ and $E_C - E_{DD}$ is the same. For example, *I-V* curves for $N_{DD} = 2\text{x}10^{17}$ cm^{-3} and $N_{DA} = 2\text{x}10^{16}$ cm^{-3} are almost the same as those for $N_{DD} = 5\text{x}10^{16}$ cm^{-3} and $N_{DA} = 2\text{x}10^{16}$ cm^{-3} shown in Fig.7 (The difference is 2 ~ 3 %). This is because when $N_{DD} > N_{DA}$, the ionized deep-donor density N_{DD}^+ becomes nearly equal to N_{DA} under equilibrium, and the space-charge region in the buffer layer consists of negative charges due to ionized deep acceptors. Therefore, we will show N_{DA} dependence of the characteristics.

Fig.8 shows calculated I_D-V_D curves of normal AlGaN/GaN HEMTs with different N_{DA} ($5\text{x}10^{15}$ cm^{-3}, 10^{17} cm^{-3}) in the buffer layer. Here, $N_{DD} = 2\text{x}10^{17}$ cm^{-3} and $E_C - E_{DD} = 1.0$ eV. The solid lines are steady-state *I-V* curves, and the dashed lines (x) are quasi-pulsed *I-V* curves with pulse width of 10^{-8} s, which are obtained from calculated turn-on characteristics, as described before. It is seen that for lower N_{DA}, the steady-state drain currents are higher, and the saturation behavior is poor. This is because a barrier between the channel and the buffer is less steep and the current component via the buffer layer becomes larger. This type of short-channel effect is recently discussed in (Uren et al., 2006). Combined with Fig.7, it is also clearly seen that the current collapse is more pronounced for higher N_{DA}. This is because the ionized deep-donor density N_{DD}^+, which acts as an electron trap, is higher for higher N_{DA}, and hence the trapping effects (or the resulting current collapse) are more

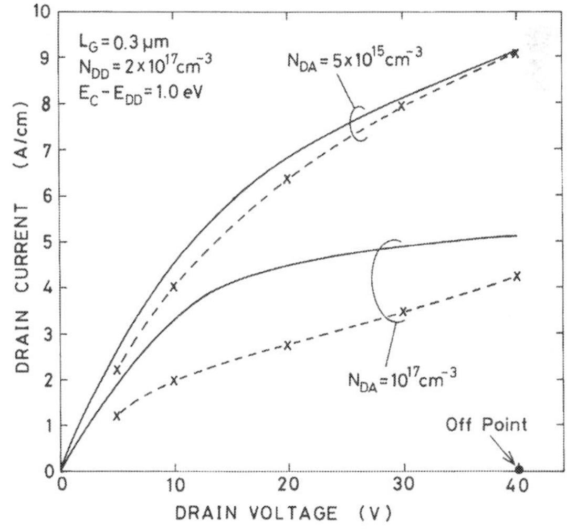

Fig. 8. Steady-state *I-V* curves ($V_G = 0$ V; solid lines) and quasi-pulsed *I-V* curves (x; $t = 10^{-8}$ s) for normal AlGaN/GaN HEMTs with different N_{DA}. Initial point is shown by (•).$L_{GD} = 1$ μm, $N_{DD} = 2\text{x}10^{17}$ cm^{-3} and $E_C - E_{DD} = 1.0$ eV.

significant for higher N_{DA}. Here, it should be mentioned that for lower N_{DA}, the current collapse could be reduced, but the threshold voltage shifts toward negative bias due to the higher buffer current, although this current should become small when L_G is long. Therefore, there may be a trade-off relationship between reducing current collapse and obtaining sharp current cutoff.

3.3.2 Dependence on off-state drain voltage

Finally, we describe dependence of current collapse on the off-state drain voltage V_{Doff}. Fig.9 shows a calculated steady-state I_D-V_D curve (solid line) and quasi-pulsed I-V curves (x; with pulse width of 10^{-8} s) which are derived from calculated turn-on characteristics. The parameter is V_{Doff}. Here, L_{GD} = 1.5 μm, N_{DD} = 2x10^{17} cm^{-3}, N_{DA} = 10^{17} cm^{-3} and E_C - E_{DD} = 1.0 eV. It is seen that for higher V_{Doff}, the drain currents in the pulsed I-V curves are lower at a given V_D. Therefore, it can be said that the current collapse is more pronounced when V_{Doff} is higher. Although some current reduction is seen when only V_G is changed (gate lag), the current reduction due to the change of V_D (drain lag) is regarded as predominant for the cases of higher V_{Doff}.

Fig.10 shows (a) electron density profiles and (b) N_{DD}^+ profiles in the off state for the two cases. The left is for V_{Doff} = 20 V (V_G = V_{th} = - 7.50 V) and the right is for V_{Doff} = 80 V (V_G = V_{th} = - 8.56 V). It is seen that for higher V_{Doff}, electron densities in the buffer layer are higher under the gate and the gate-to-drain region, because electrons are injected into the buffer layer by the applied drain voltage. These electrons are captured by the deep donors, and hence N_{DD}^+ decreases more heavily, as seen in Fig.10(b). Hence, when V_G is switched on and V_D is lowered from higher V_{Doff}, the drain current remains at a lower value. Therefore, the current collapse is more pronounced for higher V_{Doff}. The tendency shown in Fig.9 is also reported experimentally in AlGaN/GaN HEMTs (Koudymov et al., 2003).

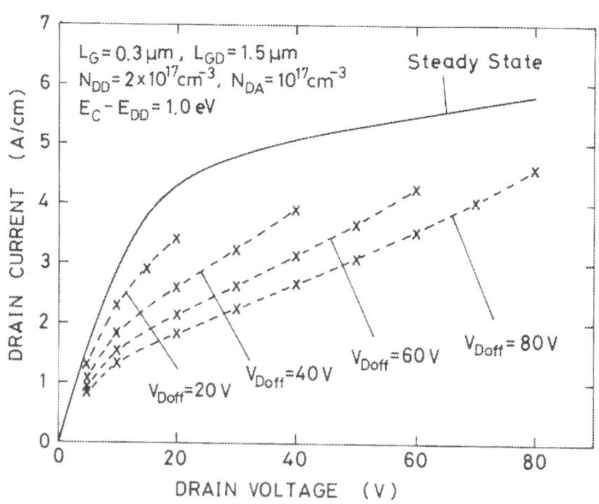

Fig. 9. Steady-state I-V curve (V_G = 0 V; solid line) and quasi-pulsed I-V curves (x; t = 10^{-8} s) for normal AlGaN/GaN HEMT, with off-state drain voltage V_{Doff} as a parameter. L_{GD} = 1.5 μm, N_{DD} = 2x10^{17} cm^{-3}, N_{DA} = 10^{17} cm^{-3} and E_C - E_{DD} = 1.0 eV.

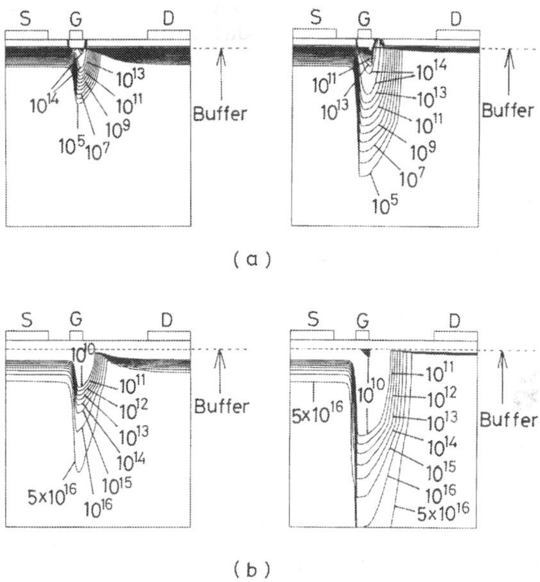

Fig. 10. (a) Electron density profiles and (b) N_{DD}^+ profiles in the off state. The left is for V_{Doff} = 20 V and the right is for V_{Doff} = 80 V. N_{DD} = 2x10^{17} cm^{-3}, N_{DA} = 10^{17} cm^{-3} and E_C - E_{DD} = 1.0 eV.

4. Effects of Field Plate

4.1 Drain lag

Next, we describe effects of a field plate. First, we discuss about the drain lag. Fig.11 shows a comparison of calculated drain-current responses of AlGaN/GaN HEMTs (N_{DD} = 2x10^{17} cm^{-3}, N_{DA} = 10^{17} cm^{-3}, E_C – E_{DD} = 0.5 eV) when V_D is lowered abruptly from V_{Dini} (40 V) to V_{Dfin} abruptly, where the gate voltage V_G is kept constant at 0 V. Fig.11(a) is for the case without a field plate (L_{FP} = 0) and Fig.11(b) is for the field-plate structure (L_{FP} = 1μm). In Fig.11, the thickness of SiN layer d is 0.03μm. In both cases with and without a field plate, the drain currents remain at low values for some periods and begin to increase slowly, showing drain-lag behavior. It is understood that the drain currents begin to increase when the deep donors in the buffer layer begin to emit electrons. This electron emission occurs because for higher V_D, more electrons are injected into the buffer layer and captured by deep donors, leading to a more negatively charged buffer layer. It is seen that the change of drain current is smaller for the cases with a field plate, indicating that the drain lag is smaller for the field-plate structure. We will discuss below why the drain lag or the trapping effect becomes smaller in the field-plate structure.

Fig.12 shows (a) electron density profiles and (b) ionized deep-donor density N_{DD}^+ profiles at V_G = 0 V and V_D = 40 V for the AlGaN/GaN HEMTs. The left is for the structure without a field plate, and the right is for the field-plate structure. In Fig.12(a), we see that for the structure without a field plate, electrons are injected deeper into the buffer layer under the

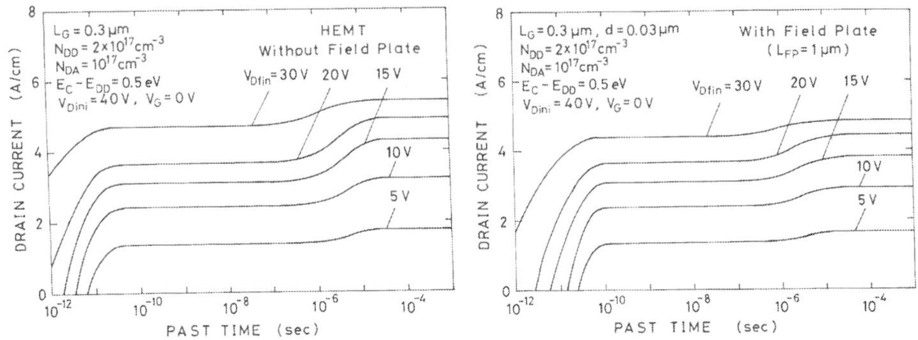

(a) (b)

Fig. 11. Calculated drain-current responses of AlGaN/GaN HEMTs when V_D is lowered abruptly from 40 V to V_{Dfin}, while V_G is kept constant at 0 V. N_{DD} = 2x10^{17} cm^{-3}, N_{DA} = 10^{17} cm^{-3} and $E_C - E_{DD}$ = 0.5 eV. (a) Without a field plate, (b) with 1 μm-length field plate.

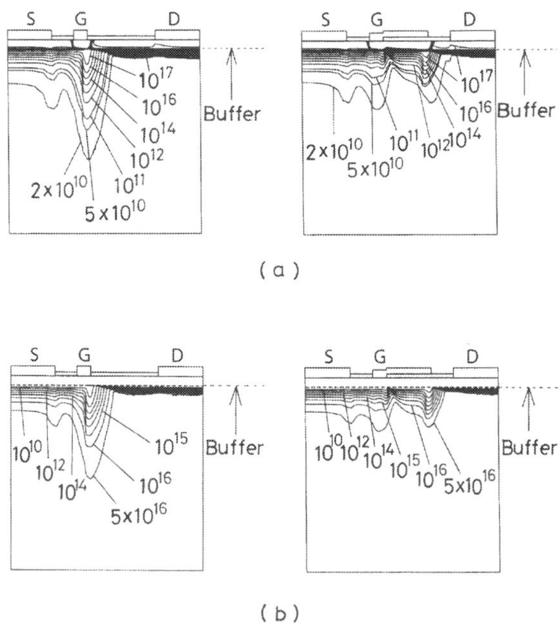

Fig. 12. (a) Electron density profiles and (b) ionized deep-donor density N_{DD}^+ profiles at V_G = 0 V and V_D = 40 V for AlGaN/GaN HEMTs. d = 0.03 μm. N_{DD} = 2x10^{17} cm^{-3}, N_{DA} = 10^{17} cm^{-3} and $E_C - E_{DD}$ = 0.5 eV. The left is for the case without a field plate, and the right is for the field-plate structure (L_{FP} = 1μm).

gate region. These electrons are captured by the deep donors, and hence N_{DD}^+ decreases there as seen in Fig.12(b). As mentioned before, when V_D is lowered abruptly, the drain currents remain at low values for some periods and begin to increase slowly as the deep donors begin to emit electrons (and N_{DD}^+ increases), resulting in the drain lag. In the case of field-plate structure, as seen in Fig.12(a), electrons are injected into the buffer layer under the drain edge of field plate as well as under the gate. But the overall injection depth is not so deep as compared to the case without a field plate. Hence, the change of N_{DD}^+ by capturing electrons is smaller for the field-plate structure as seen in Fig.12(b). This occurs because the electric fields at the drain edge of the gate become weaker by introducing a field plate, as shown in Fig.13. (Note that in the field-plate structure, the electric fields at the drain edge of the field plate can be strong for thin d.) Therefore, the drain lag becomes smaller for the field-plate structure.

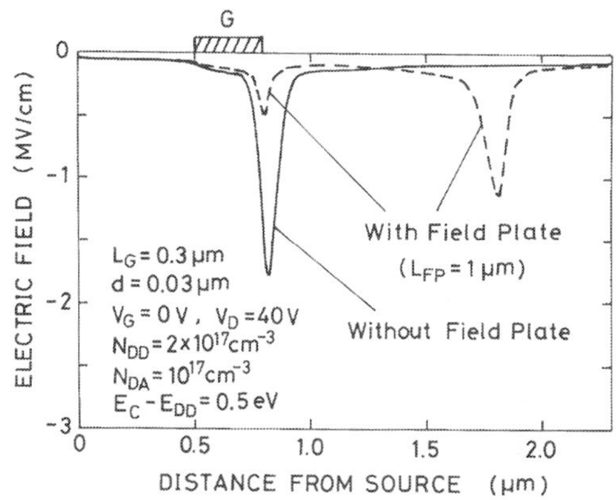

Fig. 13. Comparison of longitudinal electric field profiles at the channel side of heterojunction in AlGaN/GaN HEMTs with and without a field plate. V_G = 0 V and V_D = 40 V. d = 0.03 μm. N_{DD} = 2x10^{17} cm^{-3}, N_{DA} = 10^{17} cm^{-3} and $E_C - E_{DD}$ = 0.5 eV.

4.2 Pulsed *I-V* curves and current collapse

Next, we have calculated a case when V_G is also changed from an off point. V_G is changed from the threshold voltage V_{th} to 0 V, and V_D is changed from V_{Dini} (40 V) to V_{Don} (on-state drain voltage). The characteristics become similar to those in Fig.11, although some transients arise when only V_G is changed (gate lag). From the transient characteristics, we obtain quasi-pulsed *I-V* curves.

Fig.14 shows calculated drain current I_D - drain voltage V_D curves of AlGaN/GaN HEMTs. Fig.14(a) is for the structure without a field plate, and Fig.14(b) is for the field-plate structure (L_{FP} = 1 μm). The solid lines are steady-state *I-V* curves. In these figures, we plot by (Δ) the drain current at t = 10^{-8} s after V_D is lowered. This is obtained from the previous transient characteristics (Fig.11), and this curve (dashed line) is regarded as a quasi-pulsed *I-V* curve with pulse width of 10^{-8} s. This reduction in the drain current indicates the drain-lag behavior. In these figures, we also plot by (x) another pulsed *I-V* curve when V_G is switched

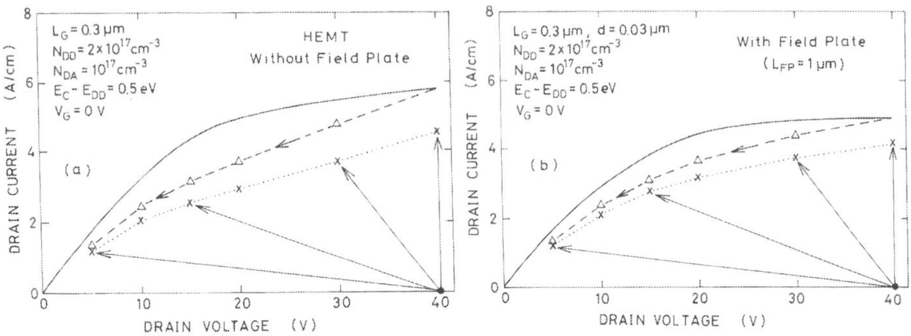

(a) (b)

Fig.14. Steady-state *I-V* curves (V_G = 0 V; solid lines) and quasi-pulsed *I-V* curves (Δ, x) of AlGaN/GaN HEMTs. (a) Without a field plate, (b) with 1 μm-length field plate. (Δ): Only V_D is changed from 40V ($t = 10^{-8}$ s), (x): V_D is lowered from 40 V and V_G is changed from V_{th} to 0 V ($t = 10^{-8}$ s).

on from V_{th} to 0 V. The drain current in this case is further reduced, indicating gate-lag and current collapse behavior. The current collapse is a combined effect of drain lag and gate lag. By comparing the cases with and without a field plate, we can definitely say that the lag phenomena (drain lag, gate lag) and current collapse become smaller for the field-plate structure.

4.3 Dependence on SiN layer thickness

Finally, we have studied how the lag phenomena and current collapse depend on the field-plate length L_{FP} and the SiN layer thickness *d*. We have found that the lag phenomena and current collapse become smaller when L_{FP} becomes longer in the range from 0 to 1 μm. This may be easily understood, because trapping effects should be reduced for longer L_{FP}. Hence, we will here focus on the dependence of lag phenomena and current collapse on the SiN layer thickness *d*.

Fig.15 shows drain-current reduction rate $\Delta I_D/I_D$ (ΔI_D : current reduction, I_D : steady-state current) due to current collapse, drain lag or gate lag as a function of *d*, where L_G = 0.3 μm and L_{FP} = 1 μm. Here *d* = 0 corresponds to a case of L_G = 1.3 μm without a field plate. The data in Fig.15 are for the field-plate structures, except for *d* = 0. When *d* is thick, the current collapse and lag phenomena are relatively large. As *d* becomes thinner, the current collapse and lag phenomena become smaller. This is because the buffer-trapping effects are reduced in the field-plate structure, as described in sections 4.1 and 4.2. However, the rates of current collapse and drain lag increase when *d* becomes very thin. This is understood that for very thin *d*, the electric field at the drain edge of the field plate becomes very strong, and electrons are injected deeper into the buffer layer under the field-plate region, contributing to the current collapse and drain lag. When *d* = 0 (L_G = 1.3 μm), that is, without a field plate, the current collapse becomes rather large. From Fig.15, we can say that there is an optimum thickness of SiN layer to minimize the buffer-related current collapse and drain lag in AlGaN/GaN HEMTs.

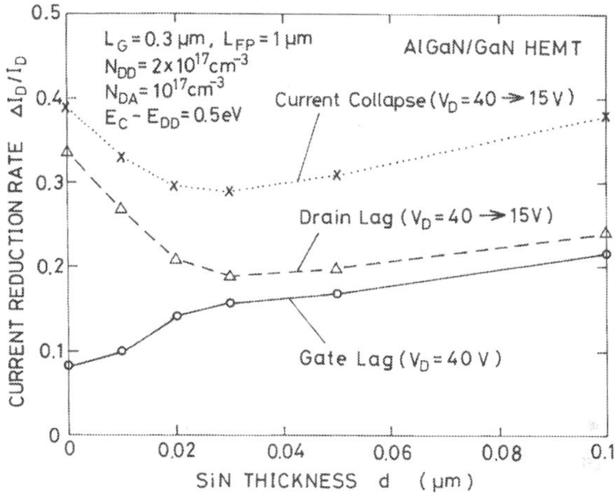

Fig. 15. Current reduction rate $\Delta I_D/I_D$ due to current collapse, drain lag or gate lag for field-plate AlGaN/GaN HEMTs as a function of SiN thickness d. L_{FP} = 1μm.

5. Conclusion

Two-dimensional transient analyses of AlGaN/GaN HEMTs have been performed in which the three level compensation model is adopted for the buffer layer, where a shallow donor, a deep donor and a deep acceptor are considered. Quasi-pulsed *I-V* curves have been derived from the transient characteristics, and compared with steady-state *I-V* curves. It has been shown that the lag phenomena and current collapse could be reproduced, as in GaN MESFETs (Horio et al., 2005).

When the drain voltage is lowered abruptly, the drain currents remain at low values for some periods and begin to increase slowly as the deep donors begin to emit electrons, showing drain-lag behavior. As compared with the case of GaN MESFETs, relatively large gate lag due to buffer traps could arise, and it is correlated to relatively high source access resistance in AlGaN/GaN HEMTs. Both the drain lag and the gate lag become causes of current collapse. The current collapse has been shown to be more pronounced when the deep-acceptor density in the buffer layer is higher and when an off-state drain voltage is higher, because the trapping effects become more significant. The drain lag could be a major cause of current collapse in the case of higher off-state drain voltage.

It is concluded that to minimize current collapse in AlGaN/GaN HEMTs, an acceptor density in the buffer layer should be made low, although the buffer current becomes large and short-channel effects become significant when the gate length is short. Some ways to reduce the short-channel effects such as using a heterojunction buffer may be needed. Here, it should be mentioned that to clarify surface-state effects in AlGaN/GaN HEMTs, which are not considered here, is an important task yet to be done.

Effects of introducing a field plate have also been studied. It has been shown that the drain lag is reduced by introducing a field plate because the electric field at the drain edge of the gate becomes weaker and electron injection into the buffer layer is reduced, resulting in the smaller buffer-trapping effects. It has also been shown that the buffer-related current collapse and gate lag are reduced in the field-plate structure. The dependence of lag phenomena and current collapse on SiN layer thickness has also been studied, indicating that there is an optimum thickness of SiN layer to minimize the buffer-related current collapse and drain lag in AlGaN/GaN HEMTs.

6. References

Ando, Y.; Okajima, Y.; Miyamoto, H.; Nakayama, T.; Inoue, T. & Kuzuhara, M. (2003). 10-W/mm AlGaN-GaN HFET with a field modulating plate. IEEE Electron Device Lett., Vol.24, No.5, pp.289-91.

Binari, S. C.; Klein, P. B. & Kazior, T. E. (2002). Trapping effects in GaN and SiC Microwave FETs. Proc. IEEE, Vol.90, pp.1048-1058.

Braga, N.; Mickevicins, R.; Gaska, R.; Shur, M. S.; Khan, M. A. & Simin, G. (2004). Simulation of gate lag and current collapse in GaN heterojunction field effect transistors. Proceedings of IEEE CSIC Symp., pp.287-290.

Daumiller, I.; Theron, D.; Gaquiere, C.; Vescan, A.; Dietrich, R.; Wieszt, A.; Leiter, H.; Vetury, R.; Mishra, U. K.; Smorchkova, I. P.; Keller, S.; Nguyen, N. X.; Nguyen, C. & Kohn, E. (2001). Current instabilities in GaN-based devices. IEEE Electron Device Lett., Vol.22, No.2, pp.62-64.

Desmaris, V.; Rudzinski, M.; Rorsman, N.; Hageman, P. R.; Larsen, P. K.; Zirath, H.; Rodle, T. C. & Jos, H. F. F. (2006). Comparison of the dc and microwave performance of AlGaN/GaN HEMTs grown on SiC by MOCVD with Fe-doped or unintentionally doped GaN buffer layer. IEEE Trans. Electron Devices, Vol.53, No.9, pp.2413-2417.

Higashiwaki, M.; Matsui, T. & Mimura, T. (2006). AlGaN/GaN MIS-HFETs with f_T of 163 GHz using CAT-CVD SiN gate-insulating and passivation layers. IEEE Electron Device Lett., Vol.27, No.1, pp.16-18.

Horio, K. & Fuseya, Y. (1994). Two-dimensional simulations of drain-current transients in GaAs MESFET's with semi-insulating substrates compensated by deep levels. IEEE Trans. Electron Devices, Vol.41, No.8, pp.1340-1346.

Horio, K.; Wakabayashi, A. & Yamada, T. (2000). Two-dimensional analysis of substrate-trap effects on turn-on characteristics in GaAs MESFET's. IEEE Trans. Electron Devices, Vol.47, No.3, pp.617-624.

Horio, K.; Yonemoto, K.; Takayanagi, H. & Nakano, H. (2005). Physics-based simulation of buffer-trapping effects on slow current transients and current collapse in GaN field effect transistors. J. Appl. Phys., Vol.98, No.12, pp.124502 1-7.

Horio, K. & Nakajima, A. (2008). Physical mechanism of buffer-related current transients and current slump in AlGaN/GaN high electron mobility transistors. Jpn. J. Appl. Phys., Vol.47, No.5, pp.3428-3433.

Karmalkar, S. & Mishra, U. K. (2001). Enhancement of breakdown voltage in AlGaN/GaN high electron mobility transistors using a field plate. IEEE Trans. Electron Devices, Vol.48, No.8, pp.1515-1521.

Khan, M. A.; Shur, M. S.; Chen, Q. C. & Kuznia, J. N. (1994). Current/voltage characteristics collapse in AlGaN/GaN heterostructure insulated gate field effect transistors at high drain bias. Electron Lett., Vol.30, pp.2175-2176.

Klein, P. B.; Freitas, Jr., J. A.; Binari, S. C. & Wickenden, A. E. (1999). Observation of deep traps responsible for current collapse in GaN metal–semiconductor field-effect transistors. Appl. Phys. Lett., Vol.75, No.25, pp.4016-4018.

Koley, G.; Tilak, V.; Eastman, L. F. & Spencer, M. G. (2003). Slow transients observed in AlGaN/GaN HFETs: Effects of SiNx passivation and UV illumination. IEEE Trans. Electron Devices, Vol.50, No.4, pp.886-893.

Koudymov, A.; Simin, G.; Khan, M. A.; Tarakji, A.; Gaska, R. & Shur, M. S. (2003). Dynamic current-voltage characteristics of III-N HFETs. IEEE Electron Device Lett., Vol.24, pp.680-682.

Koudymov, A.; Adivarahan, V.; Yang, J.; Simon, G. & Khan, M. A. (2005). Mechanism of current collapse removal in field-plated nitride HFETs. IEEE Electron Device Lett. Vol.26, pp.704-706.

Kruppa, W.; Binari, S. C. & Doverspike, K. (1995). Low-frequency dispersion characteristics of GaN HFETs. Electron. Lett., Vol.31, pp.1951-1952.

Meneghesso, G.; Verzellesi, G.; Pierobon, R.; Rarnpazzo, F.; Chini, A.; Mishra, U. K.; Canali, C. & Zanoni, E. (2004). Surface-related drain current dispersion effects in AlGaN/GaN HEMTs. IEEE Trans. Electron Devices, Vol.51, pp.1554-1561.

Mishra, U. K.; Shen, L.; Kazior, T. E. & Wu, Y.-F. (2008). GaN-based RF power devices and amplifiers. Proc. IEEE, Vol.96, No.2 , pp.287-305.

Mizutani, T.; Ohno, Y.; Akita, M.; Kishimoto, S. & Maezawa, K. (2003). A study on current collapse in AlGaN/GaN HEMTs induced by bias stress. IEEE Trans. Electron Devices, Vol.50, No.10, pp.2015-2020.

Morkoc, H. (1999). Nitride Semiconductors and Devices, Springer-Verlag.

Nakajima, A.; Itagaki, K. & Horio, K. (2009). Reduction of buffer-related current collapse in field-plate AlGaN/GaN HEMTs. phys. stat. soli (c), Vol.6, No.S2, pp.S929-S932.

Pala, N.; Hu, X.; Deng, J.; Yang, J.; Koudymov, A.; Shur, M. S. & Simin, G. (2008). Drain-to-gate field engineering for improved frequency response of GaN-based HEMTs. Solid-State Electron., Vol.52, No.8, pp.1217-1220.

Palacios, T.; Rajan, S.; Chakrabrty, A.; Heikman, S.; Keller, S.; DenBaars, S. P. & Mishra, U. K. (2005). Influence of the dynamic access resistance in the g_m and f_T linearity of AlGaN/GaN HEMTs. IEEE Trans. Electron Devices, Vol.52, No.10, pp.2117-2123.

Saito, W.; Kuraguchi, M.; Takada, Y.; Tuda, K.; Omura, I. & Ogura, T. (2005). Design optimization of high breakdown voltage AlGaN-GaN power HEMT on an insulating substrate for $R_{ON}A$-V_B tradeoff characteristics. IEEE Trans. Electron Devices, Vol.52, No.1, pp.106-111.

Tirado, J.; Sanchez-Rojas, J. L. & Izpura, J. I. (2007). Trapping Effects in the transient response of AlGaN/GaN HEMT devices. IEEE Trans. Electron Devices, Vol.54, pp.410-417.

Uren, M. J.; Nash, K. J.; Balmer, R. S.; Martin, T.; Morvan, E.; Caillas, N.; Delage, S. L.; Ducatteau, D.; Grimbert, B. & Jaeger, J. C. De (2006). Punch-through in short-channel AlGaN/GaN HFETs. IEEE Trans. Electron Devices, Vol.53, No.2 pp.395-398.

Vetury, R.; Zhang, N. Q.; Keller, S. & Mishra, U. K. (2001). The impact of surface states on the dc and RF characteristics of AlGaN-GaN HFETs. IEEE Trans. Electron Devices, Vol.48, No.3, pp.560-566.

Wu, Y.-F.; Saxler, A.; Moore, M.; Smith, R. P.; Sheppard, S.; Chavarkar, P. M.; Wisleder, T.; Mishra, U. K. & Parikh, P. (2004). 30 W/mm GaN HEMT by field plate optimization. IEEE Electron Device Lett., Vol.23, No.3, pp.117-119.

Xing, H. X.; Dora, Y.; Chini, A.; Heikman, S.; Keller, S. & Mishra, U. K. (2004). High breakdown voltage AlGaN-GaN HEMTs achieved by multiple field plates. IEEE Electron Device Lett., Vol.25, No.4, pp.161-163.

5

Study of Plasma Effects in HEMT-like Structures for THz Applications by Equivalent Circuit Approach

Irina Khmyrova
University of Aizu
Japan

1. Introduction

The growing interest to terahertz (THz) region of electromagnetic spectrum is pulled by a variety of its possible applications for free-space communications, sensing and imaging in radio astronomy, biomedicine, and in security screening for hidden explosives and concealed weapons. Terahertz imaging may also be useful for industrial processes, such as package inspection and quality control. Despite strong demand in compact solid-state devices capable to operate as emitters, receivers, photomixers of the THz radiation their development is still a challenging problem.

Plasma waves with linear dispersion law can exist in the gated two-dimensional electron gas (2DEG) (Chaplik 1972) in systems similar to field-effect transistor (FET) and high-electron mobility transistor (HEMT). At high enough electron mobility and gate length in submicrometer range 2DEG channel can serve as a resonant cavity for plasma waves with the resonant frequencies in the THz range. Experimentally observed infrared absorption (Allen,1977) and weak infrared emission (Tsui,1980) were related to plasma waves in silicon inversion layers. Excitation of plasma oscillations in the channel of FET-like structures has been proposed as a promising approach for the realization of emission, detection, mixing and frequency multiplication of THz radiation (Dyakonov,1993, Dyakonov,1996). Nonresonant (Weikle,1996) and weak resonant (Lu,1998) detection have been observed in HEMTs experimentally. Resonant peaks of the impedance of the capacitively contacted 2DEG at frequencies corresponding to plasma resonances have been revealed (Burke,2000). Terahertz detection and emission in HEMTs fabricated from different materials have been demonstrated (Knap,2002; Otsuji,2004; Teppe,2005; El Fatimy,2006; Shur,2003).

Theoretical models for plasma waves excited in the HEMT 2DEG channel are usually based on the similarity of the equations describing the behavior of electron fluid and shallow water (Dyakonov,1993; Shur, 2003; Satou,2003; Satou,2004; Veksler,2006). On the other hand, electromagnetic wave propagation in the gated 2DEG channel is similar to that in a transmission line (TL) (Yeager,1986; Burke,2000) which makes it possible to represent the gated portion of 2DEG channel by a TL model. In this chapter, we will implement distributed circuit or TL model approach to the study of plasma waves excited in the 2DEG channel of the structures similar to HEMT. Once a system is represented by an electric equivalent circuit, its performance can be simulated using a circuit simulator like SPICE (Simulation Program with Integrated Circuit Emphasis). Such an approach is less time consuming comparing to full scale computer modeling. It enables an easy variation of the system parameters during the simulation procedure provides a quick way to facilitate and improve one's understanding of the HEMT operation

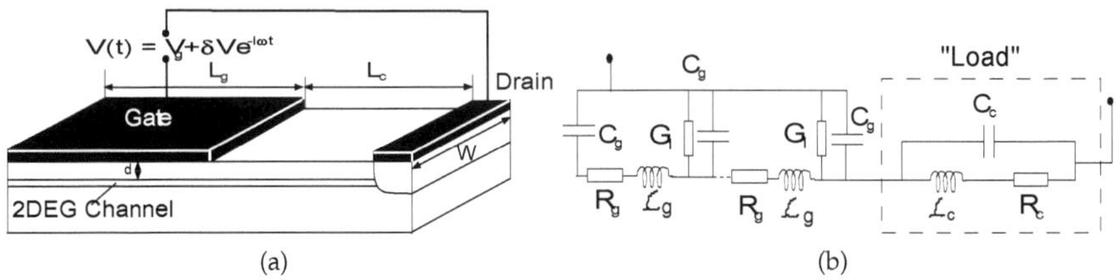

Fig. 1. Schematic structure of the HEMT (a) and its electric equivalent circuit (b). Gated portion of the 2DEG channel is represented by distributed $R\mathcal{L}C$ circuit. The rest part of the device including ungated and contact regions is enclosed in a dashed box and treated as a 'load".

in the regime of excitation of plasma wave oscillations (Khmyrova,2007). The rest of this chapter is organized as follows. In the Section 2.1 the basic electric equivalent circuit for the HEMT-like structure operating in the regime of the excitation of plasma wave oscillations is developed. Results of IsSpice simulation illustrating the dependence of resonant plasma frequency on different structure parameters are presented in Section 2.2. In the Section 2.3 load termination concept is applied to estimate reflection coefficient at the interface between gated and ungated portions of the HEMT 2DEG channel.
Section 3 focuses on the influence of the fringing effects on resonant frequency of plasma oscillations. Section 3.1 presents analytical model which allows to evaluate electric field distribution at the 2DEG surface, spatial distribution of sheet electron density, and, finally, resonant frequency of plasma oscillations in the presence of fringing effects. Section 3.2 discusses cascaded TL model and results of IsSpice simulation.
In the Section 4 results of the experiments and different analytical models are compared. Chapter summary is given in Section 5.

2. Distributed circuit approach to analysis of plasma oscillations in HEMT-like structures

2.1 Development of basic electric equivalent circuit

We consider a HEMT with schematic structure shown in Fig. 1a with a 2DEG channel formed at the heterointerface between the InGaAs narrow-gap and InAlAs wide-gap layers. Voltage $V(t) = V_g + \delta V e^{i\omega t}$ applied between the gate and drain contacts contains dc and ac components with amplitudes V_g and δV, respectively, and ac signal frequency ω. 2DEG channel beneath the gate contact of length L_g can act as a resonant cavity for the plasma waves with the fundamental resonant frequency (Dyakonov1993

$$\Omega = \frac{\pi}{2L_g}\sqrt{\frac{e^2\Sigma\, d_g}{\varepsilon_0 \varepsilon m^*}} \tag{1}$$

where e and m^* are the electron charge and effective mass, respectively, ε_0 and ε are dielectric constants of vacuum and the layer separating 2DEG channel and gate contact, thickness of the layer is d_g. 2DEG sheet electron density Σ depends on the gate bias voltage $V(t)$. In our basic equivalent circuit model we will neglect nonlinear effects and consider its gate voltage dependence in the form:

$$\Sigma = \Sigma_0 + \frac{\varepsilon_0 \varepsilon V_g}{e d_g} = \Sigma_0 (1 - \frac{V_g}{V_{th}}), \tag{2}$$

where Σ_0 is electron sheet concentration at $V_g = 0$, and V_{th} is the threshold voltage. As it follows from Eqs. (1) and (2), resonant frequencies of plasma oscillations in the 2DEG channel depend on the gate length and can be tuned by the gate bias voltage.

To express the components of electrical equivalent circuit in terms of physical parameters of the 2DEG system one may invert the Drude formula for the frequency dependent conductivity of the 2DEG and obtain its complex resistivity $\rho(\omega)$:

$$\rho(\omega) = \frac{m^*}{\Sigma e^2 \tau_{tr}} (1 + i\omega\tau_{tr}). \tag{3}$$

where $\tau_{tr} = \frac{m^*}{e}\mu$ is the transport or momentum scattering time and μ is electron mobility in the channel. The 2DEG complex resistivity contains purely resistive as well as inductive components (first and second terms in the right-hand side of Eq. (3), respectively). Therefore, to provide correct equivalent circuit representation of the system in question one should include not only the resistance of the 2DEG channel but inevitably its kinetic inductance (Burke,2000) associated with the inertia of the electrons in it. Furthermore, both resistance and inductance will depend on the gate bias voltage. For proper description of the system this fact should be also accounted for. Due to similarity of electromagnetic wave propagation in the gated 2DEG channel to that in a transmission line the distributed RC-circuit topology has been proposed for modeling of high-frequency effects in the gated 2DEG (Yeager, 1986). Later kinetic inductance has been added (Burke,2000) to distributed RC-circuit model modifying it into $R\mathcal{L}C$ distributed circuit model which we will use.

Combining Eqs. (2) and (3), for the distributed resistance R_g and kinetic inductance $\mathcal{L}_g$ of the gated portion of 2DEG channel we obtain the following expressions:

$$R_g = \frac{1}{e\mu W \Sigma_0 (1 - \frac{V_g}{V_{th}})} = \frac{R_0}{1 - \frac{V_g}{V_{th}}}, \tag{4}$$

$$\mathcal{L}_g = \frac{m^*}{e^2 W \Sigma_0 (1 - \frac{V_g}{V_{th}})} = \frac{\mathcal{L}_0}{1 - \frac{V_g}{V_{th}}}, \tag{5}$$

where W is the width of the device, R_0 and $\mathcal{L}_0$ are 2DEG channel resistance and inductance per unit length at $V_g = 0$. A distinctive feature of our equivalent circuit model is that it takes into account the dependence of the resistance R_g and inductance $\mathcal{L}_g$ on the gate bias voltage V_g (Khmyrova,2007). In this section the gate contact-2DEG channel system is considered as an ideal, i.e., fringing effects are neglected. Under this assumption its distributed capacitance can be expressed in the standard form:

$$C_g = \frac{\varepsilon_0 \varepsilon W}{d_g}, \tag{6}$$

To account for the losses due to, for example, leakage through the dielectric under the gate contact, relevant conductance G_l should be added in parallel with capacitance C_g.

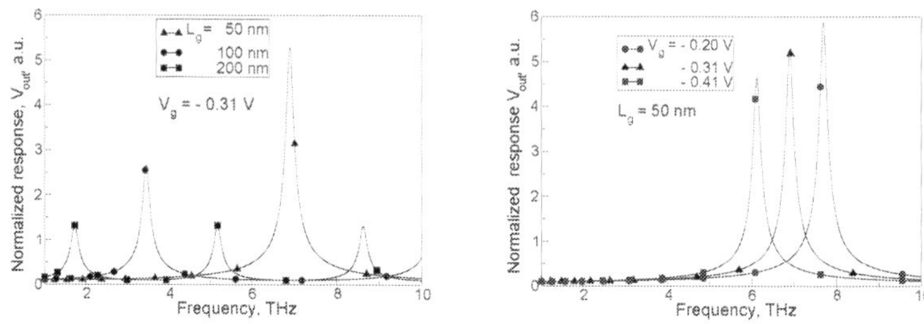

(a) (b)

Fig. 2. Normalized frequency response of the HEMT at (a) different gate lengths and $V_g = -0.31$ V; and (b) different gate bias voltages and $L_g = 50$ nm.

2.2 Results of IsSpice simulation

The developed distributed equivalent circuit of the gated 2DEG channel with $R\mathcal{L}C$-components described by Eqs. (4)-(6) is shown in Fig. 1b. It was used to simulate frequency performance of the InAlAs/InGaAs HEMT with IsSpice software which is the version of SPICE developed by Intusoft. In the simulation experiment the gated part of the 2DEG channel has been represented by a lossy transmission line component, chosen from IsSpice component library. R_g, $\mathcal{L}_g$ and C_g of the TL were calculated using Eqs. (4)-(6) and device geometrical and physical parameters listed in Table 1. First, we neglect leakage losses setting $G_l = 0$ and consider "open circuit" configuration, i.e., assume that part of the device adjacent to its gated 2DEG portion has very high resistance.

Normalized frequency response V_{out} simulated at different gate lengths L_g and $V_g = -0.31$ V is shown in Fig. 2a. Fig. 2a reveals pronounced resonant behavior with resonant peaks at frequencies corresponding to those determined by Eq. (1). The increase of the gate length L_g results in the fundamental resonant frequency reduction in line with Eq. (1). The decrease of the gate bias voltage V_g at a fixed gate length results in a decrease of the electron concentration beneath the gate contact which, in turn, leads to a resonant frequency reduction as it is shown in Fig. 2b. In other words, Fig. 2b illustrates the possibility to tune the frequency of plasma oscillations by the gate bias voltage.

Simulation using equivalent circuit approach makes it possible to evaluate quickly the influence of such factors as leakage through the dielectric separating the gate contact and 2DEG channel on the HEMT performance in the regime of excitation of plasma oscillations. Fig. 3 demonstrates the damping of oscillatory behavior of the response caused by the leakage through the dielectric separating the gate contact and 2DEG. When the conductivity across this layer increases, say, from $G_l/G_g = 0$ to $G_l/G_g = 2$ (as in Fig. 3) amplitude of plasma oscillations in the HEMT 2DEG channel decreases.

In real HEMT structures ungated regions are rather large, usually their length $L_c >> L_g$. It was assumed that such ungated regions can influence plasma oscillations excited in the 2DEG channel (Satou,2003) as well as parasitic stray capacitance and load resistors. To complete the basic equivalent circuit model we include also lumped resistance and inductance of the

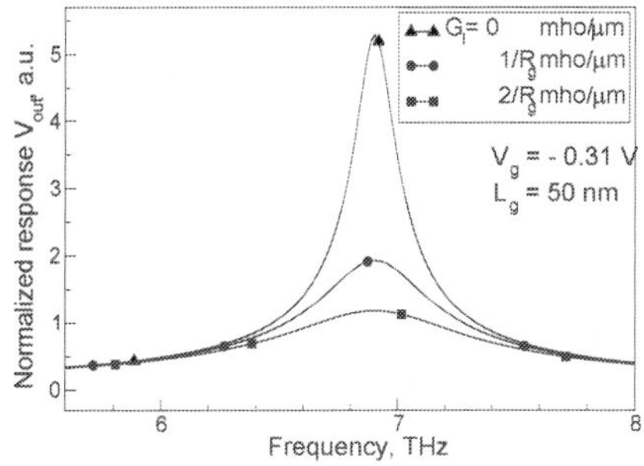

Fig. 3. Normalized HEMT frequency response at different leakage conductance G_l, $L_g = 50$ nm and $V_g = -0.31$ V.

ungated region of the 2DEG channel with the length L_c

$$R_c = R_0\, L_c, \qquad \mathcal{L}_c = \mathcal{L}_0\, L_c. \tag{7}$$

To account for parasitic capacitance a capacitor C_s has been added in parallel with resistance R_c and inductance $\mathcal{L}_c$.

Gate length	L_g	nm	50
Gate width	W	μ m	50
Thickness of the layer under the gate	d	nm	17
Length of ungated region	L_{un}	nm	100
Sheet electron density	Σ_d	cm^{-2}	5×10^{12}
Electron mobility	μ	$cm^2\ V^{-1}\ s^{-1}$	4×10^4
Threshold voltage	V_{th}	V	- 0.764
Dielectric constant	ε		12.7

Table 1. Structure parameters

2.3 Transmission line load termination concept and reflection coefficient at the gated/ungated 2DEG channel interface

Exploiting further the similarity to a transmission line, one may treat the part of the device including the ungated portion of the 2DEG channel, contacts, etc., (see Fig. 1b) as a "load" with the impedance

$$Z_L = \frac{R_c + j\omega\mathcal{L}_c}{j\omega C_s(R_c + j\omega\mathcal{L}_c + \frac{1}{j\omega C_s})} \tag{8}$$

According to the concept of a transmission line (Collin,2007) at its "load" termination reflection coefficient (the ratio of reflected and incident waves) is given by the following formula:

$\Gamma = \frac{Z_L - Z_g}{Z_L + Z_g}$, where $Z_g = \sqrt{\frac{R_g + j\omega\mathcal{L}_g}{j\omega C_g}}$ is the transmission line characteristic impedance in the absence of leakage losses $G_l = 0$. In the frequency range of interest and given structural parameters one may assume $R_g/\omega\mathcal{L}_g << 1$ and simplify characteristic impedance as

$$Z_g \simeq \sqrt{\frac{\mathcal{L}_g}{C_g}}\sqrt{1 - j\frac{R_g}{\omega\mathcal{L}_g}} \tag{9}$$

The modulus and phase of reflection coefficient Γ are shown in Figs. 4a and 4b, respectively, at different lengths of contact region L_C. At given structural parameters modulus of reflection coefficient does not approach zero, or, in other words, the "load" impedance is mismatched. In regular transmission lines any reflection at the load termination is undesirable and efforts are usually being made to eliminate it. On the contrary, for the excitation and build up of plasma waves in the system under discussion the "load" impedance mismatch seems to be beneficial.

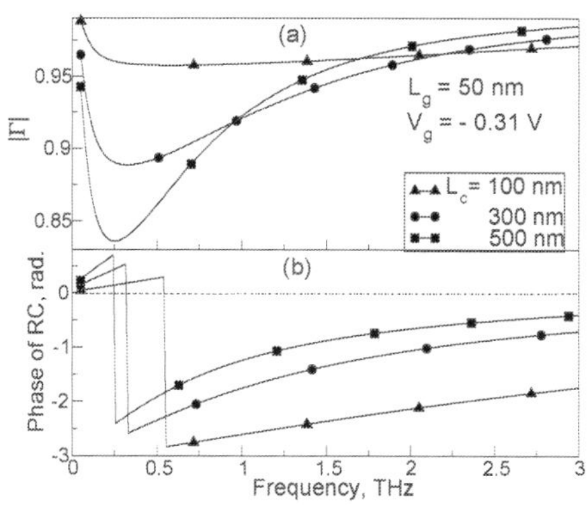

Fig. 4. Modulus (a) and phase (b) of the reflection coefficient at the load termination of the distributed circuit corresponding to the interface between gated and ungated parts of the 2DEG channel.

It was demonstrated with IsSpice simulation that the increasing length of ungated region causes the transformation of resonant peaks and shift of plasma resonances (Khmyrova,2007) although the effect was smaller comparing to that predicted by the model developed in (Satou,2003). Indeed, the situation in real devices is much more complicated. In particular, cap layer incorporated in the device structure to reduce an access source/drain contacts resistance may affect its frequency performance.

3. Impact of fringing effects on resonant frequencies of plasma oscillations

Another factor reducing the resonant plasma frequencies can be related to the nonideality of the gate contact–2DEG capacitance, i.e., fringing effects (Suemitsu,1998; Nishimura,2009). As it follows from Eq. (1) scaling down of the gate length L_g should result in the increase of the fundamental resonant plasma frequency. It seems quite natural that the gate length reduction should be accompanied by the relevant reduction of the thickness of the layer separating the gate contact and 2DEG surface. However, there are technological restrictions on the possible reduction of the thickness d_g. As a result, fringing of the electric field created by gate bias voltage may become sufficiently strong and its contribution should not be neglected.

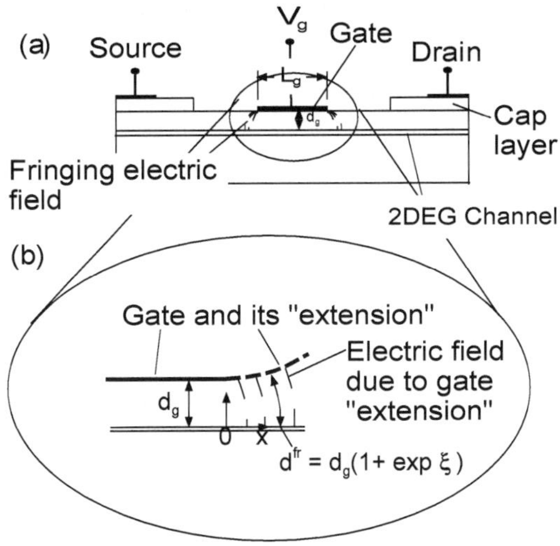

Fig. 5. Schematic of the HEMT structure (a) and model of "extended" gate (b).

3.1 Analytical model

We assume that width W of the HEMT-like structure schematically shown in Fig. 5a is large enough to provide the uniformity in the direction perpendicular to the page so that we can focus on the variation of the electric field and other related values only along x-axis indicated in Fig. 5b. Reference point $x = 0$ coincides with the edge of the gate contact, so that points with $-L_g < x < 0$ correspond to the 2DEG channel beneath the gate contact and $x > 0$ - to the ungated 2DEG channel region subjected to the fringing electric field. The thickness of the 2DEG channel is assumed to be negligibly small.

The distribution of fringing electric field at the 2DEG channel surface is similar to that in the middle plane of the fringed parallel plate capacitor with plate length L_g and separation $2d_g$. Using the conformal mapping approach elaborated in Ref. (Morse,1953) we can express in parametric form the fringing electric field distribution

$$E(\xi) = \frac{V_g}{d_g} \frac{1}{1 + \exp \xi}, \tag{10}$$

and x-coordinate

$$x = \frac{d_g}{\pi L_g}(1 + \xi + \exp \xi), \tag{11}$$

where ξ is a parameter and V_g is the gate bias voltage.

It is worth to note that electric field distribution in the structure with the gate contact extended in the direction $x > 0$ in such a way that separation between the gate contact and 2DEG surface varies as $d^{fr} = d_g(1 + \exp\xi)$ (see Fig. 5(b)) will be also described by Eq. (10) and be similar to fringing field. Furthermore, the ungated part of 2DEG channel subjected to the fringing electric field can be treated in the same way as its gated portion. For the 2DEG sheet electron density in the ungated fringed region one can write down an expression similar to Eq. (2):

$$\Sigma_{fr} = \Sigma_0^{fr} + \frac{\varepsilon\varepsilon_0 V_g}{ed_g(1+\exp\xi)}. \tag{12}$$

Here Σ_0^{fr} is sheet electron density in the fringed region at $V_g = 0$. The threshold voltage in the fringed region can be expressed as

$$V_{th}^{fr}(\xi) = -\frac{e\Sigma_0^{fr} d_g(1+\exp\xi)}{\varepsilon\varepsilon_0} = \frac{\Sigma_0^{fr}}{\Sigma_0} V_{th}^0 (1+\exp\xi), \tag{13}$$

where Σ_0 and V_{th}^0 are the surface charge density at $V_g = 0$ and threshold voltage for the 2DEG channel beneath the gate contact. On the other hand, threshold voltage is determined by physical and structural parameters (Delagebeaudeuf,1982): $V_{th} = \phi_M - \Delta E_c - V_p$, where ϕ_M is Schottky barrier height due to metalization of the gate contact, ΔE_c is conduction band discontinuity at the InAlAs-InGaAs heterointerface, and $V_p = e\Sigma_d d_d/\varepsilon\varepsilon_0$ for δ-doped HEMT (Mahajan,1998) where Σ_d and d_d are the dopant concentration and distance between the δ-doping plane and metal contact. In the fringed region the latter term is also position dependent and can be presented as $V_p^{fr}(\xi) = V_p^0(1+\exp\xi)$ resulting in the threshold voltage described by the following expression:

$$V_{th}^{fr}(\xi) = \phi_M - \Delta E_c - V_p^0(1+\exp\xi) = V_{th}^0 - V_p^0 \exp\xi, \tag{14}$$

which implies that to deplete 2DEG in the fringed region higher bias voltage should be applied to the gate contact. One can express Σ_0^{fr} from Eqs. (14) and (15), substitute it into Eq. (13) and arrive at

$$\Sigma_{fr} = \frac{\Sigma_0}{1+\exp\xi}\left(1 - \frac{V_g}{V_{th}^0}\right)\left(1 - \frac{V_p^0 \exp\xi}{V_{th}^0 - V_g}\right). \tag{15}$$

Fig. 6 shows spatial distribution of the sheet electron density Σ_{fr} for different ratios d_g/L_g at gate voltage $V_g = -0.1$ V, $\phi_M = 0.62$ V, $\Delta E_c = 0.53$ eV (Mahajan,1998), $\Sigma_d = 5 \times 10^{12}$ cm^{-2} and $d_d = 12$ nm (El Fatimy,2006).

It is seen from Fig. 6, that at small ratios $d_g/L_g << 1$ sheet electron density near the gate contact edge varies sharply, i.e., fringing of the electric field is small and can be neglected. Scaling down the gate length and gate-to-channel separation to $L_g = 50$ nm and $d_g = 17$ nm results in the ratio $d_g/L_g \simeq 0.34 < 1$. As one can see from Fig. 6 , at this ratio the fringing electric field extends at a distance L_{fr} up to several L_g. The ungated regions of length $L_c \leq L_{fr}$ being affected by the fringing field are no longer different from the gated region of the 2DEG channel and should be treated in the same way.

Thus, fringing electric field causes the extension of the 2DEG channel region controlled by the gate bias voltage V_g beyond the area covered by the gate contact. Moreover, as it is seen from Fig. 6, the fringing of the electric field influences the sheet electron density not only in the ungated regions on both sides of the gate contact, but also beneath the contact itself.

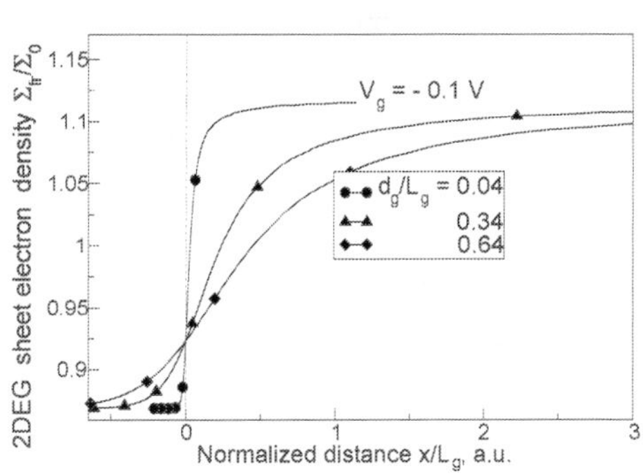

Fig. 6. 2DEG sheet electron density distribution along HEMT channel in the presence of fringing electric field at different ratios of d_g/L_g. Reference point $x = 0$ corresponds to the gate contact edge.

Due to symmetry of the HEMT structure one may consider only its half, i.e., $-0.5L_g \leq x \leq L_{fr}$ when calculating the phase change θ_{fr} of the wave traveling along the gated and fringed ungated regions (Andress,2005) of the 2DEG channel a distance $L_g + 2L_{fr}$

$$\theta_{fr} = 2\omega^{fr} \int_{x=-0.5}^{L_{fr}} \frac{dx}{s_{fr}}, \tag{16}$$

where wave's phase velocity can be expressed using Eq. (15) in the form

$$s_{fr} = \sqrt{\frac{e}{m^*}(V_g - V_{th}^0)(1 + \frac{V_p^0}{V_g - V_{th}^0} \exp\xi)} \tag{17}$$

Substitution of Eq. (17) into Eq. (16) is accompanied by the change of the variables of integration from x to ξ. For the upper limit of integration ξ_2 from Eq. (2) $L_{fr} = \frac{d_g}{\pi}(1 + \xi_2 + \exp\xi_2)$ we find $\exp(\xi_2) \simeq \frac{\pi L_{fr}}{2d_g}$. For the lower integral limit we have an equation $-0.5L_g = \frac{d_g}{\pi}(1 + \xi_1 + \exp\xi_1)$, where $\xi_1 < 0$. Substituting the modulus $|\xi_1|$ in the latter condition we get: $|\xi_1| \simeq 1 + \frac{\pi L_g}{2d_g}$. After integration from the condition for standing wave existence (Dyakonov,1996) we finally obtain the fundamental plasma resonant frequency in the presence of fringing effects

$$f^{fr} \approx \frac{f_0}{1 + \frac{4d_g}{L_g \pi R_v}\left(\sqrt{1 + \frac{\pi R_v L_{fr}}{2d_g}} - 1\right) + \chi}, \tag{18}$$

where f_0 is the fundamental frequency of plasma oscillations in the ideal case when no fringing occurs (from Eq. (1) $f_0 = \Omega/2\pi$),

$$\chi = \frac{2d_g}{\pi L_g}\left(1 + \ln\frac{4}{R_v} - 2\sqrt{\frac{2d_g}{\pi R_v L_{fr}}}\right), \tag{19}$$

and $R_v = V_p^0/(V_g - V_{th}^0)$.

3.2 Cascaded transmission line model and results of IsSpice simulation

The spanning of the 2DEG channel region controlled by the gate bias voltage beyond the area covered by the gate contact caused by fringing of the electric field requires appropriate modification of the equivalent circuit model previously developed. To represent the fringed ungated region with nonuniform sheet electron density distribution and position dependent distance d^{fr} between the "extended" gate and 2DEG channel as an equivalent electric circuit we partitioned it into small segments (Andress,2005). In each such segment sheet electron density is supposed to be uniform and distance d^{fr} constant so that they can be represented by a uniform transmission lines $TL_{1,2,...,N}$. The cascaded TL model for the gated and ungated fringed 2DEG channel regions is schematically shown in Fig. 7. Relevant $R_{fr}, \mathcal{L}_{fr}, C_{fr}$ components are derived using Eq. (15) in the form

$$R_{fr} = R_g \frac{1 + \exp\xi}{1 + \frac{V_p^0}{V_g - V_{th}^0}\exp\xi},$$

$$\mathcal{L}_{fr} = \mathcal{L}_g \frac{1 + \exp\xi}{1 + \frac{V_p^0}{V_g - V_{th}^0}\exp\xi}, \qquad C_{fr} = \frac{C_g}{1 + \exp\xi}, \tag{20}$$

whereR_g, $\mathcal{L}_g$ and C_g are determined by Eqs. (4)-(6).
In the developed cascaded equivalent circuit the ungated fringed region has been represented by 10 uniform TLs. In the IsSpice simulation TL input voltage corresponded to the gate bias voltage which contained dc and ac components. ac analysis has been conducted and response has been measured in the open circuit configuration. Test simulations performed for 5-, 10- and 15-sections model of the ungated fringed region revealed very close fundamental resonant plasma frequencies for 10 and 15 sections. Simulated frequency response of the HEMT is shown in Fig. 8. It is clearly seen from the simulation results of Fig. 8 that fringing effects cause the reduction of the resonant frequency of plasma oscillations.

4. Comparison of Theoretical and Experimental Data

Fig. 9 shows dependences of fundamental resonant plasma frequency versus gate bias voltage V_g calculated for the InAlAs/InGaAs HEMT with the gate length $L_g = 50$ nm, and structure parameters taken from Ref.(El Fatimy,2006). Curve (1) calculated using Eq. (1) corresponds to the ideal case. Curve (2) represents experimental data extracted from Ref.(El Fatimy,2006). Curve (3) has been calculated using model of Ref.(Satou,2003) which takes into account ungated regions. Curves (4) and (5) were calculated and simulated using fringing effect model. Plasma frequencies of curve 4 were calculated using Eq. (17) under the assumption based on Fig. 6 that for the ratio $d_g/L_g = 0.34$ the fringing field spreads over the whole ungated regions of length $L_c = 2 \times L_g$ on both sides of the gate contact, i.e., $L_{fr} = 2 \times 100 = 200$ nm.

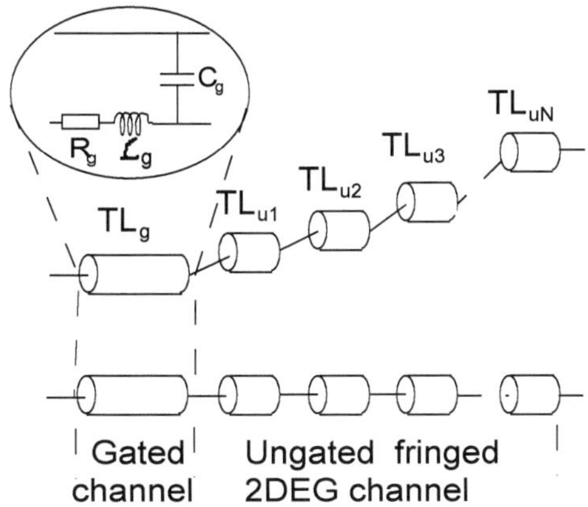

Fig. 7. Cascaded TL model for the HEMT in the presence of fringing effects. Gated 2DEG channel is represented by a single uniform TL. Ungated fringed 2DEG channel region is partitioned so that each part can be represented by uniform TL (TL_{u1}, TL_{u2},... TL_{uN}).

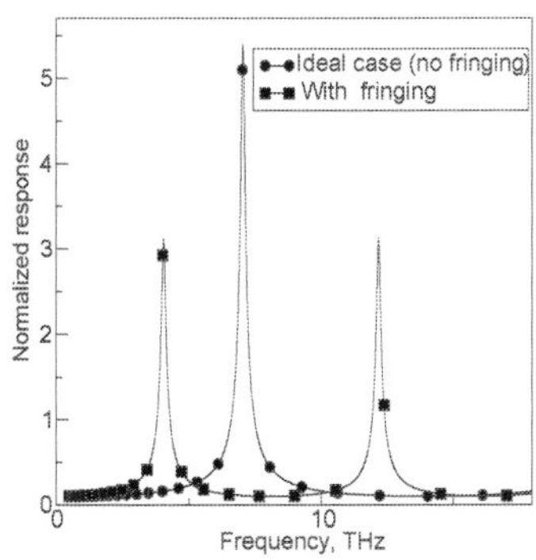

Fig. 8. Frequency response of the HEMT simulated with IsSpice software for the ideal case when fringing effects are neglected (uniform TL model for the gated channel) and for the case when ungated 2DEG channel region is fringed (cascaded TL model for the gated and fringed ungated HEMT channel regions).

As it was mentioned previously experimentally observed plasma resonant frequencies (curve 2) were much lower comparing to those predicted by Eq. (1). The model which takes into account the contribution of ungated regions (curve 3)shows, at a first glance, practically ideal agreement with experimental data. However, the incorporation of the possible contribution of some other factors, for example, cap layer may result in deviation from such ideal agreement. On the other hand, comparing curves 4 and 5 with experimental data of curve

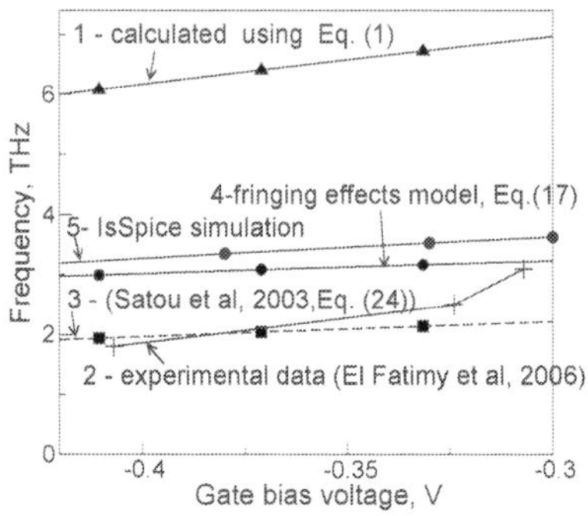

Fig. 9. Fundamental frequency of plasma oscillations: 1– Ideal case, calculated using Eq. (1); 2 – Experimental data extracted from Ref. (El Fatimy, 2006); 3 – Model for impact of ungated regions, Eq. (24) of Ref. (Satou,2003); 4 – model accounting for fringing effects, Eq. (17); 5 – Results of IsSpice simulation.

2 one can say about reasonably good agreement between them keeping in mind that further reduction of plasma frequencies is possible due to contribution of other factors, in particular, cap layer.

5. Chapter summary

In conclusion, we develop simple distributed circuit model of the HEMT-like structure to study the effects associated with the excitation of plasma oscillations in its 2DEG channel. The circuit components of the model are related to physical and geometrical parameters of the structure. Moreover, the dependence of the resistance and kinetic inductance of the gated 2DEG channel portion on gate bias voltage has been taken into account. The developed electric equivalent circuit has been used to simulate HEMT frequency performance with IsSpice software. Model accounting for the fringing effects contribution to the plasma frequency reduction is proposed. Using the concept of "extended" gate the sheet electron density distribution in the fringed ungated region of the 2DEG channel is estimated and the expression for the resonant plasma frequency in the presence of fringing effects is derived. The basic distributed circuit model has been modified into cascaded TL line model to account for the impact of fringing effects on the resonant plasma frequencies. Simulated HEMT frequency response shows the decrease of resonant frequency related to the fringing effects. The results of our model are in rather good agreement with experimental data admitting also further possible frequency reduction due to other factors, for example, the cap layer.

6. References

[1] Allen, S. J.; Tsui, D. C.; & Logan, R.A. (1977) Observation of the two-dimensional plasmon in silicon inversion layers. *Phys. Rev. Lett*, Vol. 38, 980-983.

[2] Andress, W. F. and Ham, D. (2005) Standing wave oscillators utilizing wave-adaptive tapered transmission lines, *IEEE J. Solid State Circuits*, Vol. 40, pp. 638-651.

[3] Burke, P. J.; Spielman, I. B.; Eisenstein, J. P.; Pfeiffer, L. N.; & West, K. W. (2000) High frequency conductivity of the high-mobility two-dimensional electron gas, *Appl. Phys. Lett.* Vol. 76, 745-747.

[4] Chaplik, A V. (1972) Possible crystallization of charge carriers in low-density inversion layers. *Sov. Phys. JETP*, Vol. 35, 395–398.

[5] Collin, R. E. (1992) *Foundations for Microwave Engineering* (McGrow-Hill, Inc., New York, 1992)

[6] Delagebeaudeuf, D. & Linh, N. T. (1982) Metal-(n) AlGaAs-GaAs two-dimensional electron gas FET, *IEEE Trans. Electron Devices*, Vol. ED-29, 955–960.

[7] Dyakonov, M. & Shur, M. (1993) Shallow water analogy for a ballistic field effect transistor: New mechanism of plasma wave generation by dc current *Phys. Rev. Lett.*, Vol. 71, 2465-2468.

[8] Dyakonov, M. & Shur, M. (1996) Plasma wave electronics: novel terahertz devices using two dimensional electron fluid, *IEEE Trans. Electron Device*, vol. 43, 1640-1645, 1996.

[9] El Fatimy, A.; Teppe, F.; Dyakonova, N.; Knap, W.; Seliuta, D.; Valusis, G.; Shchepetov, A.; Roelens, Y.; Bollaert, S.; Cappy, A.; & Rumyantsev, S. (2006). Resonant and voltage-tunable terahertz detection in InGaAs/InP nanometer transistors, *Appl. Phys. Lett.*, Vol. 89, 131926.

[10] Khmyrova, I. & Seijyou, Yu. (2007) Analysis of plasma oscillations in high-electron mobility transistor-like structures: Distributed circuit approach, *Appl. Phys. Lett.*, Vol. 91, 143515.

[11] Knap, W.; Kachorovskii, V.; Deng, Y.; Rumyantsev, S.; Lu, J.Q.; Gaska, R.; Shur, M.S.; Simin, G.; Hu, X.; Asif Khan, M.; Sailor, C.A.; & Brunel, L.C. (2002) Nonresonant detection of terahertz radiation in field effect transistors *J. Appl. Phys.* Vol. 91, 9346–9353.

[12] Lu,J.; Shur,M.S.; Hesler,J.L.; Sun, L.; & Weikle, R. (1998) Terahertz detector utilizing two-dimensional electronic fluid, *IEEE Electron Device Lett.*, Vol. 19, 373-375.

[13] Mahajan,A.; Arafa, M.; Fay, P.; Caneau, C.; & Adesida, I. (1998). Enhancement-mode high electron mobility transistors (E-HEMT's) lattice-matched to InP, *IEEE Trans. Electron Devices*, Vol. ED-45, 2422-2429.

[14] Morse, Ph. M. & Feshbach, H. (1953) *Methods of Theoretical Physics* (McGrow-Hill, Inc., New York, 1953)

[15] Nishimura,T.; Magome, N.; Khmyrova, I.; Suemitsu, T.; Knap, W.; & Otsuji, T. (2009). Analysis of fringing effect on resonant plasma frequency in plasma wave devices, *Jpn. J. Appl. Phys.*, Vol. 48, 04C096.

[16] Otsuji, T.; Hanabe, M.; & Ogawara, O. (2004). Terahertz plasma wave resonance of two-dimensional electrons in InGaP/InGaAs/GaAs high-electron-mobility transistors, *Appl. Phys. Lett.*, vol. 85, 2119.

[17] Satou, A.; Khmyrova, I.; Ryzhii, V.; & Shur, M. (2003) Plasma and transit-time mechanisms of the terahertz radiation detection in high-electron-mobility transistors. *Semicond. Sci. Technol.* Vol. 18, 460 .

[18] Satou, A.; Ryzhii, V.; Khmyrova, I.; Ryzhii, M.; & Shur, M. (2004) Characteristics of a terahertz photomixer based on a high-electron mobility transistor structure with optical input through the ungated regions. *J. Appl. Phys.* Vol. 95, 2084–2089.

[19] Shur, M. S. & Ryzhii, V. (2003) Plasma wave electronics. *Int. J. High. Speed Electron Syst.* Vol. 13, 575-600.

[20] Suemitsu, T.; Enoki, T.; Yokoyama, H.; & Ishii, Y. (1998) Improved recessed-gate structure for sub-01-μm-gate InP-based high electron mobility transistors, *Jpn. J. Appl. Phys.*, Vol. 37, pp. 1365-1372.

[21] Teppe, F.; Knap, W.; Veksler, D.; Shur, M. S.; Dmitriev, A. P.; Kacharovskii, V. Yu.; & Rumyantsev, S. (2005). Room-temperature plasma waves resonant detection of sub-terahertz radiation by nanometer field-effect transistor, *Appl. Phys. Lett.*, Vol. 87, 052107.

[22] Tsui, D.C.; Gornik, E. & Logan, R.A. (1980) Far infrared emission from plasma oscillations of Si inversion layers, *Solid State Commun.*, Vol. 35, 875-877.

[23] Veksler,D.; Teppe, F.; Dmitriev, A. P.; Kacharovskii, V. Yu.; Knap, W.; & Shur, M. S. Detection of terahertz radiation in gated two-dimensional structures governed by dc current. *Phys. Rev.* Vol. B 73 125328.

[24] Weikle, R.; Lu, J.; Shur, M.S.; (2006) & Dyakonov, M. (1996) Detection of microwave radiation by electronic fluid in high electron mobility transistors, *Electron. Lett.*, vol. 32, 2148-2149.

[25] Yeager, H. R. & Dutton, R. W. (1986) Circuit simulation models for the high electron mobility transistor, *IEEE Trans. Electron Devices*, Vol. ED-33, pp. 682–692.

6

Composite Right / Left Handed (CRLH) based devices for microwave applications

Stefan Simion‡, Romolo Marcelli*, Giancarlo Bartolucci#, Florea Craciunoiu‡, Andrea Lucibello*, Giorgio De Angelis*, Andrei A. Muller‡, Alina Cristina Bunea‡, Gheorghe Ioan Sajin‡

‡*National Institute for Research and Development in Microtechnologies, Bucharest, Romania*

* *CNR – Institute for Microelectronics and Microsystems, Rome Italy*

\# *University of Rome "Tor Vergata", Department of Electronics Engineering, Rome Italy*

1. Introduction

Metamaterials (MMs) became a very actual topic in the present research interest field, due to their unusual but interesting characteristics, not encountered in nature (Veselago, 1968) (Engheta & Ziolkowski, 2005). A way to obtain media having MM characteristics is to develop circuits which, under certain conditions, model the MMs properties. Using different lattice structures or periodic repetition of unit cells, different types of two-dimensional MMs have been suggested (Sievenpiper et.al., 1999; Caloz & Itoh, 2003).

A particular class of MMs consists of artificial LH (Left-Hand) transmission lines, which may be obtained using series capacitors and parallel connected inductors. With this approach and using SMDs (surface mounted devices), circuits such as branch couplers (Lin et al., 2003) and ring couplers (Okabe et al., 2004) operating up to a few GHz, have been reported. The main advantage of these circuits compared to the classical ones is the dual frequency response for any two frequency ratio.

For higher microwave frequencies, it is more convenient to replace SMD capacitors and inductors with interdigital capacitors and short-ended transmission lines, respectively, in microstrip or CPW configuration. Taking into account the series equivalent inductance of the capacitors and the parallel equivalent capacitance of the short-ended transmission lines (working as inductors, at their own resonance frequency), CRLH (Composite Right / Left Handed) cells are obtained (Lai et al., 2004). The CRLH cells open a new class of devices and applications such as backward-wave directional couplers (Caloz et al., 2004), tunable radiation angle and beamwidth antennas (Lim et al., 2004), zeroth-order resonator antennas (Sanada et al., 2004). Other new expected research directions may be found in literature (see for example Caloz & Itoh, 2005).

In the last few years, CPW (CoPlanar Waveguide) CRLH based devices were investigated and fabricated on semiconductor substrate, such as bandpass filters (C. Li et al., 2007 and W.

Tong et al, 2007), resonating antennas (Simion et al., 2007-a,b) and directional couplers (Simion et al., 2008-a,b). The main advantage of using semiconductor substrate for these devices is the possibility to integrate them monolithically in more complex circuits.
Very interesting applications can also be developed using CRLH based devices supported by magnetically biased ferrite. Recently, authors compared four related CRLH leaky-wave antennas where the CRLH structure dispersion was controlled by an applied magnetic field, for fixed frequency external tuning (Kodera & Caloz, 2008). Also, a CPW CRLH resonating antenna having a magnetically polarized ferrite as supporting substrate has been recently investigated and reported (Sajin et al., 2009).
The results presented in this chapter are focused on the results obtained by the authors in the field of CPW CRLH based devices, such as a directional coupler and a resonating antenna, both fabricated on silicon substrate, but also a resonating antenna manufactured on magnetically biased ferrite.

2. RH, LH and CRLH Transmission Lines

The transmission lines used in practice (microstrip, CPW etc.) are homogenous transmission media. They may be modeled with equivalent circuits consisting of distributed elements, which are obtained by cascading a large enough number of cells (so that the length of a cell becomes much shorter in comparison to the wavelength), each cell being a series inductance and a parallel capacitance, which will be further addressed as RH (Right-Handed)-TL Transmission Line). For the transmission line based on RH-TL cells, the phase velocity and the group velocity have positive values.
A transmission line, with a negative phase velocity (but with a positive group velocity, like for the RH-TL) would be useful in some applications. As equivalent circuit, this transmission line must consist of a large enough number of cells, each one being a series distributed capacitance and a parallel distributed inductance, further referred to as LH (Left-Handed) - TL. This type of transmission line does not exist in practice, but an artificial transmission line consisting of lumped elements can model the LH-TL behavior.
The artificial LH-TL is fabricated by using capacitors and inductors. Taking the series inductance of the capacitors and the parallel capacitance of the inductors into account, a more complicated equivalent circuit for an LH-TL cell is obtained. The structure (cell) having an equivalent circuit containing both the RH-TL equivalent circuit as well as the LH-TL equivalent circuit is known as CRLH (Composite Right / Left Handed). By cascading CRLH cells, CRLH-TLs are obtained. Depending on the frequency, the CRLH-TL may have RH-TL or LH-TL behavior.
The propagation constant and the characteristic impedance for the RH-TL, LH-TL and CRLH-TL are reviewed in the next sections (see also Caloz & Itoh, 2006).

2.1. RH-TL and LH-TL

The equivalent circuits of a cell, for RH-TL and LH-TL are shown in Fig.1 (a) and (b).

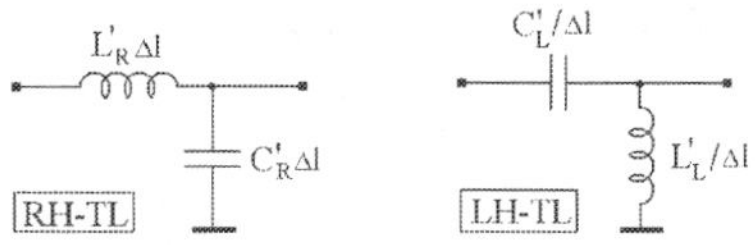

(a) (b)

Fig. 1. Equivalent circuits with distributed elements, for a cell of RH-TL (a) and LH-TL (b).

In these circuits, L'_R, C'_R and L'_L, C'_L are the distributed inductance and capacitance for RH-TL and LH-TL respectively. For these equivalent circuits, the propagation constant and the characteristic impedance can be determined with the following formulas:

$$\gamma = \sqrt{Z'(\omega)Y'(\omega)} = \alpha + j\beta \quad \text{and} \quad Z_c = \sqrt{\frac{Z'(\omega)}{Y'(\omega)}} \qquad (1\ a,b)$$

respectively, where $Z'(\omega)$ and $Y'(\omega)$ are the impedance of the series branch and the admittance of the parallel branch (see Fig.1 a,b), α is the attenuation constant and β is the phase constant.

If the line is lossless ($\alpha = 0$), then the propagation constant is pure imaginary, while the impedance is pure real. For RH-TL we can write (see Fig.1 a):

$$Z'(\omega) = j\omega L'_R \quad \text{and} \quad Y'(\omega) = j\omega C'_R \qquad (2\ a,b)$$

By inserting (2 a, b) in (1 a, b), we get:

$$\beta_{RH-TL} = \omega\sqrt{L'_R C'_R} > 0 \quad \text{and} \quad Z_{c,RH-TL} = \sqrt{\frac{L'_R}{C'_R}} \qquad (3\ a,b)$$

For LH-TL, it may be written (see Fig.1 b):

$$Z'(\omega) = \frac{1}{j\omega C'_L} \quad \text{and} \quad Y'(\omega) = \frac{1}{j\omega L'_L} \qquad (4\ a,b)$$

By inserting (4 a,b) in (1 a,b), we get:

$$\beta_{LH-TL} = -\frac{1}{\omega\sqrt{L'_L C'_L}} < 0 \quad \text{and} \quad Z_{c,LH-TL} = \sqrt{\frac{L'_L}{C'_L}} \qquad (5\ a,b)$$

The equations (3 a) and (5 a) are the dispersion equations for RH-TL and LH-TL, respectively. The qualitative graphic representation of these equations is given in Fig.2.

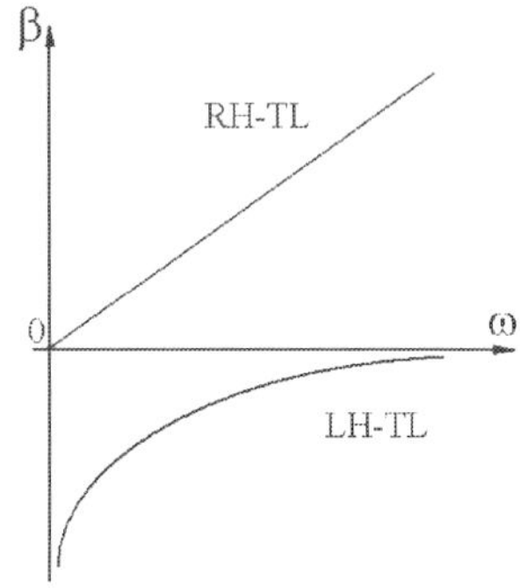

Fig. 2. Graphic representation of the dispersion equations for lossless RH-TL and LH-TL

The phase velocity, $v_f = \frac{\omega}{\beta}$ obtained from the dispersion equations (3 a) and (5 a) is positive for RH-TL and negative for LH-TL. The group velocity, $v_g = \frac{1}{\frac{d\beta}{d\omega}}$, is positive for both RH-TL and LH-TL. Therefore, the energy transport is, from generator to load in both cases, but for LH-TL, due to the fact that the phase velocity is negative, the wave is propagated backwards (from load to generator).

2.2. CRLH transmission lines

As mentioned above, the equivalent circuit of CRLH-TL is a combination of the equivalent circuits for RH-TL and LH-TL. The equivalent circuit for CRLH-TL is given in Fig.3, where, similar to RH-TL and LH-TL, Δl must be small enough compared to the wavelength.

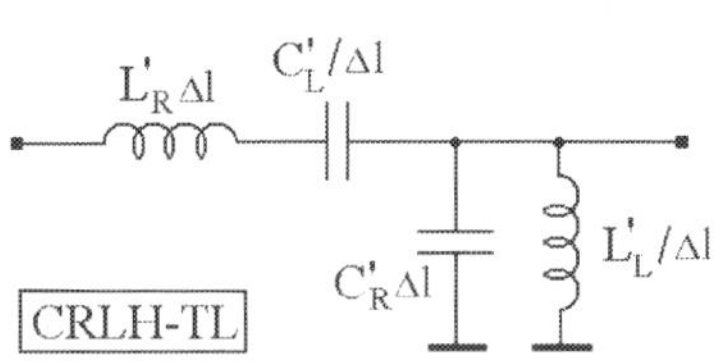

Fig. 3. The distributed equivalent circuit for a cell of CRLH-TL.

For the circuit given in Fig.3:

$$Z'(\omega) = j\left(\omega L'_R - \frac{1}{\omega C'_L}\right) \quad \text{and} \quad Y'(\omega) = j\left(\omega C'_R - \frac{1}{\omega L'_L}\right) \tag{6 a,b}$$

Using (6a,b), with the relations (1 a) and $\alpha = 0$ (lossless circuit), we get:

$$\beta_{CRLH-TL} = -\sqrt{\left(\omega L'_R - \frac{1}{\omega C'_L}\right)\left(\omega C'_R - \frac{1}{\omega L'_L}\right)} < 0 \quad \text{for } \omega < \omega_{\Gamma_1} \tag{7 a}$$

and

$$\beta_{CRLH-TL} = +\sqrt{\left(\omega L'_R - \frac{1}{\omega C'_L}\right)\left(\omega C'_R - \frac{1}{\omega L'_L}\right)} > 0 \quad \text{for } \omega > \omega_{\Gamma_2} \tag{7 b}$$

where:

$$\omega_{\Gamma_1} = \min\left[\frac{1}{\sqrt{L'_R C'_L}}; \frac{1}{\sqrt{L'_L C'_R}}\right] \quad \text{and} \quad \omega_{\Gamma_2} = \max\left[\frac{1}{\sqrt{L'_R C'_L}}; \frac{1}{\sqrt{L'_L C'_R}}\right] \tag{8 a,b}$$

For $\omega \in (\omega_{\Gamma_1}, \omega_{\Gamma_2})$, the phase constant, β, is an imaginary number, therefore the propagation constant, γ, is a real number, which means that the signal on the line is attenuated. Therefore, for $\omega \in (\omega_{\Gamma_1}, \omega_{\Gamma_2})$, the circuit behaves as a band-stop filter. If $\omega_{\Gamma_1} = \omega_{\Gamma_2}$, then there is no stop-band.

The circuit for which $\omega_{\Gamma_1} \neq \omega_{\Gamma_2}$ is called an unbalanced circuit, while the circuit for which $\omega_{\Gamma_1} = \omega_{\Gamma_2}$ is called a balanced circuit.

By simply looking at the equations (8 a,b), we get $\omega_{\Gamma_1} = \omega_{\Gamma_2}$ if:

$$L'_R C'_L = L'_L C'_R \tag{9}$$

In this case, we replace the frequencies ω_{Γ_1} and ω_{Γ_2} with ω_0 getting the following expression:

$$\omega_0 = \frac{1}{\sqrt{L'_R C'_L}} = \frac{1}{\sqrt{L'_L C'_R}} = \frac{1}{\sqrt[4]{L'_R C'_L L'_L C'_R}} \tag{10}$$

Also, for the balanced circuit, from the equations (7 a,b) and (10), $\beta = 0$.

Analyzing the dispersion equations (7 a,b), it results that, for $\omega < \omega_{\Gamma_1}$ (or $\omega < \omega_0$, for the balanced circuit), the line has a LH-TL behavior, while for $\omega > \omega_{\Gamma_2}$ (or $\omega > \omega_0$, for the balanced circuit), the line has a RH-TL behavior.

The equations (7 a,b) are the dispersion equations for CRLH-TL. The graphic representation (qualitative) of these dispersion equations, for the unbalanced and balanced circuits, are shown in Fig.4 (a) and Fig.4 (b), respectively.

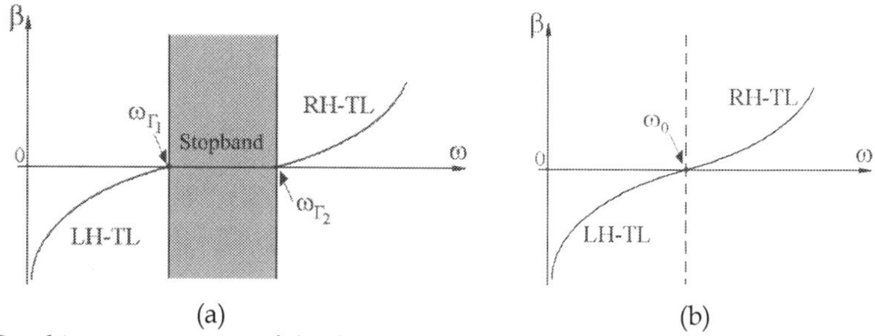

Fig. 4. Graphic representation of the dispersion equations for CRLH-TL, for unbalanced (a) and balanced (b) cases.

The characteristic impedance for CRLH-TL in an unbalanced circuit is obtained by inserting (6 a,b) in (1 b). The following formula is obtained:

$$Z_{c,CRLH-TL} = \sqrt{\frac{L'_L}{C'_L}} \cdot \sqrt{\frac{\omega^2 L'_R C'_L - 1}{\omega^2 L'_L C'_R - 1}} \tag{11}$$

If the circuit is balanced, then, with the equation (9), from (11) we get:

$$Z_{c,CRLH-TL} = \sqrt{\frac{L'_L}{C'_L}} = \sqrt{\frac{L'_R}{C'_R}}$$

or, with the values (3 b) and (5 b),

$$Z_{c,CRLH-TL} = Z_{c,RH-TL} = Z_{c,LH-TL} \tag{12}$$

It can be noticed from (12), that for a balanced CRLH-TL, the impedance matching conditions over a large frequency domain can be easily fulfilled.
If the circuit is balanced, then, using (9) in (7 a,b), we get:

$$\beta_{CRLH-TL} = \frac{\omega^2 L_R' C_L' - 1}{\omega\sqrt{L_L' C_L'}} = \omega\sqrt{L_R' C_R'} - \frac{1}{\omega\sqrt{L_L' C_L'}} \tag{13}$$

From the equation above, it can be noticed that for $\omega > \omega_0$, $\beta_{CRLH-TL} > 0$, while for $\omega < \omega_0$, $\beta_{CRLH-TL} < 0$, where ω_0 is given in (10).
Using (3 a) and (5 a), the dispersion equation (13) can be written as follows:

$$\beta_{CRLH-TL} = \beta_{RH-TL} + \beta_{LH-TL} \tag{14}$$

From (12) and (14), the equivalent circuit for a balanced CRLH-TL can be drawn by connecting in cascade the equivalent circuits for RH-TL and LH-TL, given in Fig.1 a,b.

2.3. Equivalent permittivity and permeability

For a homogenous medium, characterized by ε and μ (usually complex values, but real for lossless media), the phase constant, β, and the characteristic impedance, Z_c, have the following expressions:

$$\beta = \omega\sqrt{\varepsilon\mu}, \qquad \text{respectively} \qquad Z_c = \sqrt{\frac{\mu}{\varepsilon}} \tag{15 a,b}$$

From (15 a,b) and (1 a,b), for a lossless medium ($\alpha = 0$), we get:

$$\varepsilon = -j\frac{Y'(\omega)}{\omega} \qquad \text{and} \qquad \mu = -j\frac{Z'(\omega)}{\omega} \tag{16 a,b}$$

For a homogenous medium which may be modeled by a RH-TL equivalent circuit, inserting the equations (2 a,b) in (16 a,b), we get:

$$\varepsilon = C_R' \qquad \text{and} \qquad \mu = L_R' \tag{17 a,b}$$

For a homogenous medium which may be modeled by a LH-TL equivalent circuit, inserting the equations (4 a,b) in (16 a,b), we get:

$$\varepsilon = -\frac{1}{\omega^2 L_L'} \quad \text{and} \quad \mu = -\frac{1}{\omega^2 C_L'} \tag{18 a,b}$$

Likewise, for a homogenous medium, which may be modeled by a CRLH-TL equivalent circuit, inserting the equations (6 a,b) in (16 a,b), we get the following result:

$$\varepsilon = C_R' - \frac{1}{\omega^2 L_L'} \quad \text{and} \quad \mu = L_R' - \frac{1}{\omega^2 C_L'} \tag{19 a,b}$$

By analyzing the equations (17 a,b), (18 a,b) it is useful to observe that ε and μ are positive for RH-TL and negative for a LH-TL, respectively. The equations (17 a,b) and (18 a,b) are particular cases of the equations (19 a,b), for high enough frequencies, respectively for low enough frequencies.

The values of ε and μ for CRLH-TL can take either positive or negative values, depending on the frequency. The following conclusions can be drawn by analyzing the relations (19 a,b) and (8 a,b):

1. If $\omega < \omega_{\Gamma_1}$, then $\varepsilon < 0$ and $\mu < 0$ (20 a)

2. If $\omega > \omega_{\Gamma_2}$, then $\varepsilon > 0$ and $\mu > 0$ (20 b)

3. If $\omega \in (\omega_{\Gamma_1}, \omega_{\Gamma_2})$, then

$$\varepsilon < 0 \ \text{and} \ \mu > 0 \ \text{if} \ \frac{1}{\omega\sqrt{L_L' C_R'}} > \frac{1}{\omega\sqrt{L_R' C_L'}}$$

or (20 c)

$$\varepsilon > 0 \ \text{and} \ \mu < 0 \ \text{if} \ \frac{1}{\omega\sqrt{L_L' C_R'}} < \frac{1}{\omega\sqrt{L_R' C_L'}}$$

where ω_{Γ_1} and ω_{Γ_2} are given in (8 a,b).

For the particular case when the circuit is balanced, if $\omega_{\Gamma_1} = \omega_{\Gamma_2} = \omega_0$ (see equation (10)), the following equations can be obtained:

1. If $\omega < \omega_0$, then $\varepsilon < 0$ and $\mu < 0$ (21 a)

2. If $\omega > \omega_0$, then $\varepsilon > 0$ and $\mu > 0$ (21 b)

The propagation medium is called *metamaterial*, if $\varepsilon < 0$ and $\mu < 0$, inequalities which are valid for low enough frequencies - see (20a) for unbalanced CRLH-TL and (21a) for balanced CRLH-TL.

The analysis performed for an artificial line consisting of cascaded CRLH cells modeled by lumped components leads to similar formulas for the phase shift and for the characteristic impedance (if the length of the cell is much smaller in comparison to the wavelength), but two cut-off frequencies must be taken into account (see also Caloz & Itoh, 2006). These frequencies are due to the high-pass and low-pass behavior of the CRLH modeled with lumped elements (see section 4 for the formulas of these two frequencies) and cannot be taken into evidence for a distributed transmission line.

3. Microwave Directional Coupler with CRLH cells on Silicon Substrate

3.1. Circuit description, design and fabrication

The coupler presented here is composed of two coupled artificial CRLH-TLs, each one having two identical cascaded cells, each one consisting of series interdigital capacitors and parallel connected short-ended transmission lines.

The coupler layout, in CPW configuration, is shown in Fig.5. Four tapered lines were used for impedance matching and also to fit the probes sizes to the on-wafer measurement system.

A strong coupling is possible for this type of coupler, as compared to other kinds of couplers as, for instance, the Lange coupler (Caloz & Itoh, 2005).

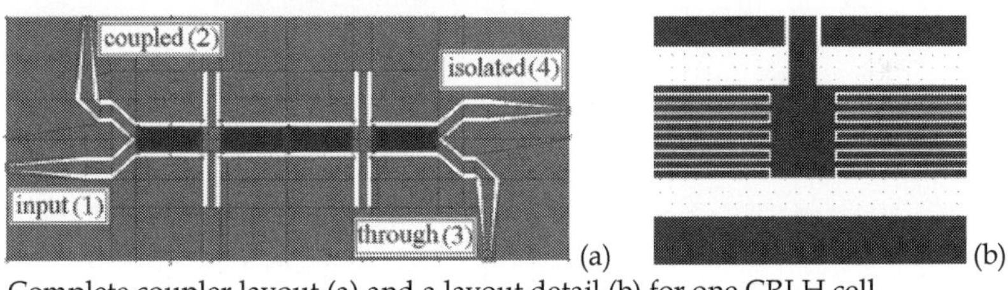

Fig. 5. Complete coupler layout (a) and a layout detail (b) for one CRLH cell.

The equivalent circuit for one CRLH cell is given in Fig.6. The series interdigital capacitor has the equivalent capacitance C_{Cs}, the series equivalent inductance L_{Cs} and the parallel equivalent capacitance (to the ground) C_{Cp}. Also, the short-ended CPW has the equivalent inductance L_{Lp} and the parallel equivalent capacitance, C_{Lp}.

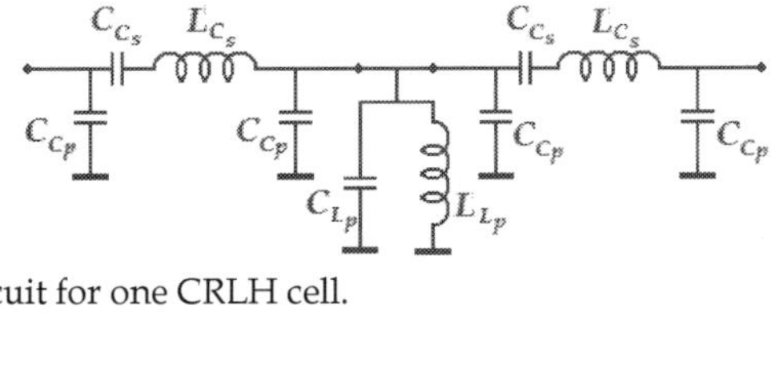

Fig. 6. Equivalent circuit for one CRLH cell.

The CRLH cell was designed such as to obtain a balanced structure for the central frequency of the coupler. This means that the series frequency of resonance has to be equal to the parallel frequency of resonance and also equal to the central frequency of the coupler (Caloz et al., 2004). Nevertheless, if two CRLH based transmission lines are coupled, the band-stop frequency due to the even/odd mode occurs around the central frequency of the coupler, due to the mutual inductances and the coupling capacitances. The band-stop frequency range is wider as the coupling is stronger.

The substrate used both in simulations and in the experimental approach was a 500 µm thick silicon wafer, with a dielectric constant of 11,9, covered by a 1 µm thick thermal SiO_2 layer, having the dielectric constant equal to 4.7. The bulk resistivity of the silicon substrate is equal to 5 kΩcm.

The design procedure starts with the interdigital capacitor. The numerical value for C_{Cs} must be chosen taking into account the central frequency of the coupler, the imposed technological resolution, the finger length (which must be much shorter than the minimum wavelength) and also the condition for the series resonance, which must be equal to the central frequency of the coupler. Taking into account these constrains, the layout of the interdigital capacitor has been optimized by using the software IE3D - Zeland. The numerical results exhibit a series resonance frequency at f_0 = 11.25 GHz, closed to the impose value. The values for the capacitor equivalent circuit elements (C_{Cs}, L_{Cs} and C_{Cp}) have been computed using the following formulas:

$$C_{Cs} = \frac{-\mathrm{Im}(Y_{12})}{2\pi f}\left[1-\left(\frac{f}{f_0}\right)^2\right]; \quad L_{Cs} = \frac{1}{(2\pi f_0)^2 C_{Cs}}; \quad C_{Cp} = \frac{\mathrm{Im}(Y_{11})+\mathrm{Im}(Y_{12})}{2\pi f}$$

where Y_{11} and Y_{12} are the admittance matrix elements of the CPW interdigital capacitor, at low frequency (f = 1 GHz). In this way, using (22), the capacitor circuit elements were computed as: C_{Cs} =0.35 pF, L_{Cs} =0.572 nH and C_{Cp} =0.082 pF.

In order to design the whole layout of the CRLH cell, the values for the equivalent circuit elements obtained for the capacitor (C_{Cs} , L_{Cs} and C_{Cp}) are used to compute L_{Lp} and C_{Lp}. The values for L_{Lp} and for the coupler characteristic impedance Z_c, may be obtained by solving the following equations system:

$$\frac{L_{Cs}}{Z_c^2} - 2C_{Cp} = \frac{L_{Lp}}{Z_{cL}^2}; \qquad \frac{L_{Cs}}{L_{Lp}Z_{cL}^{-2} + 2C_{Cp}} = \frac{L_{Lp}}{C_{Cs}} \tag{22}$$

where Z_{cL} is the characteristic impedance of the CPW stub (see Fig.5). The equations system (22) was solved for f_0 = 11.25 GHz, running Z_{cL} values to get numerical solutions for the given system. When Z_{cL} = 60 Ω, they are obtained L_{Lp} = 0.603 nH and Z_c = 41 Ω (also C_{Lp} = 0.167 pF). With the above data, the resonance frequency of the CPW stub is 11.25 GHz, the same as the resonance frequency of the interdigital capacitor (therefore, the CRLH cell is balanced). For the substrate data given above, the values for L_{Lp} and Z_c may be obtained if the length of the coplanar line is 0.8 mm, the width of the central CPW line is 100 µm and the width of the CPW slot is 100 µm. The interdigital capacitor will consist of 10 fingers, having the length equal to 1 mm and the width equal to 10 µm. Moreover, the gap between the fingers is equal to 10 µm.

The numerical results for the design coupler were obtained with IE3D - Zeland software and are presented in the next section.

The circuit was manufactured by means of a standard photolithographic process. A 500 Å Cr layer followed by 0.6 µm Au layer were evaporated onto the entire surface of the 1 µm SiO_2 layer that covers the silicon wafer. After that, the metallization pattern was defined by a usual step of wet etching. A detail of the obtained circuit is shown in Fig.7 (Simion et al., 2008-a).

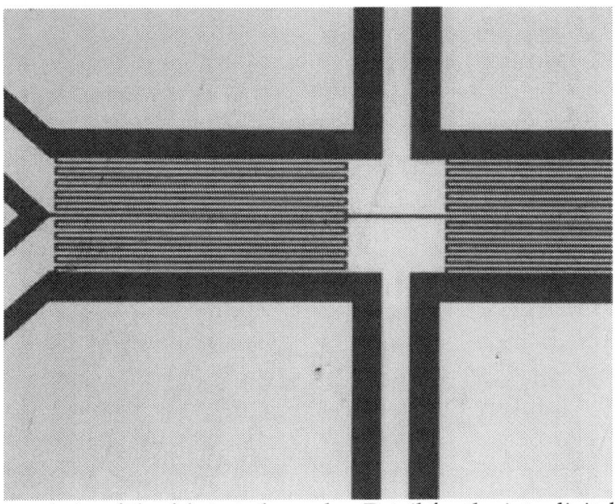

Fig. 7. Microscope photo of the fabricated coupler. Detail for the interdigital capacitors and the CPW stubs area.

3.2. Experimental characterization method and results

The fabricated coupler was experimentally characterized using a two-port vector network analyzer and an on-wafer probe heads station. In order to characterize a coupler which is a four-port circuit by using a two-port network analyzer, a similar method as proposed in (Tippet & Speciale, 1982), was employed (Simion et al., 2008-b).

For a coupler, the number of measurement sets is given by the possible combinations of two-port measurements, $C_n^2 = \dfrac{n!}{2!(n-2)!}$, which is equal to 6, if the number of ports is $n = 4$. So, the proposed method involves 6 measurement sets, performed on the coupler for each frequency. Following these two-port measurements, 6 scattering matrixes, S_{i-j}, are obtained, where i and j are the port number (see Fig.5 (a)): S_{1-2}, S_{1-3}, S_{1-4}, S_{2-3},

S_{2-4} and S_{3-4}. The scattering matrixes S_{i-j} must be re-normalized to the impedance corresponding to the loads connected to the each port, obtaining 6 scattering parameter matrixes, S'_{i-j}, according to the formula:

$$S'_{i-j} = (I_2 - \Gamma_{i-j})^{-1} \cdot (S_{i-j} - \Gamma_{i-j}) \cdot (I_2 - \Gamma_{i-j} \cdot S_{i-j})^{-1} \cdot (I_2 - \Gamma_{i-j}) \tag{23}$$

where $\Gamma_{i-j} = \begin{bmatrix} \Gamma_i & 0 \\ 0 & \Gamma_j \end{bmatrix}$, Γi and Γj are the reflection coefficients of the ports i and j, computed for the case when these ports are open-ended (the reference impedance being 50 Ω) and I_2 is the unity matrix of order two. Because each port is open, the load impedances or the reflection coefficients at the four ports are not known before. Therefore, first of all, the reflection coefficients of the all four ports, $\Gamma_1, \ldots, \Gamma_4$ must be computed. Theoretically, these reflection coefficients are equal to 1, but these values must be known more accurately. The reflection coefficients were computed by solving the following equation system:

$$\begin{aligned} S'_{1-2}[1,1] &= S'_{1-3}[1,1] \\ S'_{1-2}[1,1] &= S'_{1-4}[1,1] \\ S'_{1-2}[2,2] &= S'_{2-3}[1,1] \\ S'_{1-2}[2,2] &= S'_{2-4}[1,1] \end{aligned} \tag{24}$$

where $S'_{i-j}[1,1]$ is the element of the S'_{i-j} matrix from the first row and the first column, while $S'_{i-j}[2,2]$ is the element of the S'_{i-j} matrix from the second row and the second column.

It may be shown that imposing in (24) the conditions: $S'_{1-4}[2,2] = S'_{3-4}[2,2]$, $S'_{2-4}[2,2] = S'_{3-4}[2,2]$, $S'_{3-4}[1,1] = S'_{2-3}[2,2]$ and $S'_{3-4}[1,1] = S'_{1-3}[2,2]$ are also fulfilled.

In (24), the analytical expressions for $S'_{i-j}[1,1]$ and $S'_{i-j}[2,2]$ were developed from (23), obtaining the following formulas:

$$S'_{i-j}[1,1] = \frac{A}{B} \quad \text{and} \quad S_{i-j}[2,2] = \frac{C}{B} \qquad (25\ a,\ b)$$

where:

$$A = (S_{i-j}[1,1] - \Gamma_i) \cdot (1 - S_{i-j}[2,2] \cdot \Gamma_j) + S_{i-j}[1,2] \cdot S_{i-j}[2,1] \cdot \Gamma_j$$

$$C = (S_{i-j}[2,2] - \Gamma_j) \cdot (1 - S_{i-j}[1,1] \cdot \Gamma_i) + S_{i-j}[1,2] \cdot S_{i-j}[2,1] \cdot \Gamma_i$$

and

$$B = (1 - S_{i-j}[1,1] \cdot \Gamma_i) \cdot (1 - S_{i-j}[2,2] \cdot \Gamma_j) - S_{i-j}[1,2] \cdot S_{i-j}[2,1] \cdot \Gamma_i \cdot \Gamma_j$$

By using (25), the equation system (24) has been solved using the MATHCAD software, obtaining the reflection coefficients $\Gamma_1, \ldots, \Gamma_4$. Therefore, the all 6 matrixes S'_{i-j} were obtained numerically with (23), so, the four port matrix of the coupler having the all ports open-ended was written as follows:

$$S' = \begin{bmatrix} S'_{1-2}[1,1] & S'_{1-2}[1,2] & S'_{1-3}[1,2] & S'_{1-4}[1,2] \\ S'_{1-2}[2,1] & S'_{1-2}[2,2] & S'_{2-3}[1,2] & S'_{2-4}[1,2] \\ S'_{1-3}[2,1] & S'_{2-3}[2,1] & S'_{1-3}[2,2] & S'_{3-4}[1,2] \\ S'_{1-4}[2,1] & S'_{2-4}[2,1] & S'_{3-4}[2,1] & S'_{1-4}[2,2] \end{bmatrix}$$

Finally, the scattering matrix of the coupler is re-normalized from the load impedances corresponding to the open ports, to 50 Ω, using the formula:

$$S = (I_4 - \Gamma)^{-1} \cdot (S' - \Gamma) \cdot (I_4 - \Gamma S')^{-1} \cdot (I_4 - \Gamma)$$

where I_4 is the unity matrix of order four and $\Gamma = \begin{bmatrix} -\Gamma_1 & 0 & 0 & 0 \\ 0 & -\Gamma_2 & 0 & 0 \\ 0 & 0 & -\Gamma_3 & 0 \\ 0 & 0 & 0 & -\Gamma_4 \end{bmatrix}$

Using the experimental method proposed above, the coupler has been experimentally characterized (Simion et al., 2008-b).

For the measurements performed on the circuit, a network analyzer (HP 8510C) and onwafer probe heads station (Karl Süss PM5) have been used. These results are presented in Fig. 8 (a), for the magnitude of the scattering parameters S11, S21, S31 and S41. The simulated magnitudes of the scattering parameters S11, S21, S31 and S41 for the test circuit are shown in Fig. 8 (b). Fig. 8(c) shows the simulated and the experimental results for the phase difference between the coupled port and the through port.

By analyzing Figs. 8 (a, b) for the frequency bandwidth of 10 - 12 GHz, the experimental coupling is 5 dB ± 1dB, being in good agreement with the simulated results. The experimental input return-loss and isolation are better than 20dB for the same frequency bandwidth, a good agreement between simulated and experimental results being also observed. In Fig. 8 (c), the experimental phase is 80 - 100 deg., for frequencies between 10.25 - 11.5 GHz, the simulated results being close to the experimental ones.

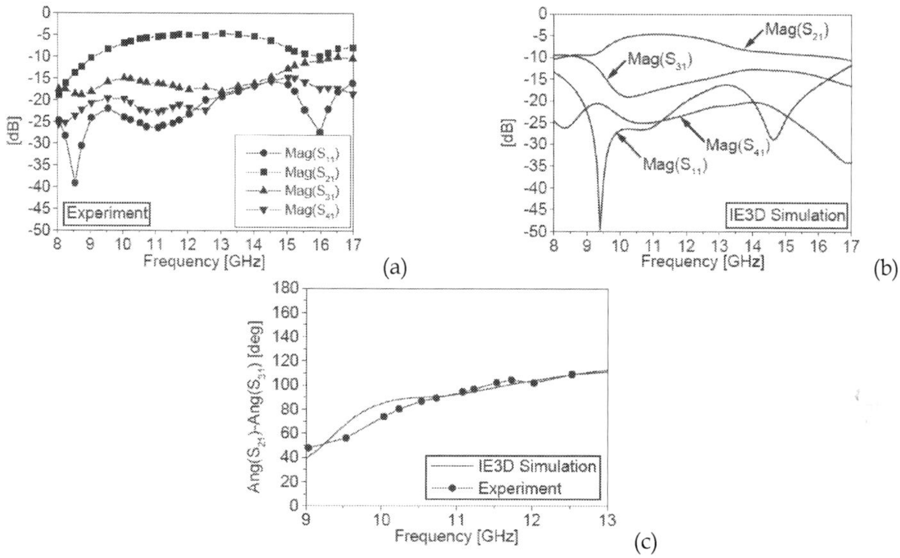

Fig. 8. (a) Experimental magnitudes of the scattering parameters, for the coupler. (b) Simulated magnitudes of the scattering parameters, for the coupler. (c) Simulated and experimental phase difference between the coupled port and the through port of the coupler.

4. CRLH Based Silicon Supported Microwave Resonating Antenna

4.1. Antenna design and layout

The antenna consists of an open-ended array of CRLH cascaded cells, each one having a T - circuit topology. Each cell consists of two series connected CPW interdigital capacitors and two parallel connected short-ended CPW transmission lines. The equivalent circuit of the CRLH cell is presented in Fig.9, where $2C_L$ and $L_R/2$ are the equivalent capacitance and the equivalent inductance of the series capacitor, while C_R and L_L are the equivalent parallel capacitance and the equivalent parallel inductance, respectively, of the two CPW transmission lines.

The parallel capacitance C_R includes the equivalent capacitance of the short-ended CPWs and the equivalent parallel capacitance of the interdigital capacitors.

Using CPW transmission lines, the circuit area could be much smaller comparing to the case when microstrip lines are used, because no large patch area is needed in order to obtain a virtual-ground capacitance used to connect the inductance L_L to the ground.

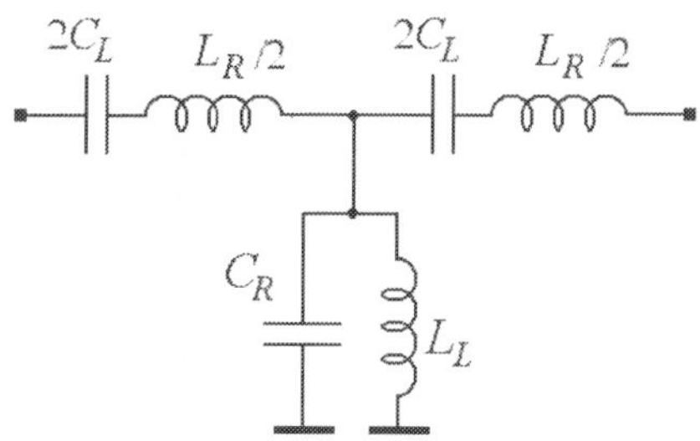

Fig. 9. Equivalent circuit of the CRLH cell used for the antenna design.

For an open-ended CRLH antenna, the zeroth-order resonance occurs at the frequency:

$$f_{sh} = \frac{1}{2\pi\sqrt{L_L C_R}} \tag{26}$$

which is the parallel resonance due to the two CPW short-ended transmission lines.
Also, there are resonance frequencies corresponding to the right-hand (RH) and the left-hand (LH) CRLH behavior (Caloz & Itoh, 2006). For the operating frequency of the zeroth-order antenna, f_{sh}, $\beta = 0$ where β is the equivalent phase constant of the CRLH cell, this frequency being the highest one for the LH frequency range.
In order to design the CRLH cell, the following formulas must also be used:

$$f_L = \frac{1}{4\pi\sqrt{L_L C_L}}; \quad f_R = \frac{1}{\pi\sqrt{L_R C_R}}; \quad f_{se} = \frac{1}{2\pi\sqrt{L_R C_L}}; \quad Z_C = \sqrt{\frac{L_L}{C_L}} \qquad (27) - (30)$$

where f_L and f_R are the cutoff frequencies for the LH and RH modes respectively, f_{se} is the series resonance of the interdigital capacitor and Z_C is the LH characteristic impedance. The frequency range for the LH mode extends from f_L to f_{sh} , while the frequency range for RH mode extends from f_{se} to f_R. The condition to be fulfilled for these frequencies is $f_R > f_{se} > f_{sh} > f_L$.
The starting point in designing the capacitor was to consider the 10 μm width of a metallic line, the length of the capacitor fingers of 0.5 mm (much smaller comparing to the operating wavelengths) and the resonance frequency f_{sh} = 14 GHz (Simion et al., 2007-a). The preliminary values for the elements of the CRLH equivalent circuit (see Fig.9) have been computed using (26) - (30), for Z_C = 50 Ω and the CPW characteristic impedance equal to 60 Ω in order to minimize the losses. After that, the layout of an elementary CRLH cell was designed and optimized. Finally, for the layout of a CRLH cell, the following results were obtained: CPWs length - 1.5 mm; CPWs central conductor width - 100 μm; width of the

CPWs slot - 100 μm; length of the interdigital capacitor at the end of the antenna - 1 mm; length of the internal interdigital capacitor - 0.5 mm; width of the metallic finger of the interdigital capacitor - 10 μm; space between two fingers of the interdigital capacitor - 10 μm; space between the interdigital capacitor and the ground planes of the CPW structure - 100 μm and the number of the metallic fingers of the interdigital capacitor - 10.

For this layout, the elements of the CRLH equivalent circuit were computed, obtaining (see Fig.9): L_L =0.55 nH, C_L = 0.18 pF, L_R = 0.3 nH and C_R = 0.23 pF, corresponding to f_L = 8 GHz, f_{sh} = 14 GHz, f_{se} = 22 GHz and f_R = 38 GHz.

The zeroth-order resonating antenna consists of three identical CRLH cells as it is shown in Fig.10. Each cell has the geometrical dimensions previously given. A CPW line of 4.5 mm length was used to connect the device to the measuring system.

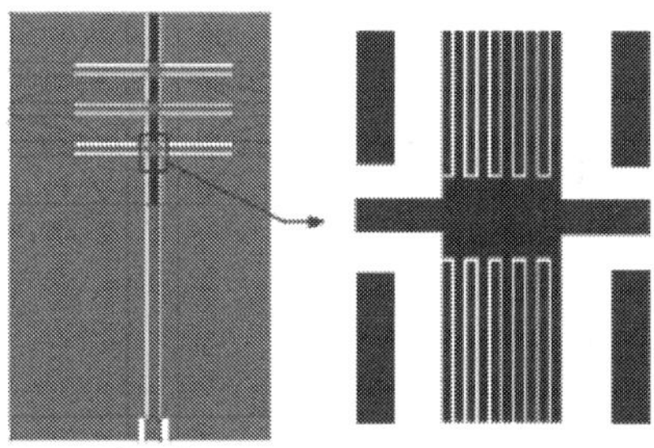

Fig. 10. The antenna layout and a detail for the area around the junction between the CPW interdigital capacitors and the inductive CPW stubs.

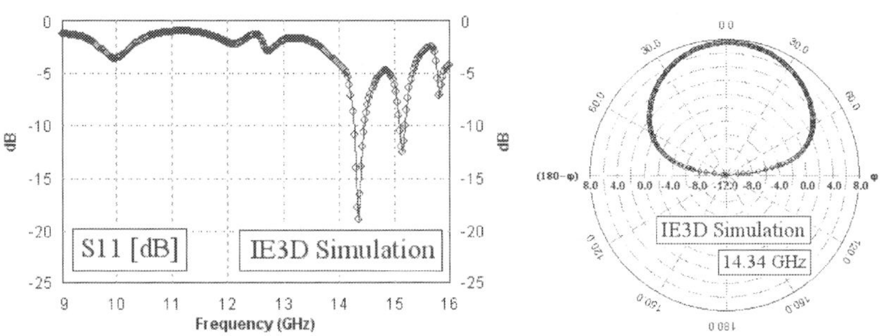

(a) (b)

Fig. 11. (a) Simulated return-loss of the CPW CRLH resonating antenna; (b) Elevation pattern gain for the CPW CRLH resonating antenna, displayed for the total electrical field.

In Fig.11 (a) and (b) the simulated parameters of the antenna - return-loss and radiation pattern - are presented.

It may be seen from Fig.11 (a) that the return-loss is close to 20 dB at 14.34 GHz which is the zeroth-order resonance frequency. Another resonance at 15.2 GHz corresponds to the series resonance of the first and the last interdigital capacitors which are longer compared to the internal ones. The antenna gain - see Fig.11 (b) - is around 7.5 dBi at the zeroth-order resonance frequency.

4.2 Antenna technological realization and experimental results

The technological process for antenna fabrication was the same as for the coupler previously presented - see section 3. The microscope photo of the antenna active area is shown in Fig.12(a). The area occupied by the antenna is 3.9×3.4 mm^2, showing a size reduction of approx. 30%, in comparison to a $\lambda/2$ patch antenna. The measurements of antenna return losses were performed using a network analyzer and an on-wafer probe heads station. In order to measure the antenna gain and pattern of the radiating beam, a mechanical system involving two antennas was prepared. Each antenna was mounted on a SMA connector test fixture, as it is shown in Fig.12(b).

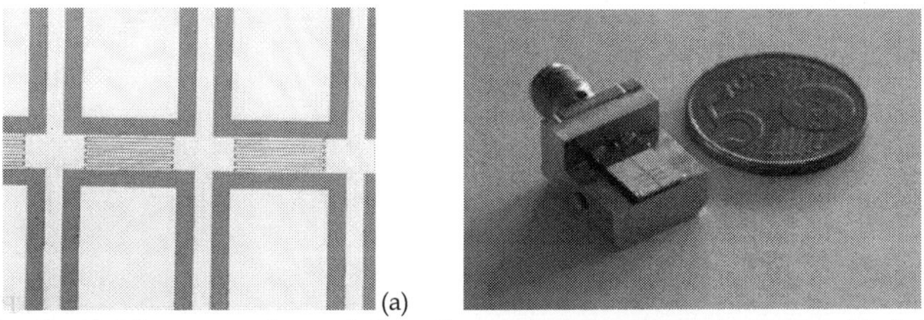

(a) (b)

Fig. 12. (a) Microscope photo of the active part of the obtained antenna; (b) Antenna mounted on a SMA test fixture for the gain and radiation pattern measurements

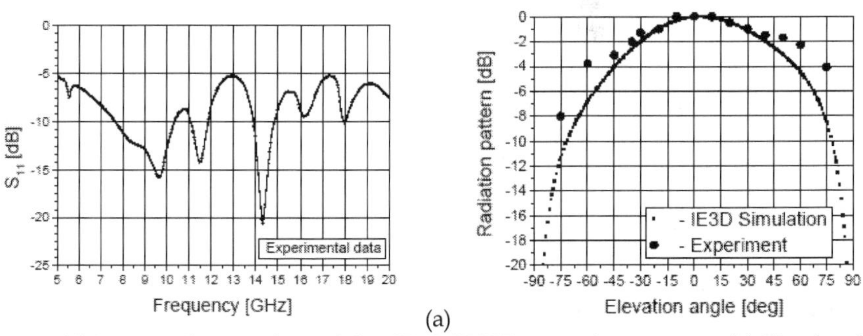

(a) (b)

Fig. 13. (a) Measured return-loss of the CPW CRLH resonating antenna; (b) Simulated and measured E - plane radiation pattern for the CPW CRLH resonating antenna.

Fig.13 (a) shows the measured return-loss of the CPW CRLH resonating antenna and Fig.13 (b) presents the simulated and measured E - plane radiation pattern for the CPW CRLH resonating antenna. It can be observed, that the resonance frequency and the return-loss at this frequency are quite the same as they were predicted by simulation. The errors are less than 0.2% for the resonance frequency and around 2 dB for the return-loss.

For the gain measurement, the two antenna method and the Friis formula, (Balanis, 1997), were used. The distance between the two antennas was $6\times\lambda$ and the magnitude of S_{21} at the resonance frequency was measured for this antenna system. Following this procedure, the measured value for the antenna gain was 6.4 dBi meaning a ~1dB difference from the simulated value.

5. Frequency Tuning of a CPW CRLH Antenna on Magnetically Biased Ferrite Substrate

5.1 CRLH antenna on magnetically biased ferrite substrate

The substrate used was a polycrystalline ferrite having the saturation magnetization Ms = 550 Gs, permittivity $\varepsilon = 13.5$, resonance linewidth $\Delta H = 16.8$ kA/m. The thickness of the ferrite substrate was 0.5 mm and the surface to be metallized was mirror polished. The antenna layout was designed for an external applied field $H_{appl} = 0$ T, namely, for the ferrite substrate in unmagnetized state, using the same method as presented in Section 4.
If a dc biasing magnetic field (H_{appl}) is applied normally to the ferrite substrate, the permeability changes its values from the unpolarized state. For the biasing magnetic field varying from $H_{appl} = 0$ T to $H_{appl} = 0{,}26$ T the effective permeability (μ_{eff}) of the chosen ferrite substrate varies from $\mu_{eff} \cong 1$ to $\mu_{eff} = 0.921$ respectively, as it is shown in Fig.14. The computation was done (Lax & Button, 1962) for a frequency f = 13 GHz.

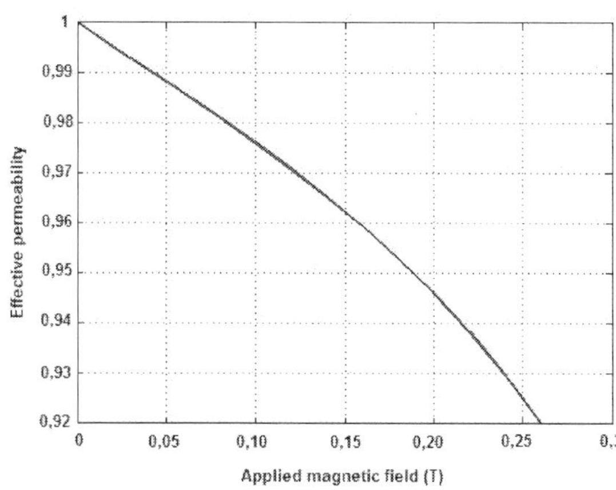

Fig. 14. Effective permeability vs. applied magnetic field

The simulation of the antenna return-loss vs. frequency was carried out for four values of μ_{eff} in the above range, corresponding to four intensities of the biasing magnetic field. The maximum value of the return-loss and the frequency for each value of the intensity of the magnetic biasing field are specified in Table 1.

Applied magnetic fied (T)	Effective permeability (μ_{eff})	Resonance frequency (GHz)	Return losses (dB)
0	1	12.88	-28.00
0.055	0.986	12.96	-20.26
0.18	0.951	13.20	-22.30
0.26	0.921	13.41	-17.70

Table 1. Computed antenna operating frequency and the corresponding return losses, when the ferrite substrate is magnetically polarized

The calculated values of S_{11} vs. frequency for the effective permeability values are shown in the graphs in Fig.15 (a)...(d). It must be noted that for this domain of the applied field the antenna structure works under the gyromagnetic resonance.

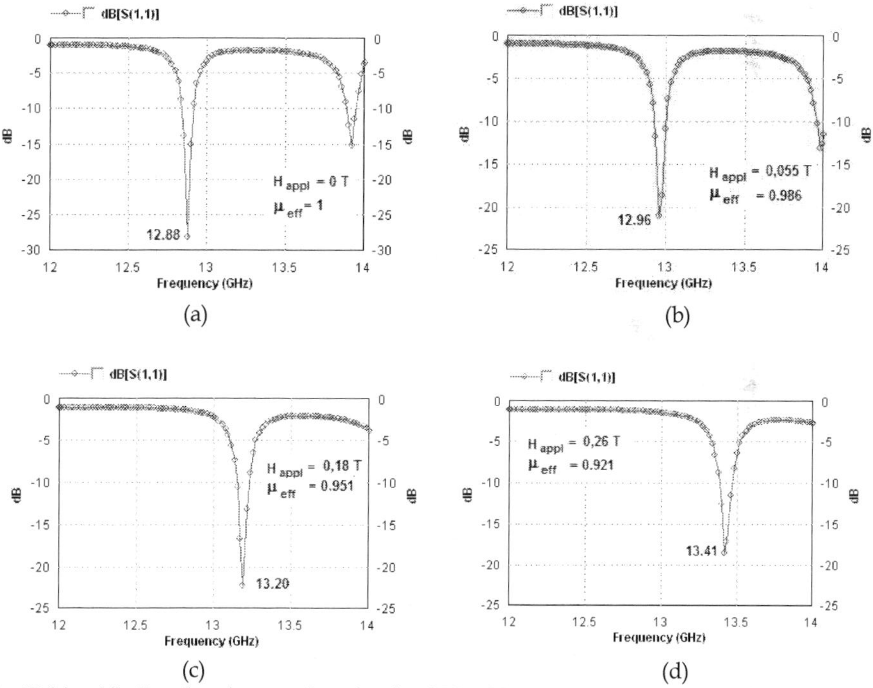

Fig.15 (a) - (d). Simulated return-loss for the CPW CRLH resonating antenna for four values of the biasing magnetic field.

It must be observed that the zeroth-order resonance frequency calculated in the absence of the magnetic biasing field was found as 12.88 GHz with a return losses level of 28 dB - see Fig.15 (a). There is another parasitic resonance at 13.9 GHz corresponding to the series resonance of the first interdigital capacitors. When the biasing magnetic field is applied, the calculated frequency for the antenna structures ranges from 12.88 GHz for H_{appl} = 0 to 13.41 GHz for H_{appl} = 0.26 T. Also, the calculated reflection losses changes from 28 dB at

$H_{appl} = 0$ to approx. 17.7 dB for $H_{appl} = 0.26$ T. Therefore a very good impedance matching is obtained for the operating frequency range.

5.2 Experimental results

The technological process for manufacturing the antenna was the same as for the coupler (see section 3).

The individual antenna structures obtained by cutting the ferrite wafer were mounted on suitable test fixtures - see Fig.16 (a) - in a measurement system able to measure microwave frequency and reflection losses while applying a biasing static magnetic field normally on the ferrite antenna substrate - see Fig.16 (b). In Fig.16 (b) number 1 denotes the antenna structure mounted on a test fixture, 2 denotes polar pieces of the electromagnet, the superior one being movable and 3 is the Hall sensor of a tesla-meter.

The experimental results concerning the measured resonating frequencies and return losses (RL) of two antenna structures, are shown in Table 2 and in Fig.17 (a) - (b).

H_{appl}. (T)	Antenna structure 1		Antenna structure 2	
	Freq. (GHz)	RL (-dB)	Freq. (GHz)	RL (-dB)
0	13.35	25	13.57	18
0.02	13.35	25	13.57	18
0.04	13.35	25	13.57	18
0.055	13.35	26	13.57	19
0.06	13.35	26	13.58	19
0.07	13.35	27	13.59	19
0.08	13.38	27	13.60	19
0.10	13.4	27	13.61	19
0.12	13.41	28	13.64	19
0.14	13.44	26	13.67	19
0.16	13.49	26	13.70	19
0.18	13.51	24	13.74	18
0.20	13.58	22	13.78	17
0.22	13.63	21	13.81	17
0.24	13.70	20	13.89	16
0.26	13.8	18	13.97	14

Table 2. The experimentally obtained resonating frequencies and return losses of the measured antennas vs. magnetic biasing field.

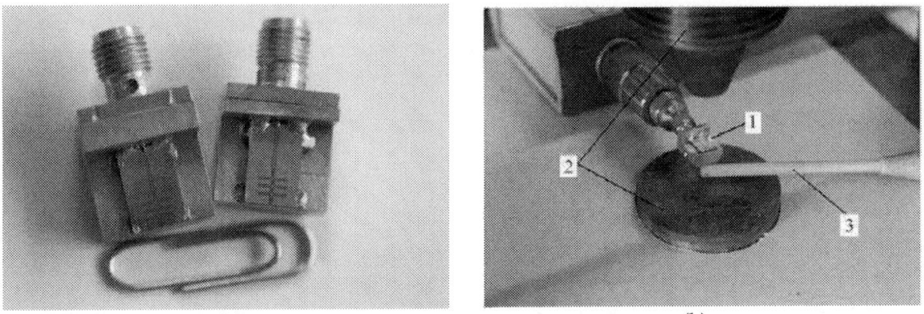

(a) (b)

Fig. 16. (a) SMA test fixture supporting the fabricated CPW CRLH zeroth-order resonating antenna on ferrite substrate; (b) Test fixture with antenna structure in an applying magnetic field facility during S_{11} measurements.

The curves showing the frequency variation of the two antennas - see Fig.17 (a) - and return losses variation - see Fig.17 (b) - as function of the applied magnetic field were obtained using the data from Table 2.

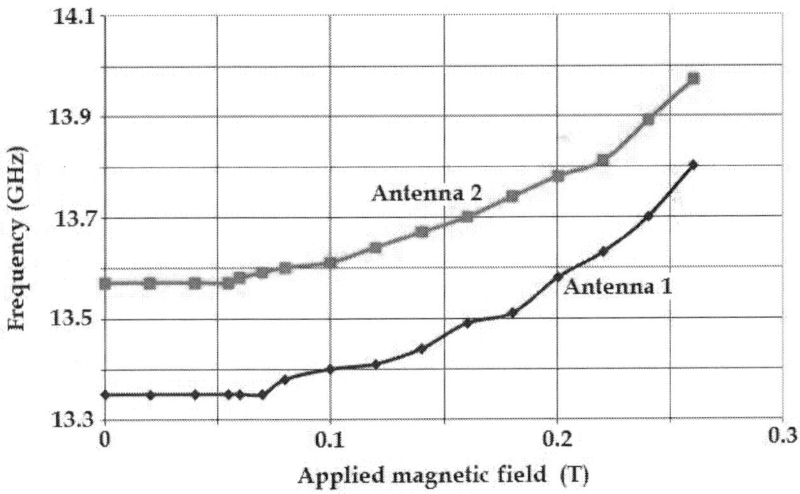

Fig. 17. (a) Frequency shift vs. applied magnetic field;

It can be observed that for a magnetic biasing field less than 0.55 T, the resonating frequencies of the two antennas were 13.57 GHz and 13.35 GHz, respectively, slightly higher than the simulated results (see Table 1). This difference between the calculated and measured resonance frequency at low H_{appl} is due to some technological limitations in obtaining the exact geometry size of each antenna structure.

If the applied magnetic field remains under the value of the saturation magnetization, (550 Gs), the working frequencies almost maintain their initial values. At higher values of H_{appl}, the resonance frequencies of the antennas tend to increase, as it may be seen in Table 2 and in Fig.17(a). The total measured frequency shift was 400 MHz for the first antenna structure and 450 MHz for the second one, when the biasing magnetic field varied from $H_{appl} = 0$ to $H_{appl} = 0.26$ T. This frequency shift fits very well with the calculated values (see Table 1).

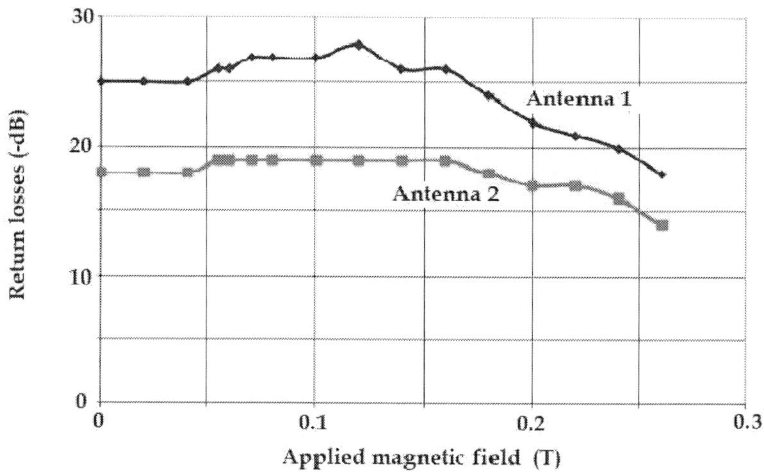

Fig. 17. (b) Return losses vs. applied magnetic field.

The measured return losses of the two antenna structures are presented in Fig.17 (b). It is observed that the experimental values for the return loss and the simulated values given in Table 1 are in very good agreement.

6. Conclusion

The chapter presents applications of CRLH structures, including a directional coupler and a resonating antenna both fabricated on silicon substrate, in order to facilitate the future device integration in a more complex monolithic integrated circuit. Also, a resonating antenna on magnetically polarized ferrite has been fabricated and experimentally tested.
CPWs were used for all these devices and a full-wave electromagnetic analysis software (IE3D - Zeland) was used for the design methods.
The experimental coupling of the coupler is 5 dB ± 1dB, for a frequency bandwidth of 10 - 12 GHz. Also, the experimental input return-loss and isolation are better than 20 dB for the same frequency bandwidth, a good agreement between simulated and experimental results being also observed. The phase difference between the ports 2 and 3 is $90^0 \pm 10^0$, for a frequency bandwidth from 10.5 GHz to 12.25 GHz. All these experimental results are in good agreement with the expected ones obtained by simulation.

The experimental operating frequency and the return-loss of the antenna fabricated on silicon substrate is ~14.5GHz and ~20dB, showing a good agreement in comparison to the simulation results (errors less than 0.2% for the resonance frequency and around 2 dB for the return-loss on the resonance frequency). Also, the experimental results for the gain and for the radiation pattern show a good agreement compared to the simulation results (for the gain, the error is around 1 dB).

Concerning the ferrite supported antenna, the experimental data shows a frequency shift of 400 MHz for one of the antenna structures and 450 MHz for the other one, due to the variation of the effective permeability of the ferrite substrate. This work demonstrates the possibility to tune a CRLH antenna made on a ferrite substrate by varying the biasing magnetic field. For this type of antenna, again, the experimental results are very close to the simulation results.

7. References

Balanis C. A. (1997). *Antenna theory - Analysis and design,* John Wiley & Sons Inc., ISBN 0-471-59268-4, New York.

Caloz C., Itoh T. (2003). Positive/negative refractive index anisotropic 2-D metamaterials, *IEEE Microwave and Wireless Components Letters*, Vol.13, No.12, pp. 547-549, ISSN 1531-1309.

Caloz C., Sanada A., T. Itoh T. (2004). A novel composite right-/left-handed coupled-line directional coupler with arbitrary coupling level and broad bandwidth", *IEEE Trans. on Microwave Theory and Techniques*, Vol.52, No.3, pp.980-992, ISSN 0018-9480.

Caloz C., Itoh T. (2005). Metamaterials for high-frequency electronics. *Proc. of the IEEE*, Vol. 93, No. 10, pp. 1744-1752, ISSN 0018-9219

Caloz C., Itoh T. (2006). *Electromagnetic metamaterials: transmission line theory and microwave applications,* John Wiley & Sons, Inc., ISBN-10: 0-471-66985-7, USA, Canada.

Engheta N., Ziolkowski R. W. (2005). A positive future for double-negative metamaterials. *IEEE Trans. on Microwave Theory and Techniques*, Vol.53, No.4, pp. 1535-1556, ISSN 0018-9480.

Kodera T., Caloz C. (2008). Comparison of various ferrite-loaded CRLH leaky-wave antenna structures. *Proc. of Asia-Pacific Microwave Conference,* Paper J3-06, ISBN 978-1-4244-2642-3, Hong Kong, Republic of China, December 2008, Hong Kong and Macau.

Lai A., Caloz C., Itoh T. (2004). Composite right/left-handed transmission line metamaterials. *IEEE Microwave Magazine,* Vol. 5, No. 3, Sept.2004, pp. 34 - 50, ISSN 1527-3342

Lax B., Button K. J. (1962). *Microwave Ferrites and Ferrimagnetics,* USA: McGraw-Hill Book Comp., Inc., New York.

Li C., Liu Y. K., Li F. (2007). Analysis of composite right/left-handed coplanar waveguide zeroth-order resonators with application to a band-pass filter. *PIERS Online*, vol.3, no.5, 2007, pp.599-602.

Lim S., Caloz C., Itoh T. (2004). Metamaterial-based electronically controlled transmission-line structure as a novel leaky-wave antenna with tunable radiation angle and beamwidth. *IEEE Trans. on Microwave Theory and Techniques*, Vol.52, No.12, pp.2678 - 2690, ISSN 0018-9480.

Lin I.-H., Caloz C., Itoh T. (2003). A branch-line coupler with two arbitrary operating frequencies using left-handed transmission lines", *IEEE MTT-S Digest*, 2003, pp.325-328. ISBN 0-7803-7695-1, Philadelphia, Pennsylvania, June 2003

Okabe H., Caloz C., Itoh T. (2004), "A compact enhanced-bandwidth hybrid ring using an artificial lumped-element left-handed transmission-line section", *IEEE Trans. on Microwave Theory and Techniques*, vol.52, no.3, pp.798-804, ISSN 0018-9480.

Sajin G., Simion S., Craciunoiu F., Marcelli R. (2007). Silicon supported microwave zeroth-order resonance antenna on metamaterial approach, *Proceedings of the 2007 Asia-Pacific Microwave Conference, APMC 2007*, pp.221–224, ISBN 1-4244-0748-6, Bangkok, Thailanda, December 2007.

Sajin G., Simion S, Craciunoiu F., Muller A., Bunea A. C. (2009). Frequency Tuning of a CRLH CPW Antenna on Ferrite Substrate by Magnetic Biasing Field. Accepted paper for European Microwave Conference, EuMW 2009, Rome, Italy, September-October 2009.

Sanada A., Kimura M., Awai I., Caloz C., Itoh T. (2004). A planar zeroth-order resonator antenna using a left-handed transmission line. *Proc. of the 34th European Microwave Conference*, pp.1341-1344, Amsterdam, The Netherlands, October 2004, Horizon House, Amsterdam.

Sievenpiper D., Zhang L., Broas R. F. J., Alexopolous N. G., Yablonovitch E. (1999). High impedance electromagnetic surfaces with a forbidden frequency band. *IEEE Trans. on Microwave Theory and Techniques*, Vol.47, No.11, pp. 2059-2074, ISSN 0018-9480.

Simion S., Sajin G., Marcelli R., Craciunoiu F., Bartolucci G. (2007-a). Silicon Resonating Antenna Based on CPW Composite Left/Right-Handed Transmission Line, *Proc. of the 37th European Microwave Conference*, pp. 478 - 481, ISBN 978-2-87487-000-2, Munchen, Germany, October 2007.

Simion S., Marcelli R., Sajin G. (2007-b). Small size CPW silicon resonating antenna based on transmission-line meta-material approach, *Electronics Letters*, Vol.43, No.17, pp.908-909, ISSN 0093-5914.

Simion S., Marcelli R., Bartolucci G., Sajin G. (2008-a). Design, Fabrication and On-Wafer Characterization of a Meta-Material Transmission Line Coupler, *International Journal of Microwave and Optical Technology - IJMOT*, Vol.3, No.3, pp. 363–369. ISSN 1553-0396.

Simion S., Marcelli R., Bartolucci G., Sajin G., (2008-b). On wafer experimental characterization for a 4-port circuit using a two-port vector network analyzer, *Proc. of the 31rst International Semiconductor Conference, CAS-2008*, pp. 223–226, ISBN 978-1-4244-2004-9; ISSN 1545-827X, Sinaia, Romania, October 2008.

Tippet J. C., Speciale R. A. (1982). A rigorous technique for measuring the scattering matrix of a multiport device with a 2-port network analyzer. *IEEE Trans. on Microwave Theory and Techniques*, Vol.30, No.5, pp. 661 - 666, ISSN 0018-9480.

Tong W., Hu Z., Chua H. S., Curtis P. D., Gibson P. A. A., Missous M. (2007). Left-handed metamaterial coplanar waveguide components and circuits in GaAs MMIC technology, *IEEE Trans. on Microwave Theory and Techniques*, vol.55, no.8, August 2007, pp.1794-1800.

Veselago V.G. (1968). The electrodynamics of substances with simultaneously negative values of ε and μ. *Sov. Physics – Usp.*, vol.47, January-February 1968, pp. 509 - 514.

7

Wide Band Gap Semiconductor Based High-power ATT Diodes In The MM-wave and THz Regime: Device Reliability, Experimental Feasibility and Photo-sensitivity

Moumita Mukherjee
Centre of MM-Wave Semiconductor Devices & Systems(CMSDS), Centre of Advanced Study in Radio Physics & Electronics, University of Calcutta, INDIA

1. Introduction

Avalanche Transit Time (ATT) Diodes which include IMPATTs, TRAPATTs, BARITTs and so on are potential solid-state sources for Microwave power. Among these devices, IMPATTs are by far the most important in view of their frequency range and power output and show great promise of increasing application in the twenty first century. During the initial phases of development of IMPATT devices in the late sixties and early seventies, Ge (Germanium) and Si (Silicon) were mainly used as semiconducting materials for IMPATT fabrication. In view of their low power capability, Ge IMPATTs have now become obsolete. In the seventies the rapid development of Si technology has made possible the emergence of Si SDR and DDR IMPATTs which can provide power at microwave and MM-wave frequency bands. GaAs (Gallium Arsenide) also emerged as a highly suitable material for fabricating IMPATT diodes in the lower microwave frequency range. Now-a-days IMPATT devices are used in microwave and MM-wave digital and analog communication systems, high power RADARs, missile seekers, and in many other defence systems.

In recent years, the development of sources for Terahertz frequency regime are being extensively explored worldwide, for applications in short-range terrestrial and airborne communications, spectroscopy, imaging, space-based communications and atmospheric sensing. To meet the rising demand of high-power, high-frequency solid-state sources, extensive research is being carried out for development of high-power IMPATT devices in MM-wave and Terahertz regime. The material parameters responsible for heat generation and dissipation in IMPATT diodes play a vital role in limiting the output power of conventional Si and GaAs IMPATT diodes at a particular frequency. Among several approaches for realizing high-power, high-frequency IMPATT sources, one option is to develop IMPATT devices based on Wide-Band-Gap (WBG) semiconductors (e.g. SiC and GaN) having high critical electric field (E_C), high carrier saturation velocity (v_S) as well as high thermal conductivity (K) *(**Table 1**)* [Trew *et al.*], since RF power output from an

IMPATT is proportional to $E_C^2 v_s^2$. Moreover, high value of K is essential to ensure good thermal stability for high-power operation of the devices. All these intrinsic material parameters of WBG semiconductors are favorable for realizing smaller transit time, an essential criterion for developing THz devices. The expected excellent performances of WBG devices can also be expressed by figures of merit (FOM).

Semiconductor	Si	GaAs	6H-SiC	4H-SiC	3C-SiC	WZ-GaN	ZB-GaN	InP	Diamond
Bandgap (E_g) (eV)	1.12	1.43	3.03	3.26	2.2	3.45	3.28	1.35	5.45
Critical Electric Breakdown field (E_C) (10^7 V.m^{-1})	3.0	4.0	25.0 (∥ to c-axis)	22.0 (∥ to c-axis)	21.2	20.0	20.0	5.0	100.0
Relative dielectric constant (ϵ_r)	11.9	13.1	9.66	9.7	9.7	8.9	9.7	12.5	5.5
Electron mobility (μ_n) ($m^2 V^{-1} s^{-1}$)	0.15	0.85	0.04 (∥ to c-axis) 0.05 (⊥ to c- axis)	0.10 (both ∥ and ⊥ to c- axis)	0.075	0.125	0.100	0.54	0.22
Hole mobility (μ_p) ($m^2 V^{-1} s^{-1}$)	0.06	0.04	0.01	0.01	0.004	0.085	0.035	0.02	0.085
Saturated carrier drift velocity (v_s) (∥ to c-axis) (10^5 ms^{-1})	1.0	1.2	2.0	2.0	2.2	2.5	2.0	2.2	2.7
Thermal Conductivity (K) (Wm^{-1} K^{-1})	150.0	46.0	490.0	490.0	320.0	225.0	130.0	69.0	2200.0

Table 1. Material properties of Si, GaAs, InP and important Wide Bandgap semiconductors.

The *Baliga* FOM is important for evaluation of high frequency application and *Johnson's* FOM considers the high-frequency and high-power capability of devices. Taking *Baliga* and *Johnson's* FOM for Si as unity, the *Baliga* and *Johnson's* FOM for GaAs are 11.0 and 7.1, respectively, while those for WBG semiconductor SiC are 29.0 and 278 and those for GaN are 77.8 and 756. Hence, SiC and GaN are found to be superior to both conventional Si and GaAs for high-frequency and high-power operation. Thus, in a bid to find single small-sized MM-wave and THz power sources, it is interesting to study the prospects of WBG semiconductor based IMPATT diodes.

In this Chapter, the DC and high-frequency characteristics of SiC and GaN based IMPATT devices at MM-wave and THz region will be presented first. This will be followed by the photo-sensitivity and experimental feasibility studies of the new-class of IMPATT devices.

2. IMPATT diode: brief history of development.

IMPATT is an acronym of *IMPact ionization Avalanche Transit Time*, which reflects the mechanism of its operation. In its simplest form, an IMPATT is a p-n junction diode reversed biased to breakdown, in which an avalanche of electron-hole pair is produced in the high-field region of the device depletion layer by 'impact ionization'. The transit of the carriers through the depletion layer leads to generation of microwave and MM-waves when the device is tuned in a suitable microwave and MM-wave cavity. These diodes exhibit negative resistance at microwave and MM-wave frequencies due to two electronic delays, viz., (i) 'avalanche build-up delay' due to 'impact ionization' leading to avalanche multiplication of charge carriers and (ii) 'transit time delay' due to the saturation of drift velocity of charge carriers moving under the influence of a high electric field.

The working principles of the device were first described by *Read* in 1958. However, the idea of obtaining a negative resistance from a reversed biased p-n junction dates back to an earlier paper (1954) by *Shockley*, in which he showed that when an electron bunch from a forward biased cathode is injected into the depletion layer of a reversed biased p-n junction a 'transit time negative resistance' is produced as the electrons drift across the high field region. The negative resistance from such early devices was found to be small and microwave power output was low. Read showed that an improved negative resistance is obtained when impact ionization is used to inject the electrons. He showed that the properties of charge carriers in a semiconductor i.e. (i) avalanche multiplication by impact ionization and (ii) transit time delay of charge carriers due to saturation of drift velocity at high electric fields, could be suitably combined in a reverse-biased p-n junction to produce a microwave negative resistance. By exploiting the time delay required to build up an avalanche discharge by impact ionization, coupled with Shockley's transit time delay, he showed that efficient microwave oscillation could be realized in his proposed p^+ n i n^+ diode. However, due to the complicated nature of the Read structure, it was not until 1965 that the first experimental Read diode was fabricated. In the early 1965 *Johnston et al.*, from Bell Laboratories, first made a successful experimental observation of microwave oscillations from a simple Si p-n junction diode. This study showed that the complicated Read structure was not essential required for generating microwave oscillations. On the basis of a small-signal analysis, *T. Misawa* showed that negative resistance would occur in a reverse biased p-n junction of any arbitrary doping profile. Since then, rapid advances have been made towards further development of various IMPATT structures, fabrication techniques as well as optimum circuit design for IMPATT oscillators and amplifiers. The frequency range of IMPATT devices can be pushed easily to MM and sub-MM wave ranges at which comparable amount of RF power generation is hardly possible by other two-terminal solid-state devices.

3. IMPATT structures and doping profiles

The typical doping profile of a Read diode makes its realization very difficult in practice. There are several other structures with simpler doping profiles which also exhibits microwave negative resistance due to IMPATT action. In practically realizable structures, the avalanche region is not very thin as was in case of Read diode and also there is no distinct demarcation between avalanche and drift regions. Single Drift Region (SDR) and Double Drift Region (DDR) IMPATTs are now commonly used belong to this category.

Single drift IMPATT (SDR) structure is based on a one-sided abrupt p-n junction of the form p^+ n n^+ or n^+ p p^+. These diodes have a single avalanche zone of finite width located at one end of the depletion layer near the junction followed by a single drift region. The doping profile at the junction and at the interface of substrate and epitaxy are approximated by use of appropriate exponential and error function. The schematic doping profile of a typical SDR diode is shown in Figure 1. Conventional SDR diodes are fabricated with Si and GaAs as base semiconductor material. SDR p^+ n n^+ IMPATT structure is better than n^+ p p^+ structure because technology of n^+ substrate is more advanced and better understood than p^+ substrate. Further, the extent of the un-depleted region between the edge of the depletion region and interface of epitaxy and substrate (un-swept epitaxy), which contributes positive series resistance and thereby dissipates microwave power, is smaller in p^+ n n^+ structure than complimentary n^+ p p^+ structure, since, compared to hole mobility, mobility of electrons in most of the semiconductors are much larger owing to its lower effective mass. The fabrication of GaAs and InP SDR IMPATTs has been mostly reported with p^+ n n^+ structure because of the advantages of better avalanche characteristics, lower loss due to un-swept epitaxy and advanced n+ substrate technology.

Double Drift IMPATT diode is another type of structure. A DDR diode is basically a p^+ p n n^+ (or its complementary) multilayer structure usually with a symmetrical step junction. A typical flat profile DDR along with its schematic doping profile and E(x) profile are shown in ***Figure 2***. The E(x) profile is characterized by a centrally located high field (> 10^7 Vm^{-1}) around the metallurgical junction along with two low field drift regions, for electrons and holes, on either side. The holes generated in the avalanche region drift through the drift region on the p-side while the generated electrons drift through the drift region on the n-side. In comparison to the SDR structure, in case of the DDR structure contribution to microwave power comes from the two drift regions. The second drift region in the DDR diode, improves the efficiency, RF power density and impedance per unit area. The impedance of an IMPATT diode can be approximated by a simple equivalent circuit which consists of a series combination of negative resistance (R_D) and reactance (X_D). In the oscillating frequency range, the magnitude of $R_D < X_D$, and thus the device reactance is approximately that of the capacitance formed by the depletion layer of the device. In the DDR structure, the added drift region increases the depletion layer width resulting in a smaller capacitance and hence a large reactance per unit area. Thus, the impedance level of a DDR diode is high as compared to that of the SDR diode. Several workers have previously suggested that the efficiency and RF power output of SDR or DDR diodes can be enhanced by modifying the epi-layer doping profile. The introduction of an impurity bump i.e. the region of high doping density, considerably improves the device efficiency. Impurity bumps can be suitably introduced in the depletion region by Molecular Beam Epitaxy (MBE) or by ion implantation to produce high-efficieny IMPATT diodes.

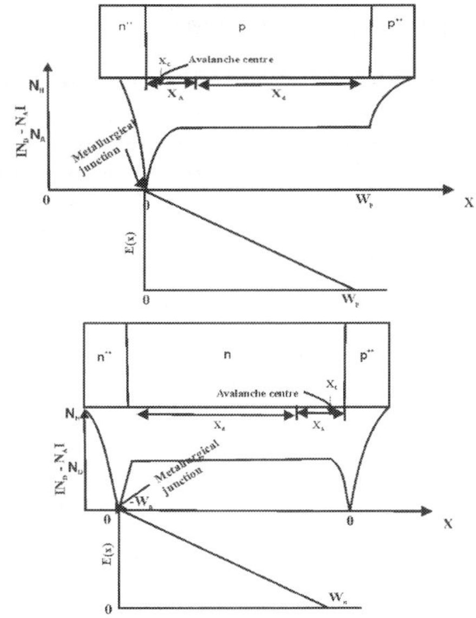

Fig. 1. Schematic diode structure, electric field and droping profiles of n^{++} pp^{++} and p^{++} n^{+} SDR diodes

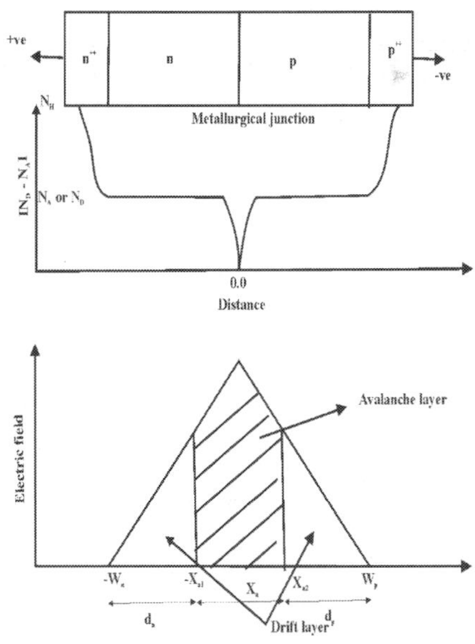

Fig. 2. The schematic diode structure, doping profile and field profile of a Double Drift flat profile diode

Two types of such modified structures are generally possible, (i) lo-hi-lo, characterized by three step doping profiles and (ii) hi-lo, characterized by two step doping profiles. Owing to some of their similarities with Read structures, such as narrow localized avalanche zone, these diodes are also called 'Quasi Read' diodes. ***Figures 3 (a-b)*** show the typical doping profile, E(x) profiles of hi-lo, lo-hi-lo SDR and DDR diodes.

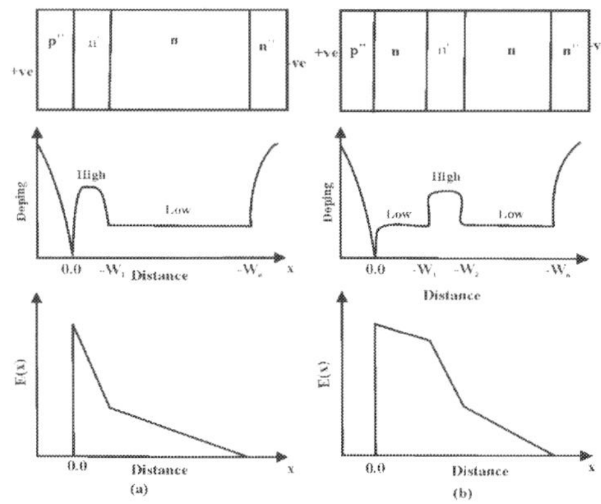

Fig. 3. (a) (i) Schematic diagram of Single Drift ,high-low' structure, doping profile and field profile

(ii) Schematic diagram of Single Drift ,low-high-low' structure, doping profile and field profile

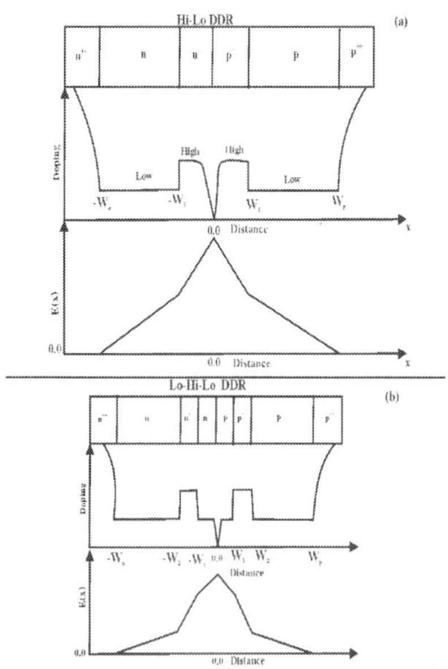

Fig. 3. (b): The schematic diode structure, doping profile and typical field profile of (i) High-Low DDR and (ii) Low-High-Low DDR IMPATT diodes

4. Basic operation principle of IMPATT diodes.

Microwave generation in an IMPATT diode can be explained on the basis of a simple Single Drift Region (SDR) structure (Read or p^+ n n^+ or p^+ p n^+). If a sinusoidal electric field is applied to the device biased to the threshold of dc breakdown, an avalanche of e-h pair is created in the avalanche region. The number of e-h pair reaches its peak after the peak of the ac field has passed. This is because the number of e-h pairs created is proportional to the product of ionization rate of an individual carrier, which is highest at the instant of the peak field, and the number density of charge carrier presents at that time. Since the number density goes on increasing as long as the applied field is added to the dc field, the peak of e-h pair generation is delayed with respect to the ac field by a phase angle of approximately 900. This delay is known as avalanche build up delay. The current pulse of carriers thus formed are injected into the drift zone, where the magnitude of the electric field is such (10^6 - 10^7 V m^{-1}) that the carriers are able to drift with saturated velocity but unable to produce additional carriers through impact ionization. This charge pulse crosses the ionization-free drift zone with saturated velocity and produces a constant induced current in the external circuit during the time of transit, W/v_S.

The external current is approximately a rectangular wave and it develops between the phase of π to 2π (***Figure 4***). The width of the drift region is so adjusted that the transit time of carriers is half the period of the ac cycle. Thus the total phase lag between applied RF voltage and external RF current is 1800, which gives rise to negative resistance. One may get the first hand idea of frequency of oscillation from the approximate equation:

$f_0 = v_s/2W$.

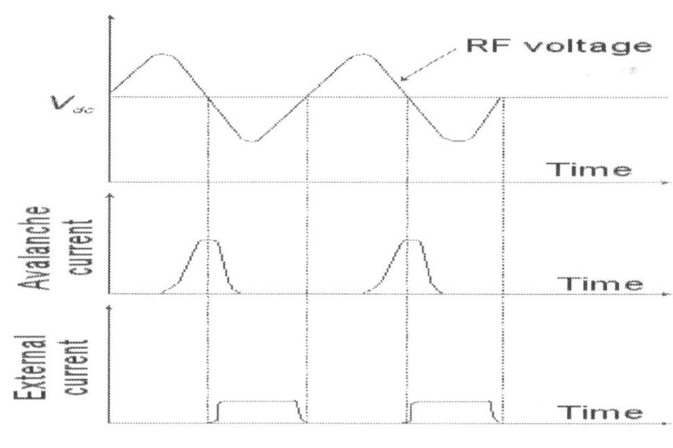

Fig. 4. Waveform of RF voltage, avalanche current and induced external current in a IMPATT diode

5. Simulation scheme for DC and high-frequency analysis of un-illuminated and iluminated IMPATT diodes of any doping profile

Numerical simulations have immense importance in producing guidelines for device design and materials research. Moreover, computer studies are essential for understanding the

properties of devices, as analytical methods do not provide accurate information regarding the dc and high frequency parameters of these devices. In the present thesis, a generalized, simple and more accurate dc computer simulation method that involves simultaneous computer solution of the nonlinear Poisson's and carrier continuity equations, as proposed by *Roy et al.* [15], has been adopted. DC modeling of the IMPATT devices has been made realistic by considering the effects of mobile space charge, inequality of ionization rates and drift velocities of charge carriers of the base materials and also their electric field and temperature dependence. The optimum depletion layer widths for a particular design frequency (f_o) are chosen from the simple transit time formula $W = 0.37\ v_{sn,sp}/\ f_o$ [16]. Here v_{sn} and v_{sp} are the saturated drift velocities of electrons and holes respectively.
DC field and carrier current profiles for various IMPATT structures can be obtained by starting the computation from the field maximum position, at the metallurgical junction. The simulation method consists of two parts: (i) DC analysis and (ii) small-signal analysis. In the dc method, Poisson and carrier continuity equations are simultaneously solved at each point in the depletion layer, subject to appropriate boundary conditions, as described elsewhere [*Roy et al* (1985), *Mukherjee et al (2007a)*]. A very small space step is considered for the accurate numerical simulation of the equations.
The DC to RF conversion efficiency (η) [*Namordi et al.* (1980)] is calculated from the semi-quantitative formula,

$$\eta\ (\%) = (V_D \times 100)\ /\ (\pi \times V_B) \qquad (1)$$

where, V_D = voltage drop across the drift region. Also, $V_D = V_B - V_A$, where, V_A = voltage drop across the avalanche region and V_B = Breakdown voltage.
The small-signal analysis of the IMPATT diode provides significant insight into the device physics and intrinsic properties of the devices. The range of frequencies exhibiting negative conductance of the diode can easily be computed by the Gummel-Blue method [*Gummel Blue* (1967)]. From the dc field and current profiles, the spatially dependent ionization rates that appear in the *Gummel-Blue* equations are evaluated and fed as input data for the high-frequency analysis. The edges of the depletion layer of the diode, which are fixed by the dc analysis, are taken as the starting and end points for the high-frequency analysis. The spatial variation of high frequency negative resistivity and reactivity in the depletion layer of the diode are obtained under small-signal conditions by solving two second order differential equations in $R(x, \omega)$ and $X(x, \omega)$.
$R(x, \omega)$ and $X(x, \omega)$ are the real and imaginary parts of the diode impedance $Z\ (x,\omega)$, such that $Z\ (x,\omega) = R(x, \omega) + j\ X(x, \omega)$. A generalized computer algorithm for simulation of the negative resistivity and reactivity in the space charge region is used in the analysis and described elsewhere [*Roy et al* (1985), *Mukherjee et al (2007a)*]. The total integrated diode negative resistance (Z_R) and reactance (Z_x) at a particular frequency (ω) and current density J_{DC}, are computed from numerical integration of the R(x) and X(x) profiles over the active space-charge layer.
The high-frequency admittance characteristics, negative resistivity profiles and device quality factor (Q) of the optimized diodes are determined by this technique after satisfying the appropriate boundary conditions for R and X, as described elsewhere [Roy *et al* (1985), Mukherjee *et al (2007a)*]. The diode quality factor (Q_P) at the peak frequency, is defined as the ratio of the imaginary part of the admittance to the real part of the admittance (at the peak frequency), i.e.,

$$-Q_p = (B_p / -G_p) \tag{2}$$

The maximum output power density (P_{output}) from the device is obtained from the expression [*Eisele et al.* (1997)]:

$$P_{output} = (V_{RF}{}^2 . \, |-G_P\,|)/2 \tag{3}$$

The diode negative conductance at the optimum frequency $|-G_P|$ is normalized to the area of the diode. V_{RF} (amplitude of the RF swing) is taken as $V_B/2$, assuming a 50% modulation of the breakdown voltage, V_B.

The value of series resistance (R_S) is determined from the admittance characteristics using a realistic analysis by Gummel-Blue [*Gummel Blue* (1967)] and Adlerstein [*Adlerstein et al (1983)*]. Under small-signal approximation, the steady state condition for oscillations is given by:

$$G_L(\omega) = |-G(\omega)| - [B(\omega)]^2 R_S(\omega) \tag{4}$$

where G_L is the load conductance. This relation provides minimum uncertainty in G_L at low power oscillation threshold. Therefore, R_S can be calculated from equation (4), considering the value of G_L as nearly equal to the diode conductance (-G) at resonance.

The leakage current (J_s), entering the depletion region of the reversed biased p-n junction of an IMPATT diode, is normally due to thermally-generated electrons and holes [$J_S = J_{ns\,(th)} + J_{ps\,(th)}$] and it is so small that current multiplication factor

$$M_{n,\,p} = J_o / [J_{ns\,(th)} \text{ or } J_{ps\,(th)}] \qquad [J_o = \text{bias current density}] \tag{5}$$

can be considered to be infinitely large. Thus the enhancement of the leakage current under optical illumination of the devices is manifested by the lowering of $M_{n,p}$. The effect of shining light from the junction side in a TM (Top Mounted) IMPATT structure, as shown in ***Figure 5(a)***, is to generate an electron-dominated photocurrent. The expression for electron current multiplication factor then changes to

$$M_n = J_o \,/\, [J_{ns\,(th)} + J_{ns\,(opt)}], \tag{6}$$

[$J_{ns\,(opt)}$ = saturation current due to photoelectrons].

Thus, the photoelectrons reduce the value of M_n, while the value of M_p remains unchanged.

Similarly, the effect of shining light from the substrate side (n^{++} edge) in a FC (Flip Chip) IMPATT structure (***Figure 5(b)***) is to generate a hole-dominated photo-current that modifies the expression for hole current multiplication factor to

$$M_p = J_o \,/\, [J_{ps\,(th)} + J_{ps\,(opt)}] \tag{7}$$

($J_{ps\,(opt)}$ = saturation current due to photo-generated holes). Thus the photo-generated holes reduce the value of M_p while the value of M_n remains unchanged.

In order to assess the role of leakage current in controlling the dynamic properties of IMPATT oscillators at MM-wave and THz frequencies, simulation experiments are carried out on the effect of M_n (keeping M_p very high ~ 10^6) and M_P (keeping M_n very high ~ 10^6) on (i) the high-frequency admittance characteristics (ii) the negative resistivity profiles, (iii) the device quality factor (Q) and (iv) of SDR and DDR diodes for both flat and SLHL structures,. The details of mathematical calculations based on modified boundary conditions due to enhancement of leakage current are described elsewhere [*Mazumder et al.* (1993)].

6. Application and State-of-the-art THz-sources

The 'terahertz gap' that lies between the infrared and millimeter regions of the electromagnetic spectrum has recently become experimentally available. Terahertz (THz) waves, or T-rays, bridge the gap between electronics and photonics, have novel properties

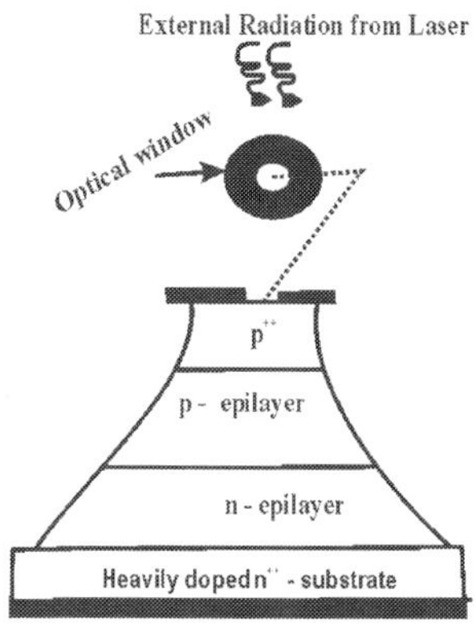

Fig. 5. (a): Schematic diagram of Top Mounted DDR IMPATT diode under optical-illumination

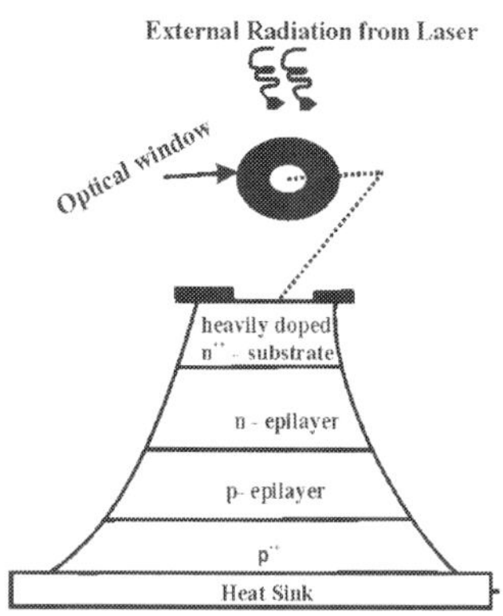

Fig. 5. (b): Schematic diagram of Flip-Chip DDR IMPATT diode under optical- illumination

and interact uniquely with many materials. The interest in THz was spawned both by researchers utilizing the microwave end of the spectrum and wants to work with shorter wavelengths, and researchers at the infrared end who saw the need for working with longer wavelengths. THz science is rapidly developing in Europe, US, Australia, Japan as well as in rest of the world. There is strong interest in the exploitation of the THz frequency range in virtually all fields of basic natural science (physics, chemistry, biology) as well as medicine [*Trew* (2005)]. Across Europe, a number of research groups at universities and in industry are working on THz science and technologies. Indeed, in the last few years the U.S. Army and the Department of Defence have focused on the advancements of THz-frequency electronic technology and on novel applications of THz-frequency sensing. Since 1999, Terahertz imaging [*Wang et al.* (2003)] has become a very important application, since it may make possible a single step removal process. This will enable improved detection rates of unhealthy tissue during surgery and should lead to a decrease in the number of repeat surgeries and in morbidity. Material spectroscopy and Biomedical sensing [*Naftaly et al.* (2005) and *Watanabe et al* (2004)] is perhaps the most rapidly developing of all THz applications. THz imaging of pathogens such as anthrax is also possible and that provides novel approaches for counter-terrorism. Terahertz imaging techniques are also used for planetary and cometary sensing as well in the earth-based studies which include monitoring of ozone depletion.

Spectroscopy was among the first applications of THz technology, for instance, in the development of basic THz fingerprints of simple molecules, such as water, carbon monoxide and ozone. Various rotational, vibrational and translational modes of complex organic molecules, including bio-molecules are within the THz range. These modes are unique to a particular molecule, and thus it is possible to obtain a 'Terahertz fingerprint' allowing for the identification of those chemical substances. The application of T-rays opens the possibilities for fast DNA analysis — in both areas of disease detection and forensics. Since THz radiation is non-ionizing, it has many potential medical applications. Apart from spectroscopic characterization, T-rays can also provide X-ray-like images. In fact, Terahertz medical imaging presents a unique solution for a variety of health-related problem, such as tissue identification through its water content, dental cavity detection and liver cancer detection [*Nishizawa et al.* (2005)]. The most important fact is that, as the photon energy of THz is much less compared to X-ray, it is not considered intrinsically harmful to living tissues as are of X-ray. It has the ability to penetrate a few millimeters of the uppermost skin layer, and thus the early detection of skin cancer is possible.

Scientists believe that the Terahertz spectrum is one of the critical technologies for defence against suicide bombers and other terrorist activities [*Karpowicz et al.* (2005)]. Now-a-days, researchers have focused their attention on the potential applications of Terahertz rays for directly detecting and imaging concealed weapons and explosives. Terahertz radiation can be transmitted through most non-metallic and non-polar mediums. When a Terahertz system is used properly, people can see through concealing barriers such as packaging, corrugated cardboard, walls, clothing, shoes, book bags, pill coatings, etc. Once the rays penetrate those materials, they can also characterize what might be hidden -be they explosives, chemical agents or others, based on a spectral fingerprint. Undoubtedly, security systems of the near future will incorporate THz technologies. It will be increasingly necessary to scan for biological, chemical and other weapons in a manner that is non-invasive and fast. Terahertz sensing provides advantages to short-range radar sensing, as they can penetrate through fog further than optical radiation. The wavelength being short enough, it provides significantly higher bandwidth than microwaves. However, the wavelength is long enough than infrared to reduce Rayleigh scattering and thus it find its application in short-range battlefield communication, where smoke prevails the infrared transmission. The advantage of THz over IR for indoor applications is that it occupies an extremely quiet band without noise or background clutter. Conventional wireless techniques for communication use microwaves at very low power. THz could increase the rate of information transfer as well as the volume. Now-a-days wireless communication technology requires more bandwidth for communication and data transfer. Although the high atmospheric attenuation at terahertz frequencies makes it difficult to have a long range mobile-communication, however a high-bandwidth, short-range and line of sight wireless link is completely realizable [*Nagatsuma et al.* (2004)]. On the other hand, atmospheric attenuation has an advantage in the reduction of coverage range of the signal in military applications to avoid communication being overheard or in frequency re-use application to avoid signal interfering.

Although all other areas of the electromagnetic spectrum are used in current technologies, development of technologies in the THz region is very difficult. The reason for this lies in the lack of suitable THz sources and receivers. Thus a critical roadblock to full exploitation of the THz band is lack of reliable, powerful (0.1W - 10.0 W CW), efficient, compact and relatively inexpensive THz radiation sources. Some of the existing THz sources are: electron beam sources, optically pumped far-infrared gas lasers, frequency multipliers, photoconductive emitters, terahertz semiconductor lasers, terahertz photo-mixers, solid-state sources, etc.

Among electron beam sources, Gyrotrons [*Flech et al.* (1999)], free electron lasers [*Krishnagopal et al.* (2004)], backward wave oscillators (BWO) [*Dobroiu et al.* (2004)] are capable of generating high-power at THz frequency region. Gyrotrons with 1 MW power at 140 GHz [*Dammertz et al.* (2002)] is feasible. Free electron lasers (FEL) are capable of operating virtually over the entire electromagnetic spectrum. A free electron laser at the University of California works at far infrared region and can generate 1 KW quasi-continuous wave signal at 300 GHz. BWOs can generate 50.0 mW of power at 300 GHz down to a few mW at 1 THz [*Schmidt et al.* (2002)]. The commercially available systems provided by Russian Company ISTOK can generate 1- 10 mW output power within the frequency range 177 GHz - 1.1 THz. Complete systems are heavy and large and need high bias voltage and water cooling systems [*Ives et al.* (2003)] , but the systems are much smaller than FEL and Gyrotrons. Electron beam devices are bulky and needs extremely high fields as well as high current densities which are main disadvantages of these devices.

Optically pumped far infrared gas lasers can produce terahertz signals. These THz sources consists of CO_2 pump laser injected into a cavity filled with a gas that help to produce THz signal [*Chao et al.* (2009)]. Semiconductor lasers show great promise for narrowband THz generation. Such lasers have many inherent limitations including low efficiency, low output power and the need for cryogenic cooling to maintain lasing conditions. The Quantum Cascade Laser (QCL) is the most promising THz semiconductor laser. *Barbieri et al.* has fabricated a continuous wave QCL that can generate 25 μW power at 4.4 THz at 52 K [*Barbieri et al.(2003)*]. Recently, the highest power THz source, pumped by an eye-safe, narrow band fiber laser system with an output of 26.4 mW, has been developed [*Leigh et al.* (2009)].

Among all two terminal solid-state sources, higher RF power levels of 23 μW at the fundamental frequency of 342 GHz and 0.6 μW at the third harmonic frequency of 1.02 THz is measured with Resonant Tunneling Diode (RTD) in the GaInAs/AlAs material systems, but these devices were operated in a "quasi CW mode" with a pulse length of 0.3 ms and a repetition rate of 300 Hz [*Orihashi et al.* (2005)]. State-of-the-art Gunn devices generate 0.2 - 5

μW power at 400 - 560 GHz frequencies [*Eisele et al.* (2005)]. Presently the maximum operating frequency range of TUNNETT devices is 355 GHz with power output of 140 μW [Eisele (2005)]. IMPact Avalanche Transit Time (IMPATT) diodes are recognized as the most powerful two terminal sources. Higher RF power and oscillation frequency were achieved from these devices by cooling he heat-sink of the diode and the waveguide circuit to 77K (liquid nitrogen) [2.11]. State-of-the-art IMPATT devices generate 2 mW- 7.5 mW power in the 300-400 GHz frequency range [*Ishibashi et al.* (1977)].
The above review shows that compact, low-cost but high-power and efficient THz sources are still lacking. Researchers have focused their attention in developing such THz sources to overcome the present limitation of THz systems. Research is continuing to increase the frequency and power level of conventional Si and GaAs based IMPATT devices to reach the THz region and also using alternate semiconductor material, such as, SiC and GaN together with improved fabrication techniques.

7. WBG semiconductors for fabricating high-power IMPATTs

The material parameters of the base semiconductors play an important role in deciding the operating frequency and output power level of IMPATT devices. So, the base semiconductor material should be chosen selectively to design high-power, high-frequency devices. The classification of WBG semiconductors is varied. Since the primary physical properties of a semiconductor scale to a certain degree with the energy gap, this parameter provides a reasonable classification scheme. However, comparison with Si and GaAs are common, because of the importance of these common materials. So, in general a WBG semiconductor is classified as a material with a bandgap at least twice the bandgap of Si. This gives a range from about 2eV (with InN and 3C-SiC) up to 6 eV (with AlN and diamond). WBG semiconductors, especially the Silicon Carbide (SiC) family and III-Nitride (GaN and its compounds) family are relatively attractive for developing new generation devices. Although the properties of these materials are very favorable, they are not as technologically mature as Si and GaAs. Rapid progress has been made in resolving the technological problems of the wide band gap semiconductors related to crystal growth, contact formation, material purity and quality.
SiC is recognized as a semiconductor of great importance in electronic applications because of its distinct properties, the possibility of easy growth on a native oxide, and the presence of numerous polytypes [*Elasser et al.* (2002)]. The SiC family of semiconductor contains the same semiconductor material grown in many polytypes. The most commonly grown SiC materials are 4H-SiC, 6H-SiC, 3C-SiC. SiC, although of varied polytypes, generally have high carrier saturation velocity and high thermal conductivity, which make them suitable for high-temperature (above 800K), high-frequency (Terahertz region) applications. *Cree Research Inc.* was the first commercial vendor of SiC wafers which are commercially available as 4-inch wafers of 4H-SiC. It is well known that SiC wafer quality deficiencies are delaying the realization of outstandingly superior 4H-SiC high-power semiconductor devices. While efforts to date have centered on eradicating micropipes, 4H-SiC wafers and epilayers also contain elementary screw dislocations in densities of the order of thousands per cm^2, nearly 100 fold micropipe densities [*Dudley et al.* (1995)]. While not nearly as detrimental to SiC device performances as micropipes, it was shown earlier that diodes containing elementary screw dislocations exhibit a 5% to 35% reduction in breakdown

voltage, higher pre-breakdown reverse leakage current, softer reverse breakdown I-V knee and concentrated microplasmic breakdown current filaments when measured under DC testing conditions. The cubic phase, 3C-SiC, however, is difficult to grow because of lack of a suitable substrate, thus it receives less interest. However, in recent years, there has been some little interest in 3C-SiC, resulting in both experimental and theoretical works. The most difficult to grow is 2H-SiC, because of its high formation energy. The most common donors in SiC are nitrogen (N) and phosphorous (P). N substitutes on C sites in the lattice, while P on Si sites. The most common acceptors are aluminum (Al) and boron (B) which substitutes on Si sites.

SiC was considered to be a promising material for fabrication of IMPATT diodes for the first time in 1973 by Keys [*Keys* (1973)]. Historically, the first simulation work on modeling and analysis of SiC IMPATT devices was done by Mehdi [*Mehdi et al.* (1988)]. They adopted the drift-diffusion method for analyzing the microwave and MM-wave characteristics of these diodes. The device operating characteristics and the power generating capabilities of the devices were studied at four different operating frequencies, 10 GHz, 35 GHz, 60 GHz and 94 GHz. Many material parameters, such as, field and temperature dependent saturation velocities and ionization coefficients of charge carriers in SiC were not available at that time and hence these were not considered in the simulation scheme. Their study however predicted that performances of SiC devices are superior to Si devices under CW mode of operation. In 1998, Meng et al. [*Meng et al.* (1998)] carried out a Read-type simulation analysis of p+n Single Drift flat profile MM-Wave IMPATT devices at 800K. The simulation demonstrates that the efficiency (DC power density) for the device is 12.4% (6.7MW cm^{-2}), 15% (4.5 MW cm^{-2}) and 15.8% (3.3MW cm^{-2}) for frequencies of 200, 100 and 50 GHz, respectively. A Read diode analysis is less accurate at the efficiency fall-off frequencies because there is no well defined avalanche region at the frequencies where efficiencies falls off. However, the study confirms the efficiency and power advantages of MM-Wave SiC IMPATT oscillators. Later, Zhao [*Zhao et al.* (2000)] have reported the Monte Carlo Particle simulation of 4H-SiC based hi-lo SDR IMPATT diode at 200 GHz. A low voltage (Vdc = 74 V) 4H-SiC IMPATT diode was designed by them to offer an efficiency of 10% at around 200 GHz with a peak output power of 11 W.

These promising theoretical results attracted the attention of experimentalists. Several research groups started the realization of 4H-SiC based IMPATT. In 1998, Konstantinov et al. fabricated epitaxial p-n diodes in 4H-SiC with uniform avalanche multiplications and breakdown [*Konstantinov et al.* (1998)]. They have performed photo-multiplication measurement to determine electron and hole ionization rates. P-n junction diodes were fabricated from p^+ -n^0 -n^+ epitaxial structures grown by vapor phase epitaxy (VPE); n0 and n+ layers were deposited on the p+ substrates. The substrates were oriented in (0001) crystal plane with a small off-orientation angle, 3.50 or lower. The photo-multiplication measurement revealed that impact ionization in 4H-SiC appears to be dominated by holes, a hole to electron ionization co-efficient ratio up to 40-50 was observed. This ionization rate asymmetry was related to band-structure effects, to the discontinuity of the conduction band or the electron momentum along the c-direction. The results had a qualitative agreement with earlier studies of impact ionization in 6H-SiC. In 6H-SiC also, electron impact ionization was strongly suppressed and that was contributed to the discontinuity of the electron energy spectrum in the conduction band. Earlier problems in SiC device development due to poor material quality and immature device processing techniques was greatly overcome with the availability of production-quality substrates and the progress made in the processing technology. Though excellent microwave performances were demonstrated in SiC MESFETs and Static Induction transistors (SIT) [*Brandt* (1998)], no experimental work was reported for SiC IMPATT devices before 2000.

First experimental success of 4H-SiC based pulsed mode IMPATT was achieved by *Yuan et al.* (2001). The DC characteristics of the high-low diodes exhibited hard, sustainable avalanche breakdown, as required for IMPATT operation. The fabricated 75 μm diameter SiC diodes were found to oscillate at 7.75 GHz at a power level of 1 mW. However, the output power level was significantly lower than the expected simulated value. They pointed out that the low-power problem is related to the measurement systems, particularly the design of the bias line. Optimization of the microwave circuit, in which the diode is embedded, is very important to properly evaluate the device performance. Any dispute in circuit optimization causes severe reduction in output power level. Thus, Yuan et al. made a comment that the measured low power, as obtained by their group, does not reflect the true power capability of SiC IMPATT. *Vassilevski et al.* (2001) also fabricated 4H-SiC based IMPATT. Microwave pulsed power of 300 mW was measured at 10 GHz. Though a comparatively higher power level was achieved, the power conversion efficiency was found to be very low ~0.3%. To increase the output power level, *Ono et al.* (2005) introduced a highly resistive guard ring that surrounds the diode periphery. The advantage of this guard ring is to reduce the electric field at the p-n junction edge of the junction periphery. A high current can thus be supplied through the diode without any destruction. Output power of 1.8W at 11.93 GHz was obtained from their fabricated diode and which is till date the highest reported output power from 4H-SiC IMPATT diodes. Nevertheless this power level is much lower than that expected. To increase the output power level, as expected from simulation studies, the residual series resistance should be minimized. No theoretical or experimental works on lo-hi-lo type 4H-SiC-based diodes are published by other workers.

To the best of author's knowledge, no experimental results are available for 6H-SiC based IMPATTs. Hence, it was established that at MM-wave region, 6H-SiC is another suitable candidate for developing high-power IMPATT devices.

The III-Nitride family of semiconductors can fill the emerging market for semiconductor optoelectronic devices. One of the important advantages of GaN over SiC is the ability to form heterojunctions. The fact that GaN together with InN and AlN, allows the formation of heterostructures provides some interesting device possibilities. The III-Nitride family consists of the binary semiconductors; InN, AlN and GaN, and the ternaries composed of them, $Al_x G_{1-x}$ N and $In_x Al_{1-x}$ N. GaN can be grown in two phases: zinc-blende (cubic) and wurtzite (hexagonal), while the remaining III-Nitride semiconductors only have the wurtzite polytype. The III-Nitride family of materials has gain interest in both opto-electronic and high-power solid-state devices. Their technological immaturity is mainly due to fabrication problems; however in recent years, advances have been made in the wurtzite-phase versions. Again as with the SiC family, wurtzite-phase materials receive most of the attention because of the relative ease of growth when compared to zinc-blende GaN. Commercial GaN based devices are grown heteroepitaxially on substrates like Sapphire and SiC. Recently, Si has been considered as a substrate for GaN growth for its low price, high crystalline quality and potential capabilities for integration with traditional Si-based electronic technology. MOCVD has become the technique of choice for the epitaxial growth of GaN material and devices [*Pearton* (2000)]. In MOCVD growth, Si and Mg are used as donor and acceptor impurities, respectively. Very recently, halide-hydride vapor phase epitaxy (HVPE) is considered as a promising technique for the fabrication of GaN based device structures, particularly for the GaN/SiC heterostructure. Reliable low-resistance ohmic contacts are essential for efficient device operation. Ohmic contact processing is still a challenging area in device technology. P- and n- type ohmic contact resistances of SiC and GaN will be discussed in the relevant chapters of the thesis.

Despite decades of study, only recently GaN-based materials have moved from research laboratories to commercial markets. This change was due to a rapid progression of improvements in epitaxial growth, demonstration of p-type conductivity and the fabrication of commercially viable devices. The fabrication of highly efficient blue and green LEDs and diode lasers is driving the development of GaN-technology. The robust and versatile properties of GaN make it an excellent candidate for high-speed and high-power electronics. Interest in GaN has exploded in past few years, leading to an expansion of its potential applications on an almost monthly basis [*Kuzuhara et al.* (2009)]. This broad spectrum of applications has led some to predict that GaN will eventually become the third most important semiconductor material, behind Si and GaAs. High-power handeling of GaN power transistors have already been demonstrated by fabrication of GaN High Electron Mobility Transistors (HEMT) and Field Effect Transistor (FET) devices [*Pearton* (2000)]. No experimental work, however, has been reported for GaN IMPATT diodes, despite the fact that the diodes are easier to fabricate than transistors. Till date, only a few simulation results on GaN based IMPATT have been reported in published journals. *Meng et al.* (1999) studied the MM-Wave performances of the wurtzite phase and zinc-blende phase GaN IMPATT devices at 800K by a Read-type modeling approach. The simulations showed that GaN wurtzite-phase p-n single drift flat-profile IMPATT oscillators at 300 GHz have an efficiency of 11% and an RF power density of 1.6 $MW.cm^{-2}$. Their studies confirm the efficiency and power density advantages of GaN IMPATT oscillators. *A. K. Panda et al.* (2001) designed and studied the performances of GaN based IMPATTs in the D-band. The maximum power that may be obtained from their designed diode was 3.775 W with an efficiency of 12.5%. Moreover, their study predicted that the wurtzite-phase GaN based IMPATT is better than its zinc- blend counterpart, as far as breakdown voltage, power output and efficiency are concerned. Later, *Reklaitis et al.* (2005) performed a Monte Carlo simulation of Wz-GaN based near-terahertz IMPATT diode. Their analysis predicted that the device may generate a RF power of ~ 3W at 0.45 THz with an efficiency of 18%. The diodes were found to be more efficient than that was designed by *Panda et al.* This study, on the other hand, predicted the possibilities of GaN based IMPATT diodes as efficient near-THz power sources. *Alekseev et al.* (2000) performed theoretical and experimental studies for the development and optimization of GaN based Gunn devices in the THz frequency region. GaN Gunn-diode oscillators at 750 GHz are expected to generate power density of $3x10^5 W\ cm^{-2}$. Before an attempt is made to fabricate GaN based IMPATT devices at Terahertz region, reliable modeling and better understanding of high-frequency properties of such devices are essential. Thus the author has studied the DC and Terahertz -frequency characteristics of the GaN based flat and lo-hi-lo types IMPATT devices at elevated junction temperature and the results will be discussed in the concerned chapters.

Unlike GaAs, wurtzite phase GaN have different ionization rates for electrons and holes ($\alpha_n \neq \alpha_P$). So from the ionization point of view, as discussed in sub-section 2.5.3, wurtzite GaN IMPATT is expected to be noisier than GaAs IMPATT. Panda et al. [2.188] showed that GaN based devices generate equal noise as Si-based IMPATTs, but higher by 6-8 dB noise values compared with GaAs based devices under the same operating conditions. However, for increased operation temperature, the noise is found to decrease [*A. K. Panda et al.* (2001)]. *Reklaitis et al.* [*Reklaitis et al.(2005)*] later studied the current voltage characteristics and the associated current noise in GaN double drift IMPATT diodes, by Monte Carlo simulations. For values of current multiplication factor greater than ten they observed a giant suppression of avalanche noise down to three orders of magnitude with respect to the standard excess noise factor. The negative feedback between fluctuations in space-charge and in number of generated e-h pairs is found to be responsible of such a giant suppression.

8. Superiority of WBG semiconductor based IMPATTs over Conventional diodes

4H-SiC based SDR (p^{++} n n^{++}) IMPATT diodes with flat and SLHL doping profiles are designed by *Mukherjee et al. (2008 a)* at around Ka-band. In order to make a comparison, Si based SDR IMPATT diode is also designed at Ka-band. The comparison reveals that 4H-SiC based SDR diodes are capable of generating a RF power of 870.0 x 10^9 Wm^{-2} with an efficiency of 20.0 %, far better than their Si counterpart [*Mukherjee et al. (2008)*]. Thereafter, the DDR IMPATT diodes are designed and studied thoroughly by *Mukherjee et al.* (2009 a) at three different window frequencies: 35 GHz (Ka-band), 140 GHz (D-band) and 220 GHz (Y-band) and the corresponding admittance plots are shown in ***Figures 6 (a-c)***. Comparative studies of SLHL and flat-profile diodes at MM-wave window frequencies by *Mukherjee et al.* (2009) reveal that the Quasi Read SLHL diodes are superior to their flat profile counterparts in terms of power output, efficiency and negative-resistance.

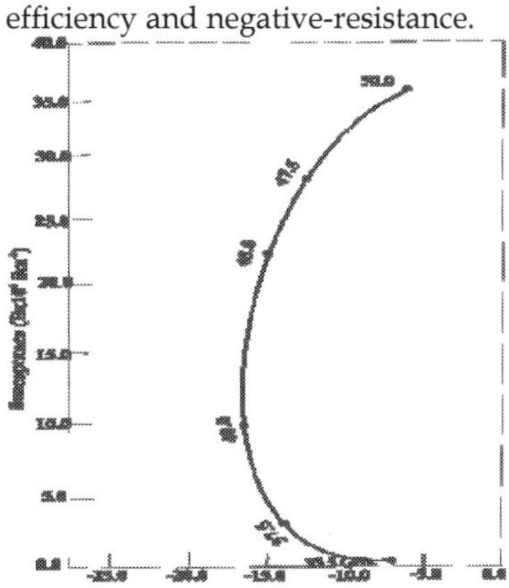

Fig. 6. (a) admittance plots of 4H-SiC DDR IMPATT at Ka--band

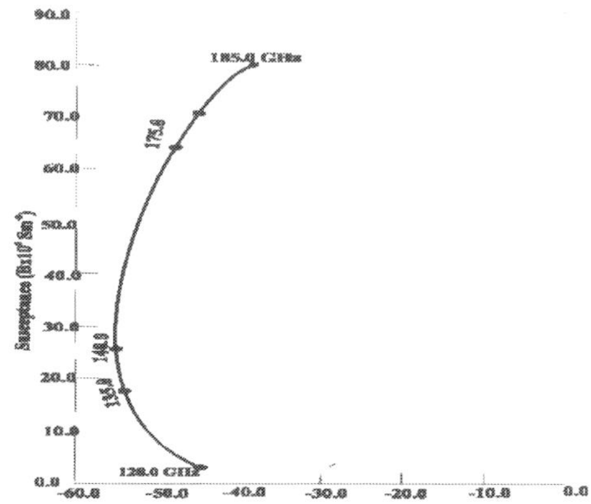

Fig. 6. (b) admittance plots of 4H-SiC DDR IMPATT at D-band

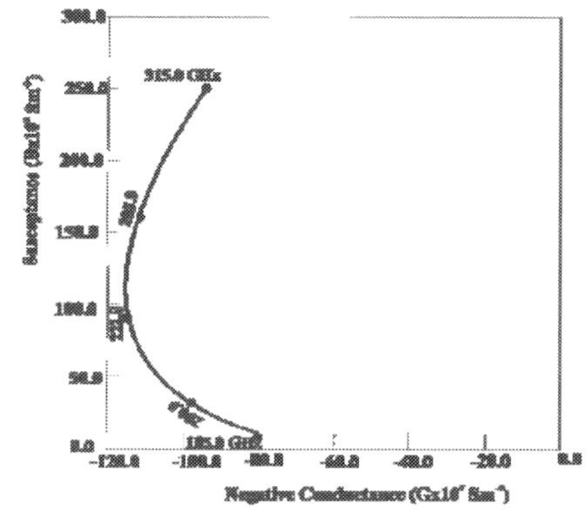

Fig. 6. (c) admittance plots of 4H-SiC DDR IMPATT at Y-band

Mukherjee et al. (2007 b) has made a systematic study on the performance of the IMPATTs designed at higher THz frequencies: 0.5 THz and 1.85 THz. The electric field profiles and admittance plots are shown in Figures 7 and 8 (a-b). It is interesting to note that even at the higher THz region (1.85 THz), 4H-SiC based diode is capable of generating a power density of 5.0x10^{11} Wm^{-2} with an efficiency of 9.0%. While estimating the power density, the effects of series resistances is considered in the analysis. The values of P_{max}, with and with-out R_S, are also studied and shown in ***Figures 9 (a-b)***.

The performances of the SLHL DDR IMPATT at THz region are further studied by *Mukherjee et al.* (2007 c). It is observed that, similar to MM-wave region, in THz region also the overall performance of SLHL diode is far better than its flat profile counterpart. It is further interesting to observe that the magnitude of R_S reduces significantly (15% - 30%) in SLHL diodes compare to that in flat profile diodes. The performances of the 4H-SiC, 6H-SiC, and 3C-SiC based THz (0.3 THz) DDR diodes are compared by *Mukherjee et al* (2008 b). The study reveals that the 4H (α)-SiC based IMPATT may yield a RF power density of 36.45 x 10^{10} Wm^{-2}, with an efficiency of 14%, which are far better than its hexagonal (6H-SiC) and cubic (3C-SiC) counterparts, under similar operating conditions. The above observations definitely establish the potential of SiC based IMPATTs at MM-wave as well in the THz region.

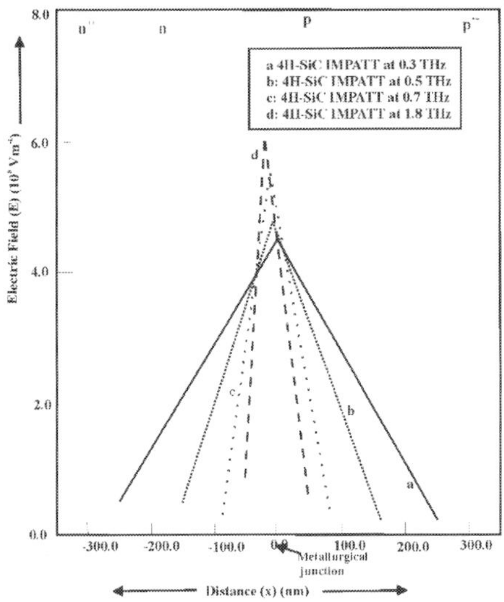

Fig. 7. E(x) profiles of 4H-SiC based Terabertz IMPATT diodes

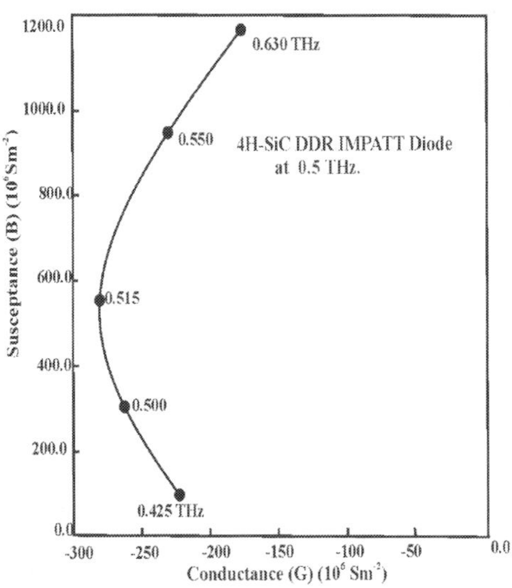

Fig. 8. (a): Admittance characteristics of 4H-SiC IMPATT at 0.5 Terahertz.

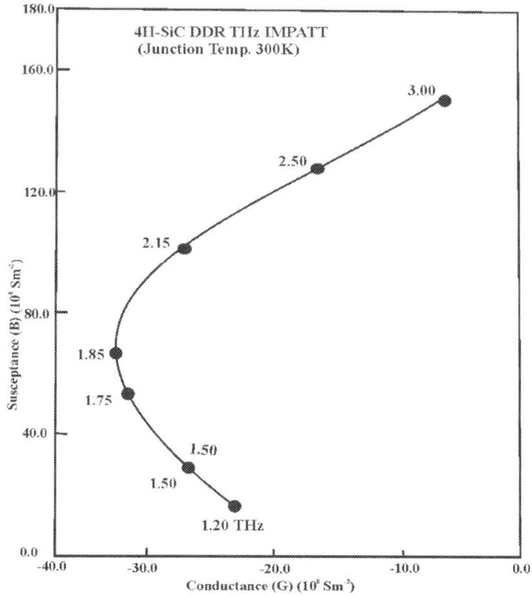

Fig. 8. (b): Admittance characteristics of 4H-SiC IMPATT at 1.8 Terahertz

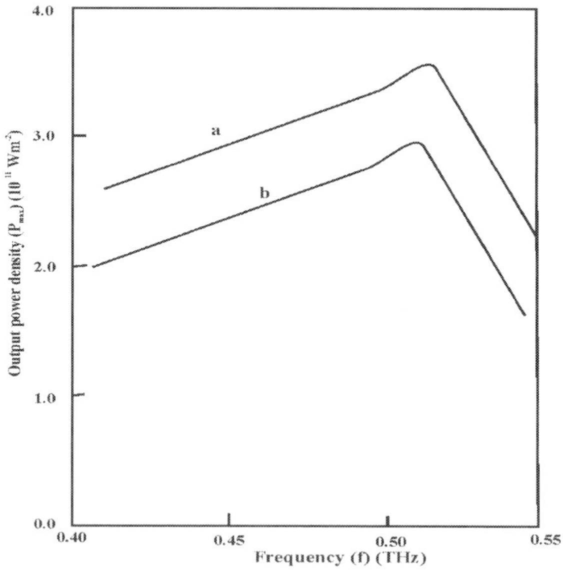

Fig. 9. (a): Effect of series reristance on output power density P_{max} of 4H-SiC 0.5 THz IMPATT diode. (a) 4H-SiC IMPATT at 300K, R_s=0.0Ω, (b) 4H-SiC IMPATT at 300K, $R_{s,total}=0.386x10^{-9}$ Ω m^2

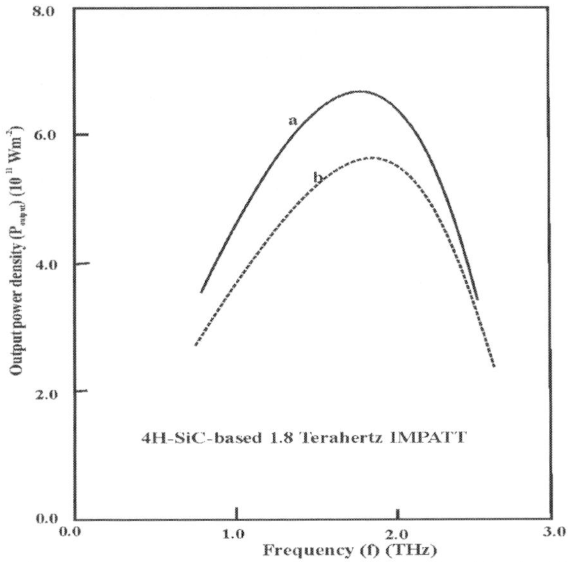

Fig. 9. (b): Effect of series resistance on output power density (P_{max}) of SiC THz IMPATT diodes. (a) 4H-SiC IMPATT at 300K, R_s=0.0Ω, (b) 4H-SiC IMPATT at 300K, $R_{s,total}$=2.2x10^{-11} Ω m^2

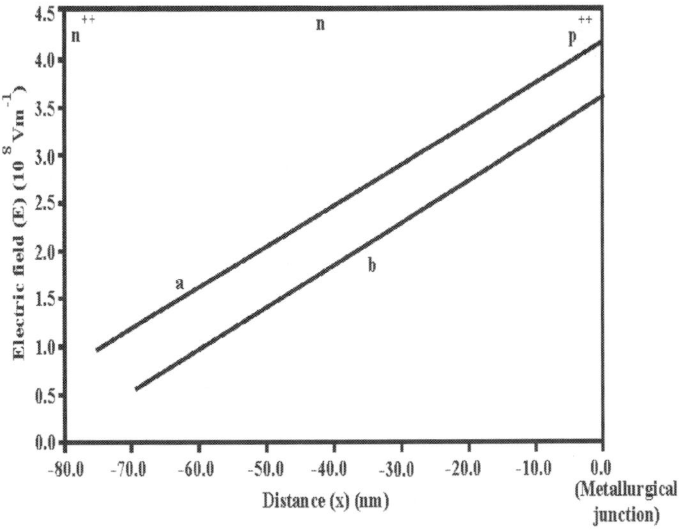

Fig. 10. Plots of electric field profile for (a) SLHL and (b) flat-type GaN (flat-profile) SDR IMPATT diodes. The distance of the n-side from the metallurgical junction has been considered as negative.

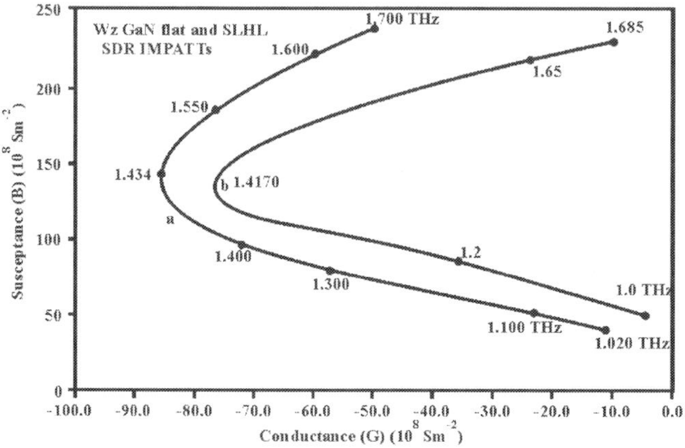

Fig. 11. Conductance (G) - Susceptance (B) plots of GaN (a) SLHL and (b) flat type SDR THz IMPATT diodes in Terahertz region.

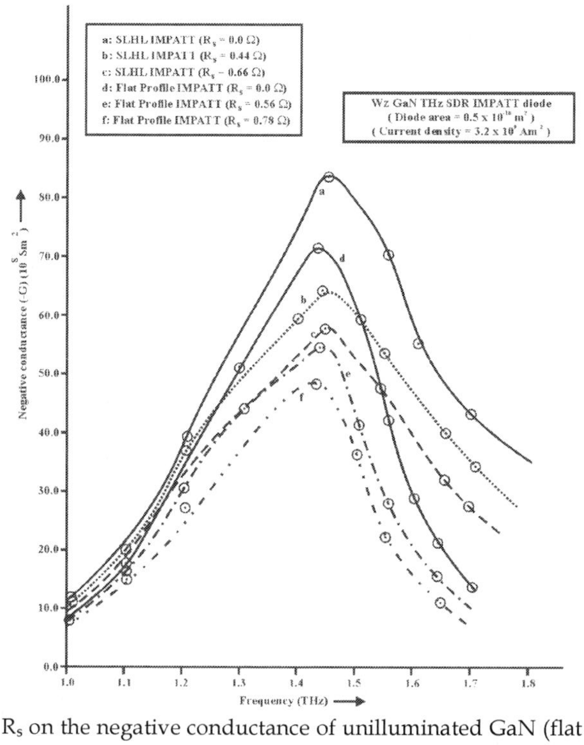

Fig. 12. Effect of R_s on the negative conductance of unilluminated GaN (flat and SLHL) SDR IMPATT diodes.

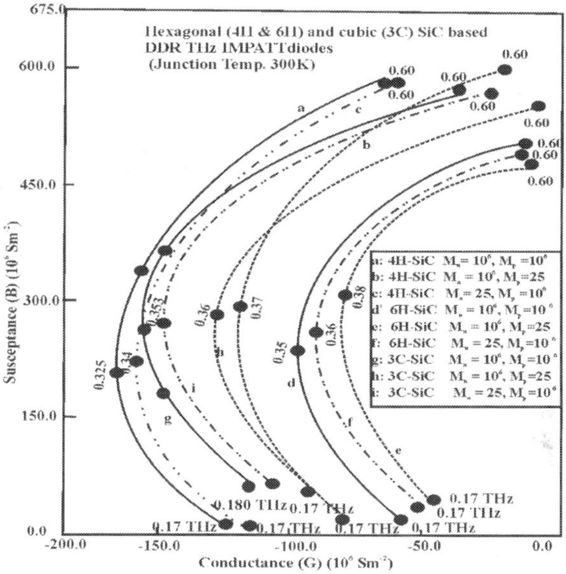

Fig. 13. Effect of photo-illumination on FC and TM illumination configurations of SiC (3C, 4H and 6H types) based terahertz DDR IMPATT diodes at room temperature (300K).

WZ-GaN based SDR (p^{++} n n^{++}) IMPATT diode is designed and studied by Mukherjee et al. at around 140.0 GHz (D-band) frequency and the performances are compared with Si IMPATT. It is found that the GaN IMPATT may generate an output-power density of 5.6×10^{10} Wm^{-2} with an efficiency of 23.5% at 145.0 GHz, far better than Si IMPATT. The prospects of SDR flat and SLHL type GaN IMPATTs at 1.45 THz are simulated and the results are reported elsewhere [*Mukherjee et al. (2007 a)*]. The studies indicate that GaN IMPATT diodes are capable of generating high power of ~2.5 W (assuming diode area = 0.5×10^{-10} m^2) at around 1.45 THz with 17% -19.0 % efficiency. The effect of series resistance on the THz performance of the device is also studied by *Mukherjee et al* (2007 a). It is interesting to note that the presence of a charge bump in flatly doped SDR structure reduces the value of parasitic series resistance in GaN-IMPATT by ~ 22%. The electric field profiles, admittance charecteristics and effects of series resistance on P_{max} of the GaN flat and SLHL diodes at 1.45 THz are shown in ***Figures 10, 11 and 12*** [*Mukherjee et al* (2007 a)].

9. Photo-sensitivity of WBG semiconductor based IMPATTs

The effects of photo-illumination on the 4H-SiC, 6H-SiC and 3C-SiC based Top Mounted and Flip Chip diodes are studied by *Mukherjee et al (2008 b)* at THz region. It is found that the optimum frequency and THz characteristics undergo sufficient variation with increase of the intensity of optical radiation. It is interesting to note that in all the three types of SiC based IMPATTs, photo-generated leakage current dominated by holes (FC illimination configuration) is more important than that dominated by electrons (TM illumination configuration), in modulating the high frequency properties of the devices. This observation has been correlated with the relative ionization coefficients of charge carriers of SiC. The admittance characteristics and negative resistivity profiles of the un-illuminated and illuminated diodes are shown in ***Figures 13 and 14***, respectively. The study reveals that the enhanced leakage current in illuminated devices, due to photo-illumination, degrades the output power density and negative resistance, but increases the tuning range of the devices. The optical illumination studies on the three types of SiC DDRs by Mukherjee et al. [*Mukherjee et al (2008 b)*] reveals that 4H-SiC based IMPATT is comparatively more photo-sensitive than its 6H- and 3C- counterparts.

The photo-sensitivity of the 4H-SiC and 6H-SiC diodes at higher THz region (> 1 THz) are studied by *Mukherjee et al*. The admittance plots are shown in ***Figures 15 and 16***. It is again found that in the higher THz region also, FC IMPATTs are more photo-sensitive than TM diodes. Compare to 4H-SiC IMPATT, 6H-SiC devices are found to be less promising in the THz region, as far as negative resistance, output power density, efficiencies and photo-sensitivity are concerned. However, it is worth mentioning that 6H-SiC based THz devices are definitely superior to conventional Si- and GaAs-based devices, since the latter categories of diodes could not reach the THz region.

The effects of photo-illumination on the MM-wave and THz performances (***Figures 17 and 18***) of the GaN based diode are studied by *Mukherjee et al.* (2007 a) and (2009 b). The study reveals that negative conductance $|-G_P|$ of the diode decrease and at the same time, the frequency range over which the device exhibits negative conductance shifts towards higher frequencies with the increasing intensity of optical radiation. As expected, the upward shift in operating frequency is found to be more (~ 16 GHz) in case of the SLHL SDR IMPATT device. Illuminated FC diodes are found to be more photo-sensitive than their TM counterparts. The inequality in the magnitudes of electron and hole ionization rates in WZ-GaN have been found to be correlated with the above results. The study indicates GaN IMPATT is a promising photo-sensitive high power THz source.

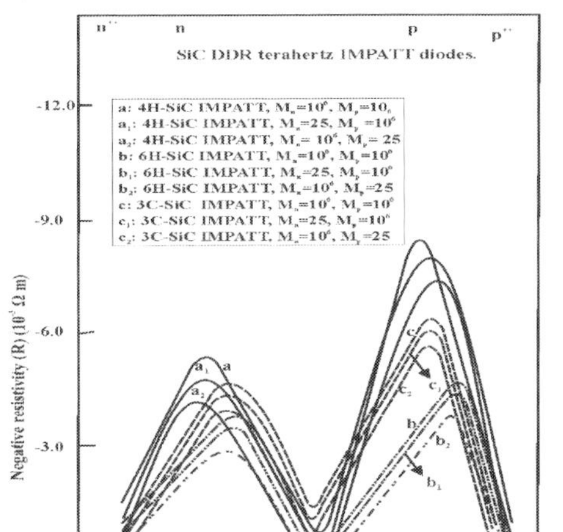

Fig. 14. Effects of radiation on negative resistivity plots of SiC IMPATT diodes

Mukherjee et al. (2008 c) have made a comparative analysis of the THz frequency performances as well as photo-sensitivity of the hexagonal GaN and SiC (4H-) based diodes, under similar operating conditions. The extensive studies establish that, though the photo-sensitivity of 4H-SiC based IMPATT is higher than its GaN counterpart, the overall Terahertz performance of un-illuminated GaN IMPATT is far better than that of a 4H-SiC based THz device, as far as power density and efficiency are concerned [*Mukherjee et al.* (2008 c)].

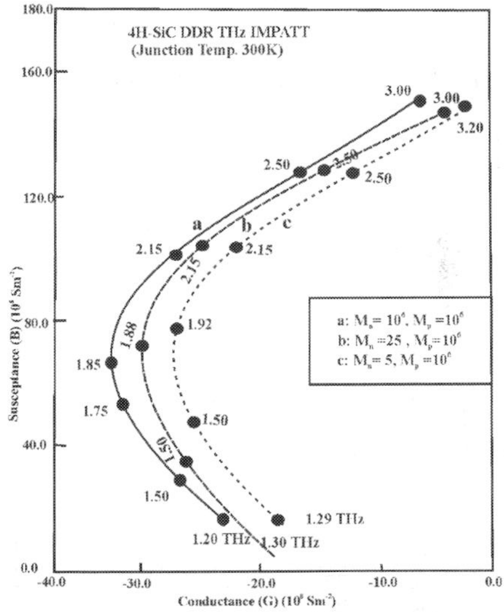

Fig. 15. (a): Effect of electron dominated photo-current (TM illumination configuration) on 4H-SiC terahertz DDR IMPATT diode at room temperature (300K).

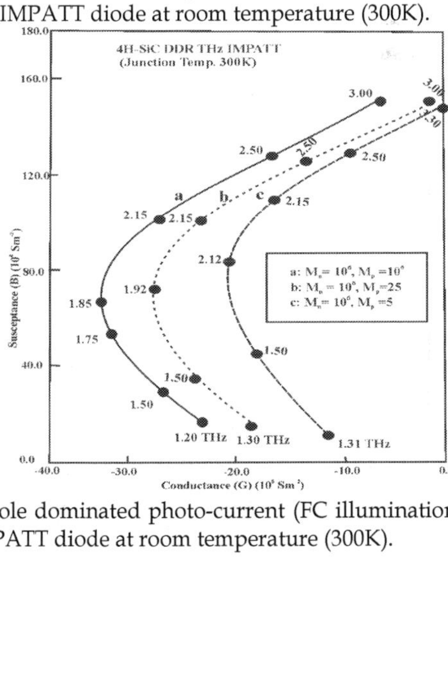

Fig. 15. (b): Effect of hole dominated photo-current (FC illumination configuration) on 4H-SiC terahertz DDR IMPATT diode at room temperature (300K).

10. Experimental feasibility of new-class of WBG IMPATT diodes

Extensive studies, as discussed in previous section, have established the superiority of WBG IMPATT diodes as high power source at THz frequency regime. However, special efforts are required in the areas of GaN and SiC material growth, doping and device processing technologies, oscillator performance characterization, etc., before the full potential of GaN and SiC IMPATT diodes are utilized at THz region. Due to lack of any experimental data on WBG semiconductor based THz IMPATTs, the simulated results could not be compared. However, it is proposed that experimental validation of these results might be possible through the following steps.

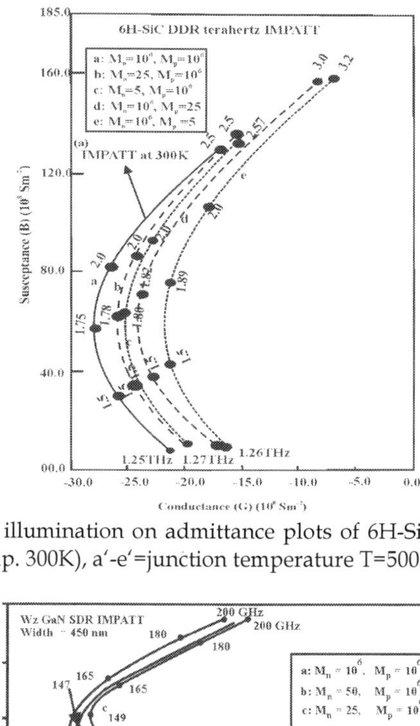

Fig. 16. Effects of optical illumination on admittance plots of 6H-SiC based DDR IMPATT diode: (a-e = Junction temp. 300K), a′-e′=junction temperature T=500K).

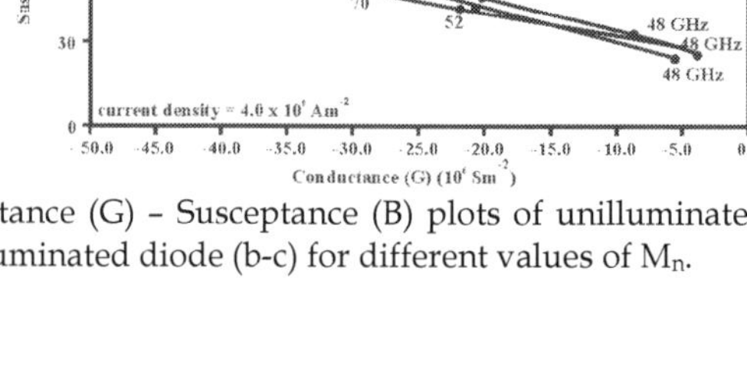

Fig. 17. (a): Conductance (G) - Susceptance (B) plots of unilluminated GaN SDR IMPATT diode (a) and the illuminated diode (b-c) for different values of M_n.

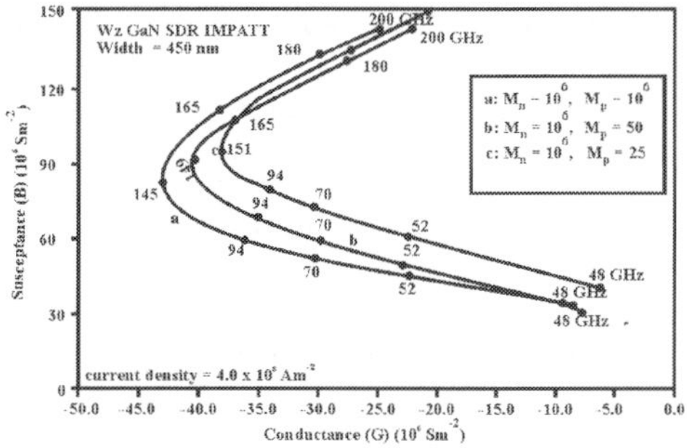

Fig. 17. (b): Comductance (G) - Susceptance (B) platos of unilluminated GaN SDR IMPATT diode (a) and the illuminated diode (b-c) for different values of M_p.

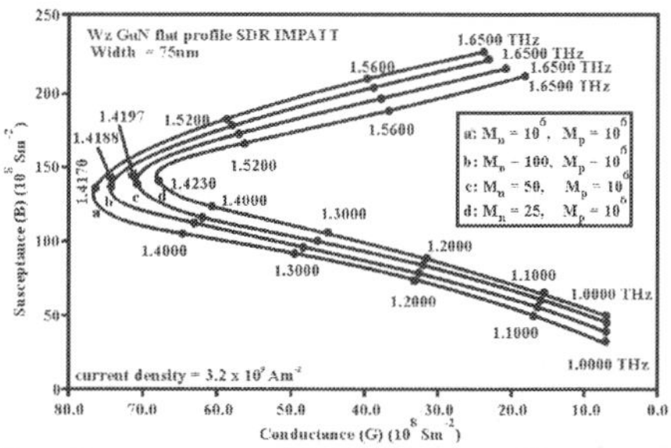

Fig. 18. Conductance (G) - Susceptance (B) plots of unilluminated GaN flat profile SDR IMPATT diode (a) and the illuminated diode (b-d) for different values of M_n.

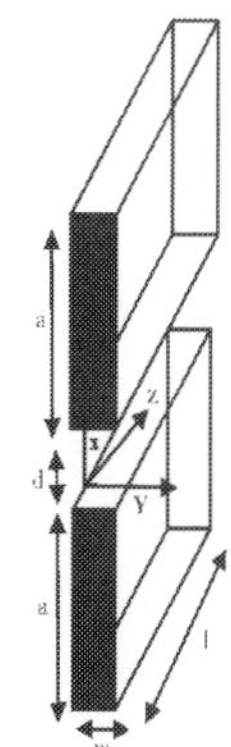

Fig. 19. Waveguide geometry: d=270 µm w=300.0 µm, and I=140.0mm

Fabrication issues related to GaN IMPATT diode:

(i) Growth of GaN p-n junction: GaN p-n junction can be obtained using metal organic chemical vapor deposition (MOCVD) technique and, to some extent, by molecular beam epitaxy (MBE). The most advanced results may be obtained by MOCVD technique, which includes the growth of GaN epitaxial layer from vapor phase. In this method, GaN may be grown from the vapor phase using metal organic gases as sources of Gallium and Nitrogen. For example, trimethylgallium (TMG) can be used as gallium source and Ammonia can be used as nitrogen source. Growth of GaN semiconductor may takes place in a reactor chamber on a substrate. During the growth, the substrate may be kept at growth temperature ranging from 800 to 1100° C. Single crystal wafers of Sapphire or SiC may serve as substrates for GaN deposition by MOCVD method. GaN THz device is found to be high power device. The large amount of current flowing in the device causes significant self-heating effect. If a very high quality GaN layer is grown on SiC substrate, SiC substrate will introduce the possibility of heat sink and that will result in a better heat-flow away from the active part of the diode.

Using MOCVD process, thin layers (usually not thicker than 5 microns) of GaN can be grown on Sapphire or SiC substrates. In order to control the type of electrical conductivity in the grown material, electrically active impurities can be introduced in the reaction chamber during the growth. Undoped GaN layers usually exhibit n-type conductivity. The value of n-type conductivity can be controlled by introducing Si impurity (in form of SiH_4 gas) in the reaction chamber during the growth. In order to obtain p-type GaN material by MOCVD method, Mg impurity can be introduced in the reactor chamber during the growth. Biscyclopentadienylmagnesium (Cp_2 Mg) may be used as a Mg source for GaN doping. In order to form a n-p junction, at first, MOCVD growth process can be carried out using Si donor impurity, to form n-type layer of GaN on an n^{++} substrate, and after that p^{++}-type cap layer can be grown on the n-layer by MOCVD process by doping high concentration of Mg impurity. However, MOCVD grown Mg-doped p^{++} layer, is highly resistive. In order to activate p-type conductivity, high temperature post-growth anneal in nitrogen atmosphere may be required. Since Mg requires large energy for ionization, in general, it is difficult to obtain heavily doped p^{++}-type GaN with Mg. Beryllium (Be) may be used to achieve p^{++}-type GaN, since the ionization energy of Be is low (~60 meV). Be-doped GaN layers may be obtained either by epitaxial growth methods, or by ion implantation. However, ion implantation may introduce severe lattice damage to the target material. Recently, *Pastor et al* (2007) assessed crystal damage of Be^+-implanted GaN by UV Raman scattering and found a correlation between implantation dose and the extent of lattice damage caused to the target.

The above mentioned procedure may be applied to form high quality GaN p-n junction. However, the MOCVD technology has a number of limitations:

(1) This is expensive method requiring complicated growth equipment. In order to form GaN p-n junctions by MOCVD process, metal organic sources, for example TMG (Ga source) must be used.

(2) Complicated chemical compounds have to be used as acceptor impurity sources, for example
biscyclopentadienylmagnesium (Cp_2 Mg) is usually used as a Mg-source.

(3) In order to form high-quality GaN material on SiC substrates, MOCVD method requires to grow a buffer layer in-between SiC and GaN, which makes impossible to fabricate devices utilizing direct contact between SiC and GaN such as GaN/SiC p-n junction.

Several experimental attempts have been taken to develop an alternative epitaxial growth technique to form GaN p-n junction. One method, which has been considered as a promising technique for the fabrication of GaN device structures, is hydride vapor phase epitaxy (HVPE). The HVPE method is convenient for mass production of semiconductor devices due to its low cost, flexibility of growth conditions, and good reproducibility. In this method, Ga metal may be used as source material, thus HVPE technology does not require expensive source material. Furthermore, since these GaN layers can be grown directly on conducting SiC wafers without insulating buffer layers, diodes with n-GaN/p-SiC heterojunctions can be fabricated by HVPE.

(ii) Formation of low resistive p^{++} and n^{++} ohmic contacts: Through photolithographic process, windows can be opened on p^{++} GaN layer for subsequent metal deposition. Using photolithography and lift-off techniques, low resistance contact metals, an alloy of Ni (20 nm)/Pd (20 nm) / Au (100 nm) [*Chu et al* (2000)], can be deposited inside the windows by an electron beam evaporator. The metal contacts should then be annealed in air, nitrogen and oxygen ambient conditions at different annealing temperatures ranging from 350 to 650 ^{0}C. The measurement of specific contact resistance may be done by circular Transmission Line Measurement (CTLM) system. A specific contact resistance of $1.0\text{x}10^{-4}\Omega\text{cm}^2$ can be achieved in this technique. If the similar metal composition is deposited on Be-implanted p^{++} type GaN layer with a carrier density of $8.1\text{x}10^{19}$ cm^{-3}, without further annealing process, the sample will show good ohmic contact with a contact resistance of $4.5\text{x}10^{-6}\Omega$ cm^2 [*Chu et al* (2000)]. A new metallization scheme has recently been developed for obtaining very low ohmic contact to n-GaN [*Burm et al* (1997)]. A composite metal layer Ti/Au/Ni/Au (150Å/2200 Å/400 Å/500 Å) can be deposited on n^{++}-GaN. Following an annealing process at 900^{0}C for 30s, a specific contact resistance (measured from TLM data) of $8.9\text{x}10^{-8}$ Ωcm^2 for a doping density of $4\text{x}10^{17}$ cm^{-3} can be obtained [*Burm et al* (1997)].

(iii) Formation of diode mesa using Reactive Ion Etching (RIE): Effective etching techniques are useful for diode fabrication. GaN has very high bond energy (8.9 eV/atom) and a wide-bandgap, which makes it almost inert to bases and acids, which are low cost and highly available wet etchants, used in Si technology. There are various methods of dry etching involving sources of external energy to initiate and sustain the break up of the high-energy bonds in GaN. A few of the dry etching techniques used for GaN include, ion milling and reactive ion etching (RIE). Ion milling relies upon physical sputtering and is not very practical for GaN because of low etch rates and extreme damage to the material caused by the purely physical process. The RIE method is a better technique of dry etching because it involves chemical etching in addition to physical etching. The etch rates for GaN using RIE with various etch chemistries range from 17 to 100 nm/min [*Adesidac et al.* (1999)]. Wet etching is an important complement to dry etching methods by providing low damage etching, low cost, and complexity. Since conventional acids and bases cannot be used to etch nitrides, a recently developed technique called photo-electrochemical (PEC) wet etching is found to etch GaN with significantly high etch rates [*Adesidac et al.* (1999)]. The PEC process, utilizes photo-generated electron-hole pairs to enhance oxidation and reduction reactions taking place in an electrochemical cell. For the PEC etching, the etch rate of about 400 nm/min for KOH solution and about 40 nm/min for HCL solution can be obtained. PEC etching is highly anisotropic etching of GaN. It is a very effective technique for forming mesa structure for design of GaN IMPATT diode.

(iv) On-Wafer DC Testing: After finishing the fabrication process, on-wafer DC testing should be performed before the diodes are packaged. DC testing will serve as the initial screening step of the device and the test results will be used for process evaluation.

(v)GaN IMPATT Device Packaging: The packaging should provide a low thermal resistance between the GaN diode chip and wave guide mount and should be mechanically rugged and hermetically sealed. The device can be bonded to a pill-type package. In pill-type configuration, the diode is bonded to a heat sink, which is usually gold plated. A ceramic or quartz ring encloses the diode and separates the heat sink from the package cap.

Fabrication issues related to SiC IMPATT diode:

SiC epiwafer (n^{++} substrate and n-type epilayer) can be procured from Cree Inc., Durham, NC, USA. The n-type doping is usually realized at Cree using nitrogen gas as the precursor. A SiC IMPATT device can be fabricated on the epiwafer following the process steps described below:

(i) Growth of p^{++} 4H-SiC layer: In order to assist p-type ohmic contact formation, the p^{++} 4H-SiC layer can be grown on top of the n-type film by Al^{2+} ion implantation; the doping concentration should be $N_A \geq 2 \times 10^{19}$ cm^{-3}. The post-implantation annealing may be performed at ≈ 1600 ° C for 45 minutes in argon atmosphere.

(ii) ***Formation of low resistive p^{++} and n^{++} contacts:*** The power dissipation of IMPATT devices strongly depend on the contact resistance. Following step (i), the samples may be cleaned by a "piranha" solution. After rinsing in DI water, the samples may be dipped in dilute hydrofluoric (HF) acid solution for 30 seconds and dried. Immediately after the cleaning, a SiO_2 layer on the p^{++} side can be grown by Plasma Enhanced Chemical Vapor Deposition (PECVD) at 285^{0}C. Through photolithographic process, windows can be opened inside the oxide layer. Using photolithography and lift-off techniques, contact metals (Al/Ti/Al) can then be deposited in the oxide windows by an electron beam evaporator. In order to obtain ohmic contacts the samples may be annealed for 3 minutes in a Rapid Thermal Anneal (RTA) furnace in nitrogen atmosphere at 950 ^{0}C. The post-deposition annealing at high temperature is generally preferred to reduce the specific contact resistance. As mentioned in the literature, Ni is the preferred material for ohmic contact on n-type SiC [*Roccaforte et al* (2006)]. Hence, for n^{++}-type contact, a Ni layer of 200 nm thickness may be evaporated on the back-side (n^{++}-side) of the wafer, followed by RTA treatment for 3 minutes at 950 ^{0}C. The choice of the metallic composition is based on the formation of Ni_2Si alloy. As mentioned in several publications [*Konishi et al* (2003)], the higher the concentration of the Ni_2Si in the contacts, the lower is the specific contact resistance. Finally, a composition of Ti/Au contact overlay is deposited on the p^{++}-side by an electron beam evaporator. The specific contact resistance can be determined from transmission line measurement (TLM) data.

(iii) The oxide layer from the p^{++}-side can then be removed by Buffered Oxide Etch (BOE).

(iv)Formation of diode mesa using Reactive Ion Etching (RIE): Si-C bonds show high chemical inertness, hence, wet etching is not efficient for reaching deep trenches. The more appropriate dry plasma etching, Reactive Ion Etching (RIE), may be used in separating diode mesa. Patterning of mesas may be carried out with photoresists: AZ 5214 (standard) and TI 35 ES (special photoresist for deep RIE Si etching) [*Lazar et al* (2006)]. A titanium/nickel (Ti/Ni) bilayer metal can be evaporated onto the sample. The metallic bilayer may be used as RIE mask that covers areas that will form mesa diodes. The dry etching can be performed in the Plasma Therm 790 reactor (say), an RIE reactor with a plasma (composed of SF_6 and O_2 gases) source generated at 13.56 MHz operating at a maximum power of 300W. The depth and etch rate of the mask may be determined by profilometry measurements with a Tencor Alpha Step.

(v) On-Wafer DC Testing: After finishing the fabrication process, on-wafer DC testing are performed before the diodes are packaged. DC testing may serve as the initial screening step of the devices and the test results can be used for process evaluation.

(vi)SiC IMPATT Device Packaging: The packaging should provide a low thermal resistance between the SiC diode chip and wave guide mount and should be mechanically rugged and hermetically sealed. The device may be bonded to a pill-type package. The details of pill-type package are described earlier.

C: Terahertz Transmission and Measurement:

It is highly desirable that a THz transmission system should have low signal attenuation, low radiation loss, and high mode-confining property. Sommerfeld wires exhibit very low attenuation and dispersion, but suffers from large field extension into the surrounding medium. Moreover, it suffers from significant radiation loss at bendings. Metallic parallel plate waveguides, on the other hand, exhibit undistorted wave propagation of broadband sub-picosecond pulses even at bendings, but the propagating wave is confined in one transverse dimension giving rise to divergence loss. Transmission of THz signals from an IMPATT device may be possible with a dielectric coplanar wave guide, the device being an integral part of the waveguide. For THz (0.1-1 THz) signal transmission, a metallic slit waveguide, fabricated by sawing a 270 μm wide slit through a 140 mm wide and 300 μm thick silicon slabs may be used. The schematic diagram of waveguide is shown in ***Figure 19***. The metallic slit waveguide shows dispersion less transmission of THz signals with very low attenuation. However, the residual edge roughness from wafer sawing [*Wachter et al* (2007)] can cause a small degree of radiation loss. For higher frequency (3 THz) power transfer, a ribbon-like structure fabricated from ceramic alumina may be utilized [*Yeh et al* (2005)]. The schematic diagrams of the structure are shown in ***Figures 20 (a-b)***.

Measurements of THz power and frequency may be done with a THz VNA (Vector Network Analyzer) or, by employing Photoconductive Method. At Terahertz regime vector network measurements are challenging because of the reduced wavelength. Also, large phase errors resulting from temperature fluctuation can occur. In addition, flexing of the cable linking the scanning waveguide probe to the measurement equipment adds to the phase uncertainty.

There are other systems for measurement and detection of THz signals for example Hot Electron Bolometer mixer- receiver and THz Time Domain Spectroscopy (THz-TDS). For detection and measurement of THz beam by photoconductive method, a THz-TDS set-up (***Figure 21***) may be used. The photoconductive setup can avoid the input and output coupling of free space Terahertz beams, giving rise to a compact and versatile setup. The details process was reported elsewhere [*Yeh et al* (2005)]. THz Time Domain Spectroscopy measurement (***Figure 21***) require multiple time delay scans, which is time consuming and can result in a systemic error caused by the intensity fluctuations of the THz beam. THz-TDS has smaller spectral range than Fourier Transform Spectroscopy (FTS) system and provide lower resolution than narrowband THz spectroscopy [*Ferguson et al* (2002)].

D: Proposed Experimental Scheme for Optical- Illumination of THz IMPATT Devices:

In reality the diode under test may be mounted in a THz package, and can be sealed in ceramic sleeve with metallic contacts on each end. To expose the diode chip for illumination, a small groove can be cut in the ceramic sleeve. The diode package may be mounted in a

THz waveguide cavity with a waveguide tuning short on one side and an output coupling probe on the other side. A small hole can be drilled on the other side of the cavity to allow the output of the modulated LASER to be coupled to the IMPATT chip via a piece of optical fiber. The intensity of optical radiation can be experimentally increased by using a convex lens between the source (UV LASER) having a particular optical power density and the THz waveguide in which the device is embedded. The diagrams, as shown in ***Figures 22 and 23***, represent proposed setup for realizing optical illumination experiment on the THz devices.

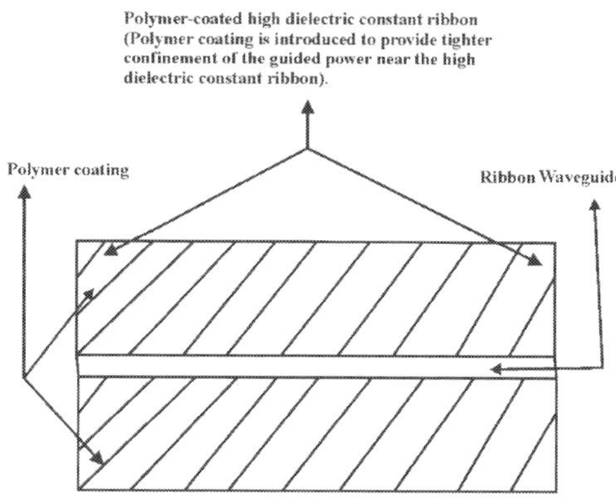

Fig. 20. (a): Longitudinal cross-sectional geometry of a polymer-coated high dielectric constant ribbon. The thickness and width of the high dielectric constant ribbon are, respectively, 0.0635λ_0. The dieletric constante of the ribbon is 10 while that of the polymer is 2.04 [35].

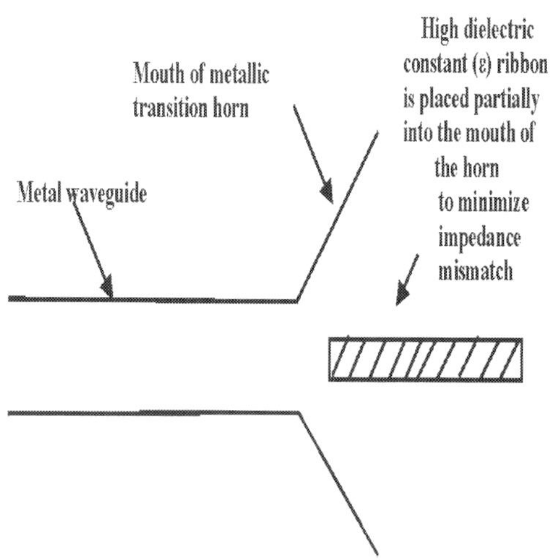

Fig. 20. (b). Rectangular metal waveguide to high dielectric constant ribbon waveguide transition [Ref.35]

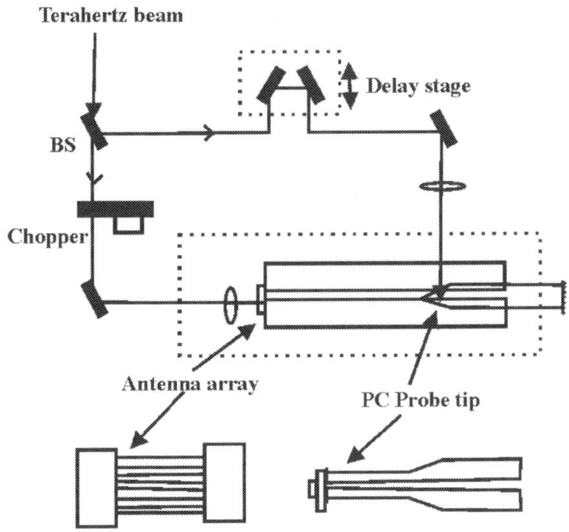

Fig. 21. Detection of Terahertz beam by photo-conductive method

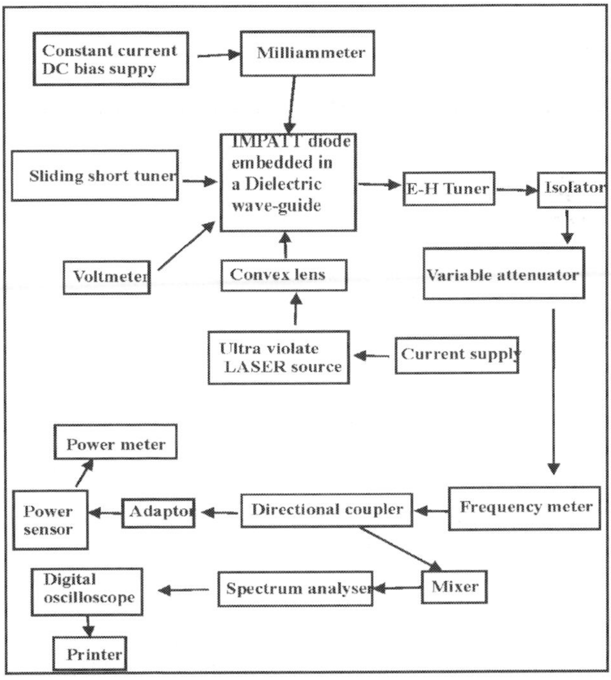

Fig. 22. Experimental setup for optical illumination experiment on 4H-SiC IMPATT diode at THz region

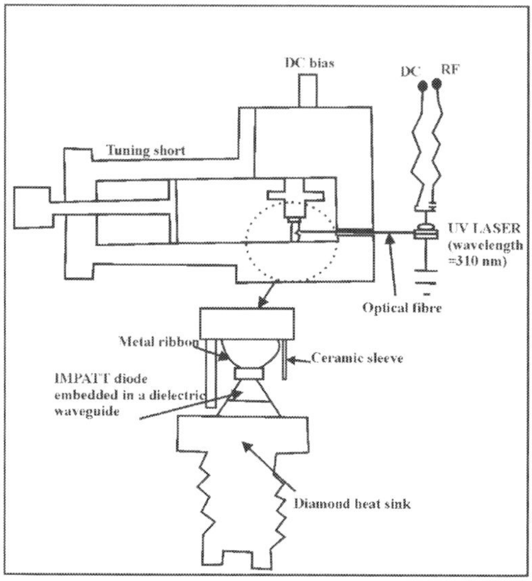

Fig. 23. Experimental setup for shining light on THz IMPATT diode

11. Conclusions

The prospects of WBG Wz-GaN and SiC (4H- and 6H-) based IMPATT devices of different structures and doping profiles were thoroughly examined both in the MM-wave and sub-millimeter wave (THz) region. The study established the potential of these WBG semiconductors in fabricating high-power and high-frequency IMPATT devices, in MM-wave as well as THz region. Si- based devices were also examined in these two regimes. The results showed that GaN and SiC based devices had superior performance over InP, in THz regime. A comparison of device properties of the 4H and 6H-SiC IMPATTs, revealed that the former is superior in terms of efficiency, output power generation capabilities and high temperature operation.

Studies were also carried out on effects of parasitic series resistance and optical illumination effects due to photo-generated carriers in the designed TM and FC diodes. The effects of optical illumination on the different semiconductor (4H- and 6H- SiC, Wz-GaN) based diodes were examined for the two different illumination configurations: TM and FC. The results indicated that device negative resistance and Q-factor degraded, resulting in decrease of power density, along with an up-shift in the frequency of operation, due to the variation of incident illumination intensity. Moreover, it was observed that the predominant hole photo-current in FC illumination configuration, had more pronounced effect in modulating the high-frequency properties of illuminated SiC and GaN devices, while illuminated Si IMPATT showed opposite behavior.

There is no doubt that, with the continuing progress in MM-wave and THz-device technology, the predictive simulation and proposed experimental feasibility of WBG semiconductor based IMPATT devices will become increasingly important for the success of research in modern high-power electronics.

12. References

Adesidac et al. (1999) "Dry and wet etching for group III nitrides", MRS Internet J. Nitride Semiconductor Res." 4S1, G1.4.

Adlerstein, M.G., Holway, L.H. ; Chu, S.L ; (1983)'Measurement of series resistance in IMPATT diodes', *IEEE Trans. Electron Devices* , vol. ED-30, pp. 179-182.

Alekseev, E ; and Pavlidis, D ; (2000) "GaN Gunn diodes for THz signal generation", *IEEE MTT-S International Microwave Symposium Digest., 2000, vol. 3, pp. 1905 - 1908.*

Barbieri, S. *et al.* ; (2003) "Continuous-Wave Operation of Terahertz Quantum-Cascade Lasers," *IEEE J. Quantum Elec.*, vol. 39, no. 4, pp. 586–591.

Brandt, R. C. et al. ; (1998) « SiC for applications in high-power electronics", in *Semiconductors and Semimetals,* Editor - Y. S. Park, New York, Academic, vol. 52, pp. 195 - 236.

Burm, J ; Chu, K ; Davis, W. A ; Schaff, W. J ; Eastman L. F ; and Eustis, T. J ; (1997)"Ultra-low resistive ohmic contacts on n-GaN using Si implantation", Appl. Phys. Lett.,vol. 70, p.464.

Chao, Q. C. et al., (2009) "Generation of intense THz pulsed lasers pumped strongly by CO2 pulsed lasers", *Chinese Physics Letters,* vol. 26, p. 064201.

Chu, C. F ; Yu, C. C ; Wang, Y. K ; Tsai, J. Y ; Lai, F. I ; and Wang, S. C ; (2000)"Low-resistance ohmic contacts on p-type GaN using Ni/Pd/Au metallization", Appl. Phys. Lett. Vol. 77, pp. 3423-3425.

Dammertz, G ; et. al., (2002) "Development of a 140-GHz 1-MW Continuous Wave Gyrotron for the W7-X Stellarator," *IEEE Trans. Plasma Sci.,* vol. 30, no. 3,pp. 808–818.

Dobroiu, A. ; Yamashita, M ; Ohshima, Y. N. ; Morita, Y ; Otani, C ; and Kawase, K. ; (2004) "Terahertz imaging system based on a backward-wave-oscillator", *Applied Optics,* vol. 43, p. 5637.

Dudley, M. ; Wang, S. ; Huang, W. ; Carter, Jr., C. H. and Fazi, C. ; (1995) *J. Phys. D* , vol. 28 pp. A63-A68.

Eisele, H. ; Naftaly, M. ; and Kamoua, R ; (2005) "Generation of submillimeter-wave radiation with GaAs TUNNETT diodes and InP Gunn devices in a second or higher harmonic mode," *Int. J. Infrared Millim. Waves,* vol. 26, no. 1, pp.1–14.

Eisele, H. ; (2005) "355 GHz oscillator with GaAs TUNNETT diode," *Electron. Letts.,*vol. 41, no. 6, pp. 55–56.

Elasser A. and Chow, T. P. (2002) "Silicon carbide benefits and advantages for power electronics circuits and systems," Proceedings of the IEEE, vol. 90, no. 6, pp. 969–86.

Eisele, H and Haddad, G.I ; (1997) 'Microwave Semiconductor Device Physics ', Ed. S. M. Sze, Wiley, New York, p. 343.

Ferguson B ; and Zhang, X ; (2002) "Materials for THz science and technology", Nat. Matr, vol. 1, pp.26-33.

Flech, K ; Danly, G ; Jory, H. R. ; Kreischer,K. E. ; Lawson, W. ; Levush, B. and Temkin, R. J. (1999) "Characteristics and Applications of Fast-Wave Gyrodevices," *Proceedings of IEEE,* vol. 87, no. 5, pp. 752–781.

Gummel, H.K. ; and Blue, J.L ; (1967) 'A small signal theory of avalanche noise in IMPATT Diodes, *IEEE Trans. Electron Devices'*, vol. ED -14, pp. 569-580.

Ishibashi T. *et al.,* (1977) "Liquid nitrogen cooled sub-millimeter wave Si IMPATT diodes", *Electronics Lett.,* vol. 13, p. 299.

Ives, R. L ; Marsden, D. ; Caplan, M. ; Kory, C ; Neilson, J ; and Schwartzkopf, S ; (2003) "Advanced terahertz backward wave oscillators," in *4th IEEE International Conference on Vacuum Electronics*, pp. 20–21.

Johnston, R. L. ; De Loach, B. C ; Cohen, B. G, (1965) "A silicon diode microwave oscillator", *Bell. Syst. Tech.* J., vol. 44, p. 369.

Karpowicz, N. ; Zhong, H. ; Zhang, C. ; Lin, K-I ; Hwang, J-S ; Xu, J ; and Zhang, X.-C. ; (2005) "Compact continuous-wave subterahertz system for inspection applications," *Appl. Phys. Lett.*, vol. 86, no. 5, pp. 54 105–3.

Keyes, R. W. ; (1973) "Silicon Carbide from the prospective of physical limits on semiconductor devices", *Proc. Silicon Carbide Conference*: 1973, pp. 534 -542.

Konstantinov, A. O. ; Waheb, Q. ; Nordell N. ; and Lindefelt, U ; (1998) "Ionization rates and critical fields in 4H-SiC junction devices", *Mat. Sci. Forum*, vol. 264 -268, pp. 513-515.

Konishi, R ; et al. (2003) "Development of Ni/Al and Ni/Ti/Al ohmic contact materials for p-type 4H-SiC" Mater. Sci. & Eng. B, vol. 98, pp. 286-293.

Krishnagopal, S ; and Kumar, V ; (2004) "Free-electron lasers," *Radiat. Phys. Chem.*, vol. 70, no. 4-5, pp. 559–569.

Kuzuhara, M. ; "Device technology based on new III-Nitride heterostructures", Proc. CS MANTECH Conference, May 18th - 21st, 2009, Tampa, Florida, USA, 2009.

Lazar, M ; et al. (2006) "Deep SiC etching with RIE", Supperlattices and Microstructures, vol. 40, pp. 388-392.

Leigh, M. A. *et al* ; (2009)"Narrowband pulsed THz source using eyesafe region fiber lasers and a non-linear crystal", IEEE Photonics Technology Letters, vol. 21, no. 1, pp. 27-29.

Mazumder, N ; and Roy, S.K ; (1993) 'Control of millimeter wave properties of high efficiency double drift region IMPATTs through enhancement of saturation current', *Phy.Status Solidi (a)*, vol. 137, pp. 267-275.

Mehdi, I ; Haddad G. I ; and Mains, R. K. ; (1998) "Microwave and millimeter-wave power generation in silicon carbide avalanche devices", *J. Appl. Phys.*, Vol. 64, No. 3, pp. 1533-1540.

Meng C. C. and Liao, G. R. (1998) "Analysis of SiC IMPATT device in Millimeter-Wave frequencies", *Microwave and Optical Technology Letters*, vol. 18, p. 167.

Meng C. C. ; Liao G. R. ; and Chen, J. W. (1999) "Analysis of Millimeter-Wave GaN IMPATT oscillator at elevated temperature", *Microwave and Optical Technology Letters*, vol. 23, p. 257.

Misawa, T ; (1966) "Negative resistance in p-n junctions under avalanche breakdown conditions: part II, *IEEE Trans. Electron Dev.*, vol. ED-13, p. 143.

Mukherjee, M ; and Mazumder, N ; (2006), 'Optically illuminated 4H-SiC THz IMPATT devices', *Egyptian J. of Solids*, vol. 30, p. 85.

Mukherjee, M ; Mazumder, N ; Roy, S. K ; and Goswami, K ; (2007a) 'GaN-IMPATT diode : a photo-sensitive high power THz source', *Semiconductor Science and Technology*, vol. 22, p. 1258.

Mukherjee, M ; and Mazumder, N ; Dasgupta , A ; (2007 b), 'Radiation effect on a high efficiency double drift region 4H-SiC Terahertz IMPATT diode', Proceedings of *IEEE 5th International Conference on Microwave and Millimeter Wave Technology (IEEE- ICMMT 2007),* April 19 -22, 2007, Tsinghua University, Guilin, China, pp. 655-658.

Mukherjee, M ; Mazumder, N ; (2008 a) 'Simulation experiment on optical modulation of 4H-SiC millimeter-wave high power IMPATT Oscillator", *Int. Journal of the European Microwave Association* (EuMA Publishing - UK), vol. 4, 2008, pp. 276-282.

Mukherjee, M ; and Roy, S. K ; (2008b) 'Design and Terahertz characteristics of hexagonal and cubic SiC based photo-irradiated IMPATT oscillators', *Proc. of IEEE Asia Pacific Microwave Conference 2008,* IEEE-APMC 2008, Hong Kong, China, 16th - 19th December, 2008, China, Paper no. 1963.

Mukherjee, M ; Mazumder, N ;, Roy, S. K ; (2008 c), 'Photosensitivity analysis of Gallium Nitride and Silicon Carbide Terahertz IMPATT oscillators: comparison of theoretical reliability and study on experimental feasibility', *IEEE Trans. Device and Materials Reliability,* vol. 8, p. 608.

Mukherjee, M ; and Roy, S.K ; (2009b) 'Optically modulated III-V Nitride based high-power IMPATT oscillator at MM-wave window frequency', *EuMA Int. J. Microwave and Wireless Technogy,* accepted for publication.

Mukherjee, M ; Mazumder, N ; (2009) 'Effect of charge-bump on high-frequency characteristics of α-SiC based double drift ATT diodes at MM-wave window frequencies", *J. IETE (India),* vol. 55, p. 118.

Nagatsuma, T.; "Millimeter-Wave Photonic Technologies for Communications and Sensor Applications," in New Photonics Technologies for the Information Age: The Dream of Ubiquitous Services, pp. 193–212, ed. by S. Sudo and K.Okamoto (Artech House, 2004).

Naftaly, M ; Foulds, A. P. ; Miles, R. E. and Davies, A. G. (2005) "Terahertz Transmission Spectroscopy of Nonpolar Materials and Relationship with Composition and Properties," *Int. J. Infrared Millim. Waves,* vol. 26, no. 1, pp. 55–64.

Namordi, M. R. ; Coleman, D. J., (1980) 'Design windows for low-high-low GaAs IMPATTs', *IEEE Trans. Electron. Devices,* vol. 27, p. 282.

Nishizawa, J.-I. ; Sasaki, T. ; Suto, K. ; Yamada, T. ; Tanabe, T. ; Tanno, T. ; Sawai, T. and Miura, Y. (2005) "THz imaging of nucleobases and cancerous tissue using a GaP THz-wave generator," *Opt. Commun.,* vol. 244, no. 1-6, pp. 469–474.

Ono, S ; Arai, M ; and Kimura, C ; (2005) "Demonstration of high power X-band oscillation in p^+ n - n^+ 4H-SiC IMPATT diodes with guard-ring termination", *Materials Science Forum,* Vol. 981-984, pp. 483-485.

Orihashi, N. ; Suzuki S. and Asada, M. ; (2005) "One Terahertz harmonic oscillation of resonant tunneling Diodes", *Appl. Phys. Lett.,* vol. 87, p. 233501-1-3.

Panda, A. K. ; Pavlidis, D ; and Alekseev, E. A ; (2001) "Noise characteristics of GaN-based IMPATTs", *IEEE Trans. Electron Devices,* vol. 48, pp.1473 -1475.

Pastor, D ; et al. (2007) "Crystal damage assessment of Be+-implanted GaN by UV Raman scattering", Semicond. Sci. Tech., vol. 22, pp. 70-73, 2007 and references therein.

Pearton, S. J. ; Ren, F ; Zang A. P. and Lee, K. P. ; (2000) "Fabrication and performance of GaN electronic devices", *Materials Science and Engineering,* vol. 250, pp. 1-158.

Read, W. T ; (1958)"A proposed high frequency negative resistance diode", *Bell. Syst. Tech. J.*, vol. 37, p. 401.

Reklaitis, A. ; and Reggiani, L. ; (2005) "Giant suppression of avalanche noise in GaN double drift impact diodes", *Solid-state Electronics*, vol. 49, pp. 405 - 408.

Roccaforte, F ; et al. (2006) "SiC Materials and Devices; vol. 1, eds: M. Shur, S. Rumyantsev, and M. Levinshtein, World Scientific Publishing, p. 40, 2006.

Roy, S.K. ; M., Sridharan ; Ghosh , M ; and Pal B.B ; (1979) 'Computer method for the dc field and carrier current profiles in the IMPATT device starting from the field extremum in the depletion layer', *Proc. 1st Conf. On Num Anal .Of Semiconductor Devices* (NASECODE I) (Dublin: Boole) (Ed. J.H. Miller), p. 266.

Roy, S. K. ; Banerjee, J. P. and Pati, S. P; (1985) "A Computer analysis of the distribution of high frequency negative resistance in the depletion layer of IMPATT Diodes", *Proc.4th Conf. on Num. Anal. of Semiconductor Devices* (NASECODE IV) (Dublin) (Dublin: Boole), pp. 494-500.

Schmidt, L. P ; Biber, S ; Rehm, G ; and Huber, K ; (2002) "THz Measurement Technologies and applications," 14th International Conference on Microwaves, *Radar and Wireless Communications*, vol. 2, pp. 581–587.

Shockley, W. ; (1954) "Negative resistance arising from transit time in semiconductor diode", *Bell. Syst. Tech.* J, vol. 33, p. 799.

Vassilevski, K ; Zorenko, K ; (2001), "4H-SiC IMPATT diode fabrication and testing", *Tech. Digest of Int. Conf. on SiC and Related Materials- ICSCRM, Tsukuba, Japan,* p. 713.

Wachter, M ; et al. (2007) "Metallic slit waveguide for dispersion-free low-loss terahertz signal transmission", Applied Physics Letters, vol. 90, pp. 061111-061111-3.

Watanabe, Y ; Kawase, K ; Ikari, T ; Ito, H ; Ishikawa, Y ; and Minamide, H. (2004) "Component analysis of chemical mixtures using terahertz spectroscopic imaging," *Opt. Commun.*, vol. 234, no. 1-6, pp. 125–129.

Wang, S. ; Ferguson, B ; Abbott, D and Zhang, X.-C. (2003) "T-ray imaging and tomography," *J. Biol. Phys.*, vol. 29, no. 2-3, p. 247.

Yeh, C ; Shimabukuro, F ; and Siegel, P. H ; (2005) "Low-loss terahertz ribbon waveguides", Applied Optics, vol.44, No.28, pp. 5937-5946.

Yuan, L ; Copper, J. A. ; Melloch M. R. ; and Webb, K. J. ; (2001) "Experimental demonstration of a Silicon Carbide IMPATT oscillator", *IEEE Electron Device Letters,* vol. 22, pp. 266 -268.

Zhao, J. H. et al., (2000) "Monte Carlo simulation of 4H-SiC IMPATT diode", *Semicond. Sci. Technol.,* vol.15, pp.1093-1100.

8

RF and microwave band-pass passive filters for mobile transceivers with a focus on BAW technology

Martha Suárez, Martine Villegas, Geneviève Baudoin
Université Paris-Est, Laboratoire ESYCOM, Groupe ESIEE Paris
France

1. Interest of RF band-pass filters in mobile communications

This chapter focuses on RF and microwave band-pass passive filters required for mobile transceiver front-ends. Band-pass filters are commonly used in wireless transceivers for communication systems such as cellular and connectivity standards or Ultra Wide Band systems. Most of the recent communications standards operate in multi-bands and have flexible frequency profiles, therefore filter banks must be implemented in order to fulfil regulation requirements in all bands.

A filter is a time-invariant linear system that can be defined by its frequency transfer function $H(f)$. It operates on an input signal $x(t)$ with respect to frequency and the resulting filtered output signal $y(t)$ has a Fourier transform equal to the product of $X(f)$ and $H(f)$: $Y(f) = H(f)X(f)$.

The ideal transfer function of a band-pass filter is equal to 1 in the pass-band and to 0 in the two stop-bands. The pass-band is defined by a range of frequencies $[f_1, f_2]$ or equivalently by its central frequency f_c and its frequency bandwidth W, with:

$$f_c = (f_1 + f_2)/2 \text{ and } W = (f_2 - f_1) \quad (1)$$

The two stop-bands are defined by the frequency ranges $[0, f_1]$ and $[f_2, +\infty]$ for the left and right stop-band respectively.

In practice, the filter transfer function is not perfect: it is not perfectly constant in the pass-band, it presents a finite attenuation (or rejection) in the stop-bands and the transitions between frequency bands are not infinitely sharp.

The different characteristics of a band-pass filter will be described in section 2. Filtering is one of the three more critical elements for the cost and size of transceivers in addition to frequency synthesizers and power amplifiers.

Many filters are present in wireless transceivers. They are distributed in the successive stages of the architectures in the baseband (BB), intermediate frequency (IF) and radiofrequency (RF) parts. This chapter focuses on RF and microwave band-pass filters.
RF band-pass filters operate on RF or microwave signals and we will use the expression RF band-pass filter to represent them regardless of whether they operate on RF or microwave signals. In the receiver, they are located just after the antenna and after the low noise amplifier (LNA). They are used to suppress out-of-band noise and blockers, to eliminate the image frequency in super-heterodyne receivers and more generally to limit the bandwidth of the received signal and the dynamic requirements of the receiver. In the transmitter, they are located before and/or after the power amplifier (PA). They are used to reject the spurious signals generated, for example by the local oscillator (LO), and to minimize power emission out of the desired frequency band that could be generated by the PA non-linearity. RF band-pass and band-reject filters are also found in the receiving and transmitting branches of duplex filters in systems using frequency division duplex (FDD) schemes. Another application of RF band-pass filters used in a filter bank is to split a large RF bandwidth into several smaller bandwidths that are easier to process. This can be useful in very wide band communications systems such as in the field of millimeter-wave 60 GHz or more generally for Ultra Wide Band (UWB) communications.
The precise role and specifications of the different filters depends on the regulation, on the standard requirements, on the architecture of the transceiver and also on the duplex scheme. Standards and regulations specify the minimum requirements for RF transceivers. They are expressed by parameters, which can take values that impose more or less stringent constraints on the RF system blocks such as filters. Among the important parameters that can influence filter (BB, IF and RF) specifications or characteristics are: frequency bands, channel and signal bandwidth, channel frequency step, duplex schemes, transmit power, output RF spectrum mask, limit on spurious emission, limit on noise, distortions, linearity, Bit Error Rate (BER), Error Vector Magnitude (EVM) and Adjacent Channel Interference expressed by the Adjacent Channel Leakage Ratio (ACLR) or the Adjacent Channel Power Ratio (ACPR).
A given standard is allocated with one or several frequency bands and these frequency bands can be split into smaller bandwidth channels allocated to different users. The RF filter is used to select the standard bands while the IF and/or BF filters select the channel bandwidths and are generally more selective than RF band-pass filters. The RF frequency bands specified by standards are usually above 50 MHz. For example, for WiMAX standards, the specified bands are between 100 and 200 MHz. Therefore, RF filters have wide pass-bands and the ratio between the bandwidth of the pass-band and the central frequency of the filter is typically of the order of a few percent.
Most mobile subscriber equipment is now multi-band, multi-mode (multi standards) and multi-radio (cellular, connectivity, FM and TV receivers, GPS, etc). They include several transmitters/receivers connected to a small number of antennas. RF low-pass, high-pass and band-pass filters are used to combine these different transceivers operating on different frequency bands that are generally quite far apart. Low-pass and high-pass filters are usually well suited for this task that most often does not necessitate very high selectivity filters with high Q resonators.

1.1 Influence of Duplex schemes on RF filter requirements

Different duplex schemes are specified in wireless communication standards to separate forward and reverse communication links in order to allow mobile equipment to share the same antenna for transmit and receive signals. These schemes are FDD (Frequency Division Duplex), TDD (Time Division Duplex), or HFDD (Half Frequency Division Duplex).

1.1.1 FDD

In the FDD method, the forward and reverse communications use different carrier frequencies separated by a frequency offset. With the FDD method, a real full duplex communication is possible, but it requires a complex RF front-end since it uses separate receiving and transmitting synthesizers. Besides, it requires two different RF filters for transmitting and receiving. The transmission must not degrade the simultaneous reception. Therefore, on the one hand, the attenuation of the transmit filter must be high enough in the receiver frequency band so that the noise introduced by the transmitter on the receiver is kept low in comparison to the noise floor of the receiver. On the other hand, the receiver RF filter must sufficiently reject the transmitter frequency band so that the transmitter does not overload the receiver. The constraints on RF filters used in duplex filters in FDD modes are usually quite stringent: The larger the frequency offset, the easier these filters. Typical values of frequency offset are 50 to 100MHz.

For example, in GSM 900 standard, the transmitter uses the uplink frequency sub-band Tx: 890-915 MHz and the receiver uses the downlink frequency sub-band Rx: 935-960 MHz.

The GSM sub-bands are separated by a frequency offset equal to 45 MHz. And each sub-band has a 25 MHz bandwidth, while the channel spacing is equal to 200 KHz. For the DCS 1800 standard, the frequency offset is 95 MHz and each sub-band has a 75 MHz bandwidth, Tx: 1710-1785 MHz and a Rx: 1805-1880 MHz.

The GSM standard (ETSI, 1999) specifies the output RF spectrum of the modulated signal for the transmitter. The spectrum mask for a class 4 mobile GSM transmitter is given in Fig. 1.

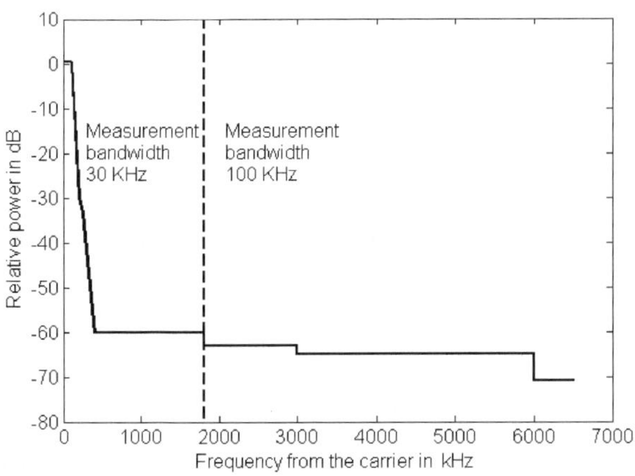

Fig. 1. GSM 900 spectrum mask of modulated signal.

Using these specifications, we can calculate the required filter attenuation (rejection) in the receive band for the GSM duplex scheme. Let's suppose that we require a transmit noise 10 dB below the noise floor of the receiver, for a receiver noise figure of 6 dB. Noting P_{out} the transmitted power in dBm, the output power spectral density in dBm/Hz for a channel bandwidth W_{ch} is:

$$PSD = P_{out} - 10\log(W_{ch}) \ [dBm / Hz] \tag{2}$$

Filter attenuation at the receive frequency must be greater than:

$$\begin{aligned} Att_{dB} &\geq P_{out} - 10\log(W_{ch}) - Mask - (-174 + NF - 10) \\ Att_{dB} &\geq 33 - 53 - 71 - (-174 + 6 - 10) = 87dB \end{aligned} \tag{3}$$

Therefore, the filter attenuation must be 87dB at 45 MHz from the carrier frequency which is a rather stringent requirement for RF filters.
The insertion losses of the filters have not been taken into account in this calculation.
The transmit/receive duplexer filters for mobile terminals must have high-performance with high out-of-band attenuation and a low in-band transmit/receive distortion and insertion loss. It is sometimes necessary to use cavity filters to fulfill the severe filter requirements of the FDD method.

1.1.2 TDD

For the TDD method, the antenna is switched alternatively between the transmitter and the receiver. The same frequency band is used for transmission and reception. The transceiver can be simplified because, even if it looks like a full duplex system for the user, the transceiver actually operates in a single mode at a time. The transmitted signal does not interfere with the received signal since transmission and reception are done at different periods of time. Therefore, the RF filter requirements are relaxed. Besides, since the transmission and reception use the same carrier frequency, a single RF filter can be used. However there are some drawbacks in TDD, e.g. the adjacent channel interference is higher than in a FDD scheme.

1.1.3 HFDD

In some standards, such as WiMAX, a Half Frequency Division Duplex is possible in order to reduce the cost and size of mobile stations. HFDD systems operate in half-duplex; the transmission and reception are done in separate bands and at separate time periods. This approach allows a single frequency synthesizer to be used and relaxes the constraints on the RF filters.

1.2 Filtering of out-of-band blockers and image frequency

RF filters are also used in the receiver to remove the RF band blockers and image frequency.

1.2.1 Blocking signals

The blocking characteristics of the receiver are specified separately for in-band and out-of-band performance. For example, for the GSM 900 standard, these bands are defined by the following frequency ranges for the mobile station: In-band: 915 MHz -980 MHz and Out-of-band: > 980 MHz-12 750 MHz.

For a small mobile station, the reference sensitivity should be met when different signals are simultaneously input to the receiver:

- a useful signal, modulated at frequency f_o, 3 dB above the reference sensitivity level or input level for reference performance,
- a continuous, static sine wave signal at a frequency f which is an integer multiple of 200 kHz and at a level of 0 dBm out-of-band and -43 dBm, -33 dBm or -23 dBm for $|f - f_o| < 1.6\,MHz$, $1.6\,MHz \le |f - f_o| < 3\,MHz$, $|f - f_o| \ge 3\,MHz$ respectively.

The FDD WCDMA standard (3GPP, 2005) specifies that the out-of-band blocking characteristics of the receiver should be such that the BER remains smaller than 10^{-3} when different signals are simultaneously input to the receiver:

- a useful signal modulated at frequency f_o with a power at -114 dBm (3 dB above the reference sensitivity) and
- a blocking signal with a power equal to -44 dBm, -30 dBm, -15 dBm, at a frequency f in the range [2050 MHz - 2095 MHz], [2025 MHz - 2050 MHz], [1000 - 2 025 MHz] respectively.

1.2.2 Image frequency

In super-heterodyne receivers, the RF filter is used to suppress the image frequency. Indeed, for a given useful RF frequency f_{RF} and a given intermediate frequency f_{IF}, it is possible to down-convert the RF frequency to the IF frequency, by mixing the RF signal with a local oscillator at a frequency f_{LO} such that:

$$f_{IF} = |f_{RF} - f_{LO}| \tag{4}$$

As the mixing generates both the difference and the sum frequencies of the input signals plus possibly some other spurious frequencies, the resulting signal is filtered after the mixing by a selective IF filter to select the down-converted signal corresponding to the desired channel. Unfortunately, not only the desired RF frequency will be down-converted to the IF frequency but also the frequency called the "image frequency" f_{im}. The image frequency satisfies the same equality and is symmetrical to f_{RF} with respect to f_{LO} and:

$$f_{IF} = |f_{im} - f_{LO}|,\ f_{LO} = \frac{f_{RF} + f_{im}}{2} \text{ and } f_{im} = 2f_{LO} - f_{RF}. \tag{5}$$

Therefore this possible image frequency has to be filtered before the mixer by an RF band-pass filter. Otherwise, if there is some undesirable signal power at the image frequency, at the receiver input, it will add as a noise to the down-converted useful signal.

1.3 Characteristics of some cellular communication and connectivity standards

Many standards exist for wireless communications, including standards for 2G, 3G and beyond 3G cellular systems (e.g. GSM, UMTS, LTE), Wireless Metropolitan Area Networks WMAN (e.g. WiMAX IEEE 802.16), Wireless Local Area Networks WLAN (e.g. Wi-Fi IEEE 802.11a/b/g/n) and Wireless Personal Area Networks WPAN (e.g. Bluetooth IEEE 802.15.1). Most of these are in the frequency range below 6 GHz but some new standards have appeared in the millimeter wave range (60 GHz radio in particular). In the first case, the data rates are in the range of several tens to several hundreds of Mbps and in the second case they can be in the range of several Gbps.

In Table 1 we consider some of the most widely used standards for wireless communications and we give some of their characteristics (for the case of a Mobile Station, uplink) that influence the design of the RF band-pass filters.

Standard	Frequency Range (MHz)	Transmission Bandwidth (MHz)	Channel Bandwidth	Duplex scheme / Frequency offset in FDD
GSM 900	890 – 915	25	200 kHz	FDD / 45 MHz
DCS 1800	1710 – 1785	75	200 kHz	FDD / 95 MHz
UMTS WCDMA (Band 1)	1920 – 1980	60	5 MHz	FDD / 190 MHz
UMTS-TDD TDCDMA	1900 – 1920 and 2010 – 2025	20 and 15	5 MHz at 3.84 Mcps	TDD
WLAN (802.11 b/g)	2400 – 2483.5	83.5	11 MHz	Half-duplex
WLAN (802.11a)	5150 – 5350	200	20 MHz	Half-duplex
WMAN Mobile WiMAX (802.16e)	2300 – 2400	100	Variable (3.5, 5, 7, 8.75, 10 MHz)	Mainly TDD
	2496 – 2690	194		
	3300 – 3400	100		
	3400 – 3600	200		
	3600 – 3800	200		

Table 1. Some parameters of mobile communications standards related to the transmitter

The relative bandwidth, i.e. the ratio between the transmission bandwidth and the central frequency of the RF band-pass filter, for these standards varies between 1 % for UMTS-TDD and 7 % for WiMAX 2496 - 2690 MHz frequency range. The value of the relative bandwidth may influence the choice of the filter technology.

It is clear from Table 1, that a single reconfigurable RF filter could not be used in a multi-radio transceiver and that the necessary RF front end filter bank is quite complex.

However tunable RF filters are necessary for reconfigurable multi-radio front-ends that can support several standards and applications. Since not all of the applications are used at the same time, it is interesting to share some RF resources between the different radios in order to reduce the hardware size of the transceiver. Reconfigurable tunable filters are one of the elements that make this possible.

A challenge is to achieve tunable filters that can be integrated on-die. For an RF band-pass filter, the characteristics that should be tunable or reconfigurable include center frequency, bandwidths, selectivity, pass-band ripple and group delay.

1.4 Case of UWB standard with an MB-OOK transceiver

As seen before, in the case of very wide or ultra wide band communications, it can be useful to split the available frequency bandwidth into several smaller bandwidths by a filter bank. An example of such an approach is the UWB architecture proposed in (Paquelet et al., 2004). UWB wireless systems based on impulse radio have the potential to provide very high data rates over short distances. Fig. 2 represents the spectral mask for UWB systems in Europe.

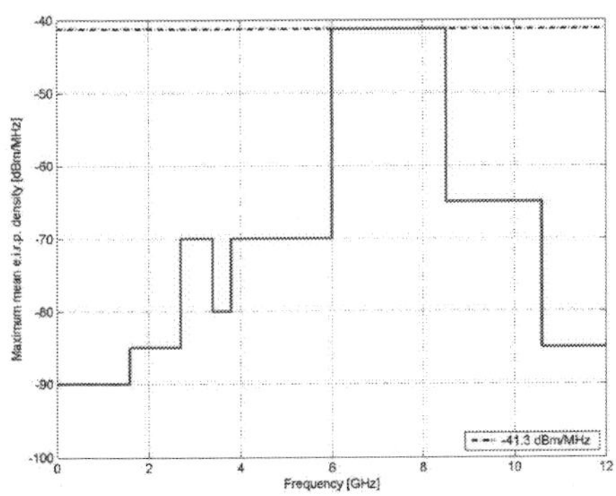

Fig. 2. Spectral mask for UWB systems in Europe

One of the possible solutions for UWB communication systems is the Multi Band On-Off Keying (MB-OOK) proposed in (Paquelet et al., 2004) which consists of an OOK modulation generalized over multiple frequency sub-bands and associated with a demodulation based on a non-trivial energy threshold comparison. Fig. 3(a) represents the architecture of the UWB MB-OOK transmitter and Fig. 3(b) shows the non-coherent processing in one sub-band of the receiver.

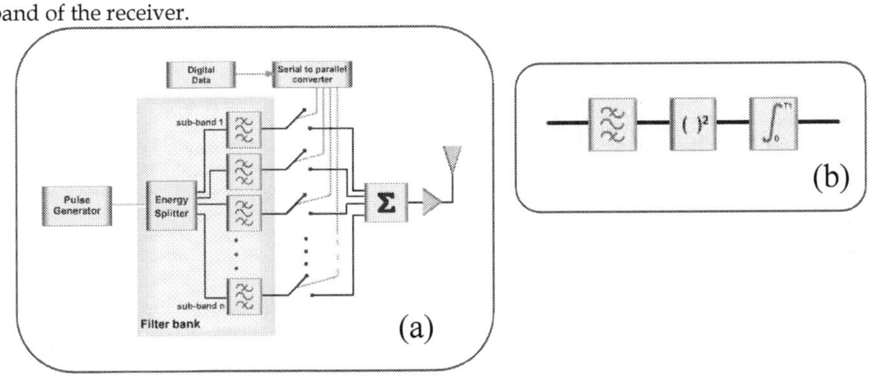

Fig. 3. (a) UWB MB-OOK transmitter architecture. (b) Non-coherent receiver: energy integration for one sub-band of the receiver

In the transmitter architecture, a pulse covering the allowed frequency band is generated with a repetition period T_r. The pulse generator is followed by a multiplexer that splits the input signal into N sub-bands. Pulses in each band are filtered and modulated by digital data at a rate $1/T_r$. Then, the modulated signals are combined and amplified before being sent through the UWB antenna.
The receiver architecture is symmetrical to that of the transmitter. It includes a low noise power amplifier (LNA), a splitter, a band-pass filter bank and then in each band, a squarer and an integrator.
Integration time in reception (T_i) and repetition time in transmission (T_r) are chosen considering the channel delay spread (T_d). To avoid inter-symbols interference, the symbol repetition period is chosen so that:

$$T_r > (T_d + T_s + T_f) \qquad (6)$$

where T_s is the duration allocated to the symbol waveform and T_f is the duration of the impulse response of one filter of the filter bank. Maximal throughput of the communication system can be estimated by multiplying its number of sub-bands and the pulse repetition rate $1/T_r$ (as long as the repetition time is long enough in comparison).
The splitter and the filter bank are common elements in the transmitter and receiver. The filter bank may be uniform or non uniform (Suarez et al., 2007a) depending on the constraints of the technology. Section 4 will present an example of a filter-bank for this architecture using BAW technology.

2. Applications and specifications of RF band-pass filters for Multi-Band reconfigurable transceiver architectures

This section considers the different characteristics of RF band-pass filters required in mobile transceivers. It's important to emphasize that the precise role and specifications of the RF band-pass filters depend on the regulation, on the standard requirements, on the architecture of the transceiver and also on the duplex scheme as detailed in the first section.
Among the important parameters that can influence RF band-pass filter's specifications and the choice of the filtering technology are: frequency bands (filter's central operation frequency), allocated bandwidth (filter's bandwidth), transmit power (filter's power handling for the transmitter case), output RF spectrum mask, limit on spurious emission and adjacent channel interference (filter's out-of-band rejection). Furthermore, low insertion loss, temperature stability and integrability are expected in mobile multi-radio filters.
The central operation frequency depends on the considered standard. As presented in the first section, wireless communication standards such as cellular and connectivity standards or Ultra Wide Band systems have specific frequency allocation. Regulation entities determine the frequency allocation chart and also the maximum output power in each frequency band. This may vary depending on the geographical region or the country. Most of the wireless communication standards are in the frequency range below 6 GHz. Allocated frequency bands determine the filter's central frequency. This is a key parameter

to choose the filtering technology which should stand a high maximal operation frequency (up to 6 GHz for multi-radio applications).

The RF filter's bandwidth is not defined by the channel bandwidth but by the allocated frequency bandwidth. Table 1 presents the different bandwidths of the RF transmission filters in a multi-radio.

In the transmitter case, since the filtering is usually carried out after the power amplification, the RF transmission filters must offer high power handling capability. Input signals may have high power dynamics (e.g. mobile WiMAX or LTE signals) and the maximum power levels may vary up to 33 dBm in the GSM case, for example.

The out-of-band rejection of the filter is generally expressed in dBc (relative power in dB to the carrier). Maximal rejection is specified at a certain frequency offset from the carrier, known as the stop bandwidth and the frequency bandwidth to reach the required attenuation is also specified as the transition bandwidth. For communications standards, the out-of-band rejection is set from the output RF spectrum mask, the limits on spurious emission and the maximal adjacent channel interference expressed by the ACLR or the ACPR (usually considering the most stringent requirements).

An example of the power spectrum mask for mobile WiMAX standard is presented in Fig. 4. This power spectrum mask has not been proposed by the WiMAX IEEE standard but by the European Telecommunications Standards Institute (ETSI, 2003).

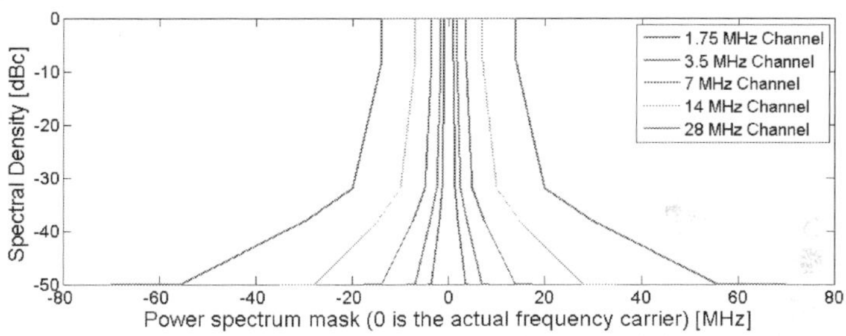

Fig. 4. WiMAX Power Spectrum mask for a high complexity modulation format.

The Power Spectrum mask is defined around the carrier and depends on the channel bandwidth. Recent standards like mobile WiMAX and LTE are very flexible and propose different channel bandwidths, number of carriers and coding and modulation formats for each carrier in order to adapt the transmission to the environment conditions (channel, network, user needs, etc). Power masks illustrate this flexibility, for example the mask of Fig. 4 is proposed just for the case of high complexity modulation format (e.g. 64 states or equivalent), which leads to the most stringent filtering constraints because of the small transition bandwidth.

In order to define the out-of-band rejection of the RF filter, a common practice is to extrapolate the power spectrum mask for a given channel bandwidth to the first and last channels in the allocated frequency band and to establish a new mask covering all the allocated frequency bandwidths.

Another important characteristic of RF band-pass filters is the Insertion Loss (IL) which should be as low as possible to increase the whole architecture power efficiency.

The group delay is another parameter to consider. For a filter's transfer function $H(s)$, at real frequencies, with $s = j\omega$:

$$H(j\omega) = |H(j\omega)| \cdot e^{j\theta(\omega)} = G(\omega) \cdot e^{j\theta(\omega)} \quad (7)$$

Where $G(\omega)$ and $\theta(\omega)$ are the gain-magnitude, or simply the gain, and the phase components respectively. Group Delay $\tau(\omega)$ is defined as:

$$\tau(\omega) = -\partial\theta(\omega) / \partial w \quad (8)$$

The group delay is expected to be constant in the whole filter's bandwidth.
EVM is typically measured at the receiver and constitutes a common indicator of signal information integrity. The maximum accepted EVM is usually given by the communications standards and in the case of WiMAX and LTE, a table with EVM values for different modulations and coding rates is established, e.g. in mobile WiMAX EVM limit is -30 dB (3.16%) for a 64-QAM (3/4) modulation (IEEE, 2005). The EVM is calculated observing all the imperfections of the transmission chain blocks. Therefore, the maximum acceptable group delay and in-band ripple of the filter depend on this EVM value and on the imperfections generated by all the other blocks of the architecture.
Finally, as size and cost are critical parameters for manufacturers, it is very often required to use a filtering technology that enables integration.

3. Available filtering technologies: advantages and trade-offs

3.1 Available technologies

The most notable RF filtering technologies include LC filters, ceramic filters, surface acoustic wave (SAW) filters, bulk acoustic wave (BAW) filters and low temperature co-fired ceramic (LTCC) filters.
LC filters can support high frequencies and can be integrated as a SoC. However, their main drawback is that they require too much area and can offer only a limited quality factor (*Q*).
Ceramic filters offer low IL (about 1.5 - 2.5 dB), high out-of-band rejection (> 35 dB) and low cost. On the other hand the large size of ceramic filters significantly penalizes the integration.
SAW filters are smaller than LC and ceramic filters, but have limitations in the frequency domain (up to 3 GHz). Depending on the application, their maximum output power rating could also be insufficient (up to 1 W). Typical IL varies between 2.5 and 3 dB and out-of-band rejection can reach up to 30 dB. The main drawback is that SAW filters are not compatible with silicon integration.
LTCC is a multi-layer technology that offers integration of high *Q* passive components along with low IL, high maximal operation frequency and acceptable out-of-band rejection. LTCC filters are smaller than LC and ceramic filters and can be integrated as SIP.
BAW filters use Film Bulk Acoustic Resonators (FBAR) that are characterized by a high quality factor *Q*. Moreover, they have low IL (1.5 - 2.5 dB), significant out-of-band rejection (≈ 40 dB) and high maximal operation frequency (up to 15 GHz). BAW filters can also deal with high output power (3 W). They are CMOS compatible and can be integrated "above IC".

CMOS-SOI technology evolution allows today to consider LC filters implementation. Indeed, the achievements in terms of quality factor are significantly improved compared to Si technologies.

3.2 SAW Technology

SAW technology is based on the use of surface acoustic waves in a piezoelectric material. Acoustic waves propagate at a speed lower than electromagnetic waves ($v_{SAW} \approx 3\,km/s$ and $v_{EM} < 3 \cdot 10^5\,km/s$ depending on the substrate used). This reduces the filter's size ($\lambda = v/f$). The Fig. 5 shows the basic structure of a SAW filter. Piezoelectric material choice, usually quartz, is important because it determines the propagation speed of the acoustic wave.

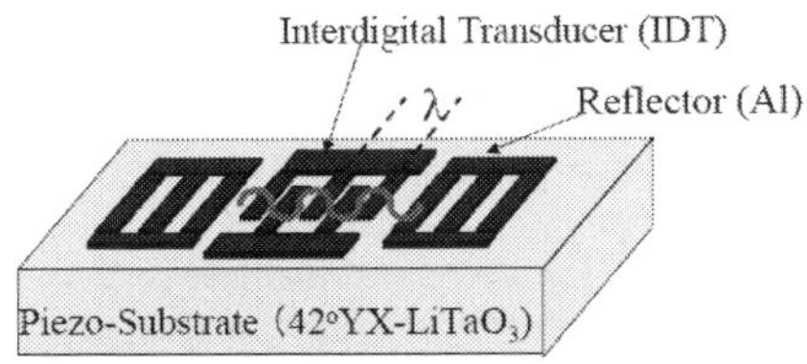

Fig. 5. Basic structure of a SAW filter.

The major drawbacks of the SAW technology are the operating frequency (<3 GHz), the significant insertion losses and the power handling (<1W). It is possible to perform filtering functions involving more complex cells, using, for example, ladder topologies:

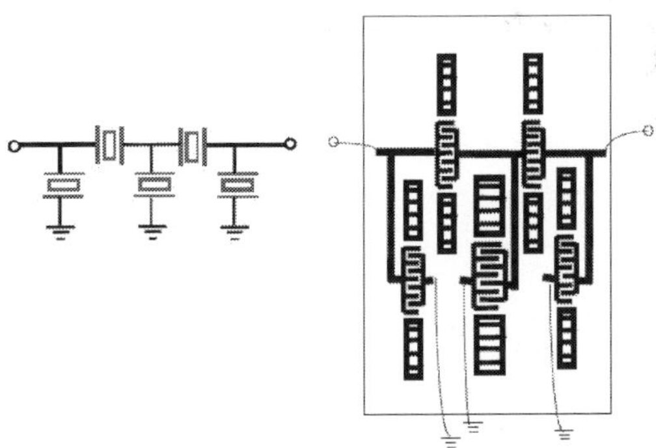

Fig. 6. (a) Ladder topology. (b) Example of a SAW filter in ladder topology

3.3 BAW Technology

3.3.1 Principle

The basic element of the BAW device is the thin film resonator which is very similar to a basic quartz crystal scaled down in size. A piezoelectric film is sandwiched between two metal films as shown in Fig. 7.

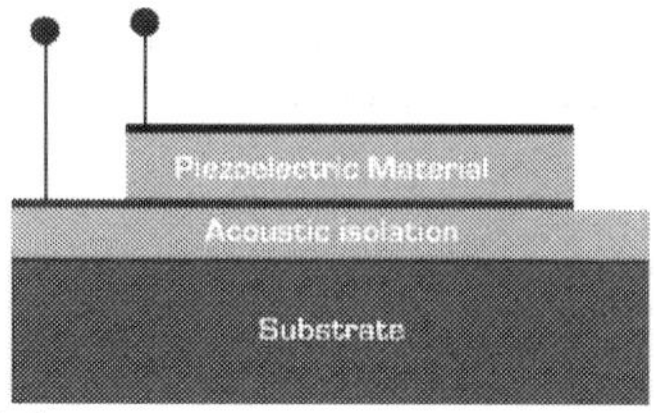

Fig. 7. BAW technology principle

The key properties of the BAW resonator are chosen to store the maximum acoustic energy within the structure, achieving a high electrical Q. The boundary conditions outside of the metal films must maintain a very high level of acoustic reflection with a vacuum being the ideal interface. The materials chosen must optimize both electrical and mechanical properties. Although there are many piezoelectric materials, Aluminium Nitride (AlN) has been established as the best balance of performance, manufacturability, and reliability.
The metal films range from Al, which offers the best performance with limited power handling, to Mo or W which offer high power handling with the cost of additional resistivity losses. The resonant frequency (f_r) is inversely proportional to the film thicknesses with both, the metal and piezoelectric dielectric, contributing to the resonant point.

$$f_r = v/2d \tag{9}$$

where v is the acoustic material velocity and d is the thickness of the piezoelectric material. BAW technology using AlN piezoelectric material allows frequency operation up to 15 GHz.

3.3.2 Resonator modeling

The Butterworth Van Dyke (BVD) model is an electric circuit model that characterizes FBAR resonators. The BVD equivalent circuit of the resonator is shown in Fig. 8.

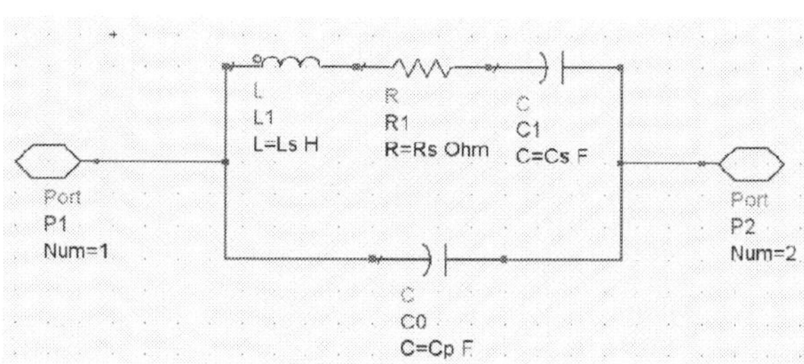

$$C_p = \frac{\varepsilon_0 \varepsilon_r A}{d}; \quad C_s = \frac{8k_t^2 C_p}{\pi^2}$$

$$L_s = \frac{\pi^3 V_a}{8 w_s^3 \varepsilon_0 \varepsilon_r A k_t^2}; \quad R_s = \frac{L_s w_s}{Q}$$

$$f_s = \frac{V_a}{2d_1}; \quad f_p = f_s\sqrt{1+\frac{8k_t^2}{\pi^2}}$$

Fig. 8. FBAR resonator - BVD model.

The resonator is in the form of a simple capacitor, having a piezoelectric material as the dielectric layer and suitable top and bottom metal electrodes. The simplified equivalent circuit of the piezoelectric resonator has two arms. C_p is the geometric capacitance of the structure. The R_s, L_s, C_s branch of the circuit is called the "motional arm," which arises from mechanical vibrations of the crystal. The series elements R_s, L_s, C_s are controlled by the

acoustic properties of the device and they cause the motional loss, the inertia and the elasticity respectively. These parameters can be calculated from equations presented in Fig. 8. ε_r is the material's relative permittivity (10.59 for the AlN), k_t^2 is the electromechanical coupling constant (6% for the AlN), V_a is the acoustic material velocity (10937 for the AlN), A is the surface area of the electrodes, d is the thickness of the piezoelectric material, and Q is the quality factor. w_s and w_p correspond to a 2π multiple of the resonance (f_s) and anti-resonance (f_p) frequencies of the resonator. Thickness of series and shunt resonators may be different; d_1 and d_2 refer to the thickness of the series and the shunt resonator respectively.
The frequency response of the FBAR resonator depends on the thickness of the thin piezoelectric film. The Fig. 9 shows an example of resonator frequency response.

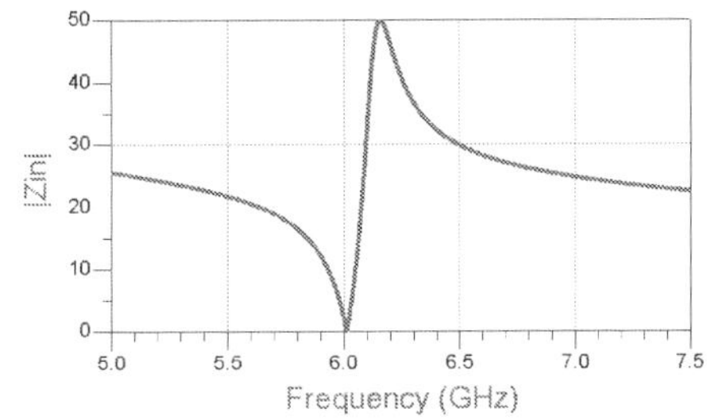

Fig. 9. FBAR resonator frequency response.

3.3.3 Filter's design principle

The filter's design is done by association of resonators. This is performed using two topologies: ladder and lattice topologies (Fig. 10). The filter's responses will be different both in terms of rejection and ripple in the band.

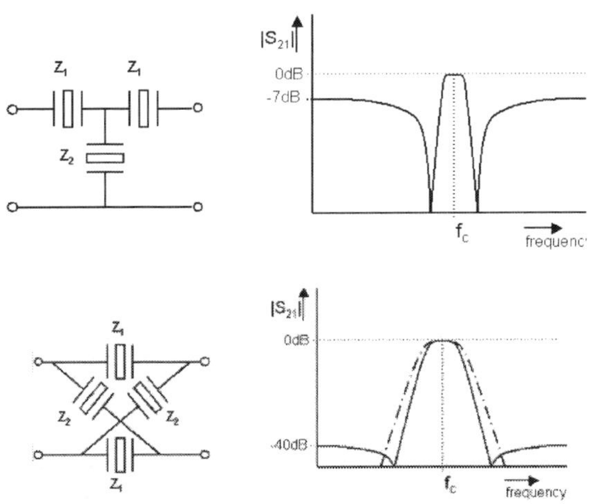

Fig. 10. Ladder and lattice topologies and filter's response.

An important parameter defining the filter bandwidth is the value of the difference between resonance frequency and anti resonance frequency. This value depends on the physical material properties.

In the case of the ladder topology design, the basic principle is to combine two resonators of different thicknesses, one serial and one parallel. The maximum bandwidth is determined by the material. The electrode's surfaces adjust impedances (Fig. 11).

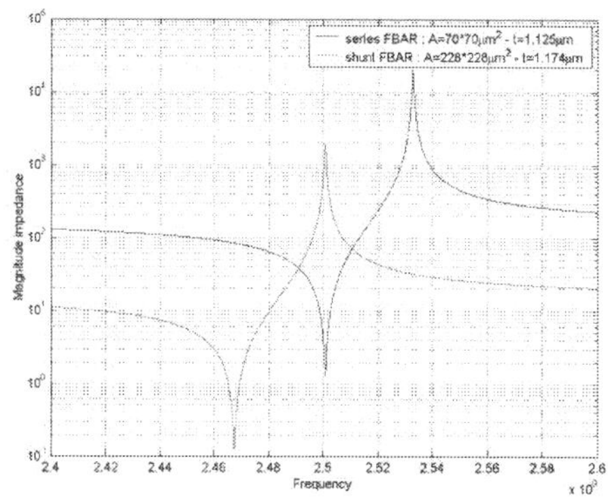

Fig. 11. Magnitude impedance of a series and a shunt FBAR resonator.

The series and the shunt resonators form a stage. In order to achieve the required frequency response and out-of-band rejection, particular stages are put together to build cascades. Each additional stage (a couple of series-shunt resonators) increases the filter order by one. Therefore, a six resonators ladder filter is a third order filter. Fig. 12 shows the evolution of the filter's frequency response according to the number of stages.

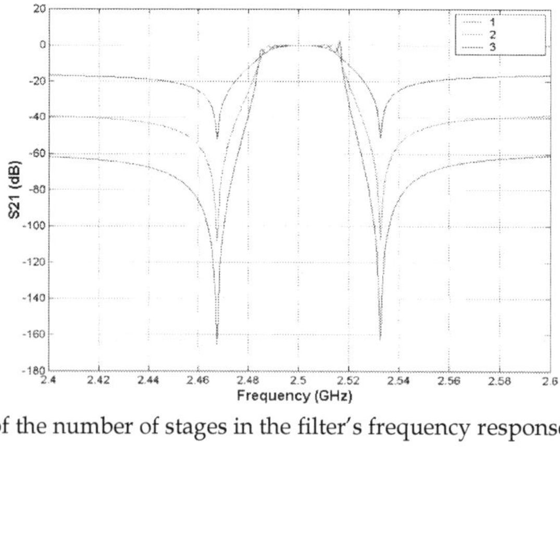

Fig. 12. Influence of the number of stages in the filter's frequency response.

Ladder filters are single ended, while lattice filters are double ended. If the filter output is at the antenna input, the ladder topology may be preferred to the lattice topology due to unbalanced signal effects.
Only two parameters need to be optimized in order to design a band-pass filter. These parameters are the area A (expressed as $l \times l$) and the resonator thickness (d_1 and d_2).
An example of a WiMAX filter in the 3.6 - 3.8 GHz frequency band was proposed in (Suarez et al., 2008). The emission filter in this case has a bandwidth of 200 MHz (Fig. 13). The out-of-band rejection is 50 dB at twice the channel bandwidth (20 MHz from the edge for a 10 MHz channel) and has been fixed from the power spectrum mask presented in Fig. 4.

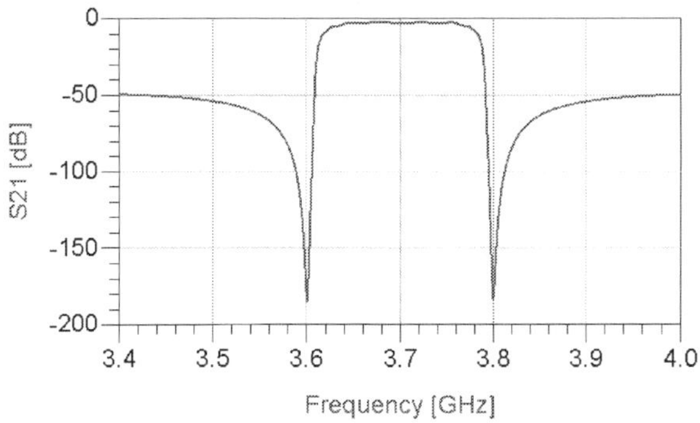

Stage	L_{serie} [µm]	L_{shunt} [µm]
1	122	87
2	102	98
3	113	107
4	124	117
5	136	128
6	150	142
7	168	132

Insertion Loses	3.06 dB
Rejection	49.03 dB

Fig. 13. [S_{21}] parameter of a WiMAX RF filter using BAW technology (7th order ladder).

3.3.4 Manufacturing technologies

There are essentially two main families of technologies for achieving the BAW filters:

- FBAR technology
- SMR technology (Solidly Mounted Resonator)

For FBAR technology, Fig. 14 shows two processes:

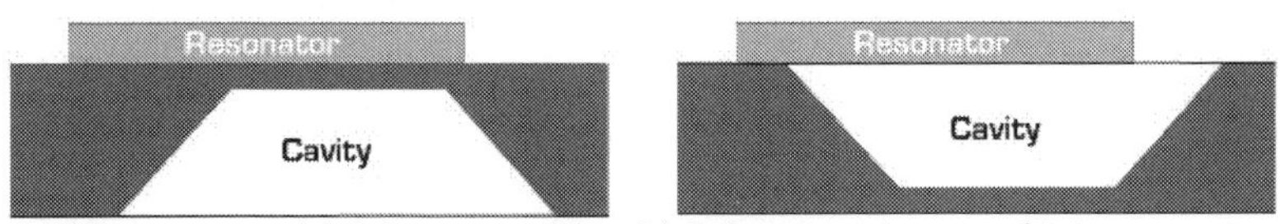

Fig. 14. FBAR processes: (a) Substrate etching (b) Substrate micro-machining.

The FBAR technology advantages are:

- a power confined to the piezoelectric material, therefore, lower losses
- low number of layers to achieve
- integration in SOC technology

The drawbacks are:

- membrane's fragility
- process complexity
- thermal dissipation

SMR technology uses a Bragg reflector as presented in Fig. 15.

Fig. 15. SMR technology - Bragg reflector.

The SMR technology advantages are:
- the possibility to process "stand alone" BAW
- good thermal dissipation in the reflector's layers

The drawbacks are:
- more important losses
- a larger number of layers

3.3.5 BAW filter's tuning

The main drawback of BAW filters (such as SAW filters) is the complexity of achieving frequency tuning. It's still possible to move the series or parallel resonance frequency using varactors. This allows a tuning of 1% of the relative bandwidth. It is also possible to use inductors (active inductors or MEMS). In this case, tunability can reach 5% of the relative bandwidth. These techniques are mostly used to achieve a post-process tuning.

4. RF band-pass filters for transmitters: implementation examples.

4.1 RF filter bank for an UWB Multi-Band On-Off Keying transceiver

This example is related to the RF band-pass filter bank in an UWB OOK architecture as described in section 1.4. In such an architecture the filter bank is a common element in transmitter and receiver. A filter bank is uniform if all the band-pass filters have the same bandwidth. In the MB-OOK architecture, the filter bank may be uniform or non uniform (Suarez et al., 2007a) depending on the constraints of the technology.
A good value of relative bandwidth (i.e. the ratio between the transmission bandwidth and the central frequency of the RF band-pass filter) for AlN BAW technology is 3%. A filter bank where all the band-pass filters keep the same relative bandwidth leads to the conception of a non uniform filter bank. The example presented in this section validates by simulations the viability of using a non-uniform filter bank (AlN BAW tehcnology) in MB-OOK UWB applications.
The filters distribution in the frequency band between 6 GHz and 8.5 GHz (Fig. 2) is calculated for a relative bandwidth of 3%. It leads to a filter bank of 10 filters. A higher number of sub-bands would allow the system to reach higher throughput. Nevertheless, the required electronic components (filters, switches, combiners, isolators) also result in increased active surface, power losses and higher power consumption. Therefore, there is a trade-off between throughput and system complexity. A ten filter configuration offers a throughput of 125 Mbps over Non Line Of Sight (NLOS) conditions, a reasonable number of sub-bands and a suitable relative bandwidth for AlN BAW filters design (Suarez et al., 2007a).

Measurements and simulations in (Diet et al., 2005) established that the Cauer or elliptic filter behavior is a good approximation to the BAW filter response. Therefore, Cauer filters were used in simulations. The band pass ripple and the out-of-band attenuation of each band-pass filter is chosen for maximal attenuation between sub-bands with the purpose of reducing the intersymbol interference in reception. Simulated filters in this filter bank are order 5 filters. Each filter has 40 dB of out-of-band rejection and less than 2.5 dB of in band ripple; as the MB-OOK architecture is based on energy detection, the values of ripple less than 3dB have been considered acceptable.
Fig. 16 presents the frequency response of the first filter of the filter bank simulated with a BVB model and without the Cauer approximation (Suarez et al., 2007b).

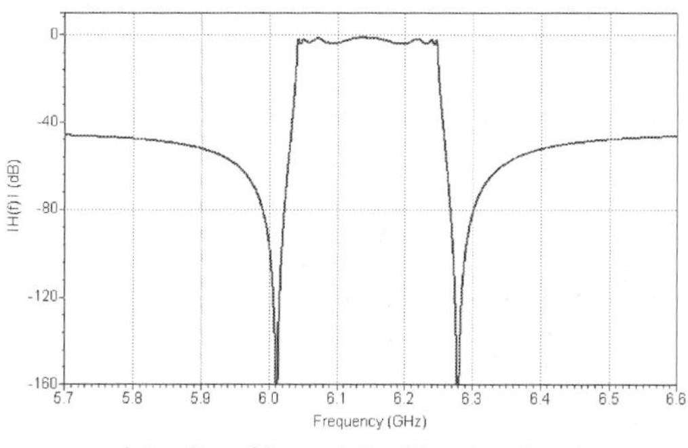

Fig. 16. Frequency response of the first filter of the filter bank.

Fig. 17 presents simulation results in time and frequency of the transmitter architecture presented in Fig. 3a including the non-uniform filter bank.

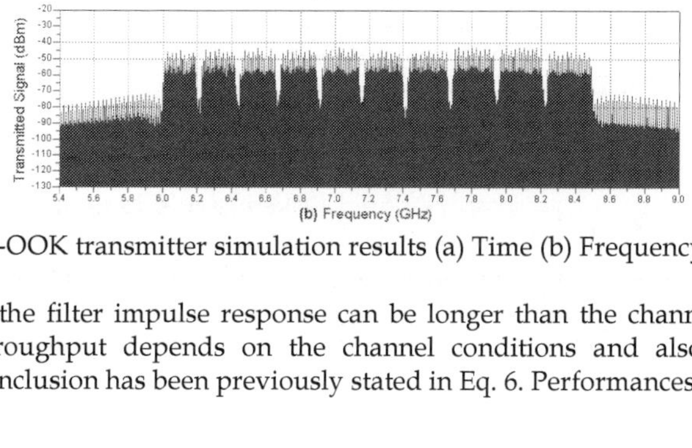

Fig. 17. UWB MB-OOK transmitter simulation results (a) Time (b) Frequency.

The duration of the filter impulse response can be longer than the channel delay spread. Thereby, the throughput depends on the channel conditions and also on the filter's response. This conclusion has been previously stated in Eq. 6. Performances of the simulated architecture validate the viability of using a non-uniform filter bank of AlN BAW tehcnology in a MB-OOK UWB transmitter: for a maximal mean error probability of 10^{-5}, the covered distance is 3.3m on the LOS case and 1.9 m on the NLOS case (Suarez et al., 2007a).

4.2 RF filter bank for a multi-radio transmitter

This example deals with a BAW RF filter bank of a multi-radio transmitter in the 800 MHz - 6 GHz frequency band. Considered communications systems include cellular phone and Wireless LANs and MANs in Europe. BAW band-pass filters are designed with AlN FBARs. Filter bank design is based on the parameters defined by the regulations (allocated frequency bands, power spectrum mask, etc.). Results presented correspond to simulations on Agilent Advanced Design System (ADS).

4.2.1 GSM

Global System for Mobile communications operates in the frequency band between 890 - 915 MHz in Uplink and 935 - 960 MHz in downlink (ETSI, 1999b). The emission filter then has a bandwidth of 25 MHz and a central frequency of 902.5 MHz. Out of band rejection must be 90dB. The transmission response of the GSM designed filter is presented in Fig. 18. The thickness of the series resonators is 6.059 µm and the thickness of the shunt resonators is 6.205µm. Fig.18 summarises the resonator's surface areas. The reached rejection is 89 dB and insertion losses are less than 2.3dB.

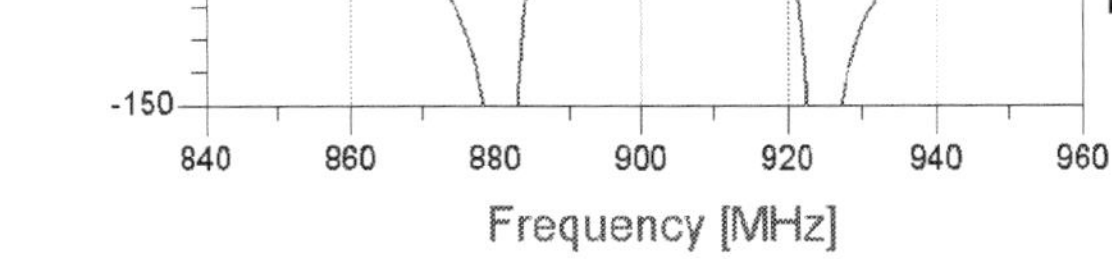

Stage	L_{serial}[µm]	L_{shunt} [µm]
1	514	548
2	280	574
3	247	573
4	243	534
5	218	544
6	281	533

Insertion Losses	2.3 dB
Rejection	89 dB

Fig. 18. Response of a 6th order ladder filter (GSM).

4.2.2 DCS 1800 MHz

Digital Cellular Systems operate in the frequency band between 1710 - 1785 MHz in Uplink and 1805 - 1880 MHz in downlink. The filter should have a bandwidth of 75 MHz and a central frequency of 1747.5 MHz. The reached rejection is 89 dB and insertion losses are less than 2.1 dB. Fig. 19 summarises the resonator's surface areas and the transmission response of the filter.

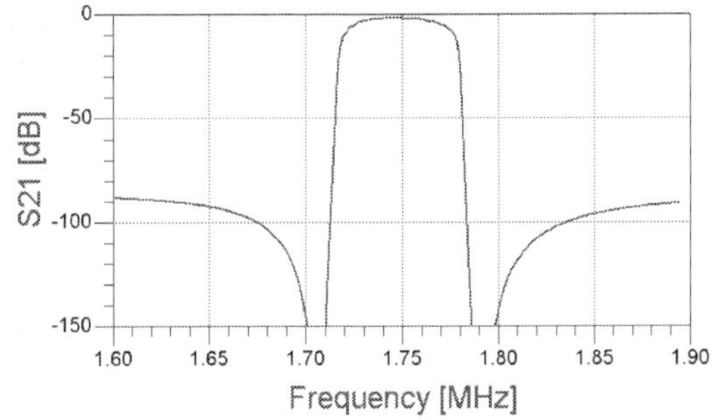

Stage	L_{serial}[µm]	L_{shunt} [µm]
1	229	286
2	175	294
3	177	288
4	177	291
5	177	288
6	177	294
7	175	286
8	229	

Insertion Losses	2.3 dB
Rejection	89 dB

Fig. 19. Response of a 7th order T ladder filter (DCS).

4.2.3 UMTS and LTE

Universal Mobile Telecommunication Systems operate in the frequency band between 1920 and 1980 MHz (3GPP, 2006). This is also one of the frequency bands allocated to the Long Term Evolution (LTE) standard. The emission filter should have a bandwidth of 60 MHz and a central frequency of 1950 MHz. Out of band rejection must be -50dB. The transmission response of the UMTS designed filter is presented in Fig. 20.
The thickness of the series and shunt resonators are 2.793 µm and 2.881µm respectively. The resonator's surface areas are summarised in Fig. 20. The reached rejection is 60 dB and insertion losses are less than 5 dB across the whole frequency band.

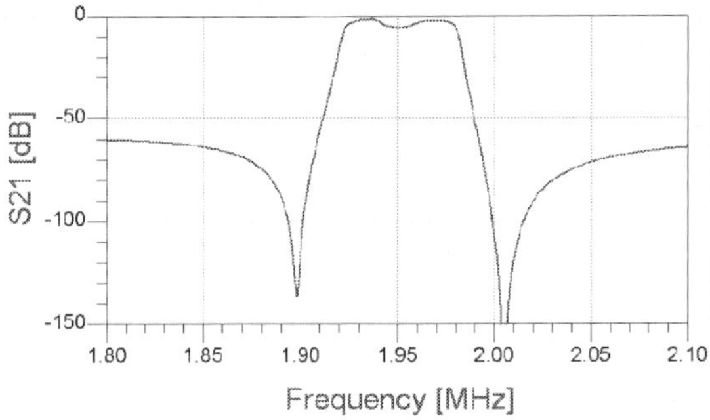

Stage	L_{serial}[µm]	L_{shunt} [µm]
1	150	300
2	125	320
3	125	300
4	148	-

Insertion Losses	5 dB
Rejection	60 dB

Fig. 20. Response of a 3rd order T ladder filter (UMTS).

4.2.4 WLAN (802.11b/g)

IEEE 802.11b and 802.11g standards establish specifications for wireless connectivity within a local area network in the 2.4 - 2.483 GHz frequency band (IEEE, 1999a and 1999b). The emission filter has the same specifications for the two standards: a bandwidth of 83.5 MHz and a central frequency of 2441.5 MHz. Out of band rejection must be -50 dB. The transmission response of the WLAN designed filter is presented in Fig. 21. The thickness of the series and shunt resonators are 2.233µm and 2.291µm respectively. Fig. 21 summarises the resonator's surface areas.

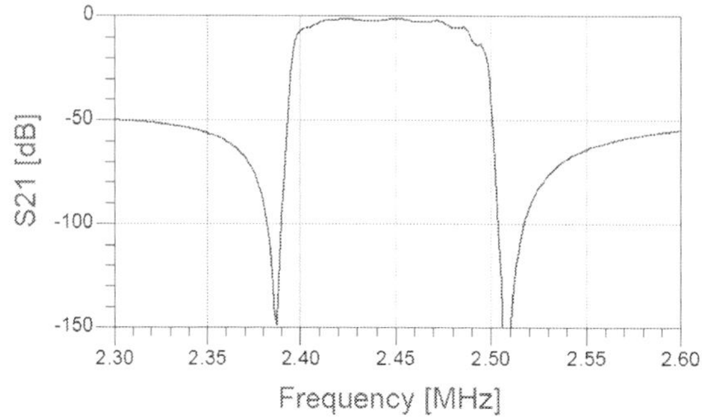

Stage	L_{serial}[μm]	L_{shunt} [μm]
1	128	135
2	129	147
3	116	134
4	116	147
5	129	135
6	128	-

Insertion Losses	2.1 dB
Rejection	50 dB

Fig. 21. Response of a 5th order T filter (2.4 GHz WLAN).

The reached rejection is 50 dB and insertion losses are less than 2.1 dB across the whole frequency band.

4.2.5 WLAN (802.11a)

The IEEE 802.11a standard establishes specifications for wireless connectivity within a local area network in the 5.150 - 5.350 GHz frequency band in Europe (IEEE, 1999c). The emission filter should have a bandwidth of 200 MHz and a central frequency of 5250 MHz. Out of band rejection must be -40 dB. The transmission response of the WLAN designed filter and the resonator's surface areas are presented in Fig. 22.
The thicknesses of the series and shunt resonators are 1.039μm and 1.070μm respectively. The reached rejection is 43 dB and insertion losses are less than 2.2 dB across the whole frequency band.

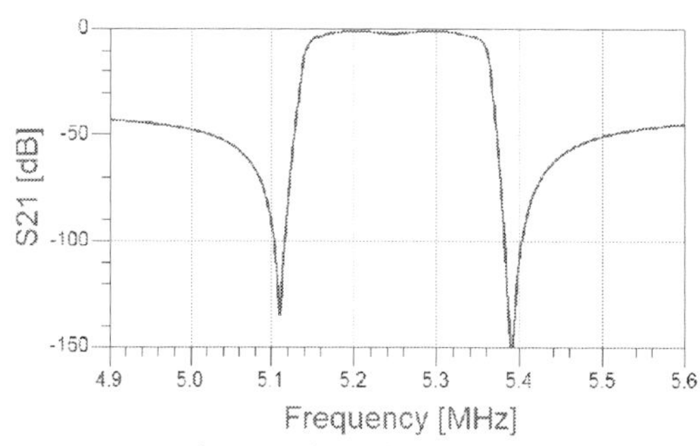

Stage	L_{serial}[μm]	L_{shunt} [μm]
1	96	94
2	67	100
3	68	100
4	67	94
5	96	-

Insertion Losses	2.2 dB
Rejection	43 dB

Fig. 22. Response of a 4th order T filter (5 GHz WLAN).

4.2.6 WMAN (802.16e)

The IEEE 802.16e standard supports operation of wireless metropolitan area networks in licensed frequency bands below 11 GHz (IEEE, 2005). In Europe, the 3.6 - 3.8 GHz frequency band is one of the bands allocated to mobile WiMAX. The emission filter, in this case, should have a bandwidth of 200 MHz and a central frequency of 3700 MHz. Out of band rejection must be -50 dB (ETSI, 2003). The transmission response of the WiMAX designed filter has already been presented in Fig. 13.

In the example presented in this section each band-pass filter has being considered independently. It is a first approach of multi-radio filter bank. The drawback of assembling 2 or more BAW filters on the board is that it requires space for chip positioning. Unitary manipulation of BAW filters is time and cost consuming. The trend is going towards a duplexer (or multiplexer) module and providing a single BAW chip with 2 or more filters (Reinhardt et al., 2009).

4.3 Other examples of RF band-pass BAW filters

This section presents some examples of band-pass filters obtained with FBAR and SMR BAW resonators for radiofrequency microelectronics applications within some European projects.

The MARTINA European project (ended in 2005) was about the design and implementation of an RF front-end for WCDMA applications using an above-IC BAW band-pass filter (SoC integration). This project validated the monolithic integration of active and passive devices on the same wafers. A filter resulting of the MARTINA European project is a stand-alone filter designed for the RX chain of a WCDMA mobile phone, and fabricated at the wafer level above BiCMOS active integrated circuits (Kerherve et al., 2006). It is a FBAR filter with 8 resonators in double lattice topology. The measured IL is -3.5dB over the 60 MHz measured bandwidth. The TX-band rejection is lower than -50 dB.

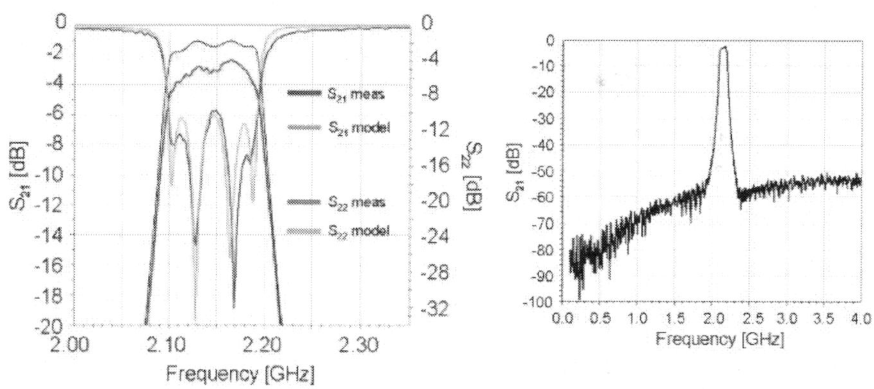

Fig. 23. FBAR Filter (a) Transmission and reflection coefficients. (b). Broadband of S_{21}

The MIMOSA European project dealt with developing a technological platform for embodiment of various RF functions, sensors and microsystems for Ambiant Intelligent applications in a mobile-phone centric approach. It proposed a SiP integration to make an ISM-band wake-up radio receiver using a selective low-noise amplifier (LNA). A modular approach was applied to an ISM-band receiver, where a stand-alone SMR-type BAW double-lattice filter was wire-bonded with CMOS LNA on the same PCB (a major differentiation to Above-IC approach). The differential BAW filter covers the whole ISM band from 2.4 to 2.48 GHz, with 3dB insertion loss and 40 dB out-of-band and image

rejection. It is a double-stage lattice SMR-type BAW filter. The fabricated filter has IL of -4dB, bandwidth of 70 MHz and rejection of -36dB (Kerherve et al., 2006).
Another European project is the MOBILIS European project. The objective of the MOBILIS project is to develop a robust and cost-effective integrated high-power RF filtering technology and demonstrate the feasibility of a mixed SoC (nanometric CMOS/system integration) and SiP (BiCMOS/power-BAW) RF power transmitter. The targeted transmitter is based on a Digital Radio transmitter architecture and addresses both the WCDMA and DCS standards. An UMTS BAW filter designed and implemented in this project is presented in Fig. 24.

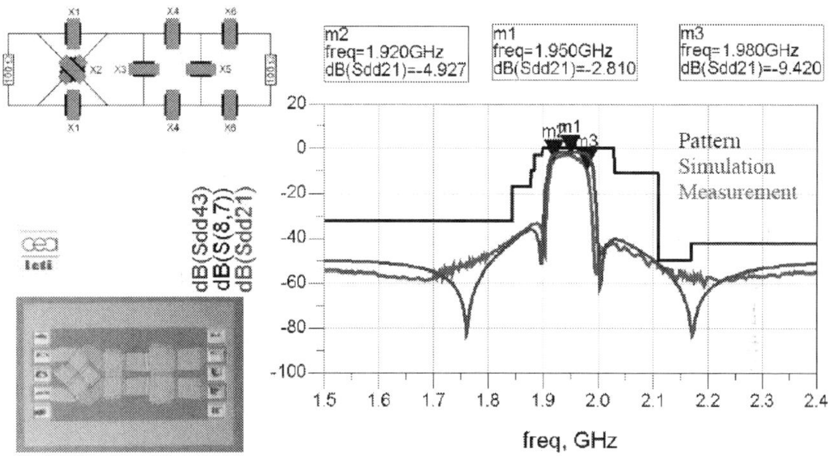

Fig. 23. UMTS band-pass filter proposed in the MOBILIS European project (Kerherve, 2009).

5. Conclusion

This chapter presented the interest of RF band-pass filters in the actual communications context. Specifically, the importance of the RF filter in the mobile transceivers and the band-pass RF filters requirements at the front-end of a mobile transmitter have been described. Some different available filtering technologies have been considered as well and their advantages and trade-offs have. The potential of BAW technology for RF band-pass filters is highlighted. Two RF BAW filters applications proposed by the authors have been described, one for an UWB application and the other one for a multi-radio system. Other examples of RF band-pass filters designed and fabricated within some European Projects have also been presented. All these implementation examples probe the interest of using BAW technology for wireless transceivers for communications. The actual trend is going towards duplexer (or multiplexer) modules and providing a single BAW chip with 2 or more filters. Filter's reconfigurability is also a research subject.

6. References

3GPP. (2005). 3rd Generation Partnership Project (3GPP), "UE Radio Transmission and Reception (FDD)," technical specification 25.101, vol. 6.8.0, 2005.

3GPP (2006). TS 25.101, 3rd Generation Partnership Project; Technical Specification Group Radio Access Network; User Equipment (UE) radio transmission and reception (FDD) (Release 7).

Diet, A.; Villegas, M.; Vasseure, C.; Baudoin, G. (2005). Impact of BAW emission filter's characteristics on 3rd Generation standards modulations. Proceedings of the International Wireless Summit, WPMC - Wireless Personal Multimedia Communications, Septembre 2005, Aalborg, Danemark.

ETSI. (1999a). European Standard, Telecommunications Series, *ETSI EN300910V8.5.1 (2000-11)*, Digital cellular telecommunications system (Phase 2+); Radio transmission and reception (GSM 05.05 version 8.5.1 Release 1999).

ETSI. (1999b). European Standard, Telecommunications Series, *ETSI TS100910V8.20.0 (2005-11)*, Digital cellular telecommunications system (Phase 2+); Radio Transmission and Reception (3GPP TS 05.05 version 8.20.0 Release 1999).

ETSI. (2003). European Standard, Telecommunications Series, *ETSI 301021 V1.6.1 (2003-07).* Fixed Radio Systems; Point-to-multipoint equipment; Time Division Multiple Access (TDMA); Point-to-multipoint digital radio systems in frequency bands in the range 3GHz to 11GHz.

Heyen, J.; Yatsenko, A.; Nalezinski, M.; Sevskiy, G.; Heide, P. (2008). WiMAX System-in-package solutions based on LTCC Technology, *Proceedings of COMCAS 2008.*

IEEE (1999a). St. 802.11b-1999, Supplement to IEEE Standard for Information technology - Telecommunications and information exchange between systems—Local and metropolitan area networks - Specific requirements - Part 11: Wireless LAN Medium Access Control (MAC) and Physical Layer (PHY) specifications: Higher-Speed Physical Layer Extension in the 2.4 GHz Band. Reaffirmed 12 June 2003.

IEEE (1999b). St. 802.11g-1999, Supplement to IEEE Standard for Information technology - Telecommunications and information exchange between systems—Local and metropolitan area networks - Specific requirements - Part 11: Wireless LAN Medium Access Control (MAC) and Physical Layer (PHY) specifications: Amendment 4: Further Higher Data Rate Extension in the 2.4 GHz Band. 27 June 2003.

IEEE (1999c). St. 802.11a-1999, Supplement to IEEE Standard for Information technology - Telecommunications and information exchange between systems - Local and metropolitan area networks Specific requirements Part 11: Wireless LAN Medium Access Control (MAC) and Physical Layer (PHY) specifications High-speed Physical Layer in the 5 GHz Band. Reaffirmed 12 June 2003.

IEEE. (2005). IEEE Standard 802.16e. Air interface for fixed and mobile broadband wireless access systems amendment 2: physical and medium access control layers for combined fixed and mobile operation in licensed bands, 2005.

Kerherve, E.; Ancey, P.; Kaiser, A. (2006). BAW Technologies: Development and Applications within MARTINA, MIMOSA and MOBILIS IST European Projects. 2006 IEEE Ultrasonics Symposium.

Kerherve, E. ; Kaiser, A. (2009). Intégration Mixte SiP/SoC d'un Transmetteur en Technologie CMOS/BiCMOS/BAW/IPD pour des Applications Multimodes WCDMA/DCS. Fetch 2009. Chexbres, Switzerland.

Kim, D.; Dong Ho Kim; Jong In Ryu; Jun Chul Kim; Chong Dae Park; Chul Soo Kim; In Sang Song. (2008). A quad-band front-end module for Wi-Fi and WiMAX applications using FBAR and LTCC Technologies, *Proceedings of APMC 2008.*

Paquelet, S.; Aubert, L.-M. & Uguen, B. (2004). An impulse radio asynchronous transceiver for high data rates. *Proceedings of IEEE Joint UWBST&IWUWBS Conf.*, ISBN 0780383737, May Kyoto, 2004, IEEE, Piscataway, New Jersey.

Reinhardt, A. (2009). BAW technology for advanced RF architectures. Workshop on BAW Technologies WSE. 2009 International Microwave Symposium. June, 2009. Boston.

Suarez, M.; Villegas, M.; Baudoin, G; Andia Montes, L. (2007a). Considering a non-Uniform filter bank in an UWB Multi Band On-Off Keying transceiver, *Journal of communications (JCM)*, Vol. 2, N° 6, Nov. 2007, pp. 24-29, ISSN 1796-2021.

Suarez, M.; Villegas, M.; Baudoin. BAW filter bank for an UWB Multi Band On-Off Keying transceiver (2007b). *Proceedings of IEEE APMC Conf.*, December Bangkok, Thailand, 2007.

Suarez, M.; Villegas, M.; Baudoin, G. (2008). Front end filtering requirements on a mobile cognitive multi-radio transmitter, *Proceedings of the 11th International Symposium on wireless Personal Multimedia Communications,* 8-11 Sept. 2008, Saariselka, Finlande.

9

DEMONSTRATION OF A POWER AMPLIFIER LINEARIZATION BASED ON DIGITAL PREDISTORTION IN MOBILE WIMAX APPLICATION

Pooria Varahram, Somayeh Mohammady,
M. Nizar Hamidon, Roslina M. Sidek and Sabira Khatun
University Putra Malaysia
Malaysia

1. Introduction

Spectrally efficient linear modulation techniques are used in the third generation systems and their performance is strongly dependent on the linearity of the transmission system, Also, the efficiency of the amplifier to be used has to be maximized, which means that it must work near saturation. Newer transmission formats, with wide bandwidths, such as multi carrier wideband code division multiple access (WCDMA), wireless local area network (WLAN), worldwide interoperability for microwave access (WiMAX), are especially vulnerable to PA nonlinearities, due to their high peak-to-average power ratio, corresponding to large fluctuations in their signal envelopes. In order to comply with spectral masks imposed by regulatory bodies and to reduce BER, PA linearization is necessary. A number of linearization techniques have been reported in recent years (Cripps, 1999;Kennington, 2000; kim & Konstantinou, 2001;Wright & Durtler 1992;Woo, et al. 2007; Nagata, 1989). One technique that can potentially compensate for power amplifier (PA) nonlinearities in such an environment is the adaptive digital predistortion technique. The concept is based on inserting a non-linear function (the inverse function of the amplifier) between the input signal and the amplifier to produce a linear output. The digital predistortion (DPD) requires to be adaptive because of variation in power amplifier nonlinearity with time, temperature and different operating channels and so on. Another limitation of predistortion is the dependence of amplifier's transfer characteristic's on the frequency content of the signal or defined as changes of the amplitude and phase in distortion components due to past signal values, that is called memory effects. The memory effects compensation is an important issue of the DPD algorithm in addition to correction of power amplifier (PA) nonlinearity especially when the signal bandwidth increases. Many studies are involved in this technique but many of them suffer from limitations in bandwidth, precision or stability (Cavers, 1990;Wright & Durtler 1992;Nagata, 1989).

In this reseach a new technique of adaptive digital predistortion that is the combination of two techniques, the gain based predistorter (Cavers, 1990) and memory polynomial model

(Ding, et al. 2004) is presented. Both previous techniques have demonstrated acceptable results but both have disadvantages. In memory polynomial predistortion the complexity of extracting the coefficients of predistortion function decrease the capability of linearization and so it needs to apply other method like (Raich, et al. 2003) for implementing it. In complex gain predistortion method (Cavers, 1990) the memory effects that cause dynamic AM-AM and AM-PM are not considered. So here the main objective is not only to demonstrate the capability of this new method to overcome for such disadvantages, but also to show that with applying this technique all the memory contents of power amplifier that is modeled with memory polynomial is compensated. for validating this technique several simulations are applied. The adaptation is based on linear convergence method in the simulations. For simplicity the effects of the quadrature modulator and demodulator and A/D and D/A is not considered. The LUT size is 10 bit and addressing the LUT is based on the input amplitude. It will be shown that with applying this method all the memory contents of the power amplifier especially the one that cause dynamic AM/AM and AM/PM are compensated. Simulation and results are examined based on Motorola's MRF1806 1.9 GHz LDMOS PA with 13 dB gain and 60W output power. To demonstrating the results several tests are shown with Mobile WiMAX signal. Finally the actual power amplifier from Mini Circuit ZVE-8G with 30 dB gain is used and tested with agilent equipments to verify the proposed algorithm.

2. Predistortion technique

Fig. 1 shows a block diagram of the adaptive digital predistortion (Cavers, 1990) . A fully adaptive digital predistortion system requires the addition of a predistortion circuit consisting of adigital predistorter and look up table (LUT) to the transmission path in addition to a feedback path consisting of a demodulator, analog to digital converter (ADC) and adaptation circuit for updating the LUT. The block diagram assumes that all components of the system except the predistorter and high power amplifier (HPA) have a linear response and hence can be ignored in the analysis. In this paper also these effects are ignored. The predistorter is equivalent to a nonlinear circuit with gain expansion response that inverse of the power amplifier gain compression AM/AM (Amplitude dependent gain) and a phase rotation that is the negative of the Power Amplifier phase rotation AM/PM (Amplitude Dependent Phase Shift). In this figure $x(n) = I + jQ$ is the quadrature modulated input signal and $v_f(n)$ is the quadrature demodulated feedback signal. These signals are sampled synchronously, and their values are used to generate a predistortion vector function $F[|x(n)|^2]$ which is stored in polar or rectangular form in a look-up table (LUT). The input signal $x(n)$ is predistorted according to $F[|x(n)|^2]$, so that the predistorted signal $v(n)$ produced the linearized output from the RF amplifier. Here the LUT is 10bit and the absolute of input signal is used for addressing it. The main objective of this paper is to study the electrical memory effects that cause dynamic memory effects (ku & . McKinley, 2002). The previous studies (Wangnyong, et al. 2004;Bosch & Gatti, 1989;Morgan, et al. 2006) were all restricted to calculation of the coefficients of the power amplifier. This way needs a lot of computation and therefore takes a lot of processor time and also never can be implemented when the number of coefficients increases.

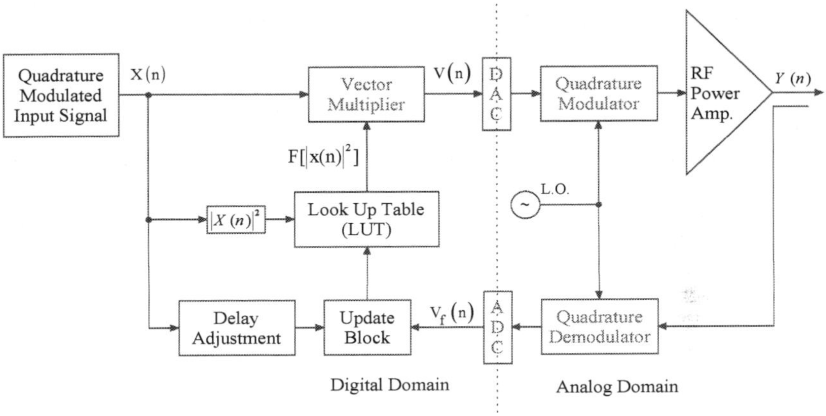

Fig. 1. Adaptive digital predistortion block

The technique that is proposed here doesn't have that drawback. It even claims that can linearized the dynamic memory effects in wideband applications. This method will be discussed in details in section III. One of the other important things in studying the predistortion method is that the predistortion attempts to add 3rd and 5th order intermodulation products to the input signals that cancels out the 3rd and 5th order intermodulation products added by the PA, thus the bandwidth of the predistorted signal must be three times greater than the bandwidth of the input signals to be able to represent up to 5th order intermodulation products. In the real world the predistorted signals are fed into a DAC and then low pass filtered at the Nyquist rate (half the input sample rate), the predistorted signal must have a sample rate of at least six times that of the original input signals. Thus in simulations the input signals are interpolated by a factor of six before being fed into the predistorter. In the next section the new technique of predistortion is discussed.

3. Complex gain predistortion

Fig. 2 shows the predistortion function $F[|x(n)|^2]$ that cascades with power amplifier that has shown with $G[|v(n)|^2]$ function. $F[|x(n)|^2]$ and $G[|v(n)|^2]$ are complex gain functions of predistortion and power amplifier.

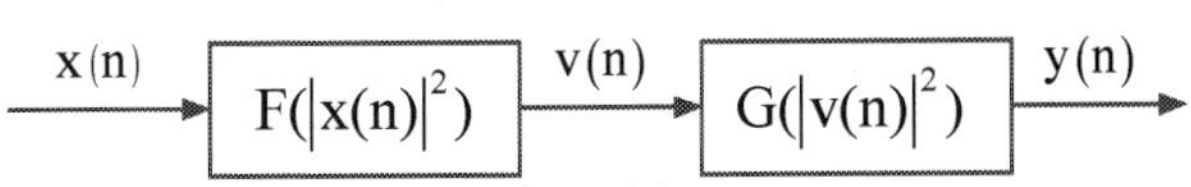

Fig. 2. Cascade of predistortion and power amplifier

As proposed in (Ding, et al. 2004) the equivalent discrete baseband PA model considering memory effects and baseband nonlinearity can be represented with a memory polynomial model which is a special case of Volterra series as below:

$$y(n) = \sum_{\substack{k=1 \\ Odd}}^{K} \sum_{q=0}^{Q} a_{kq} v(n-q) \left| v(n-q) \right|^{2(k-1)} \tag{1}$$

where $v(n)$ is the discrete input complex signal of power amplifier after predistortion block and $y(n)$ is the discrete output complex envelope signal. K is the order of nonlinearity and Q is the memory length.

This model considers only odd-order nonlinear terms due to bandpass nonlinear characteristics that cause intermodulation distortion. In (1) $v(n)$ also can be represented as below:

$$v(n) = x(n) F[\left| x(n) \right|^2] \tag{2}$$

where $x(n)$ is the discrete input complex and $F[\left| x(n) \right|^2]$ is the complex gain of the predistortion block. Equation (1) can be simplified as below:

$$y(n) = \sum_{q=0}^{Q} v(n-q) \sum_{\substack{k=1 \\ Odd}}^{K} a_{kq} \left| v(n-q) \right|^{2(k-1)} \tag{3}$$

Where the function $G_q(\left| v(n-q) \right|^2)$ can be represented as:

$$G_q[\left| v(n-q) \right|^2] = \sum_{\substack{k=1 \\ Odd}}^{K} a_{kq} \left| v(n-q) \right|^{2(k-1)} \tag{4}$$

Then (3) is as below:

$$y(n) = \sum_{q=0}^{Q} v(n-q) G_q[\left| v(n-q) \right|^2] = v(n) G_0[\left| v(n) \right|^2] + v(n-1) G_1[\left| v(n-1) \right|^2] + .. \tag{5}$$

This equation demonstrates that the memory contents of the power amplifier are not only appeared in the coefficients a_{kq} of the (1), but it also can be shown as the complex function, which means that the memory effects are appeared in the function $G_q[\left| v(n) \right|^2]$. Previous efforts only tried to extract the a_{kq} to compensate for such memory effects but here it will be shown that without having the coefficients also the memory effects can be compensated and even the compensation is better and includes all the memory (Varahram, et al. 2009).

From (2) for finding the function $F[\left| x(n) \right|^2]$, first it is assumed that Q=0 or the power amplifier is memoryless thus from (5) it can be conculded:

$$y(n) = v(n) G_0[\left| v(n) \right|^2] \tag{6}$$

Ideally the power amplifier should satisfy the below condition for having the linear output.

$$y(n) = Gx(n) \tag{7}$$

Where G is the linear gain of power amplifier.
replacing (5) in (7) then:

$$y(n) = \sum_{q=0}^{Q} v(n-q) G_q[|v(n)|^2] = Gx(n) \tag{8}$$

With assuming Q=0 and replacing the v(n) in (6) and with considering that the quadrature modulator is a perfect unity gain device the optimum predistorter characteristic, denoted by $F[|x(n)|^2]$, would satisfy:

$$x(n) F[|x(n)|^2] G_0[|x(n) F[|x(n)|^2]|^2] = Gx(n) \tag{9}$$

Then the optimum value of the predistortion complex gain is calculated from below iterative equation:

$$F_{i+1}[|x(n)|^2] = F_i[|x(n)|^2] - \frac{F_i[|x(n)|^2]}{v(n) G_0[|v(n)|^2]} V_{error}(n) \tag{10}$$

where

$$V_{error}(n) = y(n) - Gx(n) \tag{11}$$

Now assume that the power amplifier includes one memory or Q=1 then after some simplification, equation below will be generated:

$$F(|x(n)|^2) = \frac{G}{G_0[|v(n)|^2]} - \frac{v(n-1) G_1[|v(n-1)|^2]}{x(n) G_0[|v(n)|^2]} \tag{12}$$

The second fraction of (12) indicates the memory effects of the power amplifier. If Q increases then the elements in (12) also will increase.
The iterative solution for (12) is:

$$F_{i+1}(|x(n)|^2) = F_i[|x(n)|^2] - \frac{F_i[|x(n)|^2]}{v(n) G_0[|v(n)|^2]} V_{error}(n) + \frac{F_i[|x(n)|^2] v(n-1) G_1[|v(n-1)|^2]}{v(n) G_0(|v(n)|^2)} - \frac{v(n-1) G_1[|v(n-1)|^2]}{x(n) G_0[|v(n)|^2]} \tag{13}$$

This equation can be simplified as below:

$$F_{i+1}(|x(n)|^2) = F_i[|x(n)|^2] - \frac{F_i[|x(n)|^2]}{v(n)G_0[|v(n)|^2]} V_{error}(n) +$$
$$\frac{v(n-1)G_1[|v(n-1)|^2]}{G_0[|v(n)|^2]}\left(\frac{F_i[|x(n)|^2]}{v(n)} - \frac{1}{x(n)}\right) = F_{i+1}(|x(n)|^2) = F_i[|x(n)|^2] - \frac{F_i[|x(n)|^2]}{v(n)G_0[|v(n)|^2]} V_{error}(n) \tag{14}$$

The function $F[|x(n)|^2]$ in (14) is similar to (10) when the power amplifier has no memory. This formula can be extended to more memory and is still valid. Simulations and results will prove the validity of this equation later. Memory polynomial method was very complicated and it couldn't calculate all the coefficients in the volterra series and only could compensate for 2 or 3 memory length but this method proves that it can compensate all the memory contents of the power amplifier.
The important parameter in (14) is the gain factor which is the only difference between (10) and (13) and it is the $F[|x(n)|^2]$ over $v(n)G_0(|v(n)|^2)$. In the case of having memory, $G_0[|v(n)|^2]$ can not be found and also (6) can not be assumed except the case of memoryless power amplifier. As it is shown in (Ding, et al. 2004; Wright & Dutler, 1992) the gain factor can be a constant number between zero and one and it indicates the stability and convergence rate. If the gain factor sets to the larger value then the convergence is faster but the probability of convergence is low. For controlling and making the convergence slower and reach to the highest linearity especially at saturation point, (13) is as below:

$$F_{i+1}(|x(n)|^2) = F_i[|x(n)|^2] - \alpha \frac{F_i[|x(n)|^2]}{v(n)G_0[|v(n)|^2]} V_{error}(n) \tag{15}$$

In (Cavers, 1990) α is a constant between 0 and 1. This parameter indicates the convergence rate and stability and its value should be allocated with considering the linearity requirements. In (Cavers, 1990) the condition for convergence of (15) is shown. For calculating the function $F[|x(n)|^2]$ in (15) first the error vector should be calculated and then the gain factor which involves the division and then these values multiply together and with consider that $F[|x(n)|^2]$ is initially one then after some iteration the optimum value will be found.
Finding the appropriate gain factor is possible in two different ways:
Case 1:
The parameter $v(n)G_0[v(n)]$ in the gain factor is the power amplifier output without memory and it can be modeled with the block diagram in Fig. 3.

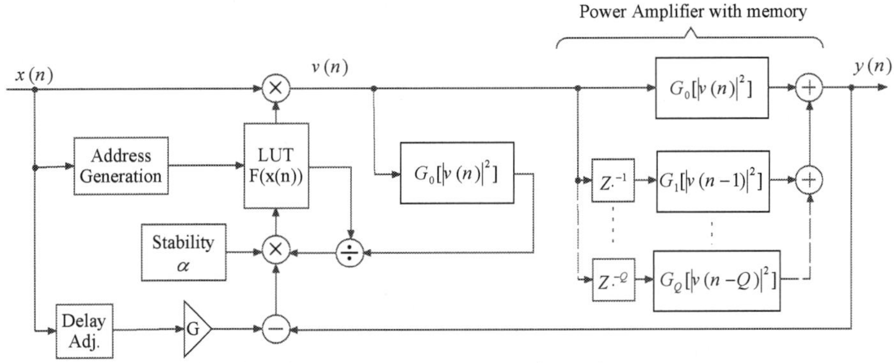

Fig. 3. predistortion block with memory compensation(case 1)

But it is still needed to know the characteristic of the power amplifier without memory. It is possible to initially calculate the coefficients of the power amplifier without memory and save it in LUT and then calculate the gain factor, but this way takes a lot of space and processing time. Because of this reason it is better to study case 2.

Case 2:

In this case the simplification could be done as below:

$$\frac{F[|x(n)|^2]}{v(n)G_0[|v(n)|^2]} = \frac{F[|x(n)|^2]}{x(n)F[|x(n)|^2]G_0[|v(n)|^2]} = \frac{1}{x(n)G_0[|v(n)|^2]} \tag{16}$$

It can be assumed that th function $F[|x(n)|^2]$ is initially equal to one then:

$$\frac{1}{x(n)G_0[|v(n)|^2]} = \frac{1}{x(n)G_0[|x(n)|^2]} = \frac{1}{y_0(n)} \tag{17}$$

Where $y_0(n)$ is the power amplifier output without memory and this is different from y(n) which includes memory. As the target is:

$$y_0(n) = Gx(n) \tag{18}$$

So the diagram in Fig. 4 will be reached which is the simplest method for linearizing the power amplifier with memory. And equation below will be the final predistortion adaptation that will be used in whole simulations:

$$F_{i+1}(|x(n)|^2) = F_i[|x(n)|^2] - \alpha \frac{1}{Gx(n)} V_{error}(n) \tag{19}$$

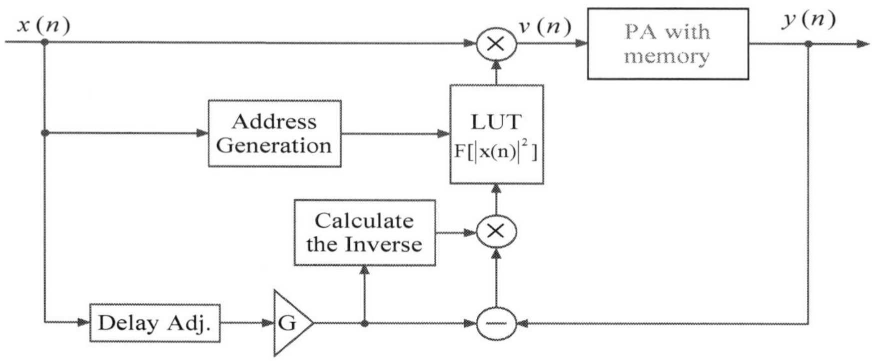

Fig. 4. predistortion block with memory compensation(case 2)

The only drawback that is remaining is to calculate the inverse of the input data which after finding it and multiplied with error vector then the LUT contents could be updated. So in implementation the main concern is the division part which the main issue in implementation and it leaves for future work.

In (15) with two or three iterations convergence is achieved and it will be shown that as compared with the indirect learning architecture method that is proposed in (Ding, 2004), the efficiency improves more and it is less complex. It will be shown that if the case 1 is applied in simulations the speed of convergence is more than case 2 which requires more iterations to convergence. The only time consuming part for implementing this method is the calculation of the gain factor which requires the inverse of the input signal. The predistorter is assumed to be implemented as a lookup table (LUT) of complex gain values (Cavers, 1990) that here is 10bit, indexed by the squared magnitude, as shown in Fig. 3. It is also possible to index by magnitude, or any other monotonic function of magnitude, depending on the regions of amplifier characteristic that need the greatest accuracy of representation. However, these considerations do not enter the analysis of the present paper. Also to help evaluate the performance of the DPD a figure for in-band distortion as well as out of band distortion which is measured with adjacent-channel-leakage-ratio (ACLR) is calculated. This involves calculating the error vector magnitude (EVM) in transmitter, which is given by the following equation

$$EVM = \frac{rms(|V_{error}(n)|)}{rms(|(x(n)|)} \quad (20)$$

where $V_{error}(n)$ is from (11) and $x(n)$ is the input signal.

4. Simulations and results

In order to validate the proposed method several simulations are done. MATLAB is applied for simulations. The power amplifier is ZVE-8G from Mini-Circuit suitable for CDMA applications. The input signal which is generated from Matlab is passed to agilent signal generator which will be upconvert the signal and pass it to the power amplifier. The power amplifier is wideband from 2 GHz to 8 GHz with 30 dB gain. Here the PA is working at 2.4 GHz. The input signal is QPSK with sample rate of 1 Mbps and the root rate cosine filter with alfa equal to 0.35. The output signal of the PA is connected to the attenuator for bring down the output power below the input power of the equipment. For receiving part of the measurement, the 89600S VXI equipment from agilent is used, which the main task is to downconvert the signal to IF frequency. The photograph of this experimental setup is shown in Fig. 5. The VSA software in PC will capture the data of the PA. The captured data can then import to Matlab for further analysis. Synchronization is used in order to achieve a complete coordination among the signal that is sent and the signal taken into Matlab.

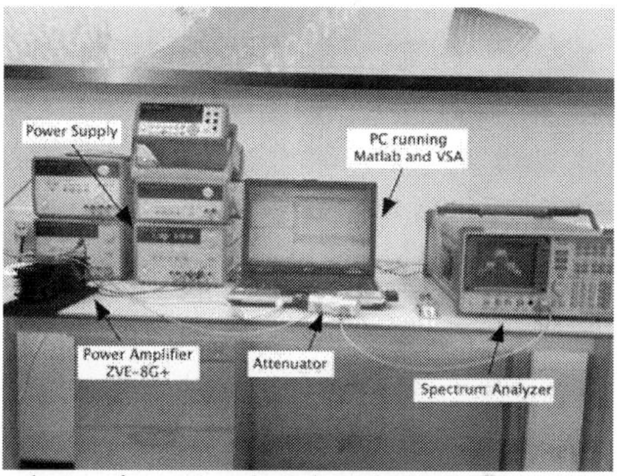

Fig. 5. the experimental setup for testing digital predistortion

These samples are used to model the power amplifier based on (1) which is the memory polynomial method. The extracted coefficients that include the memory effects are shown in table 1.

In Fig. 6 the AM-AM and AM-PM characteristics of this power amplifier are shown. It can be seen the scattering of samples in this figure that is because of the memory effects. It can be shown that when the memory effects are more these samples will be scattered more, and then the digital predistortion technique should be designed based on it. It is clear in Fig. 6a that the AM-AM characteristic is not linear when the input amplitude is increased. And also in Fig. 6b the curve bends too. This is because of the nonlinear characteristics of the power amplifier. All the input and output samples in the simulations are normalized. For modeling the memory effects of the power amplifiers authors in (Ku & Kenney, 2003) proposed a method for modeling the power amplifiers with memory. This method that is based on the spars delay taps is actually able to take into account all the memory effects of the power amplifier. The memory effect modeling ratio (MEMR) was used to show the amount of memory that this method can model. The power amplifier from mini circuit has MEMR=0.62 and the one in (Ku & Kenney, 2003) has MEMR=1 and these coefficients are shown in table 1. previous researches could present the comparison of the power amplifier with MEMR that is less than one. Here the presented method is successfully tested with these two types of PA models. In all the simulations the input back off is 3 dB. In simulations, it is avoided to reach to 1 dB compression point which increases the complexity of this method and also the effects of analog imperfections are not considered.

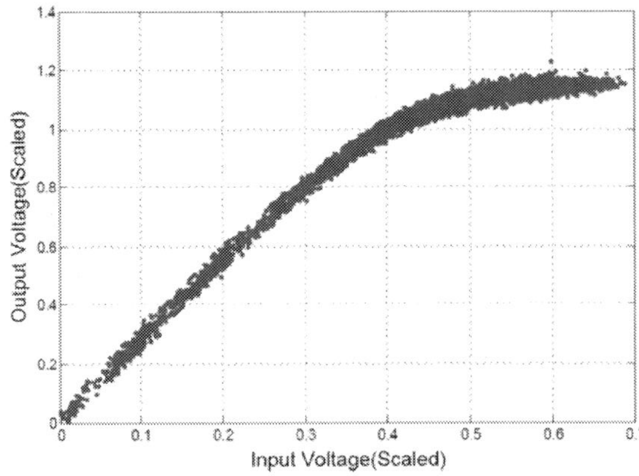

(a)

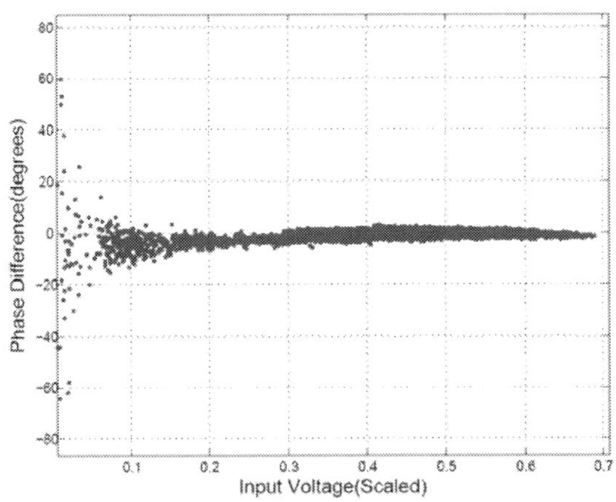

(b)

Fig. 6. AM-AM and AM-PM characteristics of the ZVE-8G PA (a) Input power versus output power. (b) Input power versus phase difference.

In table 1 the proposed method is compared with the memory polynomial method for two different power amplifiers with different memory contents for QPSK signal. The comparison is with ACLR and EVM.

Predistortion technique	Power amplifier coefficients	MEMR	ACLR(dBc) Left	ACLR(dBc) Right	EVM (%)
Memory Polynomial	a_{10}=0.9800-0.300i; a_{11}=0.06+0.03i; a_{12}=0.02+0.08i; a_{13}=-0.01+0.02i; a_{30}=-0.3+0.42i; a_{31}=-0.02+0.05i; a_{32}=-0.01-0.08i; a_{33}=0.02-0.01i;	1	-39.1	-40.2	2.55
	a_{10}=1.524-0.211i; a_{11}=0.349+0.32i;a_{12}=-0.797-0.0247i; a_{30}=-0.0355+0.72i; a_{31}=-0.010-0.012i; a_{32}=-0.0065+0.0042i; a_{50}=-0.019-0.004i; a_{51}=0.009-0.019i; a_{52}=-0.0069+0.013i;	0.62	-41.2	-42.5	2.25
Complex Gain	a_{10}=0.9800-0.300i; a_{11}=0.06+0.03i; a_{12}=0.02+0.08i; a_{13}=-0.01+0.02i; a_{30}=-0.3+0.42i; a_{31}=-0.02+0.05i; a_{32}=-0.01-0.08i; a_{33}=0.02-0.01i;	1	-42.1	-40.6	1.95
	a_{10}=1.524-0.211i; a_{11}=0.349+0.32i; a_{12}=-0.797-0.0247i; a_{30}=-0.0355+0.72i; a_{31}=-0.010-0.012i; a_{32}=-0.0065+0.0042i; a_{50}=-0.019-0.004i; a_{51}=0.009-0.019i; a_{52}=-0.0069+0.013i;	0.62	-45.1	-48.3	1.73

Table 1. Comparison of the two predistortion techniques for different power amplifiiers with QPSK signal

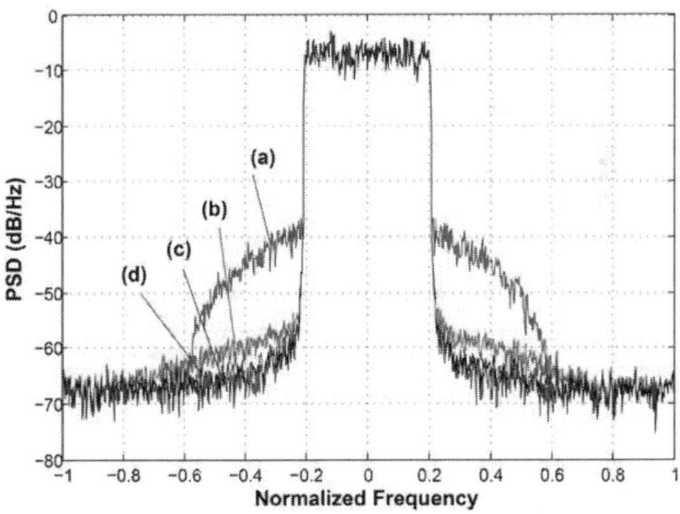

Fig. 7. Comparison of the power spectral density (PSD) between memory polynomial predistorter and gain predistortion for power amplifier without memory and Mobile WiMAX signal. (a) Output without predistortion (b) Output with memory polynomial predistortion (c) Output with gain predistortion (iteration=5) (d) Input data

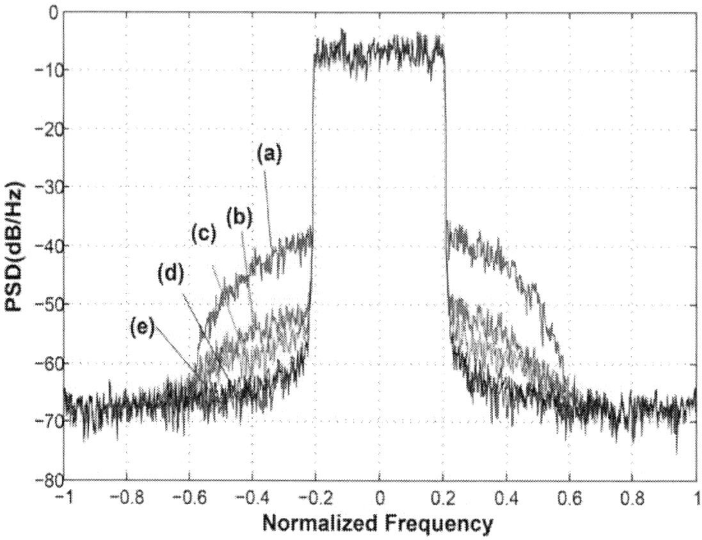

Fig. 8. Comparison of the power spectral density (PSD) between memory polynomial predistorter and gain predistortion for power amplifier wit memory and Mobile WiMAX signal. (a) Output without predistortion; (b) Output with memory polynomial predistortion (Q=0, K=3) (c) Output with memory polynomial predistortion (Q=2, K=3) (d) Output with gain predistortion (iteration=5,) (e) Input data

In Fig. 7 and Fig. 8 the Mobile WiMAX signal that is compatible with IEEE 802.16e with 10 MHz bandwidth is applied for simulations.

The power amplifier is with MEMR=0.62. In Fig. 7 the power amplifier is memoryless and has only 3 coefficients, these coefficients don't show in table 1. In Fig. 8 the power amplifier includes memory and the coefficients are presented in table 1. in this case the nonlinearity order is three and memory length is two. According to the Fig. 7 the ACLR is averagely -49.3dB when applying the memory polynomial method and with the new technique is reduced to -57.4 dB which is around 8 dB improvement in ACLR.

5. Conclusions

In this paper the new digital predistortion technique is introduced. This technique is the combination of two techniques, complex gain predistortion and memory polynomial predistortion. By applying this technique all the memory contents of the power amplifier is suppressed. Simulations and results are examined with mini circuit power amplifier with 30 dB gain that is working in 2.4 GHz and the power amplifier with MEMR=1. the QPSK signal with 1 MHz bandwidth is used for input signal. The results show the improvement of 8 dB in ACLR and improvement of 3 % in EVM. The future research should be more on the implementation of this technique using FPGA and DSP and measure the effects of analog imperfection that cause reduction in efficiency in practical implementation and add that effects in the simulations.

6. References

Bosch W. and Gatti G. (1989). "Measurement and simulation of memory effects in predistortion linearizers," *IEEE Trans. Microwave Theory Tech.*, vol. 37, pp. 1885–1890, Dec. 1989

Cavers J. K. (1990). "Amplifier Linearization Using a Digital Predistorter with Fast Adaptation and Low Memory Requirements", *IEEE Transaction on vehicular technology*, vol 39, no.4, pp. 374-382,Nov 1990.

Cripps S.C. (1999). *RF Power Amplifiers for Wireless Communications.* Norwood, MA: Artech House, 1999.

Ding L., Zhou G. T, Morgan D. R., Ma Z., Kenney J. S, Kim J., and Giardina C. R (2004). "A robust predistorter constructed using memory polynomials," *IEEE Trans. Commun.*, vol. 52, pp. 159–165, Jan. 2004.

Kenington P. B. (2000). *High-Linearity RF Amplifier Design.* Norwood, MA: Artech House, 2000.

Kim J. and Konstantinou K. (2001). "Digital predistortion of wideband signals based on power amplifier model with memory," *Electron. Lett.*, vol. 37, no. 23, pp. 1417–1418, Nov. 2001

Ku H. and Kenney J. S. (2003). "Behavioral modeling of nonlinear RF power amplifiers considering memory effects," *IEEE Trans. Microw. Theory Tech.*, vol. 51, no. 12, pp. 2495–2504, Dec. 2003.

Ku H., and McKinley M. D. (2002). "Quantifying Memory Effects in RF Power Amplifiers," *IEEE Trans. Microw. Theory Tech*, vol. 50, no. 12, December 2002.

Morgan D. R., Z. Ma, Kim J., Zierdt M. G., and Pastalan J. (2006). "A generalized memory polynomial model for digital predistortion of RF power amplifiers," *IEEE Trans. Signal Process.*, vol. 54, no. 10, pp. 3852–3860, Oct. 2006

Nagata Y. (1989). "Linear amplification techniques for digital mobile communications," in *Proc. IEEE 39th Veh. Technol. Conf.*, 1989, pp. 159–164.

Raich R., Qian H., and Zhou G. T. (2003). "Digital baseband predistortion of nonlinear power amplifiers using orthogonal polynomials," in *Proc. IEEE Int. Con f. Acoust, Speech, Signal Processing*, pp. 689–692, Apr. 2003.

Varahram P., Mohammady S., Hamidon M. N., Sidek R. M., Khatun S., (2009). "Digital Predistortion Technique for Compensating Memory Effects of Power Amplifiers in Wideband Applications," Journal of ELECTRICAL ENGINEERING, VOL. 60, NO. 3, 2009.

Wangnyong W., Miller M., and Kenney J. S. (2004). "Predistortion Linearization System for High Power Amplifiers", *IEEE MTT-S Digest*, pp. 677-680, 2004 .

Woo Y. Y., Kim J., Yi J., Hong S., Kim I., Moon J., and Kim B. (2007). "Adaptive digital feedback predistortion technique for linearizing power amplifiers," *IEEE Trans. Microw. Theory Tech.*, vol. 55, no. 5, pp. 932–940, May 2007.

Woo Y. Y., Kim J., Hong S., Kim I., Moon J., Yi J., and Kim B. (2007). "A new adaptive digital predistortion technique employing feedback technique," in *IEEE MTT-S Int. Microw. Symp. Dig.*, Jun., pp. 1445–1448.

Wright A. S. and Durtler W. G. (1992). "Experimental performance of an adaptive digital linearized power amplifier," *IEEE Trans. Veh. Technol.*, vol. 41, no. 4, pp. 395–400, Nov.

10

A Fast Method to Compute Radiation Fields of Shaped Reflector Antennas by FFT

Abolghasem Zeidaabadi Nezhad[1], Zaker Hossein Firouzeh[2] and Hamid Mirmohammad-Sadeghi[2]
[1]*Department of Electrical & Computer Engineering, Isfahan University of Technology*
[2]*Information and Communication Technology Institute (ICTI)*
IRAN

1. Introduction

Interest in reflector antennas due to attractive features including high gain, multiple beam, low side lobe level, and robustness have been growth both in military and commercial areas in microwave frequency range since World War I. In particular, low side lobe level radiation pattern, is an essential requirement for satellite communication systems. Offset reflector antennas featuring adjacent high gain beams with good isolation across the same frequency bandwidth make them good candidates for these applications. Depending on the required radiation pattern, feeds may or may not be located at the focus. A cluster of feeds can be used in the focal region of offset reflectors for multiple beams application [Skolnik, 1990; Chu and Turrin, 1973; Rudge, 1975; Janken et al., 1973; Ingerson and Wong, 1974; Tian et al., 2007].

There have been published lots of articles dealing with analyze and design of electrically small and large reflectors by analytical or numerical techniques [Love, 1978; Wood, 1986; Lo and Lee, 1988; Scott, 1990]. These methods vary from the traditional Aperture Field Method (AFM) that involves integrating the electric fields scattered by the reflector onto a projected planar aperture, to more modern hybrid methods that employ a variety of techniques each of which is applicable over different regions of the radiation pattern. Some of the aforementioned methods yield exact solutions such as, Method of Moments (MoM) but require significantly computational resources. Larger reflectors like large radio astronomy antennas need to be analysed using high-frequency techniques which use approximations based on asymptotic solutions for canonical problems. The approximate techniques have been proved to be quite successful in predicting both far and near-field patterns and are in very good agreement with the measurements. Methods of evaluating the electromagnetic fields radiated from a reflector antenna fall into two categories: exact and approximation methods. These various techniques have been explained as follows [Philips et al., 1996].

The Method of Moments is the most accurate technique in all known methods used in electromagnetic scattering analysis. The formulation of the governing equations of the problem (such as Electric Field Integral Equation (EFIE)) is exact, and highly accurate solutions can be obtained by a suitable choice of basis and testing functions. The induced current distribution on the reflector surface is obtained by MoM to calculate the electromagnetic scattering fields. The Method of Moments is well documented in the literature providing classical textbooks that describe the technique [Harrington, 1993; Moore and Pizer, 1984; Miller et al., 1992]. Using this approach, one can accurately evaluate the full radiation properties of any antenna by wire grid [Popovic et al., 1982] or patch model [Wilton & Butler, 1981]. Additionally, the antenna can be analyzed in free space or infinite dielectric half plane. The main drawback of this method is the large computational resources required. Therefore, the MoM is not a practical approach for antennas larger than a few wavelengths. Recently, Wavelet-Based Moment Method (WBMM) has been applied on large reflector antennas to calculate the current distribution accurately and fast. By serving wavelets as basis and testing functions in a wavelet expansion, a sparse matrix is generated from the previous MoM dense matrix, which may save computational cost. By use of this method, the currents on the reflector surface can be calculated faster than the conventional MoM [Lashab et al., 2007; Lashab et al., 2008; Herzberg, 2005].

Approximation methods or high frequency methods can be applied on electrically large antennas to predict far-field and near-field radiation characteristics [Rusch and Potter, 1970; Kauffman et al., 1976]. These methods are subdivided into three types; those based on the Kirchoff's approximation (such as Aperture Field Method, Scalar Radiation Integral/Projected Aperture Method, Physical Optics (PO), Gaussian Beam Mode optics (GBM)), Ray tracing methods (such as Geometrical Optics (GO), Geometrical or Uniform Theory of Diffraction (GTD/UTD)), and also corrected Kirchoff's methods (such as Physical Optics and Physical Theory of Diffraction (PTD)) [Philips et al., 1996]. The high frequency methods are well suited for the analysis of electrically large reflector antennas as well as for the modelling of secondary effects such as the interaction of the main antenna with adjacent structures similar to building or terrain obstacles.

Methods based on Kirchoff's approximation are utilized generally for the estimation of the antenna boresight, main lobe and few adjacent sidelobes. Aperture field method and projected aperture method fail to predict the cross polar pattern with sufficient degree of accuracy [Silver, 1949]. In addition, the minimum radii of curvature should be more than five wavelengths in size for validity of Geometrical Optics [Philips et al., 1996]. The most widely used method is Physical Optics which involves the vector summation of radiated fields due to individual currents induced over the structure by the illuminating fields which have been evaluated by considering the Geometrical Optics propagation due to the primary feed [Balanis, 1989]. In terms of CPU time and storage requirements, the Aperture Field Method or projected aperture method is significantly more efficient than Physical Optics. However, PO is generally more accurate than the former and correctly predicts the main beam and close in sidelobes. It also gives a better prediction of the cross polar pattern.

Methods based on the Kirchoff's approximation suffer from the inability to provide accurate predictions at angles widely displaced from the antenna's main beam direction. Although the computational requirements for the summation of the Physical Optics current are not as severe as in the case of the MoM, it is nevertheless a significant factor to consider taking into account that the sampling interval on the reflector surface needs to be of the order of a wavelength or less [Philips et al., 1996].

An alternative fast analytical way for evaluating the radiative properties of the antenna is the Gaussian Beam Mode (GBM) analysis [Goldsmith, 1982; Lamb, 1986; Bogush and Elgin, 1986; Lesurf, 1990]. The Gaussian beam mode fields are given by simple analytical expressions; hence this formulation offers advantages due to the simplicity and the speed of evaluation. The analysis is valid in both the near and the far-field points along the paraxial directions with equally rapid and straightforward evaluations. Furthermore, the expansions are valid in the transition regions. However, GBM cannot predict the far-out sidelobe region and blockage due to subreflector or struts which should be either negligible or avoided for valid results.

The wide angle and non-paraxial direction can be predicted by a ray description of the diffracted field. This is accomplished with the introduction of diffracted rays within the framework of Geometrical Theory of Diffraction [Keller, 1958; Keller, 1962], or its uniform versions such as the Uniform Theory of Diffraction [Kouyoumjian and Pathak, 1974] and Uniform Asymptotic Theory (UAT) [Ahluwalia et al., 1968], which provides diffracted field expressions valid across the shadow boundaries of the incident or reflected fields. The diffraction coefficients contain Fresnel integrals that are easily evaluated, thus providing a fast and efficient algorithm for the analysis of large reflectors. Most of the time is actually used in locating the points of reflection and diffraction on the reflector, given the source and field points. With multiple reflectors and complex geometrical shapes, this can sometimes be quite time-consuming although not nearly as much as in evaluating double integrals over large surfaces. It must be emphasised that any ray description fails at caustics and so a purely ray technique cannot predict the far-field characteristics of a large antenna near boresight. Such regions can be analysed by the Equivalent Current Method (ECM) [Ryan and Peters, 1969], which works back from the GTD solution away from caustics to obtain an equivalent current that would produce identical fields there. This current is then used to extrapolate the field at the caustics.

A PO plus UTD scheme may well be the first option for evaluating the radiated field from a large reflector antenna. The PO is used to predict the radiation characteristics in the boresight region and the UTD to evaluate the wide angle characteristics of the radiation pattern. Prediction of the antenna backlobe can be calculated by defining an equivalent edge current using the basic UTD formulation. This current is integrated in the usual far-field sense to provide the radiation pattern in the rearward direction. The implementation of a UTD scheme offers significant speed enhancement combined with minimal memory requirements. The GO plus UTD combination can also provide very rapid field calculations, but fails to predict the main beam characteristics. However, it can be used to provide full near-field predictions [Philips et al., 1996].

The problem of wide angle and non-paraxial direction prediction can be solved when the Physical Optics method is augmented with Physical Theory of Diffraction (PTD), which is a systematic extension of the PO approach, just as GTD is an extension to GO [Ufimtsev, 2007]. The PTD correction consists of a fringe current acting along the rim of the quasi-optical system. The value of these fringe currents is such that the edge condition is satisfied; hence the diffraction phenomenon is accurately described. The PO plus PTD scheme can provide accurate near and far-field information along any direction. However, this method when applied to electrically large systems may require large amounts of memory and computer time due to evaluation of surface and line integration.

After choosing the proper method to analyze the desired reflector antenna, the speed and the accuracy of the radiation pattern computation of the reflector antennas is very important. The most straightforward as well as the most time-consuming method is the direct numerical quadrature of the radiation integral for the aperture antenna. Several approaches have been proposed for the efficient numerical evaluation of the double radiation integrals. The earliest of these is the so-called Ludwig algorithm in which the double integral can be evaluated in explicit closed forms [Ludwig, 1968; Ludwig, 1988]. Alternatively, one can expand the radiation integral into a series such as the one in the Jacobi-Bessel method [Rahmat-Samii et al., 1980; Galindo-Israel and Mittra, 1977]. The Jacobi-Bessel method is most suited for computing the pattern of antennas which have a circular projected aperture. Because the Jacobi polynomials satisfy a special type of recursion relationship, they are also useful for computing the radiation pattern of parabolic reflector antennas. Another approach to the secondary pattern computation of planar or parabolic antennas has been suggested by Drabowitch [Drabowitch, 1965]. This approach is based on the two-dimensional sampling theorem. The coefficients of the interpolating functions for the secondary pattern are computed by periodically sampling the secondary pattern at intervals determined by the aperture dimensions. These coefficients are subsequently used in conjunction with the interpolating functions to compute the secondary pattern at an arbitrary observation angle.

In this chapter, an algorithm is presented to evaluate aperture numerical integration by FFT method. This coordinate system is used for all antenna configurations. The proposed algorithm can be applied to all shaped reflector antennas which has been illuminated by defocused feeds with arbitrary patterns. In this method, in order to calculate the radiation patterns, the equations of geometrical optics are used to calculate the reflected electric field using the radiation patterns of the feed and the parameters defining the reflector surface. In addition, the direction of the reflected ray and the point of intersection of the reflected ray with the aperture plane are obtained by use of geometrical optics. These fields comprise the aperture field distribution which is integrated over the aperture plane by FFT to yield the far-field radiation pattern and to calculate other antenna parameters. Shaped Reflector Antenna Design and Analysis Software (SRADAS) based on this numerical method can analyze and simulate all shaped reflector antennas with large dimensions in regard to the wavelength. SRADAS has been implemented and used in Information and Communication Technology Institute (ICTI) to analyze and simulate different practical parabolic and shaped reflector antennas [Zeidaabadi Nezhad and Firouzeh, 2005].

The organization of this chapter is as follows: Proposed fast method to calculate the radiation integral of a parabolic reflector antenna is explained in Section 2. Required mesh size in order to accomplish the optimum mesh density in calculating the radiation integrals with desired accuracy and speed is introduced in section 3. In order to confirm the validity of the proposed calculation method, in Sections 4 and 5, two types of practical antennas are analyzed by this method and the results are compared with the results achieved by the commercial software package FEKO and measurements, as well. Finally, concluding remarks are given in Section 6.

2. Calculation of the radiation integrals by FFT

Fig. 1 shows the three-dimensional geometry of a parabolic reflector antenna. A feed located at the focal point of a parabola forms a beam parallel to the focal axis. In addition, the rays emanating from the focus of the reflector are transformed into plane waves. The design is based on optical techniques, and it does not take into account any diffraction from the rim of the reflector. Since a parabolic antenna is a parabola of revolution, the equation (1) describes the parabolic surface in terms of the spherical coordinates r', θ', ϕ', where f is the focal distance. Because of its rotational symmetry, there are no variations with respect to ϕ'. The projected cross-sectional area of reflector on the aperture plane -the opening of the reflector- is S_0, and on the focal plane is S_0'.

$$r' = \frac{2f}{1+cos\theta'} = f\,sec^2\left(\frac{\theta'}{2}\right) \quad \theta' \le \theta_0 \tag{1}$$

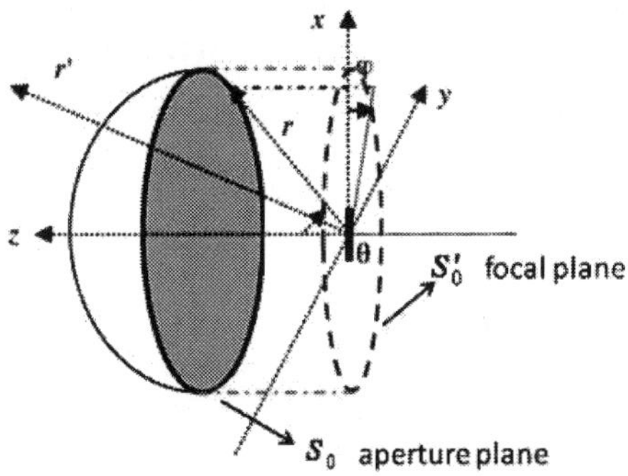

Fig. 1. Three-dimensional geometry of a parabolic reflector antenna

The total pattern of the system is computed by the sum of secondary field and the primary field of the feed element. For the majority of feeds like horn antennas, the primary pattern in the boresight direction of the reflector is of very low intensity and usually can be neglected. The advantage of the AFM is that, the integration over the aperture plane can be performed easily for any feed position and any feed pattern, whereas the double integration on current distribution over the reflector surface is time-consuming in PO method, and it becomes difficult when the feed is placed off-axis or when the feed radiation pattern has no symmetry.

The radiation integrals over S_0' computing the far fields by AFM can be written as [Balanis, 2005]:

$$E_{\theta S} = \frac{j\beta e^{-j\beta r}}{4\pi r}(1-cos\theta)\iint_{S_0'}\left(-E_{ax}\,cos\phi - E_{ay}\,sin\phi\right)$$
$$\exp(j\beta(x'sin\theta\,cos\phi + y'sin\theta\,sin\phi))\,dx'dy' \tag{2a}$$

$$E_{\phi S} = \frac{j\beta e^{-j\beta r}}{4\pi r}(1-cos\theta)\iint_{S_0'}\left(-E_{ax}\,sin\phi + E_{ay}cos\,\phi\right)$$
$$\exp(j\beta(x'sin\theta cos\phi + y'sin\theta sin\phi))dx'dy' \tag{2b}$$

$$\beta = \frac{2\pi}{\lambda}\ , \quad u = sin\theta cos\phi \quad v = sin\theta sin\phi \tag{2c}$$

In equations (2), E_{ax} and E_{ay} represent the x- and y-component of the reflected fields over S'_0. The spherical coordinates of the observation point is r, θ, ϕ. λ and β are wavelength and phase constant of the propagated wave in free space, respectively.
A feed at the focal point of a parabola forms a beam parallel to the focal axis. Therefore, the only difference between fields on S_0 and S'_0 is the constant phase because of the distance between the aperture plane S_0 and the focal plane S'_0. Additional feeds displaced from the focal point form multiple beams at angles off the antenna axis. In this case, the reflected fields from the reflector are not parallel to the focal axis resulting in a severe phase distortion between fields on S_0 and S'_0. Therefore, the integral equations (2a) and (2b) are calculated over the aperture plane S_0, not the focal plane S'_0. The phase distortion increases with the angular displacement in beamwidths and decreases with an increase in the focal length.
In order to calculate the reflected filed from the reflector, a rectangular mesh is created on the focal plane S'_0 as shown in Fig. 2. According to AFM and GO, the reflected fields out of S'_0 are vanished. Two-dimensional FFT is used to compute the integral equations (2a) and (2b) rapidly [Bracewell, 1986]. Integrals P_X and P_Y are defined as:

$$P_x = \iint_{S'_0} E_{ax} \; e^{j\beta(x'u+y'v)} dx'dy'$$
$$P_y = \iint_{S'_0} E_{ay} \; e^{j\beta(x'u+y'v)} dx'dy' \tag{3}$$

Using the equations of (3), radiation fields $E_{\theta S}$ and $E_{\phi S}$ can be calculated by:

$$E_{\theta S} = \frac{j\beta e^{-j\beta r}}{4\pi r}(1-cos\theta)\{-P_x\, cos\phi - P_y\, sin\phi\}$$
$$E_{\phi S} = \frac{j\beta e^{-j\beta r}}{4\pi r}(1-cos\theta)\{-P_x\, sin\phi + P_y\, cos\phi\} \tag{4}$$

The mesh grid is generated by the following expressions:

$$\Delta x' = \frac{d}{M-1} \;,\; x' = -\frac{d}{2} + m.\Delta x' \;,\; m = 0,1,2,\ldots,M-1 \tag{5}$$
$$\Delta y' = \frac{d}{N-1} \;,\; y' = -\frac{d}{2} + n.\Delta y' \;,\; n = 0,1,2,\ldots,N-1$$

Where, M and N are the number of points which have been distributed uniformly in x- and y-direction of S'_0 plane. The aperture diameter of the parabolic reflector is d. The relations of (3) and (5) lead to:

$$P_x = \exp\left(-j\beta u\frac{d}{2}\right)\exp\left(-j\beta v\frac{d}{2}\right)\;\Delta x'\Delta y'\sum_{m=0}^{M-1}\sum_{n=0}^{N-1} E_{ax}(m,n) e^{j\beta v\frac{nd}{N-1}} e^{j\beta u\frac{md}{M-1}} \tag{6}$$
$$P_y = \exp\left(-j\beta u\frac{d}{2}\right)\exp\left(-j\beta v\frac{d}{2}\right)\;\Delta x'\Delta y'\sum_{m=0}^{M-1}\sum_{n=0}^{N-1} E_{ay}(m,n) e^{j\beta v\frac{nd}{N-1}} e^{j\beta u\frac{md}{M-1}}$$

$E_{ax}(m,n)$ and $E_{ay}(m,n)$ are the electric fields at the mesh points (m,n) of S_0' plane. Comparison between the equations of (6) with FFT formulas, the angles of spherical coordinate of far-field radiation electric fields, θ_{kl} and ϕ_{kl} are calculated using the following expressions:

$$A = \frac{d}{\lambda}\frac{M}{M-1} \quad , \quad k = -A\, sin\theta_{kl}\, cos\phi_{kl} \quad , \quad k = 0,1,2,\ldots,M-1$$
$$B = \frac{d}{\lambda}\frac{N}{N-1} \quad , \quad l = -B\, sin\theta_{kl}\, sin\phi_{kl} \quad , \quad l = 0,1,2,\ldots,N-1 \tag{7a}$$
$$0 \le \theta_{kl} \le \pi \ , \ 0 \le \phi_{kl} \le 2\pi \ or \ -\pi \le \phi_{kl} \le \pi$$

$$\theta_{kl} = sin^{-1}\left(\left[\left(\frac{k}{-A}\right)^2 + \left(\frac{l}{-B}\right)^2\right]^{1/2}\right)$$
$$\phi_{kl} = \tan^{-1}\left(\frac{-Al}{-Bk}\right) \tag{7b}$$

Two-dimensional FFT (FFT2) formulas can be used to rewrite the relations of (6), that is:

$$P_x = \exp\left(-j\beta u\frac{d}{2}\right)\exp\left(-j\beta v\frac{d}{2}\right)\ \Delta x'\Delta y'\,FFT\,2\left(E_{ax}\right)$$
$$P_y = \exp\left(-j\beta u\frac{d}{2}\right)\exp\left(-j\beta v\frac{d}{2}\right)\ \Delta x'\Delta y'\,FFT\,2\left(E_{ay}\right) \tag{8}$$

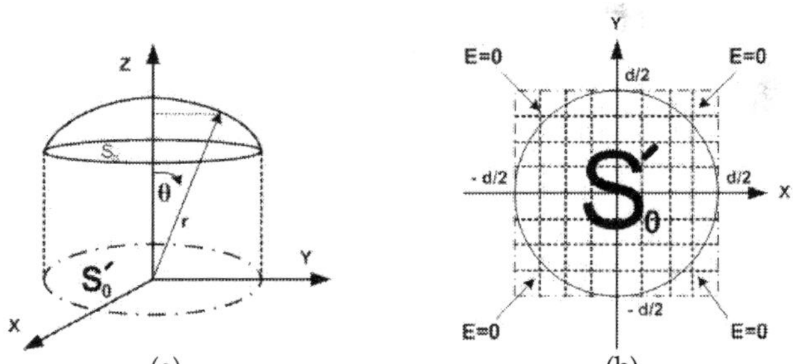

Fig. 2. (a) Plane S_0' is the projected cross-sectional area of reflector on the focal plane. (b) A rectangular mesh is created on the plane S_0'. d is the aperture diameter of the parabolic reflector.

Finally, radiation fields of $E_{\theta S}$ and $E_{\phi S}$ are computed by using FFT2. It can be written as:

$$E_{\theta S} = \frac{j\beta e^{-j\beta r}}{4\pi r}\left(1-cos\theta_{kl}\right)\left\{-cos\phi_{kl}P_x - sin\phi_{kl}P_y\right\}$$
$$E_{\phi S} = \frac{j\beta e^{-j\beta r}}{4\pi r}\left(1-cos\theta_{kl}\right)\left\{-sin\phi_{kl}P_x + cos\phi_{kl}P_y\right\} \tag{9}$$

Based on the equations (8) and (9), far fields of $E_{\theta S}$ and $E_{\phi S}$ are calculated, where θ_{kl} are the angles calculated by (7). Negative values of l and k are used to find the fields in other regions. In addition, number of main points of the major lobe found by FFT is constrained. Therefore, it is interpolated from above points to obtain E- and H-planes Half-Power beamwidths (HP_E and HP_H). The principal E- and H-plane radiation patterns can be calculated by substituting $\phi = \pi/2$ and $\phi = 0$, respectively. According to [Stutzman and Thiele, 1981], directivity D can be calculated as:

$$D = \frac{4\pi}{\Omega_A}$$
$$\Omega_A = \iint_{u^2+v^2\le 1} \left|F(u,v)\right|^2 \frac{dudv}{\sqrt{1-u^2-v^2}} \tag{10}$$

Ω_A is the antenna beam solid angle. $F(u,v)$ is the radiation pattern of the reflector antenna in terms of the variables u and v. The main equations are prepared to analyze a paraboloidal reflector and to compute the radiation characteristics. Moreover, SRADAS can be applied to all shaped reflector antennas with defocused feed elements provided that dimensions of the reflector are large in regard to the wavelength.

3. Calculation of the optimum mesh size

In general, if x(t) is a continuous function of t in the interval of [a ,b], Fourier transform pair of x(t) can be written as the following [Bracewell, 1986]:

$$x(t) = \int_{-\infty}^{+\infty} X(f) e^{+j2\pi ft} df \tag{11a}$$

$$X(f) = \int_{-\infty}^{+\infty} x(t) e^{-j2\pi ft} dt \tag{11b}$$

To calculate the Fourier integrals numerically, the interval of [a, b] is divided into N-1 segments uniformly by use of N points. The step size ΔT is:

$$\Delta T = \frac{b-a}{N-1}$$
$$t = a + n\Delta T \quad , \quad n = 0,1,2,\ldots, N-1 \tag{12}$$

After substitution of (12) in the Fourier transform pair, (11b) can be estimated using the following expression:

$$X(f) \approx \sum_{n=0}^{N-1} x[n] e^{-j\frac{2\pi}{N} f \frac{n(b-a)N}{(N-1)}} \tag{13}$$

By comparison (13) to Discrete Fourier Transform, it can be written as:

$$X(k)=\sum_{n=0}^{N-1}x[n]e^{-j\frac{2\pi}{N}kn} \qquad k=0,1,2,...,N-1$$
$$k=N\Delta Tf \Rightarrow f_k=\frac{k}{N\Delta T}\Rightarrow f_{max}=\frac{N-1}{N\Delta T} \tag{14}$$

Where, ΔT should be less than $\frac{N-1}{N}\frac{1}{f_{max}}$ until the maximum frequency component of x(t) can be detected by Discrete Fourier Transform. Since N is very large, the preceding relation can be reduced as:

$$\Delta T \le \frac{1}{f_{max}} \tag{15}$$

It is obvious that the radiation integrals are the spatial Fourier Transform of the aperture electric fields. Therefore, f_x and f_y are the spatial frequencies corresponding to x and y axes, respectively. The mesh size should satisfy the following relation based on (15):

$$\Delta x<\frac{1}{f_x}\ ,\ f_x=\frac{u}{\lambda}\Rightarrow \Delta x<\frac{\lambda}{u}\ ,\ \ Max(u)=1\Rightarrow \Delta x<\lambda \tag{16}$$

Similarly, it can be proved that $\Delta y<\lambda$. As a result, the mesh size should be less than λ until the aperture electric fields can be sampled correctly to compute the radiation integrals numerically by FFT accurately. Near-field measurement in the case of planar scanning shows the sampling interval is better to choose $\lambda/2$ or less to have more accurate phase detection [Philips et al., 1996].

In this section, in order to evaluate the effect of mesh size in calculating the radiation pattern, a typical parabolic reflector antenna excited by a feed horn has been simulated. The operating frequency of the antenna is 1.3 GHz. The diameter and the focal distance are 13.5m and 5.31m, respectively. Simulated results for different mesh sizes by SRADAS have been shown in Table 1. For mesh size greater than λ=23.08cm, the antenna parameters such as Gain and Half Power (HP) beamwidths are not accurate. However, when the mesh size is less than λ, the condition (16) is satisfied, and the radiation characteristics will be calculated correctly. When mesh points of M=128 and N=128 are chosen, the main beam is at the angle of θ=180° and the Gain is 38.91dB. HP beamwidths are 2.34° in the E-plane radiation pattern and 0.9° in the H-plane radiation pattern. Calculated side lobe levels are -35dB and -25dB in E-plane and H-plane, respectively. E-plane and H-plane radiation patterns have been depicted in Fig. 3.

In order to validate the proposed calculation method the parabolic reflector antenna has been simulated by FEKO software. FEKO results have been given in Table 2. As it can be noticed there are some discrepancies between the proposed analytical method and FEKO result. The first reason is that, for simplicity, diffraction effects have been ignored in calculation in SRADAS. Also, the interpolation method has been used to compute both of Gain and HP beamwidths. However, the consumed time of simulation by SRADAS is about one-third of FEKO simulation time. The calculation speed of SRADAS is faster than the simulation performed for the same structure by FEKO software and both results are in good agreement.

In the following sections, to evaluate the proposed calculation method, SRADAS, antenna parameters of two practical radar antennas are calculated and the results are compared both with FEKO and measurements.

M	N	dx (cm)	dy (cm)	Main beam θ (deg.)	HP_E^o	HP_H^o	Gain (dB)
32	32	43.55	43.55	180	62.89	32.42	11.06
48	48	28.72	28.72	180	24.62	14.12	18.73
56	56	24.55	24.55	180	5.24	2.32	33.30
64	64	21.47	21.47	180	2.34	0.9	38.83
100	100	13.64	13.64	180	2.34	0.9	38.90
128	128	10.63	10.63	180	2.34	0.9	38.91

Table 1. Simulated results for different meshing sizes by SRADAS

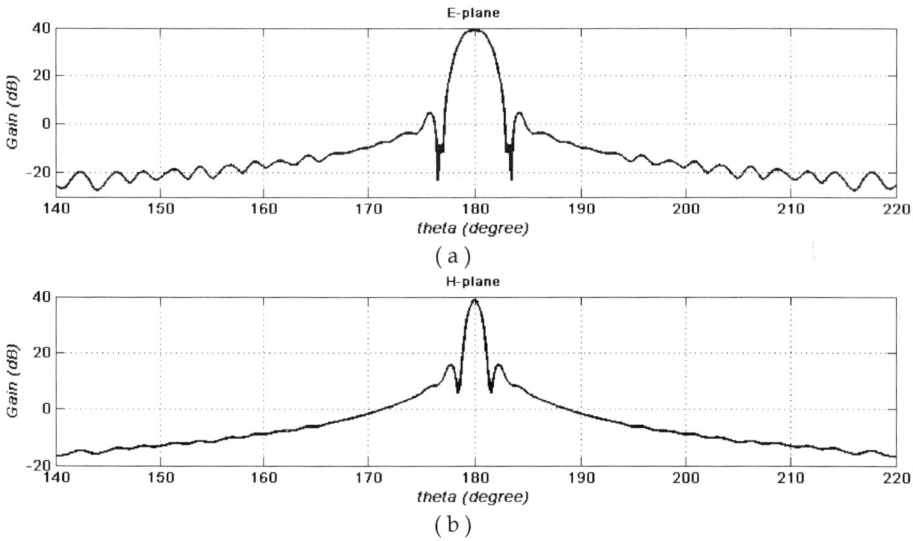

Fig. 3. (a) E-plane radiation pattern (b) H-plane radiation pattern

	Main beam θ (deg.)	HP_H^o	HP_E^o	Gain (dB)	SLL_E (dB)	SLL_H (dB)
SRADAS	180	2.34	0.9	38.91	-35	-25
FEKO	180	2.5	1.1	39.2	-33	-22

Table 2. Comparison radiation characteristics simulated by FEKO software and SRADAS

4. Analysis of a shaped reflector antenna illuminated by two displaced feed horns

A shaped reflector antenna fed by two displaced feed horns (Fig. 4) has been simulated by SRADAS. The operating frequency of the antenna is 1.4 GHz. Reflector aperture is 7.0m in height and 13.5m in width with a focal axis of 5.31m. The profile of azimuth curve is parabola and the profile of elevation curve is an unusual function. The 3-dimensional mathematical function which determines the reflector surface of the antenna is obtained by curve fitting method as:

$$z = -0.0471x^2 + 5.3100\cos^{0.6364}\left(\frac{y}{4.9267}\right) \tag{17}$$

The feed horns have been placed in y-direction symmetrically in relation to the origin. The feed horn which has been located in (0, 0,-0.185m) radiates the higher beam and the other one located in (0, 0, 0.185m) radiates lower one.

Simulated results by SRADAS have been shown in Table 3. Both of M and N for meshing the reflector aperture are 128. Results provided by FEKO have been given in Table 4. Because of large dimensions, radiation patterns of the antenna have been measured using outdoor far-field measurement method (open-site method). The gain of higher beam is 35.5dB and that of lower one is 34.5dB. Azimuth HP beamwidth ($HP_H°$) is $1.2°$ and elevation HP beamwidth ($HP_E°$) is about $10.5°$. The results obtained for this antenna using presented numerical method are in good agreement with both measurements and FEKO software. The consumed time of simulation by SRADAS is about one-third of the time consumed by FEKO.

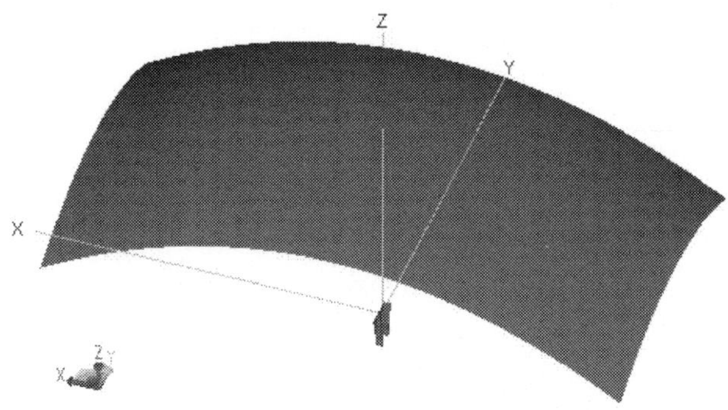

Fig. 4. A shaped reflector antenna illuminated by two horns

Beam	Y (m)	Elevation (deg.)	Azimuth (deg.)	Gain (dB)	$HP_E°$	$HP_H°$	SLL_E (dB)	SLL_H (dB)
Low	+0.185	-1.85	+90	34.1	11.1	1.0	-40	-35
High	-0.185	+1.85	+90	34.1	11.1	1.0	-40	-35

Table 3. Radiation characteristics simulated by SRADAS

Beam	Y (m)	Elevation (deg.)	Azimuth (deg.)	Gain (dB)	$HP_E°$	$HP_{XOZ}°$	SLL_E (dB)	SLL_{XOZ} (dB)
Low	+0.185	-2	+90	33.4	12.2	0.8	-34	-32
High	-0.185	+2	+90	33.4	12.2	0.8	-34	-32

Table 4. Radiation characteristics simulated by FEKO software

5. Simulation of AN/TPS-43 Antenna

The TPS-43 Radar as shown in Fig. 5 is a transportable three-dimensional air search Radar which operates in frequency range of 2.9 to 3.1 GHz with a 200 mile range. The reflector antenna is a paraboloid of revolution with elliptic cross-section from front view. Reflector aperture is 4.27m high by 6.20m wide with focal axis of 2.6m. The reflector is illuminated by 15 horn antennas which has been moved progressively back from the focal plane [Skolnik, 1990]. Feed horn 2 has been located at the focus of the reflector. The feed array features the use of a stripline matrix to form the 6 height-finding beams. Transmitting radiation pattern of Radar is fan beam for surveillance but receiving radiation pattern is stacked beam to detect height of a target.

Using SRADAS software, AN/TPS-43 antenna has been simulated and the results have been shown in Table 5 and Fig. 6. Comparing the results provided by the proposed method in Fig. 6 with results reported by M. L. Skolnik [Skolnik, 1990] for this antenna confirm the integrity of SRADAS. Some discrepancies can be noticed between them, those are, because of that the locations of feed horns of available Radar are a little different from Radar in reference [Skolnik, 1990]. In addition, diffraction effects have been neglected by use of SRADAS. The calculation speed of SRADAS is so faster than the simulation performed for the same structure by FEKO software and both results are in good agreement.

Fig. 5. AN/TPS-43 Antenna

	Elevation (deg.)	Azimuth (deg.)	Gain (dB)	HP_E (deg.)	HP_H (deg.)
Beam 1	0	0	39.05	1.80	0.9
Beam 2	4.32	90	38.12	1.98	0.9
Beam 3	7.15	90	37.06	1.98	0.9
Beam 4	12.30	90	36.10	3.78	0.9
Beam 5	17.50	90	35.03	4.30	0.9
Beam 6	23.40	90	31.20	5.22	1.26

Table 5. Radiation characteristics of final 6 beams of TPS-43 Radar simulated by SRADAS

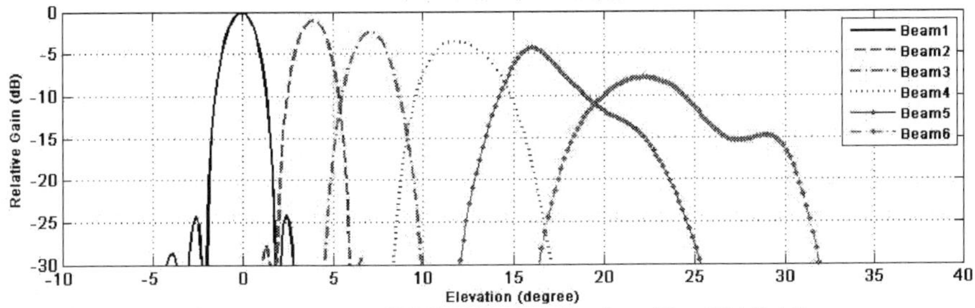

Fig. 6. Elevation radiation patterns of TPS-43 Radar simulated by SRADAS

6. Conclusion

The development and application of a numerical technique for the rapid calculation of the far-field radiation patterns of a reflector antenna excited by defocused feeds have been reported. The reflector has been analyzed by Aperture Field Method (AFM) and Geometrical Optics (GO) to predict the radiation fields in which the radiation integrals computed by FFT. The analytical and numerical results demonstrate that the maximum mesh size of the aperture plane should be less than a wavelength to compute the radiation integrals accurately. Developed Shaped Reflector Antenna Design and Analysis Software (SRADAS) based on MATLAB applied for two practical Radar antennas shows that SRADAS can be used for all shaped reflector antennas with large dimensions compared to operating wavelength. SRADAS has been implemented and used in Information and Communication Technology Institute (ICTI) to analyze and simulate different practical parabolic and shaped reflector antennas. Not only SRADAS has a library of conventional reflectors, but also is it possible to define the geometry of the desired reflector. In addition, SRADAS has the ability to simulate all of shaped reflector antennas fed by defocused feeds rapidly with good accuracy in comparison with available commercial software FEKO. The consuming simulation time performed by SRADAS is less than that of simulated by FEKO software. Consequently, SRADAS can be used as an elementary tool to evaluate the designed reflector antenna in regard to achieving the most important radiation characteristics of the reflector antenna. After that, more accurate simulation can be down by complicated and time-consuming electromagnetic softwares such as FEKO or NEC.

7. Acknowledgment

The authors would like to thank the staff of Information and Communication Technology Institute (ICTI), Isfahan University of Technology (IUT), Islamic Republic of Iran for their co-operator and supporting this work.

8. References

Ahluwalia, D. S.; Lewis, R. M. & Boersma, J. (1968). Uniform asymptotic theory of diffraction by a plane screen, *SIAM Jour. Appl. Math.*, Vol. 16, No. 4, 783-807, 0063-1399

Balanis, C. A. (1989). *Advanced Engineering Electromagnetics,* John Wiley & Sons, 0471621943, New York

Balanis, C. A. (2005). *Antenna Theory, Analysis and Design,* 3rd ed., John Wiley & Sons, 047166782X, New York

Bogush, A. & Elgin, T. (1986). Gaussian field expansion for large aperture antennas, *IEEE Trans. Antennas & Propagation*, Vol. 34, No. 2, 228-243, 0018-926X

Love, A. W. (1978). *Reflector Antennas,* John Wiley & Sons, 0471046051, New York

Bracewell, R. N. (1986). *The Fourier Transform and Its Applications,* McGraw-Hill, 0070070156, New York

Chu, T. S. & Turrin, R. H. (1973). Depolarization properties of offset reflector antennas. *IEEE Trans. Antennas & Propagation*, Vol. 21, No. 3, 339-345, 0018-926X

Drabowitch, S. (1965). Application aux antennas de la theorie du signal. *L'Onde Electrique,* Vol. LXIV, No. 458, 550-560

Galindo-Israel, V. & Mittra, R. (1977). A new series representation for the radiation integral with application to reflector antennas. *IEEE Trans. Antennas & Propagation,* Vol. 25, No. 5, 631-641, 0018-926X

Goldsmith, P. F. (1982). Quasi-optical techniques at millimeter and sub-millimeter wavelengths, In: *Millimeter and Infrared Waves,* Vol. 6, Button, K. J. (Ed.), 277-343, Academic Press, 0121477126, New York

Harrington, R. F. (1993). *Field Computation by Moment Methods,* IEEE Press, 0780310144, New York

Herzberg, T.; Ramer, R. & Hay, S. (2005). Antenna analysis using wavelet representations, *Progress In Electromagnetics Research Symposium,* pp. 48-52, Hangzhou, China, Aug. 2005

Ingerson, P. G. & Wong, W. C. (1974). Focal region characteristics of offset fed reflectors, *IEEE/Antennas and Propagation Society International Symposium,* pp. 121-123, Jun. 1974

Janken, J. A.; English, W. J. & DiFonzo, D. F. (1973). Radiation from multimode reflector antennas, *IEEE/Antennas and Propagation Society International Symposium,* pp. 306-309, Aug. 1973

Kauffman, J. F.; Croswell, W. F. & Jowers, L. J. (1976). Analysis of the radiation patterns of reflector antennas. *IEEE Trans. Antennas & Propagation,* Vol. 24, No. 1, 53-65, 0018-926X

Keller, J.B. (1958). A geometric theory of diffraction, In: *Calculus of Variations and its Applications, Proceedings of Symposia in Applied Math, Vol. 8,* Graves, L. M. (Ed.), McGraw-Hill Book Company Inc., New York

Keller, J. B. (1962). Geometrical theory of diffraction, *Jour. Opt. Soc. Amer.,* Vol. 52, No. 2, 116-130, 1084-7529

Kouyoumjian, R.G. & Pathak, P. H. (1974). A uniform geometrical theory of diffraction for an edge in a perfectly conducting surface. *Proceedings of the IEEE,* Vol. 62, No. 11, 1448-1461, 0018-9219

Lashab, M.; Benabdelaziz, F. & Zebiri, C. (2007). Analysis of electromagnetic scattering from reflector and cylindrical antennas using wavelet-based moment method. *Progress In Electromagnetics Research,* Vol. 76, 357–368, 1070-4698

Lashab, M.; Zebiri, C. & Benabdelaziz, F. (2008). Wavelet-based moment method and physical optics use on large reflector antennas. *Progress In Electromagnetics Research M*, Vol. 2, 189–200, 1937-8726

Lamb, J. W. (1986). Quasi-optical coupling of gaussian beam systems to large cassegrain antennas, *Int. Jour. Infrared Millimeter Waves*, Vol. 7, 1511-1536, 0195-9271

Lesurf, J. C. G. (1990). *Millimetre-wave Optics Devices and Systems*, Adam Hilger, 0852741294, Bristol

Lo, Y. T. & Lee, S. W. (1988). *Antenna Handbook: Theory, Applications, and Design*, Van Nostrand Reinhold Co., 0442258437, New York

Ludwig, A. C. (1968). Computation of radiation patterns involving numerical double integration, IEEE Trans. Antennas & Propagation, Vol. 16, No. 6, 767-769, 0018-926X

Ludwig, A. C. (1988). Comments on the accuracy of the Ludwig integration algorithm, *IEEE Trans. Antennas & Propagation*, Vol. 36, No. 4, 578-579, 0018-926X

Miller, E. K.; Medgyesi-Mitschang, L. & Newman, E. H. (1992). *Computational Electro-Magnetics: Frequency Domain Method of Moments*, IEEE Press, 0879422769, New York

Moore, J. & Pizer, R. (1984). *Moment Methods in Electromagnetics: Techniques and Applications*, Research Studies Press, 0863800130, Letchworth

Philips, B.; Philippakis, M.; Philippou, G. Y. & Brain, D. J. (1996). *Study of Modelling Methods for Large Reflector Antennas*, Radio Communications Agency, London

Popovic, B. D.; Dragovic, M. B. & Djordjevic, A. R. (1982). *Analysis and Synthesis of Wire Antennas*, Research Studies Press, 0471900087, New York

Rahmat-Samii, Y.; Galindo-Israel, V. & Mittra, R. (1980). A plane-polar approach for far-field construction from near-field measurements. *IEEE Trans. Antennas & Propagation*, Vol. 28, No. 2, 216-230, 0018-926X

Rusch, W. V. T. & Potter, P. D. (1970). *Analysis of Reflector Antennas*, Academic Press, 0126034508, New York

Rudge, A. W. (1975). Multiple-beam antennas: offset reflectors with offset feeds. *IEEE Trans. Antennas & Propagation*, Vol. 23, No. 3, 317-322, 0018-926X

Ryan, C. E. Jr. & Peters, L. Jr. (1969). Evaluation of edge diffracted fields including equivalent currents for caustic regions, IEEE Trans. Antennas & Propagation, Vol. 17, No. 3, 292-299, 0018-926X

Scott, C. (1990). *Modern Methods of Reflector Antenna Analysis and Design*, Artech House, 0890064199, Boston, Mass

Silver, S. (1949). *Microwave Antenna Theory and Design*, MIT Radiation Laboratory Series, Vol. 20, McGraw-Hill, New York

Skolnik, M. I. (1990). *Radar Handbook*, 2nd ed., McGraw-Hill, 007057913X, New York

Stutzman, W. L. & Thiele, G. A. (1981). *Antenna Theory and Design*, John Wiley & Sons, 047104458X, New York

Tian, Y.; Zhang, Y. H. & Fan, Y. (2007). The analysis of mutual coupling between paraboloid antennas. *Jour. of Electromagn. Waves and Appl.*, Vol. 21, No. 9, 1191–1203, 0920-5071

Ufimtsev, P. Ya. (2007). *Fundamentals of the Physical Theory of Diffraction*, John Wiley & Sons, 047009771X

Wilton, D. R. & Butler, C. M. (1981). Effective methods for solving integral and integro-differential equations, *Electromagnetics*, Vol. 1, No. 3, 289-308

Wood, P. J. (1986). *Reflector Antenna Analysis and Design*, Peter Peregrinus, 0863410596

Zeidaabadi Nezhad, A. & Firouzeh, Z. H. (2005). *Analysis and Simulation of Shaped Reflector Antennas*, Internal Report at ICTI Library, Isfahan, IRAN

11

Numerical Analysis of the ElectromagneticShielding Effect of Reinforced Concrete Walls

Gaobiao Xiao and Junfa Mao
Shanghai Jiao Tong University
P.R. China

1. Introduction

The shielding effect of buildings to electromagnetic waves is investigated by many researchers (Dalke etal., 2000), where the analysis of reinforced concrete walls has attracted special interests. In most cases, a reinforced concrete wall is treated as an infinitely extended periodic structure and periodic boundary conditions are used to get its transmission and reflection characteristics, from which the shielding effect of the wall can be evaluated. However, when analyzing indoor electro-magnetic environment, the surrounding reinforced concrete walls are not infinitely extended. Therefore, the infinitely extending plane structure model cannot provide an accurate enough prediction to the indoor electromagnetic environment, especially the fields at corners or ends of walls. Some numerical methods, such as method of moment (MoM) seem to be applicable for analyzing these kinds of problems. However, the computational cost may be too high since these kinds of electromagnetic systems usually not only have electrical sizes of tens or hundreds wavelengths, but also have very fine internal structures.

In this chapter, we present a method to circumvent the heavy burden on computing sources while adequate accuracy can still be achieved. A reinforced concrete wall with fine structures is first divided into small blocks. Each block is treated like a multi-layered scatterer and is analyzed independently by using cascaded network techniques. Then, the electromagnetic characteristics of that block is expressed by a generalized transition matrix (generalized T-matrix, GTM) that is defined on a specified reference surface containing the block, which is different from the T-matrix introduced by Waterman (Waterman, 2007). The mutual coupling effects among blocks are evaluated by using the generalized transition matrices of all blocks directly. The scattered fields of the whole system are calculated using a generalized surface integral equation method (GSIE). Furthermore, we will show that characteristic basis functions (CBF) and synthetic basis functions (SBF) can be used to accelerate the evaluation process effectively.

The following contents are included in this chapter: (1) description of the basic concept of generalized transition matrix and the generalized surface integral equation method, (2) the cascading network technique in analyzing multi-layered media, (3) the implementation of

GSIE in conjunction with CBF\SBF, and (4) as an application example, the evaluation of the shielding effect of a 2D reinforced wall with the proposed method.

2. Generalized Surface Integral Equation

The electromagnetic problem under consideration is shown in Fig. 1. A 2-D cavity is surrounded by a $10\lambda \times 10\lambda$ concrete wall. The wall is 1.0λ thick and is uniformly reinforced by 36 circular steel cylinders with radius of 0.1λ and conductivity of $\sigma = 1.1\times10^6$ Sm^{-1}. The relative permittivity of the concrete is $\varepsilon_r = 10 - j2.8$. The excitation is assumed to be a TM type plane wave with incident angle of ϕ.

In order to calculate the electromagnetic fields in the cavity, the wall is divided into 36 cells. Each cell consists of a steel cylinder and a concentric concrete cylinder with square cross section of $1.0\lambda \times 1.0\lambda$, as is sketched in Fig. 1. A cell is a multilayered scatterer and its scattering characteristics can be fully described by a generalized transition matrix, which relates the scattered tangential fields to the incident tangential fields directly, as is shown in Fig. 2.

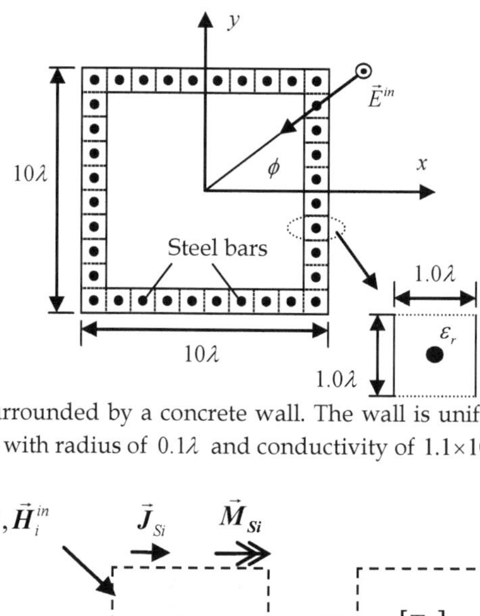

Fig. 1. A 2-D cavity surrounded by a concrete wall. The wall is uniformly reinforced by 36 circular steel cylinders with radius of 0.1λ and conductivity of $1.1\times10^6 \mathrm{Sm}^{-1}$.

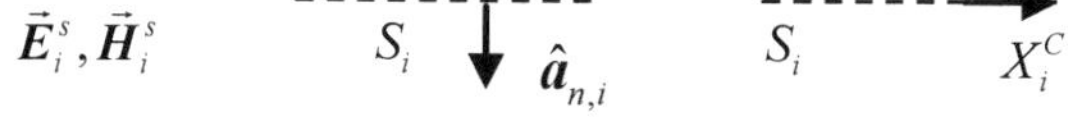

Fig. 2. A block is modelled with its associated generalized transition matrix $[T_{bi}]$. $\vec{J}_{Si}$ and $\vec{M}_{Si}$ are equivalent surface electric current and magnetic current, respectively.

A scatterer with two layers of homogeneous media is illustrated in Fig. 3(a), where V_1 denotes the medium between interface S_1 and S_2, with permittivity ε_1 and permeability μ_1. For a block of the concrete wall, S_1 is the air-concrete interface and S_2 the concrete-steel interface. The normal unit vectors of all interfaces are chosen to point outwardly. The outermost medium, denoted by V_0 is assumed to be free space.

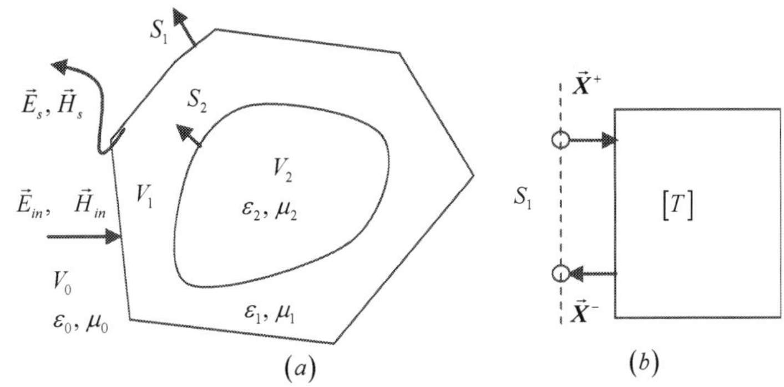

Fig. 3. (a) A scatterer with two layers. (b) One-port device model.

We examine the field scattering problem on interface S_1 at first. The incident fields may come from both sides of the interface, and being scattered to both sides of it, as is shown in Fig. 4(a). The interface may be modelled as a generalized two-port device, with its two reference surfaces being chosen as S_1^+ and S_1^-. The notation S_1^+ means approaching S_1 from outside while S_1^- means approaching S_1 from interior area.

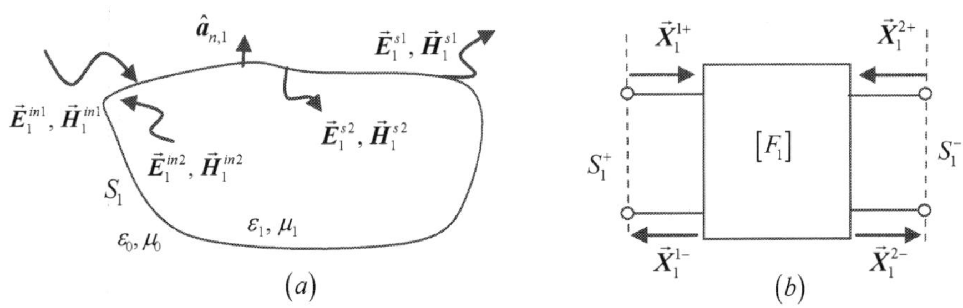

Fig. 4. (a) Scattering on interface S_1, with incident fields from both sides of it. (b) Generalized two-port device model, with its two reference surfaces approaching S_1 from outer or interior medium.

Define the rotated tangential components of incident fields and rotated tangential components of scattered fields of a block as follows (Xiao et al., 2008):

$$\left[\vec{X}_1^{1+}\right]=\begin{bmatrix}\vec{E}_1^{1+}\\ \vec{H}_1^{1+}\end{bmatrix}=\begin{bmatrix}-\hat{a}_{n,1}\times\vec{E}_1^{in1}\\ \hat{a}_{n,1}\times\vec{H}_1^{in1}\end{bmatrix}\Bigg|_{S_1^+}, \tag{1}$$

$$\left[\vec{X}_1^{2+}\right]=\begin{bmatrix}\vec{E}_1^{2+}\\ \vec{H}_1^{2+}\end{bmatrix}=\begin{bmatrix}\hat{a}_{n,1}\times\vec{E}_1^{in2}\\ -\hat{a}_{n,1}\times\vec{H}_1^{in2}\end{bmatrix}\Bigg|_{S_1^-}, \tag{2}$$

$$\left[\vec{X}_1^{1-}\right]=\begin{bmatrix}\vec{E}_1^{1-}\\ \vec{H}_1^{1-}\end{bmatrix}=\begin{bmatrix}-\hat{a}_{n,1}\times\vec{E}_1^{s1}\\ \hat{a}_{n,1}\times\vec{H}_1^{s1}\end{bmatrix}\Bigg|_{S_1^+}, \tag{3}$$

$$\left[\vec{X}_1^{2-}\right]=\begin{bmatrix}\vec{E}_1^{2-}\\ \vec{H}_1^{2-}\end{bmatrix}=\begin{bmatrix}\hat{a}_{n,1}\times\vec{E}_1^{s2}\\ -\hat{a}_{n,1}\times\vec{H}_1^{s2}\end{bmatrix}\Bigg|_{S_1^-}. \tag{4}$$

Note that the normal unit vector for S_1^- is chosen to be $-\hat{a}_{n,1}$. Based on Huygens' equivalence source principle, the scattered fields from interface S_1 can be expressed in terms of their tangential fields on the interface using the dyadic Green's function,

$$\vec{E}_1^{s1}(\vec{r})=\int_{S_1}\left[-j\omega\mu_0\overline{\overline{G}}_0\bullet\vec{H}_1^{1-}-\nabla\times\overline{\overline{G}}_0\bullet\vec{E}_1^{1-}\right]dS', \tag{5}$$

$$\vec{H}_1^{s1}(\vec{r})=\int_{S_1}\left[-j\omega\varepsilon_0\overline{\overline{G}}_0\bullet\vec{E}_1^{1-}+\nabla\times\overline{\overline{G}}_0\bullet\vec{H}_1^{1-}\right]dS', \tag{6}$$

$$\vec{E}_1^{s2}(\vec{r})=\int_{S_1}\left[-j\omega\mu_1\overline{\overline{G}}_1\bullet\vec{H}_1^{2-}-\nabla\times\overline{\overline{G}}_1\bullet\vec{E}_1^{2-}\right]dS', \tag{7}$$

$$\vec{H}_1^{s2}(\vec{r})=\int_{S_1}\left[-j\omega\varepsilon_1\overline{\overline{G}}_1\bullet\vec{E}_1^{2-}+\nabla\times\overline{\overline{G}}_1\bullet\vec{H}_1^{2-}\right]dS', \tag{8}$$

where $\overline{\overline{G}}_l=\left(\overline{\overline{I}}+\frac{\nabla\nabla}{k_l^2}\right)g_l$ is the dyadic Green's function, and g_l is the scalar Green's function in medium V_l, $k_l=\omega\sqrt{\mu_l\varepsilon_l}$ and $l=0,1$. The fields have to satisfy the following boundary conditions on interface S_1,

$$\vec{E}_1^{1+}+\vec{E}_1^{1-}=-\vec{E}_1^{2+}-\vec{E}_1^{2-}, \qquad \vec{H}_1^{1+}+\vec{H}_1^{1-}=-\vec{H}_1^{2+}-\vec{H}_1^{2-}. \tag{9}$$

Take the scattered electric field expressed by (5) as an example. If we move the observation point to surface S_1 from the outside area, and take into account the singularities of the dyadic Green's functions when source and observation points overlap, the rotated tangential scattered electric field on surface S_1^+ can be written as

$$\vec{E}_1^{1-} = -\hat{a}_{n,1} \times \vec{E}_1^{s1}\Big|_{S_1^+} = -\hat{a}_{n,1} \times \int_{S_1} \left(-j\omega\mu_0 \overline{\overline{G}}_0\right) \bullet \vec{H}_1^{1-} dS' - \hat{a}_{n,1} \times \int_{S_1} \left(-\nabla g_0 \times \vec{E}_1^{1-}\right) dS' + \frac{\Omega}{4\pi} \vec{E}_1^{1-} \ . \quad (10)$$

For smooth surface, the solid angle is $\Omega = 2\pi$, equation (10) becomes

$$\frac{1}{2}\vec{E}_1^{1-} = \hat{a}_{n,1} \times \int_{S_1} \nabla g_0 \times \vec{E}_1^{1-} dS' + \hat{a}_{n,1} \times \int_{S_1} \left(j\omega\mu_0 \overline{\overline{G}}_0\right) \bullet \vec{H}_1^{1-} dS' . \quad (11)$$

If we examine the scattered electric field $\vec{E}_1^{s2}(\vec{r})$ and move the observation point to surface S_1 from its interior area, we can show that the rotated tangential scattered electric field on S_1^- satisfies

$$-\frac{1}{2}\vec{E}_1^{2-} = \hat{a}_{n,1} \times \int_{S_1} \nabla g_1 \times \vec{E}_1^{2-} dS' + \hat{a}_{n,1} \times \int_{S_1} \left(j\omega\mu_1 \overline{\overline{G}}_1\right) \bullet \vec{H}_1^{2-} dS' . \quad (12)$$

Integral equations for rotated tangential magnetic fields can be expressed in similar way,

$$\frac{1}{2}\vec{H}_1^{1-} = -\hat{a}_{n,1} \times \int_{S_1} \left(j\omega\varepsilon_0 \overline{\overline{G}}_0\right) \bullet \vec{E}_1^{1-} dS' + \hat{a}_{n,1} \times \int_{S_1} \nabla g_0 \times \vec{H}_1^{1-} dS' \ , \quad (13)$$

$$-\frac{1}{2}\vec{H}_1^{2-} = -\hat{a}_{n,1} \times \int_{S_1} \left(j\omega\varepsilon_1 \overline{\overline{G}}_1\right) \bullet \vec{E}_1^{2-} dS' + \hat{a}_{n,1} \times \int_{S_1} \nabla g_1 \times \vec{H}_1^{2-} dS' \ . \quad (14)$$

The same process can also be applied to the incident fields. For example, according to the Huygens' principle and the distinction theorem, we have

$$-\frac{1}{2}\vec{E}_1^{1+} = \hat{a}_{n,1} \times \int_{S_1} \nabla g_0 \times \vec{E}_1^{1+} dS' + \hat{a}_{n,1} \times \int_{S_1} \left(j\omega\mu_0 \overline{\overline{G}}_0\right) \bullet \vec{H}_1^{1+} dS' \ , \quad (15)$$

$$-\frac{1}{2}\vec{H}_1^{1+} = -\hat{a}_{n,1} \times \int_{S_1} \left(j\omega\varepsilon_0 \overline{\overline{G}}_0\right) \bullet \vec{E}_1^{1+} dS' + \hat{a}_{n,1} \times \int_{S_1} \nabla g_0 \times \vec{H}_1^{1+} dS' \ . \quad (16)$$

With these integral equations, we are able to establish a set of coupled surface integral equations on interface S_1. Firstly, we consider the case that an incident field illuminates on S_1 from the outer side only, *i.e.*, $\vec{X}_1^{1+} \neq 0,\ \vec{X}_1^{2+} = 0$. Combining (11)-(16) and eliminating $\vec{E}_1^{1-}, \vec{H}_1^{1-}$ gives

$$\hat{a}_{n,1} \times \int_{S_1} \nabla(g_0 + g_1) \times \vec{E}_1^{2-} dS' + \hat{a}_{n,1} \times \int_{S_1} \left(j\omega\mu_0 \overline{\overline{G}}_0 + j\omega\mu_1 \overline{\overline{G}}_1\right) \bullet \vec{H}_1^{2-} dS' = \vec{E}_1^{1+} \ , \quad (17)$$

$$-\hat{a}_{n,1} \times \int_{S_1} \left(j\omega\varepsilon_0 \overline{\overline{G}}_0 + j\omega\varepsilon_1 \overline{\overline{G}}_1\right) \bullet \vec{E}_1^{2-} dS' + \hat{a}_{n,1} \times \int_{S_1} \nabla(g_0 + g_1) \times \vec{H}_1^{2-} dS' = \vec{H}_1^{1+} \ . \quad (18)$$

It is not difficult to check that (17) and (18) are equivalent to PMCHW formulation (Rao etal., 1982) by denoting $\vec{J}_s = \vec{H}^{2-}$, $\vec{J}_{ms} = \vec{E}^{2-}$. Symbolically, the rotated tangential scattered fields on interface S_1^- can be expressed in terms of the incident fields as

$$\left[\vec{X}_1^{2-}\right] = \left[\tau_1(\cdot)\right]\left[\vec{X}_1^{1+}\right]. \tag{19}$$

Therefore, the rotated tangential scattered fields on interface S_1^+ is obtained by

$$\left[\vec{X}_1^{1-}\right] = -\left\{\boldsymbol{I} + \left[\tau_1(\cdot)\right]\right\}\left[\vec{X}_1^{1+}\right]. \tag{20}$$

$\boldsymbol{I}$ is the identity tensor. It is natural to interpret $\left[\tau_1(\cdot)\right]$ as a transmission operator and $\left[\mathcal{T}(\cdot)\right] = -\left\{\boldsymbol{I} + \left[\tau_1(\cdot)\right]\right\}$ a reflection operator.

Secondly, we consider the case that incident field illuminates on S_1 from interior side only, i.e., $\vec{X}^{1+} = 0,\ \vec{X}^{2+} \neq 0$. It can be proven that the rotated tangential scattered fields are subject to the following coupled surface integral equations,

$$\hat{\boldsymbol{a}}_{n,1} \times \int_{S_1} \nabla(g_0 + g_1) \times \vec{E}_1^{1-} dS' + \hat{\boldsymbol{a}}_{n,1} \times \int_{S_1} \left(-j\omega\mu_0 \overline{\overline{\boldsymbol{G}}}_0 - j\omega\mu_1 \overline{\overline{\boldsymbol{G}}}_1\right) \bullet \vec{H}_1^{1-} dS' = -\vec{E}_1^{2+}, \tag{21}$$

$$\hat{\boldsymbol{a}}_{n,1} \times \int_{S_1} \left(j\omega\varepsilon_0 \overline{\overline{\boldsymbol{G}}}_0 + j\omega\varepsilon_1 \overline{\overline{\boldsymbol{G}}}_1\right) \bullet \vec{E}_1^{1-} dS' + \hat{\boldsymbol{a}}_{n,1} \times \int_{S_1} \nabla(g_0 + g_1) \times \vec{H}_1^{1-} dS' = -\vec{H}_1^{2+}. \tag{22}$$

Comparing (21) (22) with (17) (18), we can write that

$$\left[\vec{X}_1^{1-}\right] = -\left[\tau_1(\cdot)\right]\left[\vec{X}_1^{2+}\right], \tag{23}$$

$$\left[\vec{X}_1^{2-}\right] = \left\{-\boldsymbol{I} + \left[\tau_1(\cdot)\right]\right\}\left[\vec{X}_1^{2+}\right]. \tag{24}$$

In general situations, incident fields come from both sides of an interface, the total scattered fields are obtained by field superposition, which can be written in matrix form as

$$\begin{bmatrix} \vec{X}_1^{1-} \\ \vec{X}_1^{2-} \end{bmatrix} = \begin{bmatrix} -\left\{\boldsymbol{I} + \left[\tau_1(\cdot)\right]\right\} & -\left[\tau_1(\cdot)\right] \\ \left[\tau_1(\cdot)\right] & -\boldsymbol{I} + \left[\tau_1(\cdot)\right] \end{bmatrix} \begin{bmatrix} \vec{X}_1^{1+} \\ \vec{X}_1^{2+} \end{bmatrix}. \tag{25}$$

It can be re-arranged in field transfer equation,

$$\begin{bmatrix} \vec{X}_1^{1+} \\ \vec{X}_1^{1-} \end{bmatrix} = \left[\mathcal{F}_1(\cdot)\right] \begin{bmatrix} \vec{X}_1^{2+} \\ \vec{X}_1^{2-} \end{bmatrix}, \tag{26}$$

where $\left[\mathcal{F}_l(\cdot)\right]$ may be interpreted as a transfer operator.

The scattered fields $\vec{E}_1^{s2}, \vec{H}_1^{s2}$ will transfer to interface S_2 and serve as the incident fields on S_2, i.e., $\vec{E}_2^{1+}, \vec{H}_2^{1+}$, as is illustrated in Fig. 5. Similarly, the scattered fields $\vec{E}_2^{1-}, \vec{H}_2^{1-}$ from interface S_2 will transmit through the medium layer and illuminate on S_1 as the incident fields of $\vec{E}_1^{2+}, \vec{H}_1^{2+}$. From Huygens' principle, the scattered fields associated with $\vec{E}_1^{2-}, \vec{H}_1^{2-}$ are

$$\vec{E}_1^{s2}(\vec{r}) = -\int_{S_1}\left(\nabla g_1 \times \vec{E}_1^{2-}\right)dS' + \int_{S_1}\left(-j\omega\mu_1\overline{\overline{G}}_1\right)\bullet\vec{H}_1^{2-}dS', \tag{27}$$

$$\vec{H}_1^{s2}(\vec{r}) = -\int_{S_1}\left(j\omega\varepsilon_1\overline{\overline{G}}_1\right)\bullet\vec{E}_1^{2-}dS' + \int_{S_1}\nabla g_1 \times \vec{H}_1^{2-}dS'. \tag{28}$$

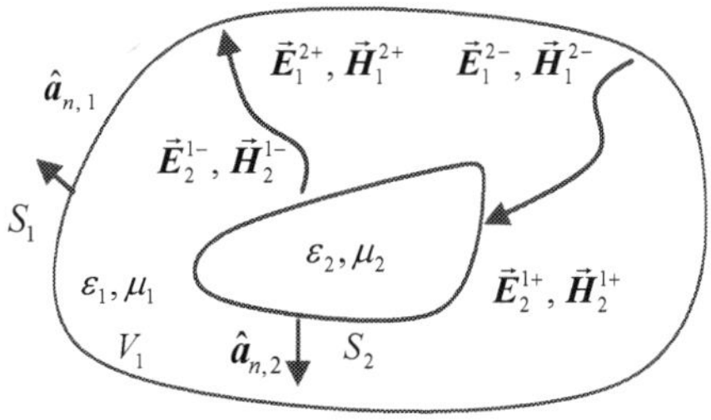

Fig. 5. Field transmission between interface S_1 and S_2.

Using the definition of rotated tangential components, we get

$$\vec{E}_2^{1+} = -\hat{a}_{n,2} \times \vec{E}_1^{s2}\left(\vec{r}_2^+\right) = \hat{a}_{n,2} \times \int_{S_1}\left(\nabla g_1 \times \vec{E}_1^{2-}\right)dS' + \hat{a}_{n,2} \times \int_{S_1}\left(j\omega\mu_1\overline{\overline{G}}_1\right)\bullet\vec{H}_1^{2-}dS', \tag{29}$$

$$\vec{H}_2^{1+} = \hat{a}_{n,2} \times \vec{H}_1^{s2}\left(\vec{r}_2^+\right) = -\hat{a}_{n,2} \times \int_{S_1}\left(j\omega\varepsilon_1\overline{\overline{G}}_1\right)\bullet\vec{E}_1^{2-}dS' + \hat{a}_{n,2} \times \int_{S_1}\nabla g_1 \times \vec{H}_1^{2-}dS'. \tag{30}$$

Equations (29) and (30) describe the field transmission from $\vec{E}_1^{2-}, \vec{H}_1^{2-}$ to $\vec{E}_2^{1+}, \vec{H}_2^{1+}$. Similarly, the relationship between the incident fields on the inner side of S_1 and the scattered fields from S_2 can be derived as

$$\vec{E}_1^{2+} = -\hat{a}_{n,1} \times \int_{S_2}\nabla g_1 \times \vec{E}_2^{1-}dS' - \hat{a}_{n,1} \times \int_{S_2}\left(j\omega\mu_1\overline{\overline{G}}_1\right)\bullet\vec{H}_2^{1-}dS', \tag{31}$$

$$\vec{H}_1^{2+} = \hat{a}_{n,1} \times \int_{S_2}\left(j\omega\varepsilon_1\overline{\overline{G}}_1\right)\bullet\vec{E}_2^{1-}dS' - \hat{a}_{n,1} \times \int_{S_2}\nabla g_1 \times \vec{H}_2^{1-}dS'. \tag{32}$$

Symbolically we denote

$$\left[\vec{X}_1^{2+}\right] = \left[\mathcal{D}_{12}(\cdot)\right]\left[\vec{X}_2^{1-}\right], \tag{33}$$

$$\left[\vec{X}_2^{1+}\right]=\left[\mathcal{D}_{21}(\cdot)\right]\left[\vec{X}_1^{2-}\right]. \tag{34}$$

(33) and (34) depict the relationship between the fields of two consecutive interfaces, with $\left[\mathcal{D}_{ij}(\cdot)\right]$ being the transmission operator. The first subscript of $\left[\mathcal{D}(\cdot)\right]$ indicates the destination interface, while the second subscript indicates the source interface. This convention is used throughout this chapter. Obviously, the homogeneous medium layer can be considered as a section of generalized transmission line, with $\left[\mathcal{D}_{21}(\cdot)\right]$ and $\left[\mathcal{D}_{12}(\cdot)\right]$ being its forward and backward transmission operators, respectively.
Since the innermost layer V_2 is homogeneous, the scattering problem on S_2 can be considered as a special case of the above scattering problem with $\vec{X}_2^{2+}=0$. The scattered fields are only related to the incident fields from the outer side of the interface by equation (20), which is rewritten as follows

$$\left[\vec{X}_2^{1-}\right]=-\left\{\boldsymbol{I}+\left[\tau_2(\cdot)\right]\right\}\left[\vec{X}_2^{1+}\right]=\left[\mathcal{T}_2(\cdot)\right]\left[\vec{X}_2^{1+}\right]. \tag{35}$$

Therefore, the two-layered scatterer is modelled as a cascaded network of two generalized devices connected by a transmission line, as is shown in Fig. 6.

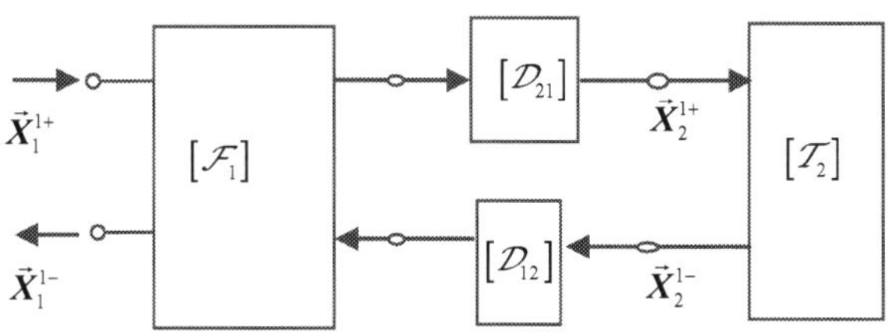

Fig. 6 A cascaded network model for a two-layered scatterer.

Assume that all field components are expanded and tested with a kind of vector basis functions on the reference surface S. The generalized transition matrix $\left[T_{bi}\right]$ of block i is defined according to

$$\left[X_i^-\right]=\left[T_{bi}\right]\left[X_i^+\right]. \tag{36}$$

$\left[T_b\right]$ can be used to describe the scattering characteristics of a scatterer with arbitrary shapes and materials. At a given frequency, $\left[T_b\right]$ is calculated independently only once for each block and can be re-used if necessary. For homogeneous media, the entries of $\left[T_b\right]$ can be found from (17), (18) by applying Galerkin's scheme directly. For scatterers with complex

structures, $[T_b]$ may be obtained using properly chosen method (Creticos & Schaubert, 2006; Lean, 2004; Polewski etal., 2004; Taskinen & Ylä-Oijala, 2006; Umashankar etal.,1986;). In this chapter, the generalized transition matrix of a two-layered block is obtained using the above described cascaded network technique.
Assume that all rotated tangential fields on the interface S_l are expanded with a set of vector basis functions $\vec{f}_{l,j}(\boldsymbol{r})$ as follows:

$$\vec{E}_l^{u\pm}(\boldsymbol{r}) = \sum_{i=1}^{N_l} e_{l,j}^{u\pm} \vec{f}_{l,j}(\boldsymbol{r}), \qquad \vec{H}_l^{u\pm}(\boldsymbol{r}) = \sum_{j=1}^{N_l} h_{l,j}^{u\pm} \vec{f}_{l,j}(\boldsymbol{r}), \tag{37}$$

where N_l is number of vector functions used on interface S_l and $u = 1,2$. Functions $\vec{f}_{l,i}(\boldsymbol{r})$ are also used as test functions. For the sake of convenience, we denote

$$\tilde{e}_{l,i}^{u\pm} = \int_{t_i} \vec{E}_l^{u\pm}(\boldsymbol{r}) \cdot \vec{f}_{l,i}(\boldsymbol{r}) dS, \qquad \tilde{h}_{l,i}^{u\pm} = \int_{t_i} \vec{H}_l^{u\pm}(\boldsymbol{r}) \cdot \vec{f}_{l,i}(\boldsymbol{r}) \, dS, \tag{38}$$

where t_i is the support of vector basis function $\vec{f}_{l,i}(\boldsymbol{r})$. It can be derived that

$$\left[\tilde{e}_l^{u\pm}\right] = [\boldsymbol{P}_l]\left[e_l^{u\pm}\right], \quad \left[\tilde{h}_l^{u\pm}\right] = [\boldsymbol{P}_l]\left[h_l^{u\pm}\right]. \tag{39}$$

Here $[\boldsymbol{P}_l]$ is a Gramm matrix formed by the inner products of the basis functions and test functions on the surface S_l. The entries of the Gramm matrix are defined as

$$P_l(i,j) = \int_{t_i} \vec{f}_{l,j}(\boldsymbol{r}) \cdot \vec{f}_{l,i}(\boldsymbol{r}) dS. \tag{40}$$

Applying Galerkin's method to (17) (18) yields

$$\begin{bmatrix} [\boldsymbol{A}_l^{ee}] & [\boldsymbol{A}_l^{eh}] \\ [\boldsymbol{A}_l^{he}] & [\boldsymbol{A}_l^{hh}] \end{bmatrix} \begin{bmatrix} [e_l^{2-}] \\ [h_l^{2-}] \end{bmatrix} = \begin{bmatrix} [\tilde{e}_l^{1+}] \\ [\tilde{h}_l^{1+}] \end{bmatrix} = \begin{bmatrix} [\boldsymbol{P}_l] & 0 \\ 0 & [\boldsymbol{P}_l] \end{bmatrix} \begin{bmatrix} [e_l^{1+}] \\ [h_l^{1+}] \end{bmatrix}, \tag{41}$$

where all elements of the coefficient matrices have their conventional form of double surface integrations on the corresponding meshes. For example, on S_1 we have

$$\boldsymbol{A}_1^{ee}(i,j) = \int_{t_i} \vec{f}_{1,i}(\boldsymbol{r}) \cdot \hat{\boldsymbol{a}}_{n,1} \times \int_{t_j} \nabla(g_0 + g_1) \times \vec{f}_{1,j}(\boldsymbol{r}) dS' dS. \tag{42}$$

A generalized transmission matrix $[\boldsymbol{\tau}_l]$ can be defined as

$$[\boldsymbol{\tau}_l]=\begin{bmatrix}[\boldsymbol{A}_l^{ee}] & [\boldsymbol{A}_l^{eh}] \\ [\boldsymbol{A}_l^{he}] & [\boldsymbol{A}_l^{hh}]\end{bmatrix}^{-1}\begin{bmatrix}[\boldsymbol{P}_l] & 0 \\ 0 & [\boldsymbol{P}_l]\end{bmatrix}\triangleq[\boldsymbol{A}_l]^{-1}[\boldsymbol{P}_{cl}]. \tag{43}$$

Although $[\boldsymbol{\tau}_l]$ is frequency-dependent, it does not depend on incident fields. The generalized transfer matrix of interface S_l can then be defined accordingly,

$$[\boldsymbol{F}_l]=\begin{bmatrix}[\boldsymbol{F}_l^{11}] & [\boldsymbol{F}_l^{12}] \\ [\boldsymbol{F}_l^{21}] & [\boldsymbol{F}_l^{22}]\end{bmatrix}=\begin{bmatrix}[\boldsymbol{\tau}_l]^{-1}-\boldsymbol{I} & [\boldsymbol{\tau}_l]^{-1} \\ -[\boldsymbol{\tau}_l]^{-1} & -[\boldsymbol{\tau}_l]^{-1}-\boldsymbol{I}\end{bmatrix}=\begin{bmatrix}[\boldsymbol{P}_{cl}]^{-1}[\boldsymbol{A}_l]-\boldsymbol{I} & [\boldsymbol{P}_{cl}]^{-1}[\boldsymbol{A}_l] \\ -[\boldsymbol{P}_{cl}]^{-1}[\boldsymbol{A}_l] & -[\boldsymbol{P}_{cl}]^{-1}[\boldsymbol{A}_l]-\boldsymbol{I}\end{bmatrix}. \tag{44}$$

The same mesh structure and vector basis functions are applied when evaluate field transmission between consecutive interfaces. We still consider the two-layered scatterer shown in Fig. 5. Applying Galerkin's method to integral equations (29)-(32) yields

$$[\boldsymbol{P}_{c1}][\vec{\boldsymbol{X}}_1^{2+}]=[\boldsymbol{D}_{12}][\vec{\boldsymbol{X}}_2^{1-}], \tag{45}$$

$$[\boldsymbol{P}_{c2}][\vec{\boldsymbol{X}}_2^{1+}]=[\boldsymbol{D}_{21}][\vec{\boldsymbol{X}}_1^{2-}], \tag{46}$$

where $[\boldsymbol{D}_{21}]$ is the field transmission matrix from interface S_2 to S_1.

Tracing the field transmission route illustrated in Fig. 5 leads to,

$$[\boldsymbol{P}_{c1}][\vec{\boldsymbol{X}}_1^{2+}]=[\boldsymbol{P}_{c1}][\boldsymbol{D}_{12}][\vec{\boldsymbol{X}}_2^{1-}]=[\boldsymbol{P}_{c1}][\boldsymbol{D}_{12}][\Gamma_2][\vec{\boldsymbol{X}}_2^{1-}]=[\boldsymbol{P}_{c1}][\boldsymbol{D}_{12}][T_2][\boldsymbol{P}_{c2}]^{-1}[\boldsymbol{D}_{21}][\vec{\boldsymbol{X}}_1^{2-}]. \tag{47}$$

Denote

$$[\Gamma_1^-]=[\boldsymbol{D}_{12}][T_2][\boldsymbol{P}_{c2}]^{-1}[\boldsymbol{D}_{21}][\vec{\boldsymbol{X}}_1^{2-}]. \tag{48}$$

The generalized T-matrix for a bock is found to be

$$[T_b]=\left([\boldsymbol{F}_1^{21}][\Gamma_1^-]+[\boldsymbol{F}_1^{22}]\right)\left([\boldsymbol{F}_1^{11}][\Gamma_1^-]+[\boldsymbol{F}_1^{12}]\right)^{-1}. \tag{49}$$

Generally speaking, a multilayered scatterer can also be handled in this recursive way. A multilayered scatterer interacts with the surrounding environment all through the outermost interface. It is natural to use the generalized transition matrix $[\boldsymbol{T}]$ defined on the outermost interface to describe the whole scattering characteristics of the scatterer. For arbitrary incident fields, the rotated tangential scattered fields can be found from the rotated tangential incident fields directly by multiplying the generalized T-matrix.

This technique is used to analyse the block of the reinforced concrete wall. The scattering problem at the interface between concrete and air is solved by using surface integral equations (17)-(22), with its transfer matrix obtained by (44). The scattering problem at the surface of the steel bar is treated with the same approach. Coordinate transform may be applied to overcome the numerical problem caused by high conductivity of the steel (Li & Chew, 2007). The generalized transition matrix $[T_2]$ of the steel bar is determined according to equation (20). The field transmission flow chart is illustrated in Fig. 7.

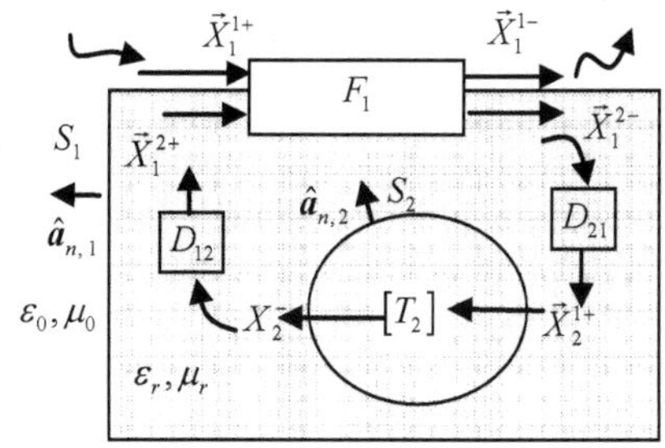

Fig. 7. Field transmission between interfaces S_1 and the surface of the steel core S_2 .

After the scattering characteristics of all blocks are represented by their generalized T-matrices, the scattered fields of the whole wall can be analysed by taking into account of the mutual couplings between all blocks, as is shown in Fig. 8, where M scatterers are considered. The generalized transition matrix of scatterer m is denoted by $[T_m]$. It is defined on the reference surface S_m , and has been calculated independently in advance.

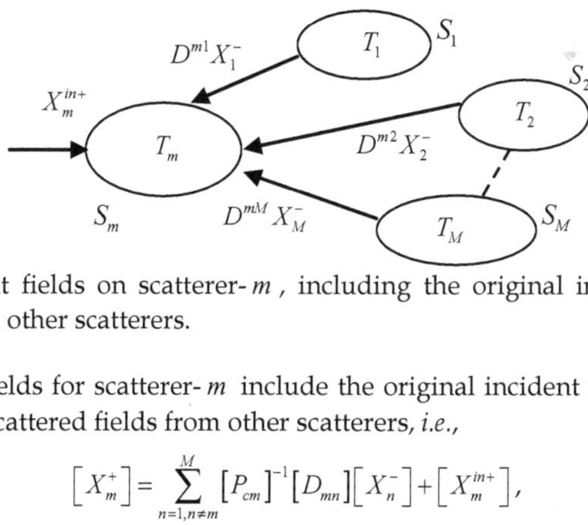

Fig.8 Total incident fields on scatterer- m , including the original incident fields and all scattered fields from other scatterers.

The total incident fields for scatterer- m include the original incident fields on its reference surface S_m and all scattered fields from other scatterers, *i.e.*,

$$\left[X_m^+\right] = \sum_{n=1,n\neq m}^{M} \left[P_{cm}\right]^{-1} \left[D_{mn}\right]\left[X_n^-\right] + \left[X_m^{in+}\right], \tag{50}$$

where $[D_{mn}]$ are field transmission matrix from scatterer- n to scatterer- m . They are defined like (29) (30), except that the integral areas are replaced by the corresponding reference surfaces of S_m and S_n . Denote

$$\left[D_{mn}\right] = \begin{bmatrix} \left[D_{mn}^{ee}\right] & \left[D_{mn}^{eh}\right] \\ \left[D_{mn}^{he}\right] & \left[D_{mn}^{hh}\right] \end{bmatrix}. \tag{51}$$

All entries are evaluated with double surface integrals. For example, we have

$$D_{mn}^{ee}(p,q) = \int_{t_p}\left[\hat{a}_{n,m} \times \int_{t_q} \nabla \times \overline{\overline{G}}_0 \cdot \vec{f}_q(\mathbf{r}')dS_n\right] \cdot \vec{f}_p(\mathbf{r})dS_m, \tag{52}$$

where $\vec{f}_q(\mathbf{r}')$ and $\vec{f}_p(\mathbf{r})$ are basis functions on the reference surface of scatterer-n (source) and scatterer-m (destination), respectively, and $\hat{a}_{n,m}$ is the normal unit vector of S_m. $[P_{cm}]$ is the Gramm matrix on the surface of scatterer-m. The total scattered fields of scatterer-m on the specified reference surface is then obtained by $[X_m^-] = [T_m][X_m^+]$. By virtue of equation (36) and the surface integral equations (17) (18), we can derive the discretized form of the surface integral equation defined on surface S_m as follows

$$[X_m^-] - \sum_{n=1,n\neq m}^{N_b} [T_n][D_{mn}][X_n^-] = [T_m][X_m^{inc}]. \tag{53}$$

$[\mathbf{X}_m^{inc}]$ is the original incident field on S_m. It can be proved that the equivalent sources satisfy a similar equation,

$$[I_{Sm}] - \sum_{n=1,m\neq n}^{N_b} [R_n][D_{mn}][I_{Sn}] = [R_n][X_n^{inc}]. \tag{54}$$

Here N_b is the number of blocks, I_{Sm} are the expanding coefficient of equivalence surface sources, where $[R_n] = [I] + [T_n]$, $[I]$ is the identity tensor. Equation (53) or (54) is the matrix form of the generalized surface integral equation formulation of a complex multi-scale system. For perfect conducting surfaces, equation (54) is more convenient to use because $\vec{\mathbf{M}}_{Sm} = 0$ in these cases. A complex multi-scale electromagnetic system may be divided into sub-blocks with proper sizes and reference surfaces. Small objects are usually grouped together and contained in one reference surface. Large bulk of inhomogeneous/homogeneous media or conducting media can be divided into smaller blocks. The generalized transition matrix of each block is then found using relevant method according to its structure; each block is assumed to be placed alone in the infinite homogeneous background media with parameters of ε_0 and μ_0. The total electromagnetic characteristics of the system can be evaluated using the generalized surface integral equations, i.e., equation (53) or (54). With this method, the unknowns involved in the final linear system are significantly reduced.

3. Field Transmission Matrices between Adjacent Blocks

Assume that block-m and block-n share a common interface S_{mn}, as is shown in Fig. 9, where $\hat{\mathbf{a}}_{n,m}$ and $\hat{\mathbf{a}}_{n,n}$ are normal unit vectors of surface S_m and S_n, respectively. The scattered fields from block-m are expressed in terms of the rotated tangential fields on surface S_m using equations (5)-(6). Taking into account the singularity of the dyadic Green's function when the source point and observatory point overlap on surface S_{mn}, we can write that

$$\vec{E}^{+}_{nm} = -\hat{a}_{n,n} \times \vec{E}^{s}_{m}(\vec{r}) = \hat{a}_{n,n} \times \int_{S_m} \left(j\omega\mu \overline{\overline{G}}_0 \bullet \vec{H}^{-}_{m} \right) dS' + \hat{a}_{n,n} \times \int_{S_m} \left(\nabla g_0 \times \vec{E}^{-}_{m} \right) dS' - \frac{\gamma}{2} \vec{E}^{-}_{m}, \tag{55}$$

$$\vec{H}^{+}_{nm} = \hat{a}_{n,n} \times \vec{H}^{s}(\vec{r}) = -\hat{a}_{n,n} \times \int_{S_m} j\omega\varepsilon \overline{\overline{G}}_0 \bullet \vec{E}^{-}_{m} dS' + \hat{a}_{n,n} \times \int_{S_m} \left(\nabla g_0 \times \vec{H}^{-}_{m} \right) dS' - \frac{\gamma}{2} \vec{H}^{-}_{m}, \tag{56}$$

where $\vec{E}^{+}_{nm}$ and $\vec{H}^{+}_{nm}$ are the rotated tangential fields on S_n created by $\vec{E}^{-}_{m}$ and $\vec{H}^{-}_{m}$ on surface S_m. $\gamma = 1$ if $\vec{r} \in S_{mn}$ and $\gamma = 0$ elsewhere. The fact that $\hat{a}_{n,n} = -\hat{a}_{n,m}$ on surface S_{mn} has been used to get equations (55) (56). Note that equations (55) and (56) also hold true if we replace $\vec{E}^{-}_{m}$ by $\vec{M}_{Sm}$ and $\vec{H}^{-}_{m}$ by $\vec{J}_{Sm}$ simultaneously.

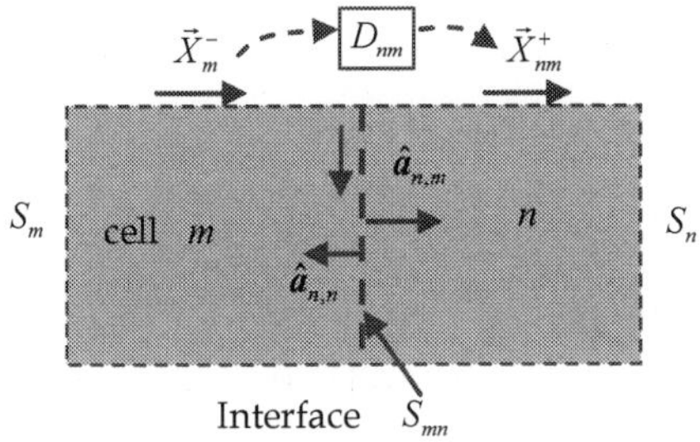

Fig. 9 Block-m and block-n share a common interface S_{mn}.

Assume that the rotated tangential fields are expanded and tested with vector basis functions $\vec{f}_{m,j}$ and $\vec{f}_{n,i}$ on surface S_m and S_n, respectively, then (55) (56) can be cast into

$$\left[X^{+}_{nm}\right] = \left[D_{nm}\right]\left[X^{-}_{m}\right] = \left(\left[\widetilde{D}_{nm}\right] - \frac{1}{2}\left[P_{nm}\right]\right)\left[X^{-}_{m}\right], \tag{57}$$

here, $\left[\widetilde{D}_{nm}\right]$ corresponds to the Cauchy principal value of the field transmission matrix from block-m to block-n, which can be calculated with numerical integrations. $\left[P_{nm}\right]$ is a Gramm matrix with entries defined by

$$P_{nm}(i,j) = \int_{S_{ti}} \vec{f}_{m,j} \bullet \vec{f}_{n,i} dS . \tag{58}$$

Apparently, $\frac{1}{2}\left[P_{nm}\right]$ corresponds to the residual term caused by the singularity of the dyadic Green's function. $P_{nm}(i,j) = 0$ for non-adjacent blocks.

4. Characteristic Basis Functions and Synthetic Basis Functions

Characteristic basis functions (CBFs) (Prakash & Mittra, 2003) are used to further reduce the unknowns in the final linear system. In this example, the CBFs are created in the following way: determine a neighbour domain for each block at first, which consists of N_{adj} nearest blocks. Since the generalized transition matrix of each block is readily available, the scattered fields of the small system consisting of an individual block and its neighbour domain can be calculated easily by solving a small linear equation system as follows,

$$\left[X_m^-\right] - \sum_{n=1,n\neq m}^{N_{adj}} [T_m][D_{mn}]\left[X_n^-\right] = [T_m]\left[X_m^{inc}\right]. \tag{59}$$

The rotated tangential scattered fields (or equivalent surface sources if equation (54) is used.) are then selected as the characteristic basis function of that block, denoted by $\left[X_m^C\right]$, and the total rotated tangential fields are expanded with

$$\left[X^-\right] = \sum_{m}^{N_b} \alpha_m \left[X_m^C\right]. \tag{60}$$

Therefore, equation (53) is cast into a linear system with only N_b unknowns,

$$[A][\alpha] = \left[X_c^{inc}\right]. \tag{61}$$

where the interaction factor between block-m and block-n is defined by

$$A_{mn} = \begin{cases} \left[X_m^C\right]^t \bullet \left([T_m][D_{mn}]\left[X_n^C\right]\right)^*, & \text{for } m \neq n \\ \left[X_m^C\right]^t \bullet \left[X_n^C\right]^*, & \text{for } m = n \end{cases}, \tag{62}$$

and $X_c^{inc}(m) = \left[X_m^C\right]^t \bullet \left[X_m^{inc}\right]^*$. The upper script t stands for transpose operation and * for conjugate.

Synthetic basis functions (Matekovits etal., 2007) can also be applied to reduce the number of unknowns involved in the generalized surface integral equation (53). In order to implement SBFs with GSIE, we have to generate a response space for a block using the method described in (Matekovits etal., 2007), where it is pointed out that the degree of freedom of the SBFs for a block is less than the total unknown number N_m for the block expressed with low-order basis functions. Therefore, based on equivalence principle, an auxiliary source space may be used to handle the effect from all other scatterers in the system except the one under consideration, and it is possible to use SBFs fewer than N_m for that block. This may be considered as an information compression process. The characteristics of the practical structure should be fully made use of in order to achieve a large compression ratio. In radiation problems, the fact that near fields evanesce rapidly from a scatterer is a key factor to use to compress the response space. Hence, the auxiliary sources should be put as far as possible away from the scatterer under consideration. On the other hand, the input condition and the connecting circuits are all key factors of the problem that may have significant effect on the response space and should be included in the compression procedure. However, the method is hard to implement in cases when sub-blocks are tightly connected, because the auxiliary sources should be put exactly on the interface of a sub-block in these cases. As a result, the compression ratio is usually found to be very low. We have supposed a modified scheme to implement SBFM, which is in somewhat similar to the method described in (Tiberi etal., 2006)). Instead of using point sources, N_s plane waves with incident angles of $\phi_s = 2\pi s/N_s$, $s = 1,\cdots,N_s$ are chosen as the auxiliary sources, as is illustrated in Fig. 10. In cases we have checked, we found that if same number of SBFs are used, this arrangement provides better accuracy than putting auxiliary sources directly on the boundary of the sub-block.

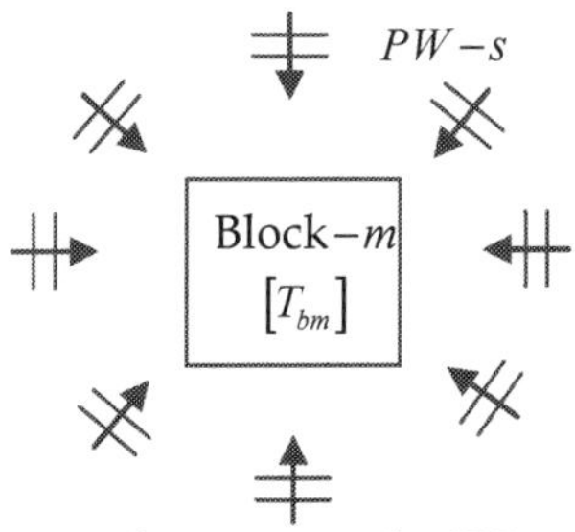

Fig. 10 Auxiliary plane waves are used to generate the SBFs of block- m .

The scattered fields of each incident plane wave are collected together to form a response space of block- m , which is a $N_m \times N_s$ matrix and denoted by $\left[X_{PWm}^{-}\right]$. Applying singular value decomposition (SVD) operation gives

$$\left[X_{PWm}^{-}\right]=[U][S][V]^{H} . \tag{63}$$

The singular values of $\left[X_{PWm}^{-}\right]$ is stored in the diagonal entries of $[S]$ in descending order, while their corresponding eigen functions are stored in the related columns of $[U]$. We use the following criterion given in (Matekovits etal., 2007) to generate the SBFs,

$$S(p,p)/S(1,1) \le \varepsilon, \quad p = 1,\cdots,N_{svd} . \tag{64}$$

The first N_{svd} eigen functions corresponding to the first N_{svd} singular values are selected as the SBFs of block- m , i.e.,

$$\left[X_{mp}^{sbf}\right]=[U]_{p}, \quad p=1,\cdots,N_{svd} , \tag{65}$$

$[U]_p$ denotes the pth column of $[U]$.

The rotated tangential fields on the reference surface of block-m are then expanded with SBFs as follows,

$$\left[X_m^-\right]=\sum_{p=1}^{N_{svd}}\alpha_{mp}\left[X_{mp}^{sbf}\right]. \tag{66}$$

Substituting (66) into (53) and testing it with SBFs (Galerkin's scheme) result in a size-reduced linear system, the unknown number of which equals to the total number of SBFs on all blocks. The interactions between block-m and block-n are also defined by equation (62), except that SBFs are used to replace CBFs.

In SBFM, because the most significant SBFs for the system are selected once for all, and the mutual coupling is treated rigorously, the numerical accuracy of SBFM is well controlled compared to CBFM.

5. Numerical Results

The field in the cavity illustrated in Fig. 1 is calculated using the above-discussed generalized surface integral equation method. The wall is divided into 36 blocks and the surface of each block is segmented to 40 segments, roughly 10 segments per wavelength. The z-axis polarized TM plane wave with unit magnitude is assumed to excite at Φ= 45°. Triangle-shaped vector basis functions $\vec{f}_{m,i}$ are used to expand all tangential electric field components, while $\hat{a}_n\times\vec{f}_{m,i}$ are used to expand all tangential magnetic field components. The generalized transition matrix of each block is calculated using cascaded network techniques. The total electric fields in the cavity are first calculated using equation (53) directly, with totally 2880 unknowns involved. The calculated electric field is plotted in Fig. 11, from which it can be seen that the fields in the cavity are attenuated from -20dB to -55dB due to the shielding effect of the walls.

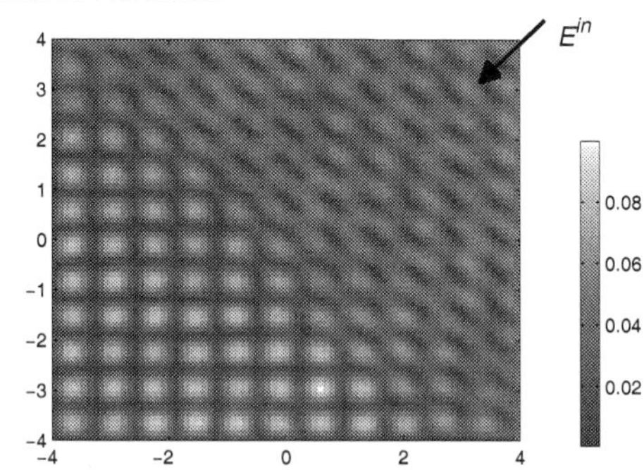

Fig.11. Calculated magnitude (Units: dB) of the total electric field in the internal area of the cavity (8λ×8λ).

Characteristic basis functions are used to re-calculate the field in the cavity. The neighbour domain of each block contains the nearest 3 blocks at one side. Therefore, a small system consisting of 7 blocks has to be solved to create a characteristic basis function. The final linear system resulting from equation (61) has only 36 unknowns. For large complex multi-scale system, this reduction rate of unknown number is really substantial. For comparison, the discrepancy of the scattered fields in the cavity calculated by using the two methods is plotted in Fig. 12 (in decibels).

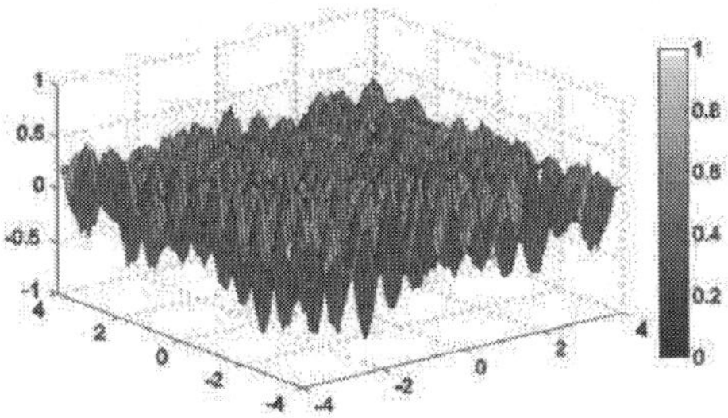

Fig. 12. The discrepancy of the electric fields calculated with the two methods. (Units: dB)

6. Conclusions

A generalized surface integral formulation for analyzing large 2-D cavity with reinforced concrete walls is presented. The wall is divided into small blocks and each block is described by an associated generalized transition matrix defined on its reference surface. Detail structure of each block is analyzed locally only once, and is replaced by a simpler reference surface. By applying the presented method in conjunction with characteristic basis functions and synthetic basis functions, the unknowns and the cost for evaluating the interactions among blocks are significantly reduced.

7. References

Dalke, R. A.; Holloway C. L.; Mckenna P.; Johnsson, M. & Ali, A. S. (2000). Effects of reinforced concrete structures on RF communications. *IEEE Trans. Electromagn. Compat.*, Vol. 42, No. 4, (Nov. 2000) 486-496, ISSN 0018-9375.

Creticos, J. P. & Schaubert, D. H. (2006). Electromagnetic scattering by mixed conductor dielectric bodies of arbitrary shape. *IEEE Trans. Antennas Propagat.*, Vol. 54, No. 8, (Aug. 2006) 2402–2407, ISSN 0018-926X.

Lean, M. H. (2004). Novel integral formulation for scattering from multilayered dielectric cylinders of arbitrary cross section. *IEEE Trans. Magnet.* Vol. 40, No. 2, (Mar. 2004) 1476-1479, ISSN 0018-9464.

Li, M. K.; & Chew, W. C. (2007). Wave-field interaction with complex structures using equivalence principle algorithm. *IEEE Trans. Antennas Propagat.*, Vol. 55, No. 1, (Jan. 2007) 130-138, ISSN 0018-926X.

Matekovits, L.; Laza, V. A. & Vecchi, G. (2007). Analysis of large complex structures with the synthetic-functions approach. *IEEE Trans. Antennas Propagat.*, Vol. 55, No. 9, (Sept. 2007) 2509-2521, ISSN 0018-926X.

Polewski, M.; Lech, R. & Mazur, J. (2004). Rigorous modal analysis of structures containing inhomogeneous dielectric cylinders. *IEEE Trans. Microw. Theory Tech.*, Vol. 52, No. 5, (May 2004) 1508-1516, ISSN 0018-9480

Prakash, V. V. S. & Mittra, R. (2003). Characteristic basis function method: A new technique for efficent solution of method of moments matrix equations. *Microwave Opt. Technol. Lett.*, Vol. 36, (Jan. 2003) 95–100, ISSN 0895-2477.

Rao, S. M.; Wilton, D. R. & Glisson, A. W. (1982). Electromagnetic scattering by surfaces of arbitrary shape. *IEEE Trans. Antennas Propagat.*, Vol. 30, No. 3, (May 1982) 409–418, ISSN 0018-926X.

Suter. E. & Mosig, J. (2000). A subdomain multilevel approach for the MoM analysis of large planar antennas. *Microwave Opt. Technol. Lett.*, Vol. 26, (Aug. 2000) 270–227, ISSN 0895-2477.

Taskinen, M. & Ylä-Oijala, P. (2006). Current and charge integral equation formulation. *IEEE Trans. Antennas Propagat.*, Vol. 54, No. 1, (Jan. 2006) 58-67, ISSN 0018-926X.

Tiberi, G. A. Manara, M. G. & Mittra, R. (2006). A spectral domain integral equation method utilizing analytically derived characteristic basis functions for the scattering from large faceted objects. *IEEE Trans. Antennas Propagat.*, Vol. 54, No. 9, (Sept. 2006) 2508-2514, ISSN 0018-926X.

Umashankar, K. R.; Taflove, A. & Rao, S. M. (1986). Electromagnetic scattering by arbitrary shaped three-dimensional homogeneous lossy dielectric objects. *IEEE Trans. Antennas Propagat.*, Vol. 34, No.6, (June 1986) 758–766, ISSN 0018-926X.

Waterman, P. C. (2007). The T -matrix revisited. *J. Opt. Soc. Am.* A, Vol. 24, No. 8, (Aug. 2007) 2257-2267, ISSN 1084-7529.

Xiao, G. B.; Mao, J. F. & Yuan, B. (2008). Generalized transition matrix for arbitrarily shaped scatterers or scatterer groups. *IEEE Trans. Antennas Propagat.*, Vol. 56, No.12, (Dec. 2008) 3723–3732, ISSN 0018-926X.

Xiao, G. B.; Mao, J. F. & Yuan, B. (2009). A generalized surface integral equation formulation for analysis of complex electromagnetic systems. *IEEE Trans. Antennas Propagat.*, Vol. 57, No. 3, (Mar. 2009) 701-710, ISSN 0018-926X.

12

52-GHz Millimetre-Wave PLL Synthesizer

Ja-Yol Lee and Hyun-Kyu Yu
ETRI (Electronics & Telecommunications Research Institute)
Korea

1. Introduction

Multiple 60-GHz WPAN (Wireless Personal Area Network) radio transceiver chips using CMOS or BiCMOS process have been developed as wireless communication service of several-gigabit rate [Reynolds, 2006] [Razavi, 2006]. In a 60GHz millimetre-wave transceiver, frequency synthesizer is a key building block. It is very difficult to design the PLL-based programmable synthesizer directly at 60GHz band without tripler or doubler. A 14.25-16-GHz programmable PLL synthesizer was presented as a frequency source for 60GHz direct-conversion receiver. Its output frequencies of 28.5-32GHz are generated using Wilkinson power combiner [Lee, 2008]. A 16-18.8GHz programmable PLL was developed as a local oscillator for 60GHz dual-conversion super-heterodyne transceiver [Floyd, 2008]. Its tripled output frequency becomes 46 to 54GHz. Also, various non-programmable PLLs have been developed as millimetre-wave frequency sources, but they are not suitable for 60GHz WPAN radio covering 57 to 64GHz range [Winkler, 2005] [Lee, 2007].

In this chapter, we present a 52-GHz PLL-based synthesizer for 60GHz dual-conversion super-heterodyne receiver. The synthesizer is composed of 26GHz programmable PLL and 52GHz frequency doubler. The 26GHz PLL consists of PFD, charge pump, loop filter, LC VCO, and four-modulus divider. The synthesizer shows a 50-53GHz locking range, and generates two channels of 50.304GHz and 52.4GHz when 262MHz reference is used. The PLL achieves phase noises of - 89dBc/Hz from 26.2GHz and - 81dBc/Hz from 52.4GHz, at 1MHz offset frequency, respectively. The synthesizer represents spurious noise level of -42dBc/Hz, and consumes 160mA at 2.5V.

In section 2 of this chapter, a 60GHz dual-conversion super-heterodyne receiver is briefly introduced. In section 3, the 52GHz PLL synthesizer including frequency doubler is described in detail. The experimental results are presented in section 4, and finally, conclusion is drawn in section 5.

2. 60GHz Dual-Conversion Super-Heterodyne Receiver

Fig. 1 shows a dual-conversion super-heterodyne receiver for 60GHz WPAN radio including RF and IF PLL. The 60-GHz dual-conversion receiver consists of RF LNA, RF Mixer, RF PLL, VGA, IF Mixer, and IF PLL. In the 60GHz receiver, the LNA amplifies the RF signals between 57GHz and 64GHz, which are down converted to 10GHz IF signal by the RF mixer. Then, the amplitude of the 10-GHz IF signal is controlled by the variable-gain amplifier (VGA). And it is fed to the double-balanced IF mixer which performs the IF-to-base-band signal conversion. The local oscillator (LO) module is configured with 52GHz RF PLL and 10GHz IF PLL. The RF PLL should provide three channels between 48.208GHz and 52.4GHz in step of 2.096GHz or between 48.576GHz and 52.8GHz by 2.112GHz step, with RF mixer down-converting the RF signals of 57-64GHz. The IF PLL provides only 10GHz or 10.032GHz fixed carrier, depending upon which reference clock of 262MHz and 264MHz is used.

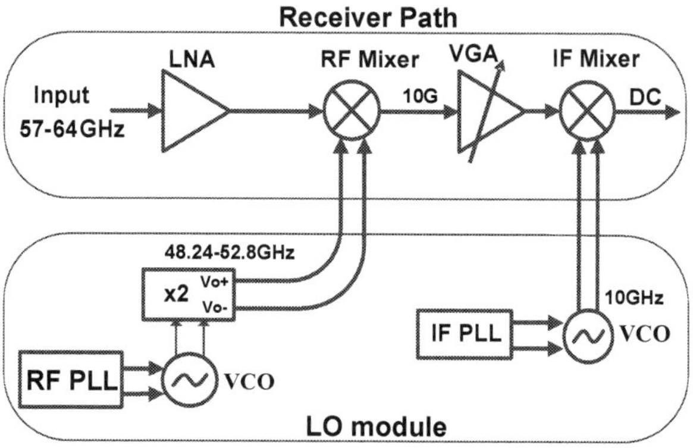

Fig. 1. 60GHz dual-conversion superheterodyne receiver with RF PLL

3. Design of 52GHz PLL Frequency Synthesizer

3.1 Frequency Synthesizer Architecture using PLL

Fig. 2 represents the 52-GHz frequency synthesizer block diagram using phase-locked loop. The frequency synthesizer consists of PFD, charge pump, loop filter, LC-tank VCO, and four-modulus divider. In the frequency synthesizer, all components including frequency doubler and loop filter are implemented by on chip. The 52-GHz PLL is designed for providing local signal to the 60-GHz receiver covering unlicensed WPAN band (57-63GHz). As shown in Fig. 2, the PLL generates 52GHz frequency through the frequency doubler making twice the output frequencies of the 26GHz VCO. The frequency doubler configured with a differential amplifier suppresses odd-mode harmonic carriers such as f_o, $3f_o$, $5f_o$, etc, but it combines the even-mode carriers ($2f_o$, $4f_o$, etc) of the differential VCO, additively. Therefore, only differential even-mode harmonic carriers are emerged from the output port of the frequency doubler whose circuit diagram is presented in 3.5.

The phase-frequency detector is designed to operate at high speed of 264MHz. The third order loop filter is implemented on chip. Since the on-chip passive loop filter occupies large chip area, each optimized values of loop-filter components should be chosen, considering its chip area, loop settling time, and spurious noise level. The LC-tank VCO is designed to generate high oscillation frequencies of 24.4-26.5GHz. In Fig.2, a four-modulus divider is designed to provide the divide number of 20 to 25 for snythesizing three-channel carriers. Also, the four-modulus divider must operate at high speed to divide the high output frequencies of the 26-GHz VCO. In a millimeter-wave PLL synthesizer, it is very difficult to design a high-speed programmable divider for synthesizing multiple channels because there are many digital logic circuits operating at low speed and some logic-signal delay in the divider block. Also, a high-speed divider consisting of many digital latches requires much amount of current, and hence, contributes to most of power consumption in the whole PLL circuit. Therefore, the programmable divider should be designed, considering trade-off between power consumption and speed. In this chapter, the 52GHz PLL synthesizer provides two channels of 50.304GHz and 52.4GHz when the reference clock of 262MHz is input. Also, the PLL generates two channels of 50.688GHz and 52.8GHz when the reference clock of 264MHz is input to the PFD.

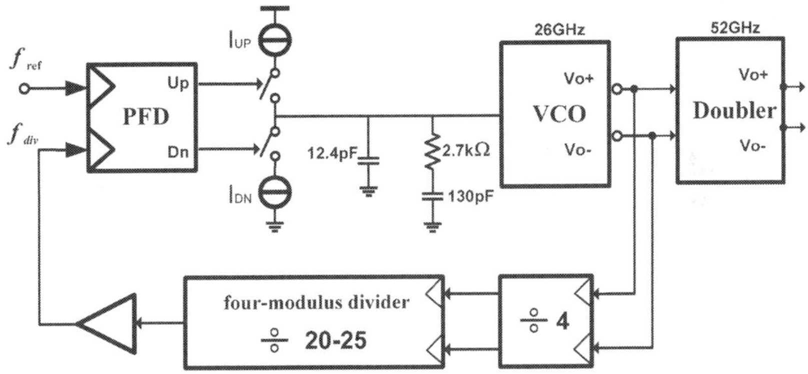

Fig. 2. 52GHz PLL frequency synthesizer

3.2 PFD and Charge Pump

Fig. 3 illustrates a common linear PFD block diagram, configured with two-edge triggered resettable DFFs and a NAND gate, and its state diagram. This PFD generates UP and Down signals that switch the current of charge pump and control its amount. The D inputs of the DFFs are fixed to logic high, and the two input signals (f_{ref} and f_{div}) of PFD trigger each DFF. The pulsewidth of both UP and Down signals is proportional to the phase difference between f_{ref} and f_{div}. As shown in Fig. 3(b), initially, both outputs are low. When one of the PFD inputs rises, the corresponding outputs becomes logic high. The state of the finite-state machine (FSM) moves from an initial state to an Up or a Down state, depending on the state of the input signal. According to the state diagram, therefore, the phase and frequency differences of the two input signals are detected by the PFD [Razavi, 2001].

Fig. 4 (a) shows the ideal graph of plotting the input-output characteristics of the tri-state PFD. Defining the output V_{out} as the difference between the average values of Up and Down signals when $f_{ref} = f_{div}$ and neglecting the effect of the narrow reset pulses, we note that the output varies symmetrically as $|\Delta\phi|$ begins from zero as shown in Fig. 4 (a). That is, the PFD ideally shows the linear characteristics for the entire range of input phase differences from -2π to 2π. The PFD including charge pump senses the transitions at the inputs, detects phase or frequency differences, and activates the charge pump accordingly. When f_{ref} is initially far from f_{div}, the PFD and the charge pump keep varying the control voltage of VCO until f_{ref} approaches f_{div} closely. When f_{ref} and f_{div} are sufficiently close, the PFD operates as a phase detector, performing phase lock. Once the phase difference is within the lock-in range and drops to zero, the cycle slipping stops and the PLL loop is locked. Here, the cycle slipping means that the phase difference changes each cycle by $2\pi\times[(T_{fref} - T_{fdiv})/\max(T_{fref}, T_{fdiv})]$. At this view point, the phase defector behaves as a linear system as shown in Fig. 4(a). However, due to the delay of the reset path, the linear range is less than 4π as shown in Fig. 4(b). The reset pulses prevent the PFD/charge pump from undergoing a deadzone around $\Delta\phi = 0$ in which the PLL loop gain drops to zero and the phase of f_{div} is locked.

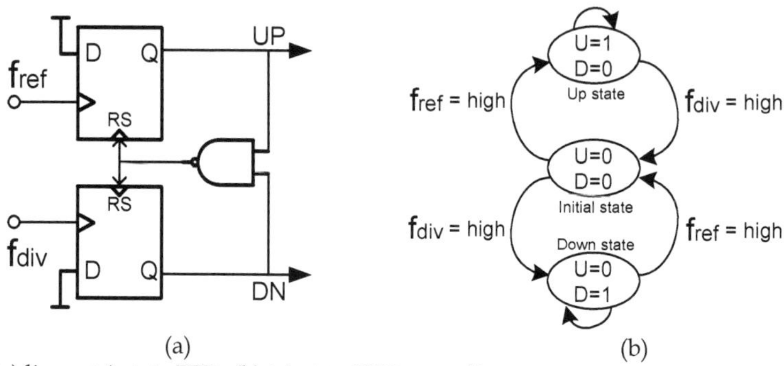

Fig. 3. (a) linear tri-state PFD, (b) tri-state PFD state diagram

Fig. 4(c) illustrates the nonideal behavior with the reference clock CKf_{ref} leading the output clock CKf_{div}, causing an Up output. As the input phase difference nears 2π, the next leading edge of CKf_{ref} arrives before the DFFs are reset due to the finite reset delay. The reset pulse overrides the new CKf_{ref} edge and then the Up signal is not activated. The subsequent CKf_{div} edge causes a Down signal. The effect appears as a negative output for phase differences higher than $2\pi-\Delta$, where $\Delta = 2\pi\times(t_{reset}/ T_{fref})$, which depends on the reset path delay t_{reset} and the reference clock period T_{fref}. Note that t_{reset} is determined by the delay of logic gates in the reset path and is not a function of input frequency. During acquisition, the frequency will not monotonically approach lock-in range because the nonideal PFD gives the wrong information periodically. The acquisition slows by how often the wrong information occurs, which depends on Δ. At an input frequency $T_{fref} = 2\times t_{reset}$ where Δ equals π, the PFD output generates the wrong information half the time and, thus, fails to acquire frequency lock unconditionally. The maximum operating frequency can be expressed as $f_{ref} \leq 1/(2\times t_{reset})$ [Mansuri, 2002].

A commonly used tri-state PFD is implemented by using NOR-based latch to build the edge-triggered resettable DFFs. The reset path includes one two-input NAND, one inverter, and three two-input NORs, which are counted twice in the reset path. Thus, the delay of the reset path is approximately equal to 10 gate delays [Razavi, 2001]. This reset pulse delay is required to remove a deadzone. However, the reset delay makes the settling time of PLL slow and disables the phase lock in the worst case of $\Delta \geq \pi$. In this PLL, the short pulse period is about 210 ps, the reference clock frequency 262MHz and Δ is about 0.11π, which can slow the loop settling time to some extent.

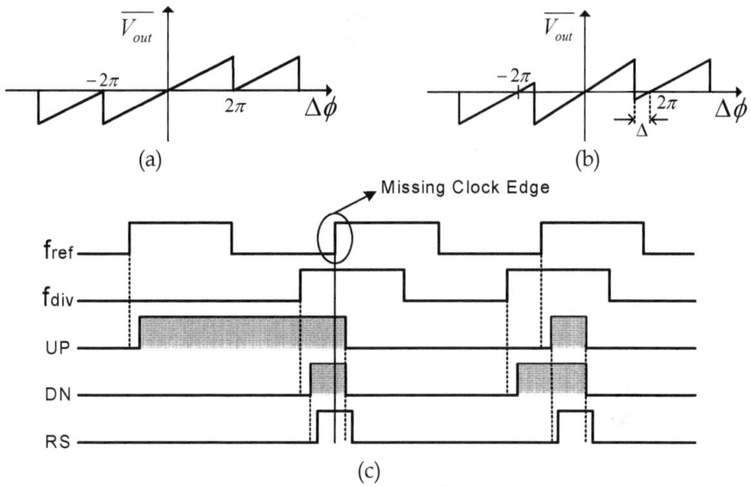

Fig. 4. (a) ideal linear characteristic of tri-state PFD (b) nonideal characteristic (c) nonideal behaviour due to nonzero reset delay

In the commonly improved PFD circuit of Fig. 5, two inverting delay stages are inserted between the clock input and the precharged PMOS (P2 and P5) [Tak, 2005]. The characteristics of the improved PFD are shown in Fig. 6. It can generate effective Up signals without missing the edge of CKf_{ref} even when the phase difference between the two input clocks is significantly close to 2π. The improved PFD represents lower power consumption and higher precision than the latch-based PFD presented by Razavi [2001] because the dynamic logic circuit has lower propagation delay and better matching, which are the advantages of the improved PFD. Therefore, the phase noise contribution of PFD can be relieved. Fig. 6(a) illustrates the operation of the improved PFD. Here, td1 is the inserted delay time between CKf_{ref} and D_CKf_{ref} (CKf_{div} and D_CKf_{div}), td2 is the pulse width for preventing the dead zone, and td3 is the time period from a rising edge of a subsequent input clock between CKf_{ref} and CKf_{div} to the falling edge of the reset signal. At the second rising edge of CKf_{ref}, the phase difference $\Delta\Phi$ is in the range of $2\pi-\Delta < \Delta\Phi < 2\pi-\delta$. At the falling edge of the following Reset signal, D_CKf_{ref} is "Low", and node X is charged to "High". Because CKf_{ref} is "High" at that moment, node Y is discharged to "Low" and Up signal becomes "High" earlier than the Down signal. Therefore, the improved PFD does not lose the edge that arrives during reset and generates the right information. At the third rising edge of CKf_{ref}, the phase difference is in the range of $\Delta\Phi \geq 2\pi - \delta$ and the Up signal remains "Low". The inserted delay td1 should be designed to be slightly smaller than td3, otherwise the PFD will fail to lock at zero input phase difference. If td1 > td3, then the delayed input clocks activate the output after the reset pulse ends. This design criterion results in a negative output voltage for $\Delta\Phi \geq 2\pi - \delta$, as illustrated in Fig. 6(b). The maximum operating frequency of the PFD is dependent on the duty ratio of each input clock. With a higher duty ratio, the PFD can operate at a higher clock frequency because the input clock should be "High" at the rising edge of the delayed input clock. Assuming half duty ratio, the maximum operating frequency of the improved PFD is 1/(2×td3), approximately equal to 1/(2×t_{reset}), while that of the latch-based PFD is about 1/t_{reset} [Mansuri, 2002] [Tak, 2005].

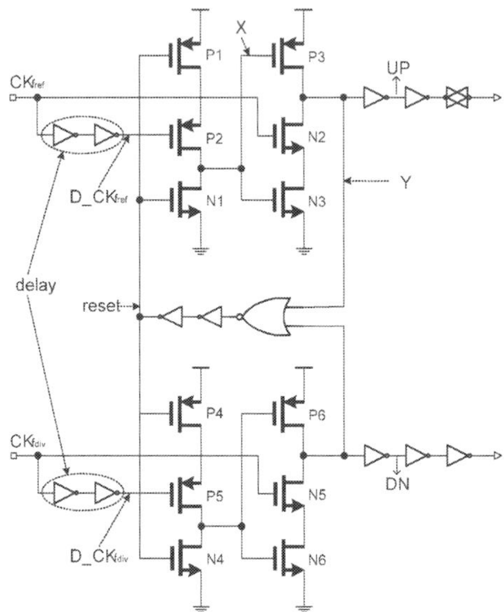

Fig. 5. PFD using edge-triggered resettable DFF

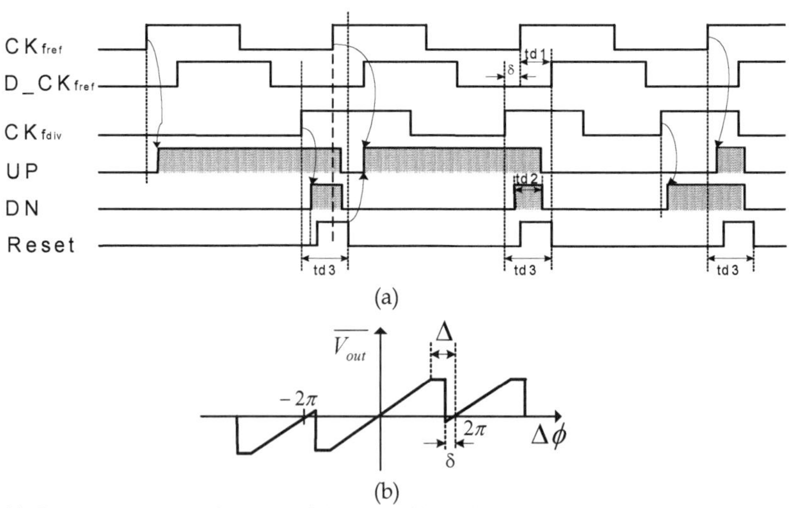

Fig. 6. (a) Operating timing diagram of the PFD (b) its characteristics

Charge pump is an important building block commonly used in PLLs for frequency synthesizers and clock generators. The dominant block causing reference spurs in a PLL is the charge pump. The reference spurs are commonly generated because there are current mismatch, leakage current, and timing mismatch in charge pump circuit. The total phase offset in PLL loop due to the non-idealities of charge pump is approximately expressed in (1) [Rhee, 1999].

$$\Delta\phi_{tot} = 2\pi\left(\frac{I_{leak}}{I_{cp}} + \frac{\Delta i}{I_{cp}}\cdot\frac{t_{on}}{T_{ref}} + \frac{\Delta t_{delay}\cdot t_{on}}{T_{ref}^2}\right) \tag{1}$$

where I_{cp} is the charge-pump current, I_{leak} is the leakage current, T_{ref} is the reference clock period, t_{on} is the PFD turn-on time, Δi and Δt_{delay} are current and timing mismatch, repectively. The resulting reference spurs for a 3rd–order PLL can be approximately given by (2)

$$P_r = 20\log\left(\frac{N\cdot f_{bw}\cdot\Delta\phi_{tot}}{\sqrt{2}f_{ref}}\right) - 20\left(\frac{f_{ref}}{f_{p1}}\right) \tag{2}$$

where N is the division ratio of the divider, f_{bw} is the loop bandwidth and f_{p1} is the frequency of the pole in the loop-filter. Eq. (2) represents that the reference spurs can be reduced by lowering loop phase error and loop bandwidth, or increasing reference frequency. The phase offset due to leakage current can be significant in sub-micron CMOS circuits, but may be reduced with a large charge-pump current. This results in high power consumption and should be hence avoided in low power application.

The use of a differential charge-pump circuit is preferable since leakage current then appears as a common mode glitch at output. The influence of current mismatch can be reduced by minimizing the turn-on time of the PFD. A small turn-on time also helps to reduce the in-band noise of the PLL. A small turn-on time requires a fast switching charge pump with minimum current mismatch. For charge pump to operate with small turn-on time and high speed, current steering techniques have to be used instead of charge pumps with drain, gate or source switching. Charge pumps with drain switching are limited by high current spikes in the first moments of the pump UP/DN operations. This is due to triode region operation of the current source transistors. These current spikes are difficult to match since the current varies with output voltage. The gate-switched charge pump eliminates the problem of current spikes, since the transistors in the current mirrors are then either off or in saturation. The main drawback is increased gate capacitance, which limits the operating speed. A higher speed is achieved with a source-switching charge pump. However, the operating speed is still limited by the time of the current mirror because the switch is connected to a low impedance node, which results in the current mirrors of either ON or in saturation [Rhee, 1999].

Fig. 7 shows the high-speed single-ended charge pump circuit using current steering technique, which improves switching time and thus allows high-speed operation. The drawback of using the current-steering charge pump switch is the mismatch between the NMOS and the PMOS transistors. This inherent mismatch can be avoided by using only NMOS switches [Magnusson, 2003]. Also, if the rise/fall times of the input signals are very small, there will not be enough time for the gate-drain capacitors to be charged or discharged by drain current. Hence, there will be very large spikes of current at the output of the charge pump and very high content of the higher harmonics in the filter output voltage [Bahreyni, 2002]. A PFD will produce logic level UP/DN signals, which are the inputs to the charge pump in Fig. 7. The large amplitude of UP and DN signals will cause the bias current of I_{up} and I_{down} to be completely switched from one branch of the corresponding differential pair to the other. For example, when both UP and DN singal are low, P2 and N1 will be ON while P1 and N2 will be OFF. Thus, the tail current, I_{up}, will be forced to flow into the loop filter. When both UP and DN signals are high, N2 and P1 will be ON while P2 and N1 will be OFF. Thus the tail current, I_{down}, will be forced to flow out from the loop filer. However, if one of the inputs is high and the other is low or vice versa, the output transistors, P2 and N2, will be on or off at the same time and no current flows into the loop filter.

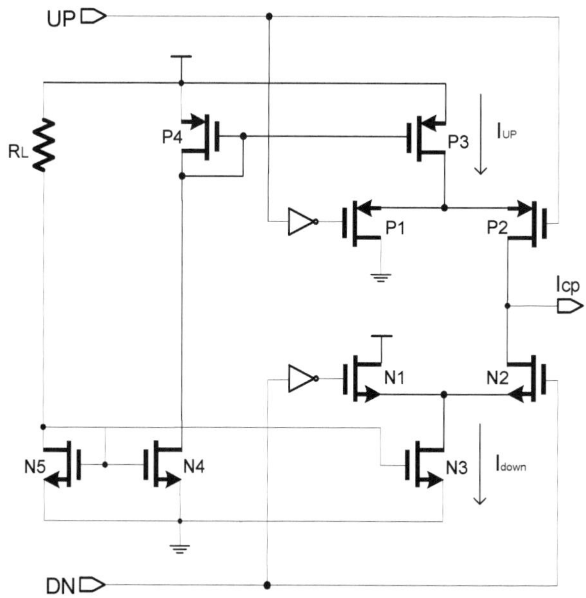

Fig. 7. High-speed current-steering single-ended charge pump

Fig. 8 shows the simulation result of the I_{up} and I_{down} current mismatch in the current-steering charge pump circuit. When the output voltage of the charge pump changes 0.2 V to 2.2 V by 0.1 V step, the I_{up} and I_{down} currents varies from 380μA to 426μA, and hence the charge pump represents about 10% current mismatch. Fig. 9 illustrates the simulation result of PFD deadzone when the input phase difference is varied across zero point. Since the curve of Fig. 9 increases linearly across zero value at Y-axis without plateau, there is no deadzone in the designed PFD circuit, which fills up the demand of td1 < td3. Here, td1 is 60ps and td3 is greater than 260 ps.

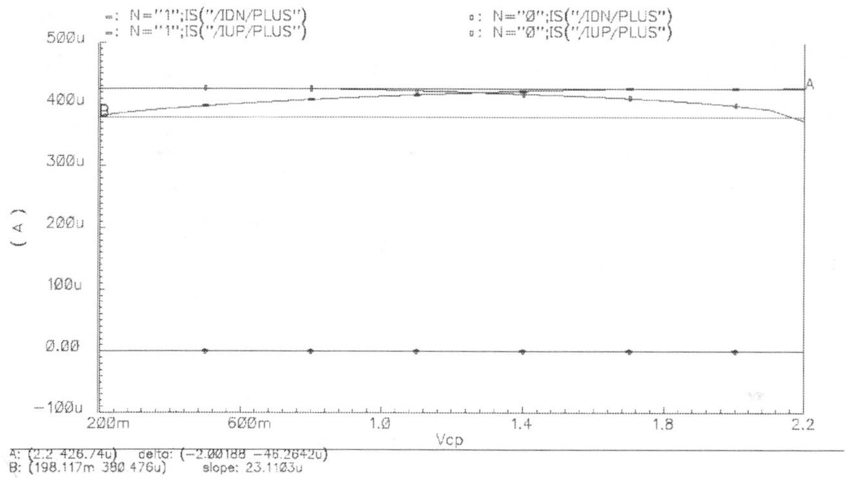

Fig. 8. Simulation result of current mismatching in charge pump

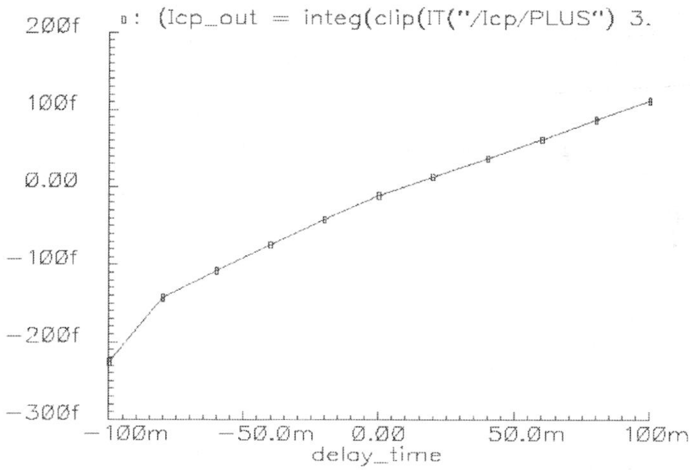

Fig. 9. Simulation result of dead zone in PFD

3.3 Programmable Divider

A multi-modulus prescaler that operates at very high frequency is one of the key building blocks for frequency synthesizers in wireless communications. The major issues for designing a multi-modulus prescaler are high operating frequency and low power consumption, while keeping low phase noise contribution to the synthesized output signal. Fig. 10 represents the block diagram of the four-modulus divider in the 52-GHz frequency synthesizer of Fig. 2. The four-modulus divider, which has four divide ratios controlled by mode control (MC) signal, is designed to simplify the hardware required for frequency synthesis. It is composed of a divid-by-4/5 dual-modulus prescaler, a divide-by-5 divider, and a control logic unit. The control logic unit generates the MC signal modulating its divide rato, and the divide ratio of the four-modulus divider can be set to be ÷ 20, ÷ 23, ÷ 24, and ÷ 25 by varying the duty ratio of the MC signal. And its duty ratio is determined by the logic values of control bits c_0 and c_1, as shown in Table of Fig. 10. For example, if c_1 is low and c_0 is high, the total divide ratio becomes ÷ 23; if c_1 is high and c_0 is low, the total divide ratio becomes ÷ 25. Fig. 11 illustrates the timing diagrams of the four-modulus divider. If the MC signal is low, the divide-by 4/5 prescaler divides the input clock signal by 5. If the signal MC is high, its divide ratio becomes ÷ 4. Therefore, if c_0 is low and c_1 is high, the dual-modulus prescaler divides the input signal of *f*vco/4 by ÷ 4 for two P_{o+} cycles and by ÷ 5 for three P_{o+} cycles, while the followed divide-by-5 divider swallows five Po+ cycles. Thus, a total divide ratio (TDR) is calculated as TDR= (÷ 4) × 2 cycles + (÷ 5) × 3 cycles = ÷ 23. The operating timing waveform of the four-modulus divider is illustrated in Fig. 11. Using the same technique as explained above, the modulus number of the four-modulus divider could be extended to more numbers of divide ratios.

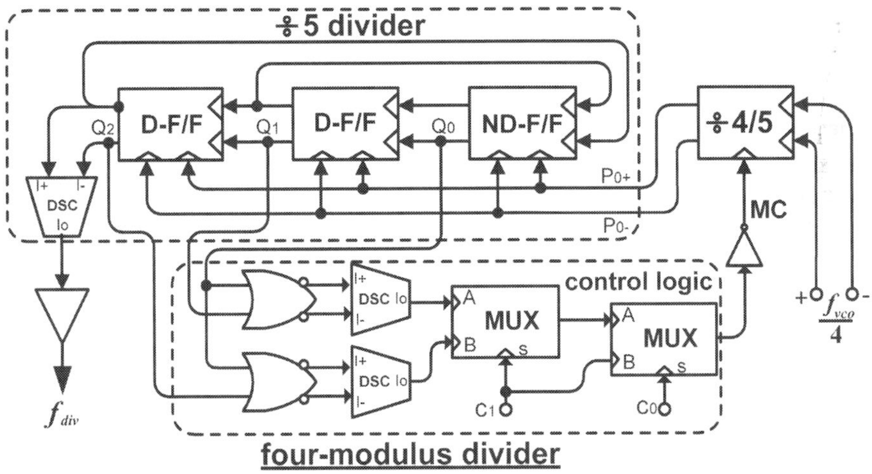

Divide ratio table

c_1 c_0	0 0	0 1	1 0	1 1
divide ratio	÷24	÷23	÷25	÷20

Fig. 10. Four-modulus divider and its divide ratio

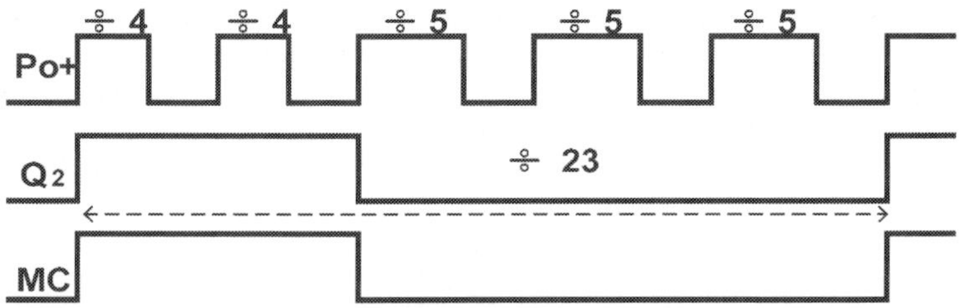

Fig. 11. Operating timing waveform of the four-modulus divider

The implementation of a high-speed prescaler in mixed-signal environment requires careful attention to certain aspects of the circuit design to contribute low noise to such sensitive analog circuit as VCO, which shares the same substrate with noisy circuits, and to the synthesized output signal. Here, both current-mode logic (CML) and ECL-like D-flipflops instead of a static CMOS logic are used to implement the four-modulus divider. The CML logic uses constant current source, which generates lower digital noise, and differential signals at both input and output, which reduce common-mode noise coupled from the power supply line and substrate because the differential circuit topology does inherently suppress the common-mode power supply and substrate noise [Park, 1998]. Another issue of the programmable divider design is reduction in power consumption at a given frequency range. Most power consumption in divider occurs in the front-end synchronous 4/5 dual-modulus prescaler because it is a part of the circuit operating at the maximum frequency of the input signal. The 4/5 synchronous dual-modulus prescaler shown in Fig. 12 contains two high-frequency fully functional ECL-like D-flipflops and one ECL-like D-flipflop with NOR logic. In the dual-modulus prescaler, the outputs of both the second D-flipflop and the third D-flipflop are feedback into the NOR D-F/F as the control inputs for generating proper division ratio. The MC signal is given to the third NOR D-F/F for modulating division ratio. The delay requirement in a critical path of the prescaler loop is severe because the 4/5 dual-modulus prescaler must operate up to a maximum of 10 GHz. The operating speed of the prescaler is limited by the delay time of each D-flipflops, and the prescaler layout. Therefore, the prescaler should be designed and laid out to achieve a delay time as small as possible and to obtain an operating frequency as high as possible.

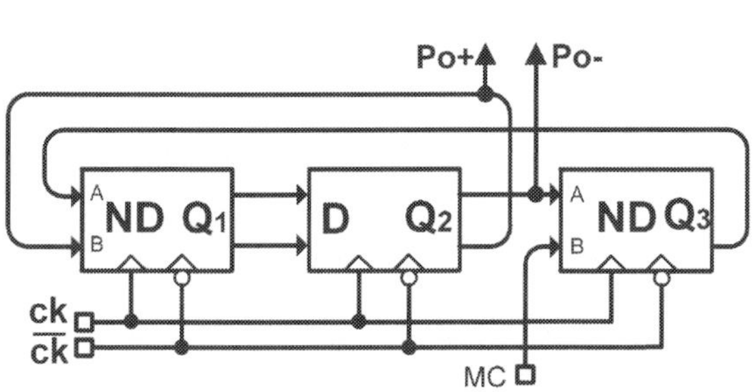

Fig. 12. 4/5 dual-modulus prescaler

Fig. 13 represents the divider circuit consisting of master-slave D-type latches. They are a rising edge-triggered E²CL D-flipflop with embedded NOR gate and a E²CL D-type flipflop, which are used in the front-end design to achieve a maximum speed and a minimum power. The master-slave D flipflop is driven by an applied clock signal (CK), and the Q of D-flipflop changes on each rising edge of the clock. Each latch consists of a differential stage (T_{r3}/T_{r4}, T_{r7}/T_{r8}) for the read-data operation and a cross-coupled stage (T_{r1}/T_{r2}, T_{r5}/T_{r6}) for the hold operation. Both load resistance R_L and bias current I_L determine logical swing. There are four distinct states that the D latch may occupy, representing state transition between latched and transparent, and on every edge of the clock the D flipflop changes state. To complete a cycle, all four-state transitions in which both master and slave latches alternate between transparent and latched states should be carried out in the divider. The maximum speed of operation of the divider circuit shown in Fig.13 can be determined by the sum of the delays of each transition. The D latches have two basic operations. The first is a current steering operation in the T_{r9}, T_{r10}/T_{r11} and T_{r12}, T_{r13}/T_{r14} differential pairs, moving between latched and transparent settings. The second is a voltage operation that can only occur after the current steering, changing the output voltage at I and Q nodes. Both of these operations introduce delay into the divider circuit and limit the maximum operating speed of the divider. Here, the delay contribution of the master's transition should be commonly improved because the master latch shows more slow cycle transition than the slave [Collins, 2005]. Also, in each latch, high-speed operation could be impaired whenever the cross-coupled stage of each latch failed to accomplish the hold-data phase. Therefore, in the master-slave D-type flip-flop, the cross-coupled pair with capacitive degeneration (C_d) is used for enhancing operation speed. In this case, it can be shown that the input conductance G(ω) of the cross-coupled pair is negative up to the frequency given by (3).

$$f_{G=0} = \frac{1}{2\pi}\sqrt{\frac{2\pi f_T\left(\frac{1}{C_d}+\frac{1}{C_\pi}\right)}{r_B}} \tag{3}$$

Here, C_π is base-emitter capacitance of transistor, r_B is base resistance, and f_T is cut-off frequency. From (3), the capacitive-degeneration cross-coupled pair has higher conductance-zero frequency point than the common cross-coupled pair, and hence there is less possibility to miss the hold-data phase at higher operating frequency. In the capacitive-degeneration divider, drawbacks such as local instabilities and unwanted oscillations could be expected. Nevertheless, a careful choice of C_d and tail current I_L results in a high free-running switching time so that oscillations do not start due to the current steering of the bottom differential pair operating at the input clock frequency [Girlando, 2005]. The method finding the optimum values of C_d and I_L is illustrated in the simulation curves of Fig. 14 through which their values are set to guarantee both operating speed as high as possible and no oscillation in the divider. First, the proper value of I_L should be set within the range of no oscillation. The Nyquist diagram of Fig. 14(a) shows the divider oscillates above 2.5mA of I_L, and then, the tail current is set by 1.5mA in this design considering power and speed. The clockwise encirclement of "1" at the horizontal axis of the Nyquist polar chart means that the transfer function of the divider circuit has poles in the right half plane i.e, it oscillates [Paul,

2001][Lee, 2002]. Second, after fixing I_L, we must check whether the divider oscillates by sweeping the value of C_d. As shown in the Nyquist diagram of Fig.14(b), the divider oscillates over 900f, and the optimum value of C_d is set to 100fF, considering process variation and speed. The value of C_d is the smaller, the higher increases the f_G of Eq. (3). When C_d is fixed to 100fF, the conductance zero frequency point of the divider is simulated by 84GHz, as shown in Fig.14(c)

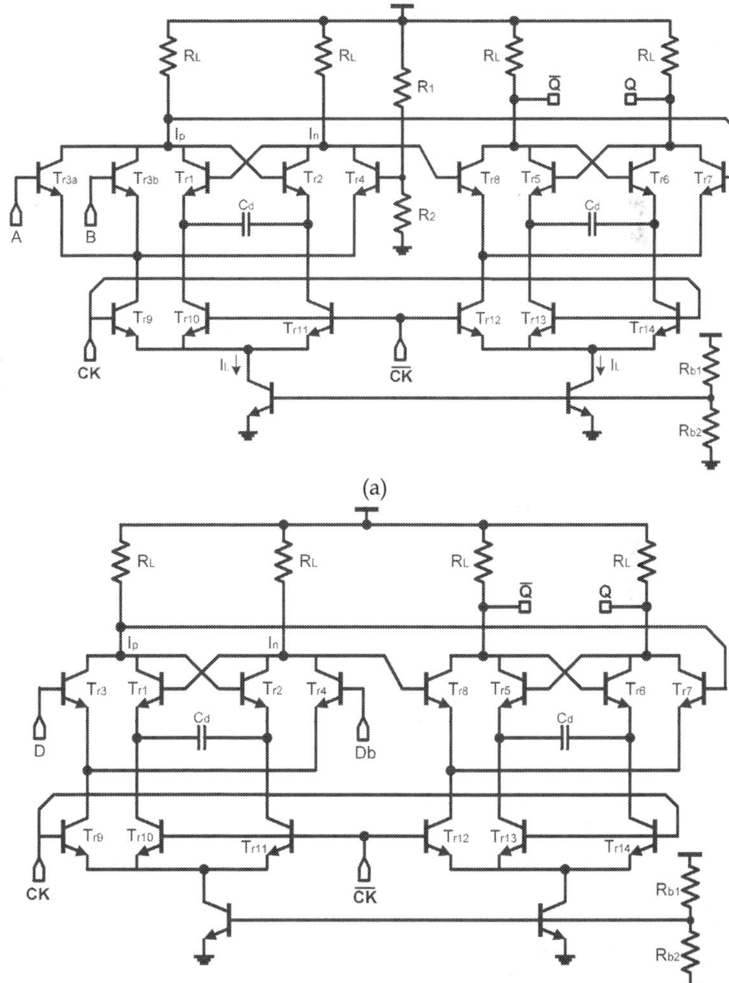

Fig. 13. (a) E²CL D-type flipflop with embedded NOR gate (b) E²CL D-type flipflop

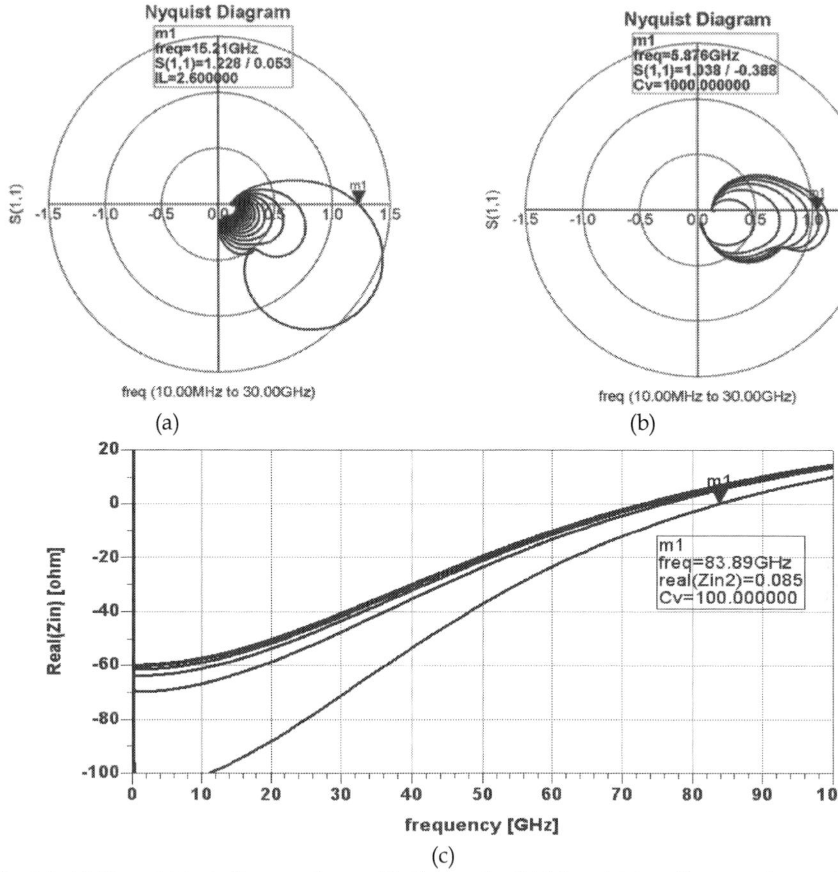

Fig. 14. (a) Nyquist test diagram for oscillation vs I_L, (b) Nyquist test diagram for oscillation vs C_d, (c) f_G simulation vs C_d, here $C_d = C_v$

The CML DFF used in the divide-by-5 circuit is made up of a cascade of a master D-latch and a slave latch with the clocks reversed in the second ones as shown in Fig. 15. The differential clocks steer the current of the current source from one side to the other side, and from the tracking mode to the hold mode. The values of load resistors are set to be as large as possible to confirm high speed at low current consumption. Transistors such as M1, M2, M3, and M4 are sized just large enough to be able to completely steer the current at worst case. Transistors M5 and M6 must be just large enough to quickly regenerate the current state during the hold mode. Finally, the current magnitude of Is must be high enough to allow a large swing at the output node and not limit switching bandwidth [Lam, 2000].

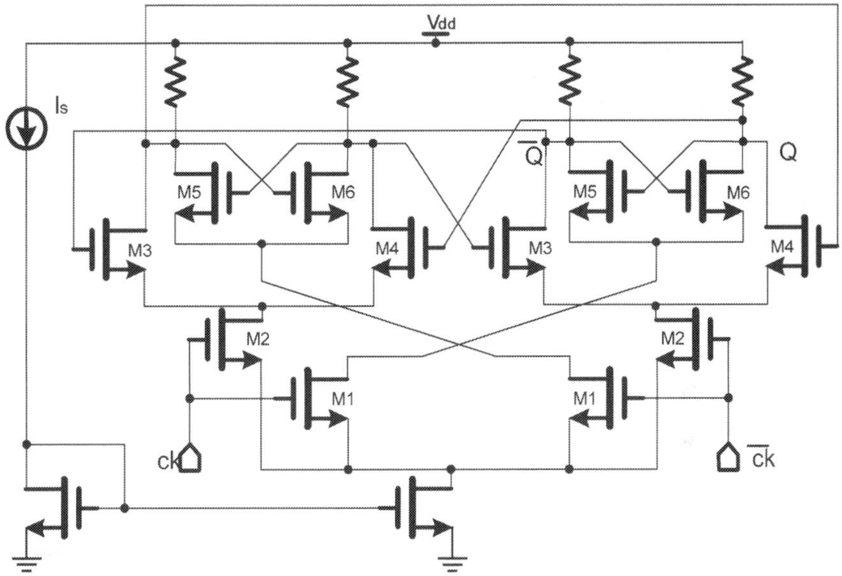

Fig. 15. CML D-type flipflop

As static logics require single-ended rail-to-rail swing, the non-rail-to-rail differential swing of the prescaler must be converted appropriately. A differential-to-single-ended signal level converter (DSC) must be inserted at the output Q_2 of the CML DFF in figure 10. The simplest circuit for this task is the four-transistor circuit shown in Fig. 16. The differential outputs of the CML DFF drive the input PMOS transistors (P1, P2), and then the single-ended output is at the drains of P2 and N2. P2 charges the output, and N2 discharges it.

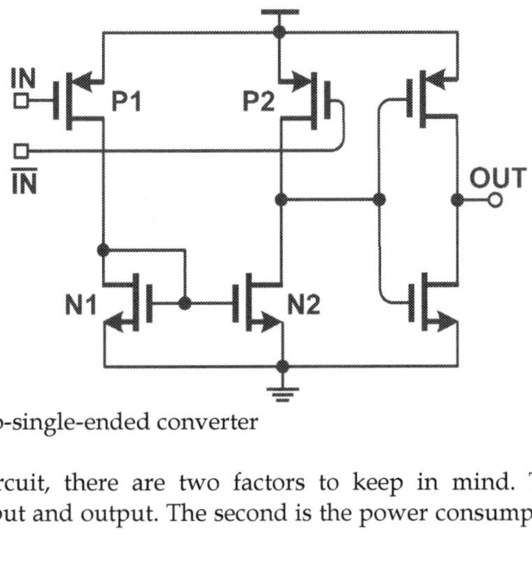

Fig. 16. Differential-to-single-ended converter

In designing this circuit, there are two factors to keep in mind. The first is the load capacitance at the input and output. The second is the power consumption since the current is not fixed and it must still be operated at a relatively high frequency. A current source cannot be used to bias this circuit because a rail-to-rail swing at the output is required. The size of input PMOS pairs must be as small as possible while providing the necessary current to charge the output node. The amount of current N2 gets to discharge the output node is determined by the current mirror configuration of N1 and N2. Thus, those transistors can almost be minimum size and still provide enough current to discharge the output node if N1 is smaller than N2. This results in a multiplication of the current through N1 to N2. In this design, N2 is 1.5 times larger than N1. For sharper rise and fall edges, one inverter is added at its output.

Fig. 17 represents a single-phase CML OR/NOR logic, which receives two single-phase inputs and then outputs complementary differential logic signals. In the CML logic, both road resistor and current-mirror transistor should be optimally sized to achieve high speed and low current at the same time. The gate voltage of inverted MOS transistor outputting Q is fixed by the voltage divider configured with R_1 and R_2, which determins the output logic level.

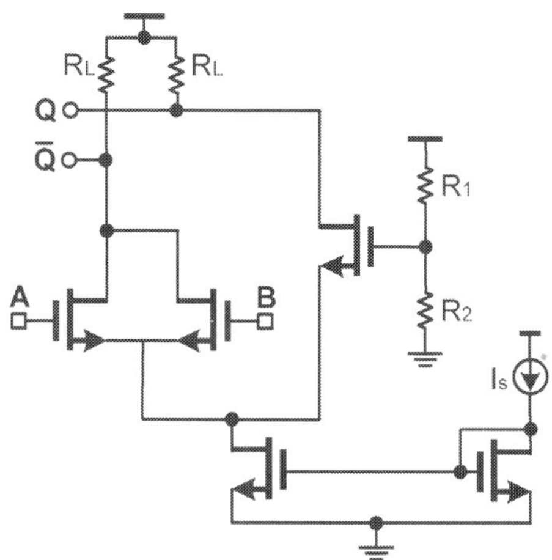

Fig. 17. Single phase complementary OR/NOR logic

3.4 Design of LC-tank VCO

In this section, a 26-GHz LC-tank VCO with 6 % tuning range is described. Here, we should design only a 26-GHz VCO because a 52-GHz frequency doubler follows it. Fig. 18(a) illustrates the circuit diagram of the 26-GHz LC-tank VCO used in the 52-GHz frequency synthesizer. It is a basic balanced differential oscillator that uses a cross-coupled differential pair. In the VCO circuit, the cross-coupled pair consisting of Q1 and Q2 generates negative conductance to compensate the LC-tank loss. In Fig. 18(a), one of three 700-pH inductors is used in the LC-tank resonator, another is connected to the collector node of oscillation transistors(Q_1 and Q_2), the remaining inductor is used as the load impedance of the common- emitter amplifier.

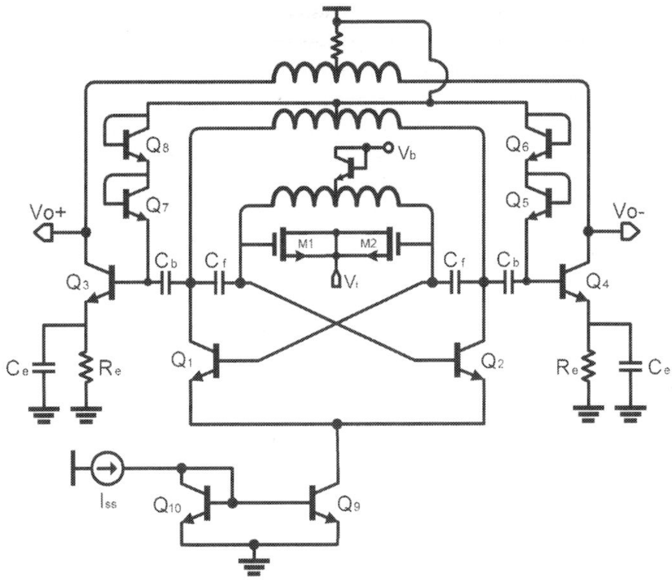

(a)

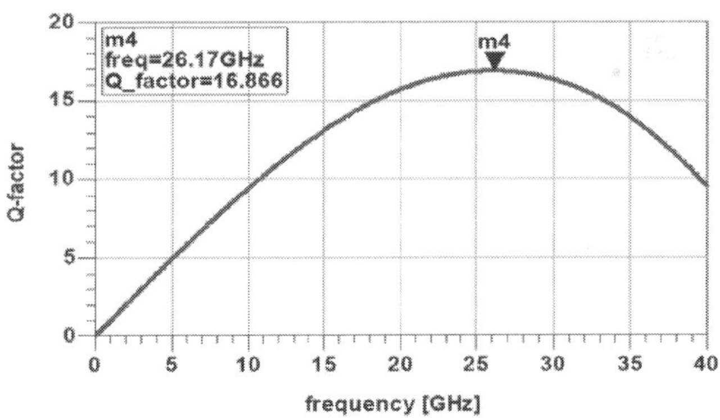

(b)

Fig. 18. (a) LC-tank VCO circuit (b) Differential Q-factor of center-tapped inductor

As shown in Fig. 18(b), the center-tapped inductor represents a quality factor of 16.8 around 26GHz. For compensating loss due to the resistance component of inductor and guaranteeing enough oscillation, a cross-coupled pair having much larger negative conductance around 26 GHz should be used. That is, since only the cross-coupled pair does not replenish enough energy causing oscillation around 26GHz due to the large loss of inductor, in Fig.18(a), the feedback capacitor C_f is inserted into the positive feedback path of the cross-coupled pair, and thus negative conductance is increased. The feedback capacitor has a role to block DC flow and couple the RF signal power. Also, C_f prevents the forward bias of the base-collector junction of the oscillation transistor, which results in high negative conductance as well as high oscillation signal amplitude. The high signal swing lowers phase noise of VCO. That is, the negative conductance is pulled up to higher frequency and increased. Both input negative resistance and effective input capacitance of the cross-coupled pair with feedback capacitor can be estimated as (4) and (5) [Veenstra, 2004][Jung, 2004].

$$R_{in} = \frac{2(r_b + r_e)^2 + 2\left[\frac{1}{\omega C_f} + \left(\frac{\omega_T}{\omega}\right)\left(\frac{1}{g_m} + r_e\right)\right]^2}{(r_b + r_e) - \left(\frac{\omega_T}{\omega}\right)^2\left(\frac{1}{\omega_T C_f} + \frac{1}{g_m} + r_e\right)} \quad (4)$$

Here, g_m is transconductance, C_f is feedback capacitance, r_e is intrinsic emitter resistance, and r_b is intrinsic base resistance. In (4), negative resistance decreases with frequency, and then the zero-point frequency negative resistance becomes zero is finally reached. Therefore, the addition of the feedback capacitor in the cross-coupled path raises the zero-point frequency upward higher frequency band. This is proved by the factor of $(1/\omega C_f)^2$ in the nominator of (4).

$$C_{in} = \frac{\left(\frac{\omega_T}{\omega^2}\right)\left(\frac{1}{\omega_T C_f} + \frac{1}{g_m} + r_b + 2r_e\right)}{2(r_b + r_e)^2 + 2\left[\frac{1}{\omega C_f} + \left(\frac{\omega_T}{\omega}\right)\left(\frac{1}{g_m} + r_e\right)\right]^2} \quad (5)$$

As shown in (5), the effective input capacitance is a function of the feedback capacitance C_f. It is noted that the effective input capacitance decreases in proportional to the factor of $(1/\omega C_f)^2$ in the denominator. As a result, the oscillation frequency can be increased due to the reduced C_{in}.

Commonly, the quality factor of LC-tank resonator in VCO is degraded by the load connected to it, and therefore, the LC-tank resonator consisting of a center-tappled inductor and two NMOS varactors is wired on the base node of the cross-couled pair. As shown in Fig.19(a), the collector and base nodes of the cross-coupled pair is separated by C_f, which has a role to protect the LC-tank resonator against the load. Additionally, it is worth noting that the negative conductance of the cross-coupled pair is different, depending upon the position looking into it from the LC-tank resonator, as illustrated in Fig.19. The curve of G_{cin} represents the input negative conductance looking into the collector node of the cross-coupled pair, and the bold line serves as the curve of G_{bin} looking into the base node of the cross-coupled pair. In Fig. 19, it is clearly apparent that G_{bin} is greater than G_{cin} about 25GHz frequency. That is, G_{bin} could be made greater than G_{cin} in some target frequency range by tuning C_f and tail current, which results in larger oscillation amplitude and lower phase noise. In summary, the feedback capacitor C_f does not only improve the loaded quality factor of the VCO, but also enlarge negative conductance at target frequency.

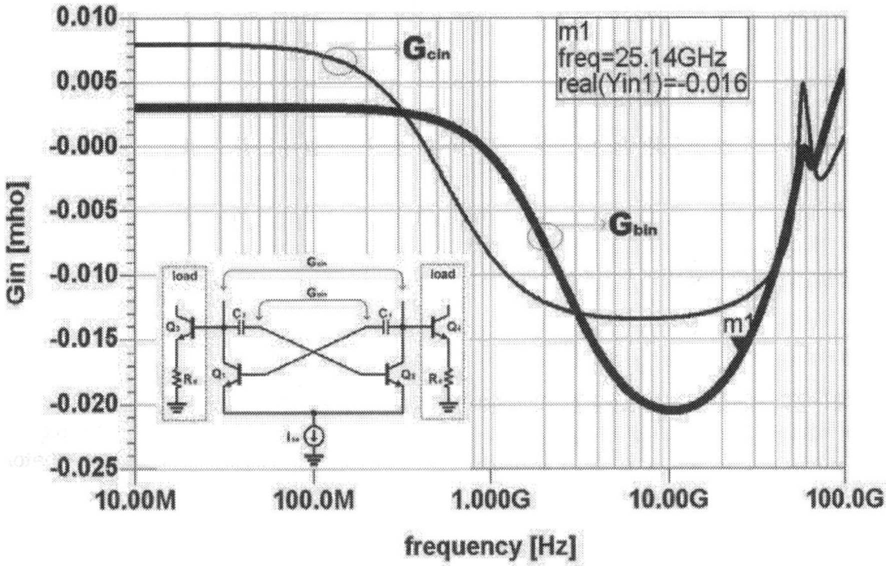

Fig. 19. Simulated input negative conductance of the cross-coupled pair

3.5 Design of 52GHz Frequency Doubler

A 52-GHz frequency doubler is presented as shown in Fig. 20. In the doubler circuit, the collector nodes of the differential amplifier configured with Q1 and Q2 are put together for extracting the even-mode signal. Also, another even-mode signal with different phase is extracted from the combined emitter node of the differential amplifier. The common-base amplifier Q3 is used for amplifying the even-mode signal extracted from the emitter node. Both Cm and Rm are used to tune the amplitude and phase difference between the signal extracted from the emitter node and the signal extracted from the collector node [Gruson, 2004]. The common-emitter amplifiers of Q4 and Q5 are used to amplify the extracted even-mode differential signals. Fig. 21 shows the simulated output spectrum of the frequency doubler, which suppresses the fundamental frequency component of 26GHz by 75dB, the third harmonic frequency component of 78GHz by 90dB, and the fourth harmonic component of 104GHz by 25dB. Since other harmonic components have been suppressed above 20dB, therefore, the second harmonic frequency component of 52GHz will show a linear sine waveform without distortion.

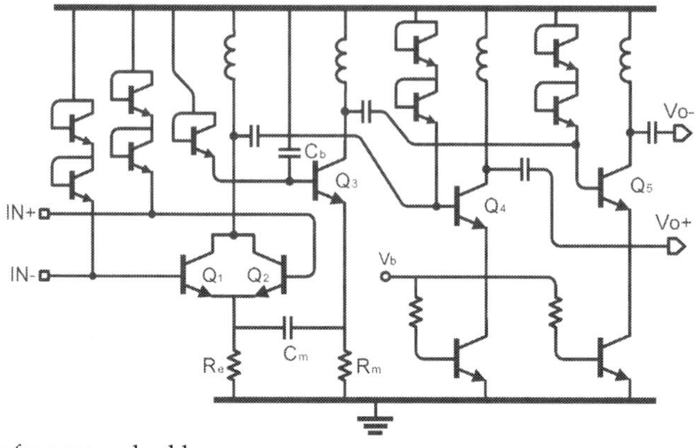

Fig. 20. 52GHz frequency doubler

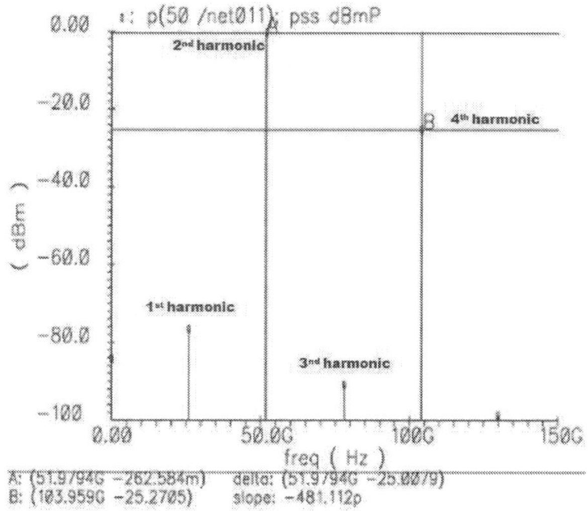

Fig. 21. Simulated output spectrum of the frequency doubler

In designing the 52GHz frequency synthesizer of Fig. 2, its full circuit is simulated using Cadence Spectre RF simulator. In the frequency synthesizer, the 3rd order loop filter is used, and is implemented by using poly resistor and MIM capacitor. In the test circuit, 262MHz reference frequency is used for close-loop simulation. Fig. 22 shows the simulated close-loop settling time of the frequency synthesizer, which is about 0.8μs.

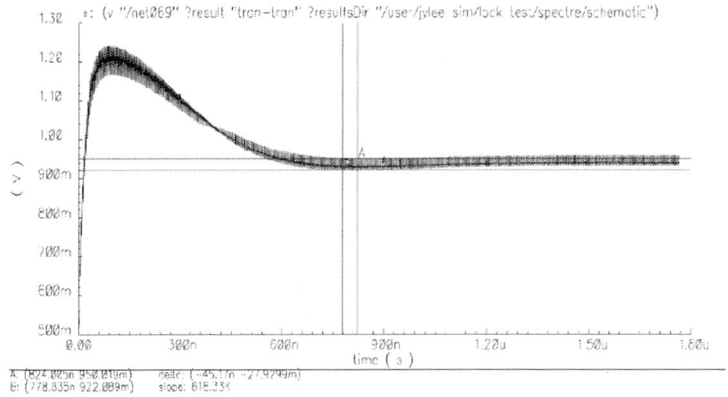

Fig. 22. Simulated close-loop settling time of the frequency synthesizer

4. Measured results

Fig.23 represents the chip microphotograph of the 52-GHz PLL synthesizer whose die area is 1.2mm² area including bonding pads. The PLL chip was designed and fabricated using 0.25-μm SiGe:C BiCMOS process technology. Both f_T and f_{max} of a HBT (hetero-junction bipolar transistor) used in this design are 180GHz and 200GHz, respectively. The PLL chip was measured using Agilent E4440A 26.5-GHz spectrum analyzer and 11970V harmonic mixer after it was mounted on probe station.

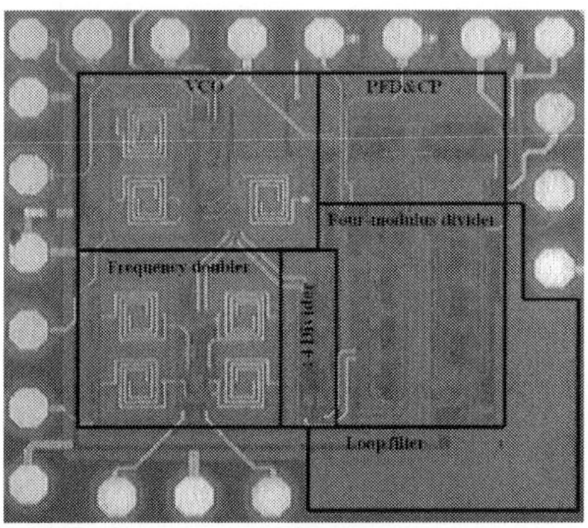

Fig. 23. Chip photograph of the frequency synthesizer

Fig. 24 shows the measured frequency tuning range of the cross-coupled differential LC VCO. Its tuning range is measured from 24.72GHz to 26.44GHz, consuming a current of 38 mA at 2.5 V. Fig. 25 shows the locked signal of 52.4GHz when 262 MHz is input to PFD and a divide ratio of ÷ 100 is selected. The PLL synthesizes two channels of 50.304GHz and

52.4GHz by 2.096GHz step. The spurious noise level is measured as - 42dBc, and this poor suppression about spurious noise is due to the small value of capacitors used in the loop filter. In this PLL chip, the size of the loop capacitors has been reduced as small as possible due to the limited chip area. The output power of the PLL is measured as - 17.6 dBm, and the decreased output power is due to both cable loss and unexpected low quality factor of the load inductor used in the amplifiers of Q_4 and Q_5. Fig. 26 represents the output power spectrum in span of 8MHz.

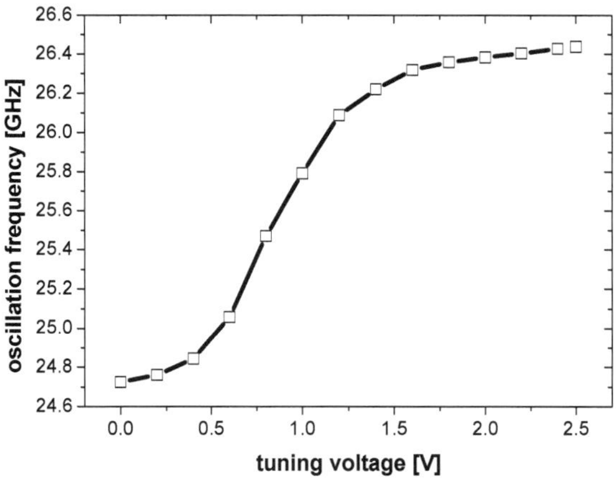

Fig. 24. Measured frequency tuning range of VCO

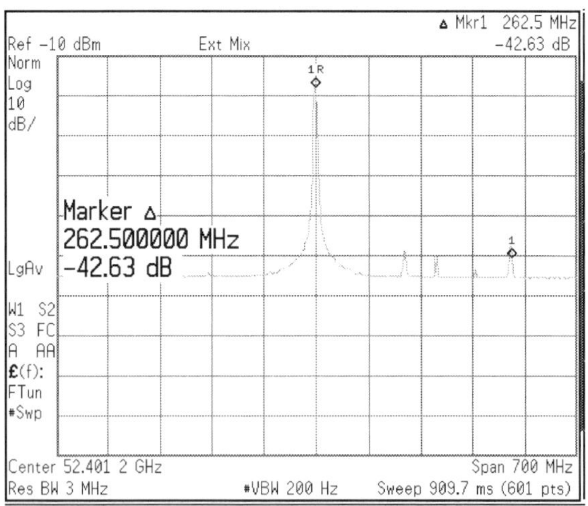

Fig. 25. Measured output spectrum of 52.4GHz locked carrier

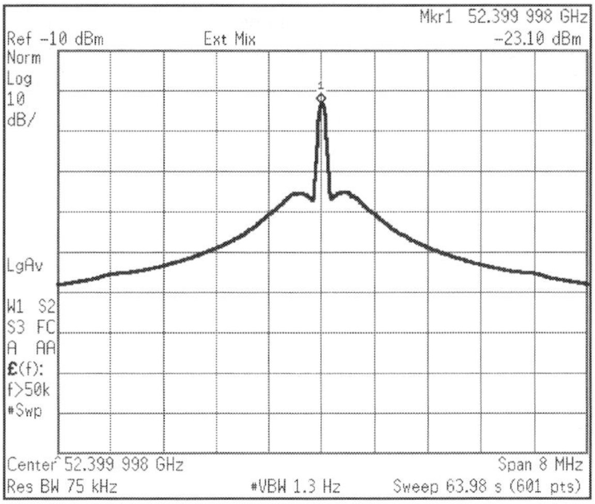

Fig. 26. Output spectrum of the PLL in span of 8MHz

Fig. 27 represents its measured phase noises, which are - 89dBc/Hz from 26.2GHz and - 81dBc/Hz from 52.4GHz, respectively, at 1MHz offset frequency. Its integrated RMS phase noise from 1MHz to 100MHz is estimated as 7.42°. The phase noise of the 52.4GHz second harmonic carrier is approximately estimated from formula (6) using the measured phase noise data of 26.2GHz first harmonic carrier in Fig. 27. Since the 26.2GHz carrier having a phase noise of - 109dBc/Hz at 10MHz offset doubles to 52.4 GHz due to the doubling operation of the frequency doubler, the phase noise of 52.4GHz carrier increases by 6 dB and becomes approximately - 102dBc/Hz due to the $(f_o)^2$ term of formula (6), which does almost fit to the measured phase noise curve of Fig. 27. That is, the phase noise of the 52.4-GHz carrier is degraded by about 7dB at offset frequency under 10MHz, compared with that of 26.2GHz fundamental carrier. This is close to the expected degradation of 6dB caused by the operation doubling frequency. Above 10MHz offset, the phase noise degradation of the doubled 52.4GHz carrier increases more and more due to the noise floor of the measurement system.

$$L(f_{off}) = 10\log\left\{\left(1+\frac{f_o}{2Qf_{off}}\right)^2\left(\frac{FkT}{2P_o}\left(1+\frac{f_{1/f^3}}{f_{off}}\right)\right)\right\} \quad (6)$$

Here, f_o is carrier frequency, Q is loaded factor, F is noise floor of active oscillator, k is Boltzman constant, P_o is signal power, f_{off} is offset frequency, and f_{1/f^3} is $1/f^3$ corner frequency [Lee, 2000][Leeson, 1966].

In Table 1, the measured results of the 52GHz PLL synthesizer are summarized. Here, the settling time of the PLL is simulated as 800ns in Fig. 22. The PLL chip consumes a total current of 160mA of which 45% is drawn by the programmable divider.

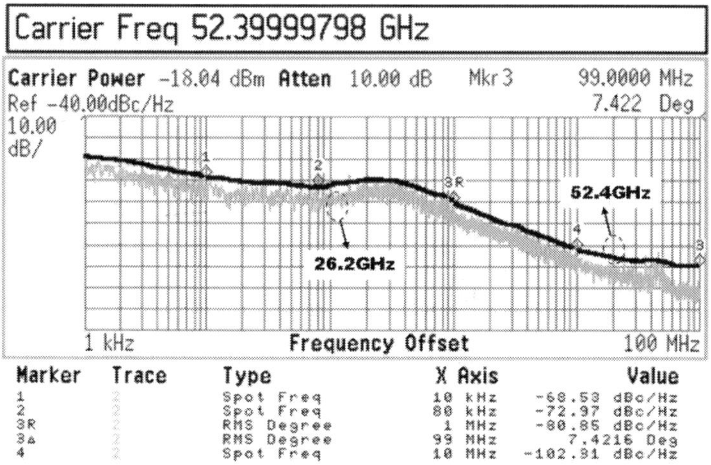

Fig. 27. Measured phase noise of the PLL

Technology	0.25µm SiGe:C BiCMOS
Supply voltage	2.5V
Reference frequency	262-264MHz
VCO tuning range	24.723 - 26.439GHz (6%)
PLL output frequency	24.9–26.50GHz
(PLL + doubler) output frequency	50.1 -52.8GHz
Loop bandwidth	600kHz-1 MHz
In-band phase noise @100kHz offset	-80 dBc/Hz from 26.2 GHz -73 dBc/Hz from 52.4 GHz
Out-band phase noise @1MHz offset	-89 dBc/Hz from 26.2 GHz -81 dBc/Hz from 52.4 GHz
Out-band phase noise @10MHz offset	-109 dBc/Hz from 26.2 GHz -102 dBc/Hz from 52.4 GHz
RMS Jitter	7.42° (from 1MHz to 100MHz)
Spurious noise level	< - 42 dBc
Settling time	< 800 ns
Current consumption	160mA at 2.5V
Chip size	1.0 mm × 1.2 mm

Table 1. Measured results of the 52GHz frequency synthesizer

5. Conclusion

In this chapter, we design and fabricate a 52GHz frequency synthesizer for 60GHz dual-conversion receiver using SiGe BiCMOS process technology. The designed PLL-based frequency synthesizer consists of a 26-GHz PLL and a 52-GHz frequency doubler. In the programmable divider, a capacitive-degeneration D-F/F is used to achieve high-speed operation. The method finding the optimum values of both degeneration capacitance and tail current is presented in order to attain high speed and guarantee no self-oscillation in the degeneration D-F/F circuit. A cross-coupled differential LC VCO with feedback capacitor is designed to generate 26GHz oscillation frequency. By tuning feedback capacitance and tail current properly, the input negative conductance at the base node of the cross-coupled pair could be enlarged at target frequency, and also, the feedback capacitance stops the loaded quality factor of VCO from being degraded by the load.

The 52GHz PLL synthesizer provides two channels of 50.304GHz and 52.4GHz by 2.096GHz step through the frequency doubler. The phase noises of the PLL are measured as -89dBc/Hz at 1MHz offset from 26.2GHz first harmonic carrier, and - 81dBc/Hz at the same offset frequency from 52.4GHz second harmonic carrier. The PLL consumes 160mA at 2.5V and takes silicon-die area of 1.2mm^2.

6. References

Schott, Reynolds et al. (2006). A Silicon 60-GHz receiver and transmitter chipset for broadband communications, *IEEE J. Solid-State Circuits,* Vol.41, No. 12, (December and 2006) (2820-2831), ISSN 0018-9200

Razavi, B. (2001). *Design of Analog CMOS Integrated Circuits,* McGraw-Hill International edition, (550-566), ISBN 0-07-237371-0, U.S.A

Razavi, B. (2006). A 60-GHz CMOS receiver front-end, *IEEE J. Solid-State Circuits,* Vol.41, No. 1, (January and 2006) (17-22), ISSN 0018-9200

Lee, J-Y et al. (2008). A 28.5-32-GHz fast setttling mutichannel PLL synthesizer for 60-GHz WPAN radio, *IEEE Trans. Microwave Theory Techniques,* Vol.56, No. 5, (May and 2008) (1234-1246), ISSN 0018-9480

Floyd, B. A. (2008). A 16-18.8-GHz sub-integer-N frequency synthesizer for a 60-GHz transceivers, *IEEE J. Solid-State Circuits,* Vol.43, No. 5, (May and 2008) (1076-1086), ISSN 0018-9200

Winkler, W. et al. (2005). A fully integrated BiCMOS PLL for 60GHz wireless applications, *Proceedings of Int. Solid-State Circuit Conf.,* pp. 406-407, ISSN 0193-6530

Lee, C. et al. (2007). A 58-to-60.4GHz frequency synthesizer in 90nm CMOS, *Proceedings of IEEE Int. Solid-State Circuit Conf.,* pp. 196-197, ISSN 0193-6530

Mansuri, M. et al. (2002). Fast frequency acquition phase-frequency detectors for Gsamples/s phase-locked loops, *IEEE J. Solid-State Circuits,* Vol.37, No. 10, (October and 2002) (1331-1334), ISSN 0018-9200

Tak, G. Y. et al. (2005). A 6.3-9-GHz CMOS fast settling PLL for MB-OFDM UWB applications, *IEEE J. Solid-State Circuits,* Vol.40, No. 8, (August and 2005) (1671-1677), ISSN 0018-9200

Rhee, W. (1999). Design of high-performance CMOS charge pumps in phase-locked loops, *Proceedings of IEEE International Sym. On Circuits and Systems (ISCAS),* vol.2, pp. 545-548, ISBN 0-7803-5471-0

Magnusson, H. ;Olsson, H. (2003). Design of a high-speed low-voltage (1V) charge-pump for wideband phase-locked loops, *Proceedings of 10th IEEE International Conf. On Electronics, Circuits and Systems (ICECS),* vol.1, pp. 14-17, ISBN 0-7803-8163-7

Bahreyni, B. (2002). A novel design for deadzone-less charge-pump with low harmonic content at the output, *Proceedings of 45th Midwest Sym. On Circuits and Systems (MWSCAS)*, vol.3, pp. 397-400, ISBN 0-7803-7523-8

Park, B. H. & Allen .P. E. (1998). A 1GHz, low-phase-noise CMOS frequency synthesizer with integrated LC VCO for wireless communications, *Proceedings of IEEE Custom Integrated Circuits Conf.*, pp. 567-570, ISBN 0-7803-4292-5

Collins, T. E. et al. (2005). Design analysis and circuit enhancements for high-speed bipolar flip-flops, *IEEE J. Solid-State Circuits,* Vol.40, No.5, (May and 2005) (1166-1174), ISSN 0018-9200

Girlando, G. et al (2005). A monolithic 12-GHz heterodyne receiver for DVB-S applications in silicon bipolar technology, *IEEE Trans. Microwave Theory Techniques,* Vol.53, No.3, (March and 2005) (952-959), ISSN 0018-9480

Paul, R. G. et al. (2001). Analysis and *Design of Analog Integrated Circuits, John Willy & Sons* 4th edition, (733-741), ISBN 0-471-32168-0, U.S.A

Lee, J-Y et al. (2003). An 1.8GHz voltage-controlled oscillator using current-current negative feedback network, *Proceedings of 6th European Conf. On Wireless Technology (EuWiT)*, pp. 113-116, ISBN 2-9600551-5-2

Lam, C. & Razavi, B. et al. (2000). A 2.6-GHz/5.2-GHz frequency synthesizer in 0.4-μm CMOS technology, *IEEE J. Solid-State Circuits,* Vol.35, No.5, (May and 2000) (788-794), ISSN 0018-9200

Veenstra, H & Heijden E. (2004). A 35.2-37.6GHz LC VCO in a 70/100GHz f_T/f_{max} SiGe technology, *Proceedings of IEEE Int. Solid-State Circuit Conf.*, pp. 394-395, ISSN 0193-6530

Jung, B. & Harjani, R. (2004). High-frequency LC VCO design using capacitive degeneration, *IEEE J. Solid-State Circuits,* Vol.39, No.12, (December and 2004) (2359-2370), ISSN 0018-9200

Gruson, F. et al. (2004). A frequency doubler with high conversion gain and good fundumental suppression, *Proceedings of IEEE International Microwave Sym. Dig. (IMS)*, vol.1, pp. 175-178, ISBN 0-7803-8331-1

Lee, T.H. & Hajimiri, A. (2000). Oscillator phase noise : a tutorial, *IEEE J. Solid-State Circuits,* Vol.35, No.3, (March and 2000) (326-336), ISSN 0018-9200

Leeson, D.B. (1966). A simple model of feedback oscillator noise spectrum, *Proceedings of the IEEE*, vol.54, No. 2, (February and 1966) (329-330), ISSN 0018-9219

13

Metamaterial Transmission Line and its Applications

Changjun Liu and Kama Huang
School of Electronics and Information Engineering,
Sichuan University, *China*

1. Introduction

Metamaterial structures have found a wide interest around the world since the properties of left-handed media were proposed for the first time in 1967 by a Soviet physicist Veselago. Besides many successful investigations on three-dimensional metamaterials, e.g. invisible cloaks and perfect lens, there are many researches on two-dimensional and one-dimensional metamaterials. Homogeneous negative index transmission lines or left-handed transmission lines are, generally speaking, metamaterial transmission lines, which belong to one-dimensional metamaterials. They do not exist in nature, and have to be approached by some artificial structures, which are usually constructed from a series of discontinuous sections operating in a restricted frequency range.

A typical realization of metamaterial transmission line is found in a quasi-lumped transmission line with elementary cells consisting of a series capacitor and a shunt inductor. As in practice, the normal shunt capacitance and series inductance cannot be avoided, the concept of the composite right/left-handed (CRLH) transmission line was developed, and a number of novel applications have been demonstrated.

In this chapter, we focus on metamaterial transmission line designs and applications. Firstly, the concept of metamaterial transmission line is introduced briefly. Secondly, the circuit models, which facilitate the analysis of metamaterial transmission lines, are discussed. The relations between composite right/left-handed transmission line and band-pass filters are analyzed. Finally, some applications of metamaterial transmission lines in microwave components, e.g. diplexers, baluns, and power dividers, are presented.

2. Basic models

2.1 Full circuit models

The equivalent circuit model of a conventional right-handed transmission line is shown in Fig. 1(a). It consists of series inductors and shunt capacitors, of which the dimensions are much less than the wavelength of the operating frequency. It is well known that many important characteristics, such as characteristic impedance, phase velocity, dispersions and so on, can be obtained from the circuit model.

The equivalent circuit model of a left-handed transmission line is shown in Fig. 1(b), which is a dual of Fig. 1(a). All series inductors in the right-handed transmission line model are replaced by capacitors in the left-handed transmission line model, and all shunt capacitors are substituted by inductors. It is an ideal model, which does not exist in nature.

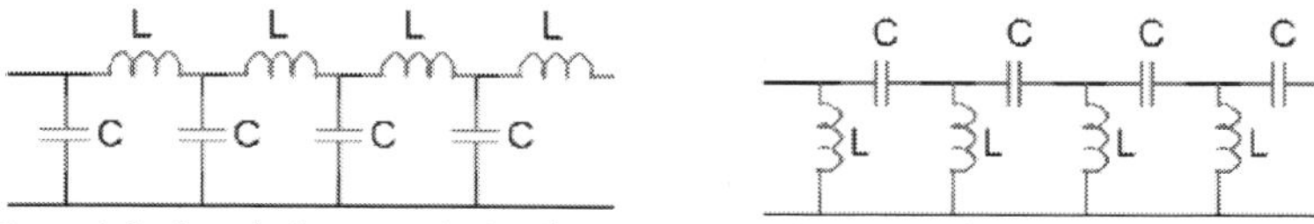

(a) Pure right-handed transmission line (b) Pure left-handed transmission line

Fig. 1. Equivalent circuit model of right-handed and left-handed transmission line

A composite right/left-handed transmission line model, as shown in Fig. 2, is more suitable than the left-handed transmission line circuit model, since the parasite series inductance and shunt capacitance cannot be avoided in nature. It consists of series resonators L_R and C_L and shunt resonators C_R and L_L, where the subscript "L" and "R" denote left-handed and right-handed, respectively. This transmission line circuit model is a combination of left-handed and right-handed transmission line. At low frequency, C_L and L_L are dominant, the transmission line shows left-handed characteristics; at high frequency, L_R and C_R are dominant, the transmission line shows right-handed characteristics.

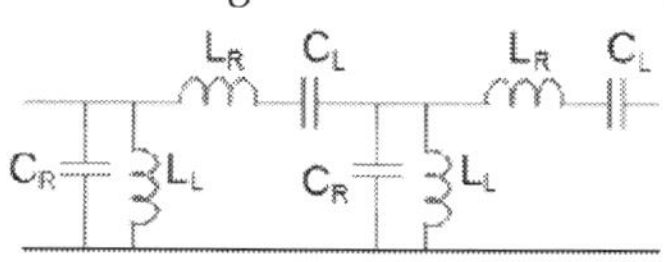

Fig. 2. Equivalent circuit model of composite right/left-handed transmission line

There is no band-gap between left-handed and right-handed regions, if a so-called balanced condition is satisfied. The balanced condition is

$$L_L C_R = L_R C_L \tag{1}$$

which implies the series and shunt LC resonators have the same resonant frequency - transition frequency - ω_0. The characteristic impedance of the composite right/left-handed transmission line is

$$Z_E = Z_L = Z_R \tag{2}$$

where $Z_L = \sqrt{\frac{L_L}{C_L}}$ and $Z_R = \sqrt{\frac{L_R}{C_R}}$ are the pure left-handed and right-handed characteristic impedances which are frequency independent with the homogenous transmission line approach. The cut-off frequencies of the composite right/left-handed transmission line is

$$\begin{cases} \omega_{cL} = \omega_R \left(\sqrt{1 + \frac{\omega_L}{\omega_R}} - 1 \right) \\ \omega_{cR} = \omega_R \left(\sqrt{1 + \frac{\omega_L}{\omega_R}} + 1 \right) \end{cases} \tag{3}$$

where $\omega_L = \dfrac{1}{\sqrt{L_L C_L}}$ and $\omega_R = \dfrac{1}{\sqrt{L_R C_R}}$ are the resonant frequencies of the left-handed and right-handed LC circuit, respectively. The transition frequency can be written as

$$\omega_0^2 = \omega_{cL}\omega_{cR} = \omega_L \omega_R = \frac{1}{\sqrt{L_L L_R C_L C_R}} \tag{4}$$

Considering the degree of freedom in the composite right/left-handed transmission line design with periodic elements, there are four parameters, namely L_L, C_L, L_R, and C_R. When the balanced condition is applied, there are only three independent parameters left. Once two cut-off frequencies and the characteristic impedance Z_E (matching to the system impedance Z_0) are fixed, a unique CRLH TL configuration is determined

$$\begin{cases} \omega_0 L_R = \dfrac{1}{\omega_0 C_L} = \dfrac{\omega_0}{\omega_{cR} - \omega_{cL}} 2Z_0 \\ \omega_0 C_R = \dfrac{1}{\omega_0 L_L} = \dfrac{\omega_0}{\omega_{cR} - \omega_{cL}} \dfrac{2}{Z_0} \end{cases} \tag{5}$$

2.2 Relation between filters and metamaterial transmission line models

The circuit model of a composite right/left-handed transmission line is very close to a high order band-pass filter, as shown in Fig. 3. The circuit model is a periodic structure, while a band-pass filter is usually not. However, for high order Chebyshev filters, the central part is a periodic structure as well. Are there some relations between circuit models of Chebyshev filters and composite right/left-handed transmission line? We review the standard band-pass filter design procedure and reveal the relation between them.

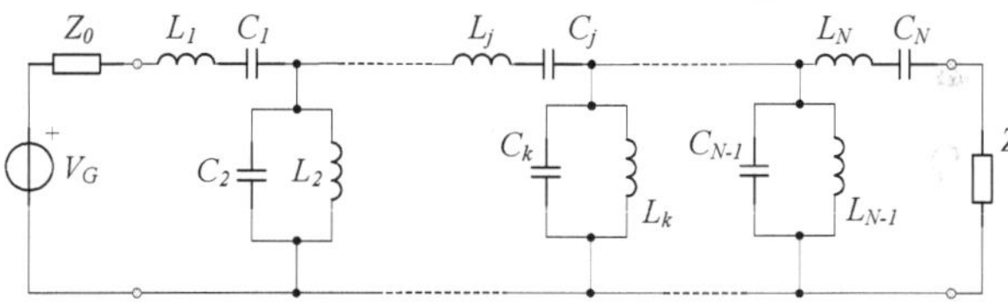

Fig. 3. Equivalent circuit of a band-pass filter

An Nth order band-pass filter in principle has the same LC circuit model. The band-pass filter design is usually achieved from the low-pass to band-pass transformation, in which a low-pass prototype filter is applied. The mapping formulas is

$$\begin{cases} \omega_0 L_j = \dfrac{1}{\omega_0 C_j} = \dfrac{\omega_0}{\omega_2 - \omega_1} g_j Z_0 & \text{for series resonators} \\ \omega_0 C_k = \dfrac{1}{\omega_0 L_k} = \dfrac{\omega_0}{\omega_2 - \omega_1} \dfrac{g_k}{Z_0} & \text{for shunt resonators} \end{cases} \tag{6}$$

where g_i is the ith element value (either the inductance or the capacitance) in a prototype low-pass filter, ω_1 and ω_2 are the lower and higher cut-off frequencies, respectively, and $\omega_0 = \sqrt{\omega_1\omega_2}$ is the central frequency of the band-pass filter. Z_0 is the system impedance.

Once these parameters are fixed, the band-pass filter is uniquely determined. From Eq. (6), it can be obtained

$$L_j C_j = L_k C_k = \frac{1}{\omega_0^2} \tag{7}$$

which means that the balanced condition of a composite right/left-handed transmission line always holds in a band-pass filter design. Since the mapping formula is a generic formula, it may be applied to other kinds of prototype LPF as well. Thus the balanced case of a composite right/left-handed transmission line is automatically realized in band-pass filters from any kind of prototype low-pass filter with series and shunt LC resonators that are built from the mapping formula Eq. (6). Butterworth, Gaussian, or Chebyshev band-pass filters with any pass-band ripple constructed from Eq. (6) will satisfy the balanced condition of a composite right/left-handed transmission line. In most prototype low-pass filters, the element values g_i usually vary in a certain range and lead to a non-periodic structure. The central section of a high order Chebyshev filter, however, has a periodic structure and is very close to a CRLH TL. Fig. 4 shows an example of a 21st order Chebyshev prototype LPF with different pass-band ripples. For elements not close to either end, the element values are periodic. It should be noted that always two adjacent filter elements form one equivalent transmission line cell, thus the central part of a Chebyshev is a periodic structure, independently of the ripple.

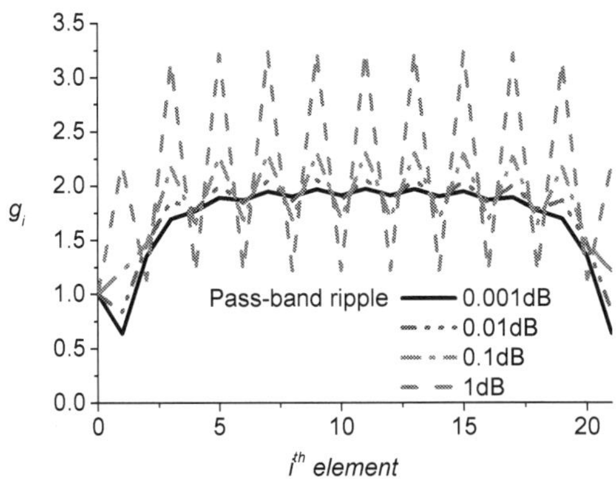

Fig. 4. Element values of 21st order Chebyshev prototype low-pass filters

As shown in Fig. 4, the lower the pass-band ripple, the smoother are the element values. In the limiting case, the values of the central elements are close to g_i=2. Moreover, the higher the filter order, the better is the approach to a periodic structure. When the element value is 2, Eq. (5) and Eq. (6) are same. Thus, when the Chebyshev band-pass filter and the composite right/left-handed transmission line have the same cut-off frequencies ($\omega_1 = \omega_{cL}$, $\omega_2 = \omega_{cH}$) and system impedance Z_0, they will own the same LC circuit determined by Eq. (5) and Eq. (6), respectively. This implies that once three parameters (two cut-off frequencies and the matching impedance) are fixed, the corresponding composite right/left-handed transmission line in the balanced case and the central part of the corresponding high order Chebyshev BPF with low pass-band ripple have identical LC configurations.

Once two cut-off frequencies and the system impedance are given, a uniform composite right/left-handed transmission line in the balanced case and a high order Chebyshev band-pass filter with low pass-band ripple can be uniquely implemented with LC circuit, respectively. The composite right/left-handed transmission line and the central part of the Chebyshev BPF own the same periodic LC structures. In other words, a uniform composite right/left-handed transmission line in the balanced case can be considered a part of a high-order low-ripple Chebyshev band-pass filter with the same cut-off frequencies and system impedance.

Therefore, it has been proved that a composite right/left-handed transmission line - in the balanced case - is identical to the central part of a high-order low pass-band ripple Chebyshev band-pass filter with the same characteristic impedance and cut-off frequencies. Theoretically the composite right/left-handed transmission line can be analyzed as a special band-pass filter.

There is an extra parameter in Chebyshev band-pass filters - the pass-band ripples. The central part of a high-ripple high-order Chebyshev band-pass filter is equivalent to a composite right/left-handed transmission line with mismatched impedance to the source and load. With similar analysis, a dual-composite right/left handed transmission line can be regarded as a part of a high order Chebyshev band-stop filter with low pass-band ripples. Thus, the relations between composite right/left handed transmission lines and filters have been presented. The negative phase velocity exists in filters for a long time. But it has not been studied as in a metamaterial transmission line. The discussion on the relations between filters and composite right/left-handed transmission lines give one more freedom in metamaterial transmission line design. It helps the impedance matching, e.g. achieving a broader pass-band with better frequency responses near cut-off frequencies. It helps the design of metamaterial transmission lines from classical filter theory.

3. Implementations of metamaterial transmission lines

3.1 Composite right/left-handed transmission line realization

A common metamaterial transmission line is a composite right/left-handed transmission line, which is shown in Fig. 5. It is a periodic structure, which consists of series capacitors, i.e. interdigital capacitors, and shunt inductors, i.e. short-ended microstrip lines. There are series inductors and shunt capacitors as well due to the parasite effects. The dimension of each unit should be much less than the guided wavelength, e.g. $p < \lambda_g/5$. Otherwise, it cannot approach a transmission line from discrete units.

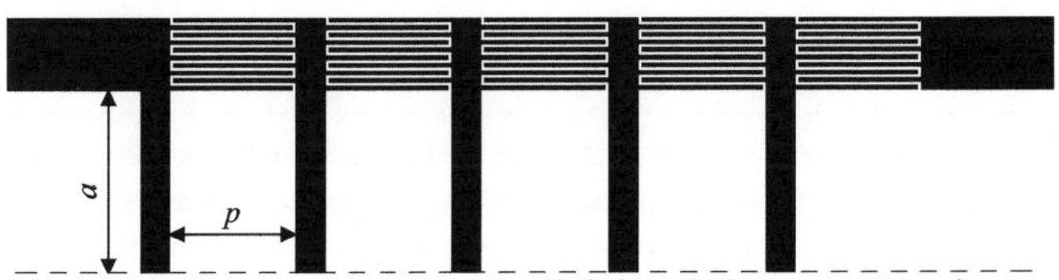

Fig. 5. Microstrip structure of a composite right/left-handed transmission line

The balanced condition can be achieved by adjusting the series and shunt resonators, respectively. The bandwidth of this structure is much limited. The pass-band of upper frequency branch becomes worse due to the self-resonance of the interdigital capacitors. The frequency response of the right-handed region is not as good as the left-handed region. The other drawback is the high insertion loss of the metamaterial transmission line from the interdigital capacitors and the radiation from shunt short-ended microstrip lines.

A microstrip directional coupler at 2.45 GHz is shown in Fig. 6. It is composed of a composite right/left-handed transmission line (the lower one) and a conventional microstrip transmission line (the upper one). Tight coupling is achieved in this coupler. It is a quasi-zero dB directional coupler, which means all incident power is coupled instead of being transmitted. The structure of the composite right/left-handed transmission line is as in Fig. 5.

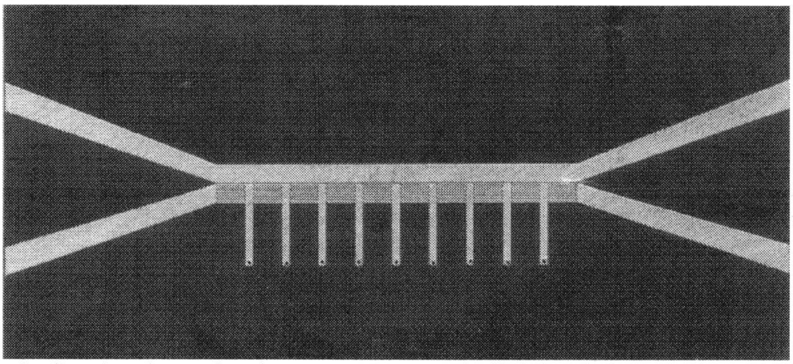

Fig. 6. A directional coupler from composite right/left-handed transmission line and conventional microstrip transmission line

At the transition frequency, there is a phenomenon called zeroth order resonance. The metamaterial transmission line share the same phase and the phase velocity goes to zero, which implies the wavelength is infinite. However, the group velocity is not zero and the incident power can be delivered.

The composite right/left-handed transmission line has found many successful applications, and extends the performance of conventional microwave components.

3.2 Composite right/left-handed transmission lines with lumped elements

Based on the theory of composite right/left-handed transmission line and the LC circuit network, a composite right/left-handed transmission line using surface mounted capacitors and short-ended microstrip lines is available. The interdigital capacitors are replaced by lumped capacitors, e.g. ATC chip capacitors. With chip capacitors, a higher series capacitance is easier to reach, and the metamaterial transmission line is more compact. Moreover, the limitation of self-resonance of interdigital capacitors has been released, and a better frequency response can be achieved. A composite right/left-handed transmission line with lumped capacitors is shown in Fig. 7. Six chip capacitors and five short-ended microstrip lines comprise a section of metamaterial transmission line.

The substrate of the metamaterial transmission line is F4B-2 with relative dielectric constant 2.65 and thickness 1 mm. Six chip capacitors are ATC600S series with 0603 package. The capacitance of the capacitor at each end is 5.6 pF, and the capacitance of the capacitors in between is 1.8 pF. The dimension of the microstrip line is: l_{RH} = 5.14 mm, w_{RH} = 2.78 mm, l_{ind} = 9.47 mm, and w_{ind} = 1 mm.

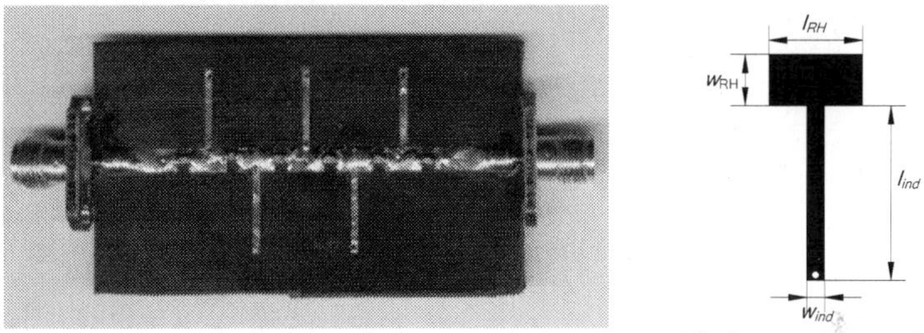

Fig. 7. A composite right/left-handed transmission line with lumped capacitors and the microstrip structure

The performance of the composite right/left-handed transmission line is fabricated and measured with Agilent N5230A vector network analyzer. The amplitude and phase response are shown in Fig. 8. Its bandwidth is from 1 GHz to 5 GHz while the center frequency is at 2.97 GHz, at which there is no phase shifting. When the frequency is between 1 GHz and 2.97 GHz, it works in the left-handed mode ($\beta < 0$); when the frequency is between 2.97 GHz and 5 GHz, it works in the right-handed mode ($\beta > 0$).

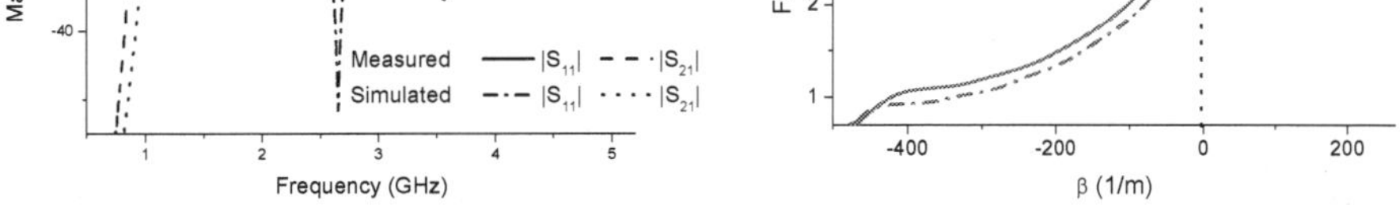

Fig. 8. Amplitude and dispersion of the metamaterial transmission line

This composite right/left-handed transmission line is compact and easy to fabricate, and the insertion loss is less than 1.3 dB, and the return loss is greater than 10 dB. Measurements agree well with simulation results. A broad pass-band is achieved with this metamaterial transmission line. The shortcoming of this metamaterial transmission line is that the performance is strongly dependent on the chip capacitors, which makes the cost higher as well.

4. Applications of metamaterial transmission lines

Metamaterial transmission lines have been applied to many fields successfully. They may extend the performance of conventional microwave components due their unique dispersion characteristics. On the other aspect, they may reduce the dimensions of some conventional microwave components greatly. In the fowling, several applications of metamaterial transmission line are presented.

4.1 Leaky-wave antenna

The kernel of the leaky-wave antenna is a metamaterial transmission line, which is built by microstrip multilayer structure. It contains two RT/Duroid 5880 substrates with relative dielectric constant ε_r=2.2 of thickness 0.254 mm and 1.58 mm, respectively. Microstrip resonators are uniformly and alternatively distributed in the top and the middle layer, as shown in Fig. 9, to form the equivalent series capacitance and shunt inductance required by the metamaterial transmission line. All microstrip resonators are identical. The unite element is a symmetric microstrip patch with length 16.8mm and width 7.4mm. Those unit elements in dash line, e.g. with number 2 and 4, are located in the middle layer, and those in solid line, e.g. with number 1, 3, and 5, are in the top layer. Therefore, a periodic microstrip structure is applied to approaching the uniform metamaterial transmission line.

The series capacitance in the metamaterial transmission line is from the overlapping between two unit elements in the top and the middle layer respectively. The overlapping region between them is 2.2mm, and the distance between them is 3.8mm. The distance between two adjacent elements is much less than the guided wavelength at the operating frequency. When the feeding is in balanced mode, the metamaterial transmission lines work in the odd mode as two parallel transmission lines. The symmetric plane of the metamaterial transmission line becomes a virtual ground due to the symmetry of the whole structure. The series inductance is from the unavoidable parasite effects and the width variation of the microstrip as well. The shunt inductance and capacitance are provided by the vertical branch connected to the virtual ground.

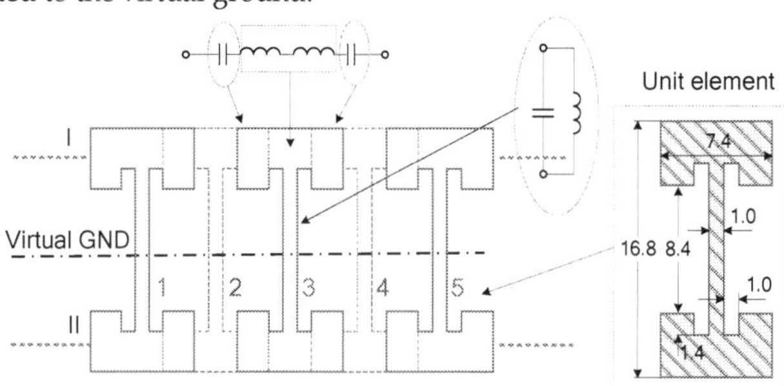

Fig. 9. Structure of the metamaterial transmission line with the equivalent circuits and the dimension of one unit element (© 2007 IEEE)

One segment of the above metamaterial transmission line is applied to build a resonator to analyze its characteristics. Two 1 mm gap capacitors are applied to feed the segment of the metamaterial transmission line at both ends, and make it resonate under the under-coupling situations. The metamaterial transmission line resonator has been simulated by the Sonnet software to obtain resonant frequencies. The simulated results of $|S_{21}|$ are shown in Fig. 10, where the number of unit elements in the metamaterial transmission line is from one to nine with only odd numbers.

The resonant frequency of one isolated unit element was 3.42 GHz, which increased to 3.95 GHz when more unit elements were coupled together. The zeorth order resonant frequency was between 3.3 GHz to 4.1 GHz. With the increase of the number of unit elements, the zeroth order resonant frequency approached to around 4 GHz, and fixed there. When there were more unit elements coupled together in the metamaterial transmission line, it was possible to obtain the resonance of the metamaterial transmission line in the left-handed or right-handed mode with dominant or high order modes. In the left-handed resonance region, the resonant frequency increased with the increase of the number of unit elements, since the guided wavelength increases with the increase of frequency in left-handed region.

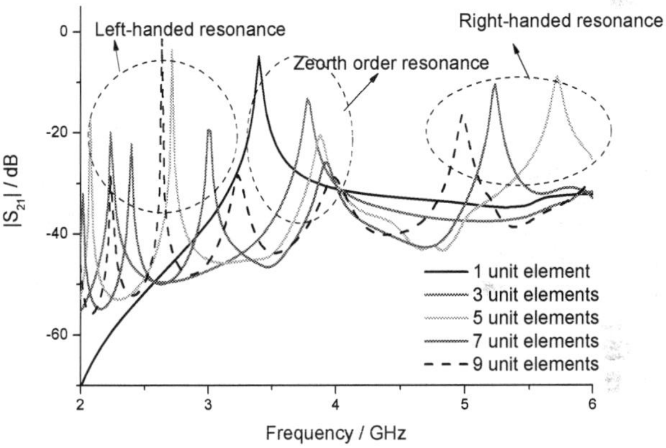

Fig. 10. Simulated resonant frequencies of the NRI TL of different unit elements

At the matched situation, in which the source and load are matched to the characteristic impedance of the metamaterial transmission line with appropriate coupling. All unit elements share the same phase at the zeroth order resonant frequency since the wavelength is infinite. The metamaterial transmission line with 21 unit elements was simulated. At $f_0 = 4.08\,\text{GHz}$ the phase difference between any unit elements was zero, as shown in Fig. 11. It was also worth noticing that: a) the phase difference between any two unit elements became greater when frequency was farther away from f_0; b) the phase difference relation reversed (from advanced to delayed or versus) when frequency changed from $f < f_0$ to $f_0 < f$.

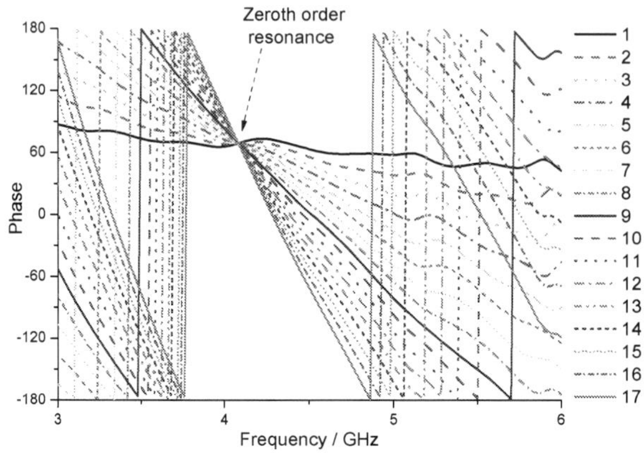

Fig. 11. Phase distribution among the NRI TL with 17 unit elements which are numerated from 1 to 17 (© 2007 IEEE)

The dispersion of the metamaterial transmission line with 17 unit elements is obtained from the simulated phase difference between the source and the load. The results are as shown in Fig. 12. In the left-handed dominant region $f < f_0$, the wavelength increases with the increase of frequency. At the centre frequency $f = f_0$, the wavelength is infinite. In the right-handed dominant region $f > f_0$, the wavelength degreases with the increase of frequency.

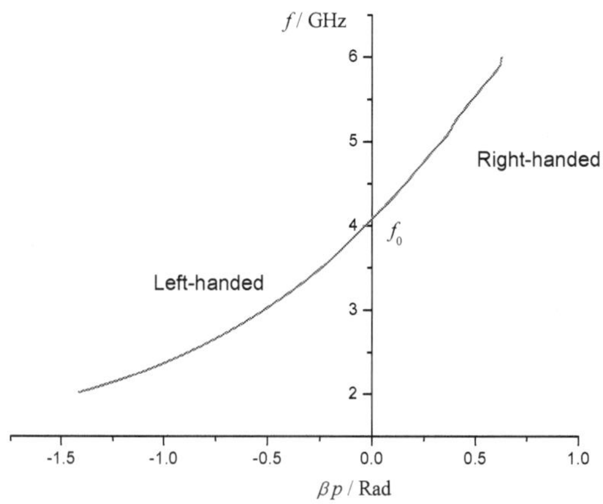

Fig. 12. Simulated dispersion of the metamaterial transmission line with 17 unit elements where p is the length of one unit element. The transition frequency between dominant left-handed and right-handed is 4.08 GHz.

The metamaterial transmission line should be fed in a balanced mode with reversed phases. Therefore, a balun is required to feed the metamaterial transmission line by coaxial lines. Since the metamaterial transmission line has a good performance and may tolerance the amplitude and phase variation, the requirement on the balun performance is low. To build a compact balun, we integrated the balun directly to the metamaterial transmission line. The balun is directly integrated into the negative index transmission lines with two attached microstrip stubs which lead to a very compact antenna feeding system.
The frequency scanned leaky-wave antenna contains the balun, the metamaterial transmission line, and the matched load, as shown in Fig. 13. When the metamaterial transmission line is fed by reversed phase microwave signals with equal amplitudes, a virtual ground occurs in the symmetry plane. The advantage of this constellation is that it does not require any vias to build the shunt inductors. Moreover, it has a broader bandwidth and potential applications in higher frequency band compared to the implementation with interdigital capacitors, since the resonance from those interdigital capacitors has been removed.

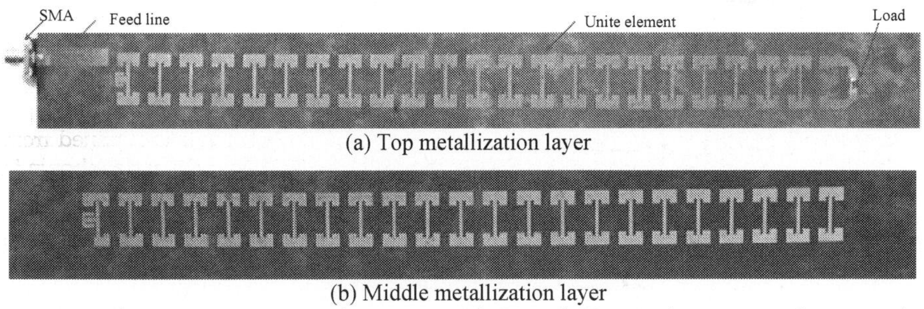

Fig. 13. Leaky-wave metamaterial antenna with 21 and 22 unit elements on the top and in the middle metallization layer, respectively. The bottom layer is the ground. The feeding port is a 50 Ω microstrip line connected to a SMA connector (© 2007 IEEE).

As shown in Fig. 14, the reflections introduced by each unit element have zero phase shifts at the center frequency and add up coherently at the feeding port, leading to an increased $|S_{11}|$ at the central frequency. On the other hand, the ripple of the reflection is obvious. It is caused by the periodic nature of the leaky-wave antenna based on metamaterial transmission lines.
The novel leaky-wave antenna is designed on the metamaterial transmission line. The antenna comprises 30 elements, and the total size with the integrated microstrip balun is about 200 mm by 20 mm. Each element contains a pair of inductors and capacitors. The antenna works at C-band from about 3.5 GHz to 5.5 GHz with a return loss smaller than 10 dB. In the operating frequency range, the main beam scanning angle is from -45° to 45°. The antenna gain is higher than 10 dB in this frequency range. All side lobe levels are more than 10 dB below the main beam at all scanning angles. Simulated antenna patterns are shown in Fig. 15.

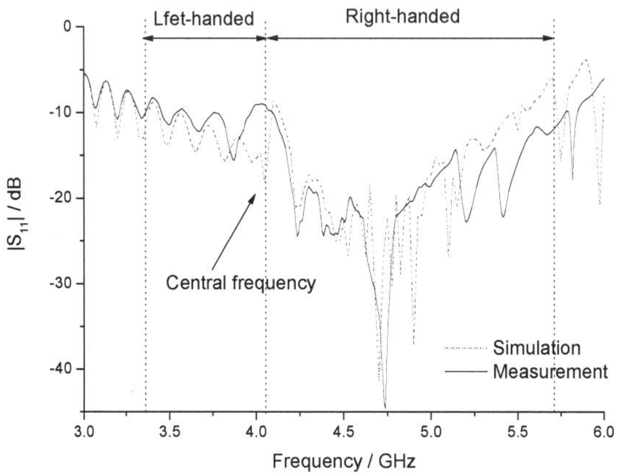

Fig. 14. Measured and simulated voltage reflection coefficients of the leaky-wave antenna. The operating band is from about 3.3 GHz to 5.7 GHz. The central frequency is 4.15 GHz, at which the reflection is slightly higher.

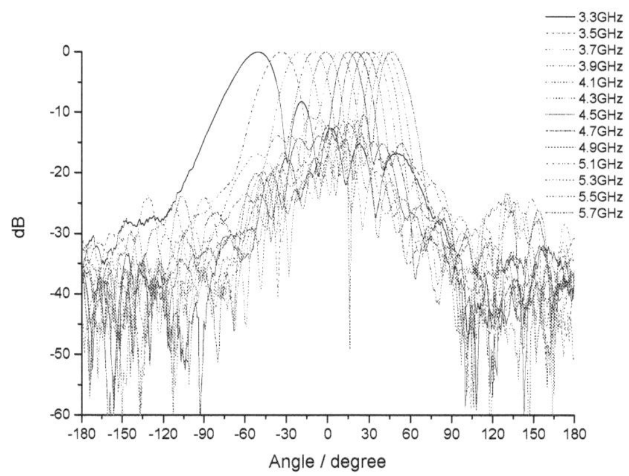

Fig. 15. Measured antenna radiation patterns on the H-plane from 3.3 GHz to 5.7 GHz, in which the main beam scans from about -40° to 40°. All side lobes are lower than -10 dB. The beam-widths at low frequencies are wider than those at high frequencies (© 2007 IEEE).

The leaky-wave frequency scanned antenna based on metamaterial transmission lines is composed of two metamaterial transmission lines with a two-layer structure, a balun and a load. It steers main beam from end-fire to back-fire from -40° to 40° with the antenna gain about 10 dB. It is a compact structure due to the integrated balun. It exhibits a good performance at C band from 3.3 GHz to 5.7 GHz. Meanwhile, it can be extended to high frequency band application since it owns no vias.

4.2 Balun

The principle structure of the proposed balun is shown in Fig. 16, in which the dimensions of one unit are presented. Each unit contains a series interdigital capacitor and a narrow shunt microstrip branch. At the dashed line in Fig. 16, the balun may be divided into two identical symmetric parts, namely the upper part and the lower part. In the example of Fig. 16, there are five units in either part.

When port I and II are fed with opposite phase and identical amplitude, and port III and IV are connected to matched loads, the odd mode will be excited in the symmetric structure. An electric wall is formed in the symmetry plane along the dashed line, forming a virtual ground. A similar metamaterial transmission line structure has been reported in antenna applications. Then, the vertical microstrip branches of length *a* are equivalent to shunt inductors. The upper part and the lower part become a pair of coupled metamaterial transmission lines. The pass-band of the balun is determined by the metamaterial transmission lines, which is between the cut-off frequencies of the left-handed and right-handed modes. The characteristic impedance of the metamaterial transmission line is equal to Z_C.

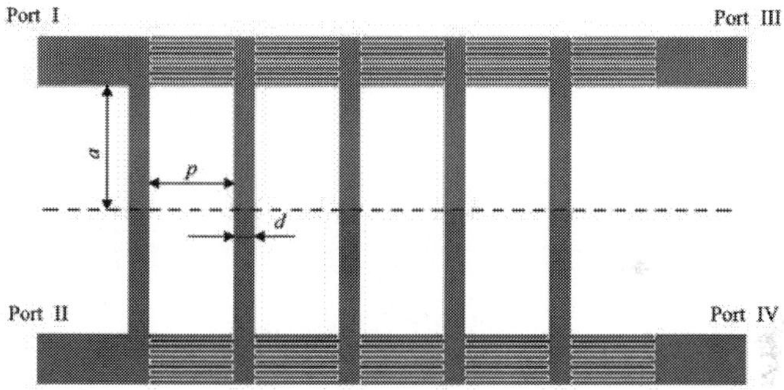

Fig. 16. Structure of the metamaterial balun based on metmaterial transmission lines. The gap between fingers of an interdigital capacitor is 0.10 mm, and the width of fingers is 0.15 mm. Other dimensions are: a = 6.05 mm, p = 4.2 mm, and d = 1 mm (© 2008 IEEE).

When port I and II are fed with identical amplitude and phase, the even mode will be excited in the balun. A magnetic wall, equivalent to open circuits, is formed in the symmetry plane. In this case, however, the vertical microstrip branches turn from short-ended to open-ended. The previous pass-band for the odd mode becomes a stop-band.

Odd and even mode equivalent circuits and their combination are shown in Fig. 17 (a), (b), and (c), respectively. In the combined case, port I is the unique input port. Port III and IV are output ports. The odd mode is supported in the pass-band of the metamaterial transmission line, while the even mode sees a stop-band and is suppressed. There is only the odd-mode left in the output resulting in identical amplitude and opposite phase at ports III and IV. Thus, the proposed design works well as a balun, with the unbalanced input at port I and the balanced outputs between ports III and IV.

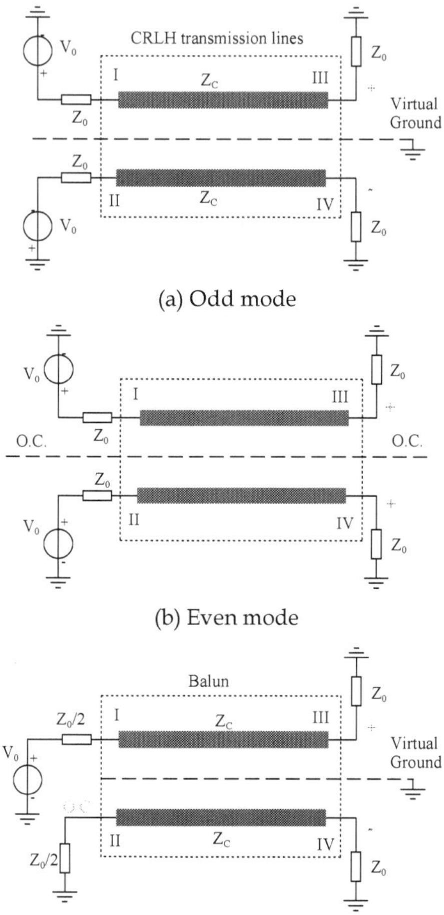

(a) Odd mode

(b) Even mode

(c) Combination of the odd and even mode

Fig. 17. Odd and even mode equivalent circuits and their combination (© 2008 IEEE)

In the proposed balun, the source impedance is required to be half of the load impedance. In applications and measurements, input/output impedances are usually Z_0. Therefore, the characteristic impedance Z_C of the metamaterial transmission line should be either $2Z_0$ or Z_0, and impedance matching circuits have to be involved in either output or input port. A trade-off is to choose the impedance Z_C between $2Z_0$ and Z_0, and apply short transition microstrip lines for impedance matching at both input and output ports.

When port III and IV are matched, port I and II are isolated against each other. When port I is the input port, port II may be an open circuit. The matching load at port II can be omitted. To keep the structure symmetric, a section of open-ended microstrip line is kept at port II. In the case of mismatched loads are applied at the output ports, a matched load at port II should be used. Moreover, based on the relation between Chebyshev band-pass filters and metamaterial transmission lines, the first and last units of the metamaterial lines are adjusted to achieve better impedance match and wider bandwidth.

Following the above steps, we have designed and optimized two baluns from metamaterial transmission lines with five and seven units, shown in Fig. 18 (a) and (b), namely balun A and B, respectively. The substrate material is RT Duroid 5880 of thickness of 1.57 mm and a relative dielectric constant of $\varepsilon_r = 2.2$. All input and output ports are matched to $Z_0 = 50\ \Omega$.

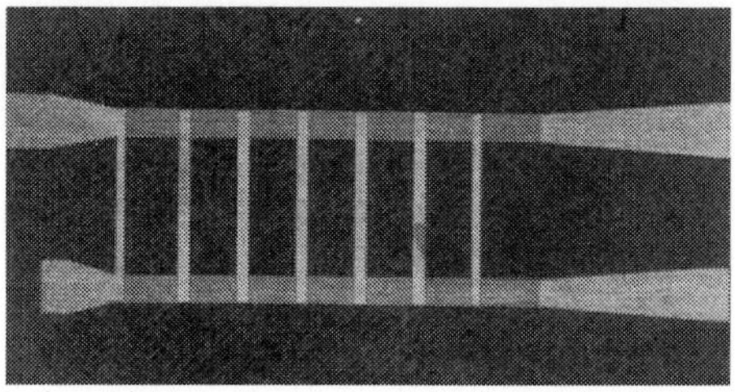

Fig. 18. Photos of a baluns based on metamaterial transmission lines. The balun is of sizes of about 65 mm × 35 mm (© 2008 IEEE).

The baluns were measured using a HP 8510C vector network analyzer. Simulated and measured results have good agreements for both amplitudes and phases, as shown in Fig. 19.

The output amplitude difference is less than 0.7 dB, and either insertion loss is smaller than 5 dB from 1.6 GHz to 4.0 GHz. The return loss is greater than 10 dB. The phase difference between the two output ports is between 178° and 184° (181° ±3°) within the frequency range from 1.5 GHz to 3.6 GHz.

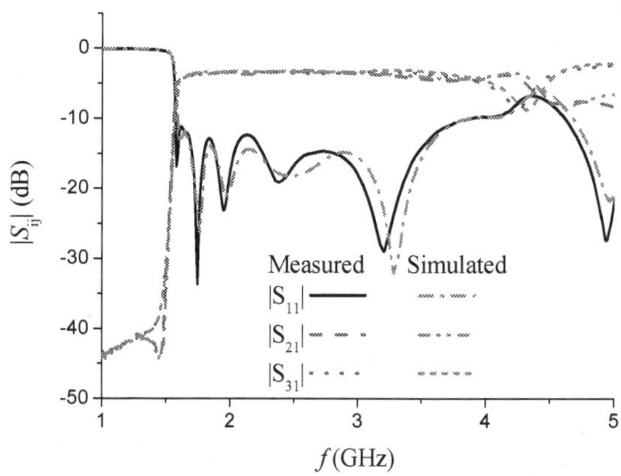

(a) Return loss and transmission performance of balun B

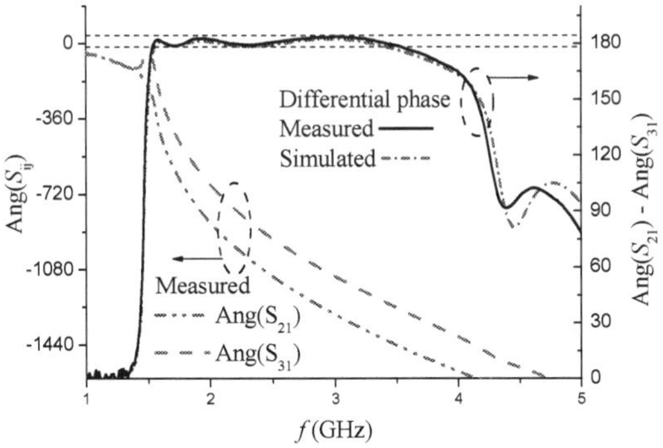

(b) Measured differential phase of balun B

Fig. 19. Simulated and measured results of baluns A and B

The novel broadband metamaterial balun design has been demonstrated. No vias and power dividers are used. Two baluns with five and seven units, respectively, have been fabricated, measured, and compared. Simulations show a good agreement with measured results. The bandwidth of the balun is from about 1.6 GHz to 3.6 GHz, while the output balance is better than 0.7 dB, and the differential phase is 181° ±3°.

4.3 Diplexer

A microstrip realization of dual-composite right/left-handed transmission line is shown in Fig. 20. The parallel LC resonator is composed of a interdigital capacitor (C_L) and a short narrow microstrip line (L_R). The series LC resonator is composed of a small patch (C_R) and a short narrow microstrip line (L_L). LC components are adjusted at both ends to achieve better impedance match and bandwidth.

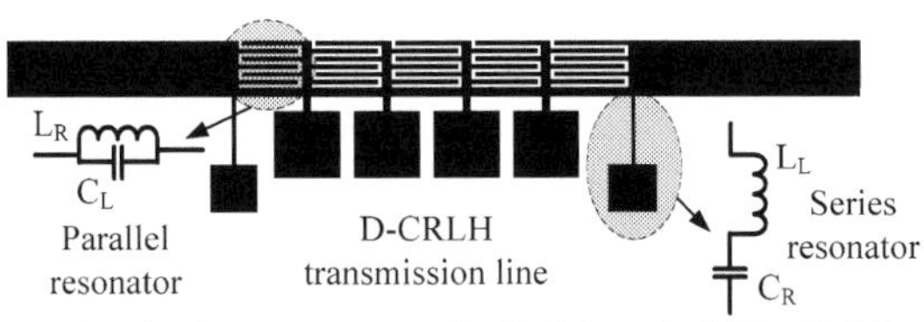

Fig. 20. Configuration of a dual-composite right/left-handed (D-CRLH) transmission line

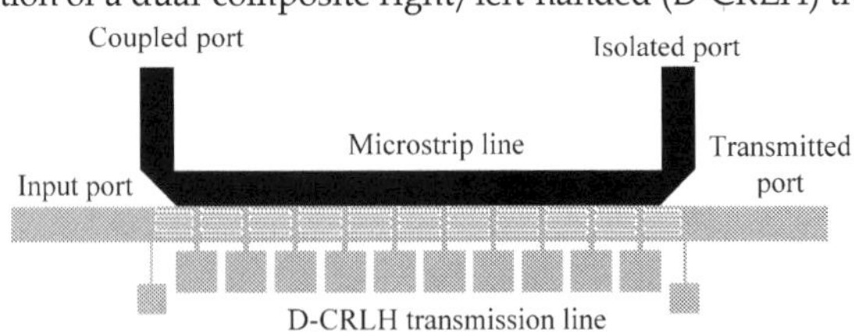

(b) Configuration of the coupler

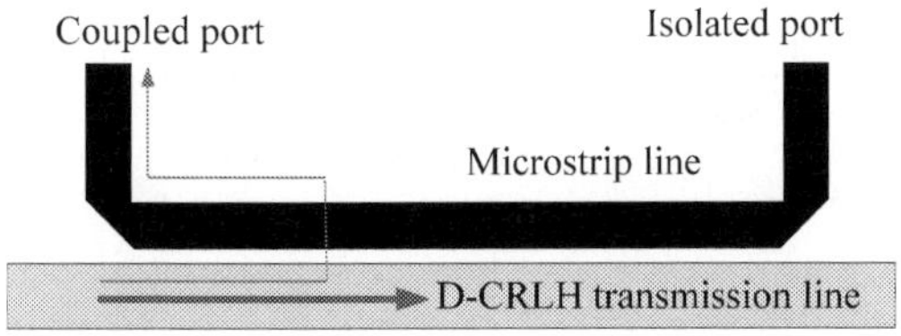

(c) Coupler in the right-handed mode

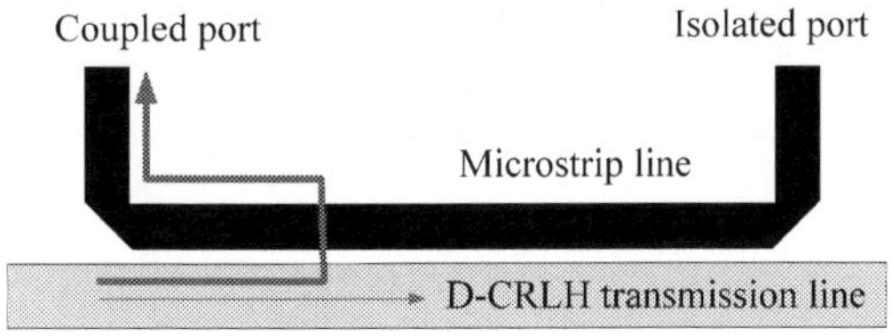

(d) Coupler in the left-handed mode

Fig. 21. Directional coupler from dual-composite right/left-handed transmission lines (© 2009 IEEE)

A conventional microstrip line and a dual-composite right/left-handed transmission line parallel to each other to form a directional coupler, as shown in Fig. 21 (a). The demonstrated D-CRLH transmission line contains twelve unit elements.

It is a dual-band directional coupler, which works at both right-handed and left-handed regions. However, the coupling coefficients are much different from each other. When the frequency is below the band-gap, it works in the right-handed mode and a weak coupling is obtained, as shown in Fig. 21 (b). On the other hand, when the frequency is above the band-gap, it works in the left-handed region and a strong coupling is achieved, as shown in Fig. 21 (c).

The coupling coefficient difference between the left-handed and right-handed modes can be higher than 20 dB in the above design. If the coupling coefficient is nearly 0 dB at the left-handed mode, it may reach 20 dB at the right-handed mode. Thus, microwave at different frequencies are transmitted to different ports dependent on its modes.

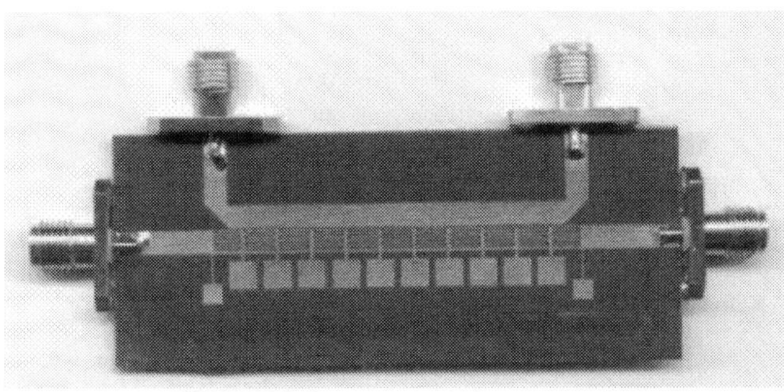

Fig. 22. Fabricated diplexer (© 2009 IEEE)

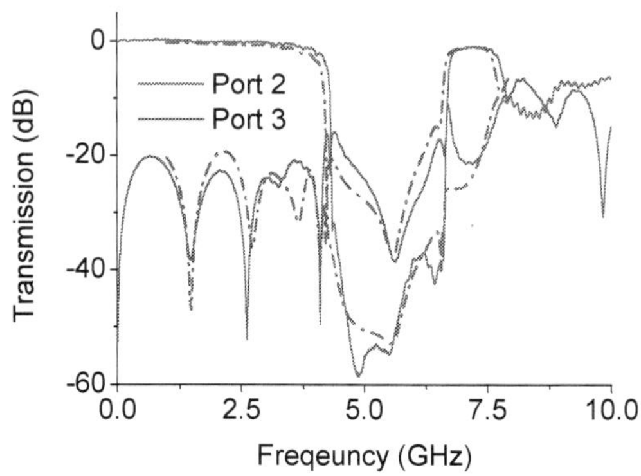

(a) Transmission

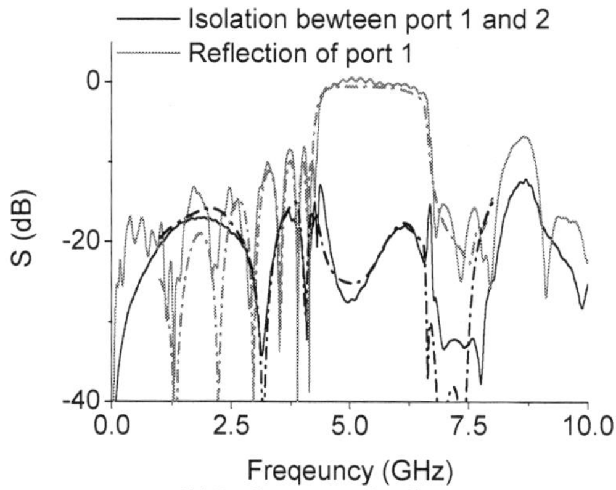

(b) Isolation and reflection

Fig. 23. Simulated and measured results. Simulated and measured results are in solid lines and dashed lines, respectively (© 2009 IEEE).

The fabricated diplexer is shown in Fig. 22, which contains a D-CRLH transmission line and a conventional microstrip line. There are twelve unit elements in the D-CRLH transmission line to achieve 0 dB coupling in the left-handed mode. The diplexer is fabricated on FB4-2 substrate with dielectric constant of 2.65 and thickness of 1 mm. The thickness of copper foils of both sides is 0.017 mm. The dimension of the diplexer is about 65 mm by 25 mm. All spacings and finger widths of interdigital capacitors are 0.2 mm, and the length of each interdigital capacitor is 3.27 mm. The inductor LR is formed by short microstrip lines with width 0.4 mm. The capacitor CR is a rectangle patch with dimension 3.4 mm by 3.4 mm. The inductor LL is formed by microstrip line with width of 0.4 mm and length of 0.8 mm. The distance between the conventional and the D-CRLH transmission lines is 0.15 mm.

Port 1 to port 4 are the input port, the transmitted port, the coupled port, and the isolated port, respectively. Measurements have been performed by Agilent E8362B vector microwave network analyzer. The measured results are shown in Fig. 23 with solid lines. Simulations have been performed by Zeland IE3D, and are shown in Fig. 23 with dashed lines. Simulated results agree well with measured results.

The transmission characteristic of the diplexer is shown in Fig. 23(a). When the frequency is below 3.7 GHz, microwave is delivered to port 2 with an insertion loss less than 1 dB. Due to the parasitic effects of interdigital capacitors, there is a upper frequency limitation for the D-CRLH transmission line. In the proposed design, it is about 7.6 GHz. Thus, when the frequency is from 6.8 GHz to 7.6 GHz, microwave is transferred to port 3 with an insertion loss less than 1.7 dB.

The rejection band of the diplexer is from about 4.4 GHz to 6.6 GHz as shown in Fig. 23(b). The isolation between port 2 and port 3 is higher than 20 dB in lower frequency band, and greater than 35 dB in higher frequency band. A quite high isolation is achieved over a wide frequency band, e.g. from DC up to 3.7 GHz.

4.4 Power divider

Fig. 24 shows a symmetrical microstrip dual-composite right/left-handed transmission line structure. This structure is a planar configuration constituted of four unit cells. Each cell includes an interdigital capacitor and several simple microstrip elements. The fundamental four LC parameters (L_R, L_L, C_R, C_L) of the metamaterial transmission line are obtained by the conditions of $Z_0 = 50\Omega$, $f_{cR} = 4.5$ GHz and f_{cL}=6.5 GHz. The capacitance C_R and the inductance L_R and L_L are implemented as simple microstrip elements, the capacitance C_L is implemented as interdigital capacitor.

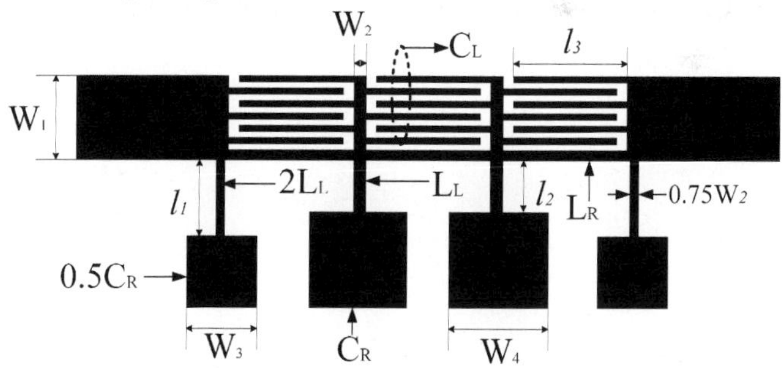

Fig. 24. Layout of the dual-composite right/left-handed transmission line

The substrate is F4B-2 with ε_r=2.65 and h =1 mm. The system impedance is Z0 = 50 Ω, and W_1 = 2.8 mm. The characteristic impedance of L_R and L_L are 125 Ω, which is a reasonable value to realize a microstrip inductor. W_2 = 0.4 mm is obtained. The width of $2L_L$ is set to 0.75 W_2, and the characteristic impedance is 136 Ω.

The proposed 3dB D-CRLH transmission line power divider is the one shown in Fig. 25, where each branch of the divider is a balanced D-CRLH line. In theory, the power divider contains the same two wide pass-band as the D-CRLH transmission line: the low frequency right-handed pass-band (from DC to cut-off frequency f_{cR}), and the high frequency left-handed pass-band (from cut-off frequency f_{cL} to unlimitedly high-frequency). But in practice, the bandwidth of the matching network cannot be made wide enough to cover the whole left-handed and right-handed pass-band, and the performance of the D-CRLH transmission line is poor at high frequency band due to radiation loss and parasitic effects. However, the performance of the D-CRLH transmission line is perfect at the low frequency right-handed pass-band. Thus, we use the right-handed pass-band of the D-CRLH transmission line for power divider design, in which a wideband fourth order Chebyshev matching transformers is applied as well. The microstrip line length of each order is $\lambda_g/4$ at the centre frequency $f_0 = \sqrt{f_{cR} \times f_{cL}}$.

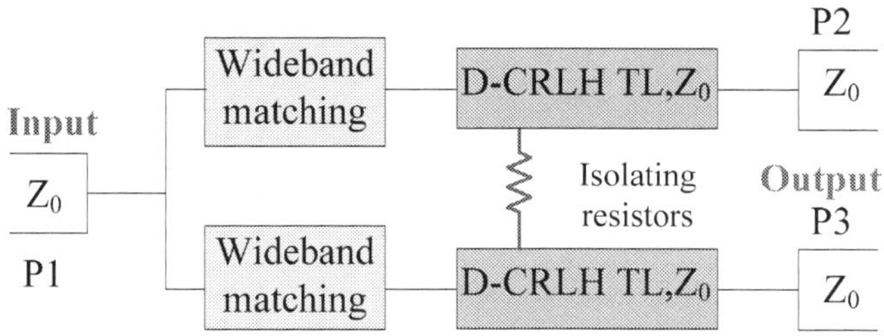

Fig. 25. D-CRLH TL power divider schematic diagram.

Fig. 26 shows the photograph of the fabricated D-CRLH transmission line power divider, in which there are five resistors with different values between the two D-CRLH transmission lines. They are simulated and optimized by IE3D to obtain the highest isolation between output ports. From left to right, the resistance values are 50 Ω, 100 Ω, 200 Ω, 400 Ω, and 800 Ω, respectively. The overall size of the power divider is about 50 mm × 40 mm.

Fig. 26. Photograph of the fabricated D-CRLH transmission line power divider

The proposed power divider was measured using an Agilent N5230A vector network analyzer. The measured S-parameters of the design are illustrated in Fig. 27 (a). It can be seen the performance of the power divider is excellent. From 0.85 GHz to 4.3 GHz, $|S_{11}|$ is below -15 dB, the isolation is higher than 15 dB, and the insertion loss is less than 1 dB. Fig. 27 (b) shows the amplitude balance is below 0.3 dB between output ports, and the phase

balance is less than 3°. The bandwidth of the D-CRLH TL power divider is about five octaves. Thus, an ultra wide band is realized.
The simulated S-parameters of the design are illustrated in Fig. 28, which agree well with the measured results.

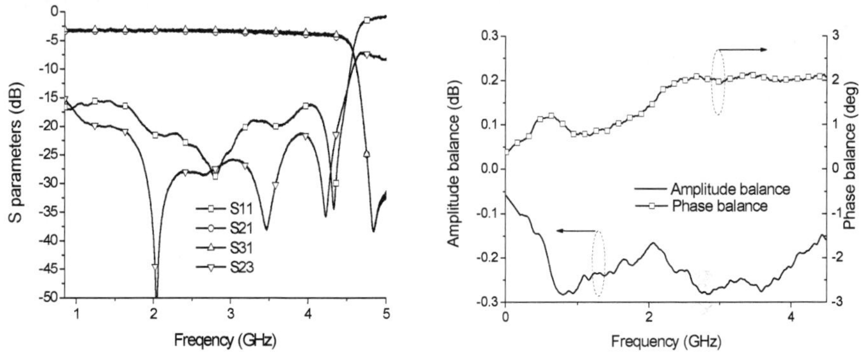

(a) S parameters (b) Amplitude balance and phase balance

Fig. 27. Measured results of the power divider

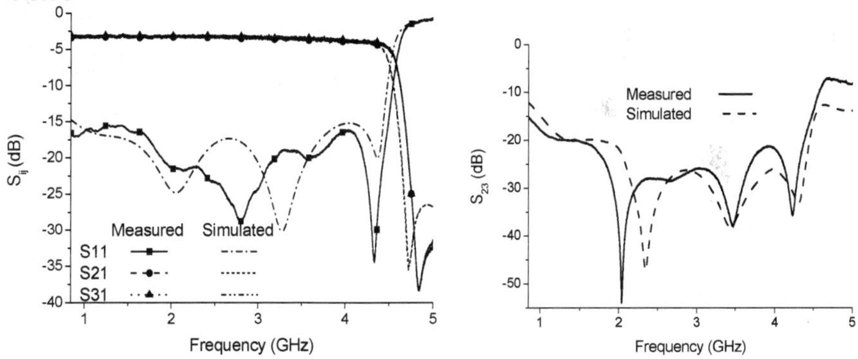

Fig. 28. Comparison between simulation and measurements

Compared with those of the conventional Wilkinson power dividers, the novel D-CRLH power divider obtains a bandwidth over 100% with reasonable return loss and isolation. Usually the bandwidth of a conventional Wilkinson power divider is about 20%. The insertion loss is less than power dividers based on CRLH transmission lines. The size of the proposed D-CRLH power divider is only about 70%. Thus, it is compact power divider with broadband width.
The novel 3 dB power divider based on dual-composite right/left handed transmission line has the feature of compact size, wide band, low insertion loss, high isolation, and good amplitude and phase balance. It can be applied to microwave signal separation, array antenna feed and so on.

Another scheme of power divider is to use the zeroth order resonance of a metamaterial transmission line. All output ports are in phase at the transition frequency. Due to the intrinsic parallel connection (the wavelength is infinite at the resonant frequency) in the power divider, the number of output ports is limited.

Multi-port power dividers based on metamaterial transmission lines are available as well. Based on the zeroth order resonance, the metamaterial transmission line shares the same phase at the transition frequency. Thus, an N-port power divider can be easily built, as shown in Fig. 29 (a) and (b). A metamaterial transmission line using ten series chip capacitors is used. The drawback of this power divider is the output impedance at each port is N times Z_0. A $\lambda_g/4$ impedance matching has to be applied. It limits the number of output ports. Meanwhile, the bandwidth of this power divider is much limited due to the zeroth order resonance.

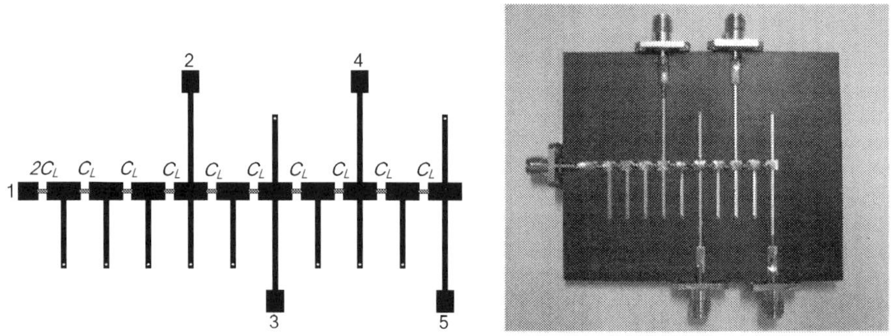

(a) Layout (b) Photo

Fig. 29. Layout and fabricated 1:4 power divider

A power divider based on a segment of metamaterial transmission line and three conventional microstrip lines, as shown in Fig. 31. Each unused conventional microstrip line end is connected with one match load. We assume that the coupling coefficients of those ideal couplers are C1, C2 and C3, respectively. The feeding power at port 1 is P1, and the output powers are P2, P3, P4, and P5 at the respective ports. When the reflection at port 1 and the insertion loss of the CRLH TL are neglected, the output powers are obtained as

$$\begin{cases} P_2 = P_1 - P_3 - P_4 - P_5 \\ P_3 = C_1 P_1 \\ P_4 = C_2 \left(P_1 - P_3\right) \\ P_5 = C_3 \left(P_1 - P_3 - P_4\right) \end{cases} \tag{8}$$

With careful design (choosing right coupler coefficients C1 to C3), the desired power division among port 2 to port 3 can be satisfied.

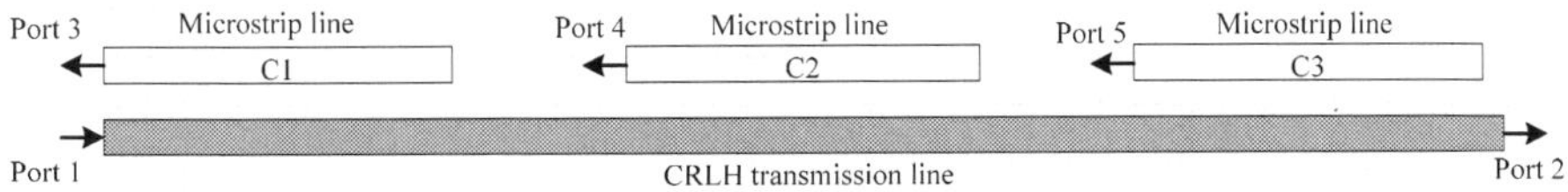

Fig. 30. Scheme of the proposed CRLH coupler. It is a 1:4 power divider, which contains one metamaterial transmission line and three conventional microstrip lines. Port 1 is input ports.

At the transition frequency of metamaterial transmission lines, there is no phase difference among their unit cells. Then, the outputs of a metamaterial coupler are in phase as well, since they have the same phase shift to the feeding port. Therefore, no phase adjusting is required in a metamaterial coupler based on the zeroth order resonance.

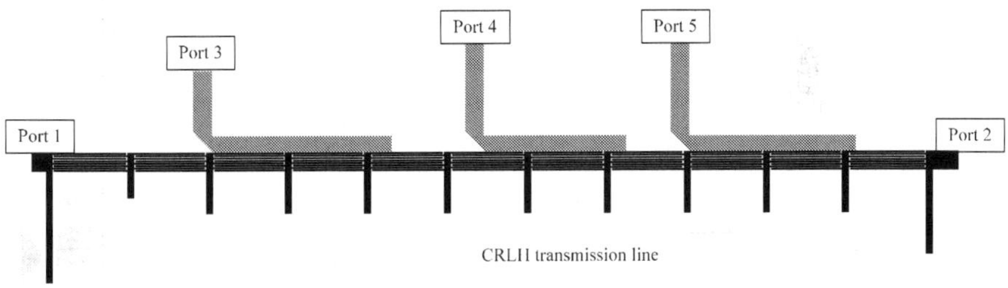

(a) Layout

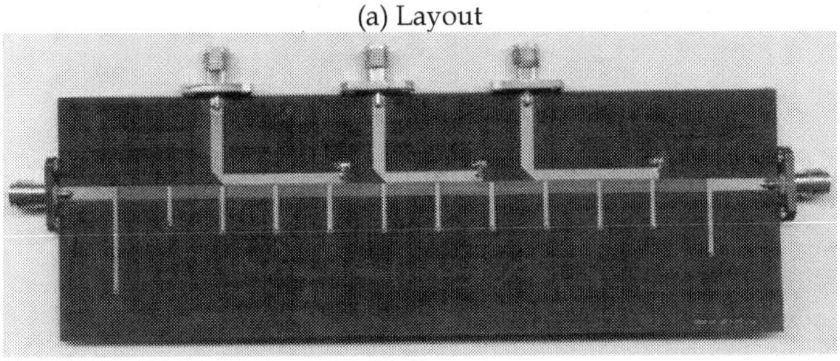

(b) Photo

Fig. 31. Layout of the proposed power divider based on CRLH couplers

The distance between the metamaterial transmission line and the coupling conventional microstrip transmission line is 0.15 mm. The layout of the proposed power divider based on metamaterial couplers is shown in Fig. 31 (a). The fabricated metamaterial power divider is shown in Fig. 31 (b). The substrate is F4B-2 with $\varepsilon_r = 2.65$ and thickness of 1 mm.

The unequal power division between port 2, port 3, port 4, and port 5 is designed to 4:2:1:1, in which the transmission from port 1 to those corresponding ports are -3 dB, -6 dB, -9 dB and -9 dB, respectively. The lengths of the coupling conventional microstrip transmission lines are adjusted to realize the desired unequal power division.

The simulated amplitude response is shown in Fig. 32 (a). At the transition frequency $f_0 = 2.4$ GHz, those transmitted amplitudes among output ports are -4.2 dB, -6.9 dB, -9.8 dB and -

9.8 dB. The reflection of port 1 at f_0 is -28 dB. The insertion loss is about 1.2 dB. For amplitude consideration, the reasonable bandwidth is about 500 MHz around the transition frequency. The problem is the coupled power at port 3 drops obviously when frequency is away from the transition frequency. A better bandwidth can be obtained with extra efforts to optimize the first coupler.

The phase shifts between output ports are shown in Fig. 31 (b). At transition frequency f0, all output ports share the same phase. Due to the dispersion characteristic of a CRLH TL, the bandwidth of ±10° is much narrower than the amplitude bandwidth. There is a phase difference of about 90° caused by the coupler. The microstrip TLs at port 3, port 4 and port 5 are extended to compensate the phase shift.

Transmissions between output ports are shown in Fig. 31 (c). The isolations between output ports are higher than 20 dB, as shown in Fig. 31 (c). Experimental results agree with the simulations well.

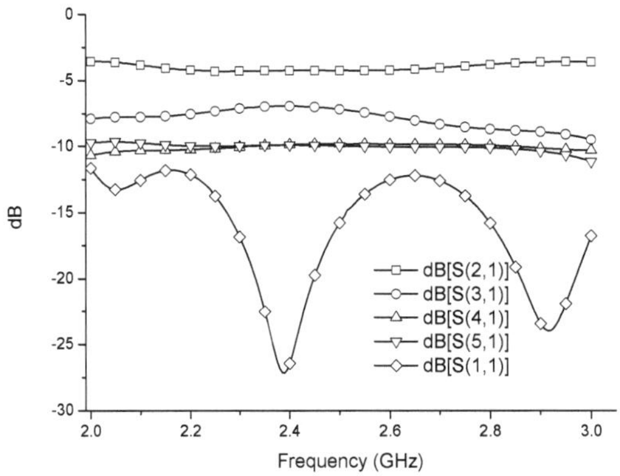

(a) Amplitude

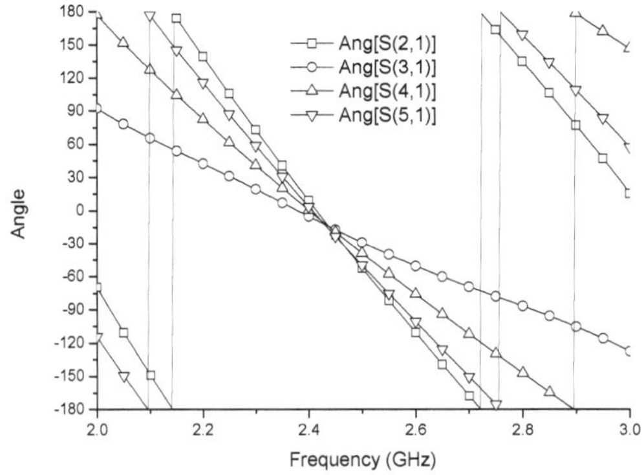

(b) Phase

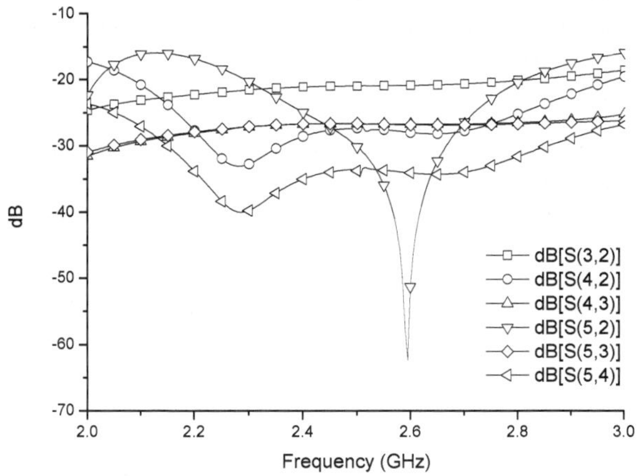

(c) Isolation

Fig. 32. Simulation results of the metamaterial power divider (© 2008 IEEE)

A novel unequal power divider based on the zeroth order resonance of a metamaterial transmission line is discussed. It is a miniaturized design along the longitudinal direction. The power divider can be easily extended to an arbitrary number of output ports. Not only even numbers but also odd numbers of output ports are suitable for the proposed power divider. Thus, the proposed power divider is a practical design.

Both equal and unequal power division are possible for the power divider. In further study, equal power divider will be considered and designed. Since the power divider is very compact along the longitudinal direction, it is suitable to realize an antenna feeding network. With desired unequal power division, an antenna array fed with the power divider may get arbitrary power supply.

The insertion loss of the metamaterial transmission at zeroth order resonance frequency is a little high. To reduce the insertion loss will make the new metamaterial power divider more reliable.

5. Conclusion

Metamaterial transmission lines are one-dimension structures. Their performances can be roughly analyzed by the circuit models, and the relation between them and band-pass filters is discussed as well. There are many applications of metamaterial transmission lines due to their excellent performance. Some typical applications, such as leaky-wave antenna, baluns, diplexers and power dividers are presented. Metamaterial transmission lines will find more and more applications of microwave components in future.

Acknowledge

This work was supported in part by the National Science Foundation of China under Grant 60971051 and the Youth Foundation of Sichuan Province under Grant 09ZQ026-016.

6. References

Caloz C. & Itoh T. (2006) *Electromagnetic metamaterials: Transmission line theory and microwave applications*. John Wiley & Sons, Inc. ISBN 0-471-66985-7; U.S.A.

Eleftheriades G.; Iyer A. & Kremer P. (2002). Planar negative refractive index media using periodically L-C loaded transmission lines, *IEEE Transaction on Microwave Theory and Technology*, Vol. 50, No. 12, 2702–2712, (Dec. 2002), ISSN 0018-9480

Eleftheriades G. & Balmain K. (2005) *Negative-Refraction metamaterials - Fundamental principles and applications*. John Wiley & Sons, Inc. ISBN 13: 978-0-471-60146-3; U.S.A.

Lai, A.; Itoh, T. & Caloz C. (2004). Composite right/left-handed transmission line metamaterial. *IEEE Microwave Magazine*, Vol. 5, No. 3, 34–50, (March 2004), ISSN 1527-3342

Liu, C. & Menzel, W. (2007). On the relation between a negative refractive index transmission line and Chebyshev filters, *Proceedings of the 37th European Microwave Conference*, pp. 704-707, ISBN 978-2-87487-001-9, October 2007, European Microwave Association, Munich, Germany

Liu, C. & Menzel, W. (2007). Frequency-scanned leaky-wave antenna from negative refractive index transmission lines, *Proceedings of the European Conference on Antennas and Propagation*, ISBN 978-0-86341-842-6, November 2007, European Association on Antennas and Propagation, Edinburg, UK

Liu, C. & Menzel, W. (2008). Broadband via-free microstrip balun using metamaterial transmission lines. *IEEE Microwave and Wireless Component Letters*, Vol. 18, No. 7, 437-439, (July 2008), ISSN 1531-1309

Wang, W.; Liu, C.; Yan, L. & Huang, K. (2009). A Novel Power Divider based on Dual-Composite Right/Left Handed Transmission Line. Journal of Electromagnetic Waves and Applications. Vol. 23, No. 8/9, 1173-1180, (Sept. 2009), ISSN 0920-5071

Pozar, D. (2004). Microwave Engineering. John Wiley & Sons, Inc., ISBN 0-471-17096-8, U.S.A.

14

Physics of Charging in Dielectrics and Reliability of Capacitive RF-MEMS Switches

George Papaioannou[1] and Robert Plana[2]
[1]*University of Athens, Greece*
[2]*Universite Paul Sabatier- LAAS*
France

1. Introduction

The dielectric charging constitutes a major problem that still inhibits the commercial application of RF MEMS capacitive switches. The effect arises from the presence of the dielectric film (Fig.1a), which limits the displacement of the suspended electrode and determines the device pull-down state capacitance. Macroscopically, the dielectric charging is manifested through the shift (Fig.1b) (Rebeiz 2003, Wibbeler et al. 1998, Melle et al. 2003, Yuan et al. 2004) or/and narrowing (Czarnecki et al. 2006, Olszewski et al. 2008) of the pull-in and pull-out voltages window thus leading to stiction hence the device failure. The first qualitative characterization of dielectric charging within capacitive membrane switches and the impact of high actuation voltage upon switch lifetime was presented by C. Goldsmith et al. (Goldsmith et al. 2001) who reported that the dependence of number of cycles to failure on the peak actuation voltage follows an exponential relationship. Particularly it was reported that the lifetime improves by an order of a decade for every 5 to 7 V decrease in applied voltage. The lifetime in these devices is measured in number of cycles to failure although experimental results have shown that this tests do not constitute an accurate figure of merit and the time the device spends in the actuated position before it fails is a much better specification to judge device reliability (Van Spengen et al. 2003).
The aim to improve the reliability of capacitive switches led to the application of different characterization methods and structures such as the MIM (Metal-Insulator-Metal) capacitors that allowed to determine the charging and discharging times constants (Yuan et al. 2004, Lamhamdi 2008) as well as to monitor the various charging mechanisms (Papaioannou 2007a), since these devices marginally approximate the capacitive switches in the pull-down state. A method that approximates more precisely the charging process through asperities and surface roughness in MEMS and allows the monitoring of the discharging process is the Kelvin Probe Force Microscopy (Nonnenmacher 1991). This method has been recently employed for the investigation of the charging and discharging processes in capacitive switches (Herfst 2008, Belarni 2008).
The charging of the dielectric film occurs independently of the actuation scheme and the ambient atmosphere (Czarnecki et al. 2006). Up to now the effect has been attributed to the charge injection during the pull-down state (Wibbeler et al. 1998, Melle et al. 2003,

Olszewski 2008, Reid 2002, Papaioannou 2006a) and dipoles orientation (Papaioannou 2005, Papaioannou 2006b), which are present in the dielectric material.
In order to minimize and control the dielectric charging and obtain devices with high capacitance aspect ratio, several materials, such as SiO_2 (Yuan 2004), Si_3N_4 (Melle 2003, Papaioannou 2005), AlN (Lisec 2004, Papaioannou 2007b, Papandreou 2009), Al_2O_3 (Berland 2003, Blondy 2007), Ta_2O_5 (Lisec 2004, Rottenberg 2002), HfO_2 (Luo 2006, Tsaur 2005), have been used. The selection has been made taking into account the maturity of low temperature deposition method and the magnitude of dielectric constant. Although these materials exhibit excellent insulating properties little attention was paid on the fact that their lattice is formed by either covalent or ionic bonds, which affect significantly the dielectric polarization/charging. It is worth noticing that among these materials, the crystalline AlN exhibits piezoelectric properties, which seems to increase significantly the device lifetime (Lisec 2004, Papandreou 2009).

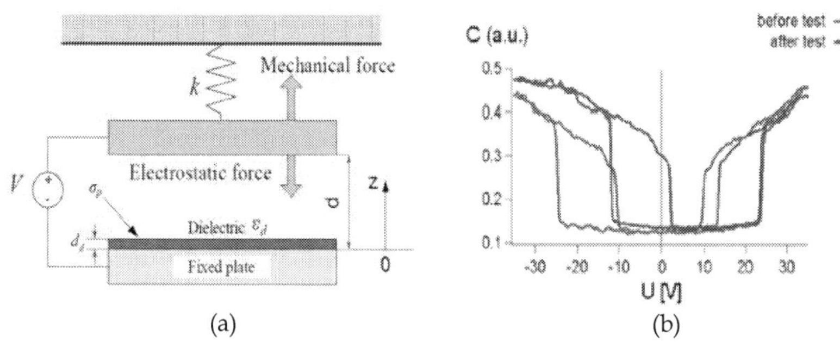

Fig. 1. (a) Simplified model of a capacitive switch based on the parallel plate model and (b) the shift of the capacitance-voltage characteristic after stress.

A key issue parameter that affects significantly the electrical properties of dielectrics and may prove to constitute a valuable tool for the determination of device lifetime is the device operating temperature. This is because temperature accelerates the charging (Papaioannou 2005, 2006, Daigler 2008) and discharging (Papaioannou 2007c) processes by providing enough energy to trapped charges to be released and to dipoles to overcome potential barriers and randomize their orientation. Finally, the presence or absence (Mardivirin 2009) of dielectric film as well as its expansion on the film on the insulating substrate (Czarnecki...) constitute a key issue parameter that influences the charging process.
The aim of the present chapter is to provide an overview and better understanding of the impact of various parameters such as the dielectric material properties, the operating temperature, etc on the physics of charging in dielectrics and reliability of capacitive RF-MEMS switches as well as to present the presently available assessment methods.
The basic polarization mechanisms in dielectrics will be presented in order to obtain a better insight on the effect of the ionic or covalent bonds of the dielectrics used in capacitive MEMS. The deviation from stoichiometry, due to low temperature deposition conditions, will be taken into account. Finally, the effect of temperature on the charging and discharging

processes will be discussed in order to draw conclusions on the possibility of identification and predict of charging mechanisms and their relation to the deposition conditions.

2. Dielectric polarization

2.1 Principles of dielectric polarization

When an electric field E is applied to an insulating material, the resulting polarization P may be divided into two parts according to the time constant of the response (Barsukov 2005):

i. An almost instantaneous polarization due to the displacement of the electrons with respect to the nuclei. This defines the high-frequency dielectric constant ε_∞ related to the refractive index.

$$\varepsilon_\infty - 1 = P_\infty / E\varepsilon_0 \tag{1}$$

The time constant of this process is about 10^{-16} s.

ii. A time-dependent polarization $\Delta P(t)$ arising from mechanisms such as the orientation of dipoles, the buildup of space charge etc in the presence of the electric field. It must be emphasized that the magnitude and sing of the time-dependent polarization is determined by the magnitude of the contributing mechanisms. If the field remains in place for an infinitely long time, the resulting total polarization P_S defines the static dielectric constant ε_S :

$$\varepsilon_S - 1 = P_S / E\varepsilon_0 \tag{2}$$

Thus the static polarization will be determined by the sum of the instantaneous and time dependent polarizations:

$$P_S = P_\infty + \Delta P(t) \tag{3}$$

The simplest assumption that allows the understanding of the response of such a system is that $\Delta P(t)$ is governed by first-order kinetics, that is, a single-relaxation time τ, such that

$$\tau \frac{\Delta P(t)}{dt} = P_S - P(t) \tag{4}$$

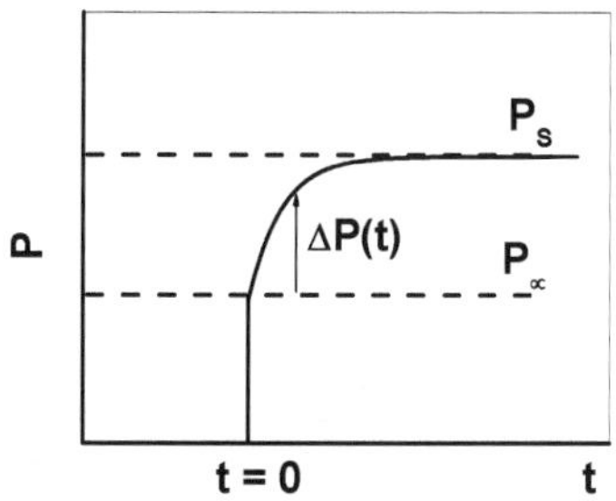

Fig. 2. Time dependence of the polarization P after the application of an electric field

This means that the rate at which P approaches P_S is proportional to the difference between them. Referring to Figure 2, on application of a unit step voltage and solving for $P(t)$, we obtain

$$P(t) = P_\infty\left[1-(P_S - P_\infty)\exp\left(-\frac{t}{\tau}\right)\right] \tag{5}$$

For most of the systems investigates, the experimental results cannot be generally described by such equation only. For this reason, it is necessary to use empirical relations that formally take into account the distribution of the relaxation times. A general form that approximates such cases is contained in the Kohlrausch-Williams-Watts (KWW) relaxation function (Kliem 2005):

$$\Delta P(t) = (P_S - P_\infty)\cdot\exp\left[-\left(\frac{t}{\tau}\right)^\beta\right] \tag{6}$$

where τ is the characteristic time constant and β the stretched factor. The KWW dielectric relaxational polarization has been found either in the time or in the frequency domain in many materials containing some degree of disorder. The list of materials is far away from being complete. Also in magnetic materials such relaxations are present. The fact that so many classes of materials exhibit the KWW behavior led to the supposition that there might be a universal law behind the experimental findings (Homann 1994). Since the observed relaxations can be distributed over more than 11 to 12 decades, the physical property causing the relaxations should be distributed in such a broad range, too. An early solution was given by H. Fröhlich (Fröhlich 1949) who reduced the broad distribution of relaxation times τ to a relatively small distribution of activation energies E_A assuming thermally activated processes with

$$\tau = \tau_0\cdot\exp\left(\frac{E_A}{kT}\right) \tag{7}$$

The linear superposition of such processes can result in the KWW relaxations. With $kT = 0.026$ eV at room temperature we find for $0.2\ \text{eV} \le E_A \le 1\text{eV}$ a distribution of τ over more than 13 decades.

2.2 Polarization/Charging mechanisms

The time dependent polarization of a solid dielectric submitted to an external electric field occurs through a number of mechanisms involving microscopic or macroscopic charge displacement. As already mentioned, according to the time scale of polarization build up we can divide the polarization mechanisms in two categories, the instantaneous and the delayed time dependent polarization. The time dependent polarization mechanisms (van Turnhout 1987, Vandershueren 1979, Barsoukov 2005, Kao 2004), which are responsible for the "dielectric charging" effects are characterized by a time constants that may be as low as 10-12 sec or as large as years, so that no relaxation is observed under the conditions of observation. These mechanisms are called slow and may occur through a number of processes involving either microscopic or macroscopic charge displacement. The slow polarization mechanisms, a summary of which is presented in Fig.3, are:

The **dipolar** or **orientational polarization** occurs in materials containing permanent molecular or ionic dipoles. In this mechanism depending on the frictional resistance of the medium, the time required for this process can vary between picoseconds to even years. The dipolar polarization of inorganic crystals may be caused by structural properties of the crystal lattice or it may be due to lattice imperfection or doping, for example in impurity vacancy dipole systems. The structural interpretation of the dielectric processes occurring in many polar materials is usually approached by assuming impaired motions or limited jumps of permanent electric dipoles. In molecular compounds for example, relaxation can be considered as arising from hindered rotation of the molecule as a whole, of small units of the molecule or some flexible group around its bond to the main chain, while in ionic crystals, it can be mainly associated with ionic jumps between neighboring sites (ion-vacancy pairs). From conventional dielectric measurements it is known that materials obeying the classical Debye treatment with a single relaxation time are rather rare.

The **space charge** or **translational polarization** is observed in materials containing intrinsic free charges such as ions or electrons or both. The space charge polarization arises from macroscopic charge transfer towards the electrodes that may act as total or partial barriers. Moreover, the charging of space-charge electrets may be achieved by injecting (depositing) charge carriers. Other methods consist in the generation of carriers within the dielectric by light, radiation or heat and simultaneous charge separation by a field. The space charge polarization causes the material to be spatially not neutral (fig.3) hence is a much more complex phenomenon than the dipolar polarization.

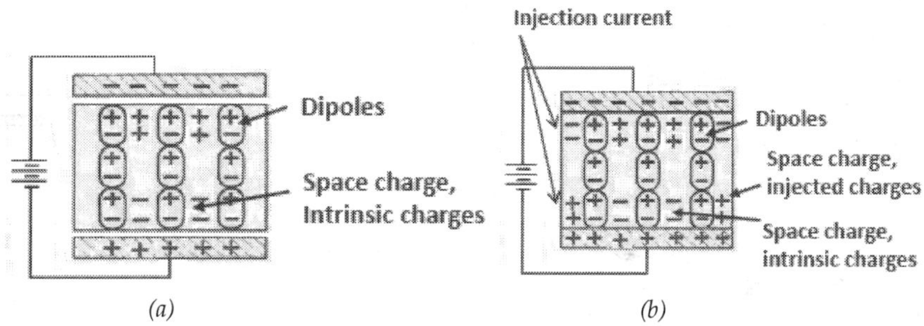

Fig. 3. Summary of polarization mechanisms under (a) non contacting and (b) contacting charging

The **interfacial polarization**, which sometimes is referred as Maxwell-Wagner-Sillars (MWS) polarization, is characteristic of systems with heterogeneous structure. It results from the formation of charged layers at the interfaces due to unequal conduction currents within the various phases. In structurally heterogeneous materials, such as complicated mixtures or semi-crystalline products, it can be expected that field-induced ionic polarization will obey more closely an interfacial model of the Maxwell-Wagner-Sillars type than a space-charge model of the barrier type. There the action of an electric field can achieve a migration charge by (a) bulk transport of charge carriers within the higher conductivity phase and (b) surface migration of charge carriers. As a consequence surfaces, grain boundaries, interphase boundaries (including the surface of precipitates) may charge. Charges "blocked" at the interface between two phases with different conductivity give a contribution to the net polarization of the body exposed to the electric field.

In most of the theoretical treatments, the polarized material is assumed to be free of charge carriers, so that the internal field and the dipolar polarization can be considered as space independent. In practice, however, dipolar and space charge polarizations often coexist and the electric field and polarization must then be considered as averaged over the thickness of the sample. Finally, the simultaneous displacement of free charges and dipoles during the polarization process may lead to a particular situation where the internal electric field is nearly zero, so that no preferred orientation of dipoles occurs.

3. Dielectric materials for RF-MEMS capacitive switches

As already mentioned the dielectric materials used in MEMS capacitive switches are as SiO_2, Si_3N_4, AlN, Al_2O_3, Ta_2O_5 and HfO_2. The charging mechanisms in each dielectric will depend on the material structure and for this reason each one will be discussed separately.

So far the dielectric charging has been intensively investigated in SiO_2 and Si_3N_4. Regarding the other materials i.e. Ta_2O_5, HfO_2 and AlN there is little information on their impact on the reliability of MEMS devices. In the case of Ta_2O_5 (Rottenberg 2002) and HfO_2 (Luo 2006, Tsaur 2005), although the materials are attractive due to their large dielectric constant, the knowledge on the charging processes is still limited and arises from the study of MIM and MIS capacitors, the latter for MOSFET gate applications. Both materials exhibit ionic conduction and in the case of Ta_2O_5 it has been shown that under high electric field space charge arises due to formation of anodic-cathodic vacancy pair, (Frenkel pair dissociation) (Duenas 2000). Moreover, isothermal current transients in chemical vapor deposited material revealed that protons are incorporated in the structure and the current transient arises from proton displacement (Allers 2003). For HfO_2 it has been shown that hole trapping produces stable charge (Afanas'ev 2004). The trapped charge density was found to be strongly sensitive on the deposition methods and the work-function of the gate electrodes. In thin layers (≤ 10nm) it was shown that charge trapping follows a logarithmic dependence on time (Puzzilli 2007). On the other hand the de-trapping rate was found to depend on the film thickness, with a power law behavior as a function of time.

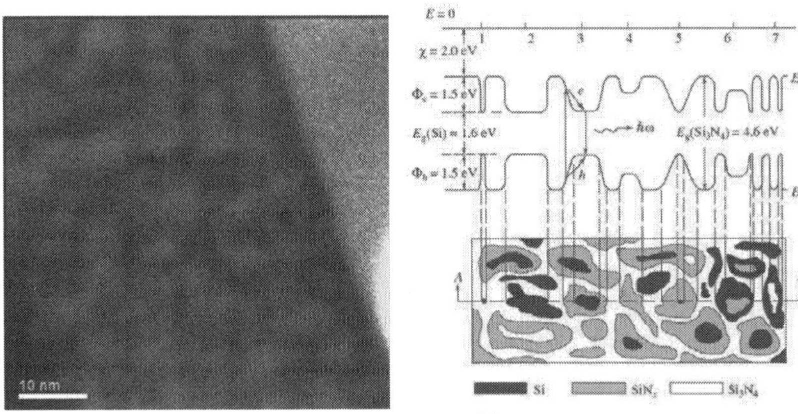

(a) (b)

Fig. 4. (a) Cross-sectional energy-filtered TEM image of Si-ncs embedded in SiNx layers deposited with a gas flow that corresponded to 21% Si excess (Carrada 2998) and (b) representation of material non-homogeneity and band gap fluctuation (Gritsenko 2004)

α-Al_2O_3 is a wide-gap insulator with a direct energy gap of about 8.3 eV (Fang 2007). The O-Al bonds in the compound exhibit highly ionic nature and theoretical calculations have shown that the valence band is well separated into two parts, with the lower part consisting of O 2s states and the upper part being dominated by O 2p states. The lower part of the conduction band is in general believed to be dominated by Al 3s states. Regarding the electrical properties and charging behavior the dc behavior of alumina has been little investigated. The experimental I(t) curves have shown that the 'quasi' steady-state current is reached for time ranging from 104 to 105 s (Talbi 2007). The transient current was reported to consist of two parts, the first one that arises mainly from the polarization of dipoles in the dielectric which dominate at short time, whereas the second part was found to correspond to the carriers transport mechanism. Moreover the conduction mechanism in the high field regime was reported to obey the space charge limited current law. The conduction mechanism high temperatures has been found to be dominated by carriers emitted from deep traps while the low temperatures one by carriers emitted from discrete shallow traps or transport in the band tails (Li 2006, Papandreou 2008). Here it must be pointed out that the characteristics of the charge traps introduced during deposition depend strongly on the deposition conditions (Papandreou 2008).
Aluminum nitride (AlN) piezoelectric thin film is very popular in RF micro-machined resonators and filters MEMS devices. The advantages arise from its high resistivity and piezoelectric coefficient, which is the largest among nitrides as well as the possibility to be deposited at temperatures as low as 500°C and patterned using conventional photolithographic techniques. AlN generally exhibits smaller piezoelectric and dielectric constant and differs from PZT materials in that it is polar rather than ferroelectric. Theoretical results have indicated that nitride semiconductors possess a large spontaneous polarization (Papandreou 2008), associated with which are electrostatic charge densities analogous to those produced by piezoelectric polarization fields. In wurtzite structure the polar axis is parallel to the c-direction of the crystal lattice that may give rise to a macroscopic spontaneous polarization, which can reach values up to 0.1 C/m^2. This macroscopic lattice polarization is equivalent to two dimensional fixed lattice charge densities with values between 10^{13} and 10^{14} e/cm^2 located at the two surfaces of a sample (Bernadini 1997). Finally, in inhomogeneous alloy layers, variations in composition are expected to create non-vanishing and spatially varying spontaneous and piezoelectric polarization fields and associated charge densities that can significantly influence the material properties. Thus in contrast to the single crystalline material, the sputtered one exhibit near-zero, positive or even negative piezoelectric response indicating a change in crystalline orientation, grain size, concentration of defects or even a complete reversal of dipole orientation (Bernadini 1997, Zorrodu 2001). Recently, AlN has been introduced in MEMS switches (Ruffenr 1999) and reliability tests have proved that under low pull-in bias or certain polarity the device degradation may be extremely low. Assessment of MIM capacitors with crystalline AlN dielectric has indicated that this behavior has to be attributed to the presence of a spontaneous polarization arising from dislocations that may induce a surface charge of the order of c.cx$10^{-7}$$Ccm^{-2}$, which is much smaller than the theoretically predicted spontaneous polarization (Papandreou 2009).
The SiO_2 and Si_3N_4 are the most important dielectrics used in modern silicon-based electronic devices. In spite of the five decades of intensive investigation, the gained knowledge has not be effectively applied in MEMS capacitive switches. The reasons behind

this deficiency lie on the fact that in MEMS capacitive switches technology the dielectric film is deposited on rough metal surfaces at low temperatures (≤ 300°C). Thus the film surface morphology is affected by the substrate and the low temperature leads to significant deviation of stoichiometry. The latter allows us to describe silicon oxide and nitride as SiO_x and and SiN_x with $x < 2$ and $x < 1.33$ respectively. The low temperature deposition gives rise to formation of silicon nanoclusters and/or nanocrystals in both materials due to the fact that Si excess is high and the phase separation mechanism is not nucleation and growth as in the case of low Si excess, but spinodal decomposition (Carrada 2008). Fig.3a shows clearly the percolation of nanocrystal after 1 min annealing at 1000 °C under Ar ambient. A simplified schematic diagram illustrating the two-dimensional structure of SiN_x (Gritsenko 2004) shows in Fig.3b (bottom) the regions of silicon phase, stoichiometric silicon nitride, and subnitrides and (top) the corresponding energy band profile. Similar is the behavior of SiO_x (Ikona 2004, Yoshida 2002).

Material	Ionic	Dipolar	Space charge	Dielectric constant
SiO_2	-	(✔)	✔	3-4.5
Si_3N_4	-	(✔)	✔	6-7.5
Al_2O_3	✔	✔	✔	8-9
AlN	✔	✔	✔	10-12
HfO_2	✔	✔	✔	11-12
Ta_2O_5	✔	✔	✔	~100

Table 1. Dielectric films for MEMS capacitive switches and charging mechanisms (✔) due to deviation from stoichiometry

Although these materials consist of covalent bonds, in substoichiometric silicon oxide the E'_{δ} defect gives rise to the formation of dipoles by trapping holes (Fleetwood 2003). Although these dipoles were observed after gamma ray irradiation, their presence in the SiO_x used in MEMS capacitive switches cannot be overruled. Moreover, the presence of such structures cannot be rejected in SiN_x.

Taking all these into account we can conclude that the charging mechanisms taking place in insulating films used in MEMS capacitive switches can be summarized in Table 1. So, in all cases the space charge polarization due to presence of free charges or injected charges as well as the dipolar polarization constitutes the major charging mechanisms. The presence of nanoclusters or nanocrystals is expected to give rise to a random distribution of dipolar polarization and in the same time is expected to give rise to interfacial polarization; a fact that needs to be experimentally demonstrated.

Presently, due to above analyzed effects, there is still no clear information on the charging of thin dielectric films used in MEMS capacitive switches. The electrical properties of these dielectrics obviously depend strongly on the deposition methods and conditions. Due to the absence of standardization of deposition methodology, the study of dielectric charging, employing MEMS and MIM devices, still leads to no concrete results. A key issue parameter, towards the solution of this problem, seems to be the dielectric film temperature since it accelerates the charging and discharging processes by providing enough energy to trapped charges to be released and to dipoles to overcome potential barriers and randomize their orientation. The effect of temperature will be analyzed in the following.

4. Assessment of dielectric charging

The dielectric charging is a complex effect, which cannot be monitored through simple test method hence requires various techniques involving specific experimental setups. The charging can be monitored using bare dielectric material or Metal-Insulator-Metal (MIM) capacitor or MEMS capacitive switch. As it will be understood from the following analysis each method provides partial information.

4.1 Kelvin Probe Force Microscopy (KPFM)

The Kelvin probe force microscopy (KPFM), also known as surface potential microscopy, is a noncontact variant of atomic force microscopy (AFM) that was invented in 1991 (Nonnenmacher 1991). The KPFM is a scanned probe method where the potential offset between a probe tip and a surface can be measured using the same principle as a macroscopic Kelvin probe.

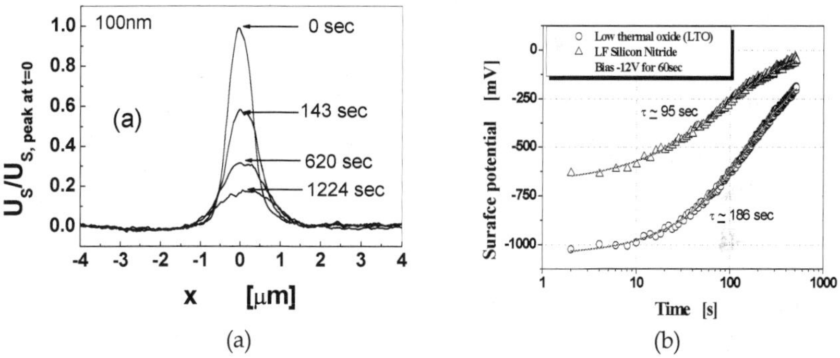

Fig. 5. (a) Dielectric film surface potential distribution vs time (Zaghloul 2008) and (b) peak charge decay (Belarni 2008).

Among the various characterization methods of dielectric charging in MEMS capacitive switches the KPFM method plays a significant role since it allows the simulation of the charging through the dielectric film and suspended electrode roughness and asperities. Presently, the Kelvin probe force microscopy is employed to provide qualitative information on dielectric free surface discharging process. The charges are injected into the dielectric with the probe conductive tip in proximity or contact to the dielectric surface. Then the tip is used to scan the charged area. In these experiments, an important result not yet fully related to switches performance, is the evolution of the deposited charges which showed that the charges are not spreading on the surface (fig.5a) (Zaghloul 2008).

The decay of the amount of charge has been attributed to the penetration and trapping into the bulk of the dielectric. The potential relaxation was reported to be exponential (fig.5b). On the other hand the surface potential induced by charges injected with the Kelvin probe tip was found to decay following the stretched exponential law. Finally, the decay time constant was reported to depend on the dielectric material and practically not affected by the tip potential (Belarni 2008).

4.2 Metal-Insulator-Metal (MIM) capacitors

The MIM capacitors, although do not substitute MEMS switches in the pull-down state, have been proved to be a valuable test structure for assessing the electrical properties of dielectric materials. The dielectric charging in MIM capacitors has been investigated through two experimental methods, the Discharge Current Transients (DCT) and the Thermally Stimulated Depolarization Currents (TSDC). Both methods are based on the application of electric field for a long time so that to produce saturation of dipole orientation and trapping of injected charges. Finally, the DCT method is better exploited if the transients are recorded at different temperatures while the TSDC method requires the current recording during the temperature sweep (Vandershueren 1979).

4.2.1 Discharge Current Transient method

The DCT method is based on the measurement of charging and discharging currents of a MIM capacitor. The discharge current transient when arises from trapped charges i.e. holes, or dipole reorientation is given respectively by

$$j = \frac{dP(t)}{dt} = -\frac{P(t)}{\tau} \tag{8}$$

were P_t is the macroscopic polarization and τ the polarization emission or relaxation time, depending on the model. Here it must be pointed out that the dielectric charging and discharging currents in principle are not equal due to the presence of external electric field during charging and the internal one during discharging (fig.6). For this reason the charging current may be masked by high leakage currents.

The DCT characterization method has been extensively applied for assessment of dielectric charging in silicon dioxide (Yuan 2005) and silicon nitride (Exarchos 2005, Lamhamdi 2008, Zaghoul 2009) films. In all cases the experimental results revealed that both charging and discharging transients are multi exponential with temperature independent time constants. On the other hand in silicon nitride the DCT method revealed the presence of thermally activated mechanisms. The decay was fitted using the stretched exponential law and the Arrhenius plot of relaxation time, Eq. 8, allowed the calculation of activation energy EA and estimation of relaxation time at room temperature $\left(\tau_{300K} \approx 6 \cdot 10^{3}\ \text{sec}\right)$ (Exarchos 2005).

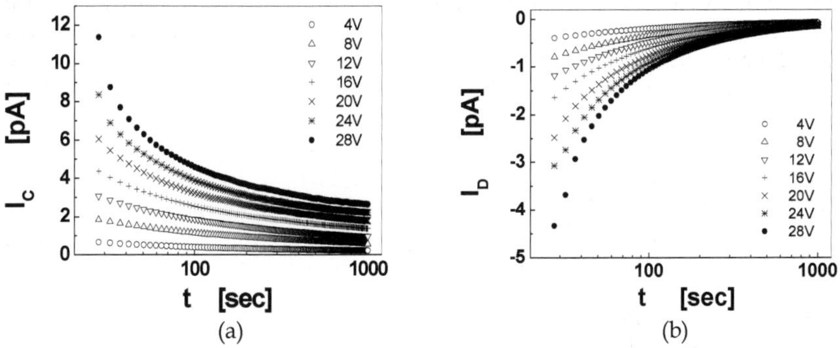

Fig. 6. Dependence of (a) charging and (b) discharging current transient on charging bias (Lamhamdi 2008)

4.2.2 Thermally Stimulated Depolarization Current method

In order to obtain a better insight on the TSDC method it is essential to take into account that in insulators, the time and temperature dependence of polarization and depolarization processes are determined by

- in the case of dipolar polarization the competition, between the orienting action of the electric field and the randomizing action of thermal motion and
- in the case of space charge polarization the processes are far more complex because several mechanisms can be involved simultaneously

The thermally stimulated depolarization current (Vandershueren 1979, Turnhout 1987) is given by:

$$j(T) \approx \frac{P_S(T_p)}{\tau_0} \cdot \exp\left(-\frac{E_A}{kT}\right) \cdot \exp\left[-\frac{1}{\gamma\tau_0} \cdot \frac{kT^2}{E_A} \exp\left(-\frac{E_A}{kT}\right)\right] \tag{9}$$

where γ is the heating rate, being kept constant during temperature scan.

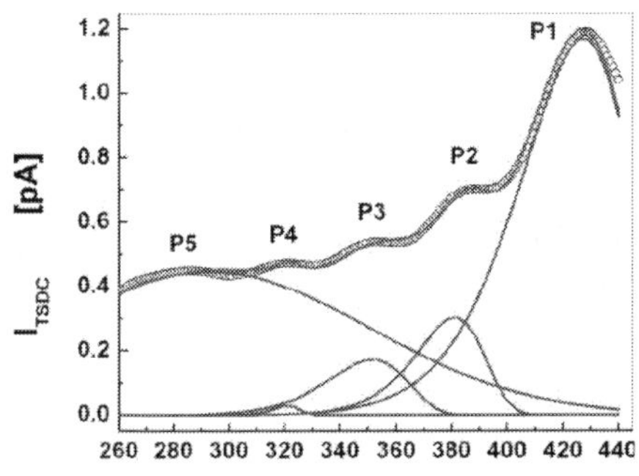

Fig. 7. Temperature dependence of TSD current of a SiN_x MIM capacitor and its analysis (Papaioannou 2007a).

The dependence of TSD current on temperature is presented in Fig.6 and analyzed by fitting Eq.9 to the experimental data. Each contribution (P1-P5) arises from a specific charging mechanism for which the activation energy E_A and τ_0 can be determined. The values of activation energy and τ_0 constitute basic parameters since they allow the calculation of relaxation times at room temperature for each contributing mechanism. Another important parameter is the range of magnitudes of relaxation times because in amorphous dielectric films the relaxation times are distributed over several decades. The difficulty of the determination of such a distribution has been minimized with the aid of Fröhlich model (Fröhlich 1949) that allowed the reduction of relaxation times to a relatively small distribution of activation energies. This model allows the extraction of the dependence of room temperature relaxation time on activation energy assuming that τ_0 is known. A simple example of the importance of this model on the prediction of MEMS reliability is plot of the dependence of room temperature relaxation time, normalized to its value at 450K (fig.8a). Since τ_0 is not known for the sake of simplicity the relaxation time has been assumed to be $\tau_{450K} = 1$. The validity of the relaxation time exponential dependence on activation energy has been confirmed through thermally stimulated depolarization current assessment of SiN_x MIM capacitors, which dielectric film was deposited with high frequency (13.6 MHz) (HF), low frequency (380 KHz) (LF) and mixed frequency (13.6 MHz + 380 KHz) PECVD method and the top contacts were extended over areas with uniform or varying stress (Papandreou 2007) (fig.8b). The results presented by Papandreou et al (Papandreou 2007) revealed that the relaxation times are distributed around specific activation energies with values of 0.17eV, 0.35eV and 0.55eV. It is interesting to point out that the distributions seem to be independent on the film deposition methods as well as the presence of uniform or non uniform stress in the nitride films. Finally it is worth noticing that presently the available data are still limited and no general rules can be extracted on the dependence of these charging mechanisms on the material "quality".

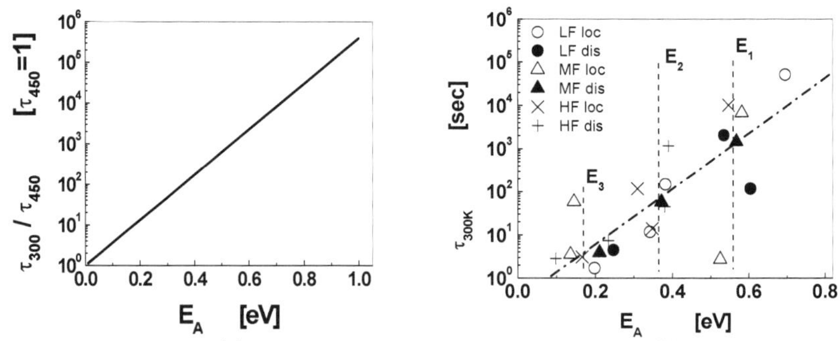

(a) (b)

Fig. 8. (a) Dependence of room temperature relaxation time vs activation energy and (b) dependence of room temperature relaxation times on the corresponding polarization mechanism activation energy (Papandreou 2007)

4.3 MEMS capacitive switches

The assessment of dielectric charging in MEMS capacitive switches constitutes an issue which still has not been established. The basic assessment methods rely on the monitoring of the shift of bias for capacitance minimum, $V_{min} = -\frac{z_d\langle\sigma\rangle}{\varepsilon_r\varepsilon_0}$ and the pull-down and pull-up voltages. The shift of the bias for capacitance minimum is a quantity that accurately provides information on the dielectric charging and does not depend on the mechanical parameters o the metal bridge or cantilever. For this reason it has been widely used (Wibbeler 1998, Papaioannou 1996, Ruan 2008 etc) to assess the charging due to cycling and ESD stress at room as well as at elevated temperatures. On the other hand the charge calculated through this method is obtained under low electric field conditions while the performance of a capacitive switch is determined by the shift of pull-down and pull-up voltages which are directly related to the device performance and occur under high electric fields. This limits the importance of V_{min}. Another method to assess the charging and discharging processes in MEMS are the pull-down and pull-up transients (Papaioannou 2005, Papaioannou 2007c). The presence of thermally activated mechanisms in dielectrics, requires the assessment of MEMS switches to be performed as a function of temperature in order to extract better information on the dielectric charging processes.

Due to the fact that these methods are routinely used for the assessment of MEMS reliability and the limited space, the values of each method will be revealed in the following paragraph.

5. Reliability of RF-MEMS capacitive switches

The reliability of MEMS capacitive switches is determined by a large number of factors. The aim of the present chapter is to present and discuss the failure mechanisms that are related to dielectric charging:

- Effect of contact roughness
- Effect of DC bias and temperature
- Influence of substrate on MEMS reliability
- Ambient effect on MEMS reliability
- MEMS reliability to ESD stress
- Reliability to RF signal power

5.1 Effect of contact roughness

The surface roughness has been recognized as a key issue performance and reliability problem in MEMS capacitive switches. The metal bridge and the dielectric film roughness constitute a technological limitation related to the deposition techniques, which does not allow the roughness better than a few tens of nanometers.

Regarding the device performance the surface roughness plays a critical role at the down position capacitance. Thus the capacitance ratio, which is given by:

$$\frac{C_{down}}{C_{up}} = \frac{\varepsilon_r t_{air} + t_{diel}}{t_{diel}} \tag{10}$$

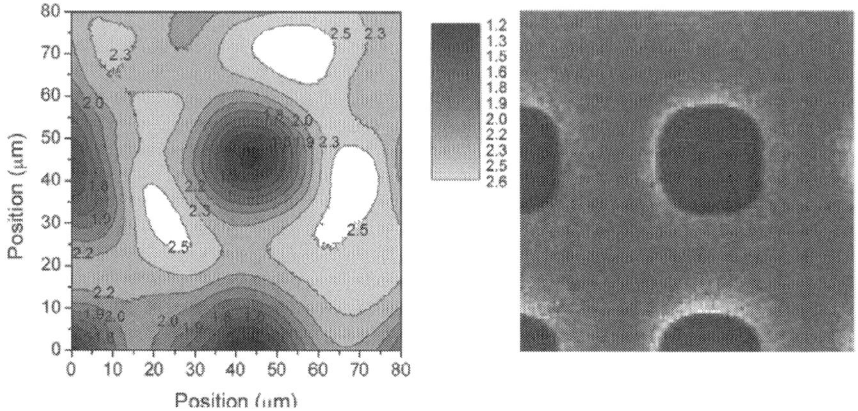

Fig. 9. Contour plot of the measured surface potential of a stressed device and the corresponding section of a SEM picture of the top electrode (Herfst 2008).

where the down state capacitance (C_{down}) must be as high as possible. In equation (11), t_{air} and t_{diel} are the thicknesses of the air and dielectric layers beneath the membrane. Due to the importance of the surface roughness on the device performance there has been an intensive effort on the modeling of the roughness using a statistical approach (Yu 2006) or AFM assessment of the bottom of switch membrane (Suy 2008). The results showed that at the up-state, the capacitance and the insertion loss increases with the RMS roughness and in the down-state, the capacitance and the isolation decreases. Moreover, it was revealed that the overall real contact area between the metal bridge and the dielectric layer surface is less than 1% of the apparent contact area, hence the down-state capacitance is mainly determined by the noncontact area between the metal bridge and the surface of the dielectric layer. The modeling suggested that the improvement of the device performance would require the RMS roughness to be kept below 10nmin order to achieve a normalized isolation of about 60% a parameter that increases with the applied hold-down voltage. Attempts to minimize this effect have been performed by adding a metal electrode, which would determine the down state capacitance, on the top of the dielectric film (Bartolucci 2008).

The surface roughness of the metal bridge and dielectric film affect directly the dielectric charging since charges are injected through the contacting areas. The effect of dielectric charging through surface roughness and asperities has been reported in several papers (Cabuz 1999, van Spengen 2002, Melle 2005, Sumant 2007, Papaioannou 2007d , Herfst 2008). Moreover charges are injected through micro gap discharge (Torres 1999, Slade 2002, Hourdakis 2006) in the proximity areas due to deviation from Pasken law and the charging is induced due to high electric field (Papaioannou 2006b) in areas where no one of the previous mechanisms can occur.

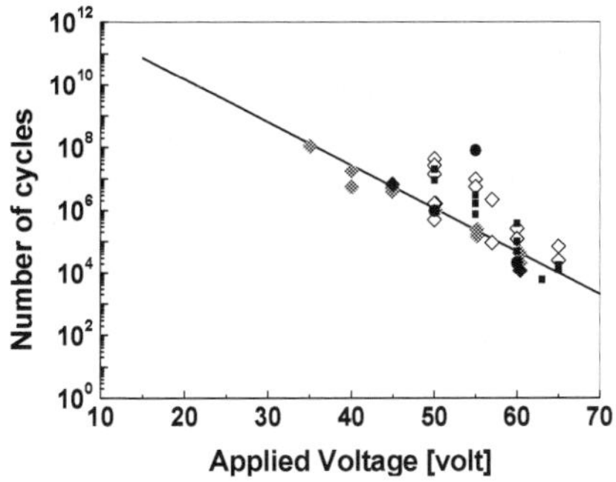

Fig. 10. MEMS lifetime dependence on actuation voltage

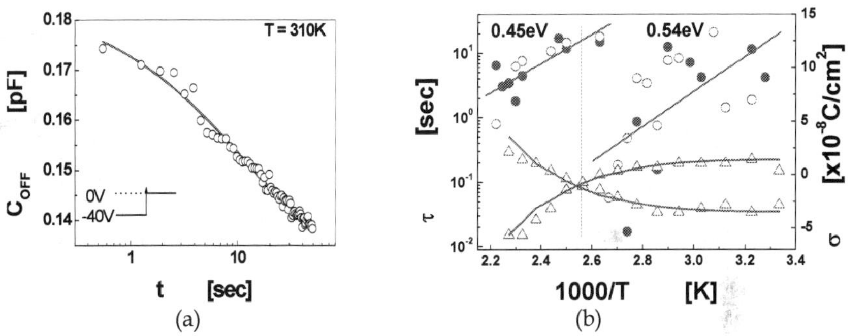

Fig. 11. (a) pull-up transient and (b) simultaneous drawing of Arrhenius plot and shift of V_{min} (Papaioannou 2007c).

These charging mechanisms in addition to the non planar dielectric film and bridge surfaces lead to a charge distribution that determines the switch behavior (Rottenberg 2008). This non uniform charge distribution has been monitored through measurements of surface potential on a capacitive switch dielectric with the aid of KPFM (fig.9) (Herfst 2008). Here it must be pointed out that the discharge data of Fig.10 allowed the determination of diffusion coefficient of electrons in SiN which was found to be of the order of $10^{-10}cm^2/sec$ (Herfst 2008).

5.2 Effect of DC bias and temperature

The first qualitative characterization of dielectric charging within capacitive membrane switches and the impact of high actuation voltage upon switch lifetime were presented by C. C. Goldsmith et al. (Goldsmith 2001). The results of evaluation of switch lifetime as a function of pull-down voltage, for all data reported in (Goldsmith 2001) are shown if Fig.10. The dependence of number of cycles on the peak actuation voltage was found to follow an exponential relationship, which deviates from Poole-Frenkel injection current relation, as C. Goldsmith et al. pointed out and plotted with a dashed line and which has been normalized at applied voltage of 15 volt. A result of technological significance was the significant improvement in lifetime as voltage decreased. Particularly it was found that the lifetime improves on the order of a decade for every 5 to 7 V decrease in applied voltage. These results definitely provided a continuing impetus to design devices of reduced switch pull-down voltage and produce wafer lots with tight pull-down voltage distributions.

Presently it is well known that the commonly quoted number of cycles to failure does not constitute a good measure of the reliability of switches suffering from charging. W.M. van Spengen et al. (van Spengen 2003) have shown that number of cycles to failure is severely affected by the actuation frequency and duty cycle. Further they have shown that since the failure is purely due to charging, the contact time (down position gives rise to charge injection) is equal for all actuation schemes.

Previously has been analyzed the effect of temperature on the charging effects in dielectrics. In the case of MEMS switches, the pull-up transient is affected by the persisting electrostatic force due to dielectric charging. Thus the fast mechanical response is followed by a slow transient which is corresponds to the dielectric film discharging process. The discharge transient was experimentally found to follow the stretched exponential relaxation law $\Delta C(t) = \Delta C_0 \exp\left[-\left(\frac{t}{\tau}\right)^{\beta}\right]$ (Papaioannou 2007c) (fig.12a). The recording of pull-up transient as a function of temperature allows us to draw the Arrhenius plot and determine the discharging process activation energy. The transient itself provides information only on the activation energy of discharging mechanism. The nature of the dominant charging mechanism i.e. dipolar or space charge polarizations, the latter arising from charge injection or intrinsic free charges, can be only obtained from the shift of the bias at minimum of capacitance-voltage characteristic, $V_{\min} = -\frac{z_d\langle\sigma\rangle}{\varepsilon_r\varepsilon_0}$, where $\langle\sigma\rangle$ is the average value of equivalent surface charge, z_d the dielectric film thickness and ε0 the vacuum dielectric constantan. Drawing both the Arrhenius and shift of V_{min} with temperature in the same figure allows us to determine the activation energy and charge origin of each contributing charging mechanism (fig.12b). Moreover, it is possible to determine the origin of the charging mechanisms, i.e if they emerge from charge injection or charge induction. Finally, the detection of thermally presence of thermally activated mechanisms may lead us to the prediction of the lifetime of capacitive switches.

5.3 Influence of substrate on MEMS reliability

The presence of two different types of charges, which are responsible for the dielectric charging, has been confirmed by experiments using either positive or negative actuation voltage (Czarnecki 2008). The conclusion was drawn from the behavior of pull-in and pull-out windows immediately after stress as well as after 1 min after stress (fig.12)

Here it must be pointed out that these results confirm the predictions of [9.39] and are supported from previous publications (Papaioannou 2005) and (Papaioannou 2007a) as well as recent ones (Olszewski 2008) where the existence of two types of charges, which contribute to dielectric charging, has been reported.

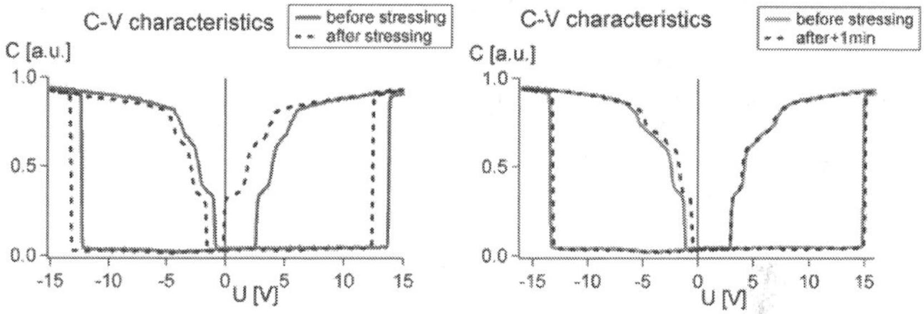

(a) (b)

Fig. 12. C-V characteristics measured (a) before and immediately after stressing with positive voltage where a negative shift is visible and (b) before and 1 minute after end of stressing with positive voltage where positive shift is visible (Czarnecki 2008).

The nowadays available information clearly proves the presence of two types of charges, which is responsible for the shift of capacitance-voltage characteristic. The presence of opposite polarity carriers is responsible for the shift C-V characteristic in voltage and capacitance domain, a behavior has been predicted (Rottenberg 2007) and shown experimentally (Papaioannou 2005, Czarnecki 2008). Moreover, the charging, when occurs with the switch in the down state, arises from the Poole-Frenkel transient current component, which gives rise to time and electric field dependent dielectric charging (Melle 2005). Here it must be emphasized that the presence of two types of charges needs further investigation since the homocharge, due to charge injection is well understood. The origin of heterocharge, which is opposite polarity charges, at the dielectric free surface needs further investigation since it is not clear whether they arise from dipole orientation or rear interface charge injection. Finally, the non symmetrical shift of pull-in and pull-out windows needs further investigation since the attribution to mobile charges cannot explain adequately the effect. This issue is highly important since the position of positive and negative charge centroids determine the magnitude and orientation dielectric polarization which in turn will directly affect the capacitance-voltage characteristic through shift of pull-in and pull-out voltages as well as the shift of capacitance minimum.

5.4 Ambient effect on MEMS reliability

The dependence of switches lifetime on the environmental gas and the gas pressure have been recently investigated (Blondy 2007, Czarnecki 2008). Although it is well understood that gasses interact with the dielectric free surface the mechanism that affects the charge storage, hence the dielectric charging, is still not understood. (fig.14)

In order to monitor the charging and discharging process the shift of V_{PI} was measured when the switch was left in the down for 10min and the discharge when was left in the up state. Humidity was removed by storing the device in high vacuum. In nitrogen ambient the charging/discharging time constants were found to be very similar. Moreover the charge trapping was found to cause the pull-voltage to drop by 7 Volts, when +50 Volts is applied for 10 minutes. When the same experiment performed in open air with typical humidity at 50%, with the devices left for 2 hours the shift of pull-in voltage was found to be completely different (fig.14). Finally, the investigation was performed assuming two charging mechanisms, the surface and bulk charging ones. According to this model when the devices was stressed in air the different behavior was attributed to change from bulk charging in the first minutes to surface charging, the latter being characterized by large time constants and degraded reliability. Finally, it is worth noticing that the effect of humidity on dielectric charging is an issue which has recently attracted attention (Peng 2009).

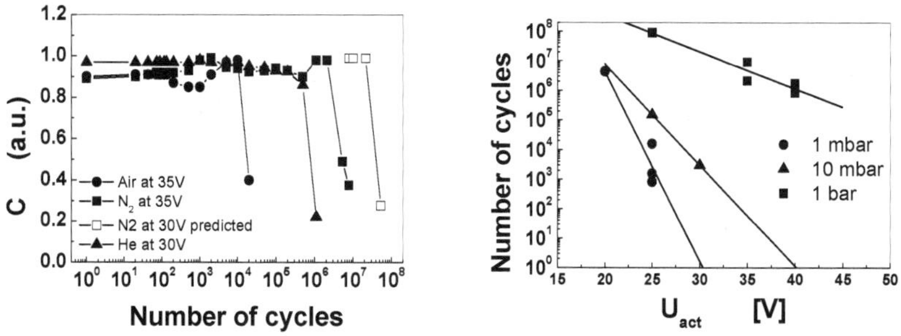

(a) (b)

Fig. 13. Effect of (a) gas ambient and (b) gas pressure on MEMS reliability (reproduced from [9.50]).

5.5 MEMS reliability to ESD stress

Electrostatic discharge (ESD) occurs when a device is improperly handled. A human body routinely develops an electric potential in excess of 1000V. Upon contacting an electronic device the discharge will create a large potential difference across the device. In MEMS electrostatic discharge and electrical overstress (EOS) damage has been identified as a new failure mode. This failure mode has not been previously recognized or addressed primarily due to the mechanical nature and functionality of these systems, as well as the physical failure signature that resembles stiction. Because RF-MEMS devices function by electrostatic actuation they are susceptible to ESD or EOS damage as well as to catastrophic failure.

Recently, ESD has started generating interest among researchers as a possible cause of failure in MEMS devices. ESD as a threat to the reliability of MEMS devices was first reported in 2000 by Walraven et al. (Walraven 2000). Reliability issues in RF MEMS switches have been reported by Tazzoli et al. (Tazzoli 2006) and Ruan et al. (Ruan 2007, 2008, 2009a, 2009b, 2009c). The robustness of MEMS capacitive switches has been assessed with Transmission Line Pulsing (TLP) (Tazzoli 2006, Ruan 2007, 2008, 2009b, 2009c) and Human Body Model (HBM) (Ruan 2009a, 2009c) setups. Due to different shape of ESD pulses the When stressing with TLP setup, from the device's mechanical point of view, the pulse is a very short (100ns) and the resulting electric field cannot induce a displacement of the membrane (switching time ≈10-20μs) (Ruan 2009b). Varying the applied pulse amplitude, ranging from 10V to 400V, J. Ruan et al (Ruan 2007, Ruan 2009b) determined the failure signature. Three different stress regions were reported (fig.14); an area where the device is still functional after the stress and two failures areas, namely soft failures area and hard failures area. The functional zone indicates the highest TLP magnitude that the component can withstand and that will not affect its behavior. Then, soft failures occur for a voltage around 350V.

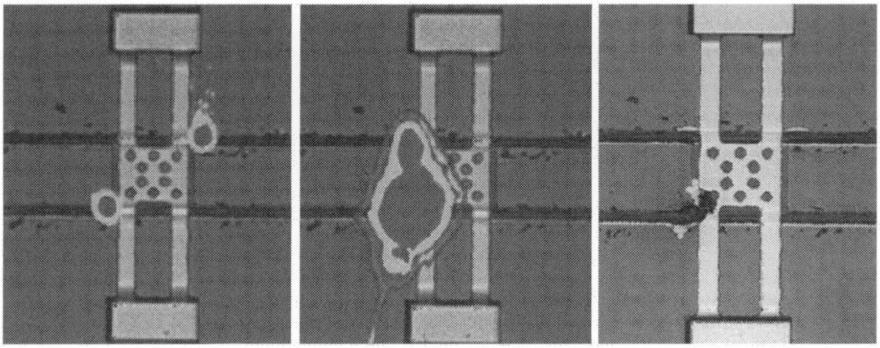

(a) (b) (c)

Fig. 14. Photo emission microscopy pictures, a) in soft failures area, b) in hard failures area and c) visible damages (after stresses) (Ruan 2009b)

In this case electric arc occurs across the air gap and can be captured using a photo emission microscopy (fig 14a). This threshold corresponds to the field emission induced breakdown level of the modified Paschen's curve. Above this threshold, hard failures happened (fig 14b) and the consequences are catastrophic (fig 14c). In the hard failures region damages were caused to the insulator layer and also to the metallic electrodes.

These three regimes can be also detected by monitoring the TLP current (fig.15a). The TLP pulses do not induce any significant current through the system up to the threshold of field emission. Above this a sharp current increase is observed and in the hard failure region the slope of the ESD current increases gradually with the pulse amplitude. It is worth noticing that in the soft failure regime the high electric field induces dielectric charging, which can be monitored through the shift of the voltage that corresponds to the minimum of capacitance. The dependence of the shift was reported to vary in all cases as a logarithmic function of the number of positive TLP pulses (Ruan 2008). Moreover, the TLP stress was found to cause narrowing of both the pull-down and the pull-up windows (Tazzoli 2006).

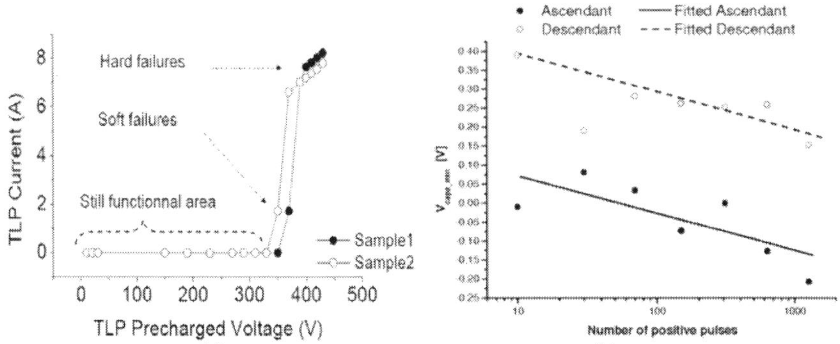

(a) (b)

Fig. 15. (a) TLP Current-Voltage failure signature of RF-MEMS capacitive switches and (b) the shift in the voltage corresponding to the minimum of capacitance as a function of the number of positive TLP pulses (Ruan 2008, 2009b).

Finally, regarding the HBM stressing it must be noticed that a MEMS capacitive switch constitutes a capacitive and non conductive load. The duration of a HBM pulse is large enough to bring the switch in the down state and cause considerable charging, much larger than the TLP stress. For this reason, recently (Ruan 2009a) proposed an accelerated stress method based on an HBM setup.

5.6 Reliability to RF signal power

In real world transmit applications the RF power that may be applied to a switch can vary over a broad range of power levels. These power levels and their effects on RF MEMS devices have not been fully explored in order to define in what applications RF MEMS

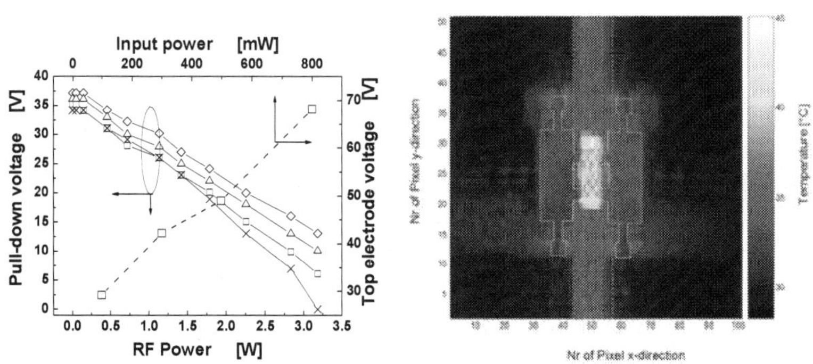

(a) (b)

Fig. 16. (a) DC pull-down voltage vs RF power level (Pillans 2002) and measured pull-up voltage vs input power (Peroulis (2004) and (b) ON-state (membrane up) temperature map for the thermal steady-state in case Pin=6W (Coccetti 2005)

switches can be used effectively. RF power signals can cause self actuation in MEMS due to high electric fields and the electrostatic force that are developed across the switch gap.
B. Pillans et al. (Pillans 2002) reported (fig. 16a) a fairly linear relationship exists between the minimum DC pull-down voltage and the RF power level applied. This result has been found to be in disagreement with the expected square relationship between the power and voltage. In the same figure (fig.16a) we included the dependence of measured pull-up voltage on input power (Peroulis 2004). Under power driving the dissipation across various parts of a switch may vary significantly. The results for a capacitive MEMS switch have shown that for 6.3W of RF incident signal at 10GHz, the temperature rise reaches the 75.5°C in case of switch in OFF state, while remains below 28°C in the ON-State for the same input signal (Coccetti 2005). Therefore, because of the linear proportionality between incident RF power and temperature, thermo-mechanical failures may be expected for the analysed device working in the OFF-State, and for input power above 10W. In this case, the hot spots temperatures are expected to reach temperatures above 120°C.

6. Conclusions

The present chapter attempts to provide a better insight on the dielectric charging and reliability of RF-MEMS capacitive switches. It has been shown that the dielectric material properties play a key issue role in the dielectric charging process. In ionic materials the ionic/dipolar polarization as well as the space charge polarization is the dominant charging mechanisms. In the case of the well established stoichiometric Si_3N_4 and SiO_2 dielectric materials the covalent bonds prevent the dipolar polarization. On the other hand, the low temperature deposition conditions, which are suitable for MEMS capacitive switches, lead to materials that are Si-rich and the significant deviation from stoichiometry gives rise to the formation of Si nanoclusters which allow the formation of defects that exhibit dipole properties, hence giving rise to dipolar polarization in addition to the space charge one.
Presently, the assessment of dielectric charging is manly based on cycling the devices between the pull-down to pull-up states. Additional assessment is performed through Metal-Insulator-Metal capacitors but these devices can provide information only on the bulk electrical properties of the dielectric film. These devices cannot be considered similar to MEMS switches since they luck free surface which interacts with ambient and accumulate surface charges. Moreover, the top electrode is in excellent contact with the top surface a fact that is not encountered in MEMS switches. A reasonable equivalent to the contacting of bridge with the dielectric film is achieved with the aid of Kelvin Probe Force Microscope (KPFM) since the contacting and charge injecting tip can reasonably simulate the surface roughness and asperities of the metal bridge. This technique has been successfully used to simulate the decay of injected charges. Moreover, it allowed the recording of charge injected in a stressed switch. The results were excellent and the preliminary evaluation of the charge decay allowed the determination of the diffusion coefficient of charges in silicon nitride. Regarding the lateral diffusion, this is still an open issue since there are no concrete results supporting such an effect. For this reason the injected charges in capacitive switches are still considered to be directly collected by the rigid electrode.
Temperature dependence of dielectric charging constitutes a key issue assessment tool. This is because the observed relaxations in disordered materials such as the dielectrics used in MEMS can be distributed over more than 11 to 12 decades. This broad distribution, as

shown by H. Fröhlich model, can be reduced to a relatively small distribution of activation energies. According to this it is highly expected to monitor thermally activated relaxation mechanisms in MEMS devices. The knowledge of the activation energy of such mechanisms provides the means to acknowledge their presence associate them with defects introduced during deposition and monitor their influence on the device performance. Regarding the bulk dielectric material this can be succeeded with both the Discharge Current Transient (DCT) and Thermally Stimulated Depolarization Current (TSDC) methods. The first requires the transient recording at different temperatures while the second one requires the temperature scan. In the case of presence of thermally activated mechanisms both methods lead to same results requiring appropriate analysis of the experimental data. Regarding the TSDC the assessment of SiN MIM capacitors revealed the distribution of time constants, normalized to room temperature, which distribution is supported by the H. Fröhlich model.
In the case of MEMS capacitive switches it has been demonstrated that the pull-up transient is reveals thermally activated mechanisms. These mechanisms have been correlated with data from TSDC measurements in MIM capacitors. Due to the fact that the mechanical performance in MEMS with metallic bridge is strongly affected by temperature, the only accurate method allows the determination of the temperature dependence of dielectric charging is the bias for capacitance minimum.
The reliability of MEMS switches is directly affected by a significant number of parameters. The failure mechanisms related to dielectric charging are the charging due to contact roughness, the DC bias and temperature, the influence of substrate, the device ambient, the ESD stress and the RF signal power. Although the nature of failure due to all these relies on dielectric charging there is still no direct connection between them. In spite of the charging and discharging acceleration observed when temperature is increased and the effect of the applied electric field intensity during when the devices is subjected to on catastrophic ESD stress and DC as well as RF power driving, there is still a significant gap of information, which would allow the unification of all these issues.

7. References

Afanas'ev, V.V., Stesmans, A. (2004), Injection induced charging of HfO2 insulators on Si, Materials Science and Engineering B Vol. 109, No. , 74–77

Allers, K.-H., Brenner, P., Schrenk, M. (2003), Dielectric reliability and material properties of Al2O3 in metal insulator metal capacitors (MIMCAP) for RF bipolar technologies in comparison to SiO2, SiN and Ta2O5, Proceedings of the Bipolar/BiCMOS Circuits and Technology Meeting, 2003, 35-38

Barsukov, E., Macdonald, J.R., Impedance Spectroscopy - Theory, Experiment, and Applications 2nd ed., John Wiley & Sons, Inc., Hoboken, New Jersey 2005

Bartolucci, G., Marcelli, R., Catoni, S., Margesin, B., Giacomozzi, F., Mulloni, V Farinelli, P. (2008), An equivalent-circuit model for shunt-connected coplanar microelectromechanical system switches for high frequency applications, Journal of Applied Physics Vol. 104, No. 084514

Belarni, M. Lamhamdi, P. Pons, L. Boudou, J. Goustavino, Y. Segui, G. Papaioannou and R. Plana, Kelvin probe microscopy for reliability investigation of RF-MEMS capacitive switches, Microelectronics Reliability 48, pp. 1232-1236, 2008

E. Berland, T. Delage, C. Champeaux, P. Tristant, A. Catherinot, and P. Blondy, "Dielectric materials in MEMS switches: a comparison between BST and Al2O3," Proceedings of the MEMSWAVE, Toulouse, 2003

Bernardini, F., Fiorentini, V., Vanderbilt, D. (1997), Spontaneous polarization and piezoelectric constants of III-V nitrides, Phys. Rev. B, Vol. 56, No., R10024-7

Blondy, P., Crunteanu, A., Pothier, A., Tristant, P., Catherinot, A., Champeaux, C. (2007), Effects of Atmosphere on the Reliability of RF-MEMS Capacitive Switches, European Microwave Integrated Circuits Conference, EuMW, pp 548-551

Mardivirin, D. Pothier, A. Crunteanu, A. Vialle, B. Blondy, P. (2009), Charging in Dielectricless Capacitive RF-MEMS Switches, IEEE Transactions on Microwave Theory and Techniques, Vol. 57, No.1, 231-236

Cabuz, C. (1999), Dielectric related effects in micromachined electrostatic actuators, IEEE Conference on Electrical Insulation and Dielectric Phenomena, 327-332

M. Carrada, M., Zerga, A., Amann, M., Grob, J.J., Stoquert, J.P., Slaoui, A., Bonafos, C., Scham, S. (2008), Structural and optical properties of high density Si-ncs synthesized in SiN_x:H by remote PECVD and annealing, Materials Science and Engineering B, Vol. 147, No. , 218–221

Chang, C.H., Qian, J.Y., Cetiner, B.A., Xu, Q., Bachman, M., Kim, H.K., Ra, Y., De Flaviis, F. Li, G.P. (2002), RF MEMS Capacitive Switches Fabricated with HDICP CVD SiNx, 2 IEEE MTT-S International Microwave Symposium Digest, Vol. 1, 231 - 234

Coccetti, F., Ducarouge, B., Scheid, E., Dubuc, D., Grenier, K., Plana, R. (2005), Thermal Analysis of RF-MEMS Switches for Power Handling Front-end, European Microwave Week, 513-516

Czarnecki, P., Rottenberg, X. Puers, R., de Wolf I. (2006), Effect of Gas Pressure on the Lifetime of Capacitive RF MEMS Switches, 19th IEEE Interna Microtional Conference on Electro Mechanical Systems, 2006. MEMS 2006 Istanbul, p. 890-893

Czarnecki, P., Rottenberg, X., Soussan, P., Ekkels, P., Muller, P., Nolmans, P., De Raedt, W., Tilmans, H.A.C., Puers, R., Marchand, L., DeWolf, I. (2008), Effect of substrate charging on the reliability of capacitive RF MEMS switches, Sensors and Actuators A in press

Daigler, R., Papaioannou, G., Papandreou E., Papapolymerou, J. (2009), Dielectric Charging in RF-MEMS Capacitive Switches - Effect of Dielectric Film Thickness, International Microwave Symposium 1275-8, 2008

De Wolf, I., van Spengen (2002), Techniques to study the reliability of Metal RF MEMS capacitive switches, Microelectronics Reliability, Vol. 42, No., 1789-1794

Duenas, S., Castan, H., Barbolla, J., Kola, R.R., Sullivan, P.A. (2000), Electrical characteristics of anodic tantalum pentoxide thin ®lms under thermal stress, Microelectronics Reliability, Vol. 40, No. , 659-662

Exarchos, M., Theonas, V., Pons, P., Papaioannou, G.J., Melle, S., Dubuc, D., Cocetti F., Plana, R. (2005), Investigation of charging mechanisms in Metal-Insulator-Metal structures, Microelectronics Reliability, Vol. 45, No. , 1782–1785

Fang, C.M., de Groot, R.A (2007), The nature of electron states in AlN and α-Al2O3, J. Phys.: Condens. Matter, Vol. 19, No. , 386223 1-6

Fleetwood, D. M. et al. (2003), Dipoles in SiO2: Border traps or not?, in Silicon Nitride, Silicon Oxide Thin Insulating Films and Other Engineering Dielectrics VII, by R. E. Sah et al. editors, 291

Fröhlich, H. (1949), Theory of Dielectrics, Oxford University Press

Goldsmith, C., Ehmke, J., Malczewski, A., Pillans, B., Eshelman, S., Yao, Z., Brank, J., Eberly, M. (2001), Lifetime Characterization of Capacitive RF MEMS Switches, IEEE MTT-S Digest pp.227-230

Gritsenko, V. A., Gritsenko, D. V., Novikov, Yu. N., Kwok R. W. M., Bello, I. (2004), Short-range order, large scale potential fluctuations and photoluminescence in amorphous SiNx, Journal of Experimental and Theoretical Physics, Vol. 98, No. 4, 760–769

Herfst, R.W., Steeneken, P.G., Schmitz, J., Mank A.J.G., van Gils M. (2008), Kelvin probe study of laterally inhomogeneous dielectric charging and charge diffusion in RF MEMS capacitive switches, 46th Annual International Reliability Physics Symposium, Phoenix, pp. 492-495

Homann, M., Kliem, H. (1994), Relaxational polarization and charge injection in thin films of silicon nitride", Microelectronics J., Vol. 25, pp. 559-566

Hourdakis, E.., Simonds, B. J., Zimmerman, Z. M. (2006), Submicron gap capacitor for measurement of breakdown voltage in air, Rev. Sci. Instrum., vol. 77, pt. 1, no. 3, 034 702

Iacona, F., Bongiorno, C., Spinella, C. (2004), Formation and evolution of luminescent Si nanoclusters produced by thermal annealing of SiO_x films, J. of Applied Physics Vol. 95, No. , 3723-

Kao, K. C. (2004), Dielectric Phenomena in Solids, With Emphasis on Physical Concepts of Electronic Processes", Elsevier Academic Press 2004

Kliem, H. (2005), Kohlrausch Relaxations: New Aspects about the Everlasting Story, IEEE Transactions on Dielectrics and Electrical Insulation Vol. 12, No. 4, 709-718

M. Lamhamdi, P. Pons, U. Zaghloul, L. Boudou, F. Coccetti, J. Guastavino, Y. Segui, G. Papaioannou and R. Plana, Voltage and temperature effect on dielectric charging for RF-MEMS capacitive switches reliability investigation, Microelectronics Reliability 48, pp 579-582, 2008

Li, C.R., Ding, L.J., Lv, J.Z., Tu Y.P., Cheng, W.C. (2006), The Relation of Trap Distribution of Alumina with Surface Flashover Performance in Vacuum, IEEE Trans. on Dielectrics and Electrical Insulation, Vol. 13, No. , 79-84

Lisec, T.; Huth, C.; Wagner, B, Dielectric material impact on capacitive RF MEMS reliability, 34th European Microwave Conference, 2004, Volume 1, Issue , 11-15 Oct. 2004 Page(s): 73 - 76

Luo, J.K., Lin, M., Fu, Y.Q., Wang, L., Flewitt, A.J., Spearing, S.M., Fleck, N.A., Milne, W.I. (2006), MEMS based digital variable capacitors with a high-k dielectric insulator, Sensors and Actuators A, Vol. 132, No. , 139–146

Melle, S., Flourens, F., Dubuc, D., Grenier, K., Pons, P., Pressecq, F., Kuchenbecker, J., Muraro, J. L., Bary, L., Plana, R. (2003), Reliability overview of RF MEMS devices and circuits, in Proc. 33rd Eur. Microwave.Conf., Oct. 2003, pp. 37–40.

Melle, S., De Conto, D., Dubuc, D., Grenier, K., Vendier, O., Muraro, J.L., Cazaux J.L., Plana R. (2005), Reliability Modeling of Capacitive RF MEMS, IEEE Tans. on Microwave Theory and Techniques, vol 53, No. , 3482-3488

Mardivirin, D., Pothier, A., Crunteanu, A., Vialle B., Blondy, P. (2009), ..., IEEE Trans. on Microwave Theory and Tech. 57, 231 (2009)

Nonnenmacher, M., O'Boyle, M. P., Wickramasinghe, H. K. (1991), Kelvin probe force microscopy, Applied Physics Letters Vol. 58, No. ,2921-

Olszewski, Z., Duane R., O'Mahony, C. (2008), A study of capacitance-voltage curve narrowing effect in capacitive microelectromechanical switches, Applied Physics Letters, Vol. 93, No. , 094101

Papaioannou, G., Exarchos, M., Theonas, V., Wang G., Papapolymerou, J. (2005),…………., IEEE Trans. on Microwave Theory and Tech. Vol.53, No. 3467-

Papaioannou G. J., Exarchos, M., Theonas, V., Psychias, J., Konstantinidis, G., Vasilache, D., Muller, A., Neculoiu D. (2006a), Effect of space charge polarization in radio frequency microelectromechanical system capacitive switch dielectric charging, Applied Physics Letters vol 89, pp 103512-1 to 4,

Papaioannou, G. J., Wang, G., Bessas D., Papapolymerou J. (2006b), Contactless Dielectric Charging Mechanisms in RF-MEMS Capacitive Switches, 1st European Microwave Integrated Circuits Conference EuMW, pp 513-516,

Papaioannou, G., Papapolymerou, J., Pons P., Plana R. (2007a), Dielectric charging in radio frequency microelectromechanical system capacitive switches: A study of material properties and device performance, Applied Physics Letters Vol.90, No. , 233507

Papaioannou G. J., Lisec T. (2007b), Dielectric charging process in AlN RF-MEMS capacitive switches, Proceedings of the 2nd European Microwave Integrated Circuits Conference, 540-543,

Papaioannou, G., Papandreou, E., Papapolymerou, J., Daigler R. (2007c), "Dielectric Discharging processes in RF-MEMS Capacitive Switches", Asia-Pacific Microwave Conference, 2007. APMC 1-4,

Papaioannou, G. J., Papapolymerou J. (2007d), Dielectric charging mechanisms in RF-MEMS capacitive switches, Proceedings of the 2nd European Microwave Integrated Circuits Conference, 1157-1160

Papandreou, E., Lamhamdi, M., Skoulikidou, C.M., Pons, P. Papaioannou, G., Plana, R. (2007), Structure dependent charging process in RF MEMS capacitive switches, Microelectronics Reliability 47, pp 1812–1817

Papandreou, E., Crunteanu, A., Papaioannou, G., Blondy, P., Dumas-Bouchiat, F., Champeaux, C., Catherinot, A. (2008), Investigation of dielecric charging mechanisms in Al2O3 RF mems capacitive switches, MEMSWAVE 2008

Papandreou, E., Papaioannou G., Lisec, T. (2009), A correlation of capacitive RF-MEMS reliability to AlN dielectric film spontaneous polarization, International Journal of Microwave and Wireless Technologies Vol. 1, No.1, 43-47

Peng Z., Palego, C. Hwang, J.C.M. Forehand, D.I. Goldsmith, C.L. Moody, C. Malczewski, A. Pillans, B.W. Daigler, R. Papapolymerou, J. (2009), Impact of Humidity on Dielectric Charging in RF MEMS Capacitive Switches, IEEE Microwave and Wireless Letters, Vol. 19, No.5, 299-301

Peroulis, D., Pacheco, S.P., Katehi, L. P.B. (2004), RF MEMS Switches With Enhanced Power-Handling Capabilities, IEEE Transactions on Microwave Theory and Techniques, Vol. 52, No. 1, 59-68

Pillans, B., Kleber, J., Goldsmith, C., Eberly, M. (2002), RF Power Handling of Capacitive RF MEMS Devices, IEEE, 329-332

Puzzilli, G., Irrera, F. (2007), Long time transients in hafnium oxide, Microelectronic Engineering, Vol. 84, No. , 2394–2397

Rebeiz, G. M., "RF MEMS," in Theory, Design and Technology. Hoboken, NJ: Wiley, 2003, pp. 185–192.

Reid, J.R., Webster, R.T. (2002), Measurements of charging in capacitive microelectromechanical switches, Electron. Lett. Vol. 38, No. 24, 1544–1545

Rottenberg, X., Jansen, H., Fiorini, P., De Raedt, W.,. Tilmans, H. A. C (2002), Novel RF-MEMS capacitive switching structures, 32nd European Microwave Conference, 1-4, 2002

Rottenberg, X., De Wolf, I., Nauwelaers, B. K. J. C., De Raedt W., Tilmans H. A. C. (2007), Analytical Model of the DC Actuation of Electrostatic MEMS Devices With Distributed Dielectric Charging and Nonplanar Electrodes, J. of Microelectromechanical Systems, vol. 16, No. , 1243-1253

Ruan, J., Nolhier, N., Bafleur, M., Bary, L., Coccetti, F., Lisec, T., Plana, R. (2007), Electrostatic discharge failure analysis of capacitive RF MEMS switches, Microelectronics Reliability Journal, Vol. 47, No. 1818–1822

Ruan, J., Papaioannou, G.J., Nolhier, N., Mauran, N., Bafleur, M., Coccetti, F., Plana, R. (2008), ESD failure signature in capacitive RF MEMS switches, Microelectronics Reliability, Vol. 48, No. , 1237–1240

Ruan, J., Nolhier, N., Papaioannou, G.J., Trnmouilles, D., Puyal, V., Villeneuve, C., Idda, T., Coccetti, F., R. Plana, R. (2009a), Accelerated lifetime test of RF-MEMS switches under ESD stress, Microelectronics Reliability in pres

Ruan, J., Nolhier, N., Trnmouilles, D, .Papaioannou, R. Plana, R. (2009b), ESD events in SiN RF-MEMS Capacitive Switches, EOS/ESD Symposium

Ruan, J., Papaioannou, G.J., Nolhier, N., Bafleur, M., Coccetti, F., Plana, R. (2009c), ESD Stress in RF-MEMS Capacitive Switches: The Influence of Dielectric Material Deposition Method, IEEE CFP09RPS-CDR 47th Annual International Reliability IEEE CFP09RPS-CDR 47th Annual International Reliability, 568-572

Ruffner, J.A., Clem, P.G., Tuttle, B.A., Dimos D., Gonzales, D.M. (1999), Effect of substrate composition on the piezoelectric response of reactively sputtered AlN thin films, Thin Solid Films, Vol. 354, No. , 256-261

Slade, P. G., Taylor, E. D. (2002), Electrical breakdown in atmospheric air between closely spaced (0.2 μm–40 μm) electrical contacts, IEEE Trans. Compon. Packag. Technol., vol. 25, no. 3, 390–396

Sumant, P.S., Cangellaris, A.C., Aluru, N. R. (2007), Modeling of dielecgtric charging in RF-MEMS capacitive switches, Microwave and Optical Technology Letters, Vol. 49, No. 12, 3188-3192

Suy, H.M.R., Herfst, R.W., Steeneken, P.G., Stulemeijer, J., Bielen, J. A. (2008), The stgatic behavior of RF MEMS capacotove switches in contact, Proc. Nanotech 2008 Vol. 3, 517 - 520

Talbi, F., Lalam F., Malec, D. (2007), DC conduction of Al2O3 under high electric field, J. Phys. D: Appl. Physics, Vol. 40, No. , 3803–3806

Tazzoli, A., Peretti, V., Zanoni, E., Meneghesso, G. (2006), Transmission line pulse (TLP) testing of radio frequency (RF) micro-machined Microelectromechanical systems (MEMS) switches, Proc. 28th EOS/ESD Symp., Tucson, AZ, Sep. 10–15, 2006, pp. 295–303.

Torres, J.-M., Dhariwal, R. S. (1999), Electric field breakdown at micrometer separations, Nanotechnology, vol. 10, no. 1, pp. 102–107

Tsaur, J. , Onodera, K., Kobayashi, T., Wang, Z.-J., Heisig, S., Maeda, R., Suga, T. (2005), Broadband MEMS shunt switches using PZT/HfO2 multi-layered high k dielectrics for high switching isolation, Sensors and Actuators A, Vol. 121, No. , 275–281

Vandershueren J. and Casiot, J., Topics in Applied Physics: Thermally Stimulated Relaxation in Solids, P. Braunlich, Ed. Berlin, Germany: Springer-Verlag, 1979, vol. 37, ch. 4.

van Spengen, W.M., Puers, R., De Wolf, I. (2002), A physical model to predict stiction in MEMS, J. Micromech. Microeng. Vol. 12, No. 5, 702–713

Van Spengen, W.M., (2002)

van Spengen, W.M., Puers, R., Mertens, R. de Wolf. I. (2003), A low frequency test-up for the reliability assessment of capacitive RF MEMS switches, J. of Micromechanics and Microengineering Vol. 13, 604-612

J. van Turnhout in: G.M. Sessler (Ed.) Topics in Applied Physics: "Electrets", vol. 33, ch. 3, pp. 81-216, Springer-Verlag, Berlin, 1987

Walraven, J. A., Waterson, B. A., De Wolf, I. (2000), Failure analysis of MEMS, Microelectronic Failure Analysis Desk Reference 2002 Supplement, pp. 75-98

Wibbeler, J., Pfeifer G., Hietschold M. (1998), Parasitic charging of dielectric surfaces in capacitive microelectromechanical systems (MEMS), Sensors and Actuators A, Vol. 71, No. , 74-80

Yoshida, K., Umezu, I., Sakamoto, N., Inada M., Sugimura, A. (2002), Effect of structure on radiative recombination processes in amorphous silicon suboxide prepared by rf sputtering, J. of Applied Physics Vol. 92, No. , 5936-

Yu, A B., Liu, A Q., Zhang, Q X, Hosseini, H M (2006), Effects of surface roughness on electromagnetic characteristics of capacitive switches, J. Micromech. Microeng. Vol. 16, No.10, 2157–2166

Yuan, X., Cherepko, S., Hwang, J., Goldsmith, C. L., Nordquist C., Dyck, C. (2004), Initial Observation and Analysis of Dielectric-Charging Effects on RF MEMS Capacitive Switches, International Microwave Symposium, 1943-1946

Yuan, X., Hwang, J. C. M., Forehand D., Goldsmith, C. L. (2005), Modeling and Characterization of Dielectric-Charging Effects in RF MEMS Capacitive Switches, International Microwave Symposium, pp 753-756

Zaghloul, U., Belarni, A., Coccetti, F., Papaioannou, G.J., Plana, R., Pons,P. (2009), U. Zaghloul, A. Belarni, F. Coccetti, G.J. Papaioannou, L. Bouscayrol, P. Pons and R. Plana, (2009), A comprehensive study for dielectric charging process in silicon nitride films for RF MEMS switches using Kelvin Probe Microscopy, 15th International Conference on Solid-State Sensors Actuators and Microsystems (Transducers09), Denver- Colorado, USA, pp. 789-793, June 21 - 25, 2009.

Zaghloul, U., Papaioannou, G., Coccetti, F., Pons, P., Plana, R. (2009), Dielectric Charging in Silicon Nitride Films for MEMS Capacitive Switches: Effect of Film Thickness and Deposition Conditions, Microelectronics Reliability, in press

Zoroddu, A., Bernardini, F., Ruggerone P., Fiorentini, V. (2001), First-principles prediction of structure, energetics, formation enthalpy, elastic constants, polarization, and piezoelectric constants of AlN, GaN, and InN: Comparison of local and gradient-corrected density-functional theory, Physical Review B, Vol. 64, No. , 045208-1 - 045208-6

RF-MEMS based Tuner for microwave and millimeterwave applications

David Dubuc[1,2] and Katia Grenier[1]
[1]*LAAS-CNRS, Toulouse and* [2]*University of Toulouse*
France

1. Introduction

This chapter sets out the basics and applications of impedance tuner for microwave and millimeterwave applications. Engineering examples, based on innovative and up-to-date Radio-Frequency MicroElectroMechanical Systems (RF-MEMS) technologies, are used to illustrate theoretical and practical principles. An explicit, comprehensive and efficient design methodology of impedance tuners is furthermore detailed. This generic design procedure is illustrated by the design of a tuner building block and followed by the description of appropriate measurements. Finally the capabilities of RF-MEMS based Impedance tuner issued from the state of the art are briefly reviewed and are followed by global conclusions.
The purposes of this chapter are then to give to the readers comprehensive informations on:

- The basics and applications of microwave and millimeterwave impedance tuners,
- The architectures of tuners,
- The implementation of tuner thanks to RF-MEMS technology,
- The design and characterization methodologies.

2. Basic definitions of Impedance Tuner

2.1 Applications and Basic definitions

Impedance matching is one of the key activities of microwave designers. Targeting maximum power transmission and/or low noise operation, impedance matching networks widely take place in all RF, microwave and millimeterwave systems. The corresponding design techniques are now well established and described in plenty of microwave books (Pozar, 2005; Collin, 2001).
For a decade, with the increase of microwave applications, requirements in term of system-reconfigurability have raised the level of complexity of circuits and especially of matching circuits. In addition to existing design constraints (detailed below), the ability of tunability without any loose of performances, and even with improved performances, has become mandatory. This is accomplished in conjunction with the use of new technologies to fulfill integration and increased frequency operation trends (Dubuc et al., 2004).

This actual trend gives rise to the development of new kinds of integrated microwave passive networks, which match/generate impedances with reconfigurable ability. Two major applications, presented in the Figure 1 and detailed below, take full benefit from these high performances and integrated circuits: (1) tunable matching networks in reconfigurable and smart RF-microsystem, and (2) impedances generators, which exhibit wide range of impedance values for devices characterization.

The Figure 1 (a) presents a reconfigurable front-end system, where impedance tuning circuits (referred in this chapter as *impedance tuner* or simply *tuner*) correspond to the key blocks in order to assure tunability under high efficiency operation (mainly high received-transmit power and low noise operation) (Rebeiz, 2003). For this application, the main features of such circuits may be listed as:

- the set of source and load impedances, which can be matched. This can be presented as the number of covered quadrants of the Smith Chart or simply by the impedance tuning range (both for the real and imaginary parts),
- the frequency bandwidth, as reconfigurability of operating frequency is concerned here,
- the insertion losses or the power efficiency of the tuner,
- the power handling capabilities,
- the DC -power consumption, as tunable-switchable elements are mandatory for tunability,
- the integration level.

As far as this last characteristic is concerned, an entire integrated system vision is considered in this chapter. This means that impedance tuners may be co-integrated with Integrated Circuits (IC). Tremendous consequences on potential applications may occur, such as the use of integrated impedance tuner for smart telecommunication systems or for millimeterwave instrumentation.

The massive integration of tuners within microsystems results in adaptative RF front-end, where functionalities can be reconfigured as well as operating frequencies (Qiao et al., 2005). The tuning capabilities of matching network gives rise to higher system efficiency and wider bandwidth. Moreover, on-wafer tuning can be also employed to compensate variations due to aging, temperature drift and unit-to-unit dispersion.

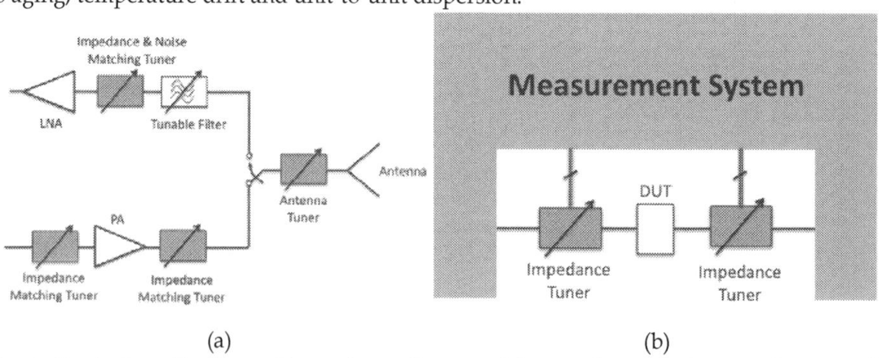

Fig. 1. Typical applications of Impedance Tuners : (a) reconfigurable front end system (b) integrated instrumentation systems

For instrumentation application, the integration of tuner circuits as close as possible to the device under test (DUT) also enlarges the measurements capabilities. The parasitic reduction in the test chain results in a rise of the maximum frequency operation: RF-MEMS tuners up to W-band have been successfully demonstrated (Vähä-Heikkilä et al., 2005). Moreover, the reduction of losses between the tuner and the DUT translates into an improvement of achievable VSWR often mandatory to provide an accurate modeling (Tagro et al., 2008). The Figure 1. (b) presents such systems : the tuners generate impedance loci featuring high impedance coverage under high frequency operation. Applications for noise or load-pull measurements can be envisioned.

2.2 Architecture of Impedance tuner

Impedance tuner architectures derive from fixed matching circuits. In this chapter, we focus on circuits able to operate in the microwave and millimeterwave domain, typically at frequencies above the X-band. Transmission lines-based circuits, which are more suitable at frequency higher than 6 GHz, are consequently discussed. Nevertheless, the next paragraph will be dedicated to semi-lumped tuner as tunability concept and expected performances can simply be introduced.

As far as lumped-elements solutions are concerned, such tuners are generally limited to 6GHz, but their associated concepts are very illustrative as they can be simply extended to all kinds of tuner. The figure 2 presents a generic reconfigurable impedance matching circuit (Pozar, 2005), which can be used as a tuner thanks to reconfigurable capacitors or inductors. Various solutions for elements' tuning are also illustrated both for inductors and capacitors. Banks of digitally commuted elements correspond to an efficient way of tuning (Papapolymerou et al., 2003).

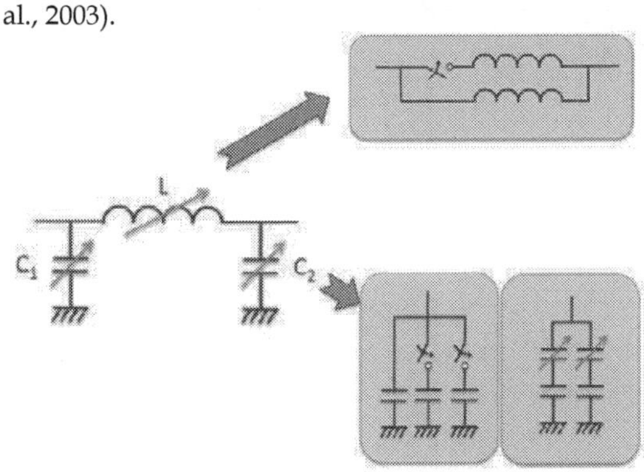

Fig. 2. Typical Lumped-Impedance Tuners

Tuning of only one element of the circuit described in figure 2 can result into a wide impedance/operating frequency tuning. The reconfigurable ability of a 4:1 impedance matching circuit has been investigated (Rebeiz, 2003). Variation of only 30 to 50% of C_2 (L and C_1 are fixed) translates into 60 to 100% of the impedance variation (for a fixed frequency) or more than 100% of fractional bandwidth (compared with 10% bandwidth for fixed elements), for a fixed set of source and load impedances.

For X-band and above (up to W-band), tuner architectures mainly involve Transmission-Lines (TL) and varactors. This type of impedance tuner is based on well-known single-double-triple-stubs impedance matching's architectures (Collin, 2001). The tuning were firstly realized using mechanical devices with either coaxial or waveguides structures, which results in cumbersome solutions requiring motors for automatic control. To integrate reconfigurable tuner, the tuning was then achieved thanks to switching elements and/or variable capacitors (diode and/or transistors), which commute or reconfigure the electrical length and/or the characteristic impedance of TL/stubs.
The figure 3. (a) presents a basic example of 3 switchable stubs featuring different electrical characteristics. To minimize the occupied space, electrical length of stubs can also be tuned with serial switches (figure 3. (b)) or shunt ones (figure 3. (c)).
Reconfigurable stub can also be realized by using switchable loading capacitor at the end of the stub or distributed along (figure 3.(d)). In specific conditions, described in the paragraph 4 of this chapter, the periodic capacitive loading translates into a equivalent TL with tunable electrical length (and characteristic impedance). The figure 4 presents such a tunable distributed transmission line, which represents the key element for multiple stubs matching network.

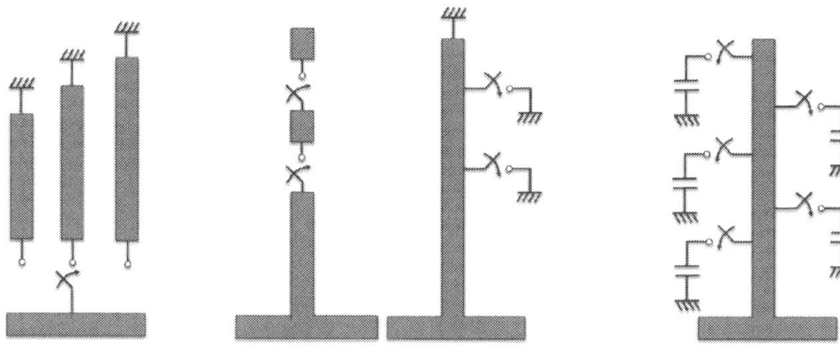

(a) (b) (c) (d)

Fig. 3. Tunable stubs, which serve as building blocs of impedance tuners.

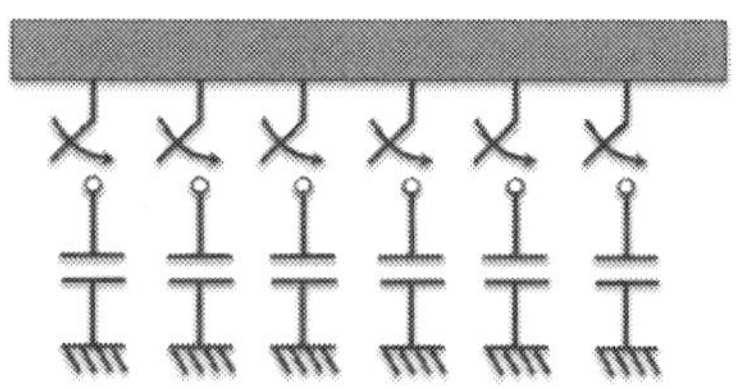

Fig. 4. Tunable Distributed Transmission Line

PIN diodes, Field Effect Transistors or switchable capacitors (also named varactor : variable capacitor) can also be exploited to tune the impedance and/or electrical length of TL. The figure 5 (a) presents a periodically loaded TL, where reactive loading elements modify the TL-phase velocity and its characteristic impedance. This topology can serve as a matching network and consequently as a tuner with limited impedance coverage. It is however suitable for power applications. The RF-current carried through switched distributed

capacitors is indeed weaker than with any other architecture (such as described in figure 3), which results in improved power handling capabilities.
More advanced impedance coverage can be achieved thanks to the use of stubs: the more the numbers of stubs take place, the wider the impedance coverage and bandwidth become (Collin, 2001). The counterpart is nevertheless an increased occupied surface and then a rise of insertion losses. This is illustrated by the schematic of figure 5. (b), which presents a reconfigurable single stub using the same principle of operation of the TL described in figure 5. (a).

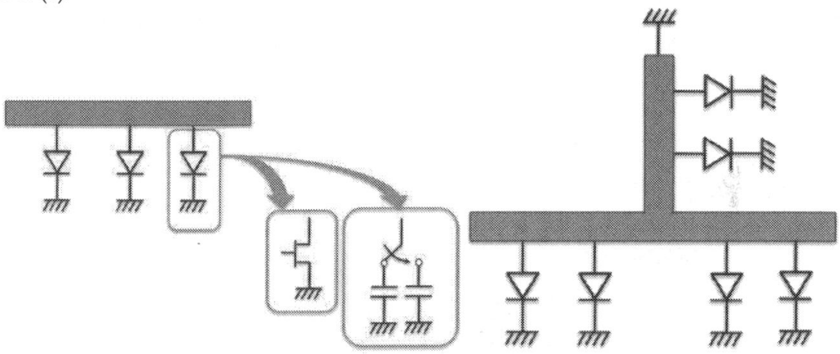

(a) (b)

Fig. 5. PIN diode, Field Effect Transistors and varactor–based impedance tuners.

The bandwidth of a tuner is also an important feature, which impacts on its architecture. The bandwidth of a lumped matching network depends on the ratio of the impedances to match. Large difference in the values of source and load impedances translates indeed into a high resonant behavior and then low circuit's bandwidth. This result can simply be pointed out with lumped circuits but is also true for distributed network.
One solution to enhance the bandwidth corresponds to use multistage transformers, for which the impedance ratio of each stage is divided by the number of stages. For TL-based-architecture, this multistage technique is built on "N-section Chebyshev impedance transformers" method for example (Collin, 2001; Pozar, 2005). As an illustration, thanks to lumped matching network as described in figure 2, the matching of impedances with a ratio of 4:1 results in a 10% fractional bandwidth for 1-stage and 30% thanks to 3-stages topology (Rebeiz, 2003). Of course, the tuning of elements can be applied in this case, not to tune the impedances to match but to improve the bandwidth (100% or more of the fractional bandwidth can be reached thanks to the tuning of the matching network). The price to pay is nevertheless an increase of the occupied surface and consequently the losses. This point limits the number of matching section to 2 or 3 stages depending on the requirements and chosen technology.
Another gain, that can be expected from multistage-tuner, corresponds to power capabilities. Increasing the bandwidth by a reduction of the resonant behavior of circuits indeed translates into a reduction of both current and voltage in the network. For fixed I-V constraints on devices and especially on RF-MEMS varactors, for which reliability highly depends on currents passing through and voltages across, the power can be raised. This explains why distributed TL, loaded with RF-MEMS varactors, corresponds to a good

candidate for high bandwidth tuner (Shen & Barker, 2005) and/or medium power applications (Lu et al., 2005).
The next paragraph presents the RF-MEMS technology, which is particularly attracting for tuner integration because of the available reconfigurable devices and the high performances they exhibit (Rebeiz, 2003).

3. RF-MEMS Technology

The selection of a technology for tuner applications is motivated by the envisioned performances expected for the circuits inside the whole system. As integration is required to address attractive applications of reconfigurable frond-end and advanced instrumentation systems, cumbersome rectangular waveguide solution with mechanical screw or ferrite for tunability is excluded. As far as high RF-performances is expected (the benefit from the integration of tuner should not be suppressed by a loose of performances), the tunability must reside in high quality components in term of:

- low losses, to reach high VSWR, high impedance coverage, low added noise and high power efficiency,
- high integration level, to assure a co-integration with active circuits and permit the integration of periodic loaded structures,
- low power consumption, as integration of tens of switches/varactors is required per matching network and tens of them per microsystems,
- high linearity to address load-pull applications as well as power amplifier matching ones.

The figure 6 presents these 4 performances for 3 different technologies which are suitable for tunability implementation: using PIN diodes or Field Effect Transistors (MMIC), using rectangular waveguide solutions which are generally based on the use of ferrite and finally the RF-MEMS technology, which corresponds to an excellent challenger for tuner application.

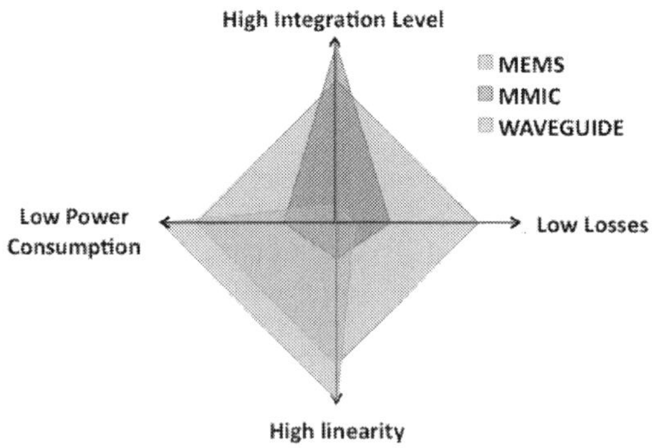

Fig. 6. Achievable tuners' performances vs technologies

One of the key features of RF-MEMS resides in their high quality factors of the resulting varactors. As already discussed, the tuner's topologies generally involve varactors and transmission lines and their losses greatly impact the overall insertion losses, more especially as the resonant behavior of circuits is intentionally high. The figure 7 illustrates the impact of capacitor quality factor and line lineic losses on the insertion losses at 20GHz for a capacitively loaded stub (see the insert of the figure 7). It is then shown that quality factor of 30 or higher is mandatory for typical transmission line losses. This, once again, highlights the RF-MEMS devices, which generally exhibit quality factors greatly higher than 30, whereas it corresponds to the maximum value obtained thanks to MMIC varactors built with FET transistors.

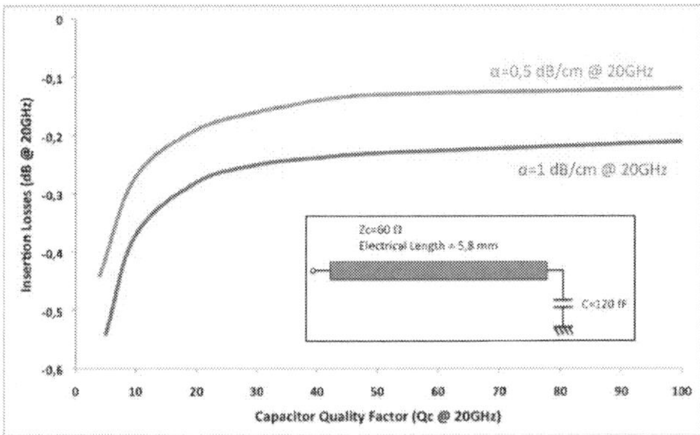

Fig. 7. Impact of capacitor's quality factor on the capacitively loaded stub insertion losses.

Numerous RF-MEMS-technologies have been developed all around the world to fulfill specific requirements (frequency operation, power handling, RF-performances, ...) (Rebeiz, 2003). The next two paragraphs present one of them, which has been developed at the LAAS-CNRS toward the integration of reconfigurable microwave passive networks over silicon active ICs (Grenier et al., 2005).

3.1 RF MEMS devices technology

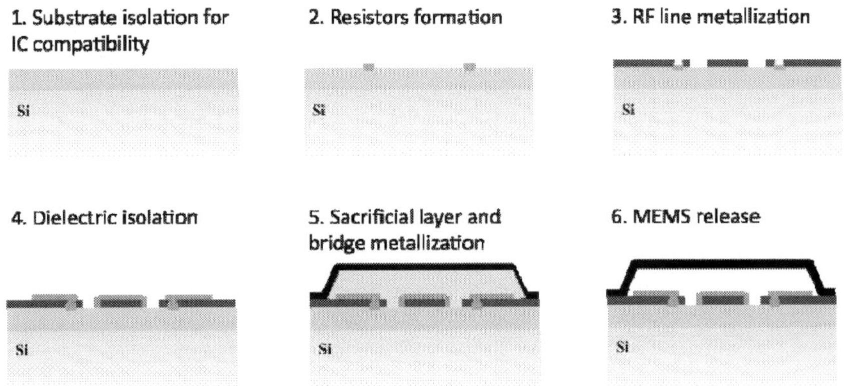

Fig. 8. Process flow of RF MEMS devices

The RF-MEMS technology described in this chapter was specifically developed in order to fulfill the previously mentioned requirements in terms of IC compatibility, low losses, high isolation as well as medium power ability. The corresponding process flow is divided in six major steps, as illustrated in Fig. 8. It includes isolation from the lossy substrate, metallization for the RF lines and the mobile membrane, as well as a thin dielectric layer and integrated resistors.

First, a polymer layer of 15 µm thick is spin-coated on top of a silicon wafer. It provides an excellent isolation between the future MEMS devices and the substrate (Grenier et al., 2004), which may include ICs for complete reconfigurable systems integration (Busquere et al., 2006). This elevation from the substrate partly confines indeed the electrical fields into a lower loss tangent material instead of the lossy silicon. The polymer "Benzocyclobuten" from Dow Chemicals, which exhibits a loss tangent close to 2.10^{-4} in the GHz range, is used. After its spin-coating, a polymerization procedure is realized at 250°C under nitrogen flow, during one hour.

An evaporated germanium layer is then patterned to realize integrated resistors. Other kinds of integrated resistors can be used such as silicon-chrome (Vähä-Heikkilä & Rebeiz, 2004-a). Nevertheless, Germanium material exhibits high value of resistivity (Grenier et al., 2007), which is in favor for low losses operation. Next step consists in the deposition of the RF lines metallization. In order to lower the metallic losses and also allow power handling through the MEMS devices, a high thickness of gold, 2 µm at least, is elaborated. Instead of an electroplating technique, which is particularly suitable for high metal thickness formation but suffers from roughness, a lift-off procedure is employed. The consequent minimization of the roughness enhances the contact quality between the metallic membrane and the MEMS dielectric and thus improves the accessible capacitive ratio.

In a fourth step, the MEMS dielectric of 0.25 µm thick is performed at 300°C by Plasma Enhanced Chemical Vapor Deposition (PECVD). After its delimitation by dry etching, a sacrificial layer is deposited and patterned. A specific care is given to this layer in order:

- to sustain the next technological steps,
- to obtain a flat MEMS bridge; several depositions and photolithographic steps are consequently required to take the RF lines relief into account,
- and to assure a good strength of the membrane anchorages.

As an air gap of 3 µm between the MEMS bridge and the central conductor of the RF line is targeted, the sacrificial layer is defined with such a thickness.

The metallic membrane is then obtained with two successive depositions: an evaporated gold layer of 0.2 µm, followed by an electroplated one of 1.8 µm. The evaporated metal, which exhibits important internal stress, is minimized to drastically decrease any risk of membrane's buckling. The gold bridge is then obtained with a classical wet etching.

The next and most critical step of the process consists in the release of the MEMS structure. It corresponds to the suppression of the sacrificial layer through chemical etching, followed by its drying.

Fig. 9 indicates the photography of a realized RF MEMS switch (which corresponds in fact to a varactor with a high capacitor ratio). This example includes a metallic membrane placed on top of a coplanar waveguide, with four membrane's anchorages and four integrated resistors. The bridge is composed of a central part, which assures the capacitor values, and two actuation electrodes located apart the line, which large surface decreases the required actuation voltage (close to 20-25 V generally) (Ducarouge et al., 2004).

Such a MEMS switch may be modeled with an RLC electrical schematic, which is embedded between two transmission lines, as illustrated in the drawing of

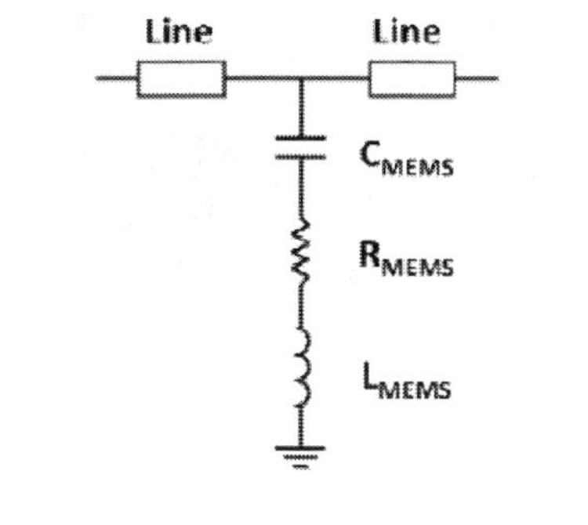

Components		Values
Lines: Z0		48Ω
$\varepsilon_{r,eff}$		3.085
Length		100µm
L_{MEMS}		28 pH
C_{MEMS}	up	70 fF
	down	2330 fF
	ratio	33
Q @ 20GHz	up	>100
	down	28
R_{MEMS}		0.12 Ω

Table 1. Electrical model of the RF MEMS switch

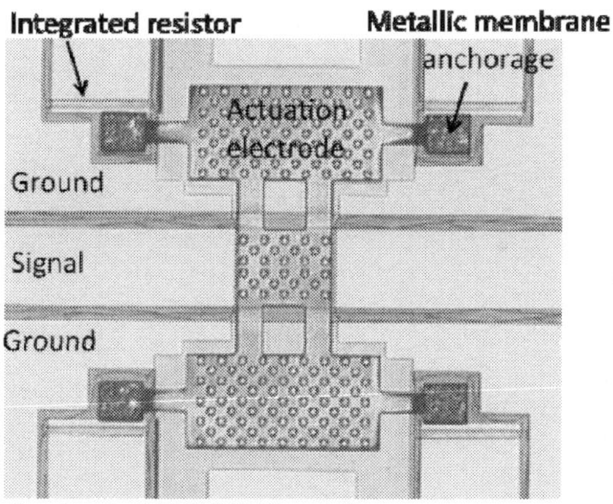

Fig. 9. Photography of a classical RF MEMS switch

The R_{MEMS} resistor corresponds to losses in the metallic membrane. The L_{MEMS} inductor is due (1) to the connecting arms between the central part of the bridge and the actuation electrodes and (2) to the suspension arms of the membrane itself. The capacitor's value C_{MEMS} depends of the state of the bridge and of the regarding surface. Its ratio in the "up" and "down" states is a good indicator about the contact quality of the MEMS switch and the corresponding technology.

As far as the access lines are concerned, they present a characteristic impedance close to 50 Ohm with an effective relative permittivity of 3 and a length of 100 µm.

The achievable RF performances of such an RF MEMS switch are presented in **Fig. 10** in both up and down states of the membrane. At 20 GHz, the insertion losses are close to 0.25dB (which include the contribution of the 200 µm long lines placed apart the membrane), whereas the isolation reaches 50 dB. The RF MEMS switch exhibits consequently very low losses and high isolation, with a capacitor ratio of 33. Power tests have demonstrated that such an RF MEMS may handle up to 1W during 30 millions of cycles in hot switching.

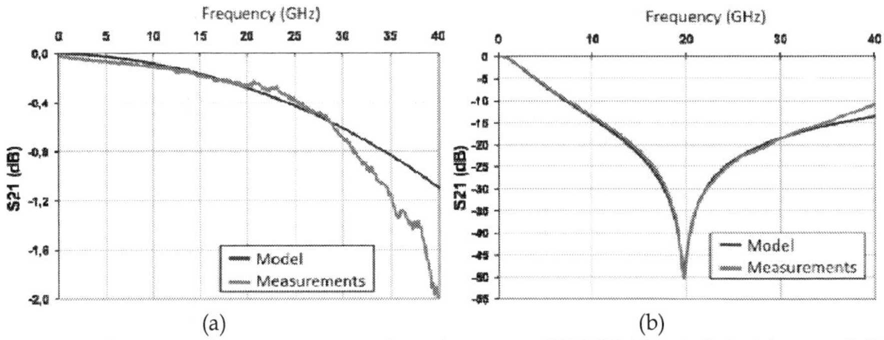

Fig. 10. Simulations and measurements of an elementary RF MEMS switch in (a) up and (b) down positions

A good agreement between modeling and measurements is achieved for both insertion losses (Fig. 10.a) and isolation (Fig. 10.b). These results validate the simple model used for the RF MEMS switch. A better fit at high frequency could however be reached if additional parasitic elements were considered, but it would highly complex the electrical model.

Depending on the technology, device architecture and targeted application, various reliability performances under low (in the milliWatt range) and medium (in the Watt range) power in hot or cold switching (the RF-power is on or off - respectively- during the MEMS switching) can be found in the literature. The reliability of RF-MEMS is actually one major concern (together with packaging issues) of the RF-MEMS researches. Considered solutions aims to optimize as much as possible the different parameters, which limits the lifetime of RF-MEMS devices/circuits such as:

(1) the actuation scheme of the devices. The frequency and the duty cycle of the biasing voltage have a high impact on the MEMS reliability (Van Spengen et al., 2002; Melle et al., 2005),
(2) the dielectric configuration, which is subject to charging. Some solutions to decrease the charging and/or enhance the discharging have already been proposed, such as adding holes (Goldsmith et al., 2007) or carbon-nanotubes (Bordas et al., 2007-b) in the dielectric for examples. In any case, dielectric charging is one major concern for high reliable RF-MEMS circuits,
(3) the thermal effects in metal lines under medium RF-power. The consequent heat induces deformation of the mobile membrane (and even buckling), which results in mechanical failure (Bordas et al., 2007-a),
(4) the electro-migration, as high current density, which is induced in metal line under medium RF-power, results in alteration of metallization and then alters the operation of the device.

As far as the elaboration of tuner is concerned, many identical MEMS structures are required to form the complete circuit. However, some technological dispersions during the fabrication of MEMS structures may not be totally avoided, especially the contact quality between the metallic membrane and the MEM dielectric. Moreover as defined previously in (Shen & Barker, 2005), capacitive ratio of 2-5:1 are required. Consequently, new MEMS varactors, which integrate Metal-Insulator-Metal (MIM) capacitors, have been developed.

3.2 RF MEMS varactor and associated technology

Based on the previous RF-MEMS devices, MIM capacitors have been added. They are placed between the ground planes and the membrane anchorages, as indicated in **Fig. 11**. They present the high advantage of being very compact, contrary to Metal-Air-Metal (MAM) capacitors (Vähä-Heikkilä & Rebeiz, 2004-a), but at the detriment of quality factor due to additional dielectric losses.

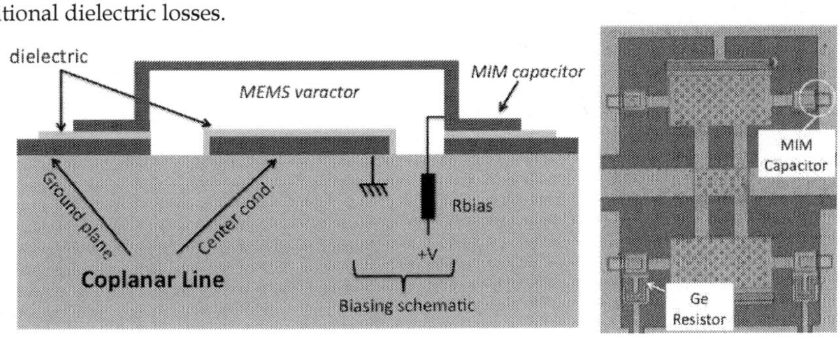

Fig. 11. Cross section view and photography of a RF MEMS switch with integrated MIM capacitors

The precedent technological process flow has consequently been modified to integrate these MIM capacitors. Two additional steps are required. After the elaboration of the RF lines, the MIM dielectric (Silicon Nitride) is deposited by PECVD and patterned. A top metallization is realized by evaporation and delimited. The MEMS process restarts then with the deposition of the MEM dielectric and continue until the final release of the structure. Because of technological limitations, MIM capacitors have to present a value equal or higher than 126fF.

The corresponding electrical model is slightly modified with the addition of a MIM capacitor, as shown in

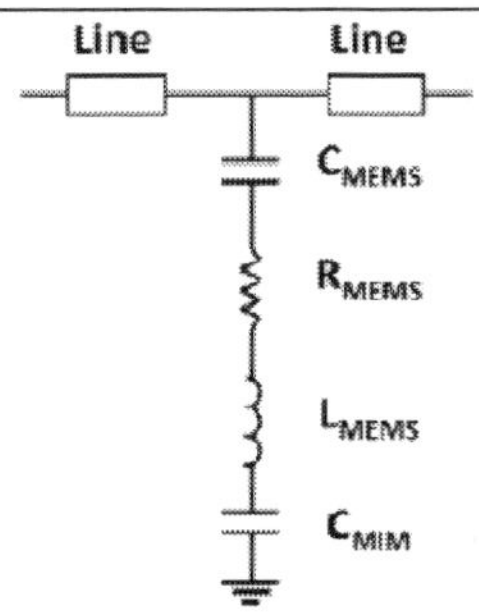

Components		Values
Line (µm)		105
LMEMS (pH)		23,5
CVAR(fF)	up	110
	down	500
RMEMS (Ω)	up	2
	down	0,15
Q@ 20GHz	up	36
	down	106
CMIM(fF)		450

Table 2. Electrical model of varactor with MIM capacitors

The MIM capacitor's value corresponds to 450fF, which leads to varactor's values (MEM and MIM capacitors in serial configuration) of 110 and 500fF in the up and down states respectively. It results in a capacitive ratio of 4.5 (Bordas, 2008).
Vähä-Heikkilä et al. have proposed another solution for the reduction and control of the capacitor ratio. They used Metal-Air-Metal (MAM) capacitors with RF-MEMS attractors (see figure 12), which results in higher quality factor, as no dielectric losses appear in the MAM device. This results in a 150% improvement in the off-state quality factor, a value of 154 was indeed obtained at 20GHz (Vähä-Heikkilä & Rebeiz 2004-a) with MAM capacitors 100 times larger than MIM ones.

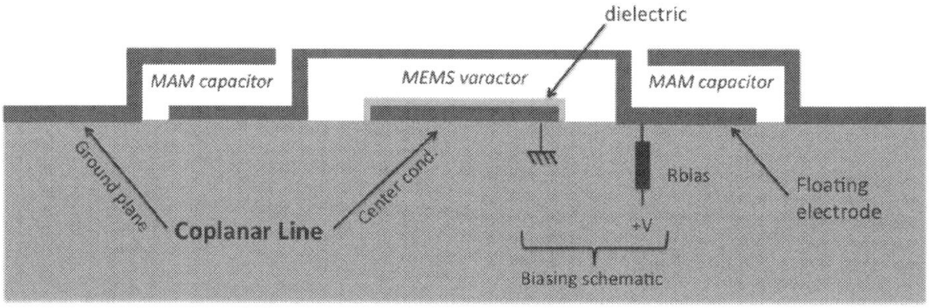

Fig. 12. Metal-Air-Metal (MAM) capacitor associated with RF-MEMS varactors used for tuning elements in tuner (Vähä-Heikkilä & Rebeiz 2004-a)

Despites these possible quality factors' improvements, quality factors higher or around 30-40 are sufficient to achieve low losses' tuners, as suggested by the figure 7. RF-MEMS devices are consequently well adapted to tuner applications (and more generally all reconfigurable applications) as they also exhibit:

(1) Controllable and predictable capacitor ratios in the range of 2-5:1,
(2) Medium power capabilities,
(3) Compatibility with a system-on-chip approach,
(4) Low intermodulation.

The next paragraph then presents an explicit method to design an RF-MEMS-based tuner.

4. RF-MEMS Tuner Design methodology: example of the design of a building block

4.1 Efficient Design Methodology

Thanks to the RF-MEMS-varactors and associated technology presented in the last paragraph, we propose to detail and illustrate an explicit design methodology of TL-based impedance tuner. The design and characterization of a basic building block of tuner: a single stub architecture, presented in the figure 13, is detailed and discussed. The investigated structure is composed of 3 TL sections: 2 input/output accesses and 1 stub. Each line is loaded by 2 switchable varactors. When the loading capacitance is increased, the line electrical length is increased and the matching is tuned. Reconfigurable varactors can be realizable thanks to a switch, which address 2 different capacitors, or by the association of fixed and tunable capacitors as illustrated in the figure 13.

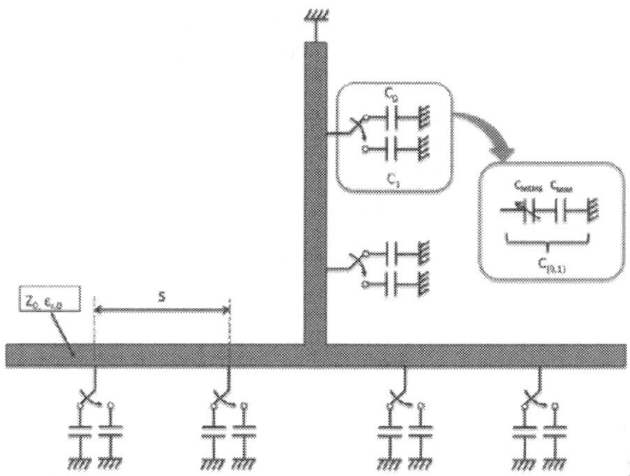

Fig. 13. Tuner's Topology

The parameters, which have to be optimized, are:

- the MIM capacitor value : C_{MIM} (we consider that the MEMS capacitor - without the MIM- is fixed by the technological constraints),
- the characteristic impedance of the unloaded line (without the varactors) : Z_0,
- the spacing s between the MEMS capacitor both for the input and the output lines and for the stub.

It follows such targets :

- an impedance coverage:
  1. as uniform as possible : target 1,
  2. providing high values of $|\Gamma|$: target 2,
  3. providing also low values of $|\Gamma|$: target 3,
- Technological feasibility (this limits some dimensions).

The target 3 is fulfilled when the characteristic impedance of the loaded line, with all MEMS in the up position (named $Z_{c,up}$) is close to 50Ω :

$$Z_{c,up} = 50\Omega \qquad (1)$$

The first target is difficult to be analytically expressed. To circumvent this difficulty, we propose to consider that this target is reached if, for each tuner's transmission line (TL), presented in the figure 14, the phase difference of the reflection scattering parameter (S_{11}) between the two MEMS states is 90°. Indeed, when a phase difference of 90° is reached for a TL, an half wise rotation is observed in the Smith Chart then leading to "a best impedance coverage".

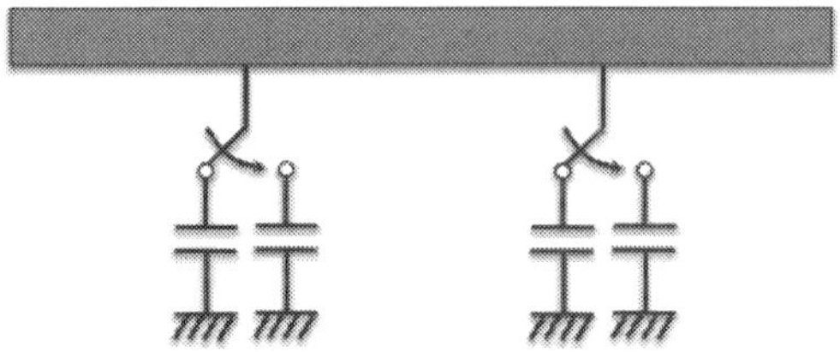

Fig. 14. TL with tunable electrical length. This element corresponds to a generic building block of complex tuner architectures.

To express this constraint, a parameter is introduced, which represents the two-states-difference of the normalized length of TL, regarding the wavelength:

$$\delta = \frac{l}{\lambda_{down}} - \frac{l}{\lambda_{up}} \tag{2}$$

The impedance coverage will then be optimally uniform if:

$$\delta = 1/4 \tag{3}$$

After some mathematical manipulations, the proposed figure of merit can be expressed as a function of the designed parameters:

$$\delta = \frac{f\sqrt{\varepsilon_{r0}}}{c} 2s\left(\sqrt{1 - R + RK_{up}}\right) - \sqrt{K_{up}} \tag{4}$$

where $K_{up}=(Z_0/Z_{c,up})^2$; R, s and ε_{r0} correspond to the capacitor ratio C_{down}/C_{up}, the spacing between varactors and the relative permittivity of the unloaded line respectively.

The design equation (4) then translates into an explicit expression of the capacitor ratio (then named R_{opt}), which permits to design the value of the MIM capacitors of the varactors:

$$R_{opt} = \frac{\left(\frac{\delta}{B} + \sqrt{K_{up}}\right)^2 - 1}{K_{up} - 1} \tag{5}$$

$$B = \frac{f\sqrt{\varepsilon_{reff}}}{c} 2s \tag{6}$$

The optimal value of the MIM capacitor is finally deduced from this optimal capacitor ratio of the varactor and the up-state value of the MEMS devices (without MIM capacitor):

$$C_{MIM}^{opt} \approx (R_{opt} - 1) \times C_{MEMS}^{up} \tag{7}$$

This last expression assumes that the MEMS capacitor ratio is large enough compared with the one of the resulting varactor.

Finally, the target 2 is fulfilled when the down-state capacitor value of the varactor is sufficiently large to 'short circuit the signal', leading to the edge of the Smith Chart. As this value is already defined by the designed equation (4), the target 2 is optimized by tuning the s value, which is -on the other side- constrained by the Bragg condition (Barker & Rebeiz, 1998) and the technological feasibility. The s value will then be a parameter to optimize iteratively in order to reach the best compromise between "wide impedance coverage (i.e. equation (1) and (4)) and "technological feasibility".
This procedure was applied to a single-stub tuner. Considering the RF-MEMS technology presented in the previous paragraph, the values summarized in the table 3 are reached after some iterations and totally defines the tuner of the figure 13.

Transmission line Characteristic Impedance		63Ω
MEMS capacitor (theoretical)	**up**	70 fF
	down	4000 fF
MIM capacitor		500 fF
Total Capacitor	**up**	60 fF
	down	450 fF
Total Capacitor Ratio		7-8

Table 3. Values of the tuner's parameters using the proposed methodology

4.2 Measured RF-Performances

The microphotography in figure 15 presents the fabricated single-stub tuner, whose electrical parameters are given in the table 3. The integration technology used has been developed at the LAAS-CNRS (Grenier et al. 2004; Grenier at al. 2005; Bordas, 2008) and, in order to integrate tuners with active circuits, the RF-MEMS devices were realized on silicon (2kΩ.cm) with a BCB interlayer of 15 µm.

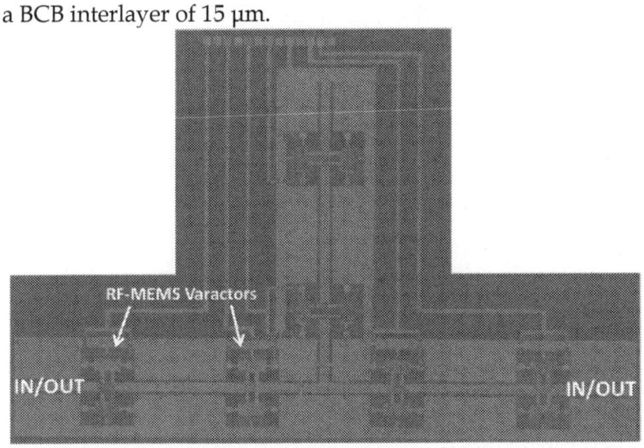

Fig. 15. Micro-photography of the fabricated RF-MEMS single stub tuner (Bordas, 2008)

The on-wafer 2-ports S parameters have been measured from 400 MHz to 30 GHz for the 2^6=64 possible states. The DC feed lines for the varactors actuation have been regrouped and connected to an automated DC -voltages supplier through a probe card (see figure 16).

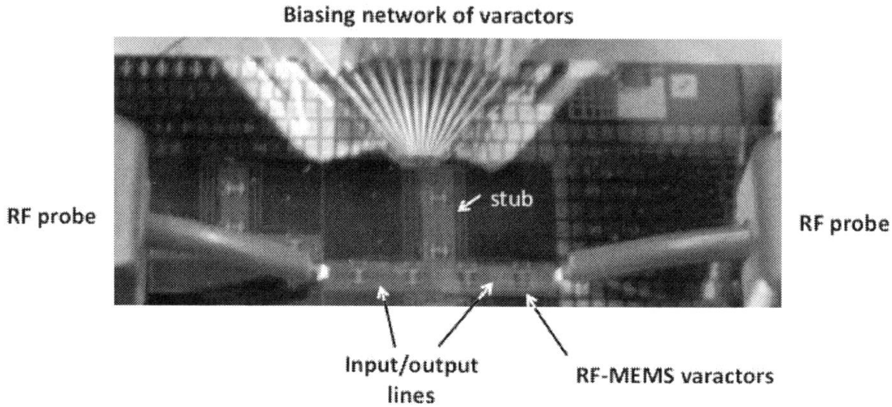

Fig. 16. Micro-photography of the fabricated tuner under testing

The measured and simulated (with Agilent ADS) S_{11} parameters vs frequency, when all the MEMS devices are in the down position, are shown in fig. 17. This demonstrates the accuracy of the RF-MEMS technologies' models over a wide frequency range.
The fig. 18 presents the measured and simulated impedance coverage at 10, 12.4 and 14GHz (64 simulated impedance values and 47 measured ones) with 50 Ω input and output terminations. There is a good agreement between the simulated and measured impedance coverage with high values of $|\Gamma_{MAX}|$ and VSWR parameters as 0.82 and 10 are respectively obtained at 14 GHz.

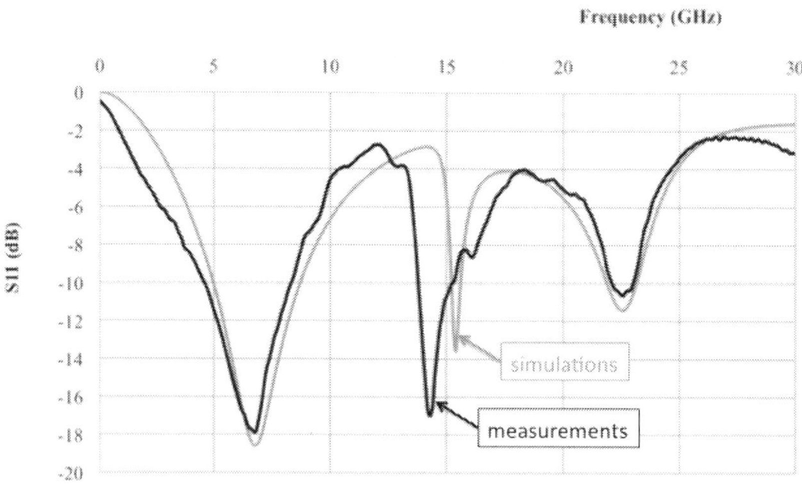

Fig. 17. Measured and simulated S11 parameter, when all MEMS devices are in the down position

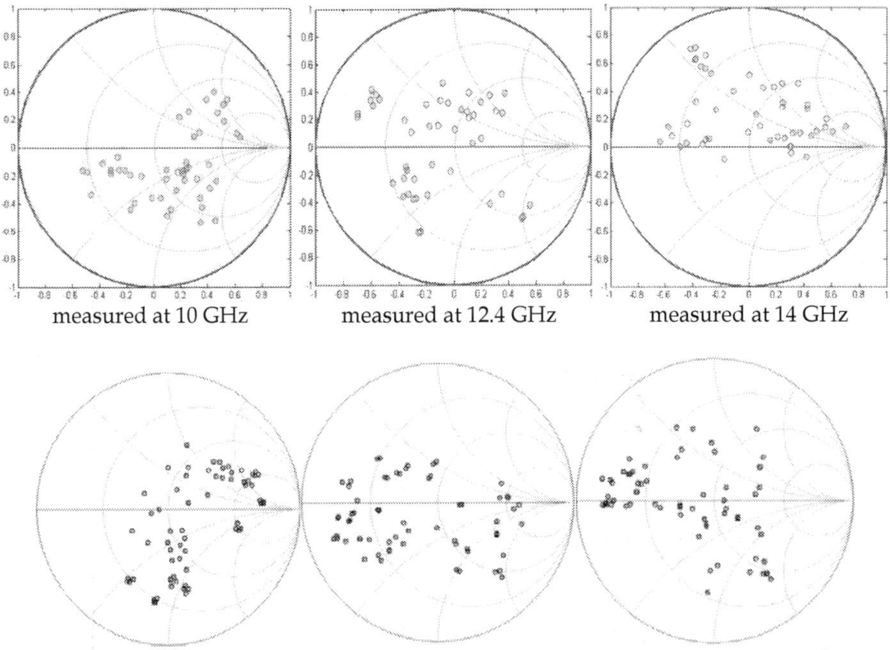

measured at 10 GHz | measured at 12.4 GHz | measured at 14 GHz

simulated at 10 GHz | simulated at 12.4 GHz | simulated at 14 GHz

Fig. 18. Measured and simulated impedances coverage of the tuner at 10, 12.4 and 14 GHz

This result then validates the proposed design methodology as a wide impedance coverage is reached after the first set of fabrication.

In term of tunable matching capability of the resulting circuit, the figure 19 presents the input impedances of the fabricated tuner, when the output is loaded by 20 Ω. The results demonstrate that the tuner is able to match 20 Ω on a 100 Ω input impedance (the 100 Ω circle is drawn in the Smith Chart of the figure 19). The corresponding impedance matching ratio of 5:1 is in the range of interest of a wide range of applications, where low noise or power amplifiers and antennas have to be matched under different frequency ranges.

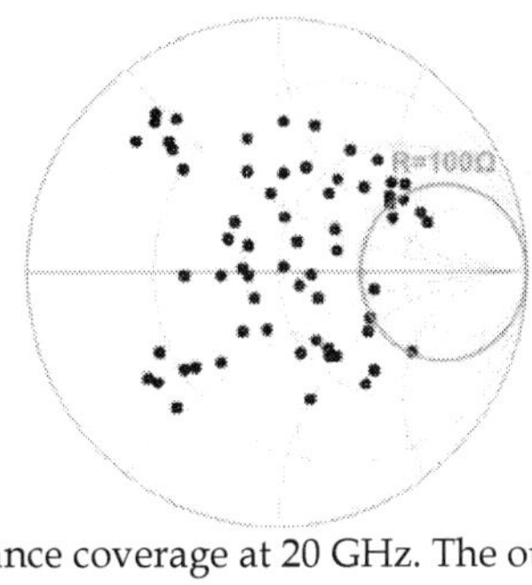

Fig. 19. Predicted input impedance coverage at 20 GHz. The output of the tuner is loaded by 20 Ω.

5. Capabilities of RF-MEMS based tuner

The previous paragraph has presented an illustration of the design of an RF-MEMS-based tuner in Ku and K-bands. Although the considered structure was quite simple (1-stub topology), the measured performances in term of VSWR and impedance coverage was very satisfactory. Of course, the presented design methodology is very generic and can also be applied for the design of more complicated tuner architecture. The figure 20 presents a double and triple stub tunable matching network.

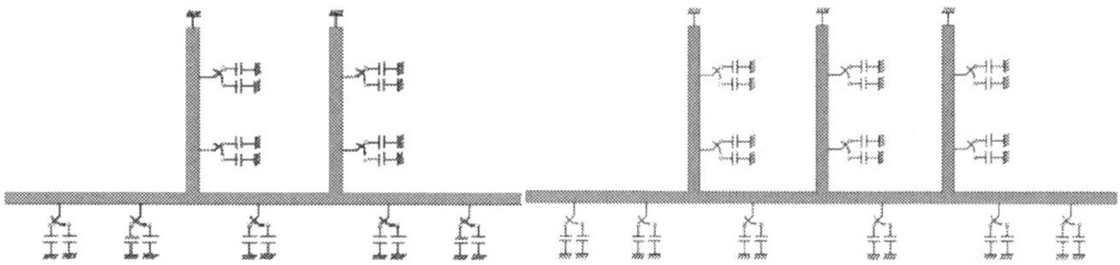

Fig. 20. RF-MEMS based tuner : double and triple stub architecture

Despites the drawbacks of such structures in terms of occupied surface and insertion losses, their impedance coverage and maximum VSWR feature improved values compare to single stub structures. The figure 21 illustrates typical results expected from double and triple stubs tuners and demonstrates the power of the design methodology presented in the paragraph 4 as well as the capabilities of RF-MEMS technologies for the implementation of integrated tuners with high performances. Excellent impedance coverage was indeed predicted as well as high value of reflection coefficient in all the four quadrant of the Smith-Chart.

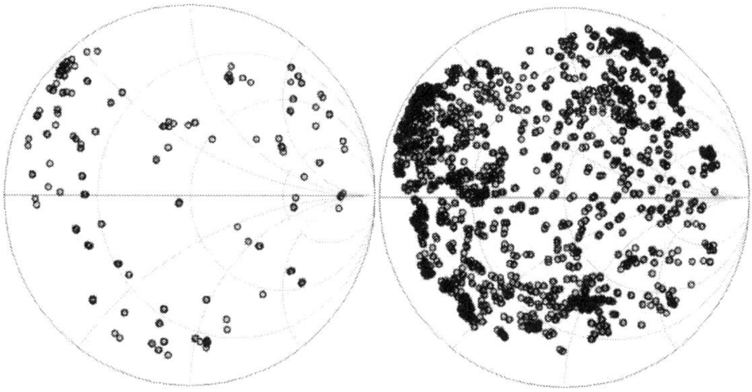

Fig. 21. Predicted impedance coverage of a 9 bits (2 stubs) and 12 bits (3 stubs) RF-MEMS tuner

The simulations predict for both architectures a $|\Gamma_{MAX}|$ value of 0.95 at 20GHz, which corresponds to a VSWR around 40. Compared with MMIC-tuner, RF-MEMS architectures clearly exhibit improvement in term of achievable VSWR. In Ka-band, the losses of FET or Diode limit the VSWR of tuner to 20 (McIntosh et al., 1999; Bischof, 1994), whereas as for RF-

MEMS-technology-based tuners exhibit values ranging from 32 (Kim et al., 2001) to even 199 (Vähä-Heikkilä et al., 2007). It clearly points out the breakthrough obtained by using RF-MEMS technologies for microwave and millimeterwave tuner applications.
Moreover, the demonstration of high RF-performances of RF-MEMS-based tuner have been successfully carried out:

1. on various architectures for

- 1-stub (Vähä-Heikkilä et al., 2004-c; Dubuc et al., 2008; Bordas, 2008; Vähä-Heikkilä et al. 2007),
- 2-stubs (Papapolymerou et al., 2003; Kim et al., 2001; Vähä-Heikkilä et al., 2005; Vähä-Heikkilä et al., 2007)
- 3-stubs (Vähä-Heikkilä et al., 2004-b; Vähä-Heikkilä et al., 2005; Vähä-Heikkilä et al., 2007)
- Distributed TL (Lu et al., 2005; Qiao et al., 2005; Shen & Barker, 2005; Lakshminarayanan & Weller, 2005; Vähä-Heikkilä & Rebeiz, 2004-a)

As anticipated (Collin, 2001), the VSWR rises when the number of stubs increases. The table 4 presents the $|\Gamma_{MAX}|$ and VSWR values for 1, 2 and 3-stubs RF-MEMS tuners. Value around 40 is achieved at 16 GHz for a 3-stub structure, which corresponds to a 100% improvement compare with a 1-stub network, but at the expense of 70% rise of the occupied surface.

Architecture	1- stub tuner	2- stub tuner	3-stub tuner
$|\Gamma_{MAX}|$ @ 16 GHz	0,91	0,93	0,95
VSWR @ 16 GHz	21	28	39

Table 4. $|\Gamma_{MAX}|$ and VSWR vs tuner architecture (Vähä-Heikkilä et al., 2007)

2. Over a wide frequency range from 4 to 115 GHz :

- C-band (Vähä-Heikkilä & Rebeiz, 2004-a),
- X-band (Vähä-Heikkilä & Rebeiz, 2004-a; Vähä-Heikkilä et al., 2004-b; Qiao et al., 2005),
- Ku-band(Papapolymerou et al., 2003; Vähä-Heikkilä et al., 2006),
- K-band (Dubuc et al., 2008; Bordas, 2008; Shen & Barker, 2005),
- Ka-band (Kim et al., 2001; Lu et al., 2005, Vähä-Heikkilä & Rebeiz, 2004-d),
- U and V-band (Vähä-Heikkilä et al., 2004-c)
- W-band (Vähä-Heikkilä et al., 2005)

One can notice that high values of $|\Gamma_{MAX}|$ and VSWR are generally achieved for high frequency operation. This is suggested by the datas reported in the table 5, which reports a tuner with an optimized impedance coverage at 16 GHz. At this frequency, a VSWR of 28 is measured, whereas at 30 GHz an impressive value of 199 is reported.

Frequency	6 GHz	8 GHz	12 GHz	16 GHz*	20 GHz	30 GHz
$\|\Gamma_{MAX}\|$	0,95	0,94	0,91	0,93	0,96	0,99
VSWR	39	32	21	28	49	199

* Optimal impedance coverage of the Smith-Chart

Table 5. $|\Gamma_{MAX}|$ and VSWR vs frequency for a 2-stubs tuner (Vähä-Heikkilä et al., 2007)

A tradeoff between impedance coverage and high value of $|\Gamma_{MAX}|$ and VSWR then exists and both features need to be considered for fair comparison.

6. Conclusions

This chapter has presented the design, technology and performances of RF-MEMS-based tuners. Various architectures have been presented in order to give a large overview of tuner-topologies. An efficient and explicit design methodology has been explained and illustrated through a practical example. The authors have moreover outlined the potential of RF-MEMS technologies for different applications (tunable impedance matching between integrated functions within smart microsystems, wide impedance values generations for devices characterization) because of their ability for IC-co-integration, low losses performances and low distortion characteristics.

7. Acknowledgements

The authors would like to specifically acknowledge Chloe Bordas, who was Ph.D student under the supervision of Katia Grenier and David Dubuc from 2005 to 2008 and worked on RF-MEMS based tuner. She was an essential backbone of the work presented in this Chapter.

We also would like to thanks Samuel Melle, Benoît Ducarouge and Jean-Pierre Busquere, Ph. D students under the supervision of David Dubuc and Katia Grenier from 2002 to 2005. Their work on RF-MEMS design, fabrication and reliability contributed to rise the knowledge of the team, and permit to envision circuits based on RF-MEMS varactors.

Katia Grenier and David Dubuc also acknowledge the support of Thales Alenia Space, the French Defense Agency (DGA) and ST-Microelectronics.

8. References

Barker, S. Rebeiz, G.M. (1998). Distributed MEMS true-time delay phase shifters and wide-band switches. *IEEE Transactions on Microwave Theory and Techniques*, Vol. 46, Issue 11, Part 2, Nov. 1998 pp:1881 - 1890

Bischof, W. (1994). Variable impedance tuner for MMIC's. *Microwave and Guided Wave Letters*, Volume 4, Issue 6, June 1994 Page(s):172 - 174

Bordas, C.; Grenier, K.; Dubuc, D.; Paillard, M.; Cazaux, J.-L.; et al. (2007-a). Temperature stress impact on power RF MEMS switches, *Microtechnologies for the new millennium 2007*, Smart sensors, actuators and MEMS, Maspalomas, Espagne. Mai 2007.

Bordas, C.; Grenier, K.; Dubuc, D.; Flahaut, E.; Pacchini, S. Paillard, M.; Cazaux, J-L. (2007-b). Carbon nanotube based dielectric for enhanced RF MEMS reliability. *IEEE/MTT-S International Microwave Symposium*, June 2007.

Bordas, C. (2008). Technological optimization of RF MEMS switches with enhanced power handling - Elaboration of a MEMS-based impedance tuner in K-band. *Ph.D. dissertation* (in French), April 2008.

Busquere, J.-P.; Grenier, K.; Dubuc, D.; Fourn, E.; Ancey, P.; et al. (2006). MEMS IC concept for Reconfigurable Low Noise Amplifier. *36th European,Microwave Conference*, 2006. 10-15 Sept. 2006 Page(s):1358 - 1361

Collin, R. E. (2001). *Field Theory of Guided Waves*, 2nd ed., IEEE Press.

Dubuc, D.; Saddaoui, M; Melle, S.; Flourens, F.; Rabbia, L.; Ducarouge, B.; Grenier, K.; et al. (2004). Smart MEMS concept for high secure RF and millimeterwave communications. *Microelectronics Reliability*, Volume 44, Issue 6, June 2004, Pages 899-907

Dubuc, D.; Bordas , C.; Grenier, K. (2008). Efficient design methodology of RF-MEMS based tuner. *European Microwave Week 2008 (EuMW 2008)*, Amsterdam (Pays Bas), 27-31 Octobre 2008, pp.398-401

Ducarouge, B.; Dubuc, D.; Melle, S.; Bary, L.; Pons, P.; et al. (2004). Efficient design methodology of polymer based RF MEMS switches. *2004 Topical Meeting on Silicon Monolithic Integrated Circuits in RF Systems, 2004*. 8-10 Sept. 2004 Page(s):298 – 301

Goldsmith, C.L.; Forehand, D.I.; Peng, Z.; Hwang, J.C.M.; Ebel, I.L. (2007). High-Cycle Life Testing of RF MEMS Switches. *IEEE/MTT-S International Microwave Symposium, 2007*. 3-8 June 2007 Page(s):1805 – 1808.

Grenier, K.; Dubuc, D.; Mazenq, L.; Busquère, J-P.; Ducarouge, B.; Bouchriha, F.; Rennane, M.; Lubecke, V.; et al. (2004). Polymer based technologies for microwave and millimeterwave applications. *50th IEEE International Electron Devices Meeting, 2004*, San Francisco, USA, Dec. 2004.

Grenier, K.; Dubuc, D.; Ducarouge, B.; Conedera, V.; Bourrier, D.; Ongareau, E.; Derderian, P.; et al. (2005). High power handling RF MEMS design and technology. *18th IEEE International Conference on Micro Electro Mechanical Systems*, 2005. 30 Jan.-3 Feb. 2005 Page(s):155 – 158

Grenier, K.; Bordas, C. Pinaud, S.; Salvagnac, L.; Dubuc, D. (2007). Germanium resistors for RF MEMS based Microsystems. *Microsystems Technologies*, DOI 10.1007/s00542-007-0448-4.

Kim, H.-T.; Jung, S.; Kang, K.; Park, J.-H.; Kim, Y.-K.; Kwon Y. (2001). Low-loss analog and digital micromachined impedance tuners at the Ka-band. *IEEE Transactions on Microwave Theory and Techniques*, December 2001, Vol. 49, No. 12, pp. 2394-2400.

Lakshminarayanan, B.; Weller, T. (2005). Reconfigurable MEMS transmission lines with independent Z0- and β- tuning. *IEEE/MTT-S International Microwave Symposium*, 2005.

Lu, Y.; Katehi, L. P. B.; Peroulis D. (2005). High-power MEMS varactors and impedance tuners for millimeter-wave applications. *IEEE Transactions on Microwave Theory and Techniques*, November 2005, Vol. 53, No. 11, pp. 3672-3678.

McIntosh, C.E.; Pollard, R.D.; Miles, R.E. (1999). Novel MMIC source-impedance tuners for on-wafer microwave noise-parameter measurements. *IEEE Transactions on Microwave Theory and Techniques*, Volume 47, Issue 2, Feb. 1999 Page(s):125 – 131

Melle, S.; De Conto, D.; Dubuc, D.; Grenier, K.; Vendier, O.; Muraro, J.-L.; Cazaux, J.-L.; et al. (2005). Reliability modeling of capacitive RF MEMS. *IEEE Transactions on Microwave Theory and Techniques*, Volume 53, Issue 11, Nov. 2005 Page(s):3482 - 3488

Papapolymerou, J.; Lange, K.L.; Goldsmith, C.L.; Malczewski, A.; Kleber, J. (2003). Reconfigurable double-stub tuners using MEMS switches for intelligent RF front-end. *IEEE Transactions on Microwave Theory and Techniques*, Volume 51, Issue 1, Part 2, Jan. 2003 Page(s):271 - 278

Pozar, D.M. (2005). *Microwave Engineering*. 3rd ed., Wiley 2005.

Qiao, D.; Molfino, R.; Lardizabal, S.M.; Pillans, B.; Asbeck, P.M.; Jerinic, G. (2005). An intelligently controlled RF power amplifier with a reconfigurable MEMS-varactor tuner. *IEEE Transactions on Microwave Theory and Techniques*, Volume 53, Issue 3, Part 2, March 2005 Page(s):1089 - 1095

Rebeiz, G. M. (2003). *RF MEMS: Theory, Design, and Technology*. New York: Wiley, 2003.

Shen, Q; Baker, N.S. (2005). A reconfigurable RF MEMS based double slug impedance tuner. *European Microwave Conference 2005*, Paris, pp. 537-540.

Tagro, Y.; Gloria, D.; Boret, S.; Morandini, Y.; Dambrine, G. (2008). In-situ silicon integrated tuner for automated on-wafer MMW noise parameters extraction of Si HBT and MOSFET in the range 60–110GHz. *72nd ARFTG Microwave Measurement Symposium*, 2008. 9-12 Dec. 2008 Page(s):119 - 122

Van Spengen, W.M.; Puers, R.; Mertens, R.; De Wolf, I. (2002). Experimental characterization of stiction due to charging in RF MEMS. *International Electron Devices Meeting, 2002*. IEDM '02. Digest. 8-11 Dec. 2002 Page(s):901 - 904

Vähä-Heikkilä, T.; Rebeiz, G.M. (2004-a). A 4-18-GHz reconfigurable RF MEMS matching network for power amplifier applications. *International Journal of RF and Microwave Computer-Aided Engineering*. Volume 14 Issue 4, Pages 356 - 372. 9 Jun 2004

Vähä-Heikkilä, T.; Varis, J.; Tuovinen, J.; Rebeiz, G. M. (2004-b). A reconfigurable 6-20 GHz RF MEMS impedance tuner". *IEEE/MTT-S International Microwave Symposium*, 2004, pp. 729-732.

Vähä-Heikkilä, T.; Varis, J.; Tuovinen, J.; Rebeiz, G.M. (2004-c). A V-band single-stub RF MEMS impedance tuner. *34th European Microwave Conference, 2004*, Volume 3, 11-15 Oct. 2004 Page(s):1301 - 1304

Vähä-Heikkilä, T.; Rebeiz, G. M. (2004-d). A 20-50 GHz reconfigurable matching network for power amplifier applications. *IEEE/MTT-S International Microwave Symposium*, 2004, pp. 717-720.

Vähä-Heikkilä, T.; Varis, J.; Tuovinen, J.; Rebeiz, G.M. (2005).W-band RF MEMS double and triple-stub impedance tuners. *IEEE/MTT-S International Microwave Symposium*, 12-17 June 2005

Vähä-Heikkilä, T.; Caekenberghe, K.V.; Varis,J.; Tuovinen, J.; Rebeiz, G.M. (2007). RF MEMS Impedance Tuners for 6-24 GHz Applications. *International Journal of RF and Microwave Computer-Aided Engineering*, 26 Mar 2007, Volume 17 Issue 3, Pages 265 - 278.

16

Broadband GaN MMIC Power Amplifiers design

María-Ángeles González-Garrido and Jesús Grajal
Departamento de Señales, Sistemas y Radiocomunicaciones, ETSIT, Universidad Politécnica de Madrid, Ciudad Universitaria s/n, 28040, Madrid, Spain

1. Introduction

Monolithic microwave integrated circuits (MMIC) based on gallium nitride (GaN) high electron mobility transistors (HEMT) have the advantage of providing broadband power performance (Milligan et al., 2007). The high breakdown voltage and high current density of GaN devices provide higher power density than the traditional technology based on GaAs. This allows the use of smaller devices for the same output power, and since impedance is higher for smaller devices, broadband matching becomes easier.

In this chapter, we summarise the design procedure of broadband MMIC high power amplifiers (HPA). Although the strategy is quite similar for most semiconductors used in HPAs, some special considerations, as well as, experimental results will be focused on GaN technology.

Apart from design considerations to achieve the desired RF response, it is essential to analyse the stability of the designed HPA to guarantee that no oscillation phenomena arises. In first place, the transistors are analysed using Rollet's linear K factor. Next, it is also critical to perform nonlinear parametric and odd stability studies under high power excitation. The strategy adopted for this analysis is based on pole-zero identification of the frequency response obtained at critical nodes of the final circuit (Barquinero et al., 2007).

Finally, to avoid irreversible device degradation, thermal simulations are required to accurately predict the highest channel temperature and thermal coupling between transistors.

2. AlGaN/GaN HEMT Technology

First of all, MMIC GaN technology has to be evaluated. High power GaN devices operate at high temperature and high-dissipated power due to the high power density of performance. Therefore, the use of substrates with high thermal conductivity like the silicon carbide (SiC) is preferred.

GaN technological process is still immature and complex. However, gate lithography resolution lower than 0.2 µm and AlGaN/GaN epi-structures on 100-mm SiC substrates are already available (Milligan et al., 2007).

Wide band gap semiconductors such as GaN and SiC are very promising technologies for microwave high power devices. The advantages of these materials over conventional semiconductors, GaAs and Si, include high breakdown field (E_g), high saturation electron velocity (v_{sat}), and high thermal conductivity. GaN/AlGaN high electron mobility transistors (HEMTs) offer even higher power performance due to the higher carrier sheet density and the higher saturation velocity of the bidimensional electron gas channel (2DEG) compared to SiC metal semiconductor field effect transistors (MESFETs). The diagram in Fig. 1 summarizes GaN-HEMT properties and its benefits.

This technology has demonstrated a power density of 30 W/mm using devices with dimensions of 0.55x246 μm^2 at 4 GHz when biased at 120 V (Wu et al., 2004).

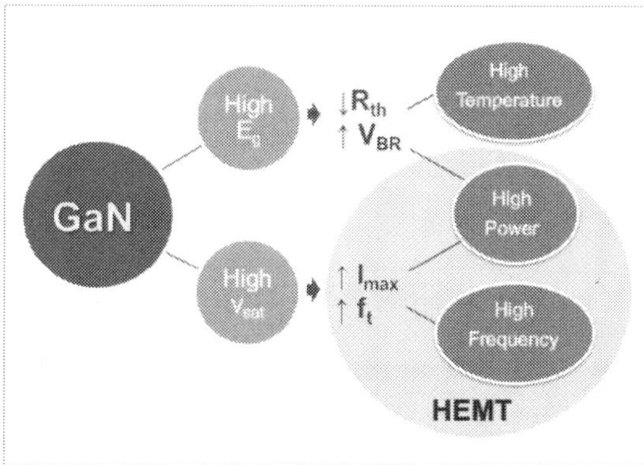

Fig. 1. GaN HEMT properties and benefits.

In a HEMT the conduction channel is confined to the interface between two materials with different band gap. This region known as 2DEG has very few ionized impurities to scatter the electrons, resulting in a very high mobility device. AlGaN/GaN heterostructures have a high sheet carrier density in the 2DEG interface without intentional doping of the structure. The spontaneous and piezoelectric polarization effects are the key factors for the charge distribution in the AlGaN/GaN HEMT (Ambacher et al., 2000).

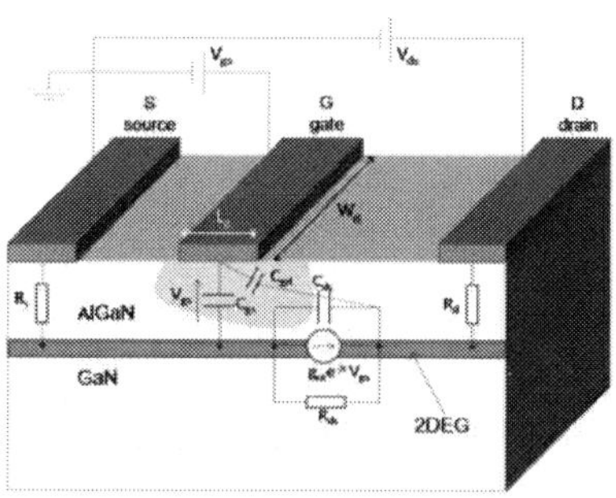

Fig. 2. AlGaN/GaN HEMT model.

The model of a HEMT that shows the small-signal parameters and the 2DEG channel is depicted in Fig. 2. Source and drain ohmic contact, as well as Schottky gate can be observed. The gate voltage (V_{gs}) controls the current (I_{ds}) that flows between the source and the drain. When V_{gs} reaches pinch-off voltage the electrons below the gate are depleted and no current can flow from drain to source.

Since AlGaN/GaN HEMTs for HPA applications work under high power conditions, non-linear models have to be used to simulate the transistor performance. The success of the design depends on the precision of the model fitting. A widely used approach is based on Angelov analytical expressions (Angelov et al., 1992). The model parameters are extracted from load pull, S-parameters, and pulse IV measurements. For instance, the nonlinear current source is characterized fitting DC and pulsed IV-measurements. The voltage controlled gate-source and gate-drain capacitance functions (C_{gs} and C_{gd}) are determined from bias dependent hot-FET S-parameter measurements. Finally, the parasitic elements of the HEMT model are extracted with cold-FET S-parameters measured from pinch-off to open channel bias conditions. The high temperature performance of GaN-HPAs demands the use of electro-thermal models (Nuttinck et al., 2003). Otherwise, power estimation will be too optimistic in CW operation.

The design methodology evaluated in this chapter is based on the experience reported by the design of several 2-6 GHz HPAs. The active devices used are 1-mm gate-periphery HEMTs fabricated using AlGaN/GaN heterostructures and gate length (L_g) technology of 0.5 µm from Selex Sistemi Integrati S.p.A foundry (Costrini et al., 2008) within Korrigan project (Gauthier et al., 2005). The HEMT cells consist of 10 fingers, each with a unit gate width (Wg) of 100 µm. The maximum measured oscillation frequency (f_{max}) of these transistors is about 39 GHz.

A considerable dispersion between wafers is still observed because of GaN technology immaturity. Table 1 shows the main characteristics (device maximum current I_{dss}, pinch-off voltage V_p, breakdown voltage V_{bgd}, C_{gs}, sheet resistance R_s and contact resistance R_c) of two wafers fabricated for the 2-6 GHz HPAs (wafer 1 and 2) and the wafer used for extracting the nonlinear electrical models (wafer 0) for the 1st-run designs. From the results in Table 1, an important deviation between the model and the measurements is expected. For instance, Cgs mismatch will produce a poor S11 fitting.

Wafer	I_{dss} mA/mm	V_p (V)	V_{bgd} (V)	C_{gs} (pF/mm)	R_s (Ω/sq)	R_c (Ωmm)
Wafer 0	972	-6.4	>70	3.2	355	0.44
Wafer 1	794	-4.7	50	2.24	456	0.47
Wafer 2	567	-2.7	71	2.88	440	0.36

Table 1. GaN wafers comparison.

Regarding passive technology, the foundries provide microstrip and coplanar models for typical MMIC components such as transmission lines, junctions, inductors, MIM capacitors and both NiCr and GaN resistors.

3. Design

The design process of a broadband HPA is described in this section. Special attention should be paid on broadband matching network synthesis and device stability.

3.1 Amplifiers Topology

The first step in an HPA design is to choose the most appropriated topology to fulfil design specifications. Single or multi-stage topology will be used depending on the gain target. In order to achieve high output power, several devices must be combined in parallel.

The classical combination topologies are the balanced and the corporative HPA. Balanced structures are made with $\lambda/4$-lines, which become quite large in designs below X-band. Furthermore, multi-stage approach for broadband design will enlarge the circuit even more. On the other hand, the corporate topology based on two-way splitters seems to be a more versatile solution for broadband designs. It can be designed with compact lumped broadband filters in frequencies below X-band while transmission lines can be used at higher frequencies.

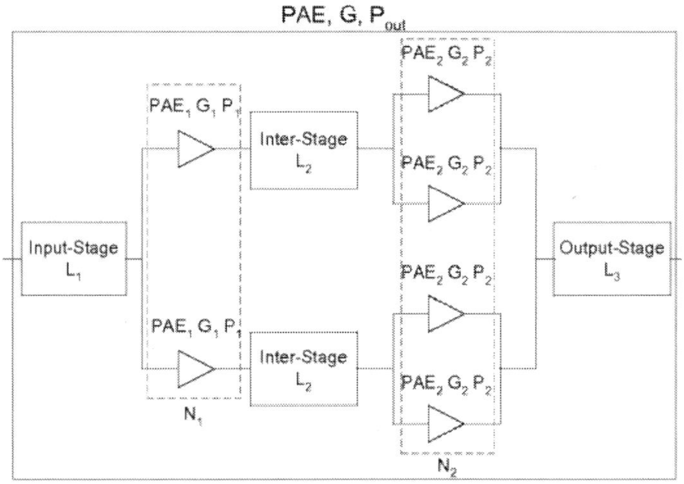

Fig. 3. Two-stage corporative topology amplifier.

The two-stage corporative topology, such as the one in Fig. 3, is widely used to design HPAs, because it offers a good compromise between gain and power. Note that the first stage consists of two unit cells which drive the output stage, composed of four equal cells. This power amplifier has three matching networks: input-, inter-, and output-stage. The labels displayed in Fig. 3 represent the loss of each matching network (L_i), the number of combined cells (N_i), the power added efficiency (PAE_i), the gain (G_i), and the output power (P_i) at the ith-stage.

Output stage loss, L_3, is critical to the HPA output power ($P_{out}=N \cdot P_{HEMT} \cdot L_3$). Besides, network loss has to be minimised mainly in the output network, because it is essential to maximise power added efficiency (PAE_{total}). Equation (1) is used to calculate PAE_{total} of a corporative topology with 2^n transistors at the output stage. The representation of equation (1) in Fig. 4 confirms the higher influence of L_3 in PAE_{total}.

$$PAE_{total} = \frac{PAE_1 PAE_2 (L_1 L_2 L_3 G_1 G_2 - 1)}{PAE_1 L_1 L_2 G_1 (G_2 - 1) + PAE_2 L_1 (G_1 - 1)} \tag{1}$$

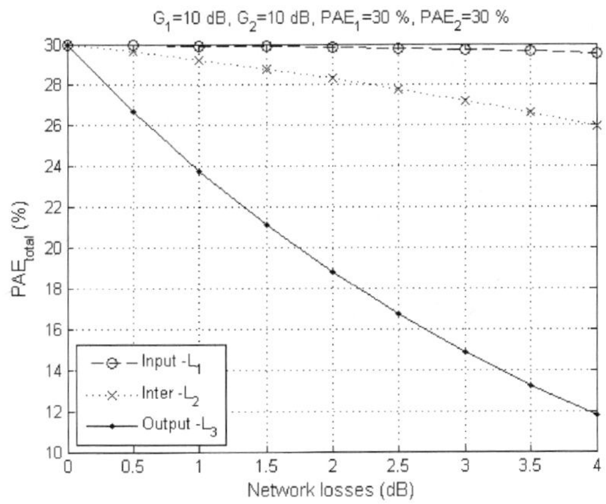

Fig. 4. PAE_{total} versus input-, inter-, and output-stage network loss. To analyse the influence of each network the loss of the other networks are set to 0 dB.

High efficiency operation is especially important for power devices, because thermal issues can degrade the amplifier performance. Dissipated power (P_{dis}) is inversely proportional to the HPA efficiency, see equation (2).

$$P_{dis} \approx P_{out} \cdot ((PAE_{total})^{-1} - 1) \tag{2}$$

P_{dis} and temperature increase (ΔT) are related in equation (3) through the thermal resistance (R_{th}). This parameter gives an idea of the thermal flow through a material or a stack of materials from a hot spot to another observation point. Therefore, R_{th} depends on the thermal conductivity of the materials and the final HPA set-up.

$$\Delta T = P_{diss} R_{th} \tag{3}$$

3.2 Unit transistor cell

The unit transistor cell size selection is based on a compromise between gain and power, because large devices have higher power, but lower gain (Walker, 1993). Moreover, input and output impedances decrease for larger devices, making the design of broadband matching networks difficult. The lack of power of small devices can be solved by combining several devices in parallel. It is worth noting that the complexity of the design increases with the number of cells to be combined.

Once the transistor size is selected, the available unit cells have to be evaluated at different bias operating conditions (Snider, 1967). The optimum operation class of a power amplifier depends on the linearity, efficiency or complexity of the design specifications. In the conventional operation classes, A, B, AB and C, the transistor works like a voltage controlled current source. On the contrary, there are some other classes, such as D, E and F, where amplifier efficiency improves by working like a switch. In the diagram of Fig. 5, different operation classes have been represented, indicating the IV-curves, the load-lines and the conduction angle (Φ) of each one. The conduction angle defines the time that the transistor is in the on-state.

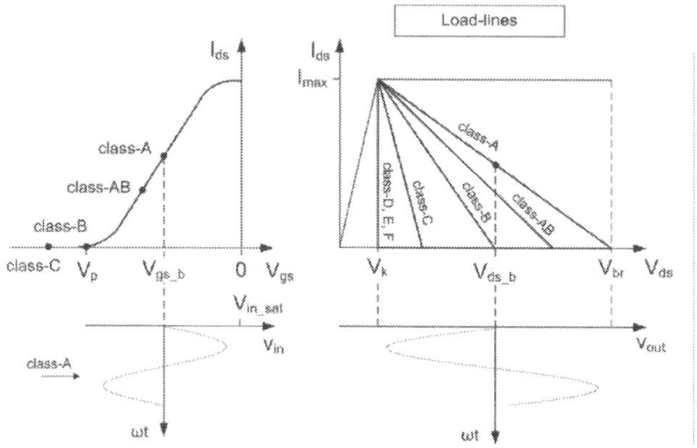

Fig. 5. HPA bias operation classes.

The maximum drain efficiency ($\eta = P_{out}/P_{dc}$) and the maximum output power of a transistor versus the conduction angle can be calculated under ideal conditions, as shown in Fig. 6, where the knee voltage (V_k) is assumed to be 0 V and RF compression is not considered. This representation shows that drain efficiency is inversely proportional to Φ, and that output power is almost constant between class AB and class A. Class-AB operation quiescent point at $30\%I_{max}$ provides simultaneously maximum power and a considerable high drain efficiency, therefore this seems to be an optimum bias point.

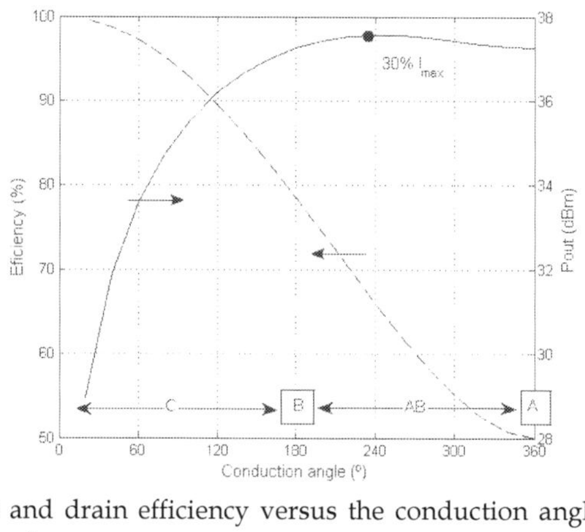

Fig. 6. Output power and drain efficiency versus the conduction angle (Φ) calculated for V_{ds}=25V and I_{max}=850mA.

Given an operation class, the RF power drive determines the actual drain efficiency. Efficiency and linearity are opposite qualities. Therefore, a compromise has to be assumed depending on the HPA design application.

3.3 Unit cell stabilization

An in-depth analysis of the stability is necessary to guarantee that no oscillation phenomena arise. Firstly, the transistors are analysed using the classical approach for linear stability based on the Rollet's (Rollet, 1962) formulas over a wide frequency band. This theorem stands that the transistor is unconditionally stable if the real part of the impedance at one port is positive ($Re(Z_{ii}) > 0$) for any real impedance at the opposite port. The Rollet's K factor as a function of the two-port network inmitance parameters ($\gamma ii=Zii=Yii$) is the following:

$$K = \frac{2Re(\gamma_{11})Re(\gamma_2) - Re(\gamma_{12}\gamma_{21})}{|\gamma_{12}\gamma_{21}|} > 1 \tag{4}$$

The easiest way to increase K to achieve K>1 is increasing the input impedance of the transistor. This can be done by adding a frequency dependent resistance (R_{stab}) at the transistor input port: $Z'_{11}=Z_{11}+R_{stab}$. Stability can also be improved with a resistor at the transistor output, but this would reduce the maximum output power. The series stabilization resistance can be calculated from equation (5).

$$R_{stab} = \frac{K|Z_{12}Z_{21}| + Re(Z_{12}Z_{21})}{2Re(Z_{22})} - Re(Z_{11}) \tag{5}$$

Another useful way to write R_{stab} is as a function of the small-signal parameters of the transistor:

$$R_{stab} = \frac{2\omega C_{ds}C_{gs}(K-1) + C_{gd}g_m K}{2\omega C_{gd}g_m(C_{gd}+C_{gs})} \tag{6}$$

Parallel RC networks in series with the transistor gate make it possible to synthesize R_{stab} in a wide frequency band, see Fig. 7.

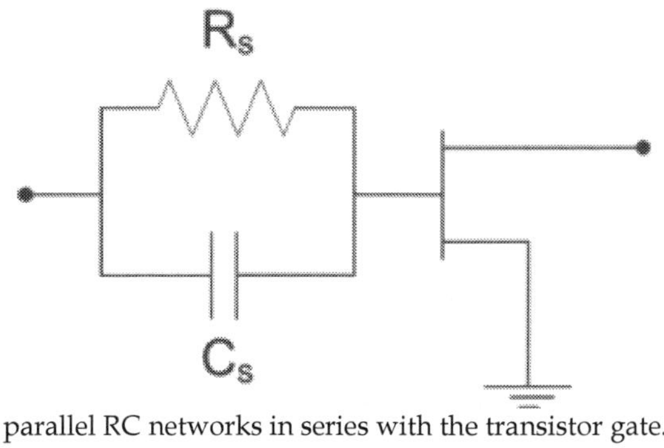

Fig. 7. Stabilization parallel RC networks in series with the transistor gate.

As an example, we take an unstable transistor (K<1) at frequencies under 12.5 GHz. The ideal stabilization resistance to make this transistor unconditionally stable (with K=1.2) at any frequency is plotted in Fig. 8 (left). In the same plot, the R_{stab} traces obtained with two different RC networks have been included. The new K factor (K_{new}) recalculated taking into account the cascade of the series RC networks and the transistors are represented in Fig. 8 (right).

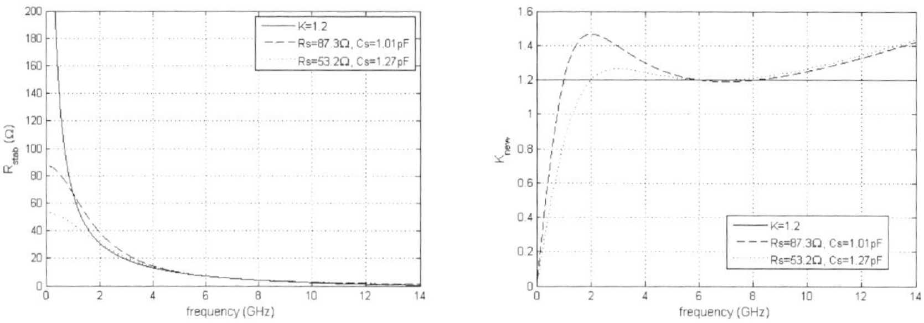

Fig. 8. R_{stab} and K_{new} calculated with an ideal network for K=1.2 and two different RC networks.

RC networks cannot be used to stabilize a transistor at low frequencies because the resulting resistor becomes too large or the capacitor too small to be feasible in MMIC technology. Therefore, off-chip stabilization networks are sometimes required to avoid the use of big components in the chip. Another solution is to add a parallel resistor (R_p) in the internal stabilization network. This resistor can be included in the gate bias path (L_b) as depicted in Fig. 9. L_b should be chosen high enough to have no influence in the frequency band of the design.

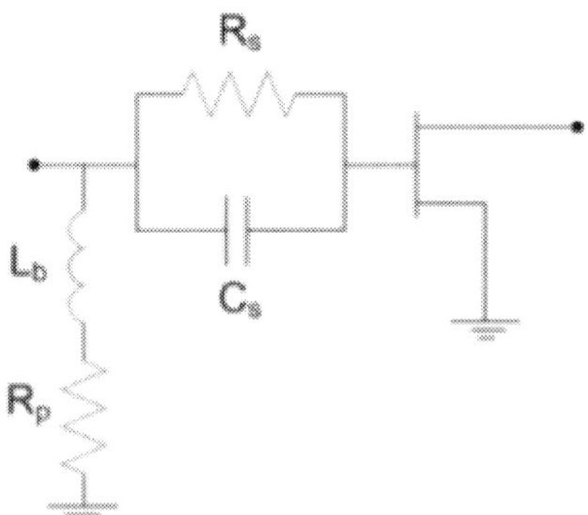

Fig. 9. Combination of a parallel RC networks in series with the transistor and a parallel resistor in the bias path to achieve unconditional stability at any frequency.

In Fig. 10 it is shown the comparison between K_{new} calculated with the series RC network of Fig. 7 and the network of Fig. 9 that combines a series and a parallel resistor. The last solution makes the transistor unconditionally stable even at low frequencies.

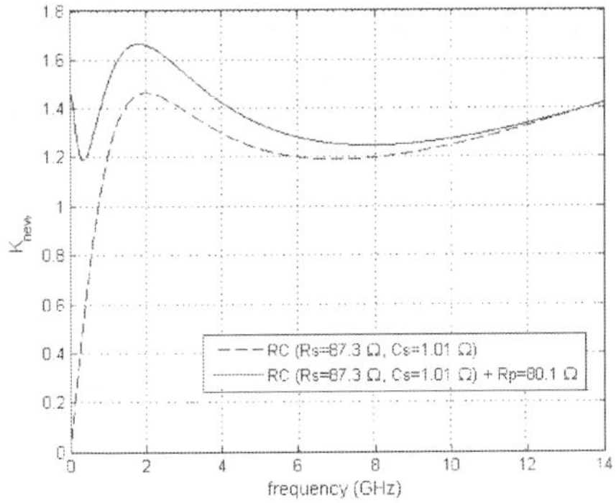

Fig. 10. K_{new} obtained with a single RC networks and with the combination of an RC network and a parallel resistor at the gate bias path.

As Fig. 8 shows, the stabilization networks introduce dissipative loss decreasing with frequency, what also contributes to compensate the device gain slope. The maximum available gain (MAG) after stabilization is obtained as:

$$MAG = \left|\frac{Z_{21}}{Z_{12}}\right| \left(K_{new} - \sqrt{K_{new}^2 - 1}\right) \tag{7}$$

The comparison between the original MAG of the transistor and the MAG obtained with the proposed stabilization networks is shown in Fig. 11.

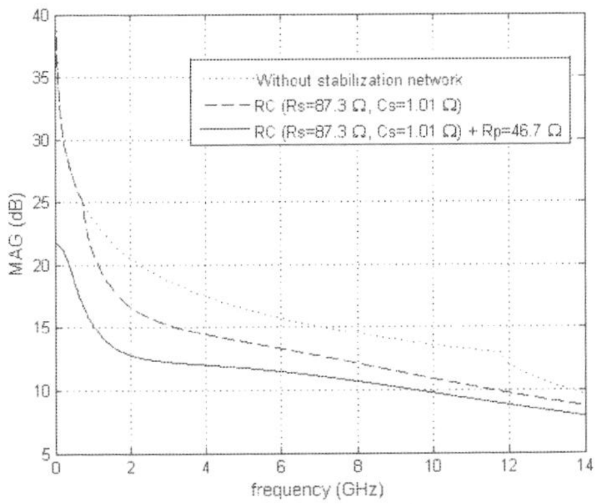

Fig. 11. Comparison of the transistor MAG without any stabilization network and with both a single RC networks and the combination of an RC network and R_p.

RC networks are also used to prevent parametric and out-of-band oscillations (Teeter et al., 1999).

3.4 Networks synthesis

The HPA matching networks (input-, inter-, and output-stage) are synthesized from filter theory and implemented with both lumped elements and transmission lines. These networks are designed to provide optimum impedances at the transistor output and conjugated matching at its input, as well as, matching the HPA input and output to 50 Ω. Stabilization networks and DC bias networks have to be included and considered in the synthesis process.

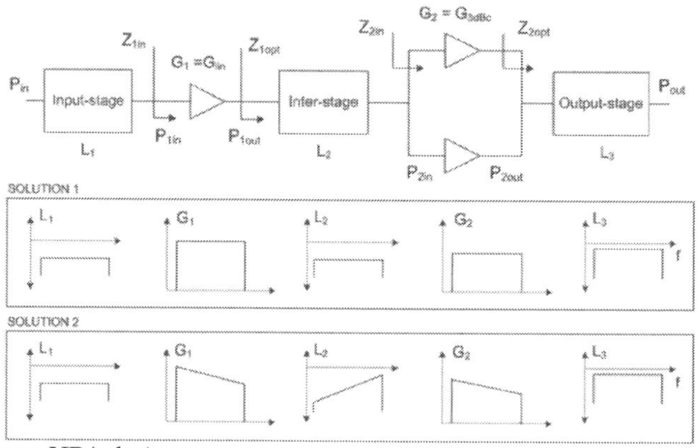

Fig. 12. Two-stage HPA design process.

Two different strategies can be followed in the HPA design as indicated in the diagram in Fig. 12; in the first solution, the transistor and the gain equalization network are considered a single block and the matching networks have to be designed with a flat frequency response. This approach is easily adopted when the stabilization network introduces frequency dependent loss. In the second solution, gain compensation is performed by the inter-stage matching network.

The following steps describe the HPA design process:

- Firstly, the transistor optimum loads for maximum power (Z_{2opt}) are calculated using load-pull techniques. The most precise method to obtain the optimum loads is by means of load-pull measurements. However, load-pull measurement equipment is expensive and, the measurement process could be tedious and long for broadband design, because many load measurements are required. If nonlinear models of the transistor are available, load-pull simulation could be done in CAD simulators. The accuracy of this option depends on the precision of the nonlinear models. When only the transistor lineal-model and the IV-curves are available, the load-pull contours have to be estimated by the Cripps method (Cripps, 1983).
- Next, the output-stage network has to be designed. This network transforms Z_{2opt} to the 50 Ω impedance of the HPA output-port and combines the power of the 2nd-stage transistors. The drain DC-bias is included in this network and it is done through a parallel inductance of a value calculated to provide the imaginary part of Z_{2opt} at the design centre frequency. Network loss (L_3) must be minimised, mainly in this output-stage because it is critical to maximise power and PAE.

- Later, the inter-stage network loads are calculated; firstly, the input impedance of the second-stage transistors loaded with the output-stage network (Z_{2in}); secondly, the optimum loads for maximum gain at the first-stage transistors (Z_{1opt}). It is important to calculate the impedances at the expected high power working conditions.
- Then, the inter-stage network is designed to synthesize the optimum loads for the first-stage (Z_{1opt}) and to match the second-stage input (Z_{2in}). The design complexity of this network is higher because two complex impedances have to be matched over a broad frequency bandwidth. Moreover, if it is required, the transistor gain roll-off should be compensated by frequency dependent losses in this network (L_2). This can be done with a RC network, as it was described in section 3.3. First-stage drain DC-bias is also done through a parallel inductance that provides the imaginary part of Z_{1opt} at the design centre frequency.
- Finally, the input-stage network is designed to match the HPA-input of 50 Ω to the first-stage transistors input (Z_{1in}). In this case, the input impedance of the transistors is also calculated loading them with the inter- and output-stage networks.

The matching networks can be designed as two-port networks. Anyhow, it is worth noting that the input impedance has to be scaled by the number of transistors to be combined (N). Then, the network can be transformed into a (N+1)-port network. The transformation is done by dividing the two-port network in different sections and scaling them depending on the branching level in the power combination (or division) network. Fig. 13 shows a diagram where the transformation process is schematized.

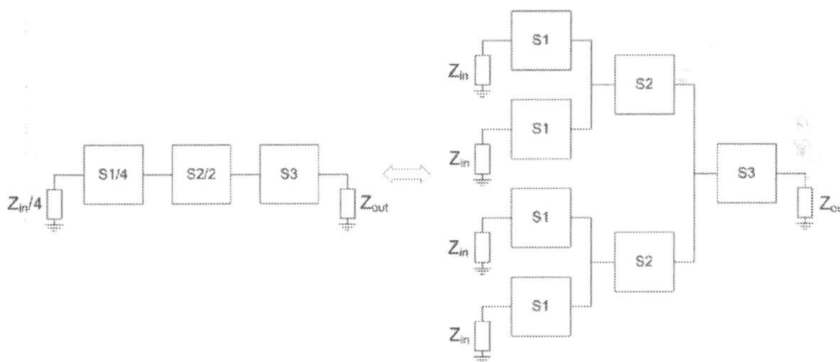

Fig. 13. Transformation of two-port to N-port networks.

4. Global stability analysis

Multistage HPAs are prone to parametric oscillations that are function of the input-power drive. The origin of these instabilities is the nonlinear capacitance of the transistor input impedance, which varies with the input-drive. Odd-mode oscillations are also frequent due to the presence of multiple active elements and the circuit symmetry. Subharmonic oscillations at $f_{in}/2$, where f_{in} is the input signal frequency, are very common in transistors due to the nonlinear capacitance nature. However, spurious oscillations at non-harmonically related frequency f_a are also observed.

Two-port network techniques cannot be applied for HPA stability analysis due to the existence of multiple feedback loops. The standard harmonic-balance (HB) simulators used for HPA design do not include tools for high-power stability analysis either. Fortunately, there are several techniques that perform a nonlinear stability study based on the circuit linearization around the large-signal steady state obtained with HB. The method proposed by Mons (Mons et al., 1999) is rigorous and complete, but it requires the verification of the Nyquist stability criteria for every nonlinear element, what becomes tedious in complex circuits. From the design process point of view, faster stability analysis is preferred. Therefore, it is proposed a technique based on the insertion of an external small-signal perturbation in a circuit node. This way, it is possible to obtain the closed-loop transfer function provided by the impedance calculated at the observation node in a certain frequency range. Pole-zero identification of the resultant transfer function is used to verify the stability of the circuit (Jugo et al., 2003). This study can be done in both small-signal and large-signal conditions. Different observation nodes must be considered to ensure the detection of masked instabilities, because of pole-zero cancellation in certain nodes. At least, an analysis per each HPA stage is required.

Parametric simulations at different working conditions are advisable to see the evolution of critical poles. If any complex conjugated poles cross to the right half plane an oscillation is detected.

In circuits with N active devices, N modes of performance coexist. There are N-1 odd-modes and one even-mode simultaneously. For instance, in the second-stage of the HPA in Fig. 3, three odd-modes and one even-mode coexist. However, due to the symmetry, two odd-modes are equivalent, so only the odd-modes [+ - + -] and [+ + - -], and the even-mode [+ + + +] have to be studied. Using different perturbation configurations, the stability of each mode can be determined by means of the pole-identification technique. Instead of a single perturbation generator, one generator at the input of each transistor is introduced and the phase of the perturbation signals is shifted 180° depending on the excitation mode (Anakabe et al. 2005).

In Fig. 14, the frequency responses of an even-mode (left) and an odd-mode [+ - + -] (right) are depicted. Both responses have been obtained at the same operating conditions (P_{in}=19dBm, f_{in}=4GHz, V_{ds}=26V and V_{gs}=-4.2V), and it can be seen that the odd-mode presents a resonance at $f_{in}/2$ that indicates the presence of a possible subharmonic oscillation. This means the transfer function may have poles with positive real part.

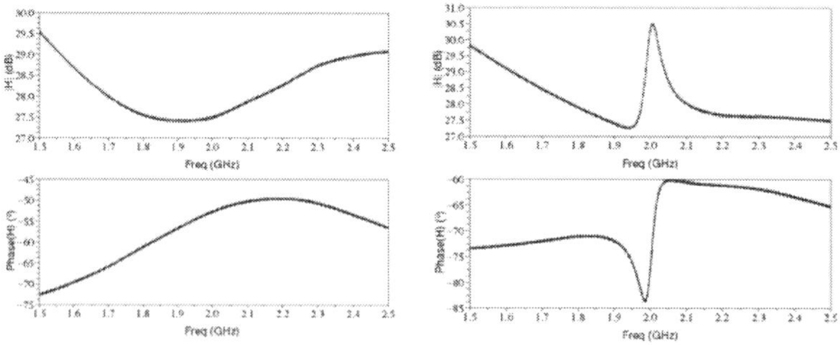

Fig. 14. Closed-loop transfer function calculated with even-mode [+ + + +] (left) and odd-mode [+ - + -] (right) excitation. The HPA operation conditions are P_{in}=19dBm, f_{in}=4GHz, V_{ds}=26V and V_{gs}=-4.2V

Once an oscillation is found, the instability margin has to be determined through a parametric study about the critical operation conditions. Frequency and power of the input signal, as well as the HPA DC-bias (V_{ds} and V_{gs}) are common parameters that affect stability. The evolution with frequency and V_{ds} of the poles at $f_{in}/2$ corresponding to the frequency response in Fig. 14 are represented in Fig. 15. The HPA is unstable between 3.97 GHz and 4.03 GHz at V_{ds} = 26 V.

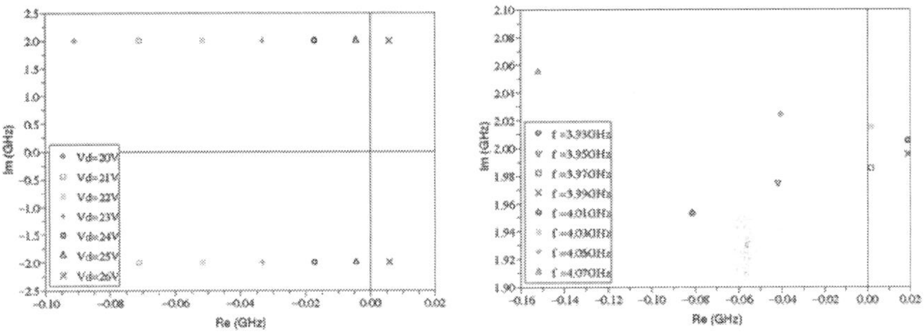

Fig. 15. Evolution of the poles at $f_{in}/2$ versus V_{ds} (left) and f_{in} (right). The HPA initial operation conditions are P_{in}=19dBm, f_{in}=4GHz, V_{ds}=25V and V_{gs}=-4.2V

Once the stability nature has been determined, the HPA circuit has to be corrected to avoid oscillations that may invalidate the design. Usually, the instability is cancelled using notch filters (like RC networks (Teeter et al., 1999)) at the oscillation frequency or resistors to add loss in the oscillation feedback loop. For instance, resistors between the transistors (R_o) can be added to prevent odd-mode oscillations, see Fig. 16.

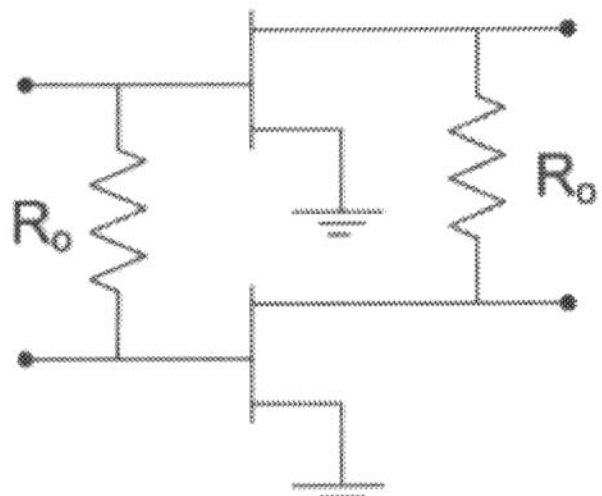

Fig. 16. Resistors to prevent odd oscillations.

5. Thermal characterization

Thermal characterization with different techniques is of crucial interest in GaN-HPAs, because it is still necessary to analyse the influence of the high power dissipated in this leading technology (Nuttinck et al., 2003).
Thermal resistances, R_{th}, at different working conditions can be calculated with commercial software like COMSOL Multiphysics (FEMLAB) or Ansys. The simulations can be performed for the unit transistor cell to obtain the maximum channel temperature (T_{chann}),

or for the final HPA to characterize the thermal coupling between the transistors. Fig. 17 shows the thermal resistance of a unit transistor cell of 1mm, and the results for an HPA with 8x1mm transistors at the output-stage. The simulation has been performed in ideal conditions and taking into account the real mounting fixture of the device on a cooper carrier.

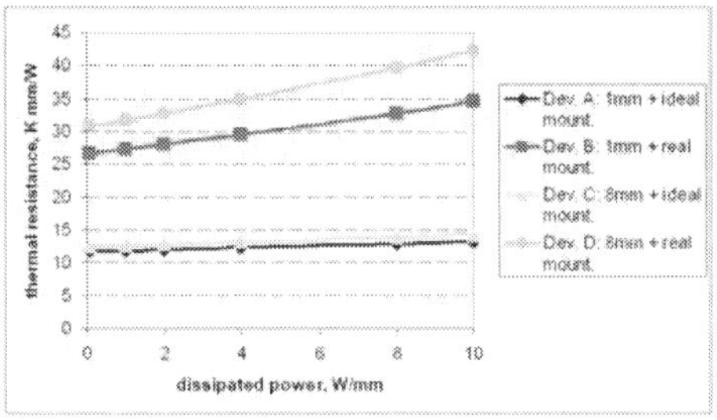

Fig. 17. Comparison between the simulated thermal resistance of a 1mm-transistor and of an HPA with 8x1mm transistors at the output-stage, in ideal and real mounting conditions.

R_{th} in the range of 13.5 °C/W has been obtained at 6W dissipated power (P_{dis}) for the ideal mounting of the 1mm-transistor. From these calculations, an estimated gradient (ΔT) around 81°C is expected between the channel and the backside temperature. However, the real assembly increases R_{th} to 32 °C/W, which means a temperature gradient of 192 °C. Thus, we see that a test fixture mounted on a cooling platform is necessary in order to provide the amplifier with a proper heat dissipation system.

6. Broadband HPA examples

Two fully monolithic broadband HPAs with an output-stage active periphery of 4 mm and 8 mm are presented in the photos of Fig. 18. They have been fabricated at Selex Sistemi Integrati S.p.A.

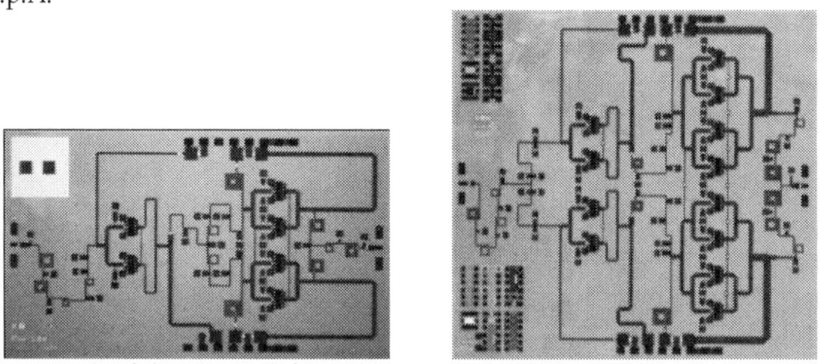

Fig. 18. Photograph of the 4 mm (left) and 8 mm (right) HPAs. The chip size is 6.6x3.7 mm^2 and 6.6x 6.0 mm^2, respectively.

Several MMIC HPAs were characterized in CW and pulsed conditions. All chips were tested at drain-source voltage, V_{ds}, from 20 to 25 V and I_d=30%Imax.

Typical measured small-signal gain and input return loss of the 4 mm-HPA are shown in Fig. 19. Over the 2-6 GHz frequency range, gain was about 18 dB and the input return loss was lower than -7 dB. Simulated results are also shown for comparison. Mismatch between simulated and measured input return loss exists because the transistor model was extracted from a previous wafer and the technological process is still in development.

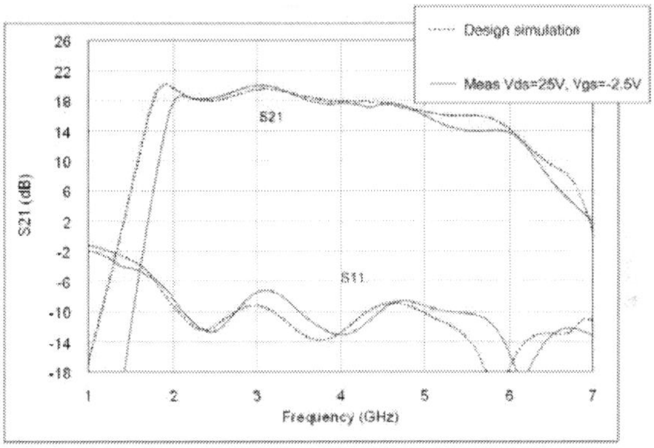

Fig. 19. Comparison of gain and input return loss measurements and simulation of the 4mm-HPA.

Pulsed and CW characterization of the 8mm-HPAs from two different wafers (see Table 1) at 4.5 GHz and V_{ds}=25V are shown in Fig. 20. The pulsed measurements were performed with short pulses of 20 µm length and 1% duty cycle. The HPA from Wafer 1 exhibited higher output power in pulsed-mode, whereas the power capacity in CW is similar. Saturation power is about 15 W in CW with better than 20% PAE and, reaching 26 W and 25% in pulsed-mode.

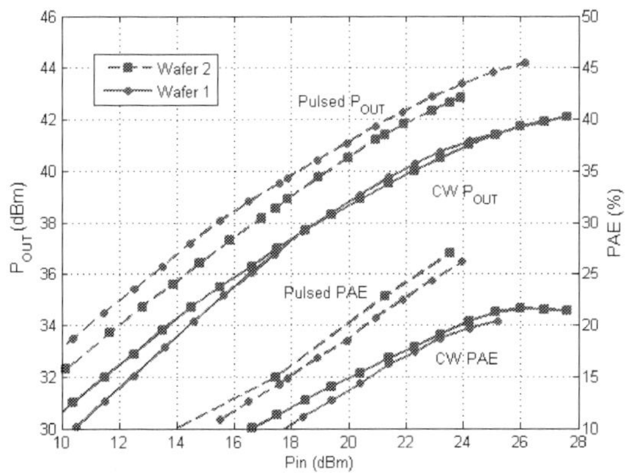

Fig. 20. CW and pulsed output power and PAE versus input power of the 8mm- HPA from Wafer 1 and Wafer 2 at 4.5 GHz, V_{ds}=25 V and 30%I_{DSS}.

Typical broadband performance of both 4mm-HPA and 8mm-HPA in CW and pulsed-mode is shown in Fig. 21. Pulsed measurements in the lower frequency band are not available because the set-up works above 3 GHz. The 4mm-HPA has greater than 40 dBm (2.5 W/mm) output power in 50% of the band in CW, and greater than 41.4 dBm (3.5 W/mm) in pulsed conditions. On the other hand, the 8mm-HPA delivers 41.2 dBm (2 W/mm) in CW and 44 dBm (3.5 W/mm) in pulsed-mode. Thermal problems are more significant for the 8mm-HPA in CW.

To characterise the power degradation due to thermal heating, the HPAs have been measured in pulsed-mode at different duty cycles and pulse lengths. The duty cycle has higher influence than the pulse length in the HPA performance. The results of this analysis with a pulse length of 100 µs are depicted in Fig. 22. The output power and PAE at V_{ds}=20 and 25 V are shown for both HPAs. As expected, the power and efficiency degradation is higher for the 8mm-HPA. This device losses approximately 40% of the power capacity from 1% to 50% duty cycle, while the 4mm-HPA falls only 30%.

There is still margin to increase CW power if the test-jig is improved to reduce its thermal resistance.

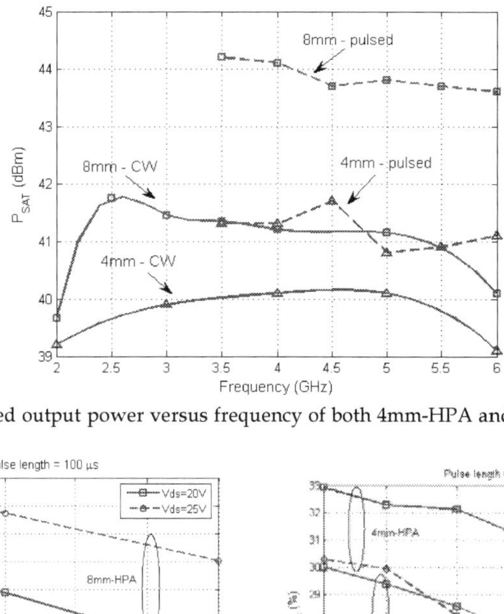

Fig. 21. CW and pulsed output power versus frequency of both 4mm-HPA and 8mm-HPA at V_{ds}=25 V and 30%I_{DSS}.

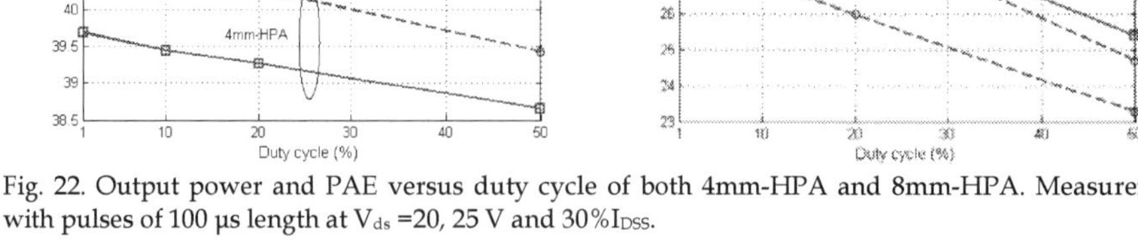

Fig. 22. Output power and PAE versus duty cycle of both 4mm-HPA and 8mm-HPA. Measurements with pulses of 100 µs length at V_{ds} =20, 25 V and 30%I_{DSS}.

7. Conclusion

This chapter makes a brief introduction of the GaN-HEMT technological process development. Based on this technology, it is established a design procedure for broadband high power amplifiers. The design is focused on the synthesis of the matching and stabilization networks of a two-stage amplifier. It is highlighted the need for nonlinear stability analysis to avoid parametric and odd-mode oscillation. Thermal characterization is also critical due to the high power dissipated in high power GaN devices. Finally, we present the analysis of results of two broadband HPA demonstrators.

8. Acknowledgment

This work has been supported by the Spanish National Board of Scientific and Technology Research under the project TEC2008-021481TEC.

9. References

Ambacher, O.; Foutz, B.; Smart, J.; Shealy, J. R.; Weimann, N. G.; Chu, K.; Murphy, M.; Sierakowski, A. J.; Schaff, W. J.; Eastman, L. F.; Dimitrov, R.; Mitchell, A.; Stutzmann, M. (2000). Two dimensional electron gases induced by spontaneous and piezoelectric polarization in undoped and doped AlGaN/GaN heterostructures. *Journal of Applied Physics*, Vol. 87, No. 1, 2000, pp. 334

Anakabe, A., Collantes, J.M.; Portilla, J.; Mons, S.; Mallet, A. (2005). Detecting and Avoiding Odd-Mode Parametric Oscillations in Microwave Power Amplifiers. *International Journal on RF and Microwave Computer-Aided Engineering*, Vol. 15, No. 5, September 2005, pp. 469-478, ISSN:1096-4290

Angelov, I.; Zirath, H.; Rosman, N. (1992). A new empirical nonlinear model for HEMT and MESFET devices, *IEEE Trans. Microw. Theory Tech.*, Vol. 40, No. 12, December 1992. pp. 2258–2266, ISSN: 0018-9480

Barquinero, C.; Suarez, A.; Herrera, A.; Garcia, J.L. (2007). Complete Stability Analysis of Multifunction MMIC Circuits. *IEEE Trans. on Microwave Theory and Tech.*, Vol. 55, No. 10, October 2007, pp. 2024-2033, ISSN: 0018-9480

Costrini, C.; Calori, M.; Cetronio, A.; Lanzieri, C.; Lavanga, S.; Peroni, M.; Limiti, E.; Serino, A.; Ghione, G.; Melone, G. (2008). A 20 watt micro-strip X-band AlGaN/GaN HPA MMIC for Advanced Radar Applications," *EUMIC 2008*. pp. 566-569, October 2008

Cripps, S. (1983). A Theory for the Prediction of GaAs FET Load-Pull Power Contours. *Proceedings of the IEEE Intl. Microwave Symp. Dig.*, Vol. 83, No. 1, 1983, pp. 221- 223

Gauthier, G.; Mancuso, Y.; Murgadella, F. (2005). KORRIGAN - a comprehensive initiative for GaN HEMT technology in Europe, *EGASS 2005*, pp. 361-363, ISBN: 8890201207

Jugo, J.; Portilla, J.; Anakabe, A.; Suárez, A.; Collantes, J. M. (2003). Closedloop stability analysis of microwave amplifiers. *Electron. Letters*, Vol. 37, Feb.y 2003, pp. 226–2228

Milligan, J.W.; Sheppard, S.; Pribble, W.; Wu, Y.-F.; Muller, StG.; Palmour, J.W. (2007). SiC and GaN Wide Bandgap Device Technology Overview. *IEEE Radar Conference*, April 2007, pp. 960-964, ISSN: 1097-5659

Mons, S.; Nallatamby, J.-C.; Quere, R.; Savary, P.; Obregon, J. (1999). A unified approach for the linear and nonlinear stability analysis of microwave circuits using commercially available tools. *IEEE Trans. Microw. Theory Tech*, 1999, pp. 2403-2409, ISSN 00189480

Nuttinck, S.; Wagner, B.K.; Banerjee, B.; Venkataraman, S.; Gebara, E.; Laskar, J.; Harris, H.M. (2003), Thermal analysis of AlGaN-GaN power HFETs. *IEEE Trans. on Microw. Theory and Tech.*, Vol. 51, No.12, Dec. 2003, pp. 2445-2452, ISSN: 0018-9480

Rollett, J. (1962). Stability and Power-Gain Invariants of Linear Twoports. *IRE Transactions on Circuit Theory*, Vol. 9, No. 1, march 1962, pp. 29-32, ISSN 0096-2007

Snider, D.M. (1967). A theoretical analysis and experimental confirmation of the optimally loaded and overdriven RF power amplifier. *IEEE Transactions on Electron Devices*, Vol. 14, No. 12, Dec 1967, pp. 851-857, I SSN: 0018-9383

Teeter, D.; Platzker, A.; Bourque, R. (1999). A compact network for eliminating parametric oscillations in high power MMIC amplifiers. *IEEE MTT-S International Microwave Symposium Digest*, Vol. 3, 1999, pp. 967-970, ISBN: 0-7803-5135-5

Walker, J. L. B. (1993). *High-Power GaAs FET amplifier*, Artech House, 1993, ISBN: 0890064792

Wu, Y.-F.; Saxler, A.; Moore, M.; Smith, R.P.; Sheppard, S.; Chavarkar, P.M.; Wisleder, T.; Mishra, U.K.; Parikh, P. (2004). 30-W/mm GaN HEMTs by field plate optimization. *IEEE Electron Device Letters*, 2004, pp. 117-119, ISSN: 0741-3106

17

Design of Multi-Passband Bandpass Filters With Low-Temperature Co-Fired Ceramic Technology

Ching-Wen Tang and Huan-Chang Hsu
National Chung Cheng University
Chiayi 621, Taiwan, R.O.C.

1. Introduction

Shared building blocks and power are required for the coexistence of a dual-band multimode wireless local area network and a mobile communication system. Therefore dual-passband bandpass filters have become key components at the front end of a concurrent dual-band receiver. There are several studies on dual-passband filters [1-9]. With sharper passband skirt, lower insertion loss and better selectivity may be resulted in the Zolotarev bandpass filters [1]. In [2], the dual-band filter is constructed with two parallel sets of filters. The frequency-selective resonators [3-6], stepped-impedance resonators [7, 8] and coupled resonator pairs [9] are also employed to design dual-passband filters.

In this article, we use low-temperature co-fired ceramic [10-15] technology to implement the three-dimensional (3D) multi-passband bandpass filters. Figure 1 shows the architecture of the proposed multi-passband bandpass filter, which is composed of multi-sectional short-circuit transmission lines and connected transmission lines. These transmission lines can be transferred individually to multilayered structure. Moreover, the short-circuit transmission lines may make more obvious isolation between passbands [16-18]. As a result, the proposed filter with controllable multiple passbands can be easily achieved by properly choosing the impedance and electrical length of each short-circuit transmission line and the connected transmission line.

2. Equivalent for filter synthesis

The immittance inverter [19] is adopted in this article to analyze the proposed filter. The transformed circuit of a transmission line shown in Fig. 2 can be utilized in the n-ordered multi-passband bandpass filter. Moreover, the transformed admittance inverter of the transmission line can be expressed as

$$J_i = Y_i \csc\theta_i \quad (1)$$

where Y_i and θ_i are the corresponding admittance and electrical length of the transmission line, respectively.

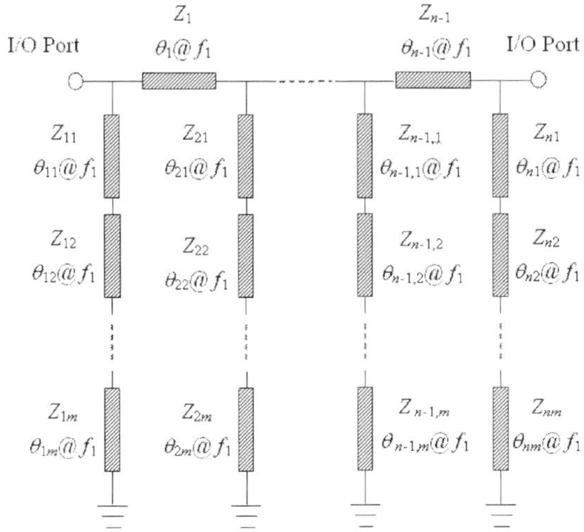

Fig. 1. Architecture of the proposed bandpass filter with multiple controllable passbands

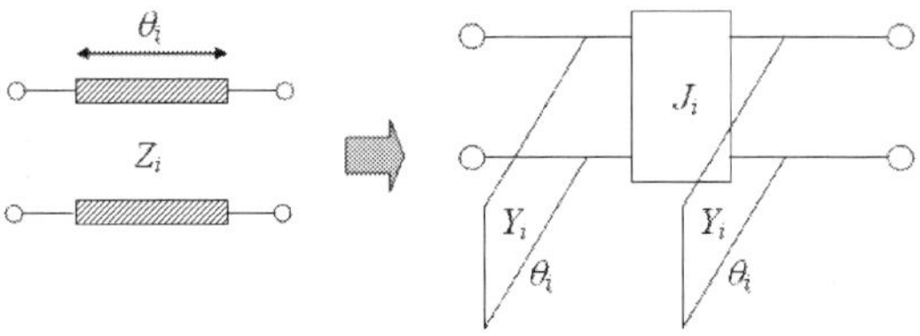

Fig. 2. Transformed circuit of the transmission line

By substituting the transformed transmission lines into the architecture in Fig. 1, the equivalent circuit of the proposed n-ordered multi-passband bandpass filter can be obtained as Fig. 3. The susceptance and its slope parameter are, respectively, given by

$$B_i(f) = -\frac{1}{X_{im}(f)} - Y_{i-1}\cot(\frac{f}{f_1}\theta_{i-1}) - Y_i\cot(\frac{f}{f_1}\theta_i), \quad \text{for} \quad i = 1,\cdots,n \tag{2}$$

$$b_i(f_r) = \frac{f_r}{2}\cdot\frac{\partial B_i}{\partial f}\bigg|_{f=f_r}, \quad \text{for} \;\; i = 1,\cdots,n \;\; \text{and} \;\; r = 1,\cdots,m \tag{3}$$

where f_r is the central frequency of the rth passband with the corresponding fractional bandwidth Δ_r, and

$$X_{ij}(f) = \begin{cases} Z_{im}\tan(\frac{f}{f_1}\theta_{im}), & \text{for } j = 1 \\ Z_{i,m-j+1}\cdot\dfrac{X_{i,j-1}(f) + Z_{i,m-j+1}\tan(\frac{f}{f_1}\theta_{i,m-j+1})}{Z_{i,m-j+1} - X_{i,j-1}(f)\tan(\frac{f}{f_1}\theta_{i,m-j+1})}, & \text{for } j = 2,\cdots,m \end{cases} \tag{4}$$

Moreover, in order to match system's impedance, 50 Ω, the input/output J-inverter J_{01} and $J_{n,n+1}$ need to be set as 0.02.

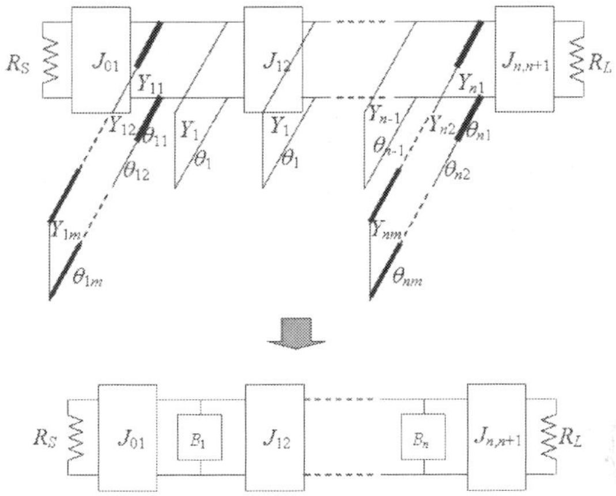

Fig. 3. Equivalent circuit of the proposed n-ordered multi-passband filter

As extremely complex procedures are required for generalized formulas to synthesize the proposed multi-passband bandpass filter, only formulas for dual-passband filter synthesis are provided in detail. On the other hand, design examples of the triple- and quadruple-passband filters are offered without detailed equations.

3. Design of the dual-passband filter

To design the dual-passband filters, $m = 2$ should be selected. Δ_1 and Δ_2 are the corresponding fractional bandwidths of the first and second passband's central frequency, f_1 and f_2, respectively. The susceptances and their slope parameters are, respectively, given by

$$B_i(f) = Y_{i1}\frac{Y_{i1}\tan(\frac{f}{f_1}\theta_{i2})\tan(\frac{f}{f_1}\theta_{i1}) - Y_{i2}}{Y_{i1}\tan(\frac{f}{f_1}\theta_{i2}) + Y_{i2}\tan(\frac{f}{f_1}\theta_{i1})} - Y_{i-1}\cot(\frac{f}{f_1}\theta_{i-1}) - Y_i\cot(\frac{f}{f_1}\theta_i), \quad \text{for} \quad i = 1,\cdots,n \tag{5}$$

$$\begin{aligned} b_i(f_r) &= \frac{f_r}{2}\cdot\frac{\partial B_i}{\partial f}\bigg|_{f=f_r} \\ &= \frac{f_r}{2f_1}\cdot\left\{\frac{Y_{i1}^2\theta_{i2}\sec^2(\frac{f_r}{f_1}\theta_{i2})\tan(\frac{f_r}{f_1}\theta_{i1})}{Y_{i2}\tan(\frac{f_r}{f_1}\theta_{i1}) + Y_{i1}\tan(\frac{f_r}{f_1}\theta_{i2})} + \frac{Y_{i1}^2\theta_{i1}\tan(\frac{f_r}{f_1}\theta_{i2})\sec^2(\frac{f_r}{f_1}\theta_{i1})}{Y_{i2}\tan(\frac{f_r}{f_1}\theta_{i1}) + Y_{i1}\tan(\frac{f_r}{f_1}\theta_{i2})}\right. \\ &\quad + \frac{Y_{i1}\left[Y_{i2} - Y_{i1}\tan(\frac{f_r}{f_1}\theta_{i2})\tan(\frac{f_r}{f_1}\theta_{i1})\right]}{\left[Y_{i2}\tan(\frac{f_r}{f_1}\theta_{i1}) + Y_{i1}\tan(\frac{f_r}{f_1}\theta_{i2})\right]^2}\cdot\left[Y_{i2}\theta_{i1}\sec^2(\frac{f_r}{f_1}\theta_{i1}) + Y_{i1}\theta_{i2}\sec^2(\frac{f_r}{f_1}\theta_{i2})\right] \\ &\quad \left. + Y_{i-1}\theta_{i-1}\csc^2(\frac{f_r}{f_1}\theta_{i-1}) + Y_i\theta_i\csc^2(\frac{f_r}{f_1}\theta_i)\right\}, \quad \text{for } i = 1,\cdots,n \text{ and } r = 1 \text{ or } 2 \end{aligned} \tag{6}$$

where $Y_0 = Y_n = 0$.
Consequently, we can obtain the following equations

$$J_{i,i+1} = \Delta_1 \sqrt{\frac{b_1^2}{g_i g_{i+1}}} = \Delta_2 \sqrt{\frac{b_2^2}{g_i g_{i+1}}}, \quad \text{for} \quad i = 1, \cdots, n-1 \tag{7}$$

$$\Delta_1 b_1 = \Delta_2 b_2 \tag{8}$$

where g_i's are the element values of the prototypical lowpass filter.
The procedures of developing the bandpass filter with controllable dual passbands are provided below.

3.1 Formula development

A. Equal bandwidth ($\Delta_1 f_1 = \Delta_2 f_2$)

With $\theta_{i1} = \theta_{i2} = \theta_i = \theta_0$ and R_f as the ratio of f_2 to f_1, the following equations are derived

$$b_2 = R_f b_1 \tag{9}$$

$$\sec(\theta_0) = -\sec(R_f \theta_0) \tag{10}$$

$$Y_i = \frac{g_0 g_1}{R_S \sqrt{g_i g_{i+1}}} \cdot \sin\theta_0 \tag{11}$$

$$Y_{11} = \frac{g_0 g_1}{R_S \Delta_1 \theta_0} - Y_{1t} \tag{12}$$

$$Y_{i2} = Y_{i1} \cdot \left[\tan^2\theta_0 - \frac{R_S \Delta_1 \theta_0 Y_{it} \sec^2\theta_0}{g_0 g_1} \right] \tag{13}$$

$$Y_{it} = Y_{i-1} + Y_i \tag{14}$$

As the fractional bandwidth of the passband is characterized by the 3 dB band-edge frequency of lower and upper bands, a steeper slope and a narrower bandwidth in the passband may be obtained with an increasing order of the filter. As a result, the formula of Δ_1 needs to be modified. With the assistance of statistical inferences, the orders, Δ_1 and R_f of the proposed dual-passband filter should be in the range of 3-8, 1-21% and 1.5-4, respectively. Therefore the modified formula can be expressed as

$$\Delta_{1m} = MF_{a1} + MF_{a2} \tag{15}$$

where

$$MF_{a1} = 0.9 + (R_f - 1.5)\left[0.5 + 0.2(N-3)\right] + \sum_{i=4}^{N}\left[0.25 + 0.08(i-4)\right] \tag{16}$$

$$MF_{a2} = \begin{cases} (\Delta_1 - 1)\left[MF + 0.01(N-1)\right], & 1\% \le \Delta_1 \le 14\% \\ (\Delta_1 - 1)MF + 0.08(N-1), & 15\% \le \Delta_1 \le 20\% \\ (\Delta_1 - 1)MF, & \Delta_1 = 21\% \end{cases} \tag{17}$$

$$MF = \begin{cases} 0.76, & N = 3 \text{ and } R_f = 1.5 \\ 0.76\left[1.29 + A(N-4)\right], & N \ge 4 \text{ and } R_f = 1.5 \\ 0.76\left[1.29 + A(N-4)\right]\left[1.31 + 0.75(R_f - 2)\right], & N \ge 4 \text{ and } R_f > 1.5 \end{cases} \tag{18}$$

$$A = \begin{cases} 0.18, & 3 \le N \le 5 \\ 0.12, & 6 \le N \le 8 \end{cases} \tag{19}$$

B. Unequal bandwidth ($\Delta_1 f_1 \neq \Delta_2 f_2$)

When the bandwidths of two passbands are unequal, the electrical lengths, θ_{i1} and θ_{i2}, are not equal, either. With $\theta_{i1} = \theta_b$ and $\theta_{i2} = \theta_{i-1} = \theta_i = \theta_a$, the following equations are obtained

$$\left(\frac{B}{C}\right)\Delta_1 = \left(\frac{D}{E}\right)R_f\Delta_2 \tag{20}$$

$$Y_i = \frac{g_0 g_1}{R_S\sqrt{g_i g_{i+1}}}\cdot\sin\theta_a \tag{21}$$

$$Y_{i1} = \frac{-Q+\sqrt{Q^2-4PS}}{2P} \tag{22}$$

$$Y_{i2} = \frac{G-H}{K} \tag{23}$$

where

$$\begin{aligned} B = {} & \left[\theta_a \tan\theta_b \sec^2\theta_a + \theta_b \tan\theta_a \sec^2\theta_b\right]\cdot\left[\cot\theta_a \tan\theta_b - \cot(R_f\theta_a)\tan(R_f\theta_b)\right] \\ & + \theta_a \tan\theta_b \csc^2\theta_a \cdot\left[\tan\theta_a \tan\theta_b - \tan(R_f\theta_a)\cdot\tan(R_f\theta_b)\right] + \theta_b \cot\theta_a \sec^2\theta_b \\ & \cdot\left[\tan\theta_a \tan\theta_b + \tan(R_f\theta_a)\tan(R_f\theta_b)\right] \end{aligned} \tag{24}$$

$$C = \tan\theta_a \sec^2\theta_b - \tan(R_f\theta_a)\tan(R_f\theta_b)\cdot\left[\tan\theta_b + \cot\theta_b\right] \tag{25}$$

$$\begin{aligned} D = {} & \theta_a \tan(R_f\theta_b)\sec^2(R_f\theta_a)\cdot\left[\cot\theta_a \tan\theta_b - \cot(R_f\theta_a)\tan(R_f\theta_b)\right] \\ & + \theta_a \tan(R_f\theta_b)\csc^2(R_f\theta_a)\cdot\left[\tan\theta_a \tan\theta_b - \tan(R_f\theta_a)\tan(R_f\theta_b)\right] \\ & + \theta_b \tan\theta_b \sec^2(R_f\theta_b)\cdot\left[\cot\theta_a \tan(R_f\theta_a) - \tan\theta_a \cot(R_f\theta_a)\right] \end{aligned} \tag{26}$$

$$\begin{aligned} E = {} & \tan(R_f\theta_a) - \tan(R_f\theta_a)\tan^2(R_f\theta_b) + \tan\theta_a \tan\theta_b \tan(R_f\theta_b) \\ & - \cot\theta_a \cot\theta_b \tan^2(R_f\theta_a)\tan(R_f\theta_b) \end{aligned} \tag{27}$$

$$G = Y_{i1}^2(\theta_a \tan\theta_b \sec^2\theta_a + \theta_b \tan\theta_a \sec^2\theta_b) \tag{28}$$

$$H = 2Y_{i1}b_1 \tan\theta_a \tag{29}$$

$$K = 2b_1 \tan\theta_b + Y_{it}\theta_b \cot\theta_a \sec^2\theta_b - Y_{it}\theta_a \tan\theta_b \csc^2\theta_a \tag{30}$$

$$P = \theta_a^2 \tan\theta_b \cot\theta_a \sec^4\theta_a + \theta_b^2 \tan\theta_a \cot\theta_b \sec^4\theta_b + 2\theta_a\theta_b \sec^2\theta_a \sec^2\theta_b \tag{31}$$

$$\begin{aligned} Q = 2\Big[& Y_{it}\theta_a^2 \tan^2\theta_b \sec^2\theta_a \csc^2\theta_a - b_1\theta_a \sec^2\theta_a \sec^2\theta_b + Y_{it}\theta_a\theta_b \tan\theta_a \tan\theta_b \csc^2\theta_a \sec^2\theta_b \\ & - b_1\theta_b \tan\theta_a \sec^2\theta_b(\tan\theta_b + \cot\theta_b) \end{aligned} \tag{32}$$

$$S = Y_{it}^2 \cot\theta_a(\theta_b^2 \cot\theta_b \sec^4\theta_b - \theta_a^2 \tan\theta_b \sec^2\theta_a \csc^2\theta_a) \tag{33}$$

Similarly, the orders, Δ_1 and R_f of the proposed dual-passband filter should be in the range of 3-6, 2-15% and 2.17-3.5, respectively. Therefore the formula of Δ_1 needs to be modified as

$$\Delta_{1m} = (MF_{a1} + MF_{a2})\cdot\Delta_1 \tag{34}$$

where

$$MF_{a1} = 1.38 - 0.031(\Delta_1 - 4) + 0.29(N-3) \tag{35}$$

$$MF_{a2} = 2(R_f - 2.17)\left[0.58 - 0.033(\Delta_1 - 4)\right] \tag{36}$$

3.2 Simulation and Experimental Results

Following are design examples of the three-ordered dual-passband bandpass filters with multilayered structure. Below are fabricated examples categorized by two passbands with equal or unequal bandwidths.

A. Two passbands with equal bandwidth ($\Delta_1 f_1 = \Delta_2 f_2$)

Figure 4 shows the three-ordered dual-passband filter. The central frequencies of two passbands are set at 2 and 5.3 GHz. The bandwidth of both passbands is chosen as 260 MHz, which is 13% of the first passband's central frequency. Moreover, the selected ripple for the prototypical Chebyshev lowpass filter is 0.01 dB. As a result, the electrical length θ_0 and the impedances Z_{11}, Z_{12}, Z_{21}, Z_{22} and Z_1 are obtained as 49.3°, 14.59, 14.64, 17.75, 27.71 and 81.9 Ω, respectively. According to these calculated parameters, theoretical predictions of the dual-passband bandpass filter are shown in Fig. 5c. Furthermore, with the assistance of full-wave electromagnetic simulator—Sonnet (Sonnet Software Inc.), these calculated parameters are converted into the multilayered structure.

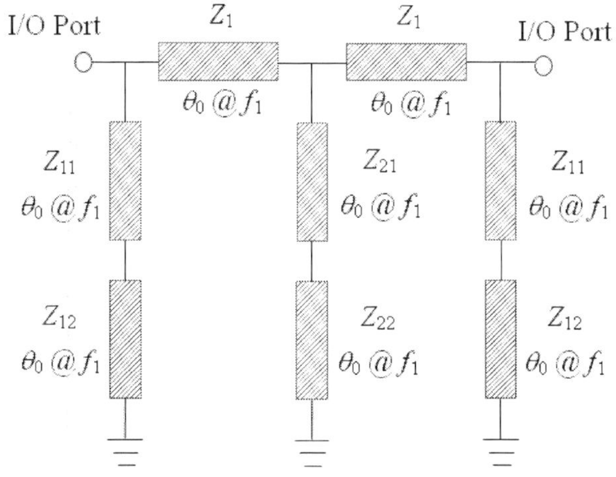

Fig. 4. Architecture of the three-ordered dual-passband filter whose passbands have equal bandwidth

The proposed multilayered dual-passband bandpass filter is fabricated on Dupont 951, whose dielectric constant and loss tangent are 7.8 and 0.0045, respectively. The multilayered 2/5.3 GHz bandpass filter is designed on 11 substrates of 0.09 mm, and its overall size is 4.98 mm × 4.01 mm × 0.99 mm. Figures 5*a* and 5*b* show the 3D architecture and the photograph of this fabricated filter; Fig. 5*c* also presents the measured results.

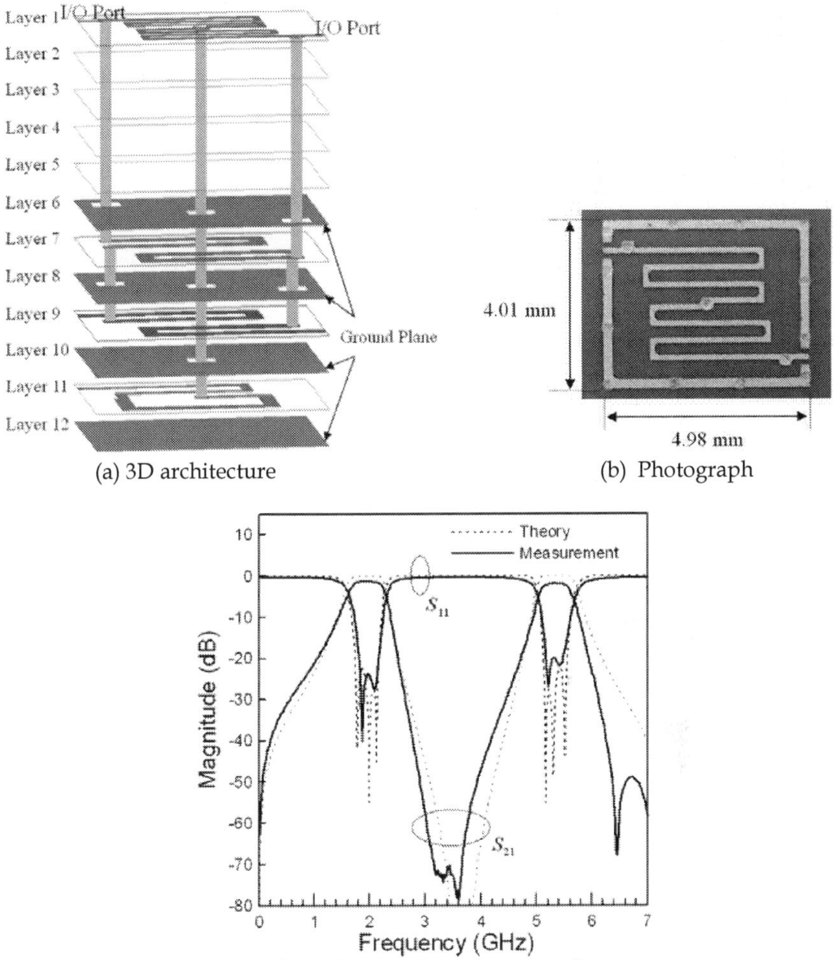

(a) 3D architecture (b) Photograph

(c) Responses of the theoretical prediction and measurement

Fig. 5. Three-ordered 2/5.3 GHz dual-passband bandpass filter whose passbands have equal bandwidth

On the one hand, within the first passband (1.8-2.2 GHz), the measured insertion loss is < 1.6 dB, whereas the return loss is > 18 dB. On the other hand, within the second passband (5.16-5.51 GHz), the measured insertion loss is < 2.2 dB, whereas the return loss is also > 18 dB.

B. First passband with greater bandwidth ($\Delta_1 f_1 > \Delta_2 f_2$)

Figure 6 shows the architecture of the three-ordered dual-passband filter. The central frequencies of two passbands are set at 2 and 5.3 GHz. The bandwidths of the first and second passband are chosen as 300 and 200 MHz, respectively, which are 15% and 10% of the first passband's central frequency. Moreover, the selected ripple for the prototypical

Chebyshev lowpass filter is 0.01 dB. With the electrical length θ_a as 35.9°, the electrical length θ_b and the impedances Z_{11}, Z_{12}, Z_{21}, Z_{22} and Z_1 are then obtained as 64.3°, 14.17, 13.44, 18.64, 28.34 and 88.24 Ω, respectively. According to these calculated parameters, theoretical predictions of the dual-passband bandpass filter are shown in Fig. 7c.

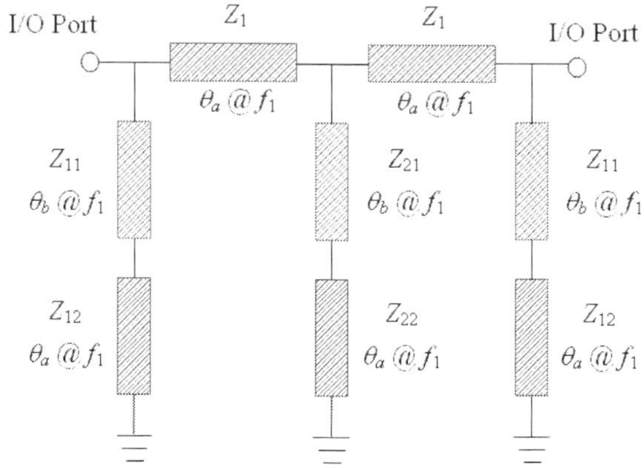

Fig. 6. Architecture of the three-ordered dual-passband filter whose passbands have unequal bandwidths

The multilayered 2/5.3 GHz bandpass filter is fabricated on 12 substrates of 0.09 mm, and its overall size is 4.98 mm × 4.06 mm × 1.08 mm. Figures 7*a* and 7*b* show the 3D architecture and the photograph of this fabricated filter; Fig. 7*c* also presents the measured results.

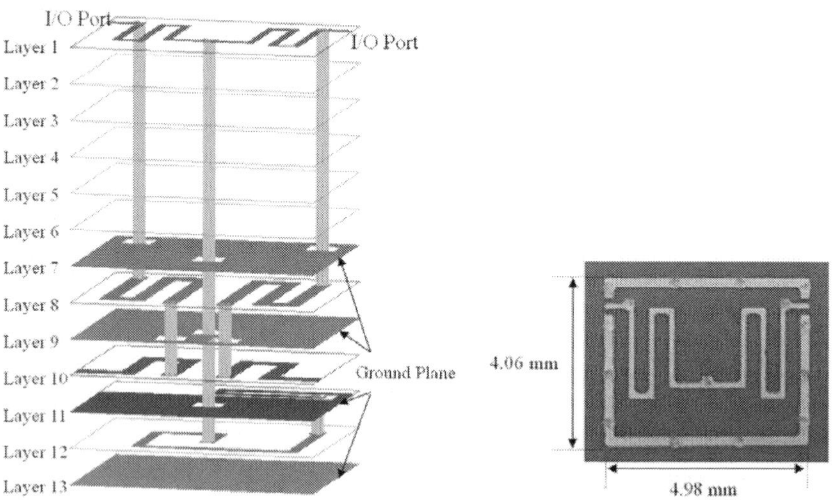

(a) 3D architecture (b) Photograph

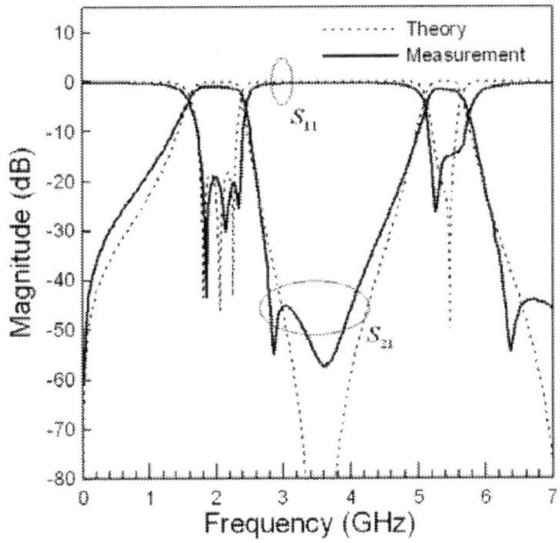

(c) Responses of the theoretical prediction and measurement

Fig. 7. Three-ordered 2/5.3 GHz dual-passband bandpass filter whose first passband's bandwidth is greater

On the one hand, within the first passband (1.79-2.35 GHz), the measured insertion loss is < 1.6 dB, whereas the return loss is > 20 dB. On the other hand, within the second passband (5.2-5.5 GHz), the measured insertion loss is < 2.1 dB, whereas the return loss is > 15 dB.

C. First passband with smaller bandwidth ($\Delta_1 f_1 < \Delta_2 f_2$)

Figure 6 also shows the architecture of the three-ordered dual-passband filter. The central frequencies of two passbands remain as 2 and 5.3 GHz. The bandwidths of the first and second passband are chosen as 200 and 300 MHz, respectively, which are 10% and 15% of first passband's central frequency. Moreover, the selected ripple for the prototypical Chebyshev lowpass filter is 0.01 dB. With the electrical length θ_a as 51°, the electrical length

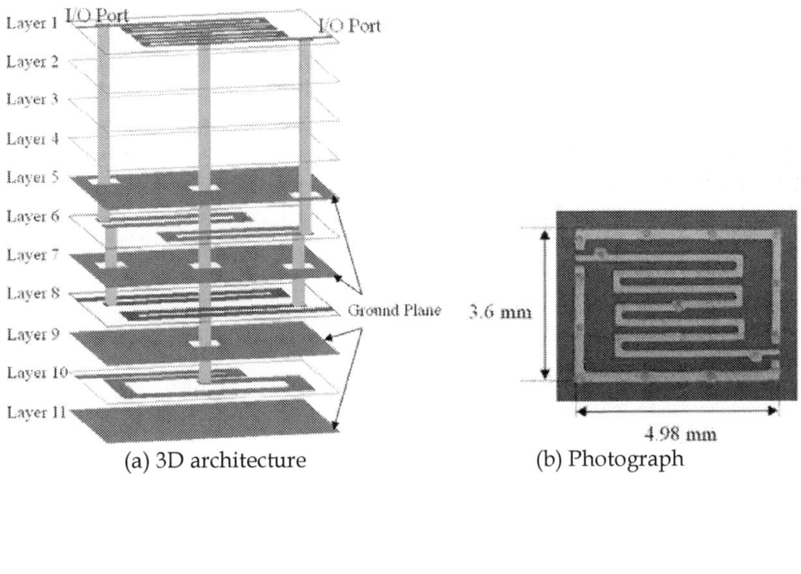

(a) 3D architecture (b) Photograph

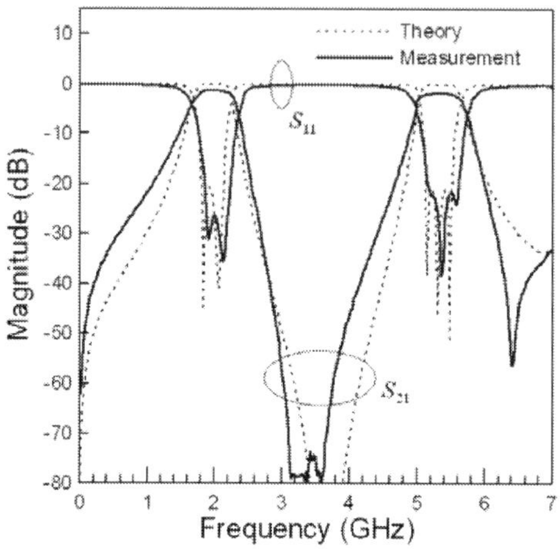

(c) Responses of the theoretical prediction and measurement

Fig. 8. Three-ordered 2/5.3 GHz dual-passband bandpass filter whose first passband's bandwidth is smaller

θ_b and the impedances Z_{11}, Z_{12}, Z_{21}, Z_{22} and Z_1 are then obtained as 47.1°, 11.89, 11.2, 11.28, 13.69 and 74.59 Ω, respectively. According to these calculated parameters, theoretical predictions of the dual-passband bandpass filter are shown in Fig. 8*c*.

The multilayered 2/5.3 GHz bandpass filter is fabricated on 10 substrates of 0.09 mm, and its overall size is 4.98 mm × 3.6 mm × 0.9 mm. Figures 8*a* and 8*b* show the 3D architecture and the photograph of this fabricated filter; Fig. 8*c* also presents the measured results.

On the one hand, within the first passband (1.85-2.22 GHz), the measured insertion loss is < 1.8 dB, whereas the return loss is > 20 dB. On the other hand, within the second passband (5.14-5.62 GHz), the measured insertion loss is < 2.5 dB, whereas the return loss is > 18 dB.

4. Multi-passband filter fabrication

The multi-passband filter can be obtained from (1) to (4). The fabricated examples of triple- and quadruple-passband filters are introduced below.

4.1 Triple-passband filter

Figure 9 shows the architecture of the three-ordered triple-passband filter. The central frequencies of three passbands are set as 2, 5 and 8 GHz. The bandwidths of the first, second and third passband are chosen as 200, 300 and 200 MHz, respectively, which are 10%, 15% and 10% of the first passband's central frequency. Moreover, the selected ripple for the prototypical Chebyshev lowpass filter is 0.01 dB. With the electrical length θ_a as 35.5°, the electrical length θ_b and the impedances Z_{11}, Z_{12}, Z_{13}, Z_{21}, Z_{22}, Z_{23} and Z_1 are then obtained as 37.2°, 20.3, 17.51, 20.34, 29.89, 13.21, 29.84 and 67.7 Ω, respectively. According to these calculated parameters, theoretical predictions of the triple-passband bandpass filter are shown in Fig. 10*c*.

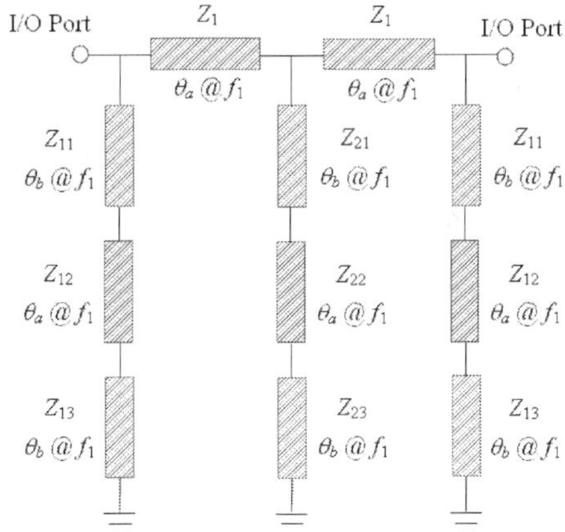

Fig. 9. Architecture of the three-ordered triple-passband filter whose passbands have unequal bandwidths

The multilayered 2/5/8 GHz bandpass filter is fabricated on 10 substrates of 0.09 mm, and its overall size is 4.72 mm × 3.7 mm × 0.9 mm. Figures 10*a* and 10*b* show the 3D architecture and the photograph of this fabricated filter; Fig. 10*c* also presents the measured results.

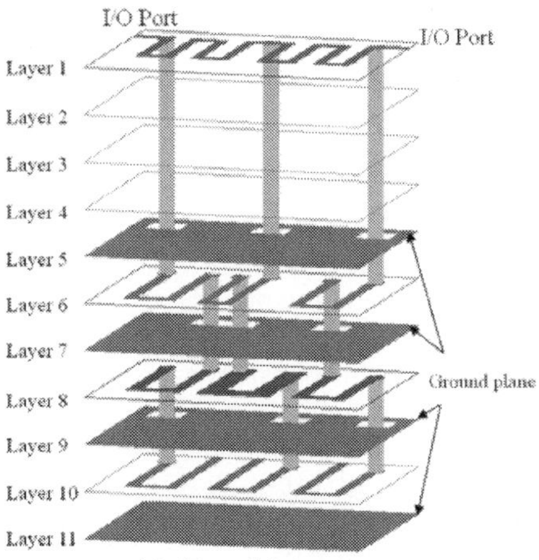

(a) 3D architecture

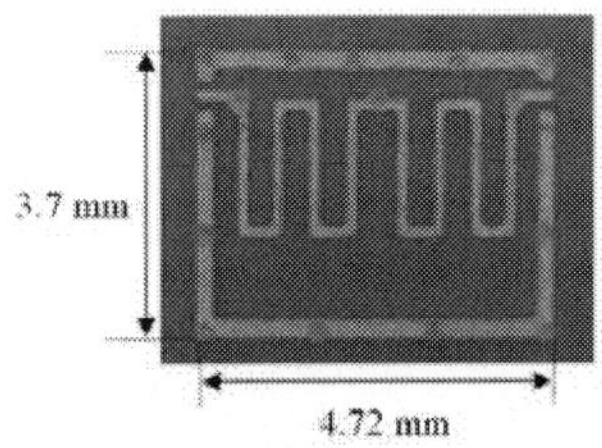

(b) Photograph

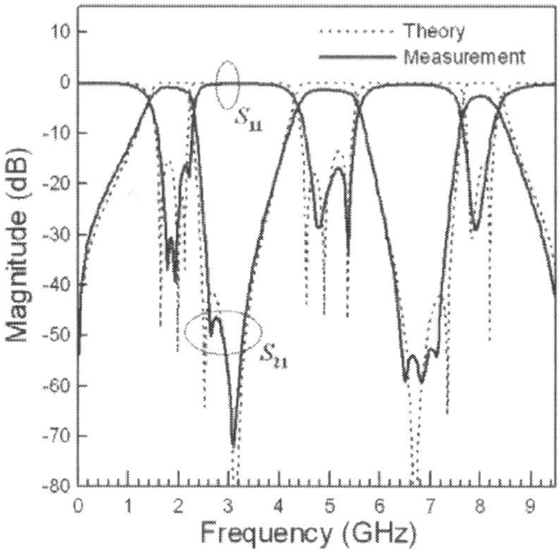

(c) Responses of the theoretical prediction and measurement

Fig. 10. Three-ordered 2/5/8 GHz triple-passband bandpass filter whose second passband's bandwidth is greater

Within the first passband (1.66-2.11 GHz), the measured insertion loss is < 1.4 dB, whereas the return loss is > 16.8 dB. Moreover, within the second passband (4.6-5.35 GHz), the measured insertion loss is < 1.8 dB, whereas the return loss is > 16 dB. Furthermore, within the third passband (7.84-8.14 GHz), the measured insertion loss is < 3 dB, whereas the return loss is > 15 dB.

4.2 Quadruple-passband filter

Figure 11 shows the architecture of the three-ordered quadruple-passband filter. The central frequencies of four passbands are set as 2, 5, 8 and 11.3 GHz. The bandwidth of four passbands is all chosen as 180 MHz, which is 9% of the first passband's central frequency. Moreover, the selected ripple for the prototypical Chebyshev lowpass filter is 0.01 dB. As a result, the electrical length θ_0 and the impedances Z_{11}, Z_{12}, Z_{13}, Z_{14}, Z_{21}, Z_{22}, Z_{23}, Z_{24} and Z_1 are obtained as 26.8°, 21.79, 19.61, 22.74, 14.54, 27.49, 16.01, 14.04, 19.84 and 71.5 Ω, respectively. According to these calculated parameters, theoretical predictions of the quadruple-passband bandpass filter are shown in Fig. 12c.

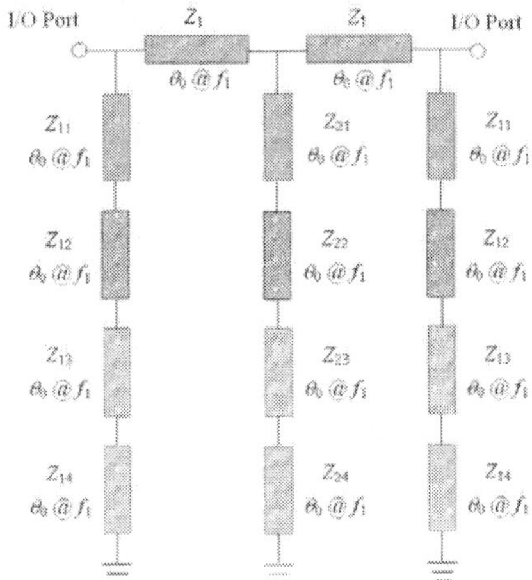

Fig. 11. Architecture of the three-ordered quadruple-passband filter whose passbands have equal bandwidth

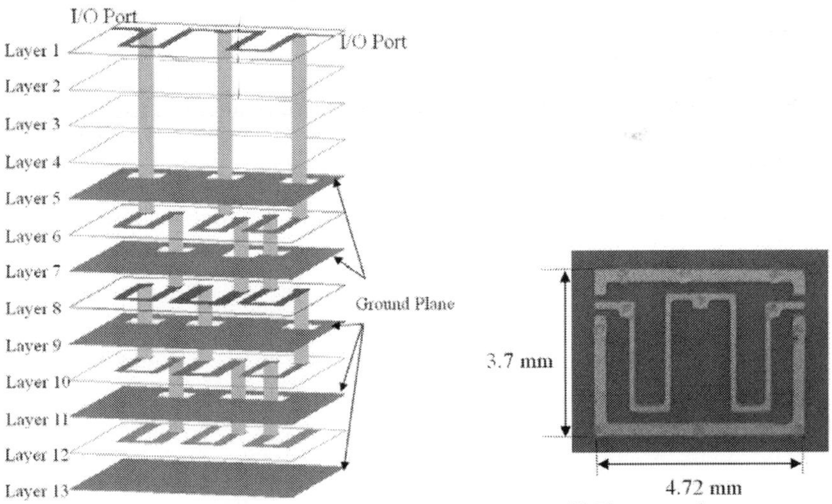

(a) 3D architecture (b) Photograph

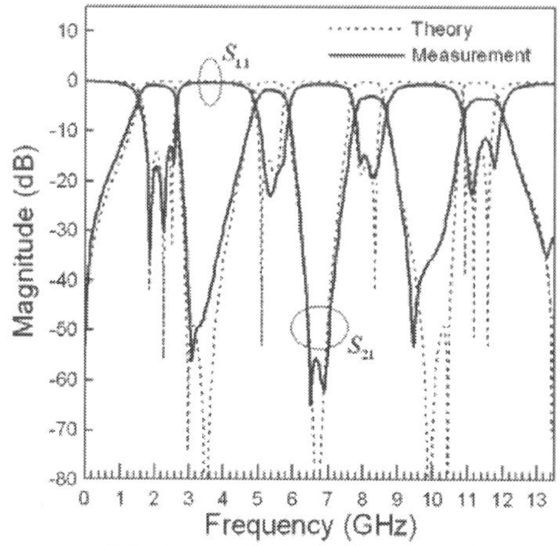

(c) Responses of the theoretical prediction and measurement

Fig. 12. Three-ordered 2/5/8/11.3 GHz quadruple-passband bandpass filter

The multilayered 2/5/8/11.3 GHz bandpass filter is fabricated on 12 substrates of 0.09 mm, and its overall size is 4.72 mm × 3.7 mm × 1.08 mm. Figures 12*a* and 12*b* show the 3D architecture and the photograph of this fabricated filter; Fig. 12*c* also presents the measured results.

Within the first passband (1.77-2.3 GHz), the measured insertion loss is < 1 dB, whereas the return loss is > 17 dB. Moreover, within the second passband (5.14-5.48 GHz), the measured insertion loss is < 1.8 dB, whereas the return loss is > 16 dB. In addition, within the third passband (8.06-8.4 GHz), the measured insertion loss is < 3 dB, whereas the return loss is > 15 dB. Furthermore, within the fourth passband (11.05-11.85 GHz), the measured insertion loss is < 4 dB, whereas the return loss is > 12.5 dB.

5. Conclusion

A novel structure of the multi-passband bandpass filters has been proposed in this article. Multi-sectional short-circuit transmission lines shunted with transmission lines are utilized. By properly choosing the proposed structure's electrical lengths and impedances, multiple passbands can be easily controlled.

In addition, the design procedures have been described in detail, and 3D architectures are provided. Because of the parasitic effect among capacitors for a physical 3D circuit, there is some slight difference between the theoretical predictions and measured results. However, generally speaking, the measured results match well with the theoretical predictions. Therefore with high integration and compact size, the proposed multi-passband filter is suitable for implementation in a multi-chip module.

By the way, because insertion loss measurement is related to the degree and bandwidth of a filter, the performance of a filter with resonators can be well indicated by translating the measurement into an unloaded Q of a resonator.

6. References

[1] Bell, H. C. (2001). Zolotarev bandpass filters, *IEEE Trans. Microw. Theory Tech.*, Vol. 49, No. 12, pp. 2357-2362

[2] Miyake, H.; Kitazawa, S.; Ishizaki, T.; Yamada, T. & Nagatomi, Y. (1997). A miniaturized monolithic dual band filter using ceramic lamination technique for dual mode portable telephones, *Proceedings of IEEE MTT-S Int. Microw. Symp. Dig.*, pp. 789-792

[3] Quendo, C., Rius, E. & Person, C. (2003). An original topology of dual-band filter with transmission zeros, *Proceedings of IEEE MTT-S Int. Microw. Symp. Dig.*, pp. 1093-1096

[4] Chang, C. H.; Wu, H. S.; Yang, H. J. & Tzuang, C. K. C. (2003). Coalesced single-input single-output dual-band filter, *Proceedings of IEEE MTT-S Int. Microw. Symp. Dig.*, pp. 511-514

[5] Tsai, C. M.; Lee, H. M. & Tsai, C. C. (2005). Planar filter design with fully controllable second passband, *IEEE Trans. Microw. Theory Tech.*, Vol. 53, No. 11, pp. 3429-3439

[6] Yim, H. Y. & Cheng, K. K. M. (2005). Novel dual-band planar resonator and admittance inverter for filter design and applications, *Proceedings of IEEE MTT-S Int. Microw. Symp. Dig.*, pp. 2187-2190

[7] Lee, J.; Uhm, M. S. & Yom, I. B. (2004). A dual-passband filter of canonical structure for satellite applications, *IEEE Microw. Wireless Comp. Lett.*, Vol. 14, No. 6, pp. 271-273

[8] Kuo, J. T.; Yeh, T. H. & Yeh, C. C. (2005). Design of microstrip bandpass filters with a dual-passband response, *IEEE Trans. Microw. Theory Tech.*, Vol. 53, No. 4, pp. 1331-1337

[9] Chen, C. C. (2005). Dual-band bandpass filter using coupled resonator pairs, *IEEE Microw. Wireless Comp. Lett.*, Vol. 15, No. 4, pp. 259-261

[10] Rong, Y.; Zaki, K. A.; Hageman, M.; Stevens, D. & Gipprich, J. (1999). Low temperature cofired ceramic (LTCC) ridge waveguide bandpass filters. *Proceedings of IEEE MTT-S Int. Microw. Symp. Dig.*, pp. 1147-1150

[11] Heo, D.; Sutono, A.; Chen, E.; Suh, Y. & Laskar, J. (2001). A 1.9 GHz DECT CMOS power amplifier with fully integrated multilayer LTCC passives, *IEEE Microw. Wireless Comp. Lett.*, Vol. 11, No. 6, pp. 249-251

[12] Leung, W. Y.; Cheng, K. K. M. & Wu, K. L. (2001). Design and implementation of LTCC filters with enhanced stop-band characteristics for bluetooth applications. *Proceedings of Asia-Pacific Microw. Conf.*, pp. 1008-1011

[13] Tang, C. W.; Sheen, J. W. & Chang, C. Y. (2001). Chip-type LTCC-MLC baluns using the stepped impedance method, *IEEE Trans. Microw. Theory Tech.*, Vol. 49, No. 12, pp. 2342-2349

[14] Tang, C. W.; Lin, Y. C. & Chang, C. Y. (2003). Realization of transmission zeros in combline filters using an auxiliary inductively-coupled ground plane, *IEEE Trans. Microw. Theory Tech.*, Vol. 51, No. 10, pp. 2112-2118

[15] Tang, C. W. (2004). Harmonic-suppression LTCC filter with the step impedance quarter-wavelength open stub, *IEEE Trans. Microw. Theory Tech.*, Vol. 52, No. 2, pp. 617-624

[16] Tang, C. W.; You, S. F. & Liu, I. C. (2006). Design of a dual-band bandpass filter with low temperature co-fired ceramic technology, *IEEE Trans. Microw. Theory Tech.*, Vol. 54, No. 8, pp. 3327-3332

[17] Sun, S. & Zhu, L. (2006). Novel design of microstrip bandpass filters with a controllable dual-passband response: description and implementation, *IEICE Trans. Electron.*, Vol. E89-C, No. 2, pp.197-202

[18] Quendo, C.; Manchec, A.; Clavet, Y.; Rius, E.; Favennec, J. F. & Person, C. (2007). General Synthesis of N-Band Resonator Based on N-Order Dual Behavior Resonator, *IEEE Microw. Wireless Comp. Lett.*, Vol. 17, No. 5, pp. 337-339

[19] Matthaei, G.L.; Young, L. & Jones, E.M. (1980). *Microwave filters, impedance-matching network, and coupling structures*, Artech House, Norwood, MA

18

The Switched Mode Power Amplifiers

Elisa Cipriani, Paolo Colantonio, Franco Giannini and Rocco Giofrè
University of Roma Tor Vergata
Italy

1. Introduction

The power amplifier (PA) is a key element in transmitter systems, aimed to increase the power level of the signal at its input up to a predefined level required for the transmission purposes. The PA's features are mainly related to the absolute output power levels achievable, together with highest efficiency and linearity behaviour.

From the energetic point of view a PA acts as a device converting supplied dc power (P_{dc}) into microwave power (P_{out}). Therefore, it is obvious that highest efficiency levels become mandatory to reduce such dc power consumption. On the other hand, a linear behaviour is clearly necessary to avoid the corruption of the transmitted signal information. Unfortunately, efficiency and linearity are contrasting requirements, forcing the designer to a suitable trade-off.

In general, the design of a PA is related to the operating frequency and application requirements, as well as to the available device technology, often resulting in an exciting challenge for PA designers, since not an unique approach is available.

In fact, PAs are employed in a broad range of systems, whose differences are typically reflected back into the technologies adopted for PAs active modules realisation. Moreover, from the designer perspective, to improve PAs efficiency the active devices employed are usually driven into saturation. It implies that a PA has to be considered a non-linear system component, thus requiring dedicated non linear design methodologies to attain the highest available performance.

Nevertheless, for high frequency applications it is possible to identify two main classes of PA design methodologies: the trans-conductance based amplifiers with Harmonic Tuning terminations (HT) (Colantonio et al., 2009) or the Switching-Mode (SM) amplifiers (Grebennikov & Sokal, 2007; Krauss et al., 1980). In the former, the active device acts as a nonlinear current source controlled by the input signal (voltage or current for FET or BJT devices respectively). A simplest schematic view of such an amplifier for FET is reported in Fig. 1a. Under this assumption, the high efficiency condition is achieved exploiting the device nonlinear behaviour through a suitable selection of both input and output harmonic terminations. More in general, the trans-conductance based amplifiers are identified also as Class A, AB, B to C considering the quiescent active device bias points, resulting in different output current conduction angles from 2π to 0 respectively.

The most famous solution of HT PA is the Class F approach (Gao, 2006; Colantonio et al, 2009), while for high frequency applications and taking into account practical limitations on

the control of harmonic impedances, several solutions have been successfully proposed (Colantonio et al., 2003).
Conversely, in the SM PA, the active device is driven by a very large input signal to act as a ON/OFF switch with the aim to maximise the conversion efficiency reducing the power dissipated in the active devices also. A schematic representation of a SM amplifier is depicted in Fig. 1b. When the active device is turned on, the voltage across its terminals is close to zero and high current is flowing through it. Therefore, in this part of the period the transistor acts as a very low resistance, ideally short circuit (switch closed) minimising the overlap between the current and voltage waveforms. In the other part of the period, the active device is turned off acting as an open circuit. Therefore, the current is theoretically zero while high voltage is present at the device terminals, once again minimising the overlap between voltage and current waveforms. If the active device shows a zero on resistance and an infinite off resistance, a 100% efficiency is theoretically achieved. The latter is of course an advantage over Class A or B, where the maximum theoretical efficiencies are 50 % and 78 % respectively. On the other hand, Class C could achieve high efficiency levels, despite a significant reduction in the maximum output power level achievable (theoretically 100% of efficiency for zero output power). Nevertheless, the HT PAs are intrinsically able to amplify the input signal with higher fidelity, since the active device is basically represented by a controlled current source (FET case) whose output current is directly related to the input voltage. Instead, in SM PAs the active device is assumed to be ideally driven in the ON and OFF states, thus exhibiting a higher nonlinear behaviour. However, this characteristic does not represent a trouble when signals with constant-envelope modulation are adopted.
On the basis of their operating principle, SM amplifiers are often considered as DC to RF converter rather than RF amplifiers.

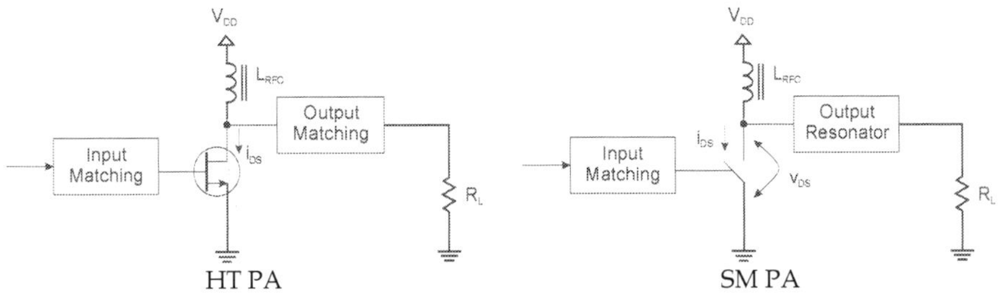

Fig. 1. Simplified view of a simple single ended HT (left) and SM (right) PA.

Different SM PA classes of operation have been proposed over the years, namely Class D, S, J, F^{-1} (Cripps, 2002; Kazimierczuk, 2008), while the most famous and adopted is the Class E PA (Sokal & Sokal, 1975; Sokal, 2001) that will be described in deep detail in the following. As will be shown, these classes are based on the same operating principle while their main differences are related to their circuit implementation and current-voltage wave shaping only.
The applications of SM PAs principles have initially been limited to amplifiers at lower frequencies in the megahertz range, due to the active device and package parasitics practically limiting the operating frequencies (Kazimierczuk, 2008). They have also been

applied to DC/DC power converters that also operate at lower switch frequencies (Jozwik & Kazimierczuk, 1990; Kazimierczuk & Jozwik, 1990). Recently, their principles of operation have been extended and applied to RF and microwave amplifier design, made possible by the high-performance active devices nowadays available based on silicon (Si), gallium arsenide (GaAs), silicon germanium (SiGe), silicon carbide (SiC), and gallium nitride (GaN) technologies (Lai, 2009).

2. Switching mode generic operating principle

The operating principle of every SM PA is based on the idea that the active device operates in saturation, thus it can be represented as a switch and either voltage or current waveforms across it are alternatively minimized to reduce overlap, so minimizing power dissipation in the device itself. If the transistor is an ideal switch, a 100% of efficiency can be achieved by the proper design of the output matching network. As reported in Fig. 1b, the output resonator can be assumed, in the simplest way, as an ideal *L-C* series resonator at fundamental frequency, terminated on a series load resistance (R_L). The role of the resonator is to shape the voltage and current waveforms across the switch in order to avoid power dissipation at higher harmonics. In fact, an ideal *L-C* series resonator shows zero impedance at resonating frequency ($\omega_0=(LC)^{-1/2}$) and infinite impedance for every $\omega\neq\omega_0$. It follows that fundamental current only is flowing into the output load and fundamental voltage only is generated at its terminals. Consequently, 100% of efficiency is obtained (being zeroed the overlap between voltage and current waveforms over the transistor, thus being nulled the power dissipated in it) and no power is delivered at harmonic frequencies in the load, being the latter not allowed to flow into the load R_L.
In actual cases, several losses mechanisms, such as ohmic and capacitive discharge or leakage, cause an unavoidable overlapping between the voltage and current waveforms, together with power dissipation at higher harmonics, thus limiting the maximum achievable efficiency levels.
The most relevant losses in SM PA are represented by:

- parasitic capacitors, such as the device drain to source capacitance C_{ds}. The presence of such capacitance causes a low pass filter behaviour at the output of the active device, affecting the voltage wave shaping with a consequent degradation in the attainable power and efficiency levels. In fact, considering the active device as the parallel connection of a perfect switch and the parasitic capacitance C_{ds}, the higher voltage harmonics are practically shorted by C_{ds} and only few harmonics can be reasonably controlled by the loading network.
- parasitic resistance, such as the drain-to-source resistance when the transistor is conducting R_{ON} (ON state). In fact, due to the non zero resistance when the switch is closed, a relevant amount of active power will be dissipated in the transistor causing a lowering in the achievable efficiency.
- non-zero transition time, due to the presence of parasitic effects, which increase the voltage and current overlap.
- implementation losses due to the components (distributed or lumped elements) employed to realise the required input and output matching networks.

The entity of the parasitic components as well as the associated losses are strictly related to the characteristics of the active device used, especially when designing RF PA (Kazimierczuk, 2008; Lai, 2008).

3. The Class E Amplifier

Firstly presented in the early 70's in (Sokal & Sokal, 1975), the Class E power amplifier recently received more attention by microwave engineers with the growing demand of high efficient transmitters in wireless communication systems.

It has been widely adopted in constant envelope based communication systems, but represents a valid alternative if combined with envelope varying technique also, like envelope elimination and restoration or Chireix's outphasing technique (Cripps, 2002).

A complete analysis of the Class E amplifier is herein presented, making the assumption of a very idealized active device switching action. The topology considered is the most common one, firstly presented in (Sokal & Sokal, 1975), although different Class E topologies have been conceived and studied during the past (Mader et al., 1998; Grebennikov, 2003; Suetsugu & Kazimierczuk, 2005). In order to clarify Class E operation, a real device-based design is also briefly presented.

3.1 Analysis with a generic duty cycle

The basic topology of a single ended Class E power amplifier is depicted in Fig. 2. The active device is schematically represented as an ideal switch and it is shunted by the capacitance C_1, which include the output equivalent capacitance of the active device also. The output network is composed by an ideal filter C_0-L_0 with a series R-L impedance.

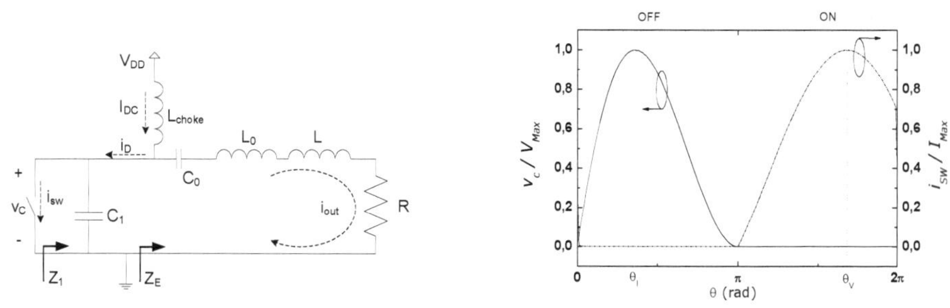

(a) (b)

Fig. 2. Basic topology of a Class E amplifier (a) corresponding ideal waveforms (b).

Such a circuit is usually analyzed in time domain, which is a straightforward but tedious process, requiring the solution of non linear differential equations. Anyway, some hypotheses can be adopted to carry out a simplified analysis useful to understand the underling operating principle.

Considering the series resonator C_0-L_0 to behave as an ideal filter, i.e. with an infinite (or high enough) Q factor, harmonics and all frequency components different from the fundamental frequency can be considered as filtered out and do not play any role in the solution of the system. As a consequence, the current flowing into the output branch of the circuit can be assumed as a pure sinusoidal, with its own amplitude I_M and its phase ϕ (Raab, 2001):

$$i_{out} = I_M \cdot \sin(\theta + \phi) \tag{1}$$

Where $\theta = \omega \cdot t$.

Consequently, from Kirchhoff laws the current i_D (see Fig. 2), which flows entirely through the switch during the ON period (i_{SW}) or entirely through the capacitance C_1 during the OFF period, can be written as:

$$i_D = I_{DC} - I_M \cdot \sin(\theta + \phi) \tag{2}$$

Assuming for simplicity a 50% of duty cycle (the analysis for a generic duty cycle is available in (Suetsugu & Kazimierczuk, 2007)), the current flowing into the switch i_{SW} can be expressed as:

$$i_{sw}(\theta) = \begin{cases} 0, & 0 \le \theta \le \pi \\ I_{DC} - I_M \cdot \sin(\theta + \phi), & \pi \le \theta \le 2\pi \end{cases} \tag{3}$$

And analogously the current in the capacitor C_1 becomes:

$$i_C(\theta) = \begin{cases} I_{DC} - I_M \cdot \sin(\theta + \phi), & 0 \le \theta \le \pi \\ 0, & \pi \le \theta \le 2\pi \end{cases} \tag{4}$$

While the voltage across the capacitance v_C can be easily inferred by integration of (4), resulting in the following expression:

$$v_C(\theta) = \begin{cases} \dfrac{1}{\omega \cdot C_1} \cdot \left(I_{DC}\theta + I_M \cdot \cos(\theta + \phi) - I_M \cdot \cos(\phi) \right), & 0 \le \theta \le \pi \\ 0, & \pi \le \theta \le 2\pi \end{cases} \tag{5}$$

The resulting theoretical current and voltage waveforms are depicted in Fig. 2b.

It can be noted that current and voltage across the switch do not overlap, thus no power dissipation exists on the active device. The unique dissipative element in the circuit is the loading resistance R, which is active at fundamental frequency only. Then, from these assumption it follows that the DC to RF power conversion happens without losses and the theoretical efficiency is 100%.

The quantities I_{DC}, I_M and ϕ have still to be determined as functions of maximum current and voltage allowed by the adopted active device, I_{Max} and V_{Max} respectively, and of operating angular frequency ω.

For this purpose, it has to note that the capacitance C_1 should be completely discharged at the switching turn on, which implies that the voltage v_C has to be null in correspondence of the instant π (see Fig. 2b):

$$v_C(\theta)\big|_{\theta=\pi} = 0 \tag{6}$$

Such condition is usually referred as Zero Voltage Switching (ZVS) condition, which implies that the capacitance C_1 should not be short circuited by the switch turn on when its voltage is still high (Sokal & Sokal, 1975).
The second condition, namely Zero Voltage Derivative Switching (ZVDS) condition, or soft-switching condition, implies that the current starts to flow from zero after the switch turn on and then increases gradually, in order to prevent worsening in circuit performance due to mistuning of the waveforms (Sokal & Sokal, 1975). This condition is written as:

$$i_C(\theta)\big|_{\theta=\pi} = \frac{d}{d\theta} v_C(\theta)\bigg|_{\theta=\pi} = 0 \tag{7}$$

Substituting (4) in the previous equations, from (6) it follows:

$$I_{DC} \cdot \pi - 2 \cdot I_M \cos(\phi) = 0 \tag{8}$$

While from (7) it follows:

$$I_{DC} + I_M \cdot \sin(\phi) = 0 \tag{9}$$

Thus the following relationships can be inferred:

$$\tan(\phi) = -\frac{2}{\pi} \tag{10}$$

$$\frac{I_{DC}}{I_M} = -\sin(\phi) = \frac{2}{\pi} \cdot \cos(\phi) \tag{11}$$

The maximum current flowing into the switch is given by:

$$I_{Max} = I_{DC} + I_M \tag{12}$$

And it occurs in correspondence of the angle

$$\theta_I = \frac{3}{2} \cdot \pi - \phi \tag{13}$$

Similarly, for the voltage across the switch its maximum value occurs in correspondence of the angle θ_V (see Fig. 2b), which can be inferred nulling the derivate of v_c given by (5).
Thus, accounting for (11), it follows:

$$\theta_V = -2 \cdot \phi \tag{14}$$

and

$$V_{Max} = -2 \cdot \phi \cdot \frac{I_{DC}}{\omega C_1} \tag{15}$$

However, the value of the capacitance C_1 is still an unknown variable. It appears in the definition of the voltage waveform, and it is convenient to use voltage constraints in order to obtain its expression. In fact, its average value must be equal to the supplied DC voltage V_{DD}; thus it follows:

$$V_{DD} = \frac{1}{2\pi} \int_0^{\pi} v_C(\theta) d\theta \tag{16}$$

Which solved lead to:

$$V_{DD} = \frac{1}{2\pi} \cdot \frac{1}{\omega \cdot C_1} \cdot \left(I_{DC} \cdot \frac{\pi^2}{2} + 2 \cdot I_M \sin(\phi) - I_M \cdot \pi \cdot \cos(\phi) \right) \tag{17}$$

from which the value of C_1 can be finally determined remembering (11)

$$C_1 = \frac{I_{DC}}{\pi \cdot \omega \cdot V_{DD}} \tag{18}$$

This also suggests a simple relationship between DC current and bias voltage.
At this point, waveforms in Fig. 2b have been completely determined in the time domain, without recurring to the frequency domain. However, the remaining elements of the circuit, DC power, output power and output impedance have still to be determined.
As stated before, all the DC power is converted to RF power and dissipated into the load resistance at fundamental frequency:

$$P_{DC} = I_{DC} \cdot V_{DD} = \frac{1}{2} \cdot I_M \cdot V_M = P_{RF} \tag{19}$$

where V_M is the amplitude of fundamental component of the voltage across R which can be obtained by (19) and replacing (11):

$$V_M = 2 \cdot \frac{V_{DD} \cdot I_{DC}}{I_M} = -2 \cdot V_{DD} \cdot \sin(\phi) \tag{20}$$

The value of the resistance R is simply obtained as the ratio between V_M and I_M:

$$R = \frac{V_M}{I_M} = 2 \cdot \frac{V_{DD}}{I_{DC}} \cdot \sin(\phi)^2 \tag{21}$$

Clearly, if a standard 50 Ohm termination is required, an impedance transformer is necessary to adapt such load to the required R value.
Finally, the inductance L is computed taking into account that its reactive energy is exchanged, at every cycle, with the capacitance C_1. Thus it follows:

$$\frac{1}{2\cdot\pi}\cdot\frac{1}{\omega C_1}\int_0^{\pi}\left[I_{DC}-I_M\cdot\sin(\theta+\phi)\right]^2 d\theta=\frac{1}{2}\cdot\omega\cdot L\cdot I_M{}^2 \tag{22}$$

where the expression in the integral represents the voltage across the capacitance C_1 during the OFF period. The value for the inductance L is therefore given by:

$$L=\frac{1}{\omega\cdot C_1}\cdot\left(\frac{\pi}{2}-\frac{4}{\pi}\cdot\cos(\phi)^2\right) \tag{23}$$

Alternatively, R and L can be found by calculation off in-phase and quadrature voltage components, as elsewhere reported (Mader et al., 1998; Cripps, 1999).
The series impedance R-L can be put together in order to obtain a more compact and useful expression for the output branch impedance (Mader et al., 1998) normalized to the shunt capacitance C_1:

$$Z_E=\frac{0.28}{\omega C_1}\cdot e^{j49°} \tag{24}$$

With reference to Fig. 2, the impedance Z_1 seen by the ideal switch is obtained by the shunt connection of the capacitance C_1 and Z_E and is herein given in its simplified formulation (Colantonio et al., 2005):

$$Z_1=\frac{0.35}{\omega C_1}\cdot e^{j36°} \tag{25}$$

Remaining reactive components, L_0 and C_0, are easily calculated by means of:

$$\omega^2=\frac{1}{L_0\cdot C_0} \tag{26}$$

Provided a high enough Q factor, the values of L_0 and C_0 are non uniquely defined and any pair of resonant element can be used.
The analysis performed here was intended for the most common case of 50% duty cycle (i.e. π conduction angle). In this case the relations are greatly simplified thanks to the properties of trigonometric functions. However, Class E approach is possible for any value of duty cycle: a detailed analysis can be found in (Suetsugu & Kazimierczuk, 2007; Colantonio et al., 2009) where all electrical properties and component values are evaluated as a function of duty cycle. It can be demonstrated that under ideal assumption the maximum output power does not occur in correspondence of a 0.5 duty cycle, but for a slightly higher value (0.511). Anyway, in terms of output power capability, this increment is extremely low (about 1‰) and a standard 0.5 duty cycle could be assumed in the design, unless differently required.

3.2 A Class E design example

In order to illustrate the application of the relations obtained in the analysis, a simple Class E design example is described, based on an actual active device, specifically a GaAs pHEMT.

The device exhibits a breakdown voltage of about 25 V and a maximum output current of 400 mA. From S-parameter simulation, an output capacitance of 0.35pF results at 2.5 GHz, the selected operating frequency. Considering this capacitance as the minimum value for the shunt capacitance C_1, the network elements can be easily calculated through the previous relationships.
From (20) and taking into account the maximum voltage, the bias voltage is set to V_{DD}=6V. Hence, from the inversion of (18), the DC component of drain current is determined, resulting in I_{DC}=105 mA.
At this point, using (21) and (23) or, alternatively, equation (24), the values of output matching network are R=33Ω and L=1.67 nH. If considering a standard output impedance of 50Ω, a transforming stage is necessary.

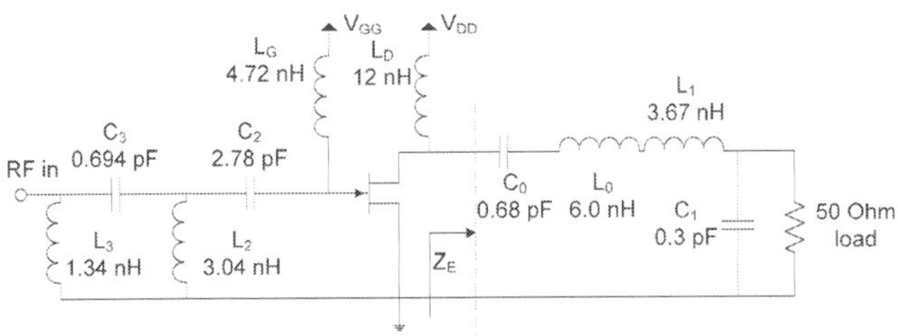

Fig. 3. Schematic of a 2.5GHz GaAs HEMT Class E amplifier.

Standing the value of optimum load, the impedance matching can be easily accomplished by a single L-C cell. A series inductance - parallel capacitance configuration has been chosen. Lumped elements for the filtering output network have then determined, selecting an inductance L_0=6nH and a resulting capacitance C_0=0.68pF. The complete amplifier schematic is depicted in Fig. 3, while the simulated output power, gain and efficiency versus input power are shown in Fig. 4.

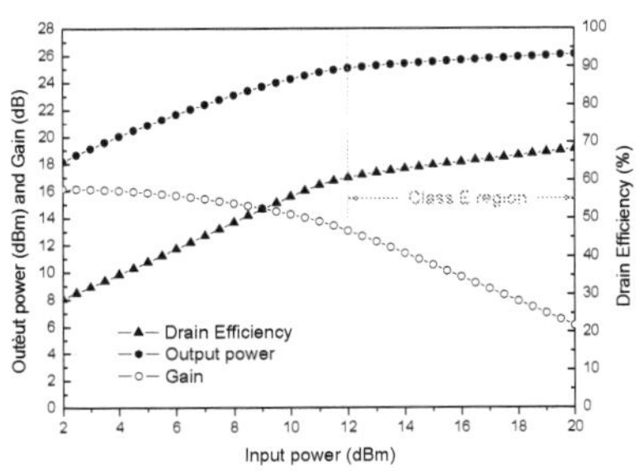

Fig. 4. Simulated performance of the 2.5GHz Class E amplifier.

It is worth to notice that, under a continuous wave excitation, Class E behavior is achieved only at a certain level of compression, i.e. when the large input sinusoidal waveform implies a "square-shaping" effect on the output current, due to active device physical limits, thus approaching a switching behavior. The output current and voltage waveforms and load line are reported in Fig. 5, showing a good agreement with the theoretical expected behavior (compare with ideal waveforms depicted in Fig. 2b).

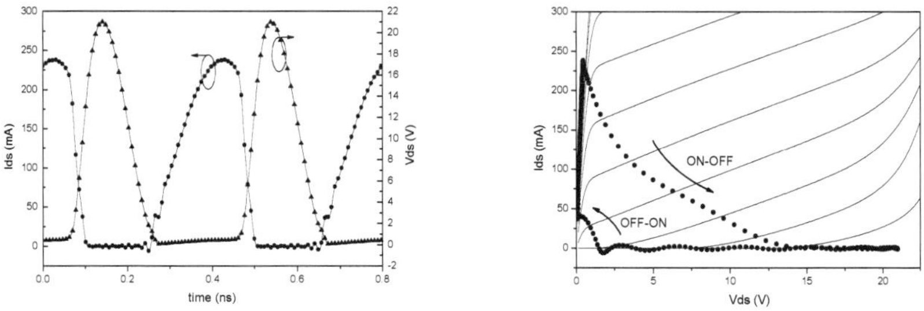

(a) (b)

Fig. 5. Output current and voltage waveforms (a) and load line (b) of the 2.5GHz Class E amplifier.

3.3 Drawbacks

As already outlined, Class E power amplifiers have some practical limitations, mainly due to their maximum operating frequency. Such limitations are partially related to the cut-off frequency of the active device, while are mainly due to the circuit topology and switching operation. In fact, as reported in (Mader et al., 1998), a Class E maximum frequency can be approximated by:

$$f_{Max} = \frac{I_{DS}}{2\pi^2 \cdot C_1 \cdot V_{DD}} \simeq \frac{I_{Max}}{56.5 \cdot C_1 \cdot V_{DD}} \tag{27}$$

Practically the lower limit of C_1 is given by the active device output capacitance C_{ds}. Consequently, the value of maximum operating frequency strongly depends on the device adopted for the design, on its size and then on the maximum current it can handle. For RF and microwave devices, the maximum frequency in Class E operation is generally included between hundred of megahertz (for MOS devices) and few gigahertz (for small MESFET or pHEMT transistors).

Additionally, at microwave frequencies higher order voltage harmonic components can be considered as practically shorted by the shunt capacitance, and the Class E behavior has to be clearly approximated. In particular, the voltage wave shaping can be performed recurring to the first harmonic components only (Raab, 2001; Mader et al., 1998), while the ZVS and ZVDS conditions cannot be longer satisfied.

Truncating the ideal voltage Fourier series at the third component, the resulting waveform is reported in Fig. 6, from which it can be noted the existence of negative values. Thus it becomes mandatory to prevent such negative values of drain voltage to respect active device physical constraint and safely operations.

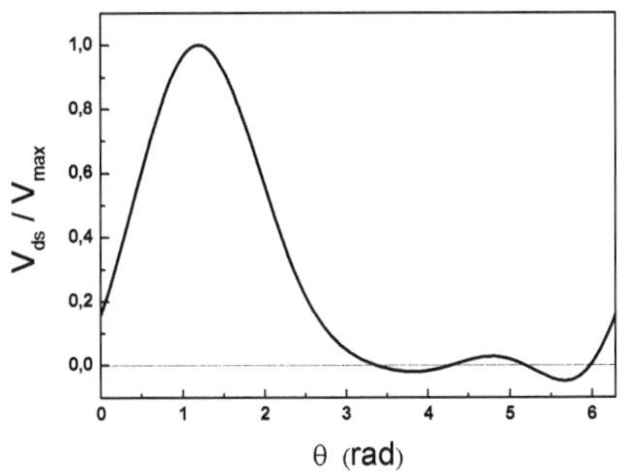

Fig. 6. Three harmonics reconstructed voltage waveform

As pointed out in (Colantonio et al., 2005), two solutions can be adopted. Obviously it is possible to increase drain bias voltage, but it would mean a non negligible increase in the DC dissipated power that in turn causes a decrease in drain efficiency levels. In addition, an increasing on peak voltage value could exceed breakdown limitations of the transistor. The other solution is based on the assumption of unaffected current harmonic components, thus optimizing the voltage fundamental component, while keeping fixed the other harmonics imposed by the network topology (i.e. the filter L_0-C_0 behavior and the device capacitance C_{ds}) (Cipriani et al., 2008). The optimization process must be implemented in a numerical form in order to reduce complexity and computing effort. The main goal is to avoid negative voltage values on drain voltage and, at the same time, maximize output power, hence efficiency. Then, for every value of frequency exceeding the maximum one, the optimum high frequency fundamental impedance, $Z_{1,HF}$, is optimized in magnitude and phase.

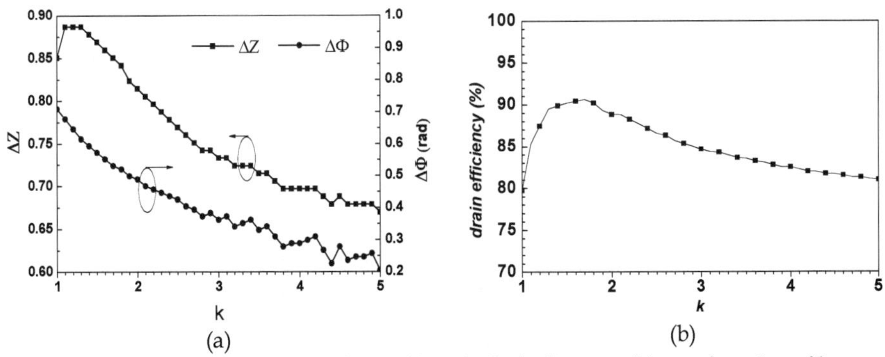

Fig. 7. Class E optimum load impedance (a) and ideal efficiency (b) as a function of k

Obtained results are expressed as normalized to ideal load impedance at maximum frequency given by (27) and depicted in Fig. 7a as a function of normalized frequency $k=f/f_{Max}$, defined as the ratio between the assumed operating frequency f and the maximum

allowed one f_{Max}, defined in (27). The plot shown in Fig. 7a can be considered as a "design chart" for high frequency Class E development, once the maximum frequency is known.
A quasi monotonic decrease of magnitude of fundamental impedance is observed, leading to a reduced voltage fundamental component and a reduced output power. At the same time, the phase decreases tending to almost purely resistive values. The related drain efficiency is reported in Fig. 7b, showing an increasing reduction with respect the ideal 100% value due to non ideal operating conditions.
A high frequency Class E PA example is shown in Fig. 8.

Fig. 8. A 2.14 GHz LDMOS Class E PA

The amplifier is designed using a medium power LDMOS transistor for base station application at 2.14 GHz. Once the bias point, maximum current and output capacitance of the transistor are fixed, the maximum frequency in Class E mode is directly derived. Considering a maximum current of 2.5 A, a bias voltage of 20 V and an output capacitance of 4.2 pF (estimated by S- parameters simulation), the maximum frequency in Class E results in 520 MHz, far below the frequency chosen for the design. If operating at 520 MHz, the load impedance would be $Z_1=25.1e^{j36°}$. At 2.14 GHz (4.1 times above f_{Max}), the load impedance is directly obtained by the design chart of Fig. 7a resulting in $Z_{1,HF}=17.5e^{j17°}$. Since a very simple equivalent model of the device is used, as Fig. 2 shows, this impedance is seen at the nonlinear current source terminals, so that eventual parasitic and package effects should be considered as belonging to the output load.

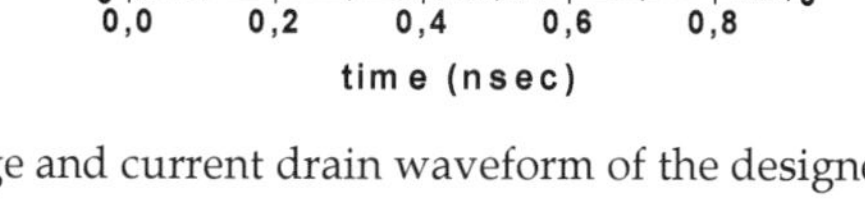

Fig. 9. Simulated voltage and current drain waveform of the designed PA

Simulated drain current and voltage waveform are depicted in Fig. 9, with reference to the internal nodes of the model. A good agreement with typical Class E waveform is remarkable, above the maximum frequency, although the perfect switching behavior cannot be satisfied.

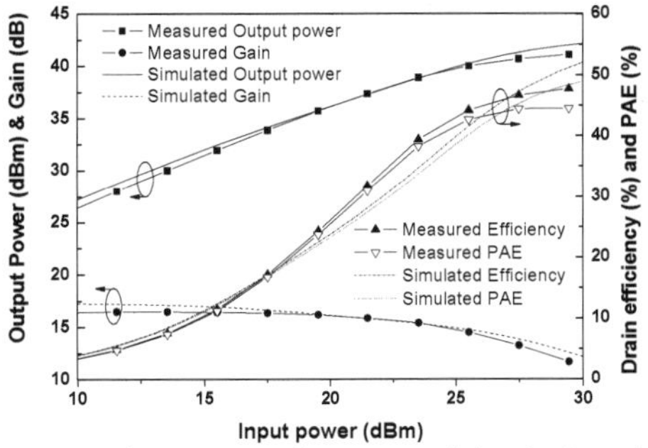

Fig. 10. Simulated and measured output performance of the designed PA

3.4 Class E matching network implementation and other topologies

Although Class E approach has basically a fixed circuit topology, different solution can be adopted for the synthesis of output matching network. Depending on the design frequency, distributed approaches are possible: proper load conditions at fundamental have to be satisfied, according to (24), and an open circuit condition must be provided at harmonics of the fundamental frequency, usually second and third harmonics (Negra et al., 2007; Wilkinson & Everard, 2001; Xu et al., 2006).

In Fig. 11a, no lumped elements are used unless block capacitors, while the 50Ω matching is synthesized through a very compact and simple structure reported in Fig. 11b (Mader et al., 1998). In order to provide harmonic suppression on the load, different quarter-wave open stubs can be used at different harmonics (Negra et al., 2007), while series transmission lines and wave impedances are properly chosen to provide the correct fundamental load.

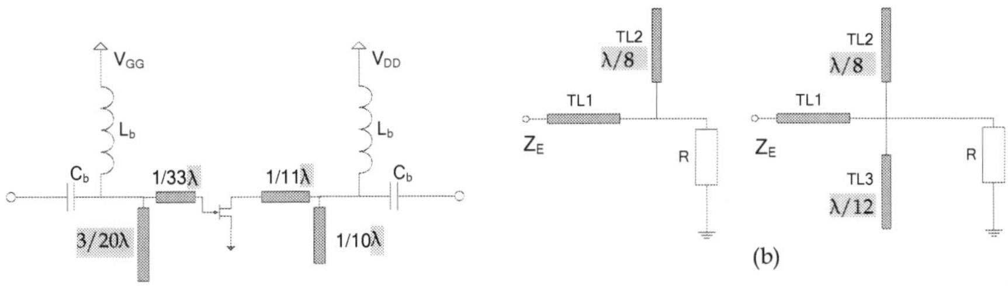

(a)

Fig. 11. Some practical examples of Class E transmission lines amplifiers.

Additionally, different circuit topologies exist that can provide the same results as the classical formulation: they have been widely investigated in (Grebennikov, 2003) and are commonly referred as parallel circuit Class E and their main characteristic is the presence of

a finite parallel inductor in the output network, required for the output device biasing supply, as reported in Fig. 12. As before assumed, the shunt capacitance C includes the transistor output capacitance C_{ds}.
The first circuit, in Fig. 12a, employs a very simple output matching network, which consist of a parallel inductor and a series blocking capacitance. Applying ZVS and ZVDS conditions on this circuit, and considering the transistor to behave as a perfect switch, the solution of the circuit is given by a second order non homogeneous differential equation, given by (28), which has to be solved in order to determine the value of all circuit parameters.

$$\omega^2 LC\frac{d^2 v_s(\omega t)}{d(\omega t)^2}+\frac{\omega L}{R}\frac{dv_s(\omega t)}{d(\omega t)}+v_s(\omega t)=V_{DD} \tag{28}$$

The values of reactive components, L and C, are:

$$C=\frac{1.025}{\omega R} \qquad L=0.41\frac{R}{\omega} \tag{29}$$

and the load resistance R is determined using the desired output power at fundamental frequency, P_1:

$$R=1.394\frac{V_{DD}^{2}}{P_1} \tag{30}$$

Due to the lack of any filtering action at the output, this circuit becomes not practical in applications - like telecommunications - which require harmonic suppression (Grebennikov, 2003). Moreover, a higher peak current value is obtained ($4.0I_{DC}$ instead of 2.862 I_{DC} for the classical topology) that has to be taken into account in the choice of the active device.
The circuit in Fig. 12b adds a series LC filter in the output branch and it is very similar to a canonic Class E amplifier using a finite DC feed inductance, unless for the absence of the "tuning" series inductance. Providing a high Q factor for the LC series filter, the current i_R flowing into the output branch can be assumed as sinusoidal: this hypothesis is used as starting point for a complete time domain analysis which is similar to what reported in paragraph 4.1. Optimum parallel capacitance C and optimum load resistance R are obtained after inferring the phase angle between in-phase and quadrature components of fundamental current:

$$\psi=\arctan\left(\frac{R}{\omega L}-\omega RC\right)=34.244° \tag{31}$$

From which:

$$C=\frac{0.685}{\omega R} \qquad L=0.732\frac{R}{\omega} \tag{32}$$

Output resistance R is derived from desired fundamental output power, P_1:

$$R = 1.365\frac{V_{DD}^{\ 2}}{P_1} \tag{33}$$

Slightly different voltage and current peak values (Grebennikov, 2003) are obtained with respect to the traditional Class E approach:

$$V_{Max} = 3.647 \cdot V_{DD} \qquad I_{Max} = 2.647 \cdot I_{DC} \tag{34}$$

In Fig. 12c the parallel inductance is replaced by a short-circuited short length transmission line: this solution is quite popular at microwave frequencies. In order to approximate the Class E optimum impedance at fundamental frequency, the electrical length and the characteristic impedance of the transmission line are determined starting from the optimum fundamental impedance and according to the relation (Grebennikov, 2003):

$$Z_0 \cdot \tan(\theta) = \omega L. \tag{35}$$

The load impedance Z_E seen at device terminals should satisfy the optimum impedance at fundamental frequency, and remembering relation (31) it is rewritten as:

$$Z_E = \frac{R}{1 - j\tan\psi} \tag{36}$$

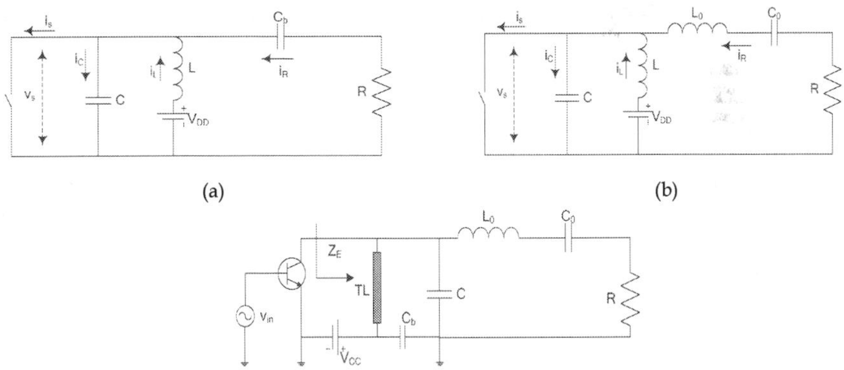

(c)

Fig. 12. Parallel circuit Class E topology (a), parallel circuit Class E with output filter (b) and transmission line parallel Class E (c).

Finally, using equation (32) to determine the optimum required parallel inductance, the electrical length of the parallel transmission line can be obtained:

$$\tan\theta = 0.732\frac{R}{Z_0} \tag{37}$$

4. The inverse Class E amplifier

The Inverse Class E amplifier, or voltage drive Class E amplifier, is commonly considered as the dual version of the Class E amplifier, in which current and voltage waveforms are interchanged, as shown in Fig. 13. It is also referred as "series-L/parallel tuned Class E", being the traditional topology defined as "parallel-C/series-tuned Class E" (Mury & Fusco, 2005; Mury & Fusco, 2007).

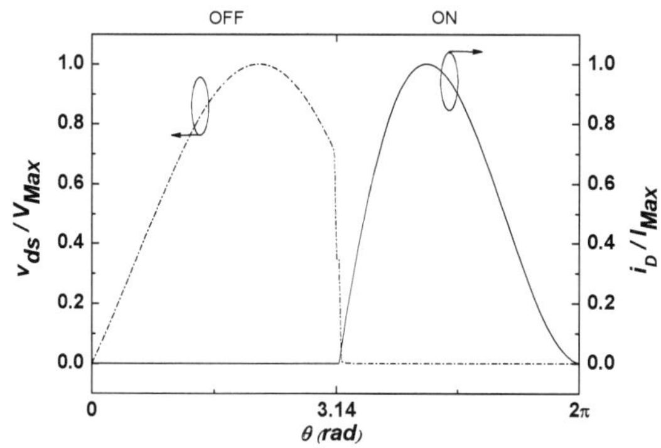

Fig. 13. Ideal inverse Class E waveforms.

A former version of the Inverse Class E amplifier was reported in (Kazimierczuk, 1981): the circuit does not have shunt capacitor, while a series tuned filter and a finite DC feed inductance is considered in the output network, as depicted in Fig. 14. Although this circuit seems to be similar to those reported in Fig. 12, it implies a different behavior, due to the different characteristic of the shunting element (an inductor instead of a capacitor). When the switch is open, and provided a high enough Q factor of the series filter, the only current flowing in the circuit is the sinusoidal output current i_R, that is the inductor current i_L. The latter causes a voltage drop across the inductor, v_L, which has a cosinusoidal form. When the switch is closed, the voltage across the inductor is instantaneously constant and equal to V_{DD}. This causes a linear increase in the current i_L. The current across the switch is calculated as the difference between i_L and i_R and assumes the typical asymmetrical shape.

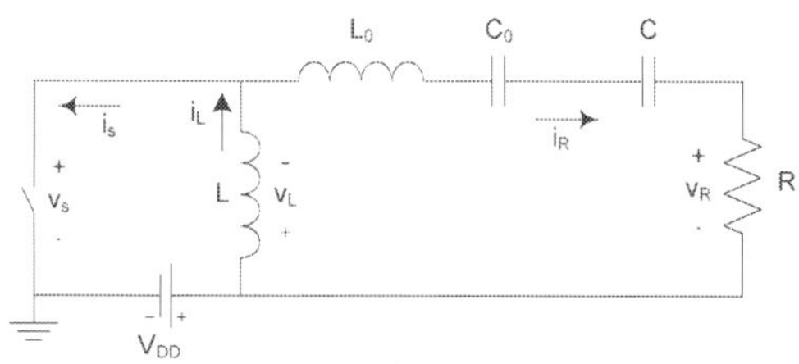

Fig. 14. Inverse Class E amplifier: no-shunt-capacitor/series-tuned topology with finite inductance.

A complete analysis of the inverse Class E amplifier is reported for the first time in (Mury & Fusco, 2005; Mury & Fusco, 2007), together with a defined topology which is shown in Fig. 15 and which is substantially different from the previous version given in (Kazimierczuk, 1981). As can be seen by a comparison of Fig. 15 and the circuit depicted in Fig. 2, each component of the traditional Class E amplifier has been replaced by its dual element in a dual configuration. A DC blocking capacitance C_b is inserted in order to prevent inductance L_0 from shorting the bias voltage.

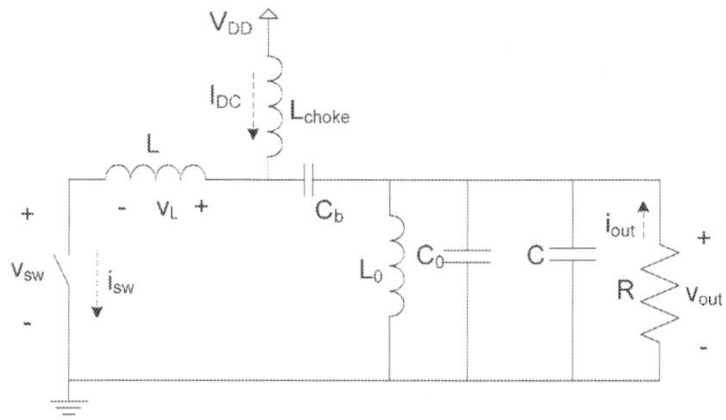

Fig. 15. Inverse Class E amplifier.

Hence, the analysis of the inverse Class E amplifier can be carried out starting from the assumption of a purely sinusoidal output voltage across the output resistance R, which produces a voltage across the inductor L given by:

$$v_L(\theta) = V_{DD} - v_o(\theta) = V_{DD} \cdot (1 - a \cdot \sin(\theta + \phi)) \tag{38}$$

This is the voltage present across the switch during the OFF time, while during the ON time the switch has no voltage across it and its current is given by integration of (38):

$$i_{SW} = \frac{1}{\omega L} \cdot \int_0^\theta v_L(\theta) d\theta \tag{39}$$

These expressions have the same form of those reported in paragraph 4.1, unless current and voltage are interchanged: the same kind of analysis as Class E can be performed on the Inverse Class E circuit. As a consequence, the same numerical results are obtained for the dual configuration, and are summarized in Table 1, referred to a 50% duty cycle operation. As can be seen, the maximum allowable voltage for the Inverse Class E operation is much smaller than for Class E: this is an unquestionable advantage of such a circuit, because the requirement on device breakdown can be drastically relaxed.
However, it is worth to notice that in Inverse Class E amplifier the output capacitance of the active device is not taken into account and set to zero in the ideal analysis: in real world circuit, this is clearly not true. Hence, some actions have to be taken in order to compensate its presence (e.g. a shunt inductance).

Circuit component or electrical value	Class E	Inverse Class E
V_{Max}	$3.562 \cdot V_{DD}$	$2.862 \cdot V_{DD}$
I_{Max}	$2.862 \cdot I_{DC}$	$3.562 \cdot I_{DC}$
C	$\frac{I_{DC}}{\pi \cdot \omega \cdot V_{DD}}$	$\frac{0.2116}{\omega^2 \cdot L}$
R	$\frac{0.1836}{\omega \cdot C}$	$\frac{1}{0.1836 \cdot \omega \cdot L}$
L	$\frac{0.2116}{\omega^2 \cdot C}$	$\frac{V_{DD}}{\pi \cdot \omega \cdot I_{DC}}$
C_0	$\frac{1}{\omega \cdot Q \cdot R}$	$\frac{Q}{\omega \cdot R}$
L_0	$\frac{1}{\omega^2 \cdot C_0}$	$\frac{1}{\omega^2 \cdot C_0}$

Table 1 - Class E and Inverse Class E comparison.

5. Class F and Class F-1 amplifier

Among the different switched mode amplifier design strategies, several authors include the Class F and inverse F (or Class F-1) schemes also.

5.1 Class F theoretical analysis

Such design strategy was introduced by Tyler at the end of fifties (Tyler, 1958) and further investigated several years later (Snider, 1967; Raab, 1996), as a simple and feasible way to improve power amplifier large-signal performances.
Following the widely accepted definition, Class-F design consists in terminating the device output with open-circuit terminations at odd harmonic frequencies of the fundamental component and short-circuiting it at even harmonics. Regarding the input network, it is typically synthesized in order to guarantee maximum power transfer at the operating frequency, neglecting or at least circumventing the device input non linear generation by using short-circuit terminations.
The theoretical output voltage and current waveforms of an ideal Class F PA are depicted in Fig. 16.

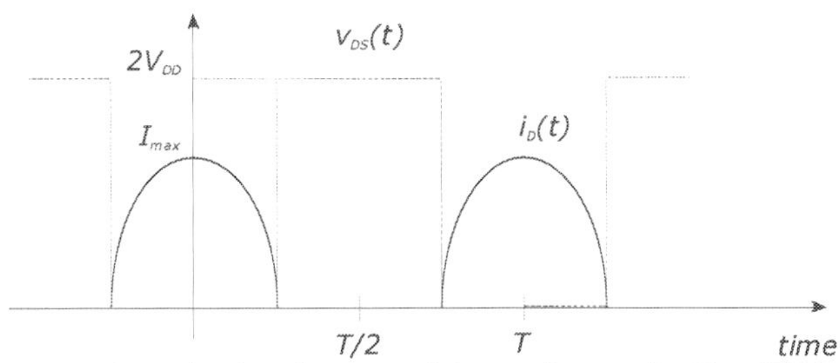

Fig. 16. Ideal output voltage (v_{DS}) and current (i_D) waveforms of a Class F PA.

The current is assumed as a truncated sinusoid waves (assuming a Class B bias condition), while the voltage is squared by the proper harmonic loading conditions. The two ideal waveforms do not overlap: no output current flows for high drain/collector voltages and maximum current occurs when drain/collector voltage waveform is at its minimum. Therefore the power dissipated in the active device is nulled ($P_{diss}=0$).
Such an ideal waveforms can be easily described by their Fourier components:

$$i_D(\theta)=\sum_{n=0}^{\infty} I_n \cdot \cos(n\theta) \qquad I_n=\begin{cases} \frac{I_{Max}}{\pi} & n=0 \\ \frac{I_{Max}}{2} & n=1 \\ \frac{2\cdot I_{Max}}{\pi}\frac{(-1)^{\frac{n}{2}+1}}{n^2-1} & n \text{ even} \\ 0 & n \text{ odd} \end{cases} \tag{40}$$

$$v_{DS}(\theta)=\sum_{n=0}^{\infty} V_n \cdot \cos(n\theta) \qquad V_n=\begin{cases} V_{DD} & n=0 \\ -\frac{4\cdot V_{DD}}{\pi} & n=1 \\ 0 & n \text{ even} \\ \frac{4\cdot V_{DD}}{\pi}\frac{(-1)^{\frac{n+1}{2}}}{n} & n \text{ odd} \end{cases} \tag{41}$$

where I_{Max} and V_{DD} the maximum output current and bias voltage, respectively.
From the previous equations it can be noted that the current and voltage Fourier components with the same order n are alternatively zeroed, thus nulling the power delivered at harmonic frequencies also ($P_{out,nf}=0$, $n>1$).
The values of the ideal terminations are inferred as the ratio between the respective Fourier components V_n and I_n, i.e.:

$$Z_n=\frac{V_n}{I_n}=\begin{cases} \frac{8}{\pi}\cdot\frac{V_{DD}}{I_{Max}} & n=1 \\ 0 & n \text{ even} \\ \infty & n \text{ odd} \end{cases} \tag{42}$$

Therefore resulting in a purely resistive terminating impedance at fundamental frequency given by $4/\pi \cdot R_{TL}$, being R_{TL} the optimum impedance of a Tuned Load scheme (i.e. short circuiting all the harmonic terminations). Conversely, short circuit condition for even harmonics and open circuit for odd ones have to be synthesized, as schematically depicted in Fig. 17.

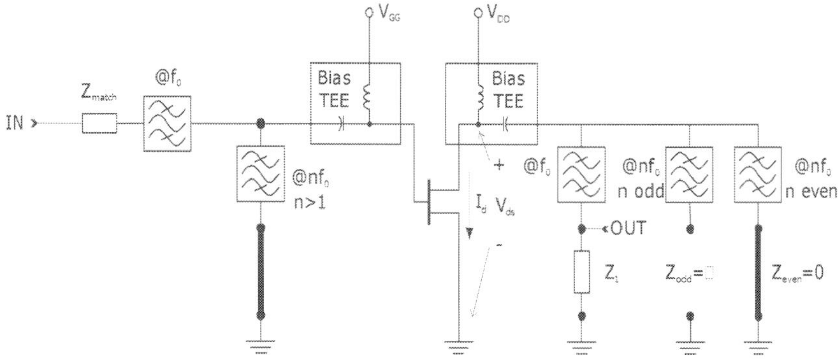

Fig. 17. Ideal structure of a Class F amplifier.

The results obtained following the Class F strategy were so interesting that, before the advent of fast switching devices, such approach was widely adopted to design PAs for amplitude modulated (AM) broadcast radio transmitters (operating at LF 30-300 KHz, MF 0.3-3 MHz and HF 3-30 MHz) or for frequency modulated (FM) broadcast radio transmitters (at VHF 30-300 MHz and UHF 0.3-3 GHz) (Wood, 1992; Lu, 1992).
Nowadays, the Class F technique is generally adopted for high frequency applications in the microwave range (i.e. up to tens of Gigahertz). Examples of Class F based on GaAs devices are available at X (9.6GHz) (Colantonio et al., 2007), Ku (14.5GHz) (Ozalas, 2005) and Ka (29.5GHz) (Reece et al., 2003) bands. For high frequency applications the active devices operate in current-mode rather than in switched-mode, and the harmonic loading conditions are implemented through lumped resonating circuits.

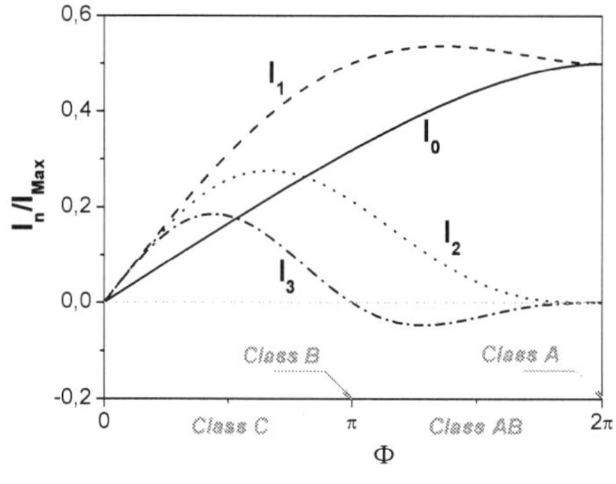

Fig. 18. Fourier components I_0, I_1, I_2, I_3 normalised to the device maximum current I_{Max}, as functions of the drain CCA Φ.

Several research efforts were focused to clarify and to implement the harmonic terminating scheme leading to the Class F optimum behavior and its experimental validation, inferring practical design guidelines also (Duvanaud et al., 1993; Blache, 1995; Colantonio et al., 1999).

The analytical results can be easily extended to bias conditions different from Class B, still assuming for the current waveform a truncated sinusoidal wave shaping, with a conduction angle (CCA) Φ larger than π (Class AB), while maintaining a square voltage waveform. In this case the corresponding current harmonic components as function of the CCA depicted in Fig. 18 result, while the estimated Class F performances are depicted in Fig. 19, compared with the corresponding theoretical one for a Tuned Load condition.

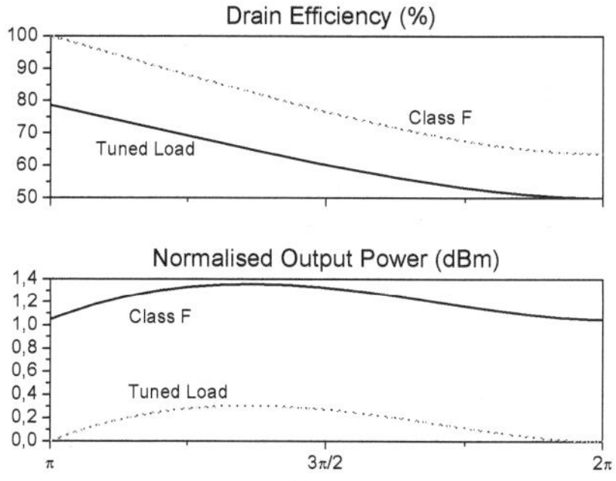

Fig. 19. Ideal performance for a Class F and reference TL amplifiers as functions of the CCA Φ. Output power normalized to the Class A output at full swing.

5.2 Class F practical limitations

Referring to the ideal solution proposed to implement the Class F loading scheme, depicted in Fig. 17, it is based on the use of resonating elements ensuring low impedance values (theoretically short-circuit conditions) for even harmonic frequencies or high values for the odd harmonics (theoretically open-circuit conditions).

Clearly, for a simple and manageable mathematical formulation, the active device model has to be strongly simplified, neglecting all parasitic and reactive elements. It implies that when dealing with an actual device, the loading condition expressed by (42) has to be fulfilled across the device intrinsic output current source, as already proposed in other similar simplified approaches (Cripps, 1983). From a practical point of view, the required ideal terminations pose significant restrictions when implementing the Class technique. In fact, while it is relatively simple to realize a short-circuit termination, compensating for the device reactive elements, the realization of an open-circuit condition becomes much more cumbersome. For example, in high frequency applications the dominant capacitive behavior shown by the active device (e.g. C_{ds}) tends to short-circuit the device output itself, thus practically not allowing the open-circuit loading condition for the higher-order odd harmonics.

At the same time, even if the internal C_{ds} capacitance can be effectively resonated by an external inductive element, the device output resistance (R_{ds}) cannot be removed, thus representing an upper limit for the impedance effectively synthesizable across the intrinsic current source. Therefore the realization of a true open termination is basically unfeasible. However, it has been demonstrated that lower-order harmonics are more effective in improving amplifier performance as compared to higher order ones (Raab, 2001). Therefore, accounting for the reduced number of harmonics can be effectively controlled, new terminating impedance values have been inferred not only at fundamental but also at harmonics, resulting in a different voltage harmonic ratio also (Raab, 1997; Colantonio et al., 2009). In fact, the new optimum voltage ratio becomes $|V_3/V_1|=1/6$ rather than $1/3$ as in the ideal case. Simultaneously, the fundamental loading impedance becomes

$$R_F = \frac{4}{\sqrt{3}} \cdot \frac{V_{DD}}{I_{Max}} \tag{43}$$

resulting in an efficiency improvement of 15% only with respect to the Tuned Load theoretical case (Colantonio et al., 2009).

A further critical point is represented by the physical mechanisms generating the harmonic components of both voltage and current waveforms. If the device output only is considered, it can be described by an independent and forcing current source, whose waveform results both from the input drive level and the device physical limitations (clipping effects), being independent on the device terminating impedances. Under this assumption, the output voltage waveform is dependent on the current one, being generated by loading each harmonic current component through the respective terminations ($V_n=Zn \cdot In$). Consequently a proper phase relationships between the output current harmonic components must be fulfilled, and in particular I_1 and I_3 must be opposite in sign. Such a condition, referring to Fig. 18 practically implies the selection of a suitable bias level close to the Class B condition, i.e. assuming a non-zero quiescent current level (deep AB bias), while leaving for instance the same harmonic terminations as derived in the ideal case. The theoretical load curve behaves as depicted in Fig. 20.

Conversely, the adoption of a Class C bias condition implies a wrong relationships among the current harmonics phase (see Fig. 18).

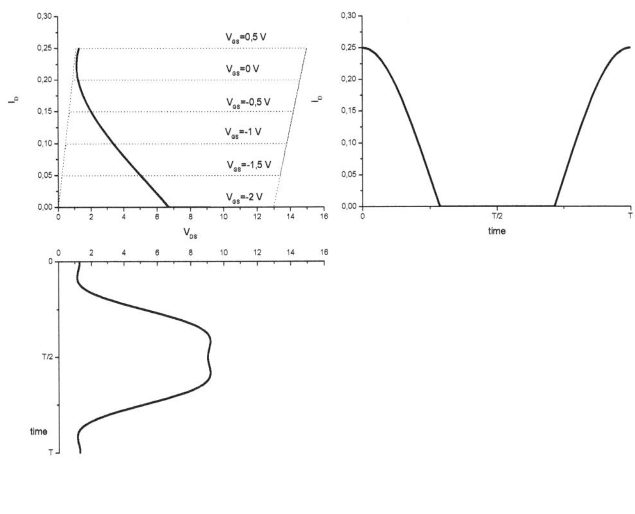

Fig. 20. Theoretical load line for a Class F amplifier.

5.3 Class F matching networks implementation

To design the Class F output section, resonating circuits are typically used as traps for the harmonic frequencies. Such resonators, sometimes referred as idlers, are limited to low-order harmonics, to reduce both circuit complexity and associated losses. Moreover, depending on the frequency range, their implementation can be performed either in lumped or distributed form, while accounting for also the DC bias.

For example, in Fig. 21 is depicted a possible implementation based on the use of a quarter-wavelength transmission line ($\lambda/4$-TL) and a parallel resonating L_0-C_0 tank to control the harmonic impedances. The optimum matching at fundamental frequency f_0 is achieved by the remaining passive components L_m and C_m (Gao, 2006).

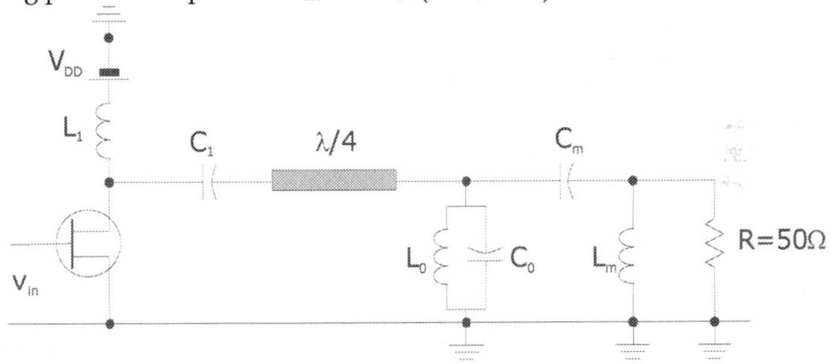

Fig. 21. Example of Class F output network design.

In practical situations, to account the biasing elements and the active device output capacitance C_{ds}, other proposed solutions are schematically depicted in Fig. 22, where the design relationships to calculate the element values can be derived evaluating the impedance loading the device output current source and then imposing the short circuit condition at $2f_0$ and the open circuit one at $3f_0$. (Trask, 1999).

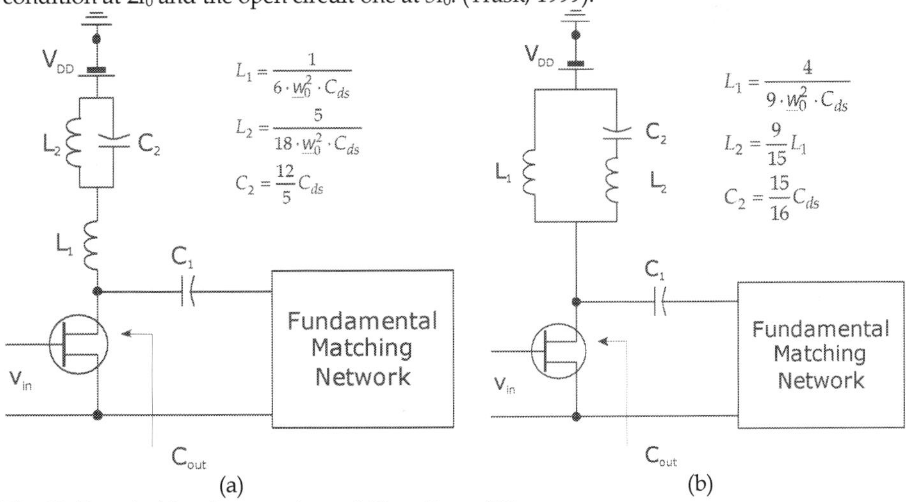

Fig. 22. Practical implementations of Class F amplifier.

In a similar way, a distributed solution can be designed and implemented by using transmission lines, as for instance reported in Fig. 23 (Grebennikov, 2000; Woo et al., 2006; Negra et al., 2007).

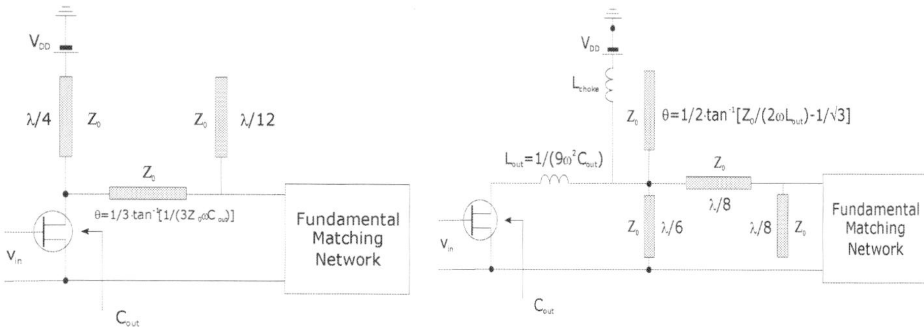

(a) (b)

Fig. 23. Distributed solutions implementing the Class F schemes.

Clearly the characteristic impedances of the TLs have to be properly selected when designing the matching network at fundamental frequency.

5.4 Class F example

An example of hybrid (MIC) Class F PA operating at 5GHz and based on GaAs device (1mm gate periphery) is reported in Fig. 24, where both the input and output matching networks were designed on Alumina substrate by using a dixtributed approach.

Fig. 24. Photo of a MIC 5GHz GaAs Class F amplifier.

The corresponding load curve and the performance as compared to a Tuned Load amplifier based on the same active device and bias point (Class B), are reported in Fig. 25 and Fig. 26, respectively.

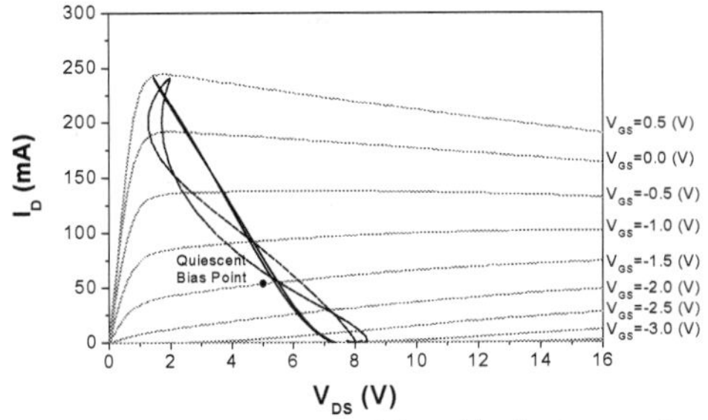

Fig. 25. Load curve of 5GHz MIC Class F amplifier (dashed curve) compared with Tuned Load case (continuous curve).

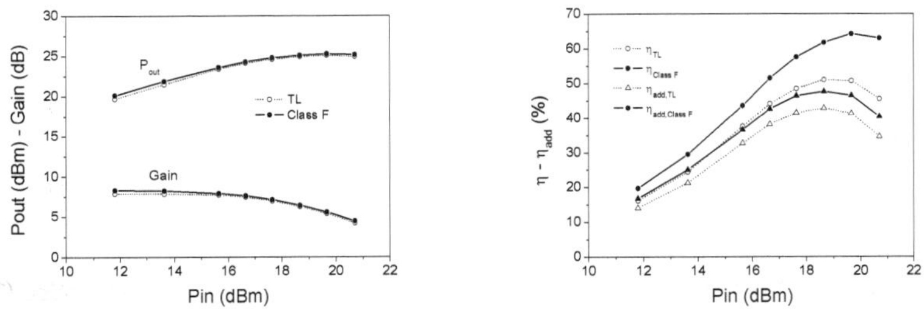

(a) (b)

Fig. 26. Output Power & Gain (a) and efficiency & power added efficiency (b) for the 5GHz MIC Class F amplifier as compared to Tuned Load amplifier.

Among the other features, it is to note that the Class F amplifier output power is higher as compared to the Tuned Load approach in the entire range of input drive. A 7-8% measured improvement (against a maximum theoretical 15% expected) is usually obtained.

It is to note that at low input power levels, a higher power gain with respect to a Tuned Load is obviously expected due to the larger output load that has been assured at fundamental frequency, as given by (43). Increasing the input drive level, the use of a Class F strategy force the output voltage (and current) to approach their maximum swings value, resulting in a proper bending of the load curve (see Fig. 25), thus improving the output power at saturation level also.

5.5 The inverse Class F amplifier

The term "inverse" in this kind of amplifier suggests somewhat inverted in the behavior: in fact, current and voltage drain waveforms are interchanged if compared to a traditional Class F amplifier (Ingruber et al., 1998; Lepine et al., 2007; Woo et al., 2006).

Inverse Class F power amplifier was introduced as a "rectangular driven Class A harmonic-controlled amplifier", in order to combine the advantage of high efficiency of a switched mode amplifier with the high gain of Class A operation (Ingruber et al., 1998).

It employs a rectangular driving voltage, which forces the active device to operate in the ohmic region or in the interdiction region, thus justifying - more than for a Class F - the classification as a switched mode amplifier. Assuming a piecewise linear simplified model for the active device, the drain current waveform can be directly inferred, resulting in a rectangular waveform. Thus, open terminations for even harmonics and short circuit terminations for odd harmonics - except for the fundamental one - give a truncated sinusoid voltage waveform. Current and voltage waveform do not overlap, thus preventing DC power consumption on the active device; additionally, thanks to the proper harmonic terminations, no power is delivered at harmonics of fundamental frequency and 100% drain efficiency is ideally achievable.

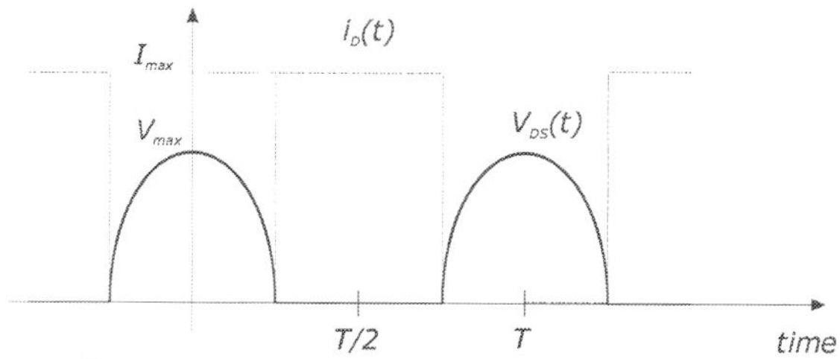

Fig. 27. Ideal output voltage (v_{DS}) and current (i_D) waveforms of an inverse Class F PA.

Referring to the ideal output current and voltage waveforms of Fig. 27, DC and fundamental frequency components can be obtained using a Fourier analysis:

$$I_0 = \frac{I_{Max}}{2} \tag{44}$$

$$I_1 = \frac{1}{\pi}\int_{\frac{\pi}{2}}^{\frac{3\pi}{2}} I_{Max}\cdot\cos(\vartheta)\,d\vartheta = \frac{2\cdot I_{Max}}{\pi} \tag{45}$$

$$V_0 = V_{DD} = \frac{1}{2\pi}\int_{-\frac{\pi}{2}}^{\frac{\pi}{2}} V_{Max}\cos(\vartheta)\,d\vartheta = \frac{V_{Max}}{\pi} \tag{46}$$

$$V_1 = \frac{1}{\pi}\int_{-\frac{\pi}{2}}^{\frac{\pi}{2}} V_{Max}\sin(\vartheta)^2\,d\vartheta = \frac{V_{Max}}{2} \tag{47}$$

Then, the expression of harmonic load impedance is given by:

$$Z_n = \frac{V_n}{I_n} = \begin{cases} \frac{\pi^2}{4}\cdot\frac{V_{DD}}{I_{Max}} & n=1 \\ \infty & n \text{ even} \\ 0 & n \text{ odd} \end{cases} \tag{48}$$

An inverse Class F amplifier requires an infinite number of resonator in order to maintain the square-shaped output current. Fig. 28 shows an immediate ideal implementation of an inverse Class F amplifier using multiple even-harmonic resonators to control the voltage and current waveforms (Grebennikov & Sokal, 2008).

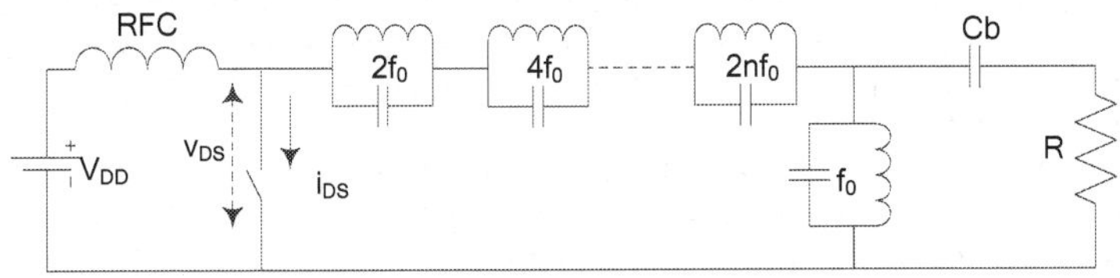

Fig. 28. Inverse Class F amplifier with multiharmonic resonator circuit.

In real world design, however, a limited number of harmonics has to be taken into account. Practical output networks implementations are often based on multiharmonic quarterwave-lenght stubs (Lepine et al., 2007; Woo et al., 2006). Input network design is the most critical aspect in inverse Class F design. In fact, if a continuous wave excitation is put on the gate, the required square wave driving voltage occurs only in deep saturation regime, thus affecting the gain. Ideal conditions would require a square wave driving voltage throughout the whole dynamic range of the amplifier. This can be obtained providing fundamental frequency and its second harmonic as a driving signal (Goto et al., 2001; Grebennikov & Sokal, 2008), so that an out of phase condition exist at their maximum amplitude. A network example which satisfied this condition is reported in Fig. 29.
Referring to switched mode finite harmonic operation, it is interesting to find a qualitative relation between the different classes of amplifier, as reported in (Raab, 2001). The transition from inverse Class F and Class F amplifiers is achieved by moving the second harmonic impedance (reactance) from zero to ∞, while decreasing the third harmonic impedance (reactance) from ∞ to zero, while in the intermediate condition, in which both second and third harmonic reactances have finite values, current and voltage waveforms assumes a resemblance to Class E amplifier.

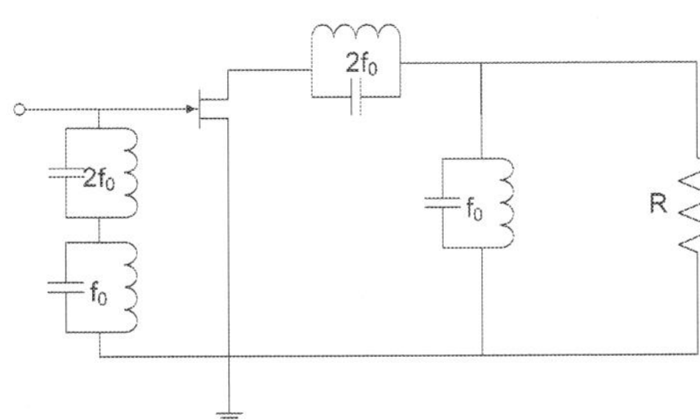

Fig. 29. Bi-harmonic inverse Class F amplifier.

6. References

Blache, F. (1995). A Novel Computerised Multiharmonic Active Load-Pull System for the Optimisation of High Efficiency Operating Classes in Power Transistors, *IEEE MTT-S Int. Microwave Symp. Dig.*, Orlando, FL, June 1995, pp.1037-1040.

Cipriani, E.; Colantonio, P.; Giannini, F. & Giofré, R. (2008). *Optimization of Class E Power Amplifier Design above Theoretical Maximum Frequency*, Proceedings of 38th European Microwave Conference, EuMC 2008, pp. 1541 - 1544, Amsterdam, October 2008.

Colantonio, P.; Giannini, F.; Leuzzi, G. & Limiti, E. (1999). On the Class-F Power Amplifier Design, *Intern. Journal of RF and Microwave Computer-Aided Engineering*, Vol. 9, No. 2, March 1999, pp.129-149.

Colantonio, P.; Giannini, F.; Leuzzi, G. & Limiti, E. (2003). Theoretical Facet and Experimental Results of Harmonic Tuned PAs, *Intern. Journal of RF and Microwave Computer-Aided Engineering*, Vol.13, No. 6, November 2003, pp. 459-472.

Colantonio, P.; Giannini, F.; Giofrè, R.; Medina, M. A. Y.; Schreurs, D. & Nauwelaers, B. (2005). High frequency class E design methodologies, *Proceedings of European Gallium Arsenide and Other Semiconductor Application Symposium*, GAAS 2005, pp. 329 - 332, Paris, October 2005.

Colantonio, P.; Giannini, F.; Giofrè R. & Limiti, E. (2007). Combined Class F Monolithic PA Design, *Microwave and Optical Technology Letters*, Vol. 49, No. 2, February 2007, pp. 360-362.

Colantonio, P.; Giannini, F. & E. Limiti (2009). *High Efficiency RF and Microwave Solid State Power Amplifiers*, John Wiley & Sons (ISBN: 978-0-470-51300-2).

Cripps, S.C. (1983). A theory for the prediction of GaAs FET Load-Pull power contours, *IEEE MTT-S Int. Microwave Symp. Dig.*, Boston, MA, May 1983, pp. 221-223.

Cripps, S.C. (1999). *RF Power Amplifiers for Wireless Communications*. Artech House, Norwood, MA (ISBN 0-89226-989-1).

Cripps, S. C. (2002). *Advanced Techniques in RF Power Amplifier Design*, Artech House, London.

Duvanaud, C.; Dietsche, S.; Pataut, G. & Obregon, J. (1993). High-Efficient Class F GaAs FET Amplifiers Operating with Very Low Bias Voltages for Use in Mobile Telephones at 1.75 GHz, *IEEE Microwave and Guided Wave Letters*, Vol.3, No. 8 , August 1993, pp.268-270.

Gao, S. (2006). High efficiency class-F RF/microwave power amplifiers, *IEEE Microwave Magazine*, Vol. 7, No. 1, February 2006, pp. 40 – 48.

Goto, S.; Kunii, T.; Ohta, A.; Inoue, A.; Hosokawa, Y.; Hattori, R. & Mitsui, Y. (2001). Effect of Bias Condition and Input Harmonic Termination on High Efficiency Inverse Class-F Amplifiers, *31st European Microwave Conference*, 2001. Oct. 2001, pp. 1 - 4.

Grebennikov, A.V. (2000). Circuit design technique for high efficiency Class F amplifiers, *IEEE MTT-S Int. Microwave Symp. Dig.*, Boston, MA, June 2000, Vol. 2, pp. 771–774.

Grebennikov, A. (2003). Switched-mode RF and microwave parallel-circuit Class E power amplifiers, *Intern. Journal of RF and Microwave Computer-Aided Engineering*, Vol. 14, No. 1, December 2003, pp. 21-35.

Grebennikov, A. & Sokal, N.O. (2007). *Switchmode RF power amplifiers*, Newnes, Elsevier, Burlington, MA, USA.

Ingruber, B.; Baumgartner, J.; Smely, D.; Wachutka, M.; Magerl, G. & Petz, F.A. (1998). Rectangularly driven class-A harmonic-control amplifier, *IEEE Transactions on Microwave Theory and Techniques*, Vol. 46, No. 11, Part 1, November 1998, pp. 1667 - 1672.

Jozwik, J. J. & Kazimierczuk, M. K. (1990). Analysis and design of class-E^2 dc/dc converter, *IEEE Trans. Ind. Electron.*, Vol. 37, No. 2, April 1990, pp. 173-183.

Kazimierczuk M.K. (1981). Class E tuned power amplifier with shunt inductor, *IEEE Journal of Solid-State Circuits*, Vol. 16, No. 1, February 1981, pp. 2 - 7.

Kazimierczuk, M.K. (2008). *RF Power Amplifiers*, Wiley, New York.

Kazimierczuk M. K. & Jozwik, J. (1990). Analysis and design of class E zero-current-switching rectifier, *IEEE Trans. Circuits Syst.*, Vol. 37, No. 8, August 1990, pp. 1000-1009.

Krauss, H. L.; Bostian, C.W. & Raab, F. H. (1980). *Solid State Radio Engineering*, Wiley, New York.

Lai, J. B. (2008). Investigation into the use of high-efficiency, switched mode Class E power amplifier for high-dynamic range, pulse-mode application, *PhD dissertation*, University of New Mexico Albuquerque, New Mexico, December, 2008.

Lepine, F.; Adahl, A. & Zirath, H. (2007). L-band LDMOS power amplifiers based on an inverse class-F architecture, *IEEE Transactions on Microwave Theory and Techniques*, Vol. 53, No. 6, Part 2, June 2005, pp. 2007 - 2012.

Lu, X. (1992). An alternative approach to improving the efficiency of high power radio frequency amplifiers, *IEEE Transaction on Broadcasting*, Vol. 38, No. 2, June 1992, pp. 85–89.

Mader, T.B.; Bryerton, E.W.; Markovic, M.; Forman, M. & Popović, Z. (1998). Switched-mode high-efficiency microwave power amplifiers in a free-space power-combiner array, *IEEE Transactions on Microwave Theory and Techniques*, Vol. 46, No. 10, Part 1, October 1998, pp. 1391 - 1398.

Mury, T. & Fusco, V.F. (2005). Series-L/parallel-tuned comparison with shunt-C/series-tuned class-E power amplifier, *IEE Proceedings Circuits, Devices and Systems*, December 2005, pp. 709 - 717.

Mury, T. & Fusco, V.F. (2007). Inverse Class-E Amplifier With Transmission-Line Harmonic Suppression, *IEEE Transactions on Circuits and Systems I: Regular Papers*, Vol. 54, No. 7, July 2007, pp. 1555 - 1561.

Negra, R.; Ghannouchi, F. M. & Bächtold, W. (2007). Study and Design Optimization of Multiharmonic Transmission-Line Load Networks for Class-E and Class-F K-Band MMIC Power Amplifiers, *IEEE Transactions on Microwave Theory and Techniques*, Vol. 55, No. 6, Part 2, June 2007, pp. 1390 - 1397.

Ozalas, M.T. (2005). High efficiency class-F MIMIC power amplifiers at ku-band, *2005 IEEE Wireless and Microwave Technology Conference*, WAMICON 2005, pp. 137 - 140.

Raab, F.H. (1996). Introduction to Class-F Power Amplifiers, *RF Design*, Vol.19, No. 5, May 1996, pp.79-84.

Raab, F.H. (1997). Class-F Power Amplifiers with Maximally Flat Waveforms, *IEEE Transactions on Microwave Theory and Techniques*, Vol. 45, No. 11, November 1997, pp. 2007-2012.

Raab, F.H. (2001). Class-E, Class-C, and Class-F power amplifiers based upon a finite number of harmonics, *IEEE Transactions on Microwave Theory and Techniques*, Vol. 49, No. 8, August 2001, pp. 1462 - 1468.

Reece, M.A.; White, C.; Penn, J.; Davis, B.; Bayne, M.Jr; Richardson, N.; Thompson, W.I.I. & Walker, L. (2003). A Ka-band class F MMIC amplifier design utilizing adaptable knowledge-based neural network modeling techniques, *IEEE International Microwave Symposium Digest*, June 2003, Vol. 1, pp. 615 - 618.

Snider, D.M. (1967). A Theoretical Analysis and Experimental Confirmation of the Optimally Loaded and Overdriven RF Power Amplifiers, *IEEE Transaction on Electron Devices*, Vol. 14, No. 6, June 1967, pp.851-857.

Sokal N. O. (2001). Class-E RF power amplifiers, *QEX* (published by American Radio Relay League, 225 Main St., Newington, CT 06111-1494, USA), January - February 2001, pp 9–20.

Sokal, N.O. & Sokal, A.D. (1975). Class E - A new class of high efficiency tuned single-ended switching power amplifiers, *IEEE Journal of Solid State Circuits*, Vol. SC-10, No. 3, June 1975, pp. 168-176.

Suetsugu, T & Kazimierczuk M.K. (2005). Voltage-clamped class E amplifier with transmission-line transformer, *IEEE International Symposium on Circuits and Systems*, ISCAS 2005, May 2005, Vol. 1, pp. 712 - 715.

Suetsugu, T. & Kazimierczuk M.K. (2007). Off-Nominal Operation of Class-E Amplifier at Any Duty Ratio, *IEEE Transactions on Circuits and Systems I: Regular Papers*, Vol. 54, No. 6, June 2007, pp. 1389 - 1397.

Trask, C. (1999). Class-F Amplifier Loading Network: A Unified Design Approach, *IEEE International Microwave Symposium Digest*, Vol. 1, June 1999, pp. 351 – 354.

Tyler, V.J. (1958). A new high efficiency high power amplifier, *Marconi Rev.* Vol. 21, No. 130, Fall 1958, pp. 96-109.

Wilkinson A. J. & Everard J. K. A. (2001). Transmission-Line Load-Network Topology for Class-E Power Amplifiers, *IEEE Transactions on Microwave Theory and Techniques*, Vol. 49, No. 6, June 2001, pp. 1202 - 1210.

Woo, Y.Y.; Yang, Y. & Kim, B. (2006). Analysis and Experiments for High-Efficiency Class-F and Inverse Class-F Power Amplifiers, *IEEE Transaction on Microwave Theory and Techniques*, Vol. 54, No. 5, May 2006, pp. 1969 - 1974.

Wood, J. (1992). The history of International broadcasting, *IET History of Technology Series 19*, 1992 (ISBN 0863413021)

Xu H.; Gao, S.; Heikman, S.; Long S. I.; Mishra, U. K. & York, R. A. (2006). A High-Efficiency Class-E GaN HEMT Power Amplifier at 1.9 GHz, *IEEE Microwave And Wireless Components Letters*, Vol. 16, No. 1, January 2006, pp. 22 - 24.

19

Developing the 150%-FBW Ku-Band Linear Equalizer

Sungtek Kahng

Abstract

This article reports the development of a linear amplitude equalizer for the linearity of the slope of the amplitude over 150% fractional bandwidth in Ku-band. The circuit model is featured by the resistor placed between each pair of a transmission-line and a stub. The design includes finding the values of resistors and stubs to have the optimal linear slope and return loss performances. The measured data show the acceptable performances of the slope variation and return loss over 2~18 GHz.

1. Introduction

The Radar Warning Receiver(RWR) for a helicopter tends to end up with the increasing overall insertion loss that is attributed to the cascaded placement of dissipative components such as Switch, Filter, Power-divider, Coupler and the like in the wide-banded RF channel. In order to reach for the target of quality performance, it is necessary to compensate for the insertion loss with flattening in the gain amplitude over the frequency band of interest. This is what is all about the gain(amplitude) equalization techniques which quite often entails the slope linearization.

To date, even though the domestic technical groups have presented that they are mature in implementing the gain equalization for commercial products operating in the relatively narrow frequency bands, they do not seem to meet the challenge of the equalization in broader bands for military applications.

As an attempt to meet the rising demands and boost the competitiveness of our technology, we have developed the linear gain equalizer working in the band as wide as over 10 GHz for the RWRs.

With a look into the current techniques of the gain equalization, it is found they can be classified to the followings : Linear and non-linear gain equalization methodologies[1-5]. The non-linear scheme is exploited to the narrow-banded sub-bands of one given broad-band, This is relatively easy to build up in the design, but it requires multitude of different stages corresponding to the sub-bands, which leads to cumbersome extra insertion losses, when the stages are electrically combined for its physical implementation. On the contrary, the linear equalizer necessitates one module, though its design seems tougher than the non-linear case. Besides, the linear equalization is advantageous in that it aims at the operation in one broad-band.

Making a noteworthy progress from what has been done previously as in [1-5], we present the linear gain equalizer working over 2 GH ~ 18 GHz by transmission lines with coupling elements.

2. Theoretical Side of Design

The gain equalizer plays a role of flattening the amplitude of the resultant insertion loss of the equalizer following the former component over the specified band[1-4].

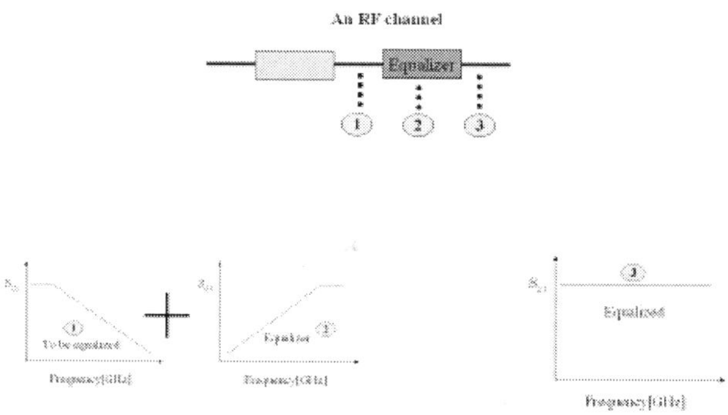

Fig. 1. Function of the linear gain equalizer

As shown in the result marked number 3, the balance is made by the ascending amplitude(marked number 1) added to the descending one(number 2). Particularly with Fig. 1, the equalizer shows the minimum loss at the high end of the band. Depending on cases, the slope of the equalizer's amplitude should be negative with the maximum loss at the upper end of the band. As a matter of course, the minimum loss of the equalizer is designed the lowest possible.

To begin, a low-ordered bandpass filter(BPF) is considered. In detail, the center frequency of the BPF is set at the end of the band(18GHz or 20GHz here), which is called the cut-off frequency in the gain equalizer design, with the ripple level of 0.1dB and the fractional bandwidth of 1. That is to say the Chebyshev filter of 1st order. Using this, it is effective to get the idea of how we get started, but falls short of satisfactory levels on return loss and linear slope. So it is inevitable to expand to a higher order circuit.

Regarding the fundamentals of the operation, the transmission line with the serial resistor is let go open at the cut-off frequency that is equivalent to the resonance frequency of series inductor and capacitor. At this point, the insertion loss becomes ideally zero. The rest of the band is designed to undergo the attenuation due to the T-network of three resistors, which determines the slope.

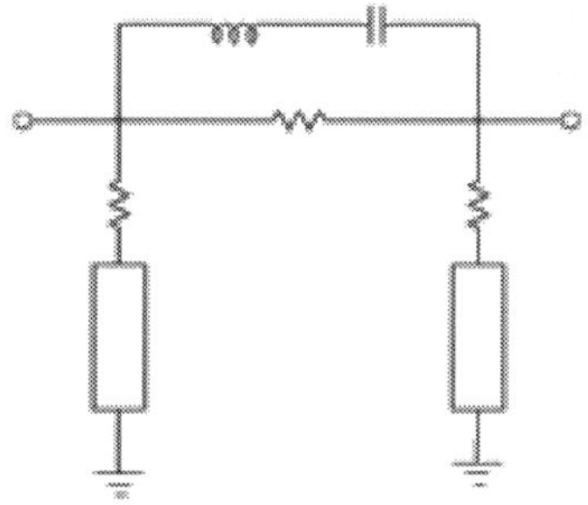

Fig. 2. Basic circuit of the linear equalizer

The design flow can be simply described as

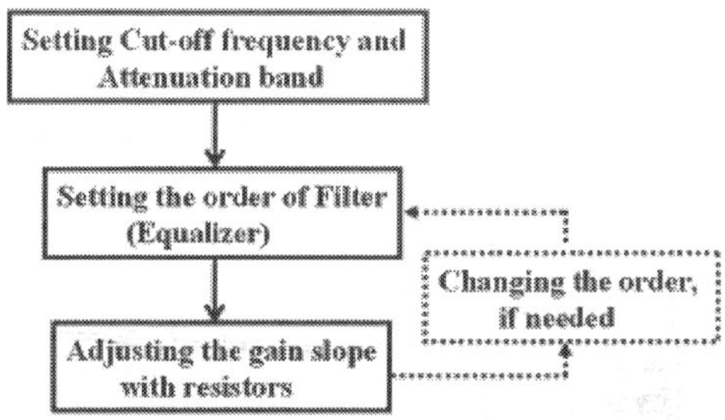

Fig. 3. Flowchart of the design

In the first place, we set the cut-off frequency of the gain equalizer at the center frequency of the nominal filter. Simultaneously, the attenuation band is defined. In the second place, the order of the filter is decided to have the slope of the amplitude as close as possible to the wanted value. And then, varying the resistors, the slope is adjusted to meet the spec. over the entire frequency band. If it is not satisfactory, return to the step where the order of the filter is determined and change to the immediate higher order.

3. Results of Design

Here comes the summary of the design specifications.

Item	Specs
fo, FBW	10GHz, 150%
Slope	10dB over the BW
Insertion loss	< 3dB at 18GHz
VSWR	< 2.0:1

Among the items, the slope is set the top priority. Maintaining the slope, the return and insertion losses are considered.

Firstly, let us start the design by changing the order of the basic circuit from 1 to 2.

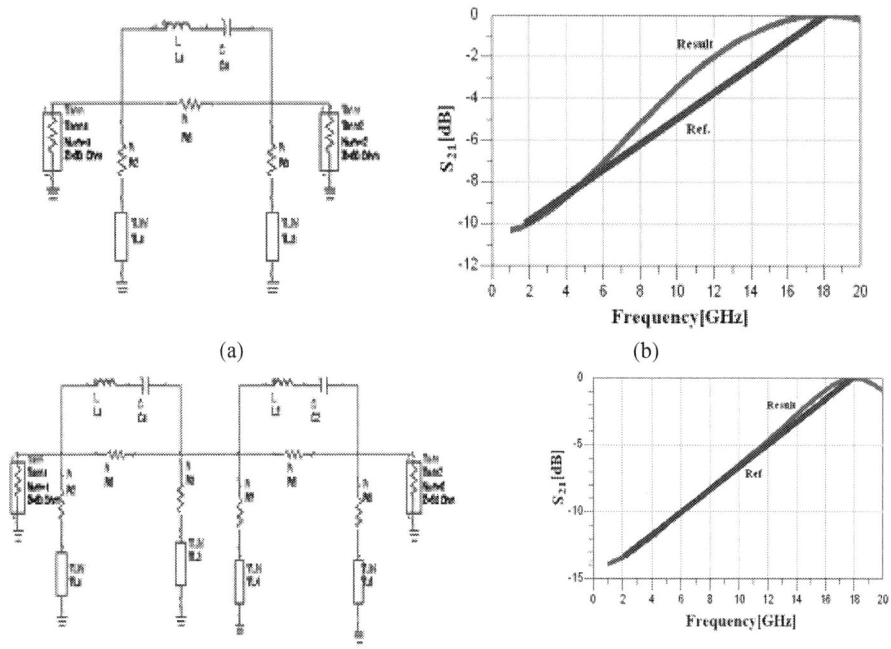

(c) (d)

Fig. 4. The 1st and 2nd order linear equalizers (a) 1st order circuit (b) Performance(1st order) (c) 2nd order circuit (d) Performance(2nd order)

The reactive elements are found by having their resonance at the cut-off frequency given in the specs. The resistors are computed, assumed that the T-networks are symmetric, to secure the gradient of the amplitude curve parallel to the given slope. Increasing the order of the equalizer, the slope performance has improved from Fig. 4(b) to Fig. 4(d). Taking into account the fabrication based upon the microstrip line, the reactive elements are replaced by the lossy transmission line(better for considering dispersion). The order of the entire circuit should be increased and the final design lends the performance in the insertion and return loss as follows.

Going through the tuning and trimming on the fabricated equalizer, the measured return and insertion losses amount to less than -10 dB and roughly 9 dB throughout the band(2GHz ~ 18GHz), respectively. Actually, the slightly non-linear behavior happens in the vicinity of 18GHz and it is believed to stem from the design ignorant of the capacitance parasitic to the resistors and transmission lines.

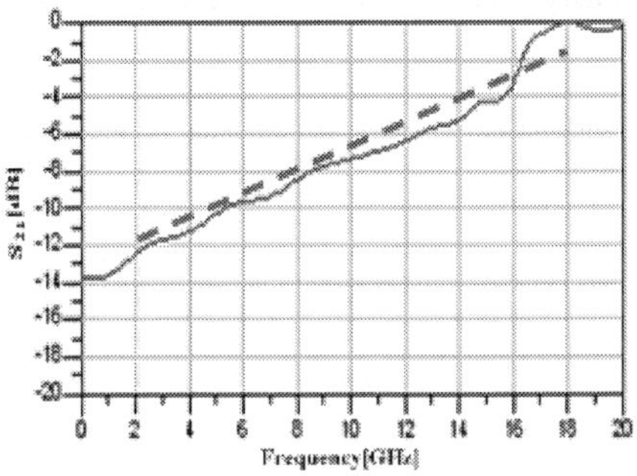

(a)

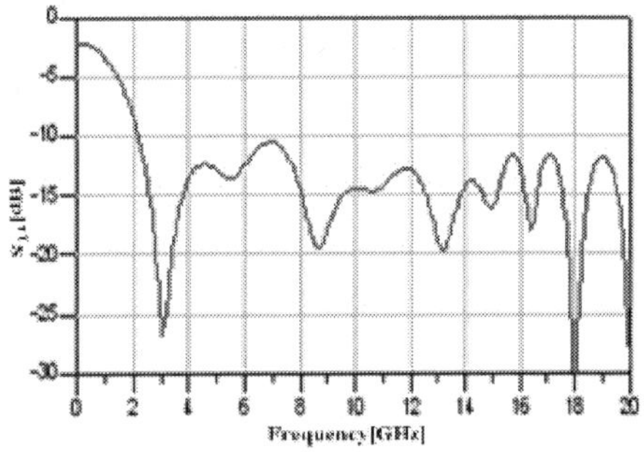

(b)

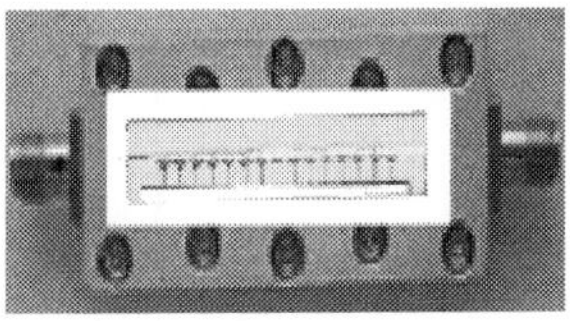

(c)

Fig. 5. The 14th order linear equalizers (a) Insertion loss (b) Return loss (c) Photo of the fabricated circuit

4. Conclusion

In this article, the design of a gain equalizer has been conceptualized to achieve the linear slope over the very wide band 2GH ~ 18GHz and good return loss performance. Besides, it has been implemented by fabrication with the microstrip transmission lines and SMT resistors. The measured data prove the realized equalizer outputs the acceptable linearity in the slope and return and insertion losses.

5. References

[1] Miodrag V. Gmitrovic et al, "Fixed and Variable Slope CATV Amplitude Equalizers," Applied Microwave & Wireless, Jan/Feb 1998, pp. 77-83.

[2] M. Sankara Narayana, "Gain Equalizer Flattens Attenuation Over 6-18 GHz," Applied Microwave & Wireless, November/December 1998.

[3] D.J.Mellor , "On the Design of Matched Equalizer of prescribed Gain Versus Frequency Profile". IEEE MTT-S International Microwave Symposium Digest , 1997, pp.308-311.

[4] Broadband MIC Equalizers TWTA Output Response . IEEE Design Feature. Oct 1993

[5] S. Kahng et al, "Expanding the bandwidth of the linear gain equalizer: Ku-band communication," KEES Journal, Vol. KEESJ18, No. 2, pp. 105-110,Feb. 2007.

[6] H. Ishida, and K. Araki, "Design and Analysis of UWB Bandpass Filter with Ring Filter," in IEEE MTT-S Intl. Dig. June 2004 pp. 1307-1310.

[7] H. Wang, L. Zhu and W. Menzel, "Ultra-Wideband Bandpass Filter with Hybrid Microstrip/CPW Structure," IEEE Microwave And Wireless Components Letters, vol. 15, pp. 844-846, December 2005

[8] S. Sun, and L. Zhu, "Capacitive-Ended Interdigital Coupled Lines for UWB Bandpass Filters with Improved Out-of-Band Performances," IEEE Microwave And Wireless Components Letters, vol. 16, pp. 440-442, August 2006.

[9] W. Menzel, M. S. R. Tito, and L. Zhu, "Low-Loss Ultra-Wideband(UWB) Filters Using Suspended Stripline," in Proc. Asia-Pacific Microw. Conf. , Dec. 2005, vol. 4, pp.2148-2151

[10] C.-L. Hsu, F.-C. Hsu, and J.-T. Kuo, "Microstrip Bandpass Filters for Ultra-Wideband(UWB) Wireless Communications," in IEEE MTT-S Intl. Dig. , June 2005, pp.675-678

[11] C. Caloz and T. Itoh, Electromagnetic Metamaterials : Transmission Line Theory and Microwave Applications, WILEY-INTERSCIENCE, John-Wiley & Sons Inc., Hoboken, NJ 2006

[12] S. Kahng, and J. Ju, "Left-Handedness based Bandpass Filter Design for RFID UHF-Band applications," in Proc. KJMW 2007, Nov.. 2007, vol. 1, pp.165-168.

[13] J. Ju and S. Kahng, "Design of the Miniaturaized UHF Bandpass Filter with the Wide Stopband using the Inductive-Coupling Inverters and Metamaterials," in Proc. Korea Electromagnetic Engineering Society Conference 2007, Nov. 2007, vol. 1, pp.5-8

[14] K. C. Gupta, R. Garg, I. Bahl, and P. Bhartia, Microstrip Lines and Slotlines, Artech House Inc., Norwood, MA 1996

20

Ultrawideband Bandpass Filter using Composite Right- and Left-Handedness Line Metamaterial Unit-Cell

Sungtek Kahng

Abstract
The design of a new UWB bandpass filter is proposed, which is based upon the microstrip Composite Right- and Left-Handed Transmission-line(CRLH-TL). In order to bring the remarkable improvement in an attempt to reduce the size, taking the features of the conventional periodic CRLH-TL, only one unit of the structure is chosen. So the component less than a quarter-wavelength is realized to achieve the ultra wide band filtering without the loss of the original advantage of the CRLH-TL. Guaranteeing the compactness in size, the interdigitated coupled lines are used to realize the strong coupling for the design that will be shown to have the size of 'guided wavelength/9.4', the fractional bandwidth over 100%, the insertion loss much less than 1 dB, and the flat group-delay with an acceptable return loss performance in the predicted and measured results.

1. Introduction

In recent years, numerous studies have been conducted to exploit the benefits of the UWB communication, since its unlicensed use was open to the public by the US FCC. As one of many such research activities, the design methods of bandpass filters have been reported[6-10].

Araki et al [6] designed the UWB bandpass filter whose bandwidth is formed by adding zeros in the sections of the transmission line. The frequency response has notches at the specific points as the very narrow regions for out-of-band suppression. H. Wang et al [7] presented the microstrip-and-CPW bandpass filter for the UWB application, which is based upon the Multi-Mode Resonator(MMR) in the form of multiples of quarter-wavelength, to broaden the bandwidth and obtain the enlarged rejection region. The idea of the MMR of the half wavelength is also used in [8] where the coupled lines of a quarter-wavelength are used as the inverter. This work shows the extension of the lower and higher stopbands owing to the increased coupling. A composite UWB filter was designed by W. Menzel et al by combining lowpass and high pass filters as a suspended stripline structure with different planes[9]. Independently, C. Hsu et al presented the composite microstrip filters for the UWB application, where seven or eight TL sections of about quarter-wavelength are sequentially connected[10]. Presently, we describe the design method of a new UWB filter on the basis of the composite right- and left-handed transmission line(CRLH-TL)[11-13].

Different from the reference [11], we take just one segment(smaller than one quarter-wavelength) from the periodic structure of the CRLH-TL to make the component very compact. Besides, instead of mixing two types, for instance, hybrid of the microstrip and CPW, the filter design is pursued with only the microstrip. Most of all, what features in our present work is that the interdigital coupled lines much smaller than a quarter wavelength and the grounded stub account for the strong capacitive coupling and the inductance for the left-handedness, respectively, and the effective inductance of the interdigital capacitor and the effective capacitance of the short-circuited inductor are used to decide the right-handedness characteristics, in order to form a ultra wideband. And then going through the implementation process, the predicted performances of the designed filter are given with the measurement of the fabricated one to validate our design methodology, where the design of the proposed BPF reveals the suitability for the UWB application, showing the size reduction to the guided wavelength/9.4, the bandwidth more than 100%, the insertion loss lower than 1 dB, the group-delay variation less than 0.5 ns with the good return loss property.

2. Design of The Crlh-Tl Type Uwb Bpf

The left-handed medium as a metamaterial has been examined theoretically and experimentally as it plays the lumped high pass filter circuit, and its unit cell in a periodic transmission line is smaller than the guided wavelength. Instead of the pure left-handedness, the CRLH-TL as a more practical circuit has been portrayed by C. Caloz et al[11]. It is represented by Fig. 2-1.

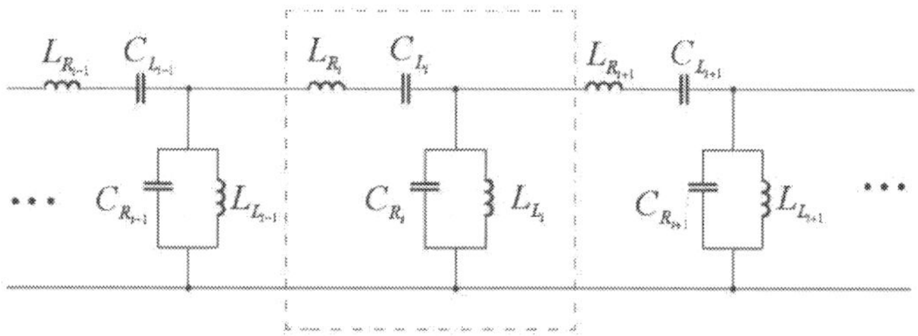

Fig. 2-1. Equivalent circuit model of the conventional periodic CRLH -TL

There are three intermediate units of the periodic CRLH-TL and the *i*-th segment is marked by the dotted line block in Fig. 1. The *i*-th segment consists of (C_{Li}, L_{Li}) for the left-handedness and (C_{Ri}, L_{Ri}) for the right-handedness property. From the standpoint of the purely left-handed unit, L_{Ri} and C_{Ri} can be considered parasitic inductance and capacitance against C_{Li} and L_{Li}, respectively. However, in our design, we use the effective inductance L_{Ri} and the effective capacitance C_{Ri} for the purpose of forming a pass-band for the UWB filter. As is addressed previously, only the basic unit, say, the *i*-th segment is taken for the present work. Its symmetric version can be expressed a Pi-equivalent circuit in Fig. 2-2.

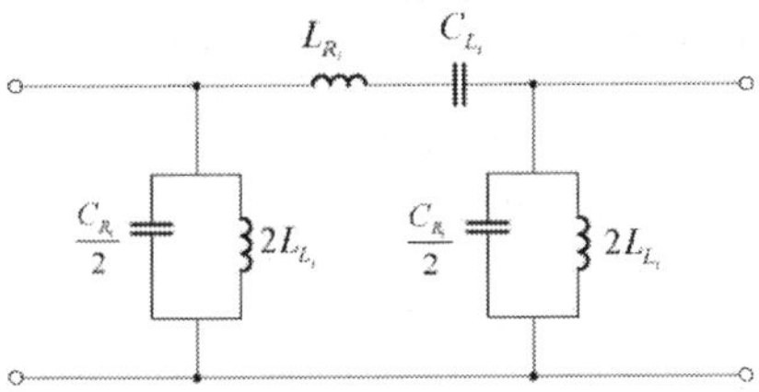

Fig. 2-2. Pi-equivalent circuit of the unit cell from the CRLH-TL

The ladder type of circuit in Fig. 1 has the exactly the same function as that in Fig. 2-2. But the difference between them is the physical configuration, and this will be shed a light on later. What is important in using the basic unit of the CRLH-TL in Fig. 2 is to determine the values of the elements (C_{Li}, L_{Li}, C_{Ri}, L_{Ri}) that produce the performances appropriate to the UWB BPF. We adopt the concept of the Balanced CRLH-TL in [11] to achieve a single broad band without any gap in between the cut-off frequencies of highpass and lowpass filtering. In the Balanced case, the three from four resonance phenomena lead to the following relations.

$$f_{Li} = \frac{1}{2\pi\sqrt{L_{Li}C_{Li}}}, \quad f_{Ri} = \frac{1}{2\pi\sqrt{L_{Ri}C_{Ri}}}$$

$$f_{sei} = f_{shi} = f_O, \quad f_O = \sqrt{f_{Li}f_{Ri}} \qquad (2\text{-}1)$$

where

$$f_{sei} = \frac{1}{2\pi\sqrt{L_{Ri}C_{Li}}}, \quad f_{shi} = \frac{1}{2\pi\sqrt{L_{Li}C_{Ri}}}$$

That f_{sei} is let equal to f_{shi} means the *balance* in the CRLH-TL, where f_{Li}, f_{Ri}, f_{sei}, f_{shi}, and f_O correspond to the lower band-edge, upper band-edge, series resonance point, shunt resonance point and center frequency, respectively. Solving the equations above, the circuit elements are identified.

In order for a BPF to have the ultra wideband, a strong coupling is essential to the implementation. In particular, the sufficient large amount of C_{Li} is required.

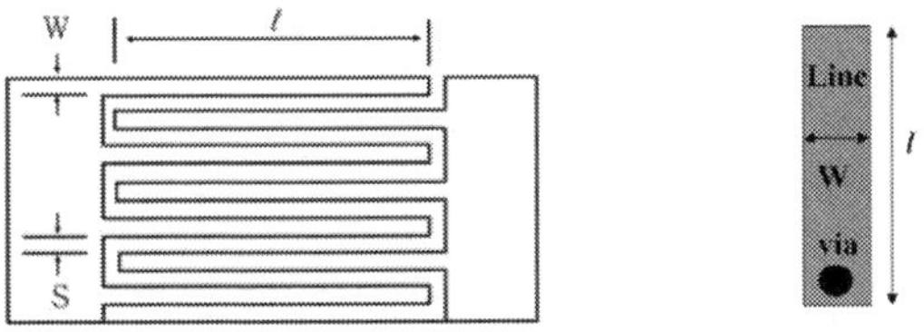

Fig. 2-3. Microstrip interdigital coupled lines and grounded stub

As explained in the introduction with other design cases where the hybrid of the microstrip/CPW or the cascaded transmissions of wavelengths are used, CLi should be large enough, as the designers' main concern. Like them, we need a strong capacitive coupling, but proceed with the microstrip interdigital coupled lines. Even if the interdigital line has been around for quite some time, as is stated before, its geometric parameters will be explored to find the desired effective inductance LRi as well as CLi in our design, different from others. Fig. 2-3 presents the typical interdigital line. The geometry of an nIDF fingered interdigital line described with W, l and S denoting the finger width, the finger length and the spacing between the two adjacent fingers, respectively. The capacitance of Fig. 2-3 is given as follows.

$$C(pF) = \frac{\varepsilon_{re} 10^{-3}}{18\pi} \frac{K(k)}{K'(k)} (n_{IDF} - 1)\ell \tag{2-2}$$

where

$$k = \tan^2\left(\frac{a\pi}{4b}\right), \quad a = \frac{W}{2}, \quad b = \frac{W+S}{2}$$

$K(\cdot)$ and $K'(\cdot)$ are the complete elliptic integral of the 1st kind and its complement. Along with the series interdigital line, the grounded shunt stub plays an important role. The expression as follows is commonly used for the inductance of the grounded stub and each finger in the interdigital line(L_{Ri}). Though it is an approximate formula, it helps us quickly approach the initial size.

$$L(nH) = 2 \times 10^{-4} l[\ln(\frac{l}{W+t}) + 1.193 + 0.224 \frac{W+t}{l}] \cdot K_g \tag{2-3}$$

where

$$K_g = 0.57 - 0.145 \ln(\frac{W}{h})$$

h and t above mean the thickness of the substrate and metallization in use. The expressions for the other circuit elements are found in [9] and used to correct the electrical behaviors based upon Eqns (2) and (3). With all these values, physical sizes are iteratively exploited until the acquisition of the desired performance.

3. Results of Implementation

Use First of all, the interdigital line's size is calculated to realize the capacitance of 0.477pF and its effective inductance of 5.53nH. Via the iterative steps using Eq's (2) and (3), the initial values are found W=0.20 mm, l =1.30mm, S=0.12 and n_{IDF} =14.

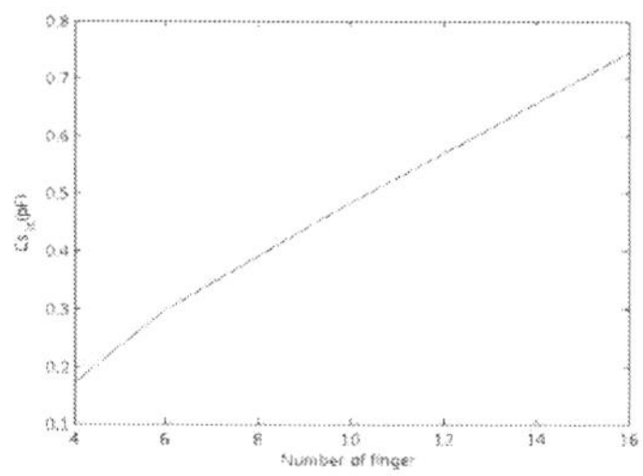

(a)

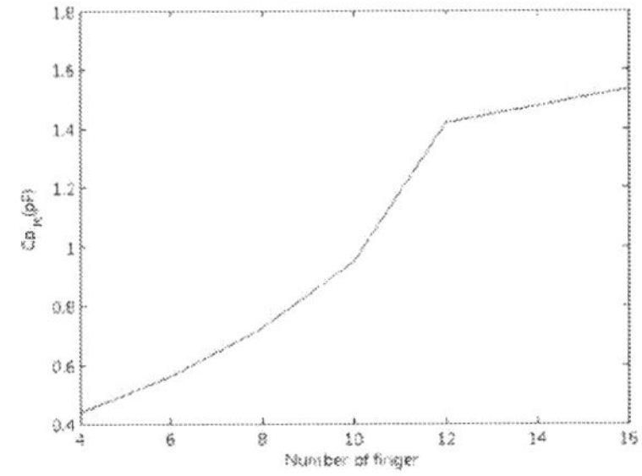

(b)

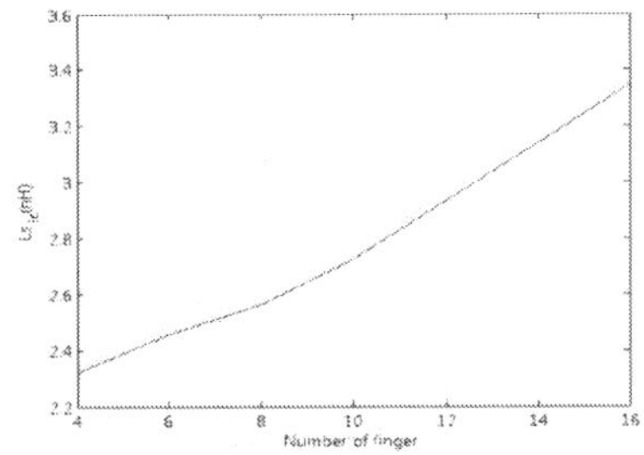

(c)

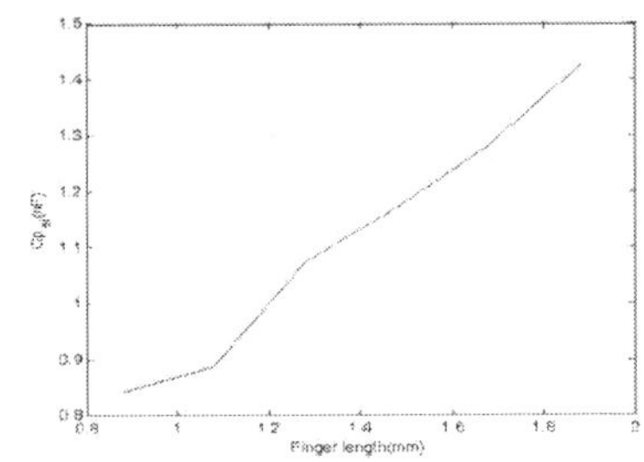

(d)

Fig. 2-4. Interdigital line's capacitance and inductance V.S. geometric changes (a) Number of fingers V.S. *Cs* (b) Number of fingers VS. *Cp* (c) Number of fingers V.S. *Ls* (d) Length of the finger V.S. *Cp*

This is followed by finding the physical dimensions of the grounded transmission line stub whose *W* and *l* are 0.5 mm and 5.0 mm with 1.13nH and 0.20pF. For the substrate, FR4(ε_r = 4.4) is used. And the circuit values result in the following dispersion diagram. Resorting to the conventional periodic CRLH-TL concept, just for convenience, we check the critical points, say, transmission and stop bands .

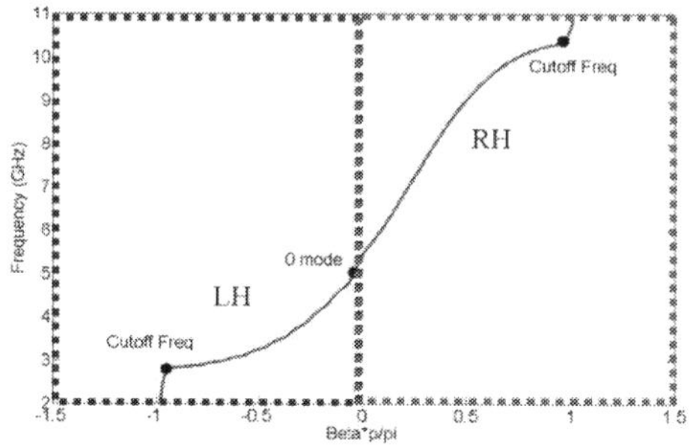

Fig. 2-5. Dispersion curve of the proposed UWB BPF

The refined physical dimensions based upon the initial values for the filter's geometry, the 3D EM full-wave simulation has been carried out.

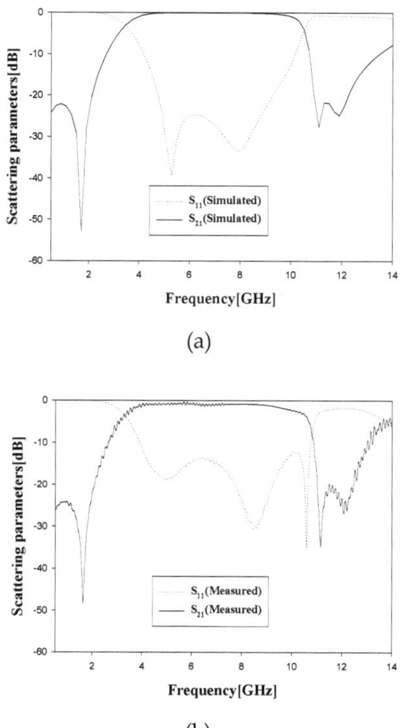

Fig. 2-6. S_{11} and S_{21} of the proposed UWB BPF (a) Simulation (b) measurement.

Fig. 2-6 plots the simulated scattering parameters S_{11} and S_{21} verified by the measurement. Excellent agreement is shown between the simulated and measured S_{21} with almost the same transmission zeros, bandwidth over 100 % and insertion loss less than 1dB. Also, good return loss is given despite the small discrepancy guessed due to the mechanical tolerance error. Next, we need to check out the group-delay of the designed filter.

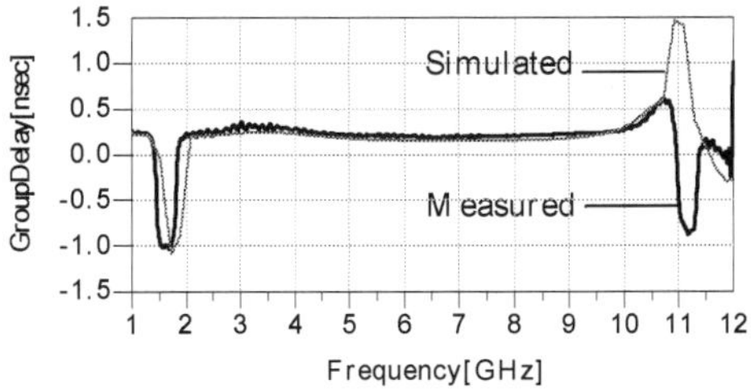

Fig.2- 7. Group-delay of the proposed UWB BPF : Simulation and measurement.

The variation of the group-delay is as small as less than 0.25 nsec over the passband. Lastly, we show the photograph of our fabricated UWB BPF.

Fig. 2-8. Picture of the designed UWB BPF

The interdigital line sandwiched by the grounded stubs composes the proposed filter which is about 4.7 mm long(far less than a quarter guided-wavelength).

4. Conclusion

The A new compact UWB BPF is proposed using the concept of the CRLH-TL. Only 1 unit of the CRLH-TL is taken for enhanced size reduction and implemented with the interdigital line and grounded stubs with their effective parasitics for the UWB. The designed BPF performs with the BW over 100%, good insertion and return loss, and flat group-delay with the overall size to the guided wavelength/9.4.

5. Acknowledgment

This work was supported by the IT R&D program of MKE/IITA. [2009-S-001-01, Study of technologies for improving the RF spectrum characteristics by using the meta-electromagnetic structure]

6. References

[1] Miodrag V. Gmitrovic et al, "Fixed and Variable Slope CATV Amplitude Equalizers," Applied Microwave & Wireless, Jan/Feb 1998, pp. 77-83.
[2] M. Sankara Narayana, "Gain Equalizer Flattens Attenuation Over 6-18 GHz," Applied Microwave & Wireless, November/December 1998.
[3] D.J.Mellor , "On the Design of Matched Equalizer of prescribed Gain Versus Frequency Profile". IEEE MTT-S International Microwave Symposium Digest , 1997, pp.308-311.
[4] Broadband MIC Equalizers TWTA Output Response . IEEE Design Feature. Oct 1993
[5] S. Kahng et al, "Expanding the bandwidth of the linear gain equalizer: Ku-band communication," KEES Journal, Vol. KEESJ18, No. 2, pp. 105-110,Feb. 2007.
[6] H. Ishida, and K. Araki, "Design and Analysis of UWB Bandpass Filter with Ring Filter," in IEEE MTT-S Intl. Dig. June 2004 pp. 1307-1310.
[7] H. Wang, L. Zhu and W. Menzel, "Ultra-Wideband Bandpass Filter with Hybrid Microstrip/CPW Structure," IEEE Microwave And Wireless Components Letters, vol. 15, pp. 844-846, December 2005
[8] S. Sun, and L. Zhu, "Capacitive-Ended Interdigital Coupled Lines for UWB Bandpass Filters with Improved Out-of-Band Performances," IEEE Microwave And Wireless Components Letters, vol. 16, pp. 440-442, August 2006.
[9] W. Menzel, M. S. R. Tito, and L. Zhu, "Low-Loss Ultra-Wideband(UWB) Filters Using Suspended Stripline," in Proc. Asia-Pacific Microw. Conf. , Dec. 2005, vol. 4, pp.2148-2151
[10] C.-L. Hsu, F.-C. Hsu, and J.-T. Kuo, "Microstrip Bandpass Filters for Ultra-Wideband(UWB) Wireless Communications," in IEEE MTT-S Intl. Dig. , June 2005, pp.675-678
[11] C. Caloz and T. Itoh, Electromagnetic Metamaterials : Transmission Line Theory and Microwave Applications, WILEY-INTERSCIENCE, John-Wiley & Sons Inc., Hoboken, NJ 2006
[12] S. Kahng, and J. Ju, "Left-Handedness based Bandpass Filter Design for RFID UHF-Band applications," in Proc. KJMW 2007, Nov.. 2007, vol. 1, pp.165-168.
[13] J. Ju and S. Kahng, "Design of the Miniaturaized UHF Bandpass Filter with the Wide Stopband using the Inductive-Coupling Inverters and Metamaterials," in Proc. Korea Electromagnetic Engineering Society Conference 2007, Nov. 2007, vol. 1, pp.5-8
[14] K. C. Gupta, R. Garg, I. Bahl, and P. Bhartia, Microstrip Lines and Slotlines, Artech House Inc., Norwood, MA 1996

Extended Source Size Correction Factor in Antenna Gain Measurements

Aleksey Solovey and Raj Mittra
L-3 Communications ESSCO, Pennsylvania State University
USA

1. Introduction

In this chapter we consider the extended source size correction factor that is widely utilized in antenna gain measurements when extraterrestrial extended radio sources are in use. Extended radio sources having an angular size that is comparable or even larger than the antenna's far-field power pattern Half Power Beam Width (HPBW) are often used to determine various antenna parameters including gain of electrically large antenna apertures (Baars, 1973). The use of such sources often becomes almost inevitable since the far-field distance of electrically large antennas can reach tens or even hundreds kilometers, which makes it impractical if not impossible to employ the conventional far-field antenna transmitter-receiver test range technique.

To illustrate the importance of the problem let's consider 30m radio astronomy antenna at 10GHz and 96GHz frequency bands. Its far-field zone distances are 60km and 580km with the HPBW equal to 4′ and 0.42′ respectively. For the comparison, among the strongest cosmic radio sources, Cassiopeia A has 4′ and Sygnus A has 0.7′ of their disk angular sizes (Guidici & Castelli, 1971). Even for the 7m communication antenna working at 20 GHz, the far-field distance is 6.5km and the far-field patterns HPBW is 1.4°, while the angular size of the Sun or the Moon disks are about of 0.5° (Guidici & Castelli, 1971).

When the radio source angular size is comparable with the antenna HPBW, the antenna radiation pattern is averaged within the solid spatial angle subtended to the source. Therefore, the measured antenna gain value appears to be less than what would be expected for the antenna's effective collecting area and the aperture illumination and the resulting gain measurements must be corrected by the extended source size correction factor to account for the convolution of the extended radio source angular size, angular source brightness distribution and the shape of the antenna's far-field radiation pattern. In this chapter two kinds of extended radio sources, having either uniform or Gaussian brightness distributions over the source disk (Baars, 1973; Kraus, 1986), along with three kinds of the most usable "Polynomial-on-Pedestal," Gaussian, and Taylor antenna aperture illuminations are examined for circular and rectangular antenna apertures.

As a result of the above considerations and based on the literature survey, the complete set of simple analytical expressions that accurately approximate the value of the extended source size correction factor have been derived and/or developed for circular and rectangular antenna apertures and for all the above combinations of extended radio source brightness distributions and antenna aperture illuminations. Those expressions eliminate the need to perform complicated and often impractical numerical integrations in order to evaluate the extended source size correction factor value for the case of particular measurement. The approximate analytical expressions for the extended source size correction factor for rectangular antenna apertures along with their tolerances for circular and rectangular apertures are obtained for the first time in literature.

Because the extended source size correction factor most conveniently can be expressed through the ratio between the extended source angular size (or its HPBW) and the antenna's far-field radiation pattern HPBW, the approximations of the antenna's HPBW for all three types of antenna aperture illuminations and for circular and rectangular antenna apertures are considered as a supplementary problem. As a result, numerous simple and accurate analytical expressions for the antenna's far-field pattern HPBW for circular and rectangular antenna apertures and for all three types of antenna aperture illuminations have been developed. This also eliminates the need to perform complicated and often impractical numerical integrations in order to evaluate the antenna's far-field pattern HPBW value for the particular antenna size(s) and aperture illumination(s). In addition, for circular and rectangular antenna apertures these expressions are shown in the form of plots.

While in this chapter, we consider the extended source size correction factor from the prospective of the antenna gain measurement, it should be noted that the same factor can also be utilized for the solution of the inverse problem: the measurement of the unknown temperature and/or flux density of a randomly polarized extended radio source using the electrically large antenna with a known antenna far-field power pattern (Ko, 1961).

2. Extended Cosmic Radio Sources and Extended Source Size Correction Factor in Antenna Gain Measurements

The IEEE Standard Std 149-1979 for the antenna test procedures (Kummer at al., 1979) defines the extended source size correction factor K by the following expression:

$$K = \frac{\iint_{\Omega_S} B_S(\Omega)d\Omega}{\iint_{\Omega_S} B_S(\Omega)F_n(\Omega)d\Omega} \qquad (1)$$

where $B_s(\Omega)$ is the angular brightness distribution of the extended radio source, $F_n(\Omega)$ is the normalized pencil beam antenna far-field power pattern, such that at the antenna boresight (direction of the peak of the main beam) $F_n(0) = 1$, and Ω_S is the solid angle subtended to the extended radio source. To obtain the correct antenna gain, the gain value measured using the extended radio source should be multiplied by K.

In order to understand the meaning and the applicability of definition (1) let's first establish the relations between the power pattern $F_n(\Omega)$ and the antenna far-field radiation power pattern $F(\Omega)$ that is normalized the way that the total power transmitted or received by the antenna within the entire 4π steradian solid angle equals to one power unit:

$$\iint_{4\pi} F(\Omega)d\Omega = \iint_{4\pi} A \times F_n(\Omega)d\Omega = 1 \tag{2}$$

where A is a constant multiplier that needs to be found. In case of the omnidirectional antenna when $F_n(\Omega) = 1$ at any direction of the solid angle Ω, condition (2) gives the following value of that multiplier A_{omni}:

$$\iint_{4\pi} A_{omni} d\Omega = 1 \Rightarrow A_{omni} = \frac{1}{4\pi} \tag{3}$$

By definition (Balanis, 2005), the maximum antenna gain G_{max} is the ratio of the maximum antenna radiation intensity, i. e. A, to the radiation intensity averaged over all directions, i. e. A_{omni}:

$$G_{max} = \frac{A}{A_{omni}} = 4\pi A \Rightarrow A = \frac{G_{max}}{4\pi} \tag{4}$$

Based on (2) and (4) one can notice that the denominator in (1) is proportional to the power $P_{extended\ source}$ received by the antenna whose maximum gain value is G_{max} in the case when its far-field pattern HPBW is comparable or greater than the size of the extended radio source spatial angle:

$$P_{extended\ source} = \iint_{\Omega_S} B_S(\Omega)\frac{G_{max}}{4\pi}F_n(\Omega)d\Omega \tag{5}$$

The numerator in (1) is proportional to the power $P_{point\ source}$ received by the same antenna in the case when its far-field antenna pattern HPBW is much bigger than the angular size of extended radio source and thus, the value $F_n(\Omega)$ in (5) can be substituted by 1 within the source spatial angle:

$$P_{point\ source} = \iint_{\Omega_S} B_s(\Omega)\frac{G_{max}}{4\pi}d\Omega \tag{6}$$

The relations (5) and (6) can be used by both ways: to find the unknown maximum antenna gain and the radiation pattern based on the measured power received by the antenna and the known source brightness distribution or, conversely, to find the unknown source brightness distribution based on the measured power received by the antenna and the known maximum antenna gain and the radiation pattern. Either way, the ratio between the received power (6) and the received power (5) is equal to the extended source size correction factor defined by (1) and represents the correction that should be made if the extended radio source, but not the point source, is used in the measurement procedure.
In order to compute the extended source size correction factor using its definition (1), both the extended source brightness $B_s(\Omega)$ and the normalized antenna power pattern $F_n(\Omega)$ as a function of the solid angle subtended to the extended source should be known. In following sections we review expressions for the extended source correction factor already existed in literature, discuss and specify the considered set of functions $B_s(\Omega)$ and $F_n(\Omega)$, its relations with actual extended radio sources and antenna configurations and disclose numerous

novel simple analytical expressions that accurately approximate the value of extended source size correction factor for those combinations of $B_s(\Omega)$ and $F_n(\Omega)$.

3. Existing Approximate Analytical Formulae for Extended Source Size Correction Factor

Based on definition (1) of the extended source size correction factor and making various simplifying assumptions about the source brightness distribution $B_s(\Omega)$ and the normalized antenna far-field power pattern $F_n(\Omega)$, several approximate analytical expressions for the extended source correction factor have been developed in literature.

For the case of a Gaussian source brightness distribution and a Gaussian antenna far-field power pattern (Guidici & Castelli, 1971) and (Baars, 1973) stated the following expression for the extended source correction factor K:

$$K = 1 + s^2 \tag{7}$$

where s is the ratio of the extended source HPBW to the antenna HPBW. For the case of a uniform source brightness distribution and a Gaussian antenna far-field power pattern (Ko, 1961) expresses value of K as:

$$K = \frac{(s\sqrt{\ln 2})^2}{1 - \exp[-(s\sqrt{\ln 2})^2]} \tag{8}$$

where s is the ratio of the disk source angular diameter to the antenna HPBW. For the case of a uniform source brightness distribution and an antenna far-field power pattern that corresponds with a uniform antenna aperture illumination (Ko, 1961) expresses the value of K as:

$$K = \frac{(1.616s)^2}{4[1 - J_1^2(1.616s) - J_0^2(1.616s)]} \tag{9}$$

where s is the ratio of the source disk diameter to the antenna HPBW.

It should be noted that expressions (7) - (9) were derived under the assumption of a small value of variable s, i. e., when the extended source angular size is noticeably less than the antenna pattern HPBW and only for the circular antenna aperture.

In Figure 1, the approximate values of the extended source size correction factor as a function of the ratio of the source diameter or the source HPBW to the antenna HPBW and given by expressions (7) - (9) are shown in comparison. As is seen from Figure 1, the values given by expressions (8) and (9) are very close to each other, while being significantly different from values given by the expression (7). Expressions (7) - (9) were selected among few others because they represent the best approximations of the extended source size correction factor available in the literature for both uniform and Gaussian extended source brightness distributions that are used the most in antenna measurements and/or calibrations.

However, there are also several problems that are associated with these approximations. First, approximations (7) and (8) are based on the shape of the antenna far-field power

pattern rather than based on the shape of the antenna aperture illumination and its edge illumination taper, which are actually known in practice. Secondly, in order to use expressions (7) - (9), the antenna HPBW as a function of the type of the aperture illumination and its edge illumination taper has to be known. Third, not all combinations of aperture illuminations and extended source brightness distributions used in practice are covered by (7) - (9). Fourth, all expressions (7) - (9) were derived for the small value of the source size to antenna HPBW ratio and it's accuracy for the case when this ratio isn't small is unknown. Fifth, these expressions were derived only for circular antenna apertures and do not cover rectangular apertures that become increasingly popular for the modern solid state antennas. Sixth, expressions (7) - (9) do not explicitly depend on the antenna aperture edge illumination taper, which is manifestly wrong since the normalized antenna power pattern $F_n(\Omega)$ and therefore, the extended source size correction factor do depend on it.
Further in this chapter we will amend and expand expressions (7) - (9) the way that eliminates all its deficiencies that were mentioned above.

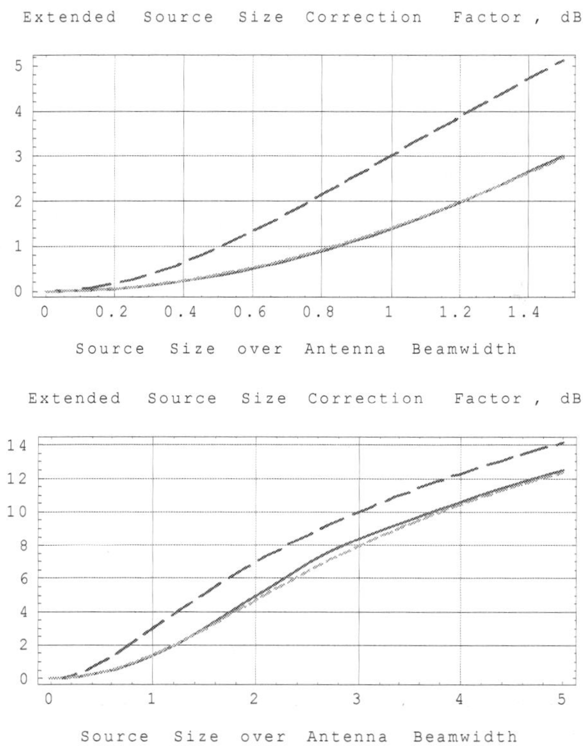

Fig. 1. Approximate expressions for extended source size correction factor in comparison. *Plot legend*: solid red - expression (9); long-dashed blue - expression (7); and short-dashed green - expression (8);

4. Brightness Distributions of Extended Cosmic Radio Sources Used in Antenna Gain Measurements

The detailed description of most cosmic extended radio sources that are used in the electrically large antenna measurements and calibrations along with the discussion of various aspects of such measurements is given, for example, in (Baars, 1973), (Guidici & Castelli, 1971) and (Kraus, 1986).
From the extended source correction factor evaluation standpoint it's enough to notice that most of these extended radio sources have circular disk shape with either axially symmetrical Gaussian or uniform brightness distributions over the solid angle that is subtended to the extended radio source disk (Baars, 1973), (Kraus, 1986). Those sources are almost entirely incoherent and that is why in (1) the antenna far-field power pattern, as oppose to the antenna far-field radiation field pattern, is in use. For example, the Cassiopeia A has a uniform brightness distribution, while the Orion A has an axially symmetrical Gaussian brightness distribution over a source disk (Baars, 1973).
Thus, for the purpose of this chapter, we will consider only these two, the uniform and the Gaussian extended radio source brightness distributions, which can be written in following forms:

$$B_{S\,uni}(\theta,\varphi)=\begin{cases}1, & \theta\le\theta_S/2\\ 0, & \theta>\theta_S/2\end{cases} \tag{10}$$

$$B_{S\,Gauss}(\theta,\varphi)=\exp\left[-\ln 2\left(\frac{2\theta}{\theta_S}\right)^2\right] \tag{11}$$

where θ_s is the angular size of the extended radio source. In the case of the uniform source brightness distribution θ_s is the physical angular size of the source disk. In the case of the Gaussian source brightness distribution, the source angular size θ_s is defined as a source HPBW, i. e., the angular size at which the brightness of the source is half of its maximum at the center of the source disk.
It should be stressed that the assumed properties of the extended radio source namely, the radiation incoherence and the types of source brightness distribution (10) and (11), are essential for the correct simulation of the extended source size correction factor.

5. Illumination Functions and Antenna Patterns for Circular and Rectangular Antenna Apertures

5.1 Circular Antenna Aperture Case

In order to calculate the value of the extended source size correction factor (1) except of the extended source brightness distribution function $B_s(\Omega)$ that was described in previous section, the antenna far-field power pattern $F_n(\Omega)$ needs to be known. Unlike the extended source brightness distribution $B_s(\Omega)$ the antenna far-field power pattern $F_n(\Omega)$ for almost all practical cases is not an analytically defined function and is rather defined through the particular illumination of the antenna aperture.

For the purpose of this chapter we chose the three most usable circularly symmetrical aperture illumination functions: "Polynomial-on-Pedestal," Gaussian, and Taylor. For convenience, we use following forms of these aperture illumination functions for the circular aperture:

$$f_{ApPoly} = B + (1 - B)\left[1 - \left(\frac{2r}{d}\right)^2\right]^2 \tag{12}$$

$$f_{ApGauss} = \exp\left[-(\sqrt{-\ln B}\,\frac{2r}{d})^2\right] \tag{13}$$

$$f_{ApTaylor} = \frac{J_0\left[j\pi\beta\sqrt{(1-(\frac{2r}{d})^2}\right]}{J_0[j\pi\beta]} \tag{14}$$

where r is the value of the radius vector from the center of the circular aperture, d is the diameter of the antenna aperture, B is the aperture edge illumination taper ($0 \le B \le 1$), and the constant β in the expression (14) can be found from the equation:

$$J_0[j\pi\beta] = \frac{1}{B} \tag{15}$$

The aperture edge illumination taper is usually expressed in dB, thus for convenience, we introduce the constant c that is associated with constant B by:

$$c = -20\log_{10} B \tag{16}$$

Here's another useful constant that is associated with constant B and will be extensively used throughout the chapter:

$$b = 1 - B \tag{17}$$

The example of all three circular aperture illumination functions (12) - (14) are shown in Figure 2 for comparison. As is seen from plots in Figure 2, expression (12) - (14) are normalized so that all aperture illuminations (12) - (14) have the same maximum value of 1 at the center of the aperture and the same minimum value at the edge of the aperture that is equal to the value of the aperture edge illumination taper B.

Knowing the antenna aperture illuminations (12) - (14) and using the aperture approach for the antenna pattern calculation, the normalized antenna power pattern for the circular antenna aperture can be expressed, according to (Johnson at all, 1993) as follows:

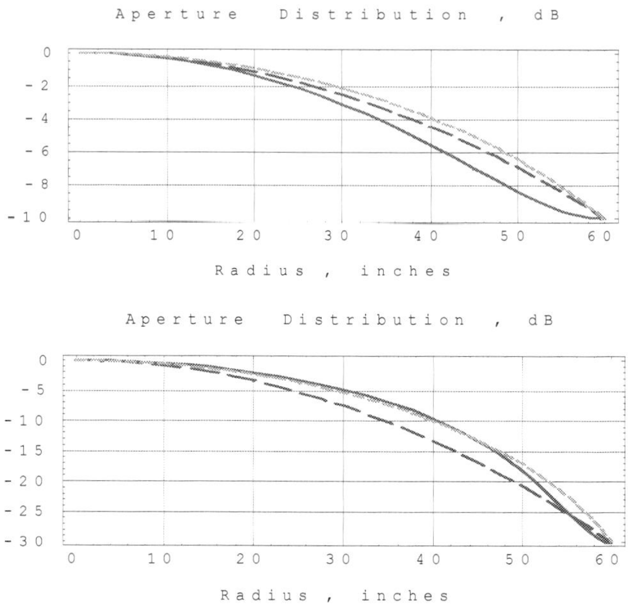

Fig. 2. Comparison between "Polynomial-on-Pedestal" (solid red), Gaussian (long-dashed blue) and Taylor (short-dashed green) aperture illuminations for 10dB (upper) and 30dB (lower) circular aperture edge illumination taper.

$$F_{in}(\theta) = \left[\frac{\int_0^{0.5d} f_i(r,B) J_0(kr \sin\theta) dr}{\int_0^{0.5d} f_i(r,B) dr} \right]^2 \tag{18}$$

where $k = 2\pi/\lambda$ is the wavenumber, θ is the antenna pattern off-boresight angle, and the index i stands for any of aperture illumination functions (12) - (14). For instance, $i = 1$ corresponds with the expression (12), $i = 2$ corresponds with the expression (13) and $i = 3$ corresponds with the expression (14).

It should be noted that in spite of well known deficiencies of the aperture approach for the antenna pattern calculation that approach is quite adequate for this particular application since in (1) only values of function (18) in the vicinity of the antenna main lobe are actually used. Plots in Figure 3 illustrate the difference between antenna patterns computed using (18) for all three aperture illuminations (12) – (14).

As is seen from these plots the difference between antenna patterns for all three aperture illuminations becomes noticeable well outside of the 3dB beamwidth off-boresight direction even for heavily tapered aperture illuminations. It should be also noticed that when the aperture edge illumination taper approaches 0dB, which means that the constant B

approaches 1, all three aperture illumination functions (12) - (14) approach the uniform illumination and thus, their antenna patterns are also converged to each other.

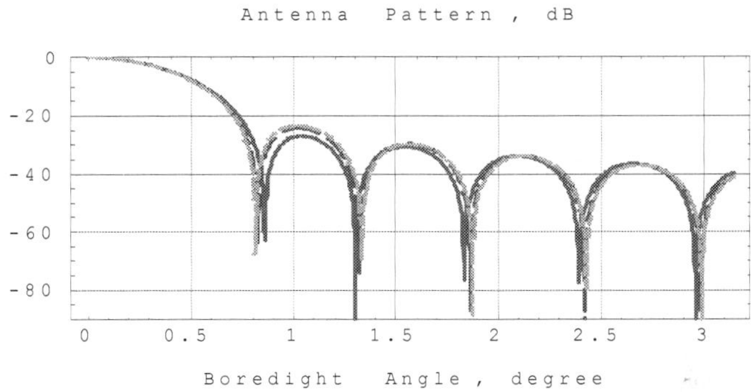

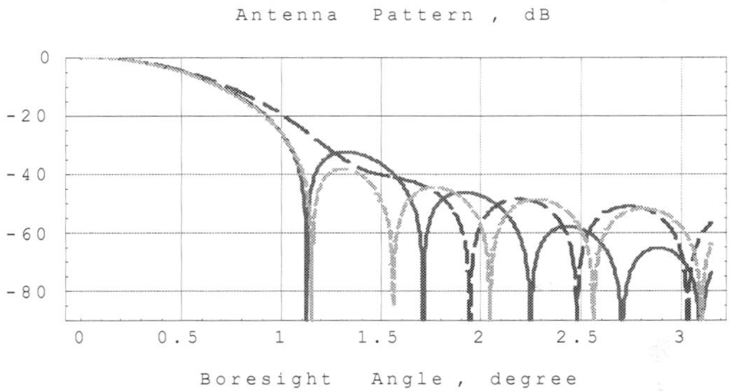

Fig. 3. Comparison between antenna patterns for "Polynomial-on-Pedestal" (solid red), Gaussian (long-dashed blue), and Taylor (short-dashed green) aperture illumination functions with 10dB (upper) and 30dB (bottom) circular aperture edge illumination taper.

5.2 Rectangular Antenna Aperture Case

It was assumed that for the case of the rectangular aperture the same aperture illumination functions (12) - (14) are applied in combination along each of the Cartesian coordinate x and y in a form of its direct product:

$$f_{ij}(x, B_x, y, B_y) = f_i(x, B_x) \times f_j(y, B_y) \tag{19}$$

where i and j stand for any of aperture illuminations (12) - (14). For example, the rectangular aperture illumination that is the "Polynomial-on-Pedestal" along the X-axis and is the Gaussian one along the Y-axis is described by the following expression:

$$f_{12}(x,B_x,y,B_y) = f_1(x,B_x) \times f_2(y,B_y) = \\ = \left\{ B_x + (1-B_x)[1-\left(\frac{2x}{d_x}\right)^2]^2 \right\} \times \exp\left[-(\sqrt{-\ln B_y}\frac{2y}{d_y})^2 \right] \tag{20}$$

where d_x and d_y are the rectangular aperture width and height, x and y are coordinates in the aperture plane with the origin in the center of the aperture, and B_x and B_y are aperture edge illumination tapers that, in general, have different values along X and Y axes.
Based on the rectangular aperture illumination function (19) and using the same aperture approach that was used for the case of the circular aperture, the normalized antenna power pattern for the rectangular aperture can be written, according to (Johnson, R. C. at all, 1993) as follows:

$$F_{ij\,n}(\theta,\varphi) = \left[\frac{\int_0^{0.5d_x}\int_0^{0.5d_y} f_{ij}(x,B_x,y,B_y)\exp[k\sin\theta(x\cos\varphi + y\sin\varphi)]dxdy}{\int_0^{0.5d_x}\int_0^{0.5d_y} f_{ij}(x,B_x,y,B_y)dxdy} \right]^2 \tag{21}$$

where the integration can be done just across the quarter of the aperture because all integrand functions are even in respect to variables x and y. Because all integrand functions are also separable in respect to variables x and y, the expression (21) can be simplified even further and present the antenna pattern of the rectangular aperture as a product of two terms one of which contains integrals only along the x coordinate axis and the other contains integrals only along the y coordinate axis:

$$F_{ij\,n}(\theta,\varphi) = \left[\frac{\int_0^{0.5d_x} f_i(x,B_x)\exp[kx\sin\theta\cos\varphi]dx}{\int_0^{0.5d_x} f_i(x,B_x)dx} \times \frac{\int_0^{0.5d_y} f_j(y,B_y)\exp[ky\sin\theta\sin\varphi]dy}{\int_0^{0.5d_y} f_j(y,B_y)dy} \right]^2 \tag{22}$$

The expression (22) is valid for the rectangular aperture antenna pattern with separable aperture illuminations along x and y axes and means, for instance, that the antenna pattern in principal plane at $\varphi = 0°$ depends only on the illumination function $f_i(x,B_x)$, while in the other principal plane at $\varphi = 90°$, it depends only on the illumination function $f_j(y,B_y)$.Plots in Figure 4 illustrate the difference between antenna patterns in principal planes computed using (22) for all three aperture illuminations (12) - (14). These plots were calculated for the rectangular aperture with the same area as the area of the circular aperture that was used to calculate plots in Figure 3. As is seen from these plots, similarly to the circular aperture case differences in antenna patterns for all three aperture illuminations (12) - (14) become noticeable well outside of the 3dB beamwidth off-boresight angles even for heavily tapered aperture illuminations. The comparison between antenna patterns for circular (Fig. 3) and rectangular (Fig. 4) antennas shows just some quantitative but not qualitative differences between them.

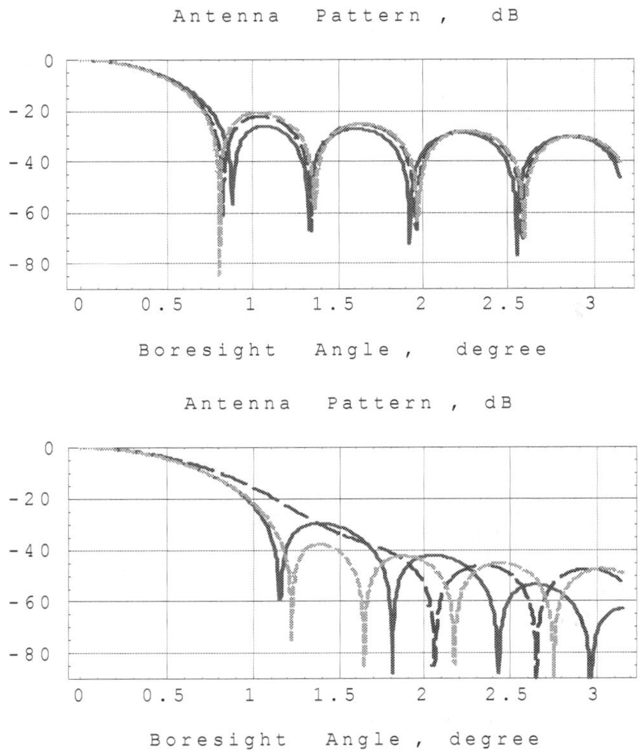

Fig. 4. Comparison between antenna patterns for "Polynomial-on-Pedestal" (solid red), Gaussian (long-dashed blue), and Taylor (short-dashed green) aperture illuminations with 10dB (upper) and 30dB (bottom) rectangular aperture edge illumination taper.

6. Simple and Accurate Approximation of Antenna HPBW for Circular and Rectangular Apertures

6.1 Circular Antenna Aperture Case

Based on the general definition of the extended source correction factor (1) and using expressions (10), (11) for the source brightness distribution and expressions (18), (22) for the circular and rectangular aperture antenna power pattern, it is possible now to calculate the exact value for the extended source size correction factor. However, in order to compare the exact value of the extended source size correction factor with its approximations given by expressions (7) - (9), the value of the antenna pattern HPBW should be known with a high degree of accuracy for each aperture illuminations (12) - (14) for circular and rectangular apertures. This is ultimately needed because the argument s of the extended source size correction factor approximate expressions (7) - (9) is defined as:

$$s = \frac{source\ diameter\ or\ source\ HPBW}{antenna\ HPBW} \tag{23}$$

According, for example, to (Johnson at all, 1993), the circular antenna HPBW can be estimated through the simple formula:

$$HPBW = \alpha \frac{\lambda}{d} \tag{24}$$

where λ is a wavelength, d is the antenna diameter and the a is the beamwidth multiplier in degrees that depends on the type of aperture illumination and edge illumination taper. The rough estimation of a as a function of the aperture edge illumination taper, without taking into account the type of the aperture illumination, was given in (Johnson at all, 1993):

$$\alpha = 55.9486 + 1.05238\, c \tag{25}$$

where c is define in (16) as an absolute value of the edge illumination taper in dB. The number of significant digits in (25) is misleading because the six digits computational accuracy implied by the expression (25) can not be achieved based solely on the value of the edge illumination taper, regardless of the type of the aperture illumination.
The numerical simulations of the circular antenna beamwidth using the expression (18) for the circular antenna radiation pattern and for all three type of aperture illuminations (12) - (14) give the values of the beamwidth multiplier a that are summarized in Table 1 and illustrated in Figure 5.

6.2 Rectangular Antenna Aperture Case

As it was mentioned in section 5.2 for the rectangular antenna aperture illuminated by the separable aperture illumination function (19), antenna pattern cuts in principal planes (along the X or Y axes) are independent of each other and, as it follows from (22), depend exclusively on its own aperture illumination function regardless of illumination function that is applied along the opposite axis in the aperture plane. Thus, to calculate the rectangular antenna HPBW in principal planes, one can still employ the expression (24) substituting the circular aperture diameter d by the width d_x or the height d_y of the rectangular aperture, respectively.
The numerical simulations of the rectangular antenna beamwidth using the expression (22) for the rectangular antenna radiation pattern and for all three type of aperture illuminations (12) - (14) give the values of the beamwidth multiplier a that are summarized in Table 2 and illustrated in Figure 6.

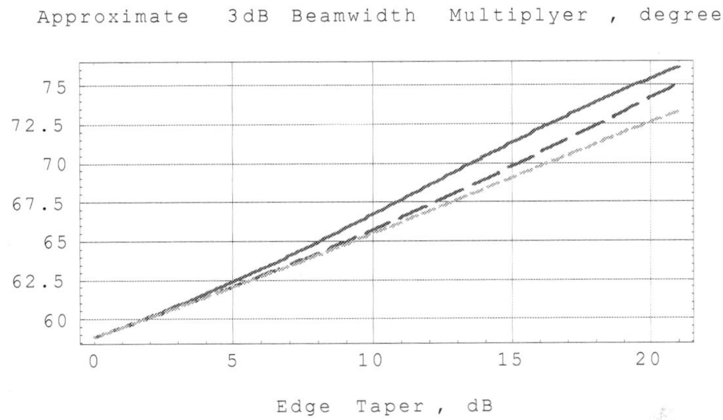

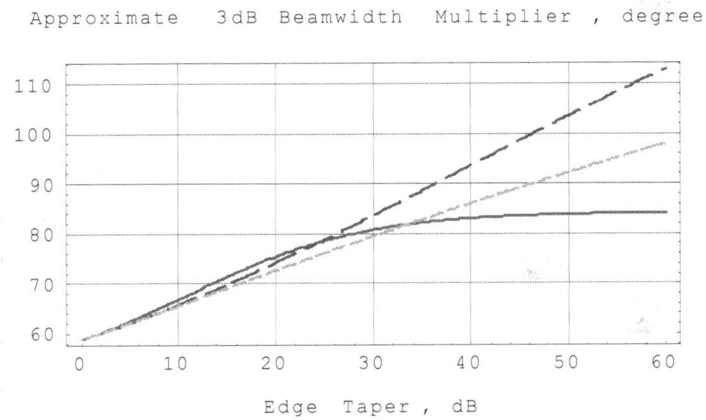

Fig. 5. Approximate 3dB beamwidth multiplier a used in (24) as a function of aperture edge illumination taper for circular aperture and "Polynomial-on-Pedestal" (solid red), Gaussian (long-dashed blue) and Taylor (short-dashed green) aperture illuminations.

Antenna Aperture Illumination	**Approximate Beamwidth Multiplier a in degrees** Circular Aperture, $c \leq$ 40dB	**Max Error,** %
Polynomial	$58.862 + 0.53523c + 0.039795c^2 - 0.001575c^3 + 0.00001562c^4$	0.13
Gaussian	$58.862 + 0.5865c + 0.01089c^2 - 0.000094c^3$	0.055
Taylor	$58.862 + 0.6247c + 0.0048c^2 - 0.000086c^3$	0.095

Table 1. Approximate formulae for antenna pattern beamwidth multiplier a used in (24) for circular aperture.

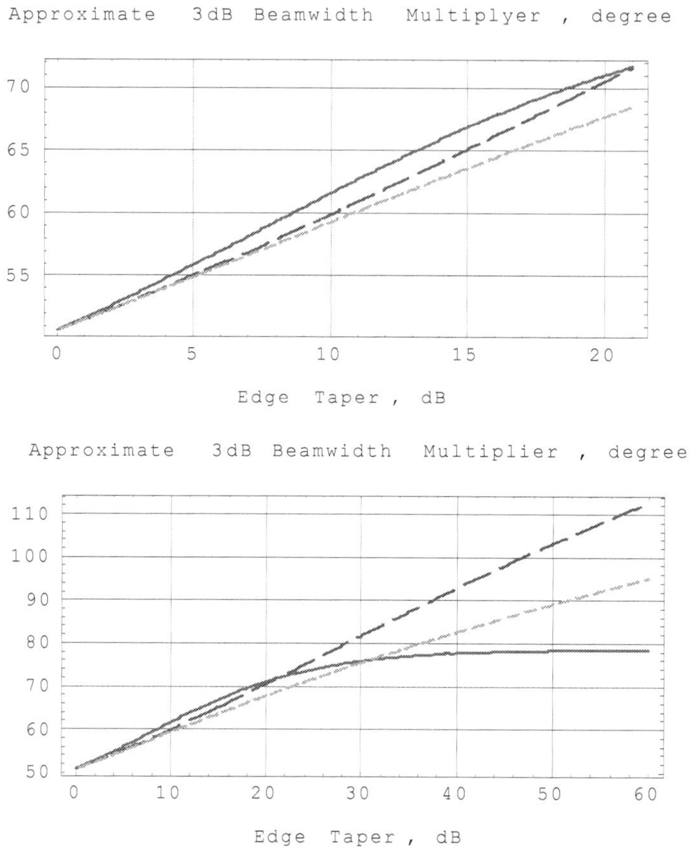

Fig. 6. Approximate 3dB beamwidth multiplier *a* used in (24) as a function of edge illumination taper for rectangular aperture and "Polynomial-on-Pedestal" (solid red), Gaussian (long-dashed blue) and Taylor (short-dashed green) aperture illuminations.

Antenna Aperture Illumination	**Approximate Beamwidth Multiplier *a* in degrees** Rectangular Aperture, $c \leq 40$dB	**Max Error, %**
Polynomial	$50.677 + 0.9449c + 0.2964c^2 - 0.001694c^3 + 0.00001971c^4$	0.26
Gaussian	$50.677 + 0.8236c + 0.011283c^2 - 0.0001402c^3$	0.037
Taylor	$50.677 + 0.863c + 0.0006c^2 - 0.000056c^3$	0.15

Table 2. Approximate formulae for antenna pattern beamwidth multiplier *a* used in (24) for rectangular aperture.

The comparison between antenna beamwidth multipliers a for circular and rectangular apertures shows that they are close to each other and have very similar behavior as a function of the aperture illumination and its edge illumination taper. As it appears from Figures 5 and 6 and Tables 1 and 2, the antenna beamwidth multiplier for rectangular aperture is slightly less than for the circular one. However, if one would compare the circular aperture with diameter d_{circle} and the square apertures with the size d_{squre} that have the same area of aperture (26), which results the smaller linear size of the square aperture than the diameter of the corresponded circular aperture, the antenna beamwidth multipliers a and therefore the antenna HPBW for circular and rectangular apertures become virtually the same.

$$\frac{\pi d_{circle}^2}{4} = d_{square}^2 \quad (26)$$

Using (24) and the appropriate value of the beamwidth multiplier a derived from the Figures 5 and 6 or Tables 1 and 2 it's easy to calculate the antenna HPBW for the respective aperture geometry and illumination with a high degree of accuracy shown in Tables 1 and 2 without the use of complicated and sometime impractical integration. It should be noted that this possibility is very instrumental by itself, regardless of the calculation of the extended source size correction factor.

7. Approximation of Exact Value of Extended Source Size Correction Factor

7.1 Circular Antenna Aperture Case

Based on some earlier studies (Solovey & Mittra, 2008) and on an extensive numerical simulations of the extended source size correction factor using its definition (1) with all combinations of the extended source brightness distributions (10) and (11) and aperture illuminations (12) - (14) for circular aperture, it was found that the most convenient way to approximate the exact value of the extended source size correction factor is to present it in the following form:

$$10\log_{10} K = 10\log_{10} K_{approx} - Corrective\ Term \quad (27)$$

where K_{approx} is given by one of the expressions (7) or (9) and the value of the *Corrective Term* should be calculated using the formulae in Tables 3 - 6 or picked up from plots in Figures 7 - 10. In those Tables and Figures the *Corrective Term* is expressed in dB and presented as a function of the ratio of the extended source size or HPBW to the antenna HPBW, the type of the aperture illumination and the illumination taper.

(a) *Uniform Source Brightness Distribution.* It was found that for the extended radio source with the uniform source brightness distribution (10) and for the circular antenna aperture, the best approximation of the extended source size correction factor is achieved when expression (9) for the K_{approx} is used in (27):

$$10\log_{10} K = 10\log_{10}\left[\frac{(1.616s)^2}{4[1 - J_1^2(1.616s) - J_0^2(1.616s)]}\right] - Corrective\ Term \quad (28)$$

where the *Corrective Term* can be found either from plots in Figures 7 and 8 or from formulae in Tables 3 and 4 and the value of s is restricted by $s < 3$, while the value of c is restricted by $c < 30$. As a reminder, the variable s in (28) and Tables 3 and 4 is defined by (23) and variables c and b in Tables 3 and 4 are defined by (16) and (17). In addition, the calculation errors of the extended source correction factor with and without the *Corrective Term* are also shown in Tables 3 and 4. As is seen from these Tables, the application of the *Corrective Term* decreases the calculation error of the extended source correction factor for the extended radio sources with the uniform source brightness distribution from the 0.038dB - 0.32dB to the 0.0051dB - 0.050dB.

By and large, the complexity of the *Corrective Term* polynomial can be increased to achieve even a smaller calculation error of the extended source correction factor. However, here and further in this chapter, the complexity of the polynomial was chosen on an ad hoc basis as a compromise between the tolerance delivered by the *Corrective Term* and the complexity of its polynomial. Another reason why one should not try to increase the complexity of the *Corrective Term* polynomial out of proportion in order to achieve increasingly lower error of the extended source correction factor approximation (27), especially for the values of s that are essentially more than *1* is this. While for the small values of s that are well within the antenna pattern main lobe, the expressions (18) and (22) for the circular and rectangular antenna patterns describe the shape of antenna pattern of real antennas fairly accurately, for the $s > 1$ the accuracy of the theoretical antenna pattern representation becomes progressively lower. The antenna pattern of real antennas outside of the main lobe essentially depends on many fine details of the particular antenna system including but not limited to the aperture blockage by the feeder(s) and struts, the fine differences between the real aperture illumination and its approximate representation by the formulae (12) - (14), etc. While those fine details have very limited influence on the shape of the main lobe of antenna pattern, they greatly affect the shape of even closest antenna sidelobes.

That is why the extended source size correction factor approximation (27) was done just within the $s < 3$ ratio and that is why it makes little sense to increase the complexity of the *Corrective Term* polynomial far more than it is necessarily to achieve a few hundredth of dB tolerance for $s < 1.5$ values and a few tenth of dB tolerance for $s < 3$ values.

(b) *Gaussian Source Brightness Distribution.* It was found that for the extended radio source with the Gaussian source brightness distribution (11) and the circular antenna aperture the best approximation of the extended source size correction factor is achieved when the expression (7) for the K_{approx} is used in (27):

$$10\log_{10} K = 10\log_{10}\left[1+x^2\right] - Corrective\ Term \tag{29}$$

where the value of the *Corrective Term* can be found either from plots in Figures 9 and 10 or from formulae in Tables 5 and 6 and the value of s is restricted by $s < 3$, while the value of c is restricted by $c < 30$. In addition, the errors in calculation of the extended source correction factor with and without the *Corrective Term* are also shown in Tables 5 and 6. As is seen from these Tables, the application of the *Corrective Term* decreases the calculation error of the extended source size correction factor for the extended radio sources with the Gaussian source brightness distribution from 0.19dB - 0.29dB to 0.025dB - 0.045dB.

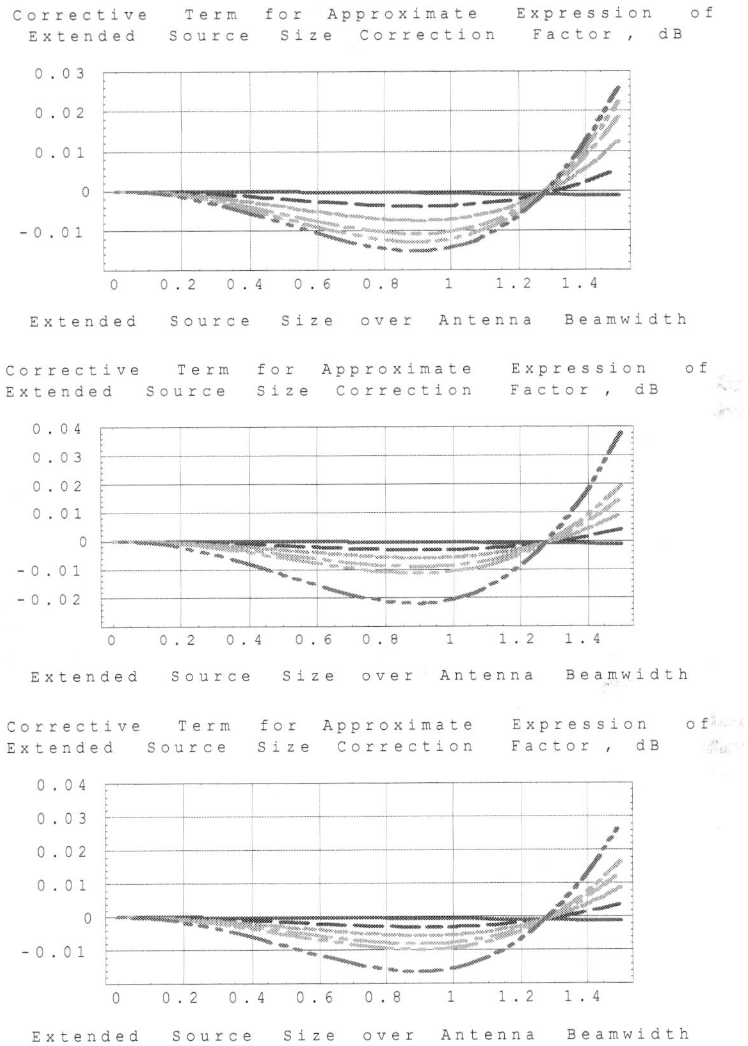

Fig. 7. *Corrective Term* for circular aperture and uniform source brightness distribution.

Plots legend for aperture illuminations: "Polynomial-on-Pedestal" (upper), Gaussian (mid) and Taylor (bottom) aperture illuminations;
Curves legend for aperture illumination tapers: 0dB (solid red), 4dB (long-dashed dark blue), 8dB (short-dashed green), 12dB (long-dashes-dotted light blue), 15dB (long-dashes-double-dotted yellow) and 30dB (long-dashes-triple-dotted purple);

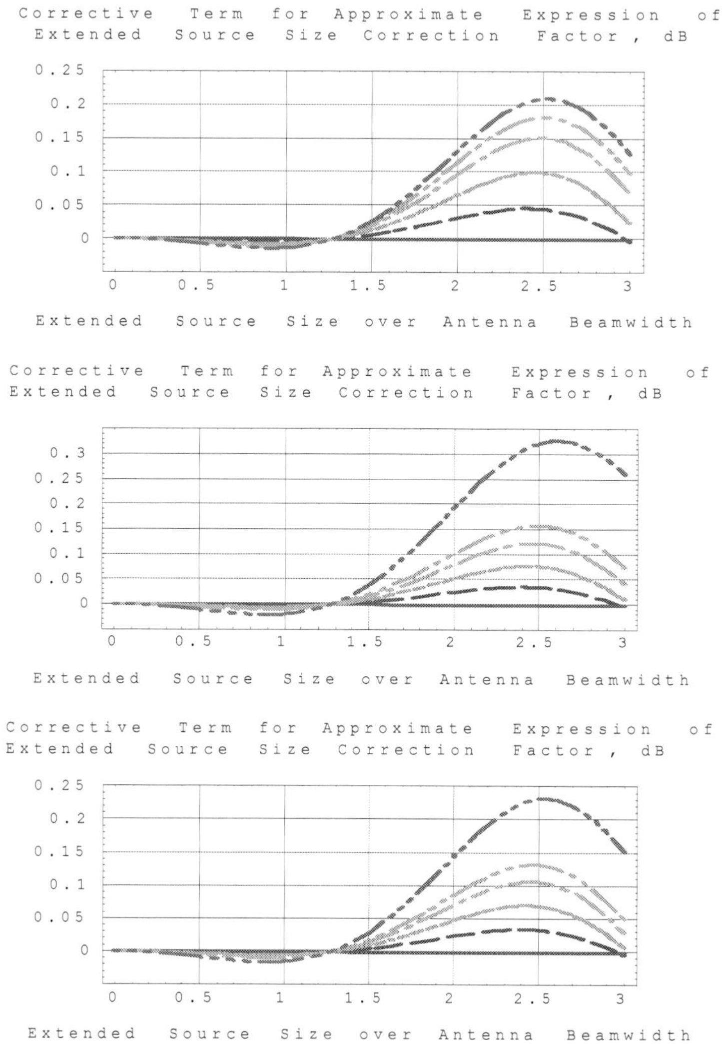

Fig. 8. *Corrective Term* for circular aperture and uniform source brightness distribution.

Plots legend for aperture illuminations: "Polynomial-on-Pedestal" (upper), Gaussian (mid) and Taylor (bottom) aperture illuminations;
Curves legend for aperture illumination tapers: 0dB (solid red), 4dB (long-dashed dark blue), 8dB (short-dashed green), 12dB (long-dashes-dotted light blue), 15dB (long-dashes-double-dotted yellow) and 30dB (long-dashes-triple-dotted purple);

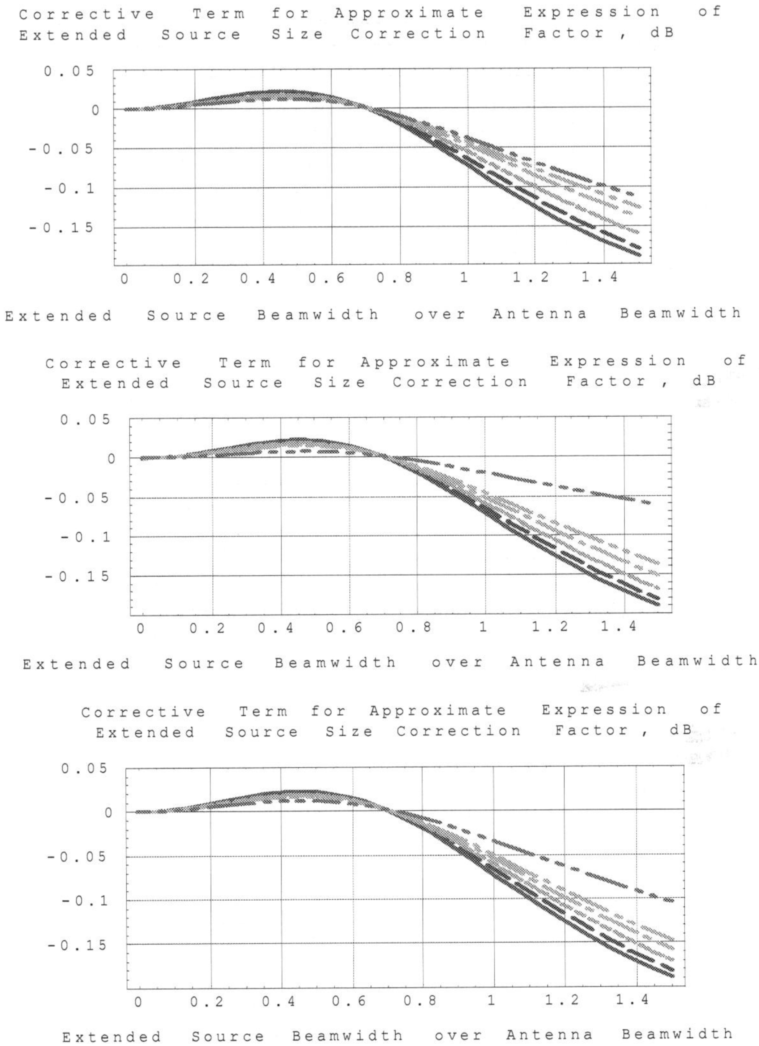

Fig. 9. *Corrective Term* for circular aperture and Gaussian source brightness distribution.

Plots legend for aperture illuminations: "Polynomial-on-Pedestal" (upper), Gaussian (mid) and Taylor (bottom) aperture illuminations;
Curves legend for aperture illumination tapers: 0dB (solid red), 4dB (long-dashed dark blue), 8dB (short-dashed green), 12dB (long-dashes-dotted light blue), 15dB (long-dashes-double-dotted yellow) and 30dB (long-dashes-triple-dotted purple);

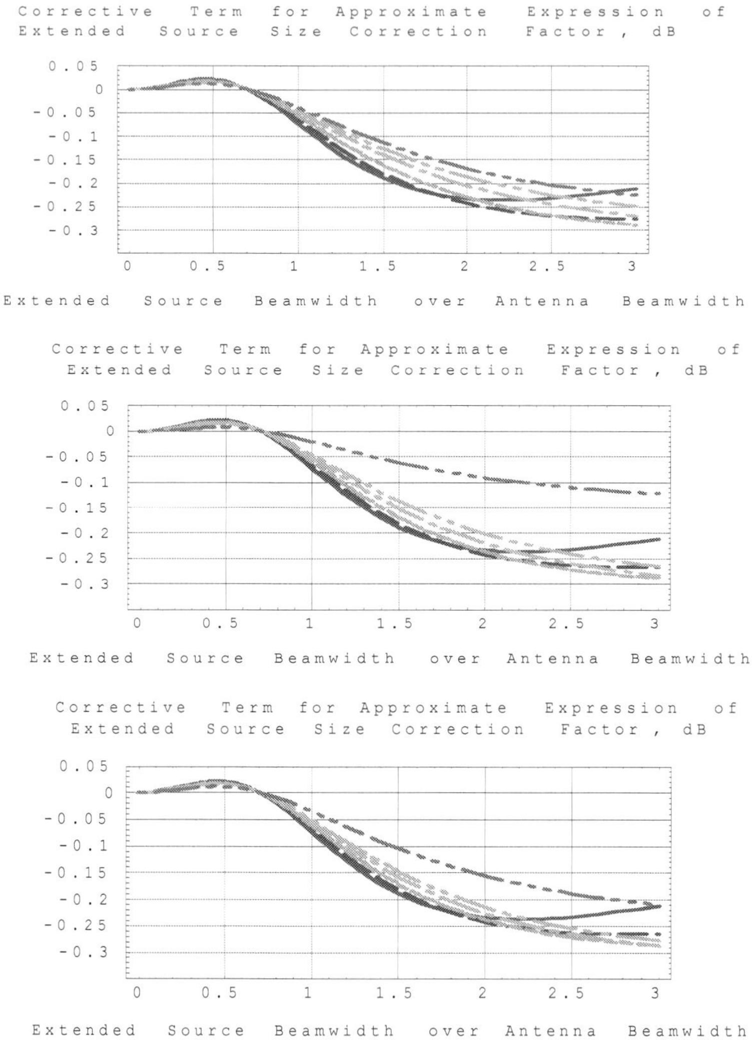

Fig. 10. *Corrective Term* for circular aperture and Gaussian source brightness distribution.

Plots legend for aperture illuminations: "Polynomial-on-Pedestal" (upper), Gaussian (mid) and Taylor (bottom) aperture illuminations;
Curves legend for aperture illumination tapers: 0dB (solid red), 4dB (long-dashed dark blue), 8dB (short-dashed green), 12dB (long-dashes-dotted light blue), 15dB (long-dashes-double-dotted yellow) and 30dB (long-dashes-triple-dotted purple);

Antenna Aperture Illumination	*Corrective Term*, dB Circular Aperture, $s \leq 1.5$, $c \leq 30$dB Maximum Error without *Corrective Term* 0.038dB	Error, dB rms max
Polynomial	$(0.00249s - 0.00516s^2 + 0.00197s^3) + b(-0.00559s - 0.01423s^2 + 0.01541s^3) + b^2(0.01411s - 0.07224s^2 + 0.04739s^3)$	0.0004 0.0017
Gaussian	$(-0.0115s + 0.0153s^2 - 0.005s^3) + b(0.0518s - 0.0504s^2 + 0.0064s^3) + b^2(-0.0318s - 0.0672s^2 + 0.0735s^3)$	0.0010 0.0051
Taylor	$(-0.0061s + 0.0079s^2 - 0.0027s^3) + b(0.0292s - 0.0401s^2 + 0.01294s^3) + b^2(-0.0155s - 0.0483s^2 + 0.0481s^3)$	0.00054 0.0027

Table 3. *Corrective Term* for Extended Radio Source with Uniform Brightness Distribution.

Antenna Aperture Illumination	*Corrective Term*, dB Circular Aperture, $1.5 \leq s \leq 3.0$, $c \leq 30$dB Maximum Error without *Corrective Term* 0.32dB	Error, dB rms max
Polynomial	$(-0.03761s + 0.03033s^2 - 0.00622s^3) + b(-0.0663s + 0.29041s^2 - 0.0705s^3) + b^2(-0.23992s + 0.20878s^2 - 0.03424s^3)$	0.0075 0.021
Gaussian	$(0.002s - 0.0007s^2 + 0.0018s^3) + b(-0.2597s + 0.0739s^2 - 0.0348s^3) + b^2(-0.5513s + 0.532s^2 - 0.0915s^3)$	0.017 0.050
Taylor	$(-0.0089s + 0.0082s^2 - 0.0009s^3) + b(-0.1152s + 0.1237s^2 - 0.0378s^3) + b^2(-0.3863s + 0.3728s^2 - 0.0675s^3)$	0.0093 0.028

Table 4. *Corrective Term* for Extended Radio Source with Uniform Brightness Distribution.

7.2 Rectangular Antenna Aperture Case

Based on the extensive numerical simulations of the extended source size correction factor using its definition (1) with all combinations of the extended source brightness distributions (10) and (11) and aperture illuminations (12) - (14) along X and Y axes for the rectangular antenna aperture, it was found that the most convenient way to approximate the exact value of the extended source size correction factor is to present it by the same formula (27) that was used in the circular aperture case. Indeed, the particular expressions for the K_{approx} and the *Corrective Term* for rectangular and circular apertures are different. Too many combinations of aperture illumination functions and its tapers along each X and Y axes for the rectangular aperture make it impractical to present the *Corrective Term* in the form of plots. Instead, for rectangular apertures, the *Corrective Term* is presented by formulae only.

(a) *Uniform Source Brightness Distribution.* It was found that for the extended radio source with the uniform source brightness distribution (10) and the rectangular antenna aperture the best approximation of the extended source size correction factor is achieved when the expressions (8) for the K_{approx} are used in (27) along X and Y axes of rectangular aperture:

$$10\log_{10} K = 10\log_{10}\left[\sqrt{\frac{(s_1\sqrt{\ln 2})^2}{1-\exp[-(s_1\sqrt{\ln 2})^2]} \times \frac{(s_2\sqrt{\ln 2})^2}{1-\exp[-(s_2\sqrt{\ln 2})^2]}}\right] - \textit{Corrective Term} \qquad (30)$$

where the *Corrective Term* can be found from formulae in Tables 7 and 8 and values of s_1 and s_2 are restricted by $s_1, s_2 < 3$, while values of c_1 and c_2 are restricted by $c_1, c_2 < 30$. Indexes 1 and 2 correspond with X and Y axes in plane of the rectangular aperture. In addition, the calculation errors of the extended source correction factor with and without the *Corrective Term* are also shown in Tables 7 and 8.

As is seen from these Tables, the application of the *Corrective Term* decreases the calculation error of the extended source correction factor for the extended radio sources with the uniform source brightness distribution from 0.076dB - 0.34dB to 0.011dB - 0.072dB.

(b) *Gaussian Source Brightness Distribution.* It was found that for extended radio sources with the Gaussian source brightness distribution (11) and the rectangular antenna aperture the best approximation of the extended source size correction factor is achieved when the expressions (7) for the K_{approx} are used in (27) along X and Y axes of rectangular aperture:

$$10\log_{10} K = 10\log_{10}\left[\sqrt{(1+s_1^2)\times(1+s_2^2)}\right] - Corrective\ Term \tag{31}$$

where the value of the *Corrective Term* can be found from formulae in Tables 9 - 11 and values of s_1 and s_2 are restricted by $s_1, s_2 < 3$, while values of c_1 and c_2 are restricted by $c_1, c_2 < 30$. In addition, calculation errors of the extended source correction factor with and without the *Corrective Term* are also shown in Tables 9 - 11. As is seen from these Tables, the application of the *Corrective Term* decreases the calculation error of the extended source correction factor for the extended radio sources with the Gaussian source brightness distribution from 0.085dB - 2.18dB to 0.011dB - 0.40dB.

7.3 Extended Source Size Correction Factor Calculation Scheme

Here is a short, step by step summary of how to calculate the extended source size correction factor using the approach discussed in this chapter:

(a) Based on the given antenna aperture illumination taper(s) and using (16), find the value of variable c for circular or values of variables c_1 and c_2 for rectangular antennas.

(b) Based on the given antenna aperture illumination taper(s) and using (17), find the value of variable b for circular or values of variables b_1 and b_2 for rectangular antennas.

(c) Based on the given antenna aperture illumination and its taper(s), find the antenna HPBW multiplier a using the appropriate formula from Table 1 or plot from Fig. 5 for circular antenna aperture. For the rectangular antenna use Table 2 or plots from Fig. 6.

(d) Using (24) find the antenna HPBW for circular or two values of the antenna HPBW in the principal planes for rectangular antennas.

(e) Based on the given source brightness distribution and its angular size, and using (23), find the value of variable s for circular or values of variables s_1 and s_2 for rectangular antennas.

(f) Based on the given source brightness distribution and antenna aperture illumination, find the approximate value of the extended source size correction factor K using either formulae (28) or (29) for circular or formulae (30) or (31) for rectangular antennas. The *Corrective Term* in those formulae can be found from either plots in Figures 7 - 10 or Tables 3 - 7 for the circular aperture. For the rectangular aperture the *Corrective Term* can be found from Tables 7 - 11.

Antenna Aperture Illumination	*Corrective Term*, dB Circular Aperture, $s \le 1.5$, $c \le 30$dB Maximum Error without *Corrective Term* 0.19dB	Error, dB rms max
Polynomial	$(0.1388s - 0.2485s^2 + 0.0433s^3) + b(0.0973s - 0.1767s^2 + 0.081s^3) +$ $b^2(-0.0171s + 0.3195s^2 - 0.1144s^3)$	0.0051 0.016
Gaussian	$(0.0575s - 0.0822s^2 - 0.0297s^3) + b(0.4417s - 0.8687s^2 + 0.3551s^3) +$ $b^2(-0.4637s + 0.9003s^2 - 0.3299s^3)$	0.0067 0.025
Taylor	$(0.0824s - 0.1285s^2 - 0.0117s^3) + b(0.3336s - 0.6783s^2 + 0.298s^3) +$ $b^2(-0.353s + 0.7089s^2 - 0.283s^3)$	0.0059 0.020

Table 5. *Corrective Term* for Extended Radio Source with Gaussian Brightness Distribution.

Antenna Aperture Illumination	*Corrective Term*, dB Circular Aperture, $1.5 \le s \le 3.0$, $c \le 30$dB Maximum Error without *Corrective Term* 0.29dB	Error, dB rms max
Polynomial	$(-0.1533s + 0.0001s^2 + 0.0097s^3) + b(0.1888s - 0.1559s^2 + 0.0186s^3)$ $+ b^2(-0.0528s + 0.1031s^2 - 0.0163s^3)$	0.0051 0.014
Gaussian	$(-0.1442s + 0.0038s^2 + 0.0086s^3) + b(0.0864s - 0.168s^2 + 0.0261s^3) +$ $b^2(0.052s + 0.1223s^2 - 0.0256s^3)$	0.018 0.045
Taylor	$(-0.1528s + 0.0063s^2 + 0.008s^3) + b(-0.011s - 0.0322s^2 - 0.002s^3) +$ $b^2(0.1657s - 0.0478s^2 + 0.0104s^3)$	0.0089 0.026

Table 6. *Corrective Term* for Extended Radio Source with Gaussian Brightness Distribution.

Antenna Aperture Illumination	*Corrective Term*, dB Rectangular Aperture, $s \le 1.5$, $c \le 30$dB Maximum Error without *Corrective Term* 0.076dB	Error, dB rms max
Polynomial Polynomial	$0.0193(s_1+s_2) + 0.0228(s_1^2+s_2^2) - 0.0083(s_1s_2) - 0.0271(s_1s_2)^2$ $- 0.0254(s_1b_1+s_2b_2) + 0.0143(s_1^2b_1 + s_2^2 b_2) + 0.00013\, s_1s_2(b_1+b_2)$	0.0014 0.0076
Polynomial Gaussian	$0.0215(s_1+s_2) + 0.0215(s_1^2+s_2^2) - 0.0087(s_1s_2) - 0.0269(s_1s_2)^2$ $- 0.0297(s_1b_1+s_2b_2) + 0.0167(s_1^2b_1 + s_2^2 b_2)$	0.0023 0.011
Polynomial Taylor	$0.02(s_1+s_2) + 0.0223(s_1^2+s_2^2) - 0.0075(s_1s_2) - 0.0274(s_1s_2)^2$ $- 0.0253(s_1b_1+s_2b_2) + 0.0143(s_1^2b_1 + s_2^2 b_2)$	0.0016 0.0077
Gaussian Gaussian	$0.0239(s_1+s_2) + 0.0202(s_1^2+s_2^2) - 0.0092(s_1s_2) - 0.0267(s_1s_2)^2$ $- 0.034(s_1b_1+s_2b_2) + 0.0192(s_1^2b_1 + s_2^2 b_2)$	0.0027 0.011
Gaussian Taylor	$0.0223(s_1+s_2) + 0.021(s_1^2+s_2^2) - 0.008(s_1s_2) - 0.0271(s_1s_2)^2$ $- 0.0296(s_1b_1+s_2b_2) + 0.0167(s_1^2b_1 + s_2^2 b_2)$	0.0024 0.012
Taylor Taylor	$0.02075(s_1+s_2) + 0.02176(s_1^2+s_2^2) - 0.00686(s_1s_2) - 0.02762(s_1s_2)^2$ $- 0.02528(s_1b_1+s_2b_2) + 0.0143(s_1^2b_1 + s_2^2 b_2)$	0.0017 0.0077

Table 7. *Corrective Term* for Extended Radio Source with Uniform Brightness Distribution.

Antenna Aperture Illumination	*Corrective Term*, dB Rectangular Aperture, $s \le 3.0$, $c \le 30$dB Maximum Error without *Corrective Term* 0.34dB	Error, dB rms max
Polynomial Polynomial	$-0.0284(s_1+s_2) + 0.0325(s_1^2+s_2^2) - 0.1447(s_1s_2) + 0.0097(s_1^3+s_2^3) -$ $0.0994(s_1s_2^2 + s_2s_1^2) - 0.0029(s_1^4 + s_2^4) + 00183(s_1s_2^3 + s_2s_1^3)$ $+ 0.0041(s_1s_2)^2 + 0.0256(s_1b_1+s_2b_2) - 0.0059(s_1b_2+s_2b_1)$	0.011 0.049
Polynomial Gaussian	$-0.0429(s_1+s_2) - 0.07451(s_1^2+s_2^2) + 0.15911(s_1s_2) + 0.14348(s_1^3+s_2^3)$ $- 0.10923(s_1s_2^2 + s_2s_1^2) - 0.03264(s_1^4 + s_2^4) + 0.02107(s_1s_2^3 + s_2s_1^3)$ $+ 0.00388(s_1s_2)^2 + 0.02566(s_1b_1+s_2b_2) - 0.00593(s_1b_2+s_2b_1)$	0.013 0.047
Polynomial Taylor	$-0.0141(s_1+s_2) + 0.0065(s_1^2+s_2^2) + 0.1502(s_1s_2) + 0.0224(s_1^3+s_2^3)$ $- 0.1007(s_1s_2^2 + s_2s_1^2) - 0.0049(s_1^4 + s_2^4) + 0.0182(s_1s_2^3 + s_2s_1^3)$ $+ 0.0046(s_1s_2)^2 + 0.0256(s_1b_1+s_2b_2) - 0.0059(s_1b_2+s_2b_1)$	0.012 0.053
Gaussian Gaussian	$-0.026(s_1+s_2) + 0.027(s_1^2+s_2^2) + 0.143(s_1s_2) + 0.014(s_1^3+s_2^3)$ $- 0.099(s_1s_2^2 + s_2s_1^2) - 0.004(s_1^4 + s_2^4) + 0.018(s_1s_2^3 + s_2s_1^3)$ $+ 0.004(s_1s_2)^2 - 0.026(s_1b_1+s_2b_2) + 0.006(s_1b_2+s_2b_1)$ $-0.014(s_1b_1^2+s_2b_2^2) + 0.06(s_1b_2^2+s_2b_1^2)$	0.016 0.072
Gaussian Taylor	$-0.02926(s_1+s_2) + 0.039(s_1^2+s_2^2) + 0.14537(s_1s_2) + 0.00598(s_1^3+s_2^3)$ $- 0.1002(s_1s_2^2 + s_2s_1^2) - 0.0023(s_1^4 + s_2^4) + 0.01853(s_1s_2^3 + s_2s_1^3)$ $+ 0.00414(s_1s_2)^2 - 0.01783(s_1b_1+s_2b_2) + 0.0042(s_1b_2+s_2b_1)$ $-0.0107(s_1b_1^2+s_2b_2^2) + 0.04706(s_1b_2^2+s_2b_1^2) - 0.00019(s_1+s_2)b_1b_2$	0.015 0.064
Taylor Taylor	$-0.02138(s_1+s_2) + 0.02502(s_1^2+s_2^2) + 0.14732(s_1s_2)$ $+0.01115 (s_1^3+s_2^3) - 0.10045 (s_1s_2^2 + s_2s_1^2) - 0.00296 (s_1^4 + s_2^4) +$ $0.01849(s_1s_2^3 + s_2s_1^3) + 0.00424(s_1s_2)^2 + 0.02554(s_1b_1+s_2b_2) -$ $0.00592(s_1b_2+s_2b_1)$	0.013 0.060

Table 8. *Corrective Term* for Extended Radio Source with Uniform Brightness Distribution.

Antenna Aperture Illumination	*Corrective Term*, dB Rectangular Aperture, $s \le 1.0$, $c \le 30$dB Maximum Error without *Corrective Term* 0.085dB	Error, dB rms max
Polynomial Polynomial	$0.0601(s_1+s_2) - 0.0371(s_1^2+s_2^2) - 0.0033(s_1s_2) + 0.0378(s_1s_2)^2$ $- 0.031(s_1b_1+s_2b_2) + 0.0315(s_1b_1^2+s_2b_2^2)$	0.0032 0.010
Polynomial Gaussian	$0.0597(s_1+s_2) - 0.037(s_1^2+s_2^2) + 0.003(s_1s_2) + 0.0356(s_1s_2)^2$ $- 0.0346(s_1b_1+s_2b_2) + 0.0369(s_1b_1^2+s_2b_2^2) - 0.0042s_1s_2(b_1+b_2)$	0.0034 0.010
Polynomial Taylor	$0.0433(s_1+s_2) - 0.0164(s_1^2+s_2^2) - 0.0028(s_1s_2) + 0.037(s_1s_2)^2$ $- 0.0016(s_1b_1+s_2b_2) - 0.0044(s_1b_1^2+s_2b_2^2)$	0.0036 0.011
Gaussian Gaussian	$0.066(s_1+s_2) - 0.0437(s_1^2+s_2^2) - 0.009(s_1s_2) + 0.0392(s_1s_2)^2$ $- 0.0334(s_1b_1+s_2b_2) + 0.044(s_1^2b_1+s_2^2b_2) + 0.0033s_1s_2(b_1+b_2)$ $- 0.0115(s_1b_1^2+s_2b_2^2) - 0.0009(s_1+s_2)b_1b_2$	0.0033 0.012
Gaussian Taylor	$0.0643(s_1+s_2) - 0.0417(s_1^2+s_2^2) - 0.0081(s_1s_2) + 0.0379(s_1s_2)^2$ $- 0.0291(s_1b_1+s_2b_2) + 0.0382(s_1^2b_1+s_2^2b_2) + 0.0032s_1s_2(b_1+b_2)$ $- 0.0094(s_1b_1^2+s_2b_2^2) + 0.0001(s_1+s_2)b_1b_2$	0.0034 0.011
Taylor Taylor	$0.044(s_1+s_2) - 0.0176(s_1^2+s_2^2) - 0.0024(s_1s_2) + 0.0363(s_1s_2)^2$ $+ 0.0007(s_1b_1+s_2b_2) - 0.0065(s_1b_1^2+s_2b_2^2)$	0.0037 0.010

Table 9. *Corrective Term* for Extended Radio Source with Gaussian Brightness Distribution.

Antenna Aperture Illumination	*Corrective Term*, dB Rectangular Aperture, $s \leq 1.5$, $c \leq 30$dB Maximum Error without *Corrective Term* 0.15dB	Error, dB rms max
Polynomial Polynomial	$-0.0216(s_1+s_2) - 0.0578(s_1^2+s_2^2) - 0.0747(s_1s_2) + 0.148(s_1^3+s_2^3)$ $- 0.067(s_1^4+s_2^4) + 0.0306(s_1s_2^3 + s_2s_1^3) + 0.0502(s_1s_2)^2$ $- 0.0033(s_1b_1+s_2b_2) - 0.0016(s_1b_2+s_2b_1)$	0.0033 0.018
Polynomial Gaussian	$0.0187(s_1+s_2) + 0.0553(s_1^2+s_2^2) - 0.0558(s_1s_2) + 0.0226(s_1^3+s_2^3)$ $- 0.0144(s_1s_2^2 + s_2s_1^2) - 0.0701(s_1^4+s_2^4) + 0.0336(s_1s_2^3 + s_2s_1^3)$ $+ 0.0551(s_1s_2)^2 - 0.0021(s_1b_1+s_2b_2) - 0.0024(s_1b_2+s_2b_1)$	0.0038 0.019
Polynomial Taylor	$0.0217(s_1+s_2) + 0.0631(s_1^2+s_2^2) - 0.0782(s_1s_2) + 0.0071(s_1^3+s_2^3)$ $+ 0.0016(s_1s_2^2 + s_2s_1^2) - 0.0642(s_1^4+s_2^4) + 0.0309(s_1s_2^3 + s_2s_1^3)$ $+ 0.0492(s_1s_2)^2 - 0.0036(s_1b_1+s_2b_2) - 0.0015(s_1b_1+s_2b_2)$	0.0034 0.017
Gaussian Gaussian	$0.017(s_1+s_2) + 0.063(s_1^2+s_2^2) - 0.07(s_1s_2) + 0.012(s_1^3+s_2^3)$ $- 0.003(s_1s_2^2 + s_2s_1^2) - 0.066(s_1^4+s_2^4) + 0.031(s_1s_2^3 + s_2s_1^3)$ $+ 0.051(s_1s_2)^2 - 0.001(s_1b_1+s_2b_2) - 0.003(s_1b_2+s_2b_1)$	0.0042 0.018
Gaussian Taylor	$0.0188(s_1+s_2) + 0.0499(s_1^2+s_2^2) - 0.037(s_1s_2) + 0.0311(s_1^3+s_2^3)$ $- 0.0302(s_1s_2^2 + s_2s_1^2) - 0.0735(s_1^4+s_2^4) + 0.0379(s_1s_2^3 + s_2s_1^3)$ $+ 0.06(s_1s_2)^2 - 0.0024(s_1b_1+s_2b_2) - 0.0023(s_1b_2+s_2b_1)$	0.0038 0.021
Taylor Taylor	$0.0179(s_1+s_2) + 0.0691(s_1^2+s_2^2) - 0.0535(s_1s_2) + 0.0055(s_1^3+s_2^3)$ $- 0.0198(s_1s_2^2 + s_2s_1^2) - 0.0647(s_1^4+s_2^4) + 0.037(s_1s_2^3 + s_2s_1^3)$ $+ 0.0553(s_1s_2)^2 - 0.0038(s_1b_1+s_2b_2) - 0.0013(s_1b_2+s_2b_1)$	0.0034 0.019

Table 10. *Corrective Term* for Extended Radio Source with Gaussian Brightness Distribution.

Antenna Aperture Illumination	*Corrective Term*, dB Rectangular Aperture, $s \leq 3.0$, $c \leq 30$dB Maximum Error without Corrective Term 2.18dB	Error, dB rms max
Polynomial Polynomial	$-0.0714(s_1+s_2) + 0.4084(s_1^2+s_2^2) - 0.4646(s_1s_2) - 0.4509(s_1^3+s_2^3)$ $+ 0.375(s_1s_2^2 + s_2s_1^2) + 0.1035(s_1^4+s_2^4) - 0.088(s_1s_2^3 + s_2s_1^3)$ $- 0.03091(s_1s_2)^2 + 0.0369(s_1b_1+s_2b_2)$	0.044 0.27
Polynomial Gaussian	$-0.073(s_1+s_2) + 0.412(s_1^2+s_2^2) - 0.452(s_1s_2) - 0.45(s_1^3+s_2^3) + 0.369(s_1s_2^2 + s_2s_1^2) + 0.103(s_1^4+s_2^4) - 0.087(s_1s_2^3 + s_2s_1^3) - 0.038(s_1s_2)^2 - 0.022(s_1b_1+s_2b_2) + 0.009(s_1b_2+s_2b_1) - 0.018(s_1b_1^2+s_2b_2^2) + 0.072(s_1b_2^2+s_2b_1^2) - 0.0008(s_1+s_2)b_1b_2$	0.048 0.36
Polynomial Taylor	$-0.0473(s_1+s_2) + 0.0121(s_1^2+s_2^2) + 0.3153(s_1s_2) - 0.1709(s_1^3+s_2^3)$ $+ 0.0852(s_1s_2^2 + s_2s_1^2) + 0.0527(s_1^4+s_2^4) - 0.062(s_1s_2^3 + s_2s_1^3) + 0.0181(s_1s_2)^2 + 0.0134(s_1b_1+s_2b_2) - 0.0133(s_1b_2+s_2b_1) + 0.0056(s_1b_1^2+s_2b_2^2)$ $+ 0.0295(s_1b_2^2+s_2b_1^2) - 0.0014(s_1+s_2)b_1b_2$	0.043 0.30
Gaussian Gaussian	$-0.0175(s_1+s_2) - 0.0526(s_1^2+s_2^2) + 0.2655(s_1s_2) - 0.1384(s_1^3+s_2^3)$ $+ 0.1023(s_1s_2^2 + s_2s_1^2) + 0.0475(s_1^4+s_2^4) - 0.0634(s_1s_2^3 + s_2s_1^3)$ $+ 0.0153(s_1s_2)^2 + 0.0546(s_1b_1+s_2b_2)$	0.077 0.40
Gaussian Taylor	$-0.0193(s_1+s_2) - 0.0554(s_1^2+s_2^2) + 0.211(s_1s_2) - 0.1505(s_1^3+s_2^3)$ $+ 0.1306(s_1s_2^2 + s_2s_1^2) + 0.0512(s_1^4+s_2^4) - 0.0694(s_1s_2^3 + s_2s_1^3)$ $+ 0.0135(s_1s_2)^2 - 0.0427(s_1b_1+s_2b_2) + 0.0142(s_1b_2+s_2b_1)$ $- 0.0223(s_1b_1^2+s_2b_2^2) + 0.093(s_1b_2^2+s_2b_1^2) - 0.0018(s_1+s_2)b_1b_2$	0.071 0.34
Taylor Taylor	$-0.0156(s_1+s_2) + 0.2169(s_1^2+s_2^2) + 0.2902(s_1s_2) - 0.2838(s_1^3+s_2^3)$ $+ 0.0916(s_1s_2^2 + s_2s_1^2) + 0.0716(s_1^4+s_2^4) - 0.0621(s_1s_2^3 + s_2s_1^3)$ $+ 0.0164(s_1s_2)^2 + 0.037(s_1b_1+s_2b_2)$	0.071 0.25

Table 11. *Corrective Term* for Extended Radio Source with Gaussian Brightness Distribution.

8. Conclusion

The calculation of the extended source size correction factor that is widely utilized in antenna gain measurements of electrically large antennas using the cosmic radio sources whose size is comparable or even larger than the antenna beamwidth was considered.
The analytical expressions for the extended source size correction factor developed in the literature were examined and their areas of applicability and associated errors were investigated for uniform and Gaussian radio source brightness distributions, "Polynomial-on-Pedestal," Gaussian and Taylor antenna aperture illuminations with up to 30dB aperture edge illumination taper(s) and for circular and rectangular antenna apertures.
Accurate analytical expressions for the extended source size correction factor for all of the above combinations of source brightness distributions, aperture illuminations and antenna shapes along with their tolerances were found for the case when the extended cosmic radio source size or its beamwidth is up to 3 times bigger than the antenna beamwidth. In addition, accurate analytical expressions for the antenna beamwidth for same combinations of source brightness distributions, aperture illuminations and antenna shapes along with their tolerances were also found.
Attained results eliminate the need to perform complicated and often impractical numerical integrations in case of the particular measurement. The approximate analytical expressions of the extended source size correction factor for rectangular antenna aperture and accurate analytical expressions for the antenna beamwidth as well as the assessment of their tolerances are obtained for the first time in literature.

9. References

Baars, J. W. M. (1973). The measurement of large antennas with cosmic radio sources. *IEEE Transactions on Antennas and Propagation,* Vol. AP-21, No. 4, (July 1973) pp. (461-474), ISSN 0018-926X

Balanis, C. A. (2005). *Antenna Theory.* John Wiley & Sons, ISBN 0-471-66782-X, Hoboken, New Jersey

Guidici, D.A. & Castelli, J.P. (1971). The use of extraterrestrial radio sources in the measurement of antenna parameters. *IEEE Transactions on Aerospace and Electronic Systems,* Vol. AES-7, No. 2, (March 1971) pp. (226-234), ISSN 0018-9251

Johnson, R. C. at all. (1993). *Antenna Engineering Handbook.* McGrow-Hill, ISBN 0-07-032381-X, New York

Ko, H. C. (1961). On the determination of the disk temperature and the flux density of a radio source using a high-gain antenna. *IRE Transactions on Antennas and Propagation,* Vol. AP-9, No. 5, (September 1961) pp. (500-501), ISSN 0096-1973

Kraus, J. D. (1986). *Radio Astronomy,* Cygnus-Quasar Books, ISBN 1-882484-00-2, Powell, Ohio

Kummer, W. H. at al. (1979). *IEEE Standard Test Procedures for Antennas,* John Wiley & Sons, ISBN 0-471-08032-2, New York.

Solovey, A. & Mittra, R. (2008). Extended source size correction factor in antenna gain measurements, *Proceedings of the 38th European Microwave Conference,* pp. 983-986, (available online only), Amsterdam, October 2008.

22

Electrodynamic Analysis of Antennas in Multipath Conditions

Eddy Luis Molina Morales and Leandro de Haro Ariet
Madrid Technological University,
Spain

1. Introduction

During the last years, a sustained growth in the number of users in mobile communication systems has led the necessity to increase their capacity. With the imminent arrival of the fourth generation and the deployment of new high data rate services, the increasing of system capacity is essential. Adaptive arrays and MIMO antennas (multiple input and multiple output) become a promising affordable solution.. The use of these techniques will not only increase the capacity but also they improve the signal quality, coverage range and they simplify the application of new services by exploiting the specific characteristics of these antenna systems. However, the introduction of these technologies involves changes in network planning and deployment, in addition, the increased complexity of both the tranceptors as the radio resource management. In these new radio technologies, antenna system devices shall be designed to get advantage from the multipath. Moreover antennas cannot be considered apart from the radio channel to define the whole structure of radio-communication system, as well as, its management and planning. This is why this chapter addresses the analysis of antennas in a multipath environment. The chapter consists of four more sections. The following one is a comprehensive and detailed analysis from the point of view of the electromagnetic theory of a receiving antenna. It will be the basis for the third section, which presents the time analysis of the output antenna signals with deep fading, oriented to a better understanding of more complex systems as the MIMO ones. In this case, the procedure is used to evaluate the spatial correlation between radiators. The fourth section provides a brief description of fading signals from a statistical point of view, deepening into the classification of radio channels. The fifth section presents some conclusions and a list of references.

2. Receiving antenna

When a wave arrives to an antenna, it will circulate a current on the antenna conductors; inducing an electromotive force (*e.m.f*) (at the input terminals of the antenna, so it excites a guided wave through the receiver input. In Figure 1 this situation is outlined, where Z_R is the input impedance of the receiver (neglecting the input transmission line, that is to say, the receiver input impedance reflected at the terminals of the line connected to the antenna). This situation is quite common using transmission lines matched to the receiver impedance.

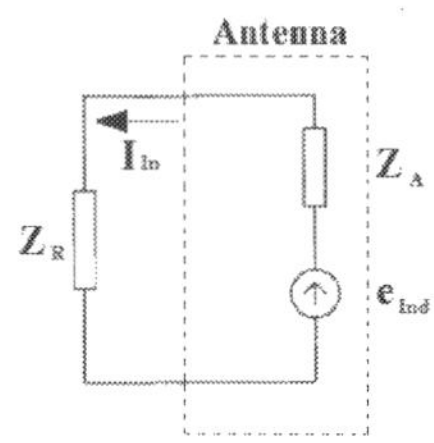

Fig. 1. Equivalent circuit of the receiving antenna

Please, note that under these conditions, the antenna behaves as a generator with an input impedance equal to the output transmitting antenna impedance. The remaining problem to solve is the ratio of effective field incident on the antenna and induced e.m.f on its terminals. In order to study this problem, the reciprocity theorem shall be applied to the situations that are shown in Figure 2.

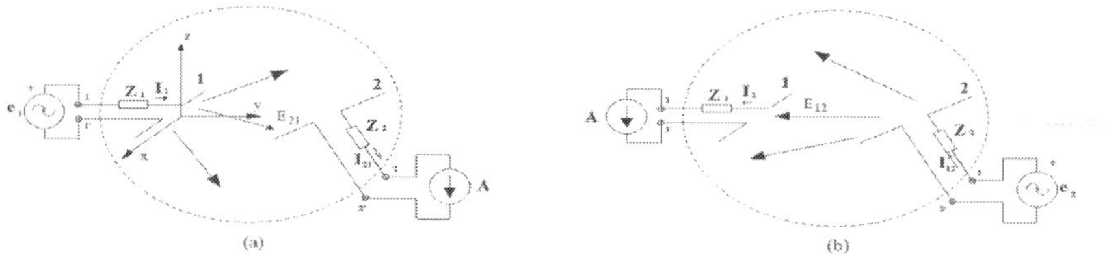

Fig. 2. Equivalent quadripole, linear and reciprocal, to study the interaction between a transmitter and a receiver antenna

Consider any two antennas, placed arbitrarily in free space mutually in far zone, in order to obtain the voltages and currents on their terminals, the previous proposed model that describes a link between two antennas as a linear and reciprocal quadripole will be used. If the terminals 1-1'of the quadripole of Figure 2a is connected to a generator the *e.m.f* e_1 and impedance Z_1, at Terminal 2-2′, connected to a load Z_2, it circulates a current I_{21}. If a generator, with an *e.m.f* e_2 and output impedance Z_2, is connected to the terminals 2-2′ (Figure 2b), and a load Z_1 is connected to terminals 1-1′, it circulates a current I_{12}. According to the reciprocity theorem (Monson, 1996):

$$\frac{e_1}{I_{21}} = \frac{e_2}{I_{12}} \tag{1}$$

and following the formula of an antenna in far field, the situation shown in Figure 2a the actual value of the field produced by the antenna 1 in the location of the antenna 2 can be determined (Dolujanov, 1965):

$$E_{21} = \frac{\sqrt{30 \cdot D_{Max1} \cdot P_{Rad1}}}{r} \cdot F_{C1}(\theta_{21}, \phi_{21}) \tag{2}$$

where the subscript 1 refers to the first antenna; θ_{21} y ϕ_{21} are the angles that define the direction in which the antenna 2, and F_{C1}. is the radiation pattern (directional characteristic). The coordinate system is placed on the antenna 1. Every one of the parameters included in the formula (2) refers to the resulting field of the antenna (superposition of the main polarization and the cross-polarization). At the input of the antenna 1, the effective value of current will be:

$$I_1 = \frac{e_1}{|Z_{A1} + Z_1|}$$

Likewise, according to the expression of the radiation resistance of an antenna (Nikolski, 1976), this is:

$$P_{Rad1} = I_1^2 \cdot R_{Rad} = \left|\frac{e_1}{Z_{A1} + Z_1}\right| \cdot R_{Rad1}$$

Substituting this result in (2) and clearing e_1, it gets:

$$e_1 = \frac{E_{21} \cdot |Z_{A1} + Z_1| \cdot r}{\sqrt{30 \cdot R_{Rad1} \cdot D_{Max1}} \cdot F_{C1}(\theta_{21}, \phi_{21})}. \tag{3}$$

When repeating the above analysis to the situation shown in Figure 2b, it gets:

$$e_2 = \frac{E_{12} \cdot |Z_{A2} + Z_2| \cdot r}{\sqrt{30 \cdot R_{Rad2} \cdot D_{Max2}} \cdot F_{C2}(\theta_{12}, \phi_{12})}. \tag{4}$$

According to the Reciprocity Theorem (1), the (3) y (4) can be related:

$$\frac{E_{21}|Z_{A1} + Z_1| \cdot r}{I_{21} \cdot \sqrt{30 \cdot R_{Rad1} \cdot D_{Max1}} \cdot F_{C1}(\theta_{21}, \phi_{21})} = \frac{E_{12}|Z_{A2} + Z_2| \cdot r}{I_{12} \cdot \sqrt{30 \cdot R_{Rad2} \cdot D_{Max2}} \cdot F_{C2}(\theta_{12}, \phi_{12})}$$

And it can also be written as:

$$\left[\frac{I_{12}}{E_{12}}\right] \cdot \frac{E_{21}|Z_{A1} + Z_1| \cdot r}{\sqrt{30 \cdot R_{Rad1} \cdot D_{Max1}} \cdot F_{C1}(\theta_{21}, \phi_{21})} = \left[\frac{I_{21}}{E_{21}}\right] \cdot \frac{E_{12}|Z_{A2} + Z_2| \cdot r}{\sqrt{30 \cdot R_{Rad2} \cdot D_{Max2}} \cdot F_{C2}(\theta_{12}, \phi_{12})} \tag{5}$$

Up to now the polarization of the incident wave has not be considered. However, the induced current in the surface of the receiving antenna is determined by those incident wave components "parallel" to the polarization of the receiving antenna[1]. That is to say, the wave field produced by the antenna 2 in the location of the antenna 1 can be expressed as:

$$\vec{E}_{12} = E_{12} \cdot \vec{e}_{2T}(\theta_{12}, \phi_{12})$$

where E_{12} is the actual value of the field and $\vec{e}_{2T}$ is normalized vector indicating the polarization of the wave transmitted by the antenna 2. When identifying the type of polarization of the antenna 1 itself in reception for the normalized vector: $\vec{e}_{2R}(\theta_{21}, \phi_{21})$, thus, the effective value of the component of the incident field that matches the type of polarization of the antenna 1 at the reception will be as (Márkov & Sazónov, 1978):

$$E_{R12} = E_{12} \cdot \left| \vec{e}_{2T}(\theta_{12}, \phi_{12}) \cdot \vec{e}_{1R}(\theta_{21}, \phi_{21}) \right| \tag{6}$$

Note that for standard polarization vectors is satisfied that:

$$\vec{e}_{2T}(\theta_{12}, \phi_{12}) \cdot \vec{e}^{\,*}_{2T}(\theta_{12}, \phi_{12}) = 1; \quad \vec{e}_{2R}(\theta_{21}, \phi_{21}) \cdot \vec{e}^{\,*}_{2R}(\theta_{21}, \phi_{21}) = 1$$

Therefore E_{R12} (6) is the real value of electric involved in the process of reception. Note that if the antennas 1 and 2 were equal and with identical directions of pointing($\theta_{12} = \theta_{21}$ y $\phi_{12} = \phi_{21}$), thus:

$$\left| \vec{e}_{2T}(\theta_{12}, \phi_{12}) \cdot \vec{e}^{\,*}_{2R}(\theta_{12}, \phi_{12}) \right| = \left| \vec{e}_{1T}(\theta_{21}, \phi_{21}) \cdot \vec{e}^{\,*}_{1R}(\theta_{21}, \phi_{21}) \right|$$

where the maximum value will happen when: $\vec{e}_{1T}(\theta_{21}, \phi_{21}) = \vec{e}^{\,*}_{1R}(\theta_{21}, \phi_{21})$, which the polarization of the transmitting antenna and the receiving one are the same but with opposite sense (seen from a common reference system), that is to say, an identical polarization seen from transmitting point of view. Therefore, we can write the expression (6) as:

$$E_{12} = \frac{E_{R12}}{\left| \vec{e}_{2}(\theta_{12}, \phi_{12}) \cdot \vec{e}^{\,*}_{1}(\theta_{21}, \phi_{21}) \right|} \tag{7}$$

[1] Polarization of an antenna is defined from transmission point of view. However, although the reception point of view is opposite to the transmission one, the polarization of an antenna is defined equally.

without distinction in the polarization vectors whether it is an antenna transmission or reception. Similarly we can write the actual value of the field incident at the antenna 2 from the antenna 1:

$$E_{21} = \frac{E_{R21}}{\left| \vec{e}_1(\theta_{21}, \phi_{21}) \cdot \vec{e}_2^{\,*}(\theta_{12}, \phi_{12}) \right|} \tag{8}$$

Given that the denominators of expressions (7) and (8) are equal, and substituting these in (5), we obtain:

$$\left[\frac{I_{12}}{E_{R12}}\right] \cdot \frac{|Z_{A1} + Z_1|}{\sqrt{R_{Rad1} \cdot D_{Max1}} \cdot F_{C1}(\theta_{21}, \phi_{21})} = \left[\frac{I_{21}}{E_{R21}}\right] \cdot \frac{|Z_{A2} + Z_2|}{\sqrt{R_{Rad2} \cdot D_{Max2}} \cdot F_{C2}(\theta_{12}, \phi_{12})} \tag{9}$$

Analysing both members; the relationship (I_{12}/E_{R12}) depend only on the characteristics of the antenna. The field incident on the antenna with the same polarization induced currents on the antenna. On the other hand the factors of the left member of (9) depend exclusively on the characteristics of the antenna 1. Similarly, it appears that all the factors of the right-hand side of (9) depend exclusively on the characteristics of the antenna 2. Since the above analysis there is anyone restriction on the type of antenna used (in general, antennas 1 and 2 are different), the obvious conclusion is:

$$\left[\frac{I_{12}}{E_{R12}}\right] \cdot \frac{|Z_{A1} + Z_1|}{\sqrt{R_{Rad1} \cdot D_{Max1}} \cdot F_{C1}(\theta_{21}, \phi_{21})} = C \tag{10}$$

where C is a constant that has the same value for all antennas. Therefore, C can be obtained by replacing in (10) the values of the parameters of any antenna; in particular the Hertz's dipole. Figure 3 shows a dipole antenna formed by a thin conductor of length *L*, with an impedance Z_R connected at its terminals, impinging a wave of linear polarization parallel to the dipole.

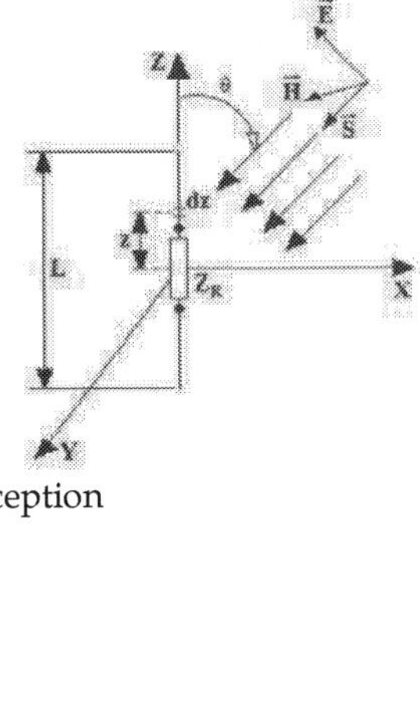

Fig. 3. Antenna type dipole in reception

This will induce a current $I(z)$ along the conductor. The $e.m.f$ de, induced a small segment of length dz will be:

$$de = E \cdot sen(\theta) \cdot dz \tag{11}$$

As for the conductor circulate the current $I(z)$, a power will be delivered to the antenna:

$$dP = de \cdot I^*(z) = E \cdot sen(\theta) \cdot I^*(z) \cdot dz$$

The total power delivered to the antenna is:

$$P = \int_L E \cdot sen(\theta) \cdot I^*(z) \cdot dz$$

In the terminals (see Figure 1) is obtained:

$$P = I_{In}^2 \cdot (Z_A + Z_R)$$

And thus:

$$I_{In}^2 \cdot (Z_A + Z_R) = \int_L E \cdot sen(\theta) \cdot I^*(z) \cdot dz$$

In the case of a Hertz dipole with length $\Delta l << \lambda$ we can presume the current uniform: $I(z) = I_{Ent}$ and, therefore:

$$I_{In}^2 \cdot (Z_A + Z_R) = E \cdot sen(\theta) \cdot I_{Ent}^* \cdot \Delta l$$

Or what is the same: $\dfrac{I_{Ent}}{E} = \dfrac{I_{12}}{E_{R12}} = \dfrac{sen(\theta) \cdot \Delta l}{|Z_A + Z_R|}$

The radiation pattern of Hertz's dipole is known $F_C(\theta) = sen(\theta)$, its directivity is of: $D_{Max} = 1.5$. Moreover, the radiation resistance is of: $R_{Rad} = 80\pi^2(\Delta l/\lambda)^2$; so, substituting these data in (10):

$$C = \frac{\lambda}{\sqrt{120} \cdot \pi}$$

where the value C obtained for the dipole Hertz is unique for all antennas, and it can be replaced in (10) to obtain the expression of the current at the terminals of the any antenna

(I_{Ind}), induced by that component of the field of the incident wave (E_R) with a polarization equal to the receiving antenna itself and arriving at the antenna in the direction defined by angles θ and ϕ:

$$I_{Ind} = E_R \cdot \frac{\lambda}{\pi \cdot |Z_A + Z_R|} \cdot \sqrt{\frac{R_{Rad} \cdot D_{Max}}{120}} \cdot F_C(\theta, \phi) \tag{12}$$

According to Figure 1 the induced *e.m.f* at the terminals of the antenna can be determined by:

$$e_{Ind} = I_{Ind} \cdot |Z_A + Z_R| = E_R \cdot \frac{\lambda}{\pi} \cdot \sqrt{\frac{R_{Rad} \cdot D_{Max}}{120}} \cdot F_C(\theta, \phi) \tag{13}$$

Taking into account the expressions (12) and (13), the values of the current and the induced *e.m.f* at the terminals of the antenna have a dependency on the direction of arrival of the incident wave, expressed by $F_C(\theta, \phi)$, which is the radiation pattern of the antenna in transmission. The expression (13) can be written like this:

$$e_{Ind}(\theta, \phi) = e_{Ind\,Max} \cdot F_C(\theta, \phi)$$

Then $e_{Ind\,Max}$ expresses the value of induced *e.m.f* when the wave arrives from the direction of maximum reception of the antenna, and $F_C(\theta, \phi)$ represents the normalized radiation pattern of the field of the antenna in reception mode, which is equal to that characteristic of their antenna transmission. On the other hand, the coefficient of directivity is function of the radiation pattern. Therefore, it confirms that its value is the same regardless of the antenna works as transmitters or receivers. Similarly, in the case of linear antennas, the effective length:

$$l_{Ef} = \frac{\lambda}{\pi} \cdot \sqrt{\frac{D_{Max} \cdot R_{Rad}}{120}} \tag{14}$$

that depends on the coefficient of maximum directivity, the radiation resistance, and will have the same value in transmission and reception. Substituting in expression (13), it becomes:

$$e_{Ind} = E_R \cdot l_{Ef} \cdot F_C(\theta, \phi)$$

where the product: $E_R \cdot l_{Ef} = e_{Ind\,Max}$ is the maximum value of the induced *e.m.f* (when $F_C = 1$). It is important to stress the significance of the effective length of the antenna; that is to say, this is a length such, when multiplied with the incident field intensity (the polarization equal to the antenna itself, incident in the direction of maximum reception), gives us the maximum value of the induced $e.m.f$. Known the induced $e.m.f$ in the antenna, with reference to Figure 1 we can determine the voltage at the input terminals of the receiver, supposed this is connected directly to the antenna, by:

$$V_R = \left| \frac{Z_R}{Z_A + Z_R} \right| \cdot e_{Ind} \tag{15}$$

Considering the presence of a transmission line (the characteristic impedance Z_0) between the antenna (impedance Z_A) and receiver (impedance Z_R), it defines the reflection coefficient at the antenna input (Pozar, 2004):

$$\rho_A = -\frac{Z_A - Z_0}{Z_A + Z_0}$$

and at the receiver input: $\rho_R = \dfrac{Z_R - Z_0}{Z_R + Z_0}$; the expression (15) will transformed:

$$V_R = \frac{|1+\rho_R| \cdot |1+\rho_A| \cdot \exp(-\alpha \cdot L)}{2 \cdot |1+\rho_R \cdot \rho_A \cdot \exp(-j2 \cdot \Gamma \cdot L)|} \cdot e_{Ind} \tag{16}$$

where $\Gamma = \alpha + j\beta$ is the propagation constant of the transmission line (which includes the attenuation constant α and the phase constant β), and L is the length of the line. In case of low frequency receivers ($f < 30\,MHz$) the condition: $|Z_R| >> |Z_A|$ is usually applied; thus (replacing in 15) is obtained: $V_R \approx e_{Ind}$; so, maximum voltage to the receiver's input. Moreover, matching the transmission line to the antenna ($\rho_A \approx 0$) and $|Z_R| >> Z_0$, $\rho_R \approx 1$ and expression (16) also gives us the maximum input voltage of the receiver: $V_R = \exp(-\alpha \cdot L) \cdot e_{Ind}$, where the factor: $\exp(-\alpha \cdot L) < 1$ takes into account transmission losses along the line. On the other hand, in case of higher frequency it is more difficult to provide high power amplifiers, so the purpose is to maximize the real power delivered by the antenna to receiver. If the receiver is directly connected to the antenna, the power supplied to the receiver is:

$$P_R = I_{Ind}^2 \cdot R_R = E_R^2 \cdot \frac{\lambda^2 \cdot R_R}{\pi^2 \cdot |Z_A + Z_R|} \cdot \frac{R_{Rad} \cdot D_{Max}}{120} \cdot F_P(\theta, \phi)$$

where R_R is the real part of the input impedance of the receiver. Maximum transmitting power forces to an impedance matching between the receiver and antenna ($Z_R = Z_A^*$). Applying this condition in the above expression and using the classical expression of the efficiency of an antenna: $R_{Rad} = \eta_A \cdot R_{In}$, we obtain (Balanis, 1982):

$$P_R = E_R^2 \cdot \frac{\lambda^2}{4 \cdot \pi^2} \cdot \frac{\eta_A \cdot D_{Max}}{120} \cdot F_P(\theta,\phi)$$

When using a transmission line of low losses between the receiver and the antenna and impedance matching between the receiver and the line, the power supplied by the antenna to the line will be:

$$P_L = E_R^2 \cdot \frac{\lambda}{\pi^2} \cdot \frac{Z_0 \cdot R_A}{|Z_A + Z_0|^2} \cdot \frac{\eta_A \cdot D_{Max}}{120} \cdot F_p(\theta,\phi) = E_R^2 \cdot \frac{\lambda^2}{4\pi^2} \cdot \left(1-|\rho|^2\right) \cdot \frac{\eta_A \cdot D_{Max}}{120} \cdot F_p(\theta,\phi)$$

A part of power will flow to the receiver:

$$P_R = P_L \cdot \exp(-2 \cdot \alpha L) = E_R^2 \cdot \frac{\lambda^2}{4\pi^2} \cdot \left(1-|\rho|^2\right) \cdot \exp(-2 \cdot \alpha L) \cdot \frac{\eta_A \cdot D_{Max}}{120} \cdot F_p(\theta,\phi) \quad (17)$$

and the other part: $P_{Cons.L} = P_L \cdot [1 - \exp(-2\alpha L)]$, will be consumed by the line. In expression (17) we can see that through the appropriate orientation of the antenna, by matching the direction of maximum reception of the antenna with the direction of arrival of the wave we get $F_P = 1$ and the received power reaches its maximum value by adjusting the direction:

$$P_{RMax} = \left(1-|\rho_A|^2\right) \cdot \exp(-2\alpha L) \cdot \eta_A \cdot \frac{E_R^2}{120\pi} \cdot \frac{\lambda^2}{4\pi} \cdot D_{Max} \quad (18)$$

Considering this expression; factor $E_R^2/120\pi$ represents the module of the Poynting vector of the incident wave to the antenna. If we multiply this factor by the physical area of the antenna, the power incident on the antenna is obtained:

$$P_{Inc} = \frac{E_R^2}{120\pi} \cdot A_{Geom} \quad (19)$$

The antenna is not able to fully grasp the incident power that really is:

$$P_{Cap} = \frac{E_R^2}{120\pi} \cdot \left(\frac{\lambda}{4\pi} \cdot D_{Max}\right) = \frac{E_R^2}{120\pi} \cdot A_{Ef} \quad (20)$$

where:

$$A_{Ef} = \frac{\lambda^2}{4\pi} \cdot D_{Max} \quad (21)$$

It will be called the effective area of the antenna; namely the area through which the antenna fully captures the incident power density. The relationship:

$$\xi_A = \frac{P_{CAP}}{P_{Inc}} = \frac{A_{Ef}}{A_{Geom}} \tag{22}$$

is called the coefficient of the utilization of the surface of the antenna. The expression (18) now we can write:

$$P_{RMax} = \left(1-|\rho_A|^2\right)\cdot \exp(-2\alpha L)\cdot \eta_A \cdot \xi_A \cdot P_{Inc} = \left(1-|\rho_A|^2\right)\cdot \exp(-2\alpha L)\cdot \eta_A \cdot P_{Cap}$$

Note that the antenna does not use all the power captured, but a fraction called useful captured power:

$$P_{CapUse} = \eta_A \cdot P_{Cap} \tag{23}$$

the other part: $(1-\eta_A)\cdot P_{Cap}$ is the power losses in the antenna as heat during the reception. Between the antenna and the transmission line is not always met the condition of impedance matching, therefore only part of the useful captured power useful is delivered to the line:

$$P_L = \left(1-|\rho_A|^2\right)\cdot P_{CapÚtil} \tag{24}$$

and, only the fraction given by (17) is delivered to the receiver. In Figure 4 shows schematically the flow of power from the wave that propagates in free space and arrives to the antenna, to the receiver. From the analysis we can summarize the conditions for optimal reception: coincidence of the polarization itself of the antenna with the polarization of the incident wave (This ensures E_R maximum); orientation the antenna to the direction of arrival of the wave ($F_P = 1$); effective area (or effective length) maximum of the antenna (which depends on its physical characteristics); high efficiency ($\eta_A = 1$); proper impedances matching between the antenna and feed line ($\rho_A \approx 0$); low losses of the feed line ($\alpha_L \approx 0$). In practice these conditions are met satisfactorily, so the power at the receiver is close to optimal value:

$$P_{Opt} = \frac{E_R^2}{120\pi}\cdot \eta_A \cdot A_{Ef} = \frac{E_R^2}{120\pi}\cdot\frac{\lambda^2}{4\pi}\cdot G_{Max} \tag{25}$$

This expression clearly reveals the role of the maximum gain at the reception. The value of G_{Max} of any antenna indicates either the number of times that the power delivered to the receiver exceeds that delivered by an isotropic radiator ($G_{Max} = 1$) under the same conditions of external excitation, coupling and losses of the transmission line. In a similar way we can say that the coefficient of directivity D_{Max} of an antenna (as was noted earlier, has the same value in transmission and reception), at the receiving antenna indicates the number of times that the power captured by the antenna exceeds that delivered by an isotropic radiator. Finally, keep in mind that the presence of induced current in the receiving antenna also determines an effect known as secondary radiation. We must emphasize the fact that, in general, the directional characteristic of the secondary radiation does not match the directional characteristic for the transmitting antenna.

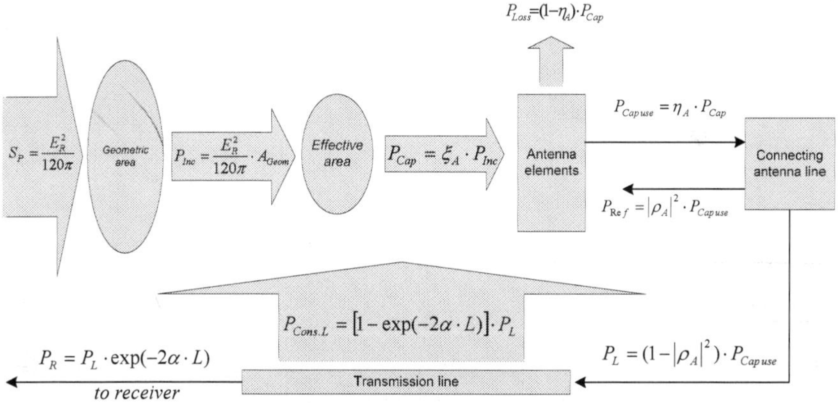

Fig. 4. Power flow from the wave in free space to the receiver

This is explained by that shape of the induced current distribution on the elements of the antenna is not equal to that found when the excitation takes place at the terminals of the antenna. However, the total power of secondary radiation can be calculated from:

$$P_{Sec\,Rad}=I_{Ind}^2\cdot P_{Rad}=E_R^2\,\frac{\lambda^2}{\pi^2}\cdot\frac{R_{Rad}^2}{|Z_A+Z_R|^2}\cdot\frac{D_{Max}}{120}\cdot F_p(\theta,\varphi) \tag{26}$$

In this expression the factor: $F_P(\theta,\phi)$ defines the directional pattern of the antenna during the reception; while $P_{Sec\,Rad}$ is the total power of secondary radiation in all directions of space. This phenomenon has a special interest in antennas that act as passive elements, where generally Z_R is the impedance of a ($Z_R=0$) or a pure reactive element ($Z_R=jX_R$). Particularly this treatment can be extended to objects that do not really fulfil the mission of the antennas, that serving (intentionally or not) as reflectors of radio waves. In these cases, as can be shown easily from (26), it is possible to reach a power of secondary radiation which is 4 times larger than the optimum power of reception given by the formula (25). The analysis of antennas in reception mode, leads to a set of conclusions of great importance. First we establish that many of the properties of the antennas are the same as transmission as reception, which simplifies its research, since it is not necessary to determine these properties in both regimes. Thus, the impedance of the antenna, its directional pattern, its directivity, efficiency, and gain are the same in both schemes of work. The expressions obtained (mainly induced *e.m.f*) in the receiving antenna (13), are useful in tasks of calculation and design of antennas in general. There are two parameters that are used in the study of the receiving antennas (aperture antennas mainly); the coefficient of utilization of the surface of the antenna and the effective area.

3. Antennas receiving mode in multipath conditions

It is said that an antenna operates under multipath conditions when in it impinge radio waves arriving from different directions. Figure 5 show the multipath phenomenon.

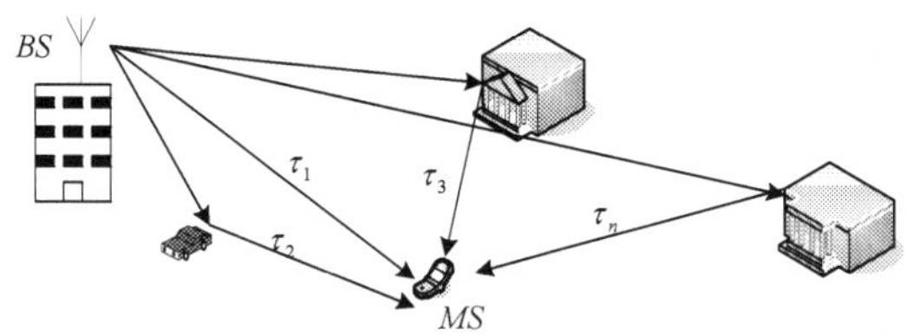

Fig. 5. Phenomenon of multipath propagation

In Figure 5 it can be seen: transmitter and receiver antennas, rays that define the different propagation paths from transmitter to receiver antenna, and the scattering elements (buildings and cars), which are called scatterer. Propagation environments, together with the communications system can be divided into: indoor and outdoor. The theory of radio channels is a rather broad topic not covered here, but from point of view of the antenna, we will present (only from the point of view spatial) similar to the patterns that characterize the radiation of the antennas (Rogier, 2006). This way, according to the angular distribution of power that reaches the antenna, we can present them as omnidirectional, and with some directionality. Then, the shape of the angular distribution of power that characterizes the channel depends on the position of the antenna inside the environment of multipath propagation. Figure 6 shows some examples.

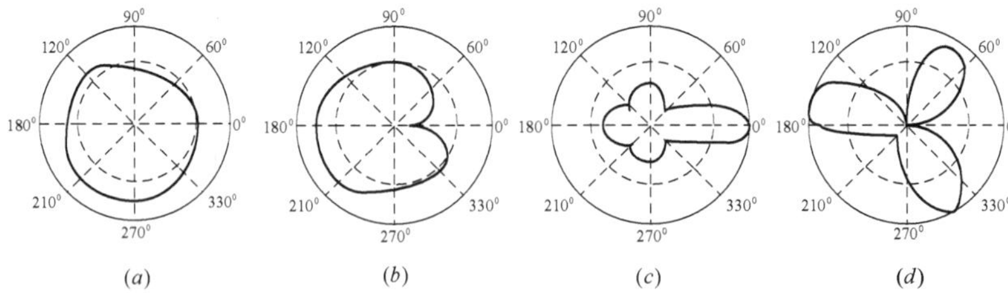

Fig. 6. Angular distribution of the power reaching the antenna by multiple pathways, a) omni directional channel, b) dead zone channel, c) directional channel, d) multidirectional channel.

Figure 6 shows patterns corresponding to the measured power at the terminals of a high directivity antenna used to sample the channel performance in time. The graphs that are shown, they correspond themselves only to the azimuthal plane, often with higher importance as in case of mobile communications.. Figure 6a shows some omni-directional angular distribution, where the incoming waves reach the antenna with a similar intensity from all directions from statistical point of view. Figure 6b presents a channel with an angular distribution indicating some directional properties, in which the waves impinge the antenna from all directions, except from one sector, normally called dead zone. Figure 6c shows the case where the waves reach the antenna from a defined direction. Figure 6d is a typical situation when waves reach the antenna from some well defined directions, (in this case three), in most cases are caused by discrete clusters of scatterer, as in the mobile communications enabling the use of smart antenna systems. In practice there may be all kinds of Multipath described in Figure 6 on a single antenna. This is the true of the antenna is in a mobile terminal that changes its spatial position over time. Induced *e.m.f* at the antenna terminals, which has been described in terms of the angular power spectrum in the plots of Figure 6, is the statistical average of the amplitude of the signal at the antenna terminals. In fact, the resulting signal has a fading performance, due to the fasorial summation of all the waves arriving to the antenna with different amplitudes and phases, due to the difference in the delay associated with each propagation path (Blaunstein et al., 2002. Under this situation induced *e.m.f* in the terminals of an antenna has a fading nature, as it is shown in Figure 7.

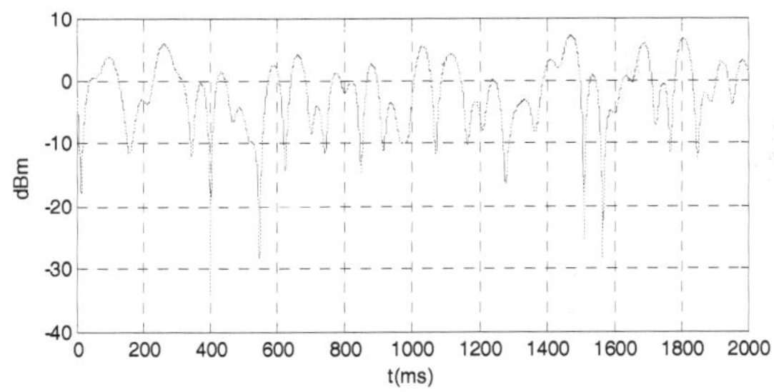

Fig. 7. Signal at the antenna terminals under multipath

The fading performance of the signal can be, explained by the multipath phenomenon using a ray model at the plane. In a first approximation, one considers n waves coming through L_n different paths to a Q- point, in which there is no antenna. The spreading angle associated with each beam is zero, so $\sigma_n = 0$, so it is valid to propose that the resulting signal $u(t)$ in Q point is given by (27):

$$u(t) = \sum_{n=1}^{L_n} a(\phi_n) \cdot s_n(t - \tau_n) + N(t) \tag{27}$$

where $a(\phi_n) = \exp(j \cdot K \cdot \cos(\phi_n))$, is the phase that comes the n signal to the Q point for the ϕ_n direction, $K = 2\pi/\lambda$ is the constant propagation wave to the working frequency, whose wavelength is λ, τ_n is the delay associated to S_n signal. It has also introduced additive noise $N(t)$ in the point. The expression (26) accurately describes the fading nature associated to the multipath. However, the expression (26) does not include the antenna. A computational procedure based on (26), that considers the presence of the antenna and its parameters to obtain the fading signals at their terminals, when it does interact with virtual radio channels is described out below (Molina & De Haro, 2007). The philosophy is based on the creating an effect similar to that described by the equation (27), and simultaneously introduce in the equation that describes the induced $e.m.f$ in an antenna in the reception mode (equation (13) of the previous section). Using as antenna a half wavelength dipole, whose radiation pattern is as shown in Figure 8.

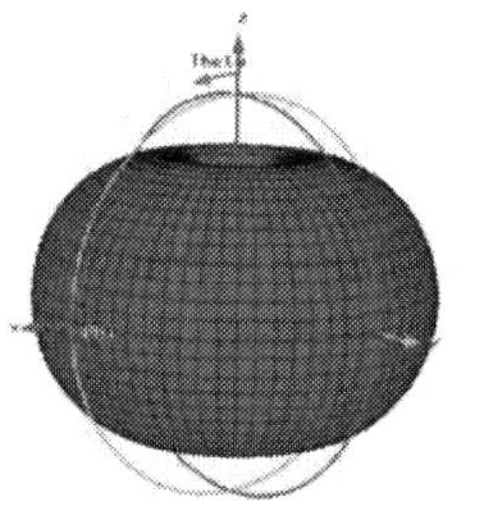

Fig. 8. Radiation pattern of half-wave dipole

The radiation pattern of dipole is displayed using a two dimensional plot of the directional characteristics of amplitude, phase and polarization. The polarization follows the definitions of the main polarization and cross polarization given by (Ludwig 1973 and Markov & Sazonov, 1978). Figure 9 shows two-dimensional pattern of amplitude and phase in the main and crossed polarizations of the dipole. This form of plot allows a faster execution due to the use of matrices in the procedure.

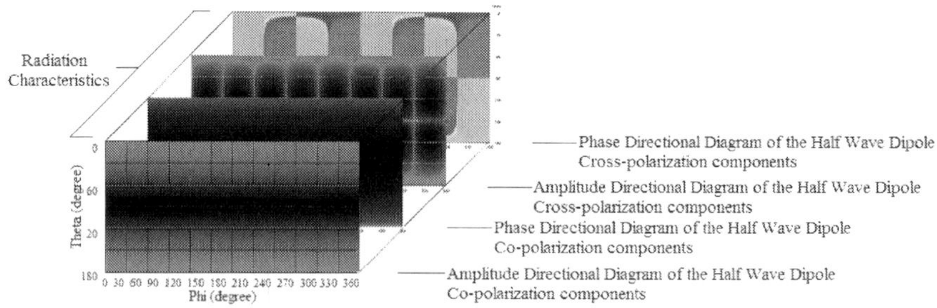

Fig. 9. Radiation characteristics of the half-wave dipole in 2-D.

Moreover, one defines the multipath radio channel as a 360 x 180 matrix that represents all possible directions of space from where can reach the waves to the antenna. It is generated dynamically by the pseudo random number of waves, their amplitudes, their phases, and the coordinates of its angular direction of arrival. In the same way you generate the angular spread associated to each ray.. Figure 10 shows the results of a simulation of the virtual channel multipath where the probability distribution associated with the generation of the angles of arrival was uniform in the 4π radians of the sphere and the probability distribution of the amplitude of the signals was uniform too

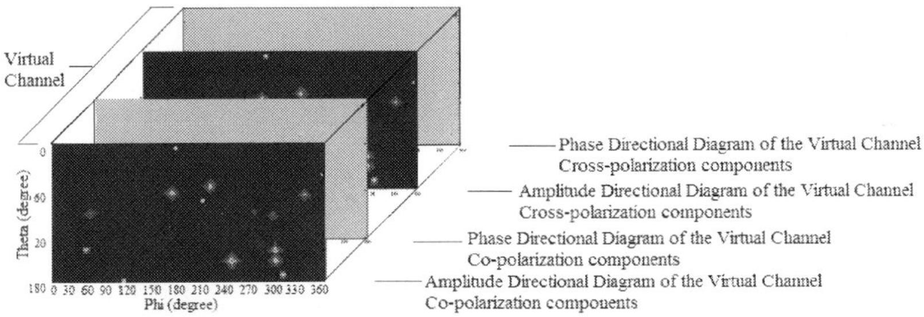

Fig. 10. Virtual multipath radio channel in 2-D.

Note in Figure 10, that as in the case of the antenna, the colour scale indicates the intensity with which the signal arrives. The channel has been defined by analogy with the antennas: amplitude and phase pattern in the main and cross polarization, corresponding to the signals that incident in the antenna. Now, the $\vec{H}(\theta,\phi,t)$ function, models the dynamic performance of space time channel, which, by analogy with the antennas, are contained all the characteristics of amplitude, phase and polarization of the channel:

$$\vec{H}(\theta,\phi,t) = H_\theta(\theta,\phi,t)\vec{i}_\theta + H_\phi(\theta,\phi,t)\vec{i}_\phi \tag{28}$$

where H_θ and H_ϕ are the functions in the principals planes of channel (in main and cross polarization). They, $\vec{i}_\theta$, $\vec{i}_\phi$ are unit vectors indicating the orientation of the electric field incident. Each one of the functions that are part of the right-hand side of (28) is defined as follows:

$$\vec{H}_\theta(\theta,\phi,t) = h_\theta(\theta,\phi,t)\cdot e^{j\psi_\theta(\theta,\phi,t)}\vec{i}_\theta \tag{29a}$$

and

$$\vec{H}_\phi(\theta,\phi,t) = h_\phi(\theta,\phi,t)\cdot e^{j\psi_\phi(\theta,\phi,t)}\vec{i}_\phi \tag{29b}$$

They, h_θ and h_ϕ are the directional pattern associated to the amplitude component of the main and cross polarization in the channel. ψ_θ and ψ_ϕ are also the patterns of the phase associated to the main and cross polarization components of the channel. With this description, and adapting the equation (13) the previous section to the present situation; induced *e.m.f* at the terminals of the dipole can be raised by the following expression in scalar form:

$$u(t) = l_{Ef} \cdot \int_0^{2\pi}\int_0^{\pi} H(\theta,\phi,t) \cdot F(\theta,\phi) \cdot sen(\theta) d\theta d\phi \quad [V] \tag{30}$$

The components for the main and cross polarization, equation (30) are:

$$\begin{aligned} u(t) = l_{Ef} \cdot & \int_0^{2\pi}\int_0^{\pi} h_\theta(\theta,\phi,t) \cdot e^{j\Psi_\theta(\theta,\phi,t)} \cdot f_\theta(\theta,\phi) \cdot e^{j\Phi_\theta(\theta,\phi)} sen(\theta) d\theta d\phi + \\ & + l_{Ef} \cdot \int_0^{2\pi}\int_0^{\pi} h_\phi(\theta,\phi,t) \cdot e^{j\Psi_\phi(\theta,\phi,t)} \cdot f_\phi(\theta,\phi) \cdot e^{j\Phi_\phi(\theta,\phi)} sen(\theta) d\theta d\phi \end{aligned} \quad [V] \tag{31}$$

where we recall that the functions f_θ and f_ϕ are the directional characteristics amplitude at the main and cross polarization and the functions Φ_θ and Φ_ϕ are the phase patterns of the antenna in these polarizations. The geometrical interaction between the antenna and the multipath radio-channel can be shown in Figure 11.

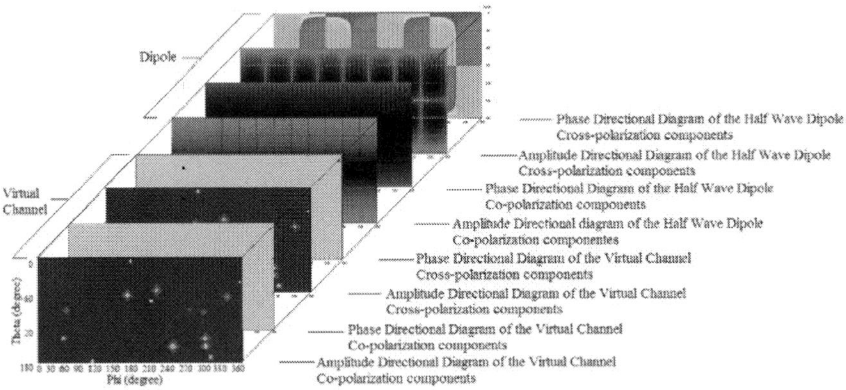

Fig. 11. Geometry of the interaction between the antenna and the channel

Last expressions can be treated in a discrete way, tabulating the functions required by the equation (31):

$$u(t) = l_{Ef} \cdot \sum_{\phi=0}^{N} \sum_{\theta=0}^{M} h_\theta(\theta,\phi,t) \cdot e^{j\Psi_\theta(\theta,\phi,t)} \cdot f_\theta(\theta,\phi) \cdot e^{j\Phi_\theta(\theta,\phi)} sen(\theta) \Delta\theta\Delta\phi + \\ + l_{Ef} \cdot \sum_{\phi=0}^{N} \sum_{\theta=0}^{M} h_\phi(\theta,\phi,t) \cdot e^{j\Psi_\phi(\theta,\phi,t)} \cdot f_\phi(\theta,\phi) \cdot e^{j\Phi_\phi(\theta,\phi)} sen(\theta) \Delta\theta\Delta\phi \quad [V] \qquad (32)$$

In (32) equation, $\Delta\theta = \pi/N$ y $\Delta\phi = 2\pi/M$, represent the angular resolution step in coordinates θ and ϕ respectively, with a value of one degree, so $N = 180$ and $M = 360$. The induced *e.m.f* obtained on the terminals of the dipole due to its interaction with the virtual channel is shown below:

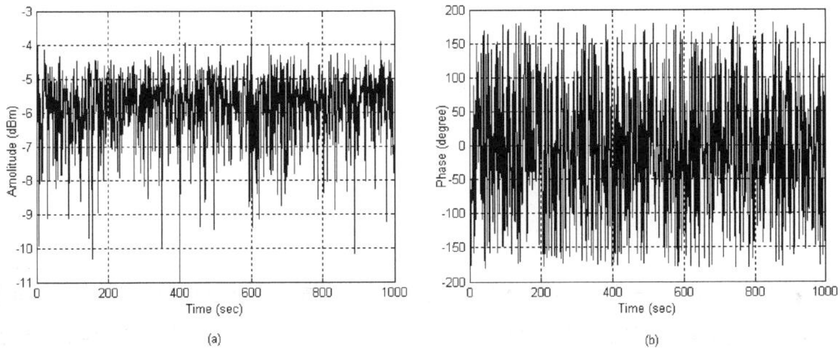

Fig. 12. Signal at the antenna terminals. a) Amplitude fading, b) phase fading.

As seen in Figure 12, the induced *e.m.f.* in the antenna terminals, obtained by the procedure has a fading performance. Moreover, amplitude fading have a statistical distribution of *Rayleigh* type and the phase is uniformly distributed in the range $[0-2\pi]$, (Molina & De Haro, 2008). Figure 13 shows the statistical adjustment of the amplitude and phase fluctuations of this signal.

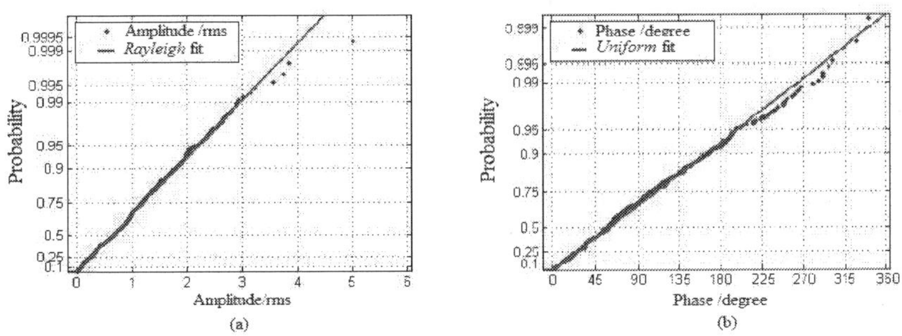

Fig. 13. Statistical adjustment of signal fading. a) Amplitude distribution of type *Rayleigh*, b) phase *uniform* in the interval $[0-2\pi]$.

An important outcome of the simulation is the fact that the signal statistics obtained corresponds closely to real signal channels (Blaunstein et al., 2002). This also provides an explanation of the operation of the receiving antennas under multipath environments, the idea can be reused for practical purposes. So, it is possible the measurement of spatial correlation in systems multi-antennas, as Kildal & Rosengren (2002) has proposed to evaluate the performance of. The evaluation of multi-antennas systems from the point of view of their spatial correlation is based on the implementation of the virtual radio channel which interact between various antennas simultaneously. The correlation between the envelopes of the induced *e.m f.* between the different antennas is calculated including the interaction with the virtual radio channel. Measured of the radiation patterns of the different antennas are obtained in an anechoic chamber, or they are simulated by means of electromagnetic simulation. It's important to emphasize that it takes into account the mutual coupling between antennas, which are implicit in the tabulated measured of the radiation characteristics (radiation patterns of antennas, directivity, and radiation resistance). The equation of the correlation coefficient between the envelopes associated with the signal terminals of any two antennas of the system is as follows (Hill, 2002):

$$\rho = \frac{s_A(t) \cdot s_B^*(t)}{\sqrt{|s_A(t)|^2 + |s_B(t)|^2}} \quad [V] \tag{33}$$

Figure 14 shows the problem of evaluation is a system of two separate and parallel dipoles, please note that both measures are related to the origin of the same coordinate system.

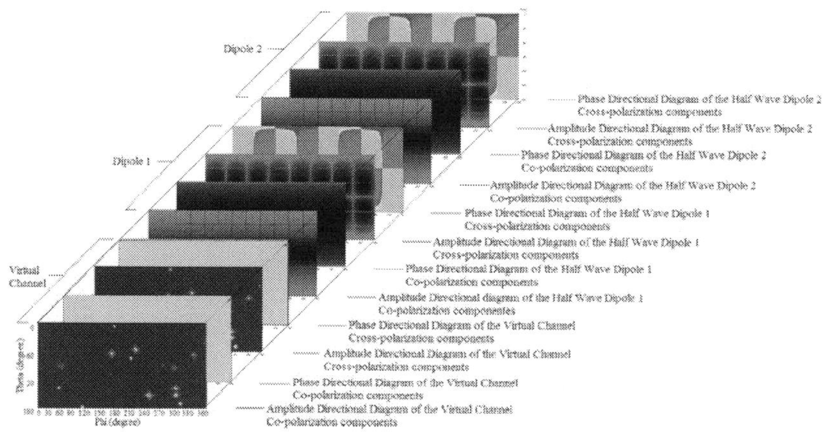

Fig. 14. Geometry of the problem with two antennas under test.

As an example the spatial correlation coefficient as a function of separation between the electric dipoles has been computed and the obtained results are shown in Figure 15.

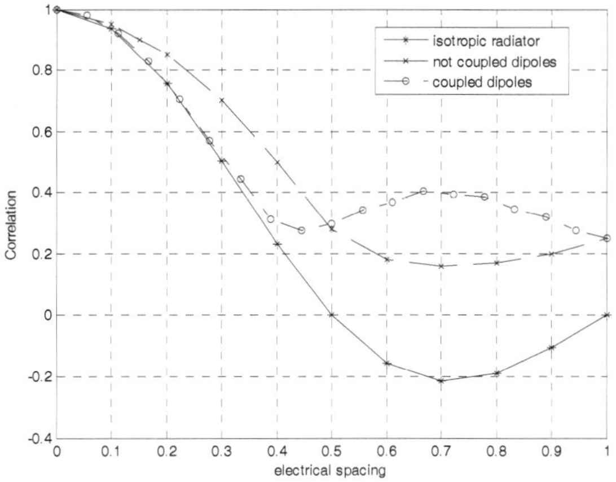

Fig. 15. Correlation coefficients as a function of spatial separation between antenna elements.

In this figure, there are three graphs. The first one shows the spatial correlation function between two isotropic radiators, without mutual coupling; The second one presents two dipoles without mutual coupling; and third plot shows the correlation between two real dipoles taking into account mutual coupling. The first case of isotropic radiators, shows a *sinc shape*, which corresponds to the well known theoretical studies. For dipole antennas, which were not taken into account the mutual coupling, the plot is nearly a *sinc*. The changes are due the radiation pattern of dipoles are not isotropic, but toroidal, with zones with high radiation and zones where the radiation is entirely null. The plot does not reach the minimum at zero or negative but the general shape of the graph is still approaching the *sinc*. The third plot is linked to the situation of coupled dipoles, is the real case and the explanation of how they are less defined exponentially decreasing is in two fundamental reasons: The first is that the radiators are not isotropic, as explained in the previous situation; the second is that when taking into account the mutual coupling, radiation patterns of the two dipoles are changing because of the electromagnetic interaction between them to vary the separation, so changing the directivity D_{Max}, the radiation resistance R_{Rad} of both dipoles, and their effective lengths l_{Ef} (14). The computations have been performed using a sampled radiation pattern matrix of dimension of 360 x 180. This matrix can include not only simulated values but measured patterns providing an early measurement of the correlation coefficients. Moreover it can evaluate systems built without having to measure them in a reverberation chamber with problems that this implies (time, complexity of the measures, special tools, etc).

4. Radio channels classification

Experience has shown that information about their average values is not sufficient to ensure the quality of performance of radio-communications systems, which takes into account a number of parameters of great importance to the design, operates and manage the radio system. Moreover, some concepts have been used in the previous sections but they need deep explanations..:

Fading is the sudden variation and reduction of signal received power with respect to its nominal value. This is due to the superposition of waves that arrive by different path. The phenomenon has a basically spatial nature, but the spatial variations of the signal are experienced as temporal variations when the receiver or transmitter they move through the dispersive channel. Figure.16 shows the parameters of the interest for the signal characterization, as: the nominal power received $P_N(dBm)$, the depth of the fading $P_F(dBm)$, and the duration of the fading τ .

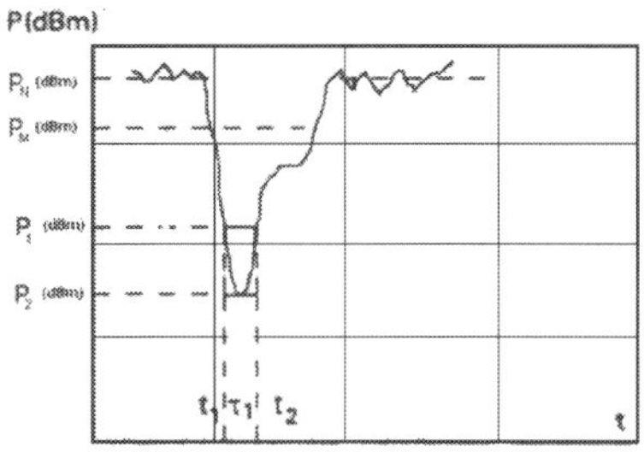

Fig. 16. Fading parameters

Fading performances produces changes in the the spectral characteristics, probability distributions of radio channel. Table 1 shows a classification according to the fading parameters according to the mentioned parameters.

Characteristics	Fading Type	
Depth	Deep	Very Deep
Duration	slow	Fast
Spectral characteristics	flat	Selective
Generating mechanism	*k* factor	Multipath
Probability distribution	Gaussian	Rayleigh, Rice

Table 1. Fading classification

Table 1, provides various kinds of fading in two columns, within which there is some relationships. Deep fade is usually selective and caused by multipath interference. A flat fading plane appears normally in case of narrow bandwidth producing the same distortion along the carrier spectrum. On the other hand, the selective fading produces different distortions along the spectrum of the modulated signal. Time variation of the desired signal and interference, plays a crucial role in the reliability analysis of a system imposing requirements to the type of modulation, transmission power, protection ratio against interference, diversity techniques, and coding method. That is why; output signal from a radio channel is studied as a random process using statistical methods to characterize them. The radio channels are classified taking the name of the statistical distribution function that describes the signal obtained. In this way, some typical channels are: *Normal, Gaussian, Rayleigh, Rice, and Nakagami*. In the case of the fading signals at the terminals of the antennas that operate under multipath, Practice has shown that the probability distributions that best fit are: *Rayleigh*, and *Nakagami-Rice*. Then will analyze these distributions (Rec. ITU-R P. 1057-1, 2001).

Rayleigh, when several multipath components with an angle of arrival that are uniformly distributed in the in the range $[0-2\pi]$, the *Rayleigh* distribution describes the fading fast of the signal envelope, both spatial and temporal. Therefore, it can be obtained mathematically as envelope limit of the sum of two noise signals in quadrature with Gaussians distributions. The probability density function, *PDF*, is expressed as follows:

$$PDF = \begin{cases} \frac{x}{\sigma^2}\exp\left\{-\frac{x^2}{2\sigma^2}\right\} & x \geq 0 \\ 0 & x < 0 \end{cases} \tag{34}$$

Equation (34), x is the random variable and σ^2 variance or average voltage of the envelope of the received signal. Its maximum value is $\exp(-0.5)/\sigma = 0.6065/\sigma$ and it corresponds to the random variable $x = \sigma$. The cumulative distribution function *CDF* which is given by:

$$CDF = \Pr(x \leq X) = \int_0^X PDF(x)dx = 1 - \exp\left\{-\frac{X^2}{2\sigma^2}\right\} \tag{35}$$

The average value x_{mean} of the *Rayleigh* distribution can be obtained from the condition:

$$x_{mean} \equiv \int_0^\infty x \cdot PDF(x)dx = \sigma \cdot \sqrt{\frac{\pi}{2}} \approx 1.253 \cdot \sigma \tag{36}$$

while the variance or average power of the signal envelope of the *Rayleigh* distribution can be determined as:

$$\sigma_x^2 = \int_0^\infty x^2 \cdot PDF(x)dx - \frac{\pi\sigma^2}{2} = \sigma^2\left(2 - \frac{\pi}{2}\right) \approx 0.429\sigma^2 \tag{37}$$

The *rms* value of the envelope signal is defined by the square root of $2\sigma^2$, is :

$$rms = \sqrt{2} \cdot \sigma = 1.414 \cdot \sigma \tag{38}$$

The median of the envelope of this signal is defined from the following condition:

$$\frac{1}{2} = \int_0^{x_{median}} PDF(x)dx \tag{39}$$

and it is obtained:

$$x_{median} = 1.177\sigma \tag{40}$$

All these parameters are presented in the x-axis of Figure 17.

Rice distribution appears when several multipath components and a line of sight component are added between the antennas of the transmitter and receiver. A parameter, known as *K factor*, is introduced, which is the rate between the following components:

$$K = \frac{\text{Power of the dominant component}}{\text{Power of the multipath components}} \tag{41}$$

Usually, the *PDF* and *CDF* functions of this distribution are expressed in terms of the *K factor*, as shown:

$$PDF(x) = \frac{x}{\sigma^2}\exp\left\{-\frac{x^2}{2\sigma^2}\right\}\cdot\exp(-K)\cdot I_0\left(\frac{x}{\sigma}\sqrt{2K}\right) \tag{42}$$

and

$$CDF(x) = 1 - \exp\left\{-\left(K + \frac{x^2}{2\sigma^2}\right)\right\}\cdot\sum_{m=0}^{\infty}\left(\frac{\sigma\sqrt{2K}}{x}\right)^m \cdot I_m\left(\frac{x}{\sigma}\sqrt{2K}\right) \tag{43}$$

The *K factor* is represented by the following ratio $K = A^2/2\sigma^2$, where A is the peak voltage of the power or envelope of the dominant component, $I_o(\bullet)$ and $I_m(\bullet)$ are the modified Bessel function of first kind and zero order and m respectively. Figure 17 shows the *Rayleigh PDF* graph and some *PDF* graphs of the several K values. See in this graph (figure 17a) that is asymmetrical bell-shaped, and that with increasing the K value (figure 17b), the graphics are changing so. In the case where $K = 0$, the *Rice* distribution becomes a *Rayleigh* distribution, this is perfectly understandable, because for this value is not presence of dominant component of signal, and is only in the presence of multipath components. When K is increasing the graphics are starting to be tighten and tend to a *Gaussian* distribution. This is a result of the increase in signal level associated with the dominant component. This is the real situation when exist the line-of-sight between the antennas of the transmitter and receiver.

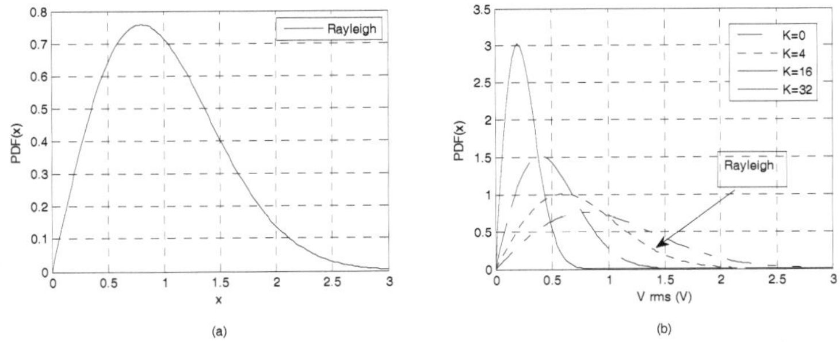

Fig. 17. a) Rayleigh density function and b) *Rayleigh PDF* for the several K values

In practice usually identified mixed distributions. Identification of the type given channel fluctuations at the terminals of an antenna probe is of great importance. Knowledge of the channel parameters associated allows selecting the most suitable antenna system, both in the transmitter as the receiver system, allows the configuration of the all signal processing. Knowing the parameter of the channel allow to define the all the structure of the radio system.

5. Conclusions

In this chapter an analysis on the behaviour of antennas in the reception system has been performed. The conclusions are the basis for the approach to the problem of antennas in receiving mode operating in multipath conditions. It was shown that many properties of the antennas are the same in transmission and in reception, which simplifies its study, since it is not necessary to determine these properties in both regimes. Thus, the impedance of the antenna, directional properties, its directivity, efficiency, and gain are the same in both ways of work. Several expressions (mainly the induced *e.m.f*), at the receiving antenna has been obtained. An analysis of the flow of electromagnetic power from the wave travelling in free space and incident at the antenna until you arrive to the receiver, several causes of loss that can occur in this tract were shown, allowing to define conditions to be met to achieve optimal reception: coincidence of the polarization of the antenna with of the incident wave, orientation of the antenna with the direction of arrival of the wave, maximum effective area antenna (length), which depends on its building characteristics, high efficiency, impedances matching between the antenna and feed line. Once the theory and basis on the antenna in reception mode has been stated, the operation under multipath conditions has been covered. A computational procedure that allows see the behaviour an antenna when waves impinge from all directions of space has been presented. The presented procedure has been proved to be useful not only from an educational point of view, but also for assessment multi-antennas system from the viewpoint of spatial correlation, which generalizes its range of applicability. Finally, the importance of classification and identification of radio channels has been explained through a statistical analysis of the fading signal at its terminals. *Rice* and *Rayleigh* channels models were used that best describe the main essence of the multipath. It was observed that when decreasing the signal level of the dominant path this one tends to *Rayleigh*. Knowing the behaviour of antennas under multipath conditions is of significant importance to the engineers who design, plan and operate radio systems, in order to configure, optimize and select proper system elements, which are ultimately defined by the set-antenna channel.

6. References

Balanis, C. A. (1982). *Antennas Theory: analysis and design,* Edit. Harper & Row, ISBN 0-471-59268-4, New Jersey

Blaunstein, N & Andersen, J. B. (2002). *Multipath Phenomena in Cellular Network,* Edit. Hartech House Inc, ISBN-158053-185-7, New York

Hill, D. A & Ladbury, J. M. (2003). Spatial-Correlation Functions of Fields and Energy Density in a Reverberation Chamber, IEEE Transaction on electromagnetic compatibility, Vol.44, No. 1. , February 2002. pp. 95-101, ISSN 0018-9375

Kildal, P-S & Rosengren, K. (2003). Electromagnetic analysis of effective and apparent diversity gain of two parallel, *IEEE Antennas and Wireless Propagation Letters,* Vol. 2, No. 1, pp 9-13, ISSN 1536-1225

Ludwig, A.C. (1973). The Definition of Cross Polarization. *IEEE Transaction on Antennas and Propagation,* pp. 116-119, Jan 1973

Márkov, G.T. & Sazónov, D. M. (1978). *Antenas.* Edit. MIR, ISBN 9785884170797, URSS

Molina, E. L & De Haro, L. A. (2007). Antenna and Wireless Multipath Virtual Channel Interaction, Proceedings *of IEEE-MTT,* pp. 1-3, ISBN 978-1-4244-0748-4, December 2007, Bangkok

Molina, E. L & De Haro, L. A. (2008). Statistical characterization of the antenna and wireless multipath virtual channel interaction, Proceedings *of IEEE-APS,* pp. 1-4, ISBN 978-1-4244-2041-4, September 2008, San Diego. CA

Monson, J.C. (1996). A new reciprocity theorem, *IEEE Transaction on Antennas and Propagation,* Vol.44, No. 1. , February 1996. pp. 10-14, ISSN 0018-9480

Nikolski, V. V. (1976). *Electrodinámica y propagación de las ondas de radio,* Edit. URSS, ISBN-9785884170551, Moscú

Rec. ITU-R P. 1057-1 (2001), Probability distributions relevant to radiowave propagation modelling, Edit. ITU

Rogier, H. (2006). Phase-mode construction of a coupling matrix for uniform circular arrays with a center element, *Microwave Optical Technology Letters,* Vol. 48, No. 2, February 2006. pp. 291-298, ISSN 0895-2477

23

Foamed Nanocomposites for EMI Shielding Applications

Isabel Molenberg, Isabelle Huynen, Anne-Christine Baudouin, Christian Bailly, Jean-Michel Thomassin and Christophe Detrembleur
Université Catholique de Louvain, Université de Liège
Belgium

1. Introduction

The addition of nanoparticles having specific properties inside a matrix with different properties creates a novel material that exhibits hybrid and even new properties. The nanocomposites presented in this paper combine the properties of foamed polymers (inexpensive, lightweight, easy to mould into any desired shape, etc.) with those of carbon nanotubes (CNTs). The addition of any conductive nanoparticles to an otherwise insulating matrix leads to a significant increase of the electrical conductivity. But CNTs have a very high aspect ratio; a much lower content of CNTs is therefore required to get the same conductivity increase as the one obtained with more compact nanoparticles.

This is especially interesting for EMI shielding materials since, as will be explained in further details in this chapter, it is desirable for such materials to have a high conductivity but a low dielectric constant, in order to minimize the electromagnetic power outside the shield casing but also to minimize the power reflected back inside the casing, as is explained in section 2. In particular, two parameters of interest when comparing shielding materials are detailed and discussed.

The polymer/CNTs nanocomposites were fabricated and characterized using a two-step diagnostic method. They were first characterized in their solid form, i.e. before the foaming process and the most interesting polymer matrices (with embedded CNTs) could be selected. This way, only the promising blends were foamed, therefore avoiding the unnecessary fabrication of a number of foams. These selected blends were foamed and then characterized. The samples, both solid and foamed, are described and their fabrication processes are briefly explained in section 3 while the characterization methods are shown in section 4.

A simple electrical model is given and explained in section 5 and an optimized topology for the foams is also proposed in the second part of the same section.

The measurement results for the solids and for the mono-layered and multi-layered foams are summarized and discussed in section 6. They are then compared to results obtained using the electrical model presented in the previous section and they are also correlated to rheological characterizations.

2. EMI shielding considerations

There are two main parameters that are used to characterize the quality of a shielding material in terms of electromagnetic power; the Shielding Effectiveness (SE) and the Reflectivity (R). The former relates the power that is transmitted through the material (P_{out}), cf. Fig. 1(left), to the incident power (P_{in}): SE = 10 log (P_{in}/P_{out}). The latter relates the power reflected back from the material (P_{ref}) to the incident power (P_{in}): R=10 log(P_{ref}/P_{in}).
The incident power is either reflected back, or transmitted through the material to the outer world, or absorbed inside the material, $P_{in} = P_{ref} + P_{out} + A^2$, where A^2 is the power absorbed inside the material. A is called the attenuation and increases as the conductivity of the material increases. The power reflected at the interface air-material is higher when the difference between the dielectric constants on both sides of the interface is more important. Since air has a dielectric constant of 1, the Reflectivity increases with the dielectric constant of the shielding material.
It must be noted that the above discussion does not take into account the reflection at the second interface, material to air. To be exact we should consider this extra reflection but its effect becomes negligible when the attenuation is sufficiently high, cf. (Huynen et al., 2008).
It is not always enough for a shielding material to exhibit a good SE, i.e. stopping power transmission to the outside world. For example, metallic materials have a high SE at high frequency but almost all the incident power is reflected back inside the shield casing (high R), possibly interfering with other inner elements (or even with the emitter itself). Materials that combine a high SE with a low R at microwave frequencies are called microwave absorbers, because the power is absorbed inside the material, cf. Fig. 1(right). For a material to have a high SE, it must exhibit a high conductivity and, in order to have a low R, it must have a low dielectric constant.

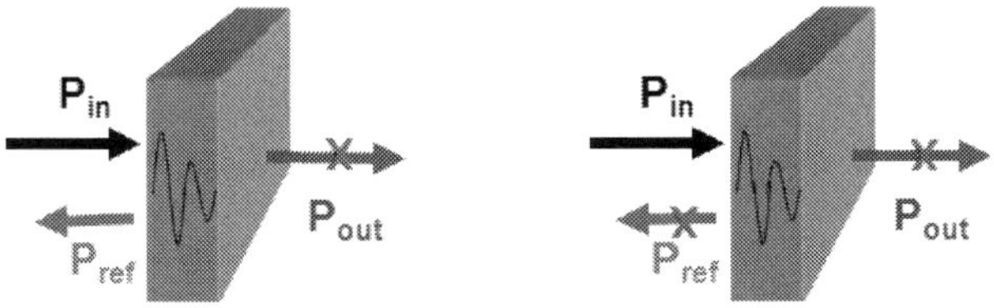

Fig. 1. Schematic diagrams for a conventional metallic shielding material (left) and for a microwave absorber (right).

Finding a material that combines a high electrical conductivity and a low dielectric constant is not trivial. This is the idea behind the use of foamed nanocomposites. In theory, foaming a material would decrease its dielectric constant (because of the porosity and because air has dielectric constant of 1). On the other hand, the addition of conductive nanoparticles to a material should theoretically increase its conductivity, cf. (Huynen et al., 2008), (Thomassin et al., 2008).
Besides having a potentially low dielectric constant, polymer foams are inexpensive, lightweight and easy to mould into any desirable shape. And, the addition of conductive nanoparticles would also reinforce the polymer matrix, improving its electrical conductivity but also its mechanical properties and its thermal conductivity, cf. (Saib et al. 2006). Carbon

nanotubes have a very high aspect-ratio (≥1000). They can therefore form extensive regular conductive networks with a much lower content than other nanoparticles having a more spherical or compact shape, cf. Fig. 2.

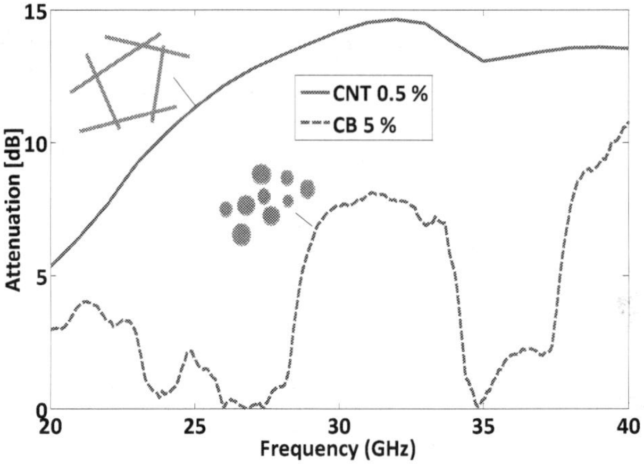

Fig. 2. Influence of the type of nanoparticles, CBs or CNTs, on the attenuation of a nanocomposite, from (Saib et al. 2006).

The applications of such foamed nanocomposites are numerous and varied, from electronics packaging to bioneural matrices, in aeronautics, automotive, environmental applications, and many more...

But the most promising application at the moment is certainly as shielding materials, more precisely as microwave absorbers. They could be made into thin, large, flexible, lightweight panels to be used as EMI shielding materials anywhere, and for little cost. Those panels could also have interesting fire retardancy or protection again electrostatic discharges properties.

3. Nanocomposite samples description

The chemical processes involved in the fabrication of foamed samples are rather complex, cf. section 3.2 for more details. It is therefore interesting to make a first selection on the nanocomposite blends before they are actually foamed. A two-step diagnostic method was developed; solid samples of polymer-CNTs blends were first characterized (the so-called screening tests), the best candidate blends were then used to fabricate foams and finally the foamed samples were also characterized.

3.1 Solid thin-film samples (screening tests)

The CNTs used for the fabrication of the nanocomposites are commercially available thin MultiWalled NanoTubes (MWNTs), with an average outer diameter of 10 nm and a purity over 95%. They were produced using Catalytic Carbon Vapour Deposition (CCVD) by Nanocyl SA (Belgium).

The polymer/CNTs composites were fabricated using two different techniques, cf. (Thomassin et al., 2008). The first one consists in melt-blending the polymer matrix, poly(ε-caprolactone) (PCL), with CNTs using a DSM microextruder. The second one, called the "coprecipitation" technique, is the solubilization of PCL in an organic solvent (tetrahydrofurane, THF, especially well suited to PCL) in the presence of the required amount of CNTs. After 30 seconds of ultrasonic treatment in order to break the CNTs bundles, the mixture is poured into heptane, which is a poor solvent for PCL. The polymer then instantly precipitates and the CNTs are trapped in it.
At this stage the samples are in solid form, cf. FIG. 3, they have not yet been foamed, they are simply referred to as "solid samples" throughout the chapter.

Fig. 3. Picture of a nanocomposite solid sample.

3.2 Foamed samples

Solid samples prepared by the methods described in the previous section were foamed using supercritical CO_2. They were first pressurized at about 200 bars at 60°C for 3h in order to saturate the sample with CO_2. The pressure was then rapidly released in a few seconds leading to the foaming of the sample. A picture of a foamed nanocomposite fabricated this way is shown on FIG. 4 (already cut into pieces prior to its characterization), cf. (Thomassin et al., 2008)

Fig. 4. Picture of a nanocomposite foamed sample (already cut into pieces prior to its characterization).

4. Characterization methods

A two-step diagnostic method to find the polymer-CNTs nanocomposites best fitted for shielding applications was developed, cf. (Molenberg et al., 2009); solid samples of polymer-CNTs blends were first characterized using a microstrip one-line method, cf. section 4.2, the best candidates were used to fabricate foams that were then characterized using a waveguide line-line method, cf. section 4.1.
Even if different methods were used to characterize the various samples, they were all based on the measurement of their scattering parameters (S_{ii}) using a Vector Network Analyzer (VNA). The dielectric constant and conductivity of the samples were then extracted from these parameters. Those measurements were made in the 8-40 GHz frequency band.
The foams were measured using waveguides, while the solid samples were measured using microstrips. This is due to their respective geometries, thin flat solid samples would not fill the waveguides enough to make precise measurements while foams are too porous and thick to be reliably measured in a microstrip configuration.

4.1 Line-line waveguide configuration technique

The Line-Line (LL) method is based on the extraction of the propagation constant (γ), cf. FIG. 5, from the measurement of the scattering parameters (S_{ij}) of two transmission lines of different lengths, here two waveguides having their inner volumes entirely filled with the sample under test, cf. (Huynen et al., 2001), (Saib et al., 2006) and (Pozar, 2005).

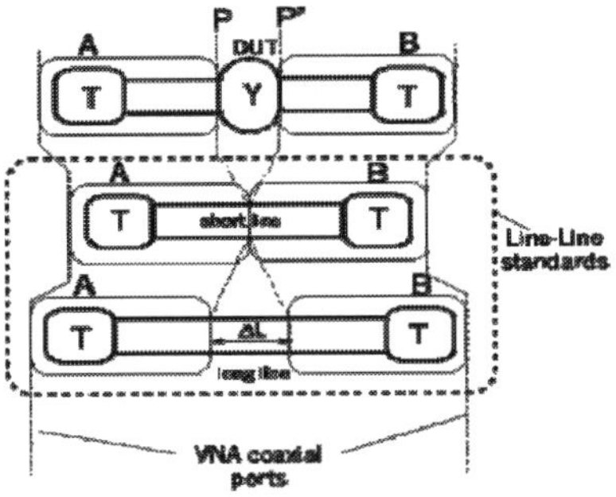

Fig. 5. Line-line method, from (Huynen et al., 2001) and experimental set-up for the Line-Line method, waveguide, transitions and coaxial cables to the VNA.

From the scattering parameters of both lines, their transfer matrices T_{L1} and T_{L2} can be extracted. After a few mathematical operations, a matrix $T_{\Delta L}$ can be calculated. It is diagonal and has the form

$$T_{\Delta L} = \begin{pmatrix} e^{-\gamma\Delta L} & 0 \\ 0 & e^{-\gamma\Delta L} \end{pmatrix} \quad (1)$$

And from the propagation constant, the dielectric constant and conductivity can be easily determined

$$\varepsilon_r = \varepsilon' - j\varepsilon'' = \left(\frac{c_0\gamma}{j\omega}\right)^2 \text{ and } \sigma = -\omega\varepsilon_0\varepsilon'' \tag{2}$$

It should be noted that the magnitude of diagonal element $e^{-\gamma\Delta L}$ in equation (1) corresponds to the attenuation undergone by the signal over a thickness ΔL of material. As γ is proportional to σ, it further confirms that a high conductivity is required for obtaining a good absorption.
Only a simple coaxial SOLT (short-return-through-line) calibration of the VNA is required and the final values of permittivity and conductivity depend only on the length difference ΔL. It is therefore especially well suited to waveguide measurements, because there is no precision waveguide calkit available in our laboratory. Such a calkit would have been necessary for more complex and accurate calibrations. The comparison between the simple SOLT calibration method and a LRM (Line-Reflect-Match) method, a more precise technique using a reference calkit, is illustrated on Fig. 6. With an LRM calibration, the reference planes of the VNA are brought after the coaxial-microstrip transitions, so that the VNA measures the scattering parameters of the line under test only. With a SOLT calibration, the VNA reference planes are placed before these transitions and their influence on the measurements of each line is not eliminated. Nevertheless, the SOLT calibration is sufficient to make sure that the $T_{\Delta L}$ matrix is diagonal, therefore ensuring the validity of the LL method.

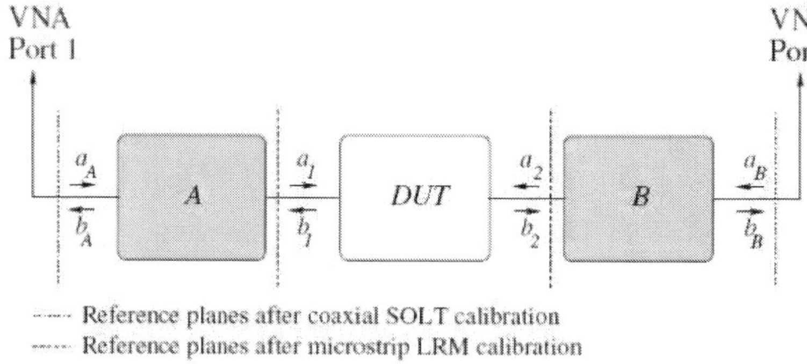

Fig. 6. Schematics of the measurement setup where A and B correspond to the coaxial-to-microstrip transition, and DUT to the device-under-test, from (Saib et al., 2004).

4.2 One-line characterization in microstrip configuration technique

The One-Line method is based on the measurement of the scattering parameters (S_{ii}) of only one line (different from the LL method described in the previous section), a microstrip line in this case. The sample to be characterized is used as substrate to fabricate a microstrip transmission line. A thin copper ribbon is glued to the top of the sample, serving as microstrip and a piece of aluminum tape is glued on the bottom of the solid sample, to form the ground plane, cf. Fig. 7(right). From the S matrix, the ABCD chain matrices can be calculated for a transmission line of length L and of characteristic impedance Z_C, cf. (Saib, 2004)

$$ABCD = \begin{pmatrix} ch(\gamma L) & Z_C sh(\gamma L) \\ \frac{1}{Z_C} sh(\gamma L) & ch(\gamma L) \end{pmatrix} \quad (3)$$

The propagation constant γ can then be extracted and the dielectric constant and conductivity can be determined, using equation (2).
This method is valid only if a precise LRM calibration of the VNA has been done. The reference planes must be put after the transitions, and the reference impedances in these planes must be set to 50 Ω, which is ensured using an LRM calibration. An Anritsu precision microstrip calkit and the corresponding 3680K Anritsu sample holder (including the transitions) were used for the microstrip measurements, cf. Fig. 7.

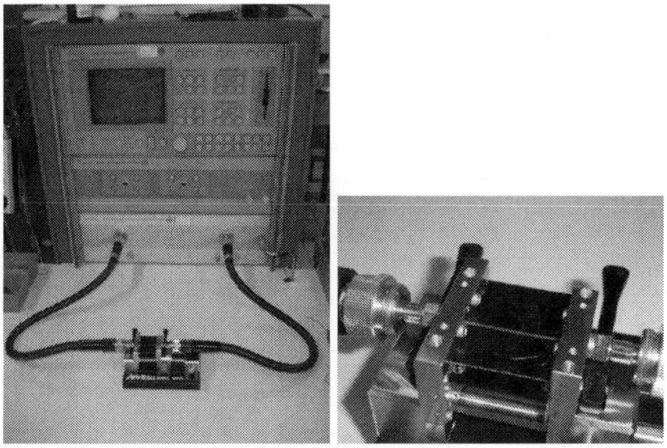

Fig. 7. Experimental set-up for the One-Line method applied to a microstrip topology, entire set-up (left), Anritsu sample holder (right) with the microstrip visible on top of the sample (substrate).

5. Modelling, design and optimization

5.1 Simple electrical model

It must first be noted that Carbon Black (CB) nanoparticles, i.e. relatively spherical carbon platelets, are considered here instead of CNTs to simplify geometrical considerations.
As can be seen on conductivity-versus-frequency plots resulting from nanocomposites measurements, cf. section 6 and (Saib et al., 2006), the measured conductivity tends to 0 for very low frequencies. This means that the conductive nanoparticles do not form a direct conductive pathway for the electrons from one side of the sample to the other side. However, at relatively high frequency the conductivity becomes significant, indicating the presence of capacitive couplings between the nanoparticles. Taking these observations into

account, a simple electrical model was developed, cf. (Saib et al., 2006). This model is represented on Fig. 8. The nanocomposite is placed between a ground plane and a microstrip line, to form a microstrip transmission line, the actual configuration of the (solid) samples during the measurements, cf. section 5.2 and 6.1.

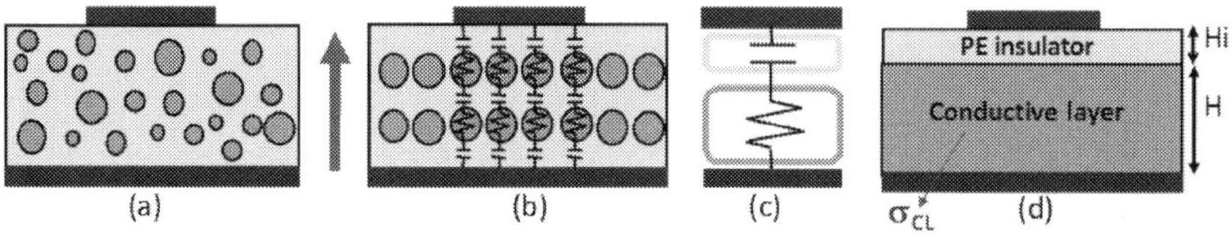

Fig. 8. Simple electrical model explaining the frequency dependence of the conductivity of the nanocomposites, from (Saib et al., 2006): (a) first approximation, the CB nanoparticles are spherical, well dispersed inside the polymer matrix and not in physical contact with one another. The arrow indicates the dominant direction of the electric field. (b) Second approximation, the CB particles form a regular network of conductive grains inside an insulating polymer matrix. (c) Equivalent admittance per unit length of microstrip line. (d) Corresponding two-layer microstrip transmission line.

The CB nanoparticles are supposed to be spherical and well dispersed inside the matrix. There is also supposed to be no direct physical contact between them, i.e. they do not touch one another, cf. Fig. 8(a). This random distribution is approximated by a regular homogeneous network of purely conductive grains inside a completely insulating polymer matrix. Therefore, the electrical equivalent of each grain is a simple resistor and there are strong capacitive couplings between all the grains. Since the electric field distribution between the top and bottom conductors of the microstrip line is quasi uniform and perpendicular to those conductors, only capacitors and resistors that are parallel to that direction have to be considered, cf. Fig. 8(b). This corresponds to a two-layered microstrip transmission line, cf. Fig. 8(d), with an insulating top layer and a conductive bottom layer (conductivity = σ_{CL}). This line has an electrical equivalent admittance per unit length that corresponds to a simple resistor-capacitor (R_T and C_T) series circuit, cf. Fig. 8(c). The simulated results obtained using this simple model are confronted with measurement results in section 6.3.

5.2 Optimization of the topology of the samples

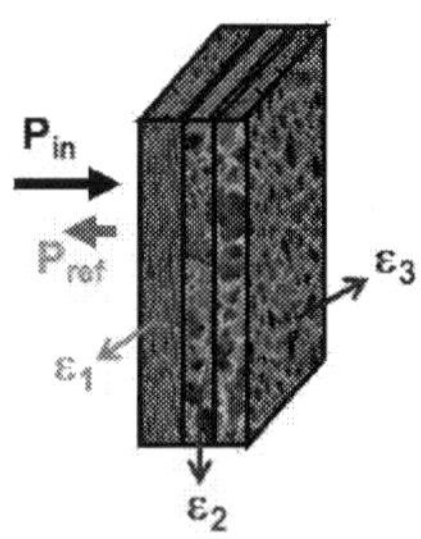

Fig. 9. Schematic diagram the three-layered foam, with $\varepsilon_1 < \varepsilon_2 < \varepsilon_3$.

In order to improve the reflectivity of the shielding material, it would be interesting to create a foam having a gradient of dielectric constant inside the material, analogous to the stealth paint protecting some spy aircrafts from radar detection. A way to achieve this gradient is to fabricate multilayered foams, each layer having a different, increasing dielectric constant, cf. Fig. 9.

The measurement results for a three-layered nanocomposite fabricated as previously described are shown and discussed in section 6.4.

6. Results and discussion

6.1 Solid samples

Microstrip transmission lines were fabricated, using the nanocomposite thin films as substrates. By connecting these lines to a 2-port vector network analyzer (VNA), their scattering parameters were measured and the dielectric constant and conductivity of the samples were then extracted, using the One-Line method described in section 4.2.

The results presented in this section were obtained for composites based on different polymer materials: PS (polystyrene), PCL (poly(ε-caprolactone)), PVC (polyvinyl chloride) and PMMA (polymethyl methacrylate), they all had a 2% CNT content. As can be seen on Fig. 10 and Fig. 11, the PMMA-based sample exhibited the lowest dielectric constant but not highest conductivity, while the PCL-based composite had the highest conductivity but also the highest dielectric constant.

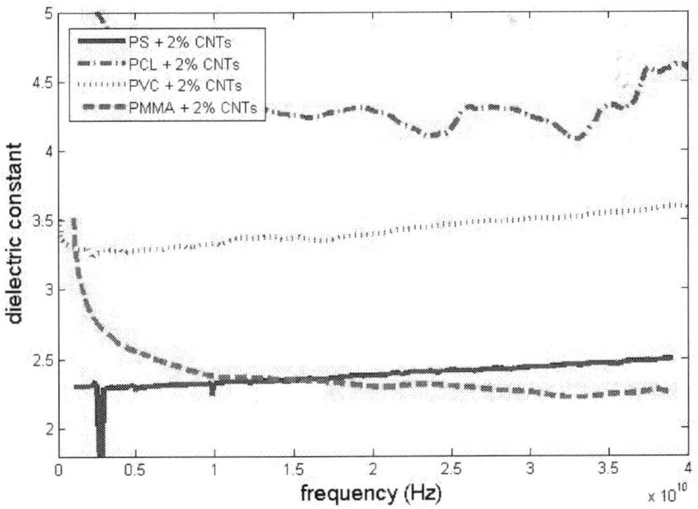

Fig. 10. Measured dielectric constant of four solid nanocomposite samples versus frequency (composites based on different polymer matrices).

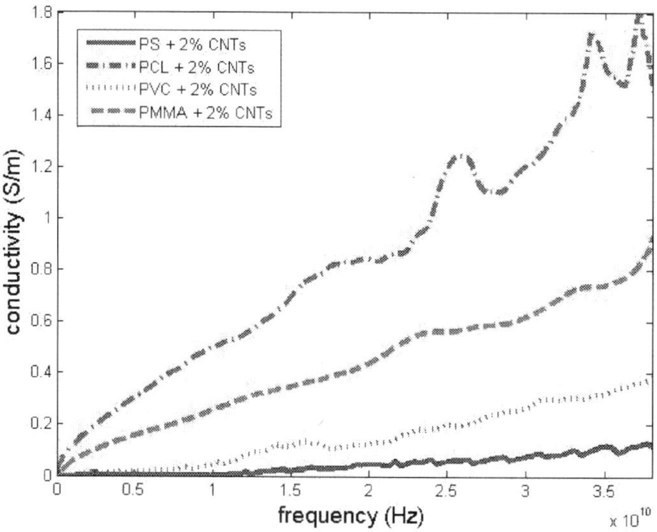

Fig. 11. Measured conductivity of the solid nanocomposite samples versus frequency (composites based on different polymer matrices).

Once a polymer matrix is selected, an optimal content of CNTs to add into the matrix must be determined. It can be shown that the addition of CNTs increases the conductivity of the nanocomposite but also increases its dielectric constant, which is not desired. Since foamed samples have lower dielectric constants than solid ones with the same composition, the exact content of CNTs should be chosen after measurement of foamed samples rather than solid ones, cf. section 6.2.

It should also be noted that other chemical process parameters have a non-negligible influence. Those are beyond the scope of the present chapter, but can be found in (Thomassin et al., 2007).

6.2 Foamed samples

PCL was selected as polymer matrix because it exhibited the highest conductivity in its solid form, cf. Fig. 11. Its dielectric constant was also the highest one, cf. Fig. 10, but foaming the nanocomposite should significantly decrease it. The foamed samples were inserted in four waveguides of different dimensions, covering the 8 to 40 GHz band, cf. Fig. 5. Their scattering parameters were measured using the Line-Line method described in section 4.1.

As mentioned in the previous section, a foamed material has a lower dielectric constant than a solid one of the same chemical composition. This is shown on Fig. 12, the black and red curves correspond to PCL samples with no CNT content, and the dielectric constant of the solid sample is twice that of the foam. It can be explained by the porosity of the foams and the subsequent presence of air inside the sample, the air having a dielectric constant of 1.

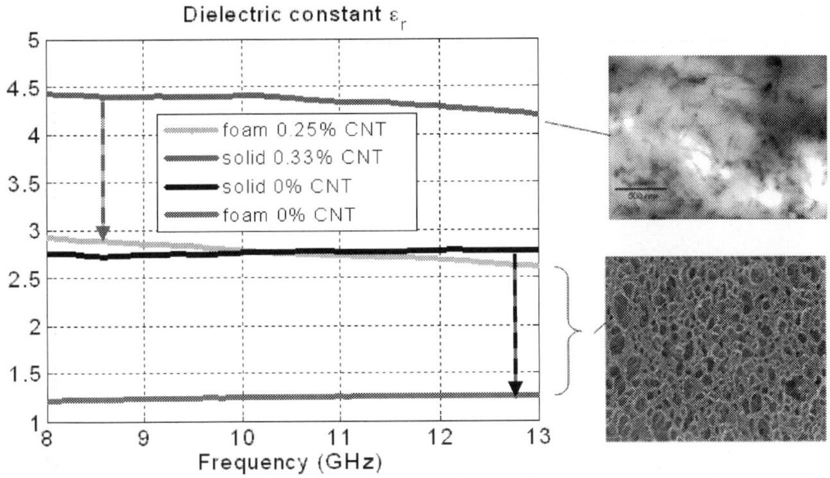

Fig. 12. Measured dielectric constant of solid and foamed PCL nanocomposite samples, all having a similar CNT content.

On the other hand, the conductivity of foams is usually higher than that of solid samples, mainly because the CNT are forced into the side walls of pores in the foamed material and this way form a more regular network. This effect is shown on Fig. 13 for a foam and a solid sample having a close but not equal CNT content, both PCL composites.

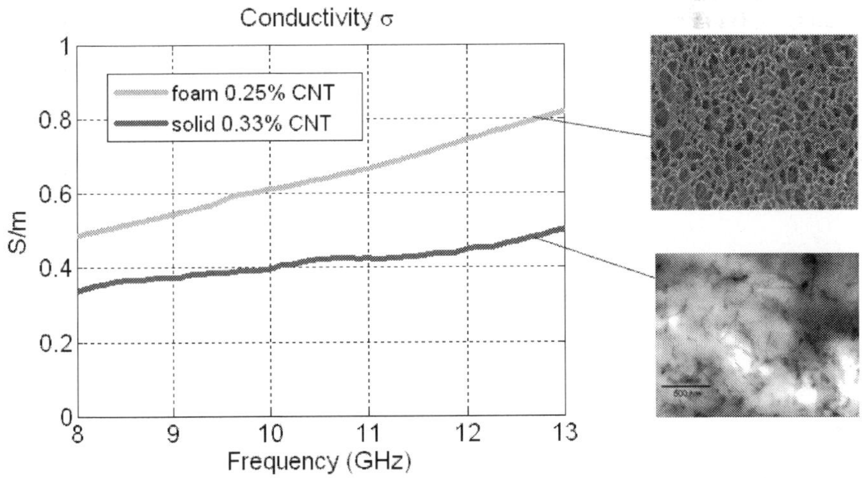

Fig. 13. Measured conductivity of solid and foamed PCL/CNTs nanocomposite samples, both having a similar CNT content.

The conductivity of foamed nanocomposites increases significantly with the CNT content, as can be seen on Fig. 14(left) and the Shielding Effectiveness also increases in the same proportions, cf. Fig. 14(right). Therefore, in order to have a good SE, it appears, from these results, that the CNT content must be as high as possible.

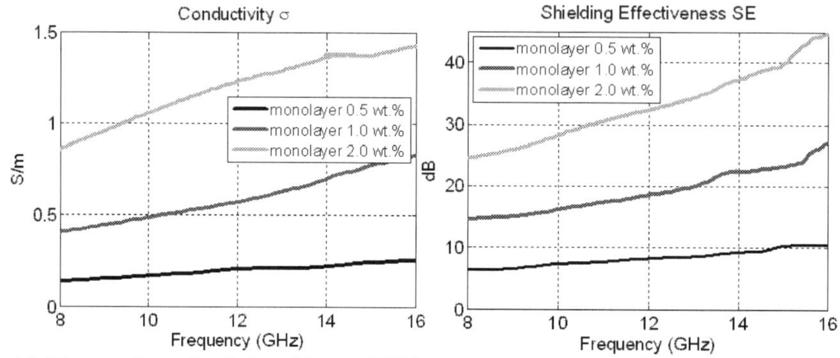

Fig. 14. Measured conductivity of foamed PCL nanocomposite samples with increasing CNT contents (left) and the corresponding Shielding Effectiveness (right).

But the addition of CNTs has an adverse effect on the Reflectivity because the dielectric constant increases significantly with CNT content, cf. Fig. 15. To measure R, a metallic sheet (or Perfect Electrical Conductor - PEC) is added on the output interface, cf. Fig. 9, and the sample with the PEC sheet is then characterized (Line-line method, same as samples without PEC). Part of the incident power is absorbed inside the material, the remaining power then reaches the PEC where it is totally reflected and, another part of the power is absorbed on the way back. Because of the PEC, the SE in this configuration is theoretically infinite. The power reflected back to the port 1 of the VNA, cf. Fig. 6, is the combination of the power directly reflected back at the input interface and the power reaching the metallic plate and being reflected back (the remaining power that has not been absorbed by the material on the way back).

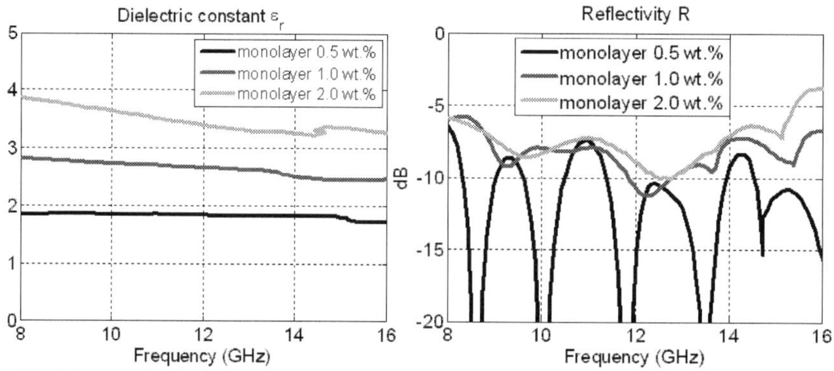

Fig. 15. Measured dielectric constant of foamed PCL nanocomposite samples with increasing CNT contents (left) and the corresponding Reflectivity (right).

In conclusion, foamed nanocomposites have both a higher conductivity and a lower dielectric constant than solid samples. The first are therefore better suited for shielding applications than the latter. The optimum CNT content has to be the result of a compromise between Shielding Effectiveness and Reflectivity, i.e. between conductivity and dielectric constant.

6.3 Comparison between results using the electrical model and from measurements

According to the electrical model developed in section 5.1, at low frequency, since there are no physical contacts between the grains, the equivalent capacitor, cf. Fig. 8(c), is a virtual 'open-circuit'; therefore there is no current flowing through the composite. At high enough frequency, the equivalent capacitor is a virtual short circuit and the conductivity becomes constant. Its value depends then only on the conductivity of the conductive layer of Fig. 8(d). In other words, at low frequency when $(\omega C_T)^{-1} >> R_T$, the conductivity is very low. At high frequency, when $(\omega C_T)^{-1} << R_T$, the conductivity is constant. In between, the conductivity increases with frequency. This corresponds to a transition frequency, f_T, equal to $f_T = (2\pi R_T C_T)^{-1}$.
The dielectric constant and conductivity extracted using this model and the values from measurements are plotted on Fig. 16 and Fig. 17 for CNT contents of 50% and 0.35 %, respectively. The conductivity of the conductive bottom layer (σ_{CL}) was arbitrarily chosen to be equal to 25 S/m. The height of the top layer (H_i) was adjusted in order to obtain the best fit between model and measurement results.

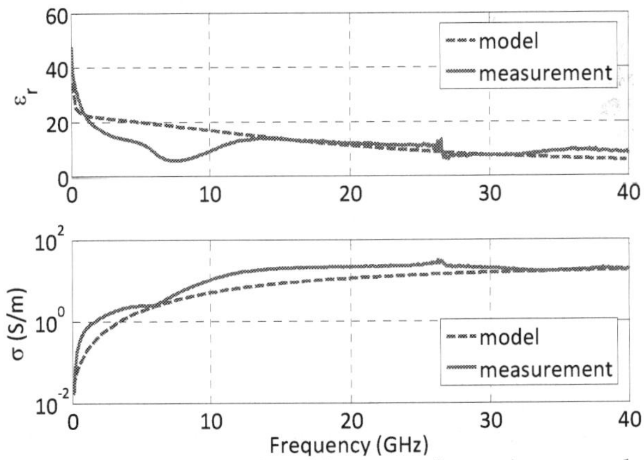

Fig. 16. Dielectric constant (top) and conductivity (bottom) extracted using the simple electrical model and from actual measurements, for a CNT content of 50% (model parameters: $H/H_i = 93/7$ with $H + H_i = 0.8$ mm and $\sigma_{CL} = 25$ S/m), from (Saib et al., 2006).

As can be seen on Fig. 16(bottom), the results obtained using our model and those from measurements show a very good adequacy. At low frequency the conductivity tends to zero. At high frequency there is a plateau, i.e. a constant conductivity. The transition frequency is of about 10 GHz. At this frequency, a behaviour change can also be observed in the dielectric constant-versus-frequency plot, cf. Fig. 16(top).

A CNT (weight) content of 50% is very high. The conductivity and dielectric constant were simulated and measured for a much more realistic CNT content of 0.35%. The results are shown on Fig. 17, the saturation plateau in the conductivity-versus-frequency plot is not reached at 40 GHz but it is expected that with a CNT content as low as 0.35%, the transition frequency would be well over 100 GHz. Nevertheless, the results obtained using the electrical and those from measurements are still in very good adequacy. It must be noted that the H/H_i ratio is very different for the 0.35 and 50% CNT content models. It can be easily understood considering that the average distance between the conductive particles decreases as the concentration of these particles is increased.

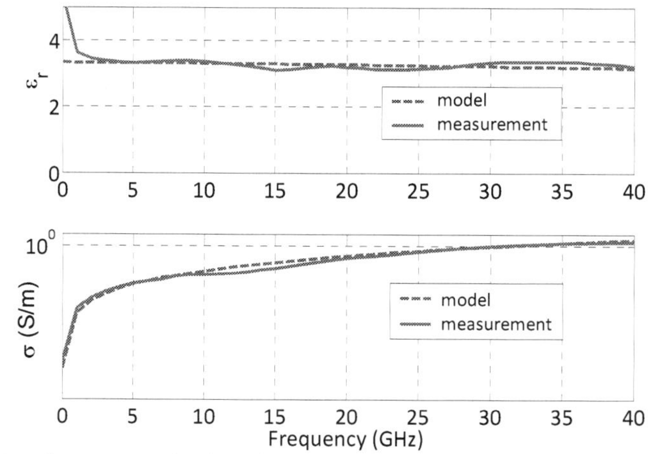

Fig. 17. Dielectric constant (top) and conductivity (bottom) extracted using the simple electrical model and from actual measurements, for a CNT content of 0.35% (model parameters: $H/H_i = 40/60$ with $H + H_i = 0.8$ mm and $\sigma_{CL} = 25$ S/m), from (Saib et al., 2006).

In conclusion, the simple electrical model developed in section 5.1 for CB particles can explain and fit very well the frequency dependence of the conductivity and dielectric constant of polymer/CNTs nanocomposites. The model shows that different frequency behaviours of nanocomposites are related to different CNT contents, or different volumetric H/H_i ratios in the model, hence different mean distances between particles.

6.3 Correlation between the electrical and rheological characterizations

As explained in sections 5.1 and 6.3, a good dispersion of the CNTs inside the polymer matrix is the key parameter for the conductivity of the nanocomposites. The mean distance between two CNTs should also be as small as possible. But raw CNTs, before they are incorporated in the polymer matrix, are entangled and tend to stick to one another, as confirmed by the Transmission Electron Microscope (TEM) picture shown on Fig. 18.

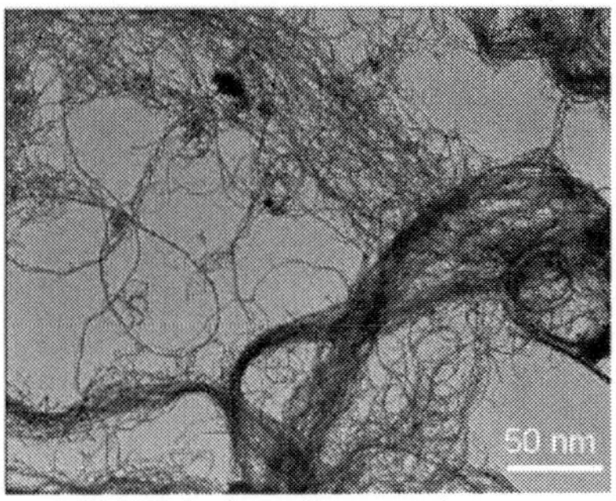

Fig. 18. TEM picture of raw MWNT before they are incoporated into the polymer matrix, from (Saib et al., 2006).

Rheological measurements can also be used to assess the dispersion of the CNTs, in particular the (oscillatory) frequency dependence of the dynamical complex modulus G*, which is the ratio of stress to strain when a sinusoidal stress or strain is applied to the sample (Dynamical Mechanical Analysis or DMA), cf. (Saib et al., 2006). For elastic materials, G* is real; the response to the sinusoidal strain is also sinusoidal and in-phase with the stimulus. For visco-elastic materials like the polymer presented in this paper, G* is complex; the response is also sinusoidal but with a phase shift. The real part (G′) of the dynamical modulus is the in-phase, elastic, response of the material, while the imaginary part (G′′) is the out-of-phase, viscous response. In polymers such as PCL or PS, an increase of G′ with CNT content at low frequency confirms the presence of mobile and flexible chains that are partially bound together by bridging structures, here the CNTs.

The real part of G* is plotted versus frequency for two sets of PS and PCL based nanocomposites, with and without CNTs, on FIG. 19. The presence of saturation at low frequency for the PCL samples containing CNTs is characteristic of the existence of a percolation threshold, after which the CNTs form a network throughout the polymer matrix, cf. Fig. 19(left).

On the contrary, there is little difference between the G′-versus-frequency curve for the PS sample with 1% CNTs and the one for pure PS, cf. Fig. 19(right). It means that the nanoparticles do not form an extended network inside the PS matrix.

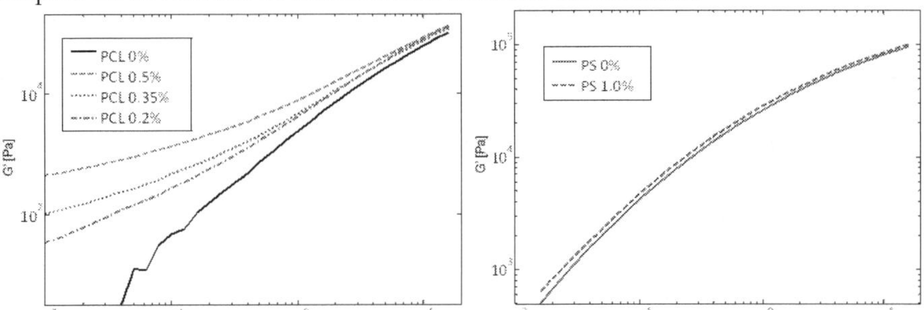

Fig. 19. Dynamical modulus real part (G′) versus frequency for PCL and PS based nanocomposites, with and without CNTs, from (Saib et al., 2006).

These conclusions based upon rheological considerations are confirmed by the measurement of the electrical conductivity. The conductivities of the PCL nanocomposite with a 0.35% CNT content and of the PS one with a concentration of 1% are plotted versus frequency on Fig. 20. The electrical model explained how the value of the electrical conductivity was linked to the mean distance between the particles and therefore to the presence of a regular, well-dispersed nanoparticles network, cf. (Saib et al., 2006). The PCL composite has a much higher conductivity than the PS one, confirming the presence of a conductive network in the former and not in the latter. This is further confirmed by TEM pictures, cf. Fig 21. The CNTs in the PCL sample seem well dispersed forming an extensive network while those in the PS composite are regrouped in distinct aggregates, over 1µm apart.

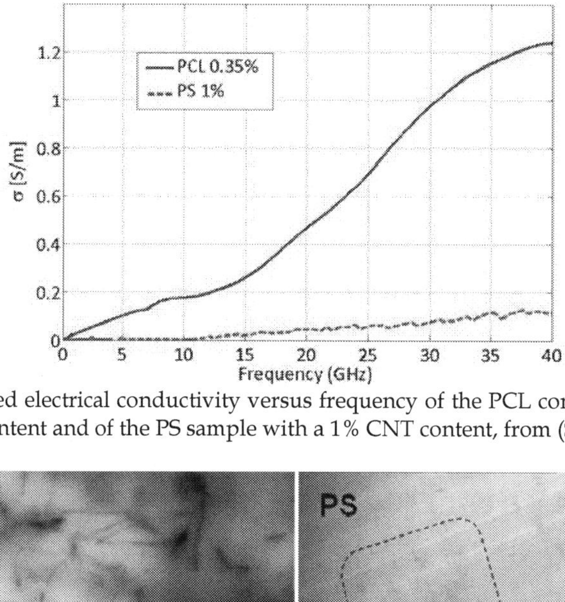

Fig. 20. Measured electrical conductivity versus frequency of the PCL composite containing a 0.35% CNT content and of the PS sample with a 1% CNT content, from (Saib et al., 2006).

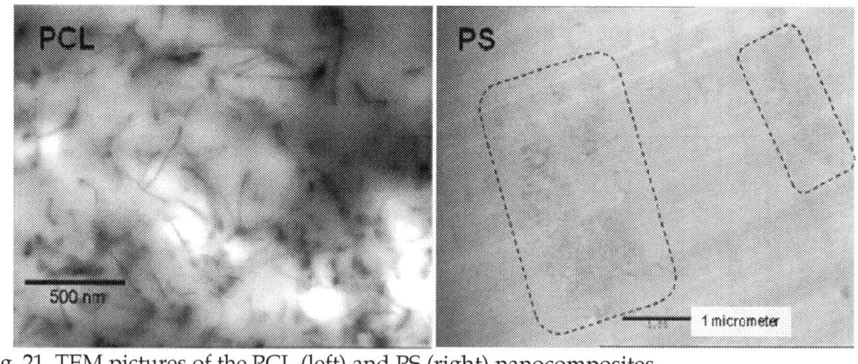

Fig. 21. TEM pictures of the PCL (left) and PS (right) nanocomposites.

6.4 Optimized multilayered foamed samples

As explained in section 5.2, a multilayered foam sample with graded concentrations of CNTs should exhibit simultaneously a high Shielding Effectiveness with a low dielectric constant for a lower CNT content.

A multilayered foam consisting of three layers of PCL with graded CNT contents (0.5%, 1% and 2% for layers 1.2, 1.1 and 0.7 cm thick, respectively) was fabricated and characterized. The total thickness of the sample is therefore equal to 3 cm and the average CNT concentration is equal to 1.03%.
The conductivity and dielectric constant were measured between 8 and 16 GHz. The Shielding Effectiveness and Reflectivity were then extracted. The results are shown on Fig. 22, along with the results for the mono-layered samples of section 6.2. As can be seen on this figure, the multi-layered foam has the same SE as the mono-layered one with the same CNT concentration as the average CNT content of the multi-layered sample. But its Reflectivity is much lower than that of all the mono-layered foams!
The graded concentration achieved by the cascade of layers with increasing CNT concentration, combined with foaming, ensures a progressive and moderate increase in dielectric constant, meaning that reflection P_r at input interface of each layer is minimized, power entering the composite is maximized, and reflection is significantly reduced over a broad frequency range. The progressive increase in conductivity yielded by graded concentration contributes to the absorption of the most part of the power over the depth of the structure by conductive dissipation.

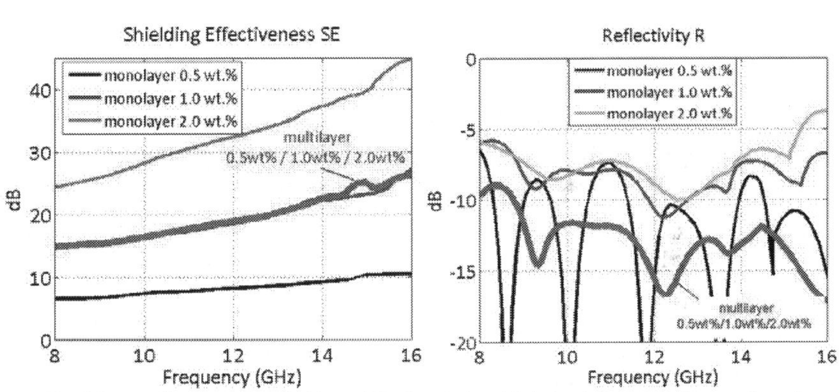

Fig. 22. Shielding Effectiveness (left) and Reflectivity (right) of the multilayered and mono-layered samples, from (Huynen et al., 2008).

The measurements therefore confirm the high potential of multilayered foams for EMI shielding applications.

7. Conclusion

Throughout this chapter, we have seen that foamed polymer/CNTs nanocomposites make very good EMI shielding materials. They exhibit a high conductivity and a relatively low dielectric constant, leading to a high Shielding Effectiveness and a relatively low Reflectivity, although a compromise in CNT content must be found between a high SE and a low R. This was confirmed by measurement results and rheological measurements but also using a simple electrical equivalent model.

Some EMI shielding considerations were first presented. In particular, the notions of SE and R were defined and the advantage of microwave absorbers on conventional metallic shielding materials was discussed. The samples were briefly described, both foamed and in their solid form, along with their fabrication methods. The electrical characterization techniques were presented and their adequacy to the samples under test was briefly discussed. An optimized multi-layered foam topology with increasing CNT contents in each subsequent layer was proposed.

The measurement results of the solid samples, mono- and multi-layered foams, were detailed and discussed. Physical explanations were proposed. These results were also compared to those obtained using the equivalent model and rheological characterization techniques, with a very good adequacy. The three-layered foam exhibited a SE similar to that of a mono-layered sample having the same average CNT content but the R of the multi-layered foam was significantly decreased.

To conclude, nanocomposite foams have good EMI properties but a compromise on the CNT content must be found in order to have a high SE and also a low R. On the other hand, multi-layered foams with graded concentrations of CNTs are especially well suited to EMI shielding applications, having a high SE with a lower R than mono-layered samples.

8. References

Huynen I., Steukers C. & Duhamel F. (2001). A wideband line-line dielectrometric method for liquids, soils, and planar substrates. *IEEE Transactions on Instrumentation and Measurement*, Vol. 50, No. 5, (October 2001) pp. 1343-1348, 0018-9456

Huynen I., Bednarz L., Thomassin J.-M., Pagnoulle C., Jerome R. & Detrembleur C. (2008). Microwave absorbers based on foamed nanocomposites with graded concentration of carbon nanotubes, *Proceedings of the 38th European Microwave Conference*, pp. 5-8, 978-2874870064, Amsterdam, April 2008, IEEE, Piscataway

Molenberg I., Thomassin J.-M., Ferain E., Detrembleur C. & Huynen I. (2009). RF Diagnostic Methods for Two Nanostructured Polymer Applications, pp. 1045-1048, 978-2874870118, Rome, *Proceedings of the 39th European Microwave Conference*, Roma, September 2009, IEEE

Pozar D. (2004). *Microwave Engineering*, Wiley, 978-0471448785, New York

Saib A. (2004). *Modeling and design of microwave devices based on ferromagnetic nanowires*. PhD Thesis, Presses Universitaires de Louvain, 978-2930344776, Louvain la Neuve

Saib A., Bednarz L., Daussin R., Bailly C., Lou X., Thomassin J.-M., Pagnoulle C., Jerome R. & Huynen I. (2006). Carbon nanotubes composites for broadband microwave absorbing materials. *IEEE Transactions on Microwave Theory and Techniques*, Vol. 54, No. 6, (June 2006) pp. 2745-2752, 0018-9480

Thomassin J.-M., Pagnoulle C., Bednarz L., Huynen I., Jerome R. & Detrembleur C. (2008). Foams of polycaprolactone/MWNT nanocomposites for efficient EMI reduction. *Journal of Materials Chemistry*. Vol. 18, No. 7, (January 2008) pp. 792-796, 0959-9428

24

Pseudo-Bessel Beams in Millimeter and Sub-millimeter Range

Yanzhong Yu[1,2] and Wenbin Dou[1]
[1] *State Key Lab of Millimeter Waves, Southeast University, Nanjing,*
[2] *School of Science, Quanzhou Normal University, Quanzhou,*
P. R. China

1. Introduction

In 1987, Durnin firstly discovered a class of novel solutions of the free-space scalar wave equation for beams that are diffraction-free (Durnin, 1987). This means that the time-averaged intensity pattern is unchanged in the cross-section when such beam propagates in free space (McGloin & Dholakia, 2005). It is Bessel beams that are the one most interesting family of diffraction-free beams. The transverse intensity distributions of ideal Bessel beams can be highly localized, and therefore they have many unique properties, such as large depth of field, propagation invariant and reconstruction (MacDonald et al., 1996; Bouchal et al., 1998) and so on. Unfortunately, the ideal Bessel beams can not be exactly generated, due to their infinite lateral extent and energy (Monk et al., 1999). Only their approximations known as the near or pseudo-Bessel beams can be obtained physically (Bouchal, 2003), but they can still propagate over extended distances in a diffraction-free manner (Arlt & Dholakia, 2000). In optics they could have prospective applications, such as optical alignment, interconnection, and promotion of free electron laser gain (Li et al., 2006), and they may be useful in power transmission, communications and imaging applications (Mahon et al., 2005) in millimeter and sub-millimeter range. Therefore, much attention has been paid to this subject, and numerous papers have been devoted to the generation and applications of Bessel beams.

More recently, the studies of Bessel beams at millimeter and sub-millimeter wavelengths have been carried out in our group. The main aim of this chapter is to present our investigation results comprehensively, including their theories, generation, propagation and potential applications. The relevant contents are organized as follows. Section 2 gives the scalar and vector analyses of Bessel Beams. How to produce pseudo-Bessel beams is described in Section 3 and 4. The comparison of propagation distance between apertured Bessel and Gaussian beams is made in Section 5. Lots of potential applications are discussed in the last Section 6.

2. Scalar and Vector Analyses of Bessel Beams (Yu & Dou, 2008a; Yu & Dou, 2008b)

2.1 Scalar analysis

In free space, the scalar field is governed by the following wave equation

$$\nabla^2 E(\vec{r},t) - \frac{1}{c^2}\frac{\partial^2}{\partial t^2}E(\vec{r},t) = 0 \tag{1}$$

where ∇^2 is the Laplacian operator, c is the velocity of light in free space, $\vec{r}$ is the position vector. Assuming that the angular frequency is ω, the field $E(\vec{r},t)$ can be written as

$$E(\vec{r},t) = E(\vec{r})\exp(-i\omega t) \tag{2}$$

Substituting (2) into (1), we have the homogeneous Helmholtz wave equation

$$\nabla^2 E(\vec{r}) + k^2 E(\vec{r}) = 0 \tag{3}$$

where $k = \omega^2\mu_0\varepsilon_0$, is the wave number in free space. Applying the method of separation of variables in cylindrical coordinates, we can derive the following solution from (3)

$$E(\vec{r},t) = E_0 J_n(k_\perp\rho)\exp(in\varphi)\exp(i(k_z z - \omega t)) \tag{4}$$

where E_0 is a constant, J_n is the nth -order Bessel function of the first kind, $\rho = \sqrt{x^2+y^2}$, $x = \rho\cos\varphi$, $y = \rho\sin\varphi$, $k_\perp^2 + k_z^2 = k^2$, $k_\perp$ and k_z are the radial and longitudinal wave numbers, respectively. Thus the time-average intensity of (4) can be given by

$$I(\rho,\varphi,z \ge 0) = I(\rho,\varphi,z=0) = | E_0 J_n(k_\perp\rho)|^2 \tag{5}$$

It can be seen from (5) that the intensity distribution always keeps unchanged in any plane normal to the z-axis. This is the characteristic of the so-called nondiffracting Bessel beams. When $n = 0$, (4) represents the zero-order Bessel beams (i.e. J_0 beams) presented by Durnin in 1987 for the first time (Durnin, 1987). The central spot of a J_0 beam is always bright, as shown in Figs. 1(a) and 1(b). The size of the central spot is determined by $k_\perp$, and when $k_\perp = k$, it reaches the minimum possible diameter of about $3\lambda/4$, but when $k_\perp = 0$, (4) reduces to a plane wave. The intensity profile of a J_0 beam decays at a rate proportional to $(k_\perp\rho)^{-1}$, so it is not square integrable (Durnin, 1987). However, its phase pattern is bright-dark interphase concentric fringes, as shown in Fig. 1(c). An ideal Bessel beam extends infinitely in the radial direction and contains infinite energy, and therefore a physically generated Bessel beam is only an approximation to the ideal. Experimentally, the generation of an approximate J_0 beam is reported firstly by Durnin and co-workers (Durnin et al., 1987). The geometrical estimate of the maximum propagation rang of a J_0 beam is given by

$$Z_{\max} = R[(k/k_\perp)^2 - 1]^{1/2} \tag{6}$$

where R is the radius of the aperture in which the J_0 beam is formed. We can see from (6) that when $R \to \infty$, then $Z_{\max} \to \infty$, provided that $k/k_\perp$ is a fixed value.

But for $n > 0$, (4) denotes the high-order Bessel beams (i.e. J_n beams, n is an integer). The intensity distribution of all the higher-order Bessel beams has zero on axis surrounded by concentric rings. For example, when $n = 3$, the J_3 beam has a dark central spot and its first bright ring appears at $\rho = 4.201/k_\perp$, as illustrated in Figs. 2(a) and 2(b). However, the phase pattern of the J_n beam is much different from that of the J_0 beam. It has $2n$ arc sections distributed evenly from the innermost to the outermost ring, as shown in Fig. 2(c).

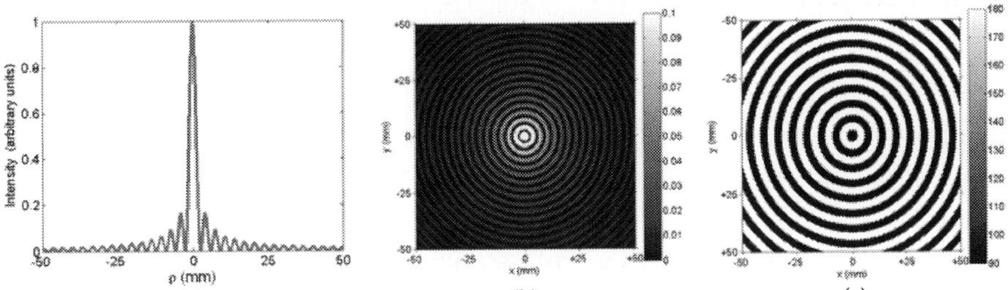

(a) (b) (c)

Fig. 1. A J_0 beam. (a) One-dimensional (1-D) intensity distribution. (b) 2-D intensity distribution plotted in a gray-level representation. (c) Phase distribution ($t=0$, $z=0$). The relevant parameters are incident wavelength of $\lambda = 3mm$, and aperture radius of $R = 50mm$, $k_{\perp} = 0.962mm^{-1}$.

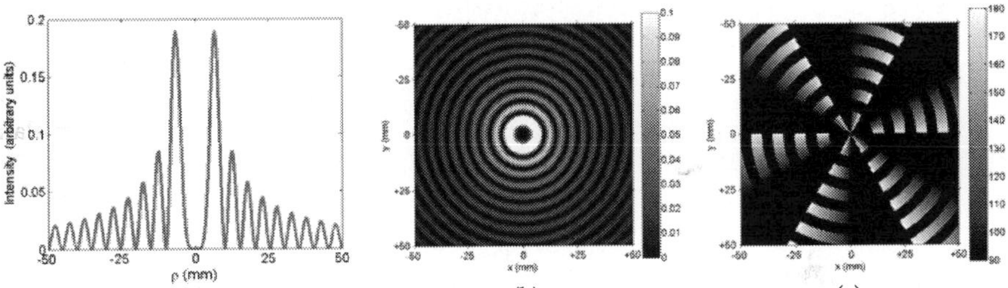

(a) (b) (c)

Fig. 2. A J_3 beam. (a) 1-D intensity distribution. (b) 2-D intensity distribution. (c) Phase distribution ($t=0$, $z=0$). The relevant parameters are the same as in Fig. 1, except $k_{\perp} = 0.638mm^{-1}$.

2.2 Vector analysis

2.2.1 TM and TE modes Bessel beams

In order to discover more characteristics of Bessel beams, the vector analyses should be performed. By using the Hertzian vector potentials of electric and magnetic types $\vec{\Pi}_e$, $\vec{\Pi}_m$, respectively, the fields are expressed as

$$\vec{E}_e = \nabla \times \nabla \times \vec{\Pi}_e = \nabla\nabla \cdot \vec{\Pi}_e + k^2 \vec{\Pi}_e, \quad \vec{H}_e = i\omega_0 \mu_0 \nabla \times \vec{\Pi}_e \tag{7}$$

$$\vec{E}_m = -i\omega_0 \mu_0 \nabla \times \vec{\Pi}_m, \quad \vec{H}_m = \nabla \times \nabla \times \vec{\Pi}_m = \nabla\nabla \cdot \vec{\Pi}_m + k^2 \vec{\Pi}_m \tag{8}$$

where $\vec{\Pi}_e$ and $\vec{\Pi}_m$ are the solutions to vector Helmholtz wave equation. In a source-free region, they satisfy the homogeneous vector Helmholtz equation, respectively. When the choice of $\vec{\Pi}_e = \Pi_e \vec{z}$ and $\vec{\Pi}_m = \Pi_m \vec{z}$, they are reduced to scalar Helmholtz equation

$$\nabla^2 \Pi_e + k^2 \Pi_e = 0, \qquad \nabla^2 \Pi_m + k^2 \Pi_m = 0 \tag{9}$$

From (3) and (4), we have deduced that the Π_e and Π_m can take the form of $J_n(k_\perp \rho)\exp(in\varphi)\exp(i(k_z z - \omega t))$. Thus, $\vec{\Pi}_e$ and $\vec{\Pi}_m$ can be written in the form

$$\vec{\Pi}_e = \Pi_e \vec{z} = P_e J_n(k_\perp \rho)\exp(in\varphi)\exp[i(k_z z - \omega t)]\vec{z} \tag{10a}$$

$$\vec{\Pi}_m = \Pi_m \vec{z} = P_m J_n(k_\perp \rho)\exp(in\varphi)\exp[i(k_z z - \omega t)]\vec{z} \tag{10b}$$

where P_e and P_m are the electric and magnetic dipole moment, respectively. By substituting (10) into (7) and (8) respectively, we finally obtain the TM and TE modes Bessel beams.

TM_n mode: TE_n mode:

$$\left.\begin{aligned}
E_{\rho e} &= iP_e k_\perp k_z J_n'(k_\perp \rho)\exp(in\varphi)\exp[i(k_z z - \omega t)] \\
E_{\varphi e} &= -\frac{n}{\rho} P_e k_z J_n(k_\perp \rho)\exp(in\varphi)\exp[i(k_z z - \omega t)] \\
E_{ze} &= P_e k_\perp^2 J_n(k_\perp \rho)\exp(in\varphi)\exp[i(k_z z - \omega t)] \\
H_{\rho e} &= \frac{n}{\rho} P_e \omega\varepsilon J_n(k_\perp \rho)\exp(in\varphi)\exp[i(k_z z - \omega t)] \\
H_{\varphi e} &= iP_e k_\perp \omega\varepsilon J_n'(k_\perp \rho)\exp(in\varphi)\exp[i(k_z z - \omega t)] \\
H_{ze} &= 0
\end{aligned}\right\} \tag{11a}$$

$$\left.\begin{aligned}
E_{\rho m} &= -\frac{n}{\rho} P_m \omega\mu J_n(k_\perp \rho)\exp(in\varphi)\exp[i(k_z z - \omega t)] \\
E_{\varphi m} &= -iP_m k_\perp \omega\mu J_n'(k_\perp \rho)\exp(in\varphi)\exp[i(k_z z - \omega t)] \\
E_{zm} &= 0 \\
H_{\rho m} &= iP_m k_\perp k_z J_n'(k_\perp \rho)\exp(in\varphi)\exp[i(k_z z - \omega t)] \\
H_{\varphi m} &= -\frac{n}{\rho} P_m k_z J_n(k_\perp \rho)\exp(in\varphi)\exp[i(k_z z - \omega t)] \\
H_{zm} &= P_m k_\perp^2 J_n(k_\perp \rho)\exp(in\varphi)\exp[i(k_z z - \omega t)]
\end{aligned}\right\} \tag{11b}$$

From (11), their instant field vectors and intensity distributions for the TM or TE modes Bessel beams can be easily obtained. Two examples for TM_0 and TE_0 modes Bessel beams are illustrated in Figs. 3 and 4, respectively. From (11a), we can see that the transverse electric field component of the TM_0 mode is only a radial part and thus it is radially polarized. This can also be seen from Fig. 3(a). Similarly, the TE_0 mode is only an azimuthal component of the electric field and thus is azimuthally polarized. Its field vectors at $t = 0$ are shown in Fig. 4(a).

2.2.2 Polarization States

To analyze the polarization states of Bessel beams, (11) in cylindrical coordinates are transformed into rectangular coordinates. Applying the relationships: $\vec{x} = \vec{\rho}\cos\varphi - \vec{\varphi}\sin\varphi$, and $\vec{y} = \vec{\rho}\sin\varphi + \vec{\varphi}\cos\varphi$, we have the following representations for the electric fields.

$$\left.\begin{aligned}
E_{xe} &= [ik_\perp J_n^{'}(k_\perp \rho)\cos\varphi + \frac{n}{\rho} J_n(k_\perp \rho)\sin\varphi]\times \\
&\quad P_e k_z \exp(in\varphi)\exp[i(k_z z - \omega t)] \\
E_{ye} &= [ik_\perp J_n^{'}(k_\perp \rho)\sin\varphi - \frac{n}{\rho} J_n(k_\perp \rho)\cos\varphi]\times \\
&\quad P_e k_z \exp(in\varphi)\exp[i(k_z z - \omega t)] \\
E_{ze} &= P_e k_\perp^2 J_n(k_\perp \rho)\exp(in\varphi)\exp[i(k_z z - \omega t)]
\end{aligned}\right\} \tag{12a}$$

$$\left.\begin{aligned}
E_{xm} &= -[\frac{n}{\rho} J_n(k_\perp \rho)\cos\varphi - ik_\perp J_n^{'}(k_\perp \rho)\sin\varphi]\times \\
&\quad P_m \omega\mu \exp(in\varphi)\exp[i(k_z z - \omega t)] \\
E_{ym} &= -[\frac{n}{\rho} J_n(k_\perp \rho)\sin\varphi + ik_\perp J_n^{'}(k_\perp \rho)\cos\varphi]\times \\
&\quad P_m \omega\mu \exp(in\varphi)\exp[i(k_z z - \omega t)] \\
E_{zm} &= 0
\end{aligned}\right\} \tag{12b}$$

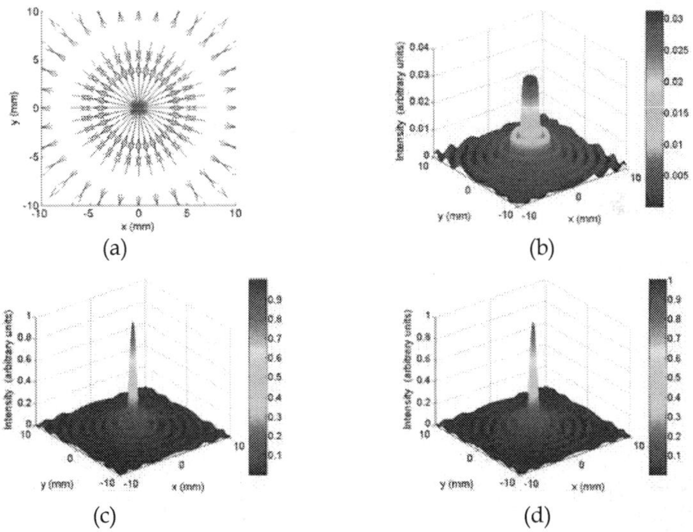

Fig. 3. TM_0 mode Bessel beam. (a) Instant vector diagram for the transverse component of the electric field ($t = 0$, $z = 0$). (b) The transverse electric field intensity ($I_\perp = |E_{\rho e}|^2 + |E_{\varphi e}|^2$). (c) The longitudinal electric field intensity ($I_z = |E_{ze}|^2$) and (d) the total electric filed intensity ($I = I_\perp + I_z$). The color bars illustrate the relative intensity. The relevant parameters are $\lambda = 3mm$, $k_\perp = 2.004mm^{-1}$, $k_z = 0.608mm^{-1}$, and $R = 10mm$.

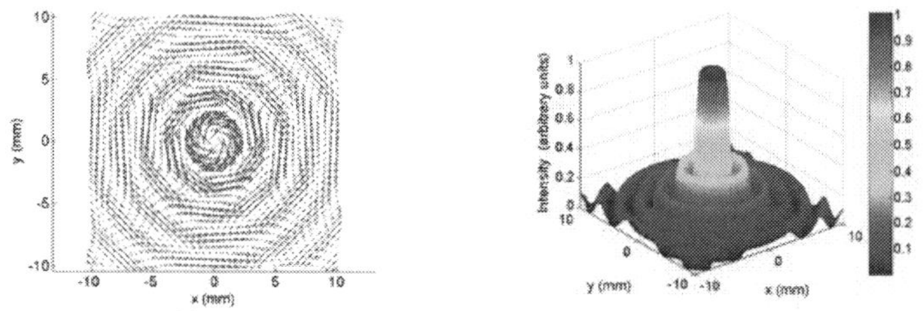

(a) (b)

Fig. 4. TE_0 mode Bessel beam. (a) Instant vector diagram for the transverse component of the electric field ($t = 0$, $z = 0$). (b) The transverse electric field intensity. The relevant parameters are the same as in Fig. 3, except $k_\perp = 1.503mm^{-1}$, and $k_z = 1.459mm^{-1}$.

The total electric fields of E_x and E_y are given by, respectively

$$E_x = A_1 E_{xe} + A_2 E_{xm}, \qquad E_y = A_1 E_{ye} + A_2 E_{ym} \tag{13}$$

where A_1 and A_2 are the proportional coefficients. Let $P_e = 1$, then $P_m = i\sqrt{\varepsilon/\mu}$. Substituting (12) into (13), we can deduce the following representations:

$$E_x = E_{xA}\exp(i\theta_1)\exp(in\varphi)\exp[i(k_z z - \omega t)], \qquad E_y = E_{yA}\exp(i\theta_2)\exp(in\varphi)\exp[i(k_z z - \omega t)] \tag{14}$$

where

$$E_{xA} = \sqrt{(B_1 \sin\varphi)^2 + (B_2 \cos\varphi)^2}, \qquad E_{yA} = \sqrt{(B_1 \cos\varphi)^2 + (B_2 \sin\varphi)^2},$$

$$B_1 = \frac{A_1 n k_z}{\rho} J_n(k_\perp \rho) - A_2 k k_\perp J_n^{'}(k_\perp \rho), \qquad B_2 = A_1 k_\perp k_z J_n^{'}(k_\perp \rho) - A_2 \frac{nk}{\rho} J_n(k_\perp \rho),$$

$$\theta_1 = \arctan(\frac{B_2 \cos\varphi}{B_1 \sin\varphi}), \qquad \theta_2 = \arctan(-\frac{B_2 \sin\varphi}{B_1 \cos\varphi}).$$

The polarization states of Bessel beams are discussed as follows:

Case 1) $\theta_2 - \theta_1 = K\pi$, where $K = 0,1,2...$ is an integer. The Bessel beam is linearly polarized. To satisfy this case and assume that $n = 0$, it is demanded from (14) that $A_1 = 0$ and $A_2 \neq 0$, or $A_1 \neq 0$ and $A_2 = 0$. Under these conditions, we can acquire the zero-order Bessel beam with linear polarization, as shown schematically in Figs. 5 and 6.

Case 2) $\theta_2 - \theta_1 = +\pi/2$ and $E_{xA} = E_{yA}$. The Bessel beam is left-hand circularly polarized. To satisfy these requirements, the demand of $A_1/A_2 = +k/k_z$ can be derived from (14). The left-hand circularly polarized Bessel beam is illustrated in Fig. 7.

Case 3) $\theta_2 - \theta_1 = -\pi/2$ and $E_{xA} = E_{yA}$. The Bessel beam become right-hand circularly polarized. Similarly, the demand of $A_1/A_2 = -k/k_z$ is needed. Fig. 8 shows the right-hand circularly polarized Bessel beam.

Case 4) In other cases, the Bessel beam is elliptically polarized.

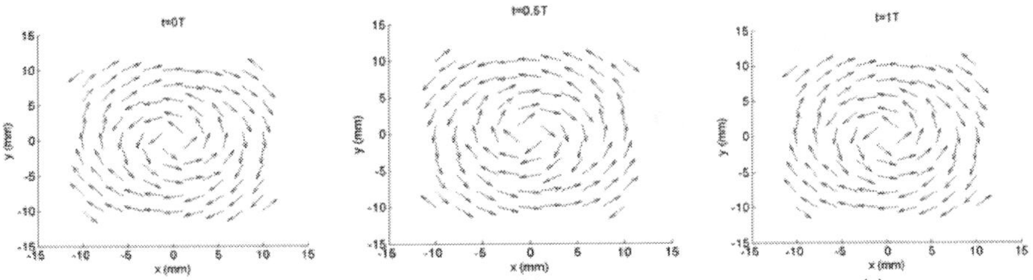

(a) (b) (c)

Fig. 5. Linearly polarized Bessel beam. (a)-(c) Vector diagrams of the transverse component of the electric field at three different instants: $t = 0$, $t = 0.5T$, $t = T$, $T = 2\pi/\omega$, respectively. The parameters used in Fig. 5 are $k_{\perp}/k = 0.25$, $n = 0$, $A_1 = 0$, and $A_2 \neq 0$.

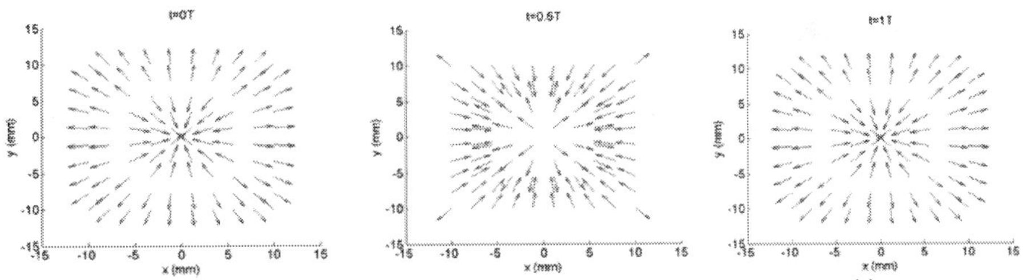

(a) (b) (c)

Fig. 6. Linearly polarized Bessel beam. (a)-(c) Vector diagrams of the transverse component of the electric field at three different instants: $t = 0$, $t = 0.5T$, $t = T$, respectively. The parameters used in Fig. 6 are the same as in Fig. 5, except $A_1 \neq 0$, and $A_2 = 0$.

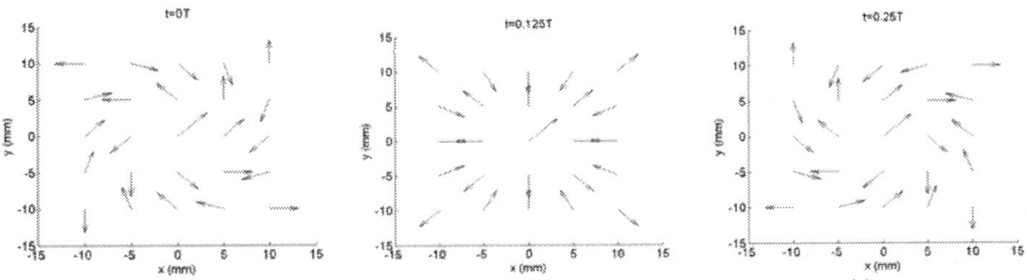

(a) (b) (c)

Fig. 7. Left-hand circularly polarized Bessel beam. (a)-(c) Vector diagrams of the transverse component of the electric field at three different instants: $t = 0$, $t = 0.125T$, $t = 0.25T$, respectively. The relevant parameters are $k_{\perp}/k = 0.4$, and $A_1/A_2 = +k/k_z$.

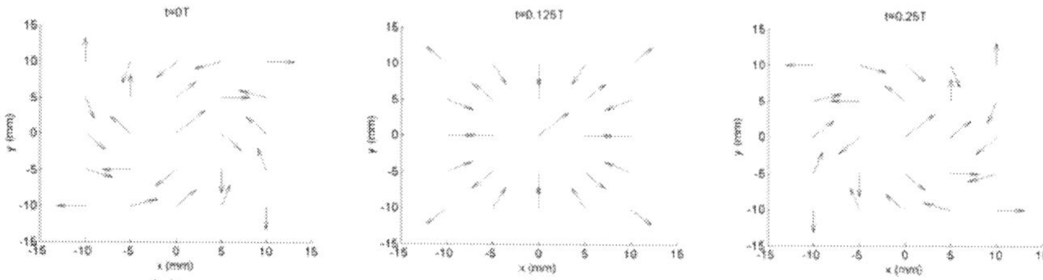

(a) (b) (c)

Fig. 8. Right-hand circularly polarized Bessel beam. (a)-(c) Vector diagrams of the transverse component of the electric field at three different instants: $t=0$, $t=0.125T$, $t=0.25T$, respectively. The relevant parameters are $k_\perp/k=0.4$, and $A_1/A_2=-k/k_z$.

2.2.3 Energy Density and Poynting Vector

Using the above equations (11), the total time-average electromagnetic energy density for the transverse modes, TE or TM, is calculated to be

$$\overline{w}=\frac{1}{4}\varepsilon|\vec{E}|^2+\frac{1}{4}\mu|\vec{H}|^2=\frac{1}{4}\varepsilon\{(k_\perp J_n)^2+(k^2+k_z^2)[(\frac{nJ_n}{\rho})^2+(k_\perp J_n^{'})^2]\} \tag{15}$$

And the time-average Poynting vector power density is given by

$$\overline{\overline{S}}=\frac{1}{2}\mathrm{Re}(\vec{E}\times\vec{H}^*)=\omega\varepsilon k_z[(\frac{nJ_n}{\rho})^2+(k_\perp J_n^{'})^2]\vec{z}+\frac{n\omega\varepsilon}{\rho}(k_\perp J_n)^2\vec{\varphi} \tag{16}$$

From (15) or (16), it can immediately be seen that neither $\overline{w}$ nor $\overline{\overline{S}}$ depends on the propagation distance z . This means the time-average energy density does not change along the z axis, and our solutions clearly represent nondiffracting Bessel beams. In addition, from (16), we note that $\overline{\overline{S}}$ has the longitudinal and transverse components, which determine the flow of energy along the z axis and perpendicular to the z-axis, respectively. However, when n=0, corresponding to TM_0 or TE_0 mode, $\overline{\overline{S}}$ is directed strictly along the z-axis and is proportional to J_1^2 .

3. Generation of pseudo-Bessel Beams by BOEs (Yu & Dou, 2008c; Yu & Dou, 2008d)

In optics, lots of methods for creating pseudo-Bessel Beams have been suggensted, such as narrow annular slit (Durnin et al., 1987), computer-generated holograms (CGHs) (Turunen et al., 1988), Fabry-Perot cavity (Cox & Dibble, 1992), axicon (Scott & McArdle, 1992), optical refracting systems (Thewes et al., 1991), diffractive phase elements (DPEs) (Cong et al., 1998) and so on. However, at millimeter and sub-millimeter wavebands, only two methods of production Bessel beams have been proposed currently, i.e., axicon (Monk et al., 1999) and computer-generated amplitude holograms (Salo et al., 2001; Meltaus et al., 2003). Although the method of using axicon is very simple, only a zero-order Bessel beam can be generated. The other method relying on holograms can produce various types of diffraction-free beams, but their diffraction efficiencies are only around 45% (Arlt & Dholakia, 2000) owing to using amplitude holograms. In order to overcome these limitations mentioned above, in our work, binary optical elements (BOEs) are employed and designed for producing pseudo-Bessel Beams in millimeter and sub-millimeter range for the first time. The suitable design tool is to combine a genetic algorithm (GA) for global optimization with a two-dimensional finite-difference time-domain (2-D FDTD) method for rigours eletromagnetic computation.

3.1 Description of the design tool

3.1.1 FDTD method computational model

The electric field distribution of the nth -order Bessel beam in the cylindrical coordinates system is rewritten as :

$$E(\rho,\varphi,z) = E_0 J_n(k_\perp \rho)\exp(in\varphi)\exp(ik_z z) \tag{17}$$

All Bessel beams are circularly symmetric, thus our calculations are concerned only with radically symmetric system. The feature sizes of BOEs are on the order of or less than a millimeter wavelength, the methods of full wave analysis are needed to calculate the diffractive fields of BOEs. The 2-D FDTD method (Yee, 1966) is employed to compute the field diffracted by the BOE in our work. The Computational model of the FDTD method is shown schematically in Fig. 9, in which the BOE is used to convert an incident Gaussian-profile beam on the input plane into a Bessel-profile beam on the output plane. z_1 is the distance between the input plane and the BOE, and z_2 is the distance between the BOE and the output plane; the aperture radius of the BOE, which is represented by R, is the same as that of the input and output planes; n_1 and n_2 represent the refractive indices of the free space and the BOE, respectively; and z is the symmetric axis and the magnetic wall is set on it to save the required memory and computing time.

When a Gaussian beam is normally incident from the input plane onto the left side of the BOE, its wave front is modulated by the BOE, and a desired Bessel beam is obtained on the output plane. It is worthy to point out that our design goal is to acquire a desired Bessel beam in the near field (i.e. the output plane). If one wants to obtain a desired field in the far field, an additional method, like angular-spectrum propagation method (Feng et al.,2003), should be employed to determine the far field.

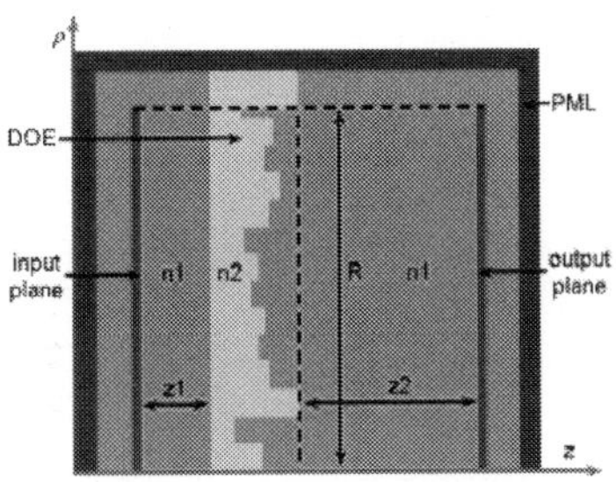

Fig. 9. Schematic diagram of 2-D FDTD computational model

3.1.2 Genetic Algorithm (GA)

To fabricate conveniently in technics, the DOE, with circular symmetry and aperture radius R, should be divided into concentric rings with identical width δ but different depth x, as shown in Fig. 10. The width δ equals R/K, K is a prescribed positive integer. The maximal depth of a ring is $x_{\max} = \lambda/(n_2 - 1)$, in which n_2 is the refractive index of the BOE. In BOEs design, the depth x of each ring can take only a discrete value. Provided that the maximal depth of a ring is quantified into M -level, in general case, $M = 2^a$, where a is a integer, the minimal depth of a ring is $\Delta x = x_{\max}/M$. Therefore, the depth x of each ring can take only one of the values in the set of $\{\Delta x, 2\Delta x, ..., M\Delta x\}$. Thus, the different combination of the depth x of each ring, i.e., $X = \bigcup_{k=1}^{K} \{x_k\}$, where $x_k \in \{\Delta x, 2\Delta x, ..., M\Delta x\}$, represents the different BOE profile. To obtain the BOE profile which satisfies the design requirement , the different combination X should be calculated, and the optimum combination is gained finally. In fact, this is a combinatorial optimization problem (COP). The GA (Haupt, 1995; Weile & Michielssen, 1997) is adopted for optimizing the BOE profile. It operates on the chromosome, each of which is composed of genes associated with a parameter to be optimized. For instance, in our case, a chromosome corresponds to a set X which describes the BOE profile, and a gene corresponds to the depth x of a ring.

The first step of the GA is to generate an initial population, whose chromosomes are made by random selection of discrete values for the genes. Next, a fitness function, which describes the different between the desired field E^d and the calculated field E^c obtained by using 2-D FDTD method, will be evaluated for each chromosome. In our study, the fitness function is simply defined as:

$$fitness = \sum_{u=1}^{U} (|E_u^c| - |E_u^d|)^2 \quad (18)$$

in which E_u^c and E_u^d are the calculated field and the desired field at the *uth* sample ring of the output plane, respectively. Then, based on the fitness of each chromosome, the next generation is created by the reproduction process involved crossover, mutation, and selection. Last, the GA process is terminated after a prespecified number of generations $Gen_{\max}$. The flow chart of the GA procedure is shown in Fig. 11.

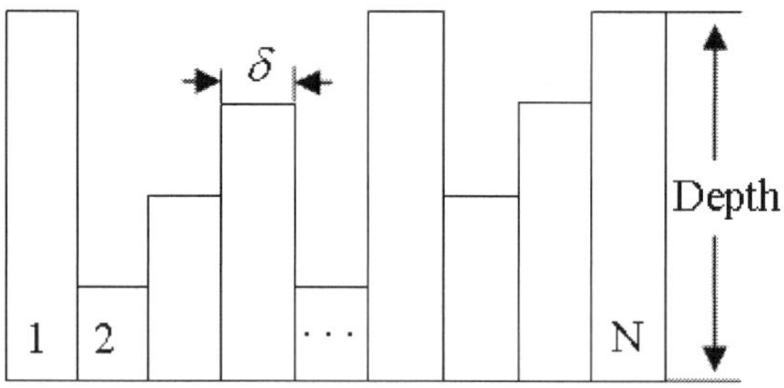

Fig. 10. Division of the BOE profile into the rings with identical width δ but different depths x

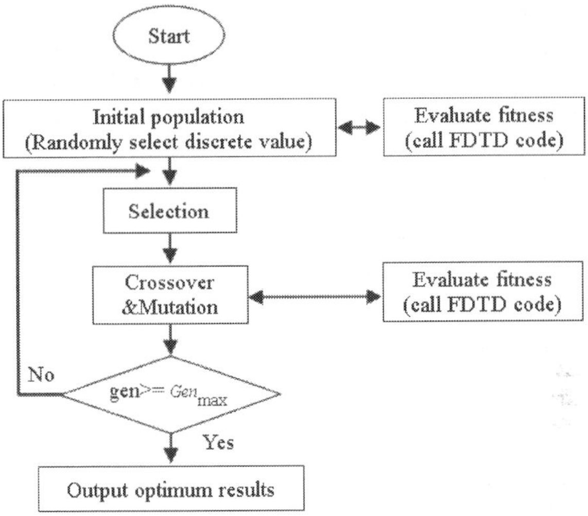

Fig. 11. The flow chart of the GA procedure

3.2 Numerical simulation results

In order to evaluate the quality of the designed BOE, we introduce the efficiency η and the root mean square (RMS) describing the BOE profile error (Feng et al.,2003), which are defined as, respectively.

$$\eta = \frac{\sum_{u=1}^{U} |E_u^c|^2 S_u^c}{\sum_{v=1}^{V} |E_v^i|^2 S_v^i} \tag{19}$$

$$RMS = \left[\frac{1}{U-1} \sum_{u=1}^{U} (|E_u^c|^2 - |E_u^d|^2)^2 \right]^{1/2} \tag{20}$$

where S_v^i and S_u^c are the areas of the vth and uth sample ring of the input and output planes, respectively; E_v^i is the incident field at the vth sample ring of the input plane, and E_u^c and E_u^d are the calculated field and the desired field at the uth sample ring of the output plane. To demonstrate the utility of the design method, we present three examples herein in which an incident Gaussian beam is converted into a zero-order, a first order and a second order Bessel beam respectively. The same parameters in three examples are as follows: an incident Gaussian beam waist of $w_0 = 4\lambda$ $n_1 = 1.0$, $n_2 = 1.45$, $z_1 = 2\lambda$, $z_2 = 6\lambda$, $\delta = \lambda/18$, $R = 8\lambda$, $K = 144$, $M = 8$, $U = V = K$. From three cases, it is clearly seen that the fields diffracted by the designed BOE's on the output plane agree well with the desired electric field intensity distributions.

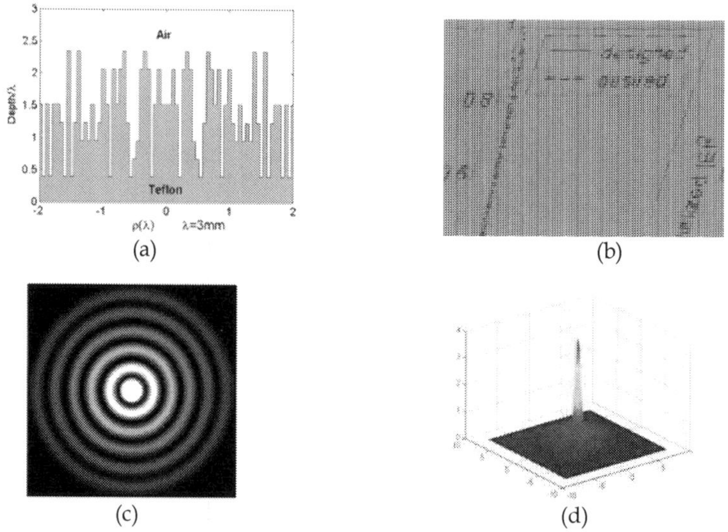

Fig. 12. Generation of a J_0 beam on the output plane. $\lambda = 3mm$, $k_{\perp} = 0.7635mm^{-1}$, $\eta = 94.494\%$ and $RMS = 5.562\%$. (a) Part of the optimized BOE profile. (b) The desired and the designed transverse intensity distribution on the output plane. (c) The 2-D transverse intensity distribution plotted in a gray-level representation, and (d) the 3-D transverse intensity distribution.

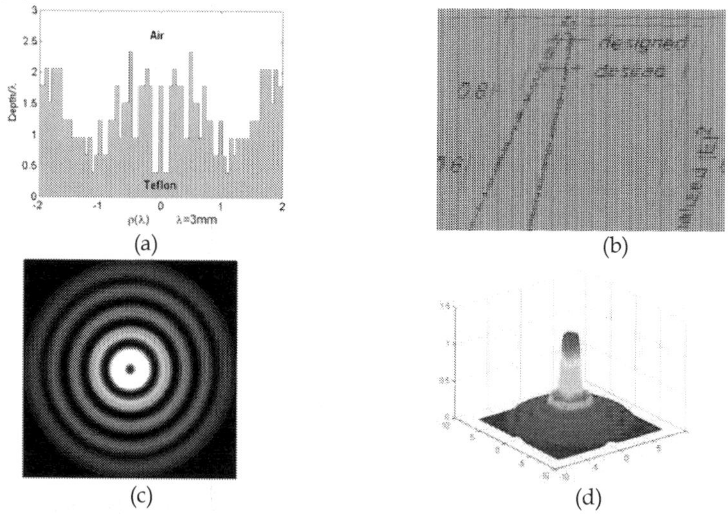

Fig. 13. Production of a J_1 beam on the output plane. $\lambda = 3mm$, $k_{\perp} = 0.6911mm^{-1}$, $\eta = 96.283\%$ and $RMS = 2.806\%$. (a) Part of the optimized BOE profile. (b) The desired and the designed transverse intensity distribution on the output plane. (c) The 2-D transverse intensity distribution, and (d) the 3-D transverse intensity distribution.

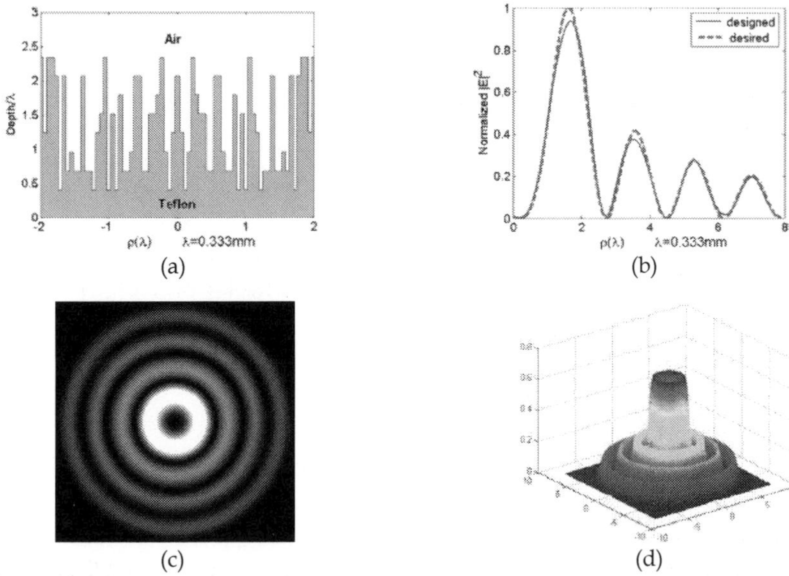

Fig. 14. Creation of a J_2 Bessel beam on the output plane. $\lambda = 0.333mm$, $k_{\perp} = 5.6406mm^{-1}$, $\eta = 97.263\%$ and $RMS = 1.845\%$. (a) Part of the optimized BOE profile. (b) The desired and the designed transverse intensity distribution on the output plane. (c) The 2-D transverse intensity distribution, and (d) the 3-D transverse intensity distribution.

4. Production of approximate Bessel beams using binary axicons (Yu & Dou, 2009)

Currently, numerous ways for generating pseudo-Bessel beams have been proposed, among which using axicon is the most popular method, owing to its simplicity of configuration and easy realization. However, at millemter and sub-millimter wavebands, classical cone axicons are usually bulk ones and therefore have many disadvantages, like heavy weight, large volume and thus increased absorption loss in the material. These limitations together make them extremely difficult in miniaturizing and integrating in millemter and sub-millimter quasi-optical systems. To overcome these problems, binary axicons, based on binary optical ideas, are introduced in our study and designed for producing pseudo-Bessel beams at sub-millimter wavelengths. The designed binary axicons are more convenient to fabricate than holographic axicons (Meltaus et al., 2003; Courtial et al., 2006) and, become thinner and less lossy in the material than classical cone axicons (Monk et al., 1999; Trappe et al., 2005; Arlt & Dholakia, 2000). In order to analyze binary axicons accurately when illuminated by a plan wave in sub-millimter range, the rigorous electromagnetic analysis method, that is, a 2-D FDTD method for determining electromagnetic fields in the near region in conjunction with Stratton-Chu formulas for obtaining electromagnetic fields in the far region, is adopted in our work. Using this combinatorial method, the properties of approximate Bessel beams generated by the designed binary axicons are analyzed.

4.1 Binary axicon design

A classical cone axicon, introduced firstly by McLeod in 1954 (McLeod, 1954), is usually a bulk one, as illustrated in Fig. 15(a), in which D is the aperture diameter and γ is the prism angle. Based on binary optical ideas, the profile of a binary axicon, whose performance required is equivalent to that of a bulk one, can be easily formed. Assuming straight-ray propagation through the bulk axicon, the relation between the phase retardation $\varphi(\rho)$ and the surface height $h(\rho)$ is given as (Feng et al., 2003)

$$h(\rho) = \varphi(\rho)/[(n_2 - n_1)k] \tag{21}$$

where k is the free space wave number, n_1 and n_2 are the refractive indexes of the air and the axicon, respectively. To generate the continuous profile of the binary axicon, the equivalent transformation can be used by (Hirayama et al., 1996)

$$h(\rho) = [\varphi(\rho) \bmod 2\pi]/[(n_2 - n_1)k] \tag{22}$$

The continuous profile of the binary axicon produced by (22) is shown in Fig. 15(c). For the multilevel axicon, the profile is quantized into equal height step Δ . The quantized height is given by

$$h_q(\rho) = \text{int}[h(\rho)/\Delta]\Delta \tag{23}$$

where $\Delta = h_{\max}/M$, $h_{\max} = \lambda/(n_2 - n_1)$ and M is the number of levels. Eq. (23) generates the multilevel profiles of the binary axicon. The schematic diagram of the 4-level binary axicon is illustrated in Fig. 15 (d). It is known that the larger the number of levels is, the higher the diffraction efficiency is, however, the higher the difficulty of manufacture becomes. Therefore, the compromise between the diffraction efficiency and the difficulty of manufacture should be considered when determining the number of levels. In our work the selection of the 32-level binary axicon is made. From Figs. 15(c) and 15(d), we can see easily that the designed binary axicon is not only more compact than the classical cone axicon, but also simpler to fabricate than the holographic axicon.

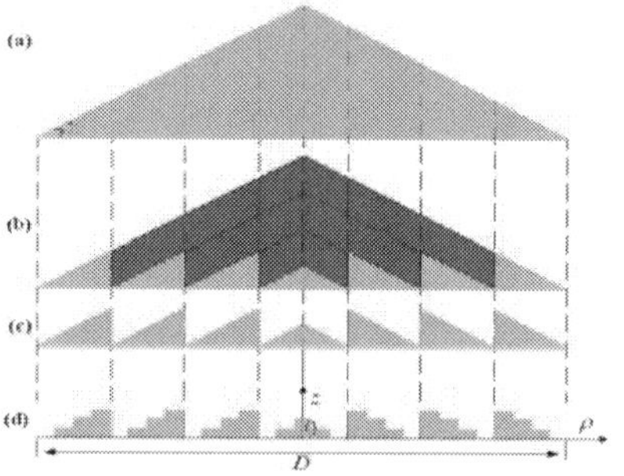

Fig. 15. The design process of a binary axicon. (a) A bulk axicon. (b) An axicon removed the unwanted material (red part). (c) An equivalent binary axicon with continuous profile. (d) An equivalent binary axicon quantized into four levels.

4.2 Rigorous electromagnetic analysis method

Because of rotational symmetry of the binary axicon, a 2-D FDTD method is applied to evaluate the electromagnetic fields diffracted by the binary axicon in the near region. The computational model of the 2-D FDTD method is shown schematically in Fig. 16, in which the binary axicon is utilized to convert an incident beam into a pseudo-Bessel beam. To stimulate the entire 2-D FDTD grid, a total-scattered field approach is applied to introduce a normally incident plane wave. In this approach the connecting boundary serves to connect the total and the scattered field regions, and is the location at which the incident field is introduced. Because of the limitation of computational time and memory, the computational range of the 2-D FDTD method is truncated by using perfectly matched layer (PML) absorbing boundary conditions (ABCs) in the near region. Therefore, in order to accurately determine the electromagnetic fields in the far region, Stratton-Chu integral formulas are applied and given by (Stratton, 1941)

$$\begin{aligned}\vec{E}(\vec{r}) &= \int_L \left\{-i\omega\mu\left[\vec{n}\times\vec{H}(\vec{r}\,')\right]G_0(\vec{r},\vec{r}\,') + \left[\vec{n}\times\vec{E}(\vec{r}\,')\right]\times\nabla' G_0(\vec{r},\vec{r}\,') + \left[\vec{n}\bullet\vec{E}(\vec{r}\,')\right]\nabla' G_0(\vec{r},\vec{r}\,')\right\}dL' \\ \vec{H}(\vec{r}) &= \int_L \left\{i\omega\varepsilon\left[\vec{n}\times\vec{E}(\vec{r}\,')\right]G_0(\vec{r},\vec{r}\,') + \left[\vec{n}\times\vec{H}(\vec{r}\,')\right]\times\nabla' G_0(\vec{r},\vec{r}\,') + \left[\vec{n}\bullet\vec{H}(\vec{r}\,')\right]\nabla' G_0(\vec{r},\vec{r}\,')\right\}dL'\end{aligned} \tag{24}$$

where $\vec{r} = (\rho, z)$ and $\vec{r}\,' = (\rho', z')$ denote an arbitrary observation point in the far region and an source point on the output boundary of the 2-D FDTD model, respectively; unit vector $\vec{n}$ is the outer normal of the closed curve, L, of the output boundary; $G_0(\vec{r},\vec{r}\,') = -iH_0^{(2)}(k|\vec{r}-\vec{r}\,'|)/4$, is the 2-D scalar Green's function in free space, $H_0^{(2)}$ is the zero-order Hankel function of the second kind and k is the wave number in free space; ω is the angular frequency; ε and μ are the permittivity and permeability, respectively.

In short, our electromagnetic analysis method is to join the 2-D FDTD method for computing the fields diffracted by the binary axicon in the near region with the Stratton-Chu integral formulas for obtaining its diffractive fields in the far region. It can be implemented by the following procedure: First, we carry out the 2-D FDTD computation and obtain the near fields, $\vec{E}(\vec{r}\,')$ and $\vec{H}(\vec{r}\,')$, on the output boundary. Then, these fields can be regarded as secondary sources and substituted into (24) to calculate the fields, $\vec{E}(\vec{r})$ and $\vec{H}(\vec{r})$, at arbitrary observation point in the far region. Note that the integral herein is over the closed curve, L, of the output boundary.

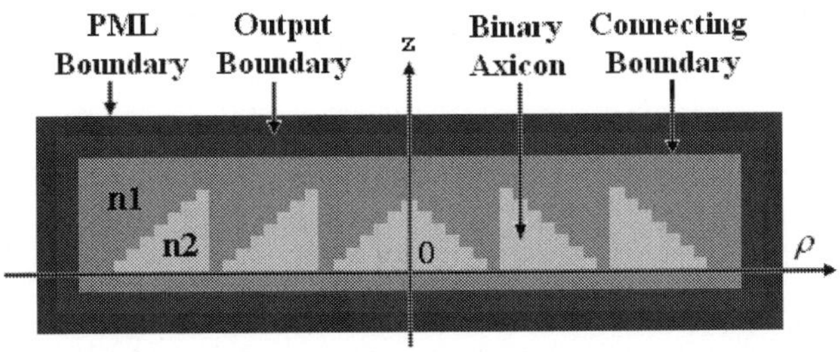

Fig. 16. Schematic diagram of 2-D FDTD computational model, where the 8-level binary axicon is embedded into FDTD grid.

4.3 Demonstration of equivalence

To demonstrate the equivalent performance between the bulk axicon and the designed binary axicon, Fig. 17 shows the on-axis intensity distributions for both the bulk axicon and the 32-level binary axicon. In this case both axicons, with the same aperture diameter $D = 40\lambda$ and prism angle $\gamma = 10^0$, are normally illuminated by a plane wave of unit amplitude. Other parameters used in Fig. 17 are as follows: an incident wavelength is $\lambda = 0.32mm$ ($f = 0.94\text{THz}$), the refractive indexes of the axicon and the air are $n_2 = 1.4491$ (Teflon) and $n_1 = 1.0$, respectively. Two distributions exhibit some differences in the near region ($z < 75\lambda$). The reason is that the binary axicon suffers more from edge diffraction and truncation effects (Trappe et al., 2005). The effects can also be seen from Fig. 18(b), which has more burr than Fig. 18(a) in the near region. However, two curves show a good agreement in the region ($z > 75\lambda$), where the propagating beam can be best approximated by the Bessel beam in terms of its intensity profile. Thus, the performance of the designed binary axicon is equivalent to that of the bulk one. In order to further demonstrate the equivalent effect between two axicons, we extend our 2-D FDTD calculated region to 200λ along z-axis, and display their electric-field amplitudes in a pseudo-color representation in Fig. 18. It can also be seen that the designed binary axicon has the same performance as the bulk one.

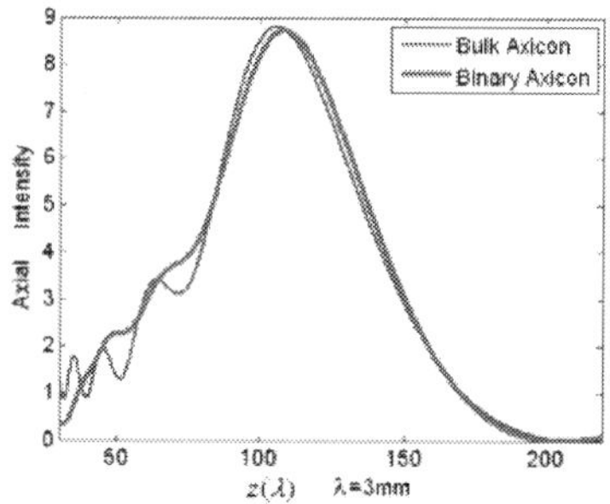

Fig. 17. The axial intensity distributions for the designed binary axicon and the bulk one.

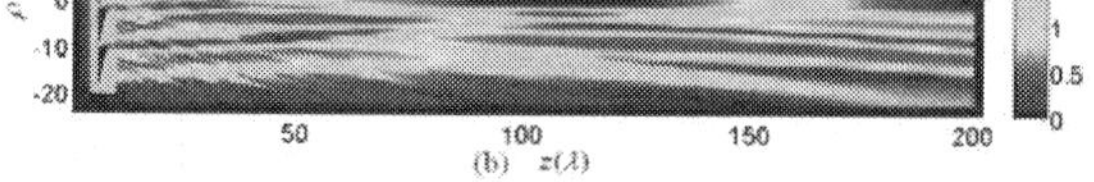

Fig. 18. Electric-field amplitude patterns plotted in a pseudo-color representation. (a) For the bulk axicon. (b) For our designed binary axicon.

4.4 Properties of pseudo-Bessel beam

In order to study the properties of a pseudo-Bessel beam, the other 32-level binary axicon having aperture diameter $D = 44\lambda$ and prism angle $\gamma = 12^0$, are examined. Other parameters used in this example are the same as in Fig .17. When this axicon is normally illuminated by a plane wave of unit amplitude, its axial and transverse intensity distributions at three representative values of z : $z = 0.8Z_{max}$, Z_{max} and $1.2Z_{max}$ are shown in Figs. 19(a)-19(d), respectively. It can be seen clearly from Fig. 19(a) that the on-axis intensity increases with oscillating, and reaches its maximum axial intensity then decreases quickly, as the propagation distance z increases. The maximum value of on-axis intensity in Fig. 19(a) is 10.297, located at $Z_{max} = 125.4\lambda$. As shown in Fig. 19(a), if L_{max} is defined as the maximum propagation distance of a pseudo-Bessel beam, we can obtain $L_{max} \approx 230\lambda$. In addition, according to geometrical optics (Trappe et al., 2005), a limited diffraction range $L = 227\lambda$ is estimated by: $L = D/(2\tan\beta)$ and $\sin(\beta+\gamma) = n_1 \sin\gamma$. We discover two results almost coincide. From Figs. 19(b)-19(d) we can observe that their transverse intensity distributions are approximations to Bessel function of the first kind. The radii of their central spot are only about 3.5λ. This indicates that the transverse intensity distribution of pseudo-Bessel beam is highly localized. It is also interesting to point out that the radius size of 3.5λ is very close to the value of 3.4λ, which is determined roughly from the first zero of the Bessel function ($2.4048\lambda/(2\pi\sin\beta)$) (Trappe et al., 2005).

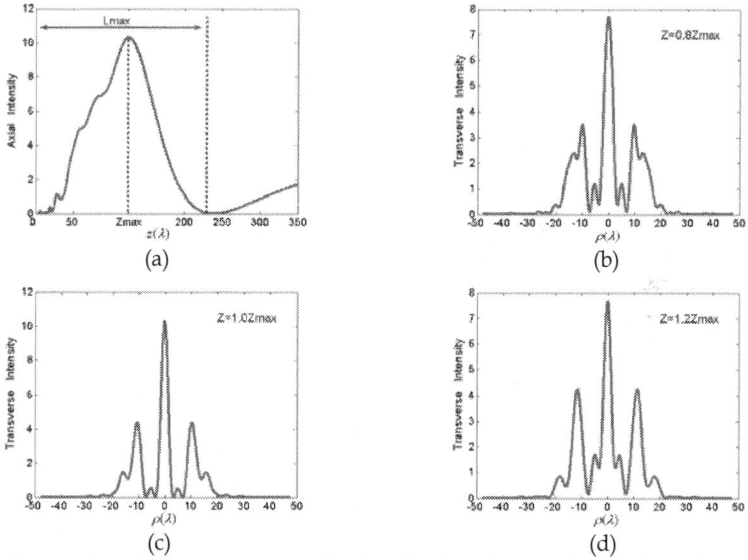

Fig. 19. The axial and transverse intensity distributions for the designed binary axicon. (a) The on-axis intensity versus propagation distance z. (b) The transverse intensity distribution at $z = 0.8Z_{max}$ plane. (c) $z = 1.0Z_{max}$. (d) $z = 1.2Z_{max}$.

5. Propagation characteristic (Yu & Dou, 2008e)

The most interesting and attractive characteristic of Bessel beam is diffraction-free propagation distance. In optics, the comparisons of maximum propagation distance had been done between apertured Bessel and Gaussian beams by Durnin (Durnin, 1987; Durnin et al., 1988) and Sprangle (Sprangle & Hafizi, 1991), respectively. However, the completely contrary conclusions were derived by them, owing to the difference between their contrast criteria. Because Bessel beams have many potential applications at millimeter and sub-millimeter wavebands, therefore, it is necessary and significant that the comparison is carried out at these bands. A new comparison criterion in the spectrum of millimeter and sub-millimeter range has been proposed by us. Under this criterion, the numerical results obtained by using Stratton-Chu formulas instead of Fresnel-Kirchhoff diffraction integral formula are presented; and a new conclusion is drawn.

5.1 Reviews of comparisons of Durnin and Sprangle

In this Subsection, the comparisons done by Durnin and Sprangle respectively are reviewed at first. Because of the circular symmetries of Bessel and Gaussian beams, thus our calculations are concerned only with circularly symmetric system. Let $(\rho',0)$ and (ρ,z) be the coordinates of a pair of points on the incident and receive planes, respectively. In optics, it is well known that scalar diffraction theory yields excellent results when the wavelength is small compared with the size of the aperture and the propagation angles are not too steep (Durnin, 1987). In the Fresnel approximation the amplitude $A(\rho,z)$ at a distance z can be obtained from Fresnel-Kirchhoff diffraction integral formula (Jiang et al., 1995)

$$A(\rho,z)=\exp(ikz+\frac{ik\rho^2}{2z})(\frac{k}{iz})\times\int_0^R A(\rho',0)J_0(\frac{k\rho'\rho}{z})\exp(\frac{ik\rho'^2}{2z})\rho'\mathrm{d}\rho' \tag{25}$$

where $\rho=\sqrt{x^2+y^2}$, $\rho'=\sqrt{x'^2+y'^2}$, R is the aperture radius of incident plane, and

$$A(\rho',0)=\begin{cases}J_0(k_\perp\rho') & \text{Bessel beam}\\ \exp(-\rho'^2/w_0^2) & \text{Gaussian beam}\end{cases} \tag{26}$$

for all $\rho'<R$, and zero for all $\rho'>R$, where $k_\perp$ is the radial wave number and w_0 is the waist radius of Gaussian beam. When $\rho=0$ in (25), the axial intensity distribution $I_A(0,z)$ can be given by

$$I_A(0,z)=\left|A(0,z)\right|^2=\left(\frac{k}{z}\right)^2\times\left|\int_0^R A(\rho',0)\exp(\frac{ik\rho'^2}{2z})\rho'\mathrm{d}\rho'\right|^2 \tag{27}$$

According to Durnin's comparison criterion (Durnin, 1987; Durnin et al., 1988) : $\rho_0=w_0$, that is, on the incident plane ($z'=0$), the central spot radius ρ_0 of zero-order Bessel beam (i.e. J_0 beam) is equal to the waist radius w_0 of Gaussian beam, as displayed in Fig. 20(a), where $\rho_0=w_0=100um$, $k_\perp=2.405/\rho_0$, $R=2mm$, $\lambda=0.6328um$, we calculate the $I_A(0,z)$ versus z curves by using (27), which are shown in Fig. 20(b). It can be seen clearly from Fig. 20(b) that the Bessel beam propagates farther than the Gaussian beam.

However, according to Sprangle's comparison criterion (Sprangle & Hafizi, 1991): $w_0=R$, and a J_0 beam has at least one side-lobe on the incident plane, as illustrated in Fig. 21(a), we can obtain the results given in Fig. 21(b). The converse conclusion that the Bessel beam propagates no farther than the Gaussian beam can be easily drawn from Fig. 21(b).

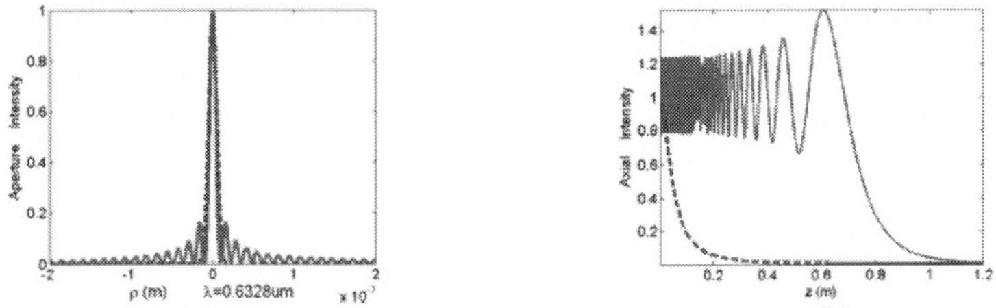

(a) (b)

Fig. 20. The comparison of Durnin. (a) Intensity distributions for a J_0 beam (—) and a Gaussian beam (----) on the incident plane where the beams are assumed to be formed. (b) Axial intensities $I_A(0,z)$ versus propagation distance z.

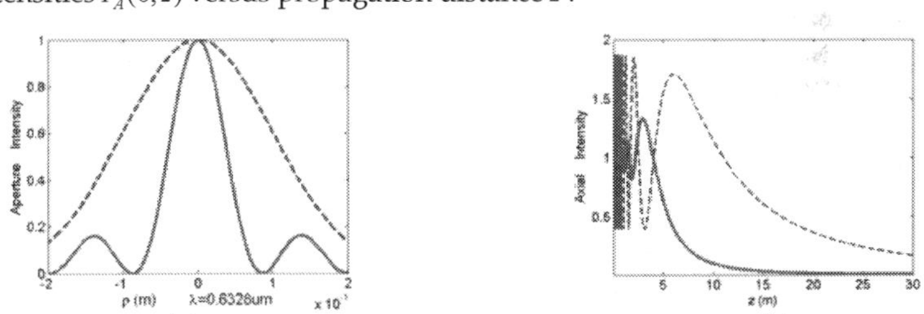

(a) (b)

Fig. 21. The comparison of Sprangle. (a) Intensity distributions for a J_0 beam and a Gaussian beam on the incident plane. (b) Axial intensities $I_A(0,z)$ versus propagation distance z.

The reason why the converse conclusions were obtained by Durnin and Sprangle respectively was that the criteria taken by them were very different. This fact can be seen from Fig. 20(a) and Fig. 21(a). Moreover, the key problem of their criteria is not objective and fair. Under Durnin's criterion, the utilization ration of aperture for the Gaussian beam is very low. In fact, we should not utilize so large aperture to eradiate a Gaussian beam with so small waist radius. However, under Sprangle's criterion, the powers carried by two beams on the incident plane are not equal. Therefore, we propose a new comparison criterion at millimeter wavelengths, which is discussed in the next Subsection.

5.2 Our comparison criterion and results

At millimeter wave bands, it is known that Fresnel-Kirchhoff diffraction integral formula based on scalar theory is not suitable for calculating the diffractive field. The Stratton-Chu formulas are one of the most powerful tools for the analysis of electromagnetic radiation problems. So, they can be credibly used to determine the diffractive field, and rewritten as (Stratton, 1941)

$$\begin{aligned}\vec{E}(\vec{r}) &= \iint_S \left\{-i\omega\mu\left[\vec{n}\times\vec{H}(\vec{r}')\right]G_0(\vec{r},\vec{r}') + \left[\vec{n}\times\vec{E}(\vec{r}')\right]\times\nabla' G_0(\vec{r},\vec{r}') + \left[\vec{n}\bullet\vec{E}(\vec{r}')\right]\nabla' G_0(\vec{r},\vec{r}')\right\} dS' \\ \vec{H}(\vec{r}) &= \iint_S \left\{i\omega\varepsilon\left[\vec{n}\times\vec{E}(\vec{r}')\right]G_0(\vec{r},\vec{r}') + \left[\vec{n}\times\vec{H}(\vec{r}')\right]\times\nabla' G_0(\vec{r},\vec{r}') + \left[\vec{n}\bullet\vec{H}(\vec{r}')\right]\nabla' G_0(\vec{r},\vec{r}')\right\} dS'\end{aligned} \tag{28}$$

where $\vec{r} = (\rho, z)$, $\vec{r}\,' = (\rho', 0)$, $G_0(\vec{r}, \vec{r}') = e^{-ik|\vec{r}-\vec{r}'|} / (4\pi|\vec{r} - \vec{r}'|)$, is the scalar Green's function. Let us assume that on the incident plane ($z' = 0$) we have a J_0 beam and a Gaussian beam, polarized in the x direction and propagating in the z direction. They are expressed in the following forms, respectively.

$$\begin{cases} \vec{E}(\vec{r}\,') = \vec{x} J_0(k_\perp \rho'), & \vec{H}(\vec{r}\,') = \vec{y} J_0(k_\perp \rho')/\eta \quad \text{Bessel beam} \\ \vec{E}(\vec{r}\,') = \vec{x} \exp(-\rho'^2/w_0^2), & \vec{H}(\vec{r}\,') = \vec{y} \exp(-\rho'^2/w_0^2)/\eta \quad \text{Gaussian beam} \end{cases} \tag{29}$$

where $\eta = \sqrt{\mu/\varepsilon}$. When $\rho = 0$ in (28), the axial intensity distribution $I_A(0, z)$ can be obtained by substituting (29) into (28)

$$I_A(0, z) = \left|\vec{E}(\vec{r})\right|^2 = \left|\vec{E}(0, z)\right|^2 \tag{30}$$

In addition, at a certain plane $z = z_f$, the transverse intensity distribution $I_T(\rho, z_f)$ can also be calculated by

$$I_T(\rho, z_f) = \left|\vec{E}(\vec{r})\right|^2 = \left|\vec{E}(\rho, z_f)\right|^2 \tag{31}$$

Now, we propose our contrast criterion: the same initial total power and central peak intensity on the same initial aperture. In order to compare conveniently, we also defined the propagation distance as the value of z-axis at which the axial intensity falls to $1/2$. Three cases are presented herein, where the same parameters are $\lambda = 3mm$ and $R = 10\lambda$. In the first example, their intensity distributions for a J_0 beam and a Gaussian beam on the initial aperture are shown in Fig. 22(a). Using (30), we get the axial intensity distributions along z-axis, as illustrated in Fig. 22(b). For the purpose of observing the propagation process, Figs. 22(c)-22(f) display the transverse intensity distributions at $z = 10\lambda, 20\lambda, 30\lambda, 40\lambda$, respectively. From this instance, A conclusion can be easily reach that the propagation distance of the J_0 beam is greater than that of the Gaussian beam, under the condition of the same initial total power and central peak intensity on the same initial aperture. In order to further confirm our conclusion, the other two examples are presented in Fig. 23 and Fig. 24. Apparently, a similar conclusion can be drawn from Figs. 23 and 24.

From Figs. 22(b), 23(b) and 24(b), we can also observe that the axial intensity distributions of Bessel beams oscillate more acutely than those of Gaussian beams. This is because the initial field distributions of Bessel beams near the edges of the aperture are much larger than those of Gaussian beams, and as a result, Bessel beam will suffer more diffraction on the sharp edges of the aperture than Gaussian beams.

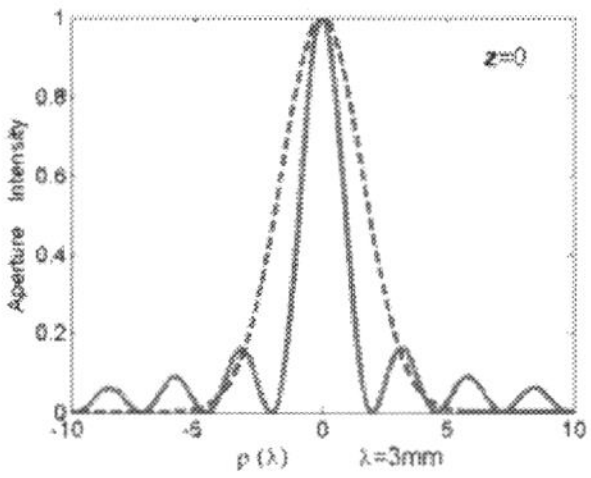

(a)

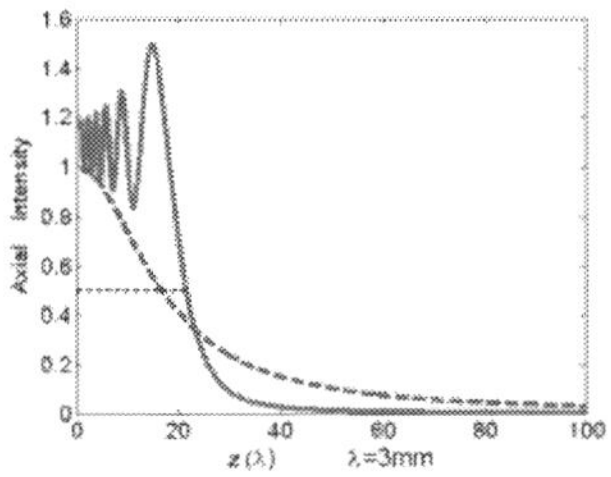

(b)

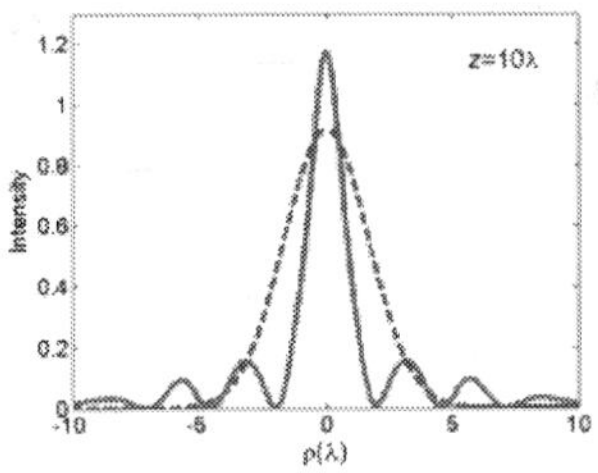

(c)

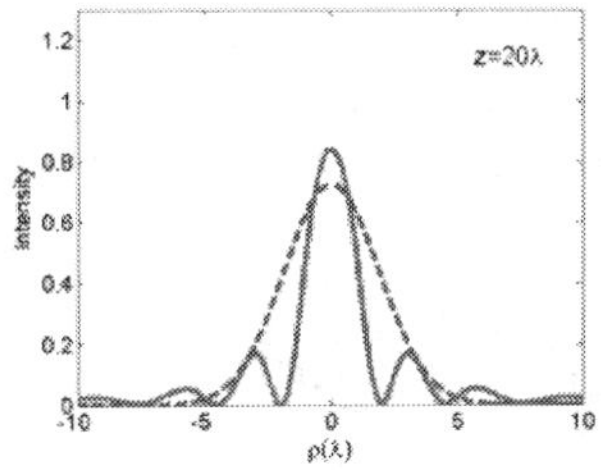

(d)

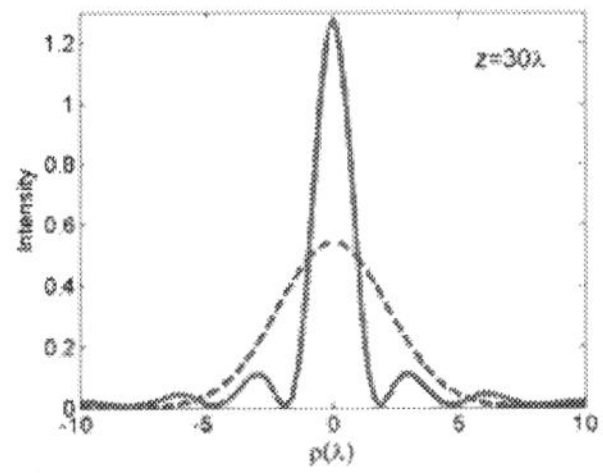

(e)

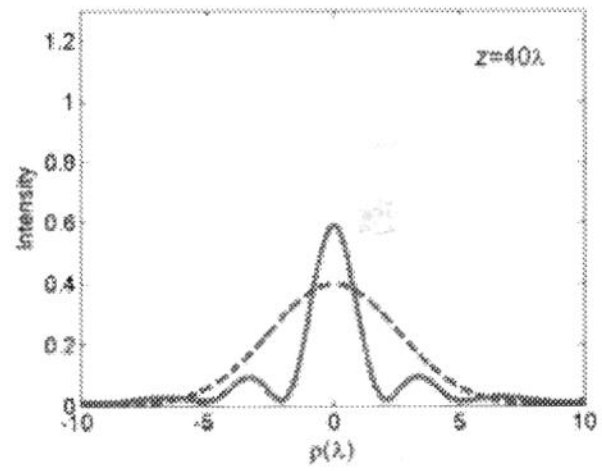

(f)

Fig. 22. The first case. (a) Intensity distributions for an apertured Bessel beam (−) and an apertured Gaussian beam (----) on the incident plane. (b) Axial intensities $I_A(0,z)$ versus propagation distance z . (c)-(f) Transverse intensity distributions at $z = 10\lambda, 20\lambda, 30\lambda, 40\lambda$, respectively.

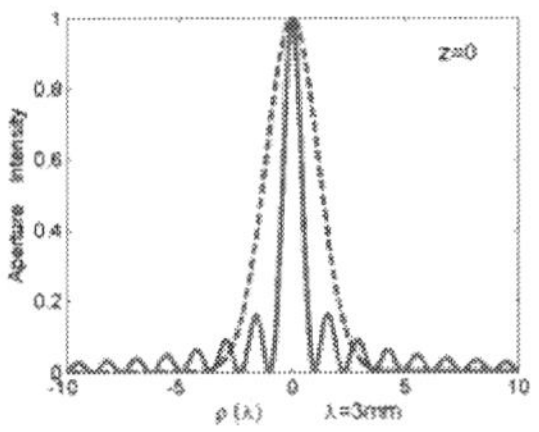

(a)

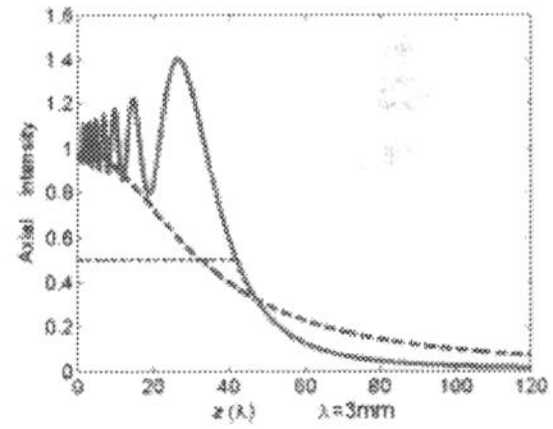

(b)

Fig. 23. The second case. (a) Initial Intensity distributions on the incident plane. (b) Axial intensities $I_A(0,z)$ versus propagation distance z .

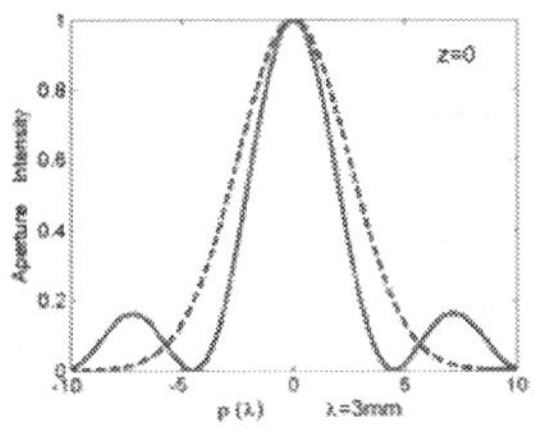

(a)

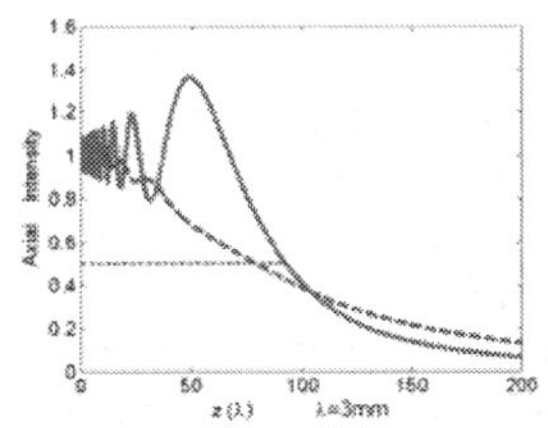

(b)

Fig. 24. The last case. (a) Intensity intensity distributions on the apture. (b) Their propagation distance.

6. Applications and Conclusions

The novel properties of the diffraction-free Bessel beams have many significant applications (Bouchal, 2003). In optics, due to the propagation invariance and extremely narrow intensity profile, Bessel beams are applicable in metrology for scanning optical systems. These beams are also suitable for large-scale straightness and measurements (Wang et al., 2003), since they can stand the atmospheric turbulence more than other beams. The imaging applications of the diffraction-free Bessel beams are also presented in (Li & Aruga, 1999), and it has been demonstrated that the imaging produced by Bessel beams can provide a longer focal depth when compared with Gaussian beams. Bessel beams can also be applied to optical interconnection and promotion of free electron laser gain (Li et al., 2006). An increasing attention is devoted to the applications of Bessel beams in nonlinear optics. The third-harmonic generation using Bessel beams was proposed by Tewari and co-workers (Tewari et al., 1996). In addition, Cerenkov second-harmonic generation by nondiffracting Bessel beams in bulk optical crystals was also suggested in (Pandit & Payne, 1997). The application of Bessel beams to increase the Z-scan sensitivity in measurement was demonstrated in (Hughes & Burzler, 1997). The radially polarized Bessel beams were applicable to accelerate the particles of the electron beam (Tidwell et al., 1992). Recently, Bessel beams are used to manipulate micrometer-sized particles. Using the self-reconstruction property of Bessel beam, it is possible to manipulate tiny particles simultaneously in multiple planes (Hegner, 2002; Garces-Chavez et al., 2002). We believe that Bessel beams are prospective for improving the resolution of images in millimeter wave imaging system (Monk, 1999). These beams may also be useful for measurements and power transmission at millimeter and sub-millimeter wavebands.

In optical region of the spectrum, diffraction-free Bessel beams have attracted much interest over the years and have been widely investigated. However, in the millimeter and sub-millimeter wave regions, exploratory work devoted to this field is much less. So, a great deal of contribution should be made to this field in the future.

7. Acknowledgments

This work is supported by NSFC under grant 60621002, the Natural Science Foundation of Fujian Province of China (No.A0610027) and the Key Project of Quanzhou City Science and Technology Program (No.2008G13).

8. References

Arlt, J. & Dholakia, K. (2000). Generation of high-order Bessel beams by use of an axicon. *Opt. Commun.*, Vol. 177, pp. 297-301

Bouchal, Z.; Wagner, J. & Chlup, M. (1998). Self-reconstruction of a distorted nondiffracting beam. *Opt. Commun.*, Vol. 151, No. 4-6, pp. 207–211

Bouchal, Z. (2003). Nondiffracting optical beams: physical properties, experiments, and applications. *Czech. J. phys.*, Vol. 53, No. 7, pp. 537-624

Cox, A. J. & Dibble, D. C. (1992). Nondiffracting beam from a spatially filtered Fabry-Perot resonator. *J. Opt. Soc. Am. A*, Vol. 9, No. 2, pp. 282–286

Cong, W. X.; Chen, N. X. & Gu, B. Y. (1998). Generation of nondiffracting beams by diffractive phase elements. *J. Opt. Soc. Am. A*, Vol. 15, No. 9, pp. 2362–2364

Courtial, J.; Whyte, G.; Bouchal, Z.; et al. (2006). Iterative algorithms for holographic shaping of non-diffracting and self-imaging light beams. *Opt. Express*, Vol. 14, No. 6, pp. 2108–2116

Durnin, J. (1987). Exact solutions for nondiffracting beams. I. The scalar theory. *J. Opt. Soc. Am. A*, Vol. 4, No. 4, pp. 651–654

Durnin, J.; Miceli J. J., Jr. & Eberly, J. H. (1987). Diffraction-free beams. *Phys. Rev. Lett.*, Vol. 58, No. 15, pp. 1499-1501

Feng, D.; Yan, Y. B.; Jin, G. F.; et al. (2003). Rigorous electromagnetic design of finite-aperture diffractive optical elements by use of an iterative optimization algorithm. *J. Opt. Soc. Am. A*, Vol. 20, No. 9, pp. 1739-1745

Garces-Chavez, V.; Mcgloin, D. & Melvill, E. H. (2002). Simultaneous micromanipulation in multiple planes using a self-reconstructing light beam. *Nature*, Vol. 419, No. 6903, pp. 145–147

Haupt, R. L. (1995). An introduction to genetic algorithms for electromagnetics. *IEEE Antennas Propag. Mag.*, Vol. 37, No. 2, pp. 7-15

Hirayama, K.; Glytsis, E. N. & Gaylord, T. K. (1996). Rigorous electromagnetic analysis of diffractive cylindrical lenses. *J. Opt. Soc. Am.* A, Vol. 13, No. 11, pp. 2219–2231

Hughes, S. & Burzler, J. M. (1997). Theory of Z-scan measurements using Gaussian-Bessel beams. *Phys. Rev. A*, Vol. 56, No. 2, pp. R1103–R1106

Hegner, M. (2002). Optics: The light fantastic. *Nature*, Vol. 419, No. 6903, pp. 125–127

Jiang, Z. P.; Lu, Q. S. & Liu, Z. J. (1995). Propagation of apertured Bessel beam. *Appl. Opt.*, Vol. 34, No. 31, pp. 7183-7185

Li, S. W. & Aruga, T. (1999). Long focal depth imaging over a long range. *J. Commun. Res. Lab.*, Vol. 46, No. 3, pp. 309–310

Li, D.; Imasaki, K.; Miyamoto, S.; et al. (2006). Conceptual design of Bessel beam cavity for free-electron laser. *Int. J. Infrared Millim. Waves*, Vol. 27, No. 2, pp. 165-171

McLeod, J. H. (1954). The axicon: a new type of optical element. *J. Opt. Soc. Am*, Vol. 44, No. 8, pp. 592–597

MacDonald, R. P.; Boothroyd, S. A.; Okamoto, T.; et al. (1996). Interboard optical data distribution by Bessel beam shadowing. *Opt. Commun.*, Vol. 122, No. 4-6, pp. 169–77

Miceli J. J., Jr. & Eberly, J. H. (1988). Comparison of Bessel and Gaussian beams. *Opt. Lett.*, Vol. 13, No. 2, pp. 79-80

Monk, S.; Arlt, J.; Robertson, D. A.; et al. (1999). Generation of Bessel beams at millimetre-wave frequencies by use of an axicon. *Opt. Commun.*, Vol. 170, pp. 213-215

Meltaus, J.; Salo, J.; Noponen, E.; et al. (2003). Millimeter-wave beam shaping using holograms. *IEEE Trans. Microwave Theory Tech.*, Vol. 51, No. 4, pp.1274-1279

Mahon, R. J.; Lanigan, W.; Murphy, J. A.; et al. (2005). Novel techniques for millimeter wave imaging systems operating at 100GHz. *Proc. SPIE. Int. Soc. Opt. Eng.*, Vol. 5789, pp. 93-100

McGloin, D. & Dholakia, K. (2005). Bessel beams: diffraction in a new light. *Contemp. Phys.*, Vol. 46, No. 1, pp. 15–28

Pandit M. K. & Payne, F. P. (1997). Cerenkov second-harmonic generation by nondiffracting Bessel beams in bulk optical crystals. *Opt. Quantum Electron.*, Vol. 29, No. 1, pp. 35–51

Stratton, J. A. (1941). *Electromagnetic Theory*. McGraw-Hill, NewYork

Sprangle P. & Hafizi, B. (1991). Comment on nondiffracting beams. *Phys. Rev. Lett.*, Vol. 66, No. 6, pp. 837

Scott, G. & McArdle, N. (1992). Efficient generation of nearly diffraction-free beams using an axicon. *Opt. Eng*, Vol. 31, No. 12, pp. 2640-2643

Salo, J.; Meltaus, J.; Noponen, E.; et al. (2001). Millimetre-wave Bessel beams using computer holograms. *Electron. Lett.*, Vol. 37, No. 13, pp. 834–835

Tidwell, S. C.; Ford, D. H. & Kimura, W. D. (1992). Transporting and focusing radially polarized laser beams. *Opt. Eng.*, Vol. 31, No. 7, pp. 1527–1531

Tewari, S. P.; Huang, H. & Boyd, R. W. (1996). Theory of third-harmonic generation using Bessel beams, and self-phase-matching. *Phys. Rev. A*, Vol. 54, No. 3, pp. 2314–2325

Turunen, J.; Vasara, A. & Friberg, A. T. (1988). Holographic generation of diffraction-free beams. *Appl. Opt.*, Vol. 27, No. 19, pp. 3959–3962

Thewes, K.; Karim, M. A. & Awwal, A. A. S. (1991). Diffraction free beam generation using refracting systems. *Opt. Laser Technol.*, Vol. 23, No. 2, pp. 105–108

Trappe, N.; Mahon, R.; Lanigan, W.; et al. (2005). The quasi-optical analysis of Bessel beams in the far infrared. *Infrared Phys. Technol.*, Vol. 46, pp. 233–247

Weile, D. S. & Michielssen, E. (1997). Genetic algorithm optimization applied to electromagnetics: A review. *IEEE Trans. Antennas Propag.*, Vol. 45, No. 3, pp. 343-353

Wang, K.; Zeng, L. & Yin, Ch. (2003). Influence of the incident wave-front on intensity distribution of the nondiffracting beam used in large-scale measurement. *Opt. Commun.*, Vol. 216, No. 1-3, pp. 99–103

Yee, K. S. (1966). Numerical solution of initial boundary value problems involving Maxwell equations in isotropic media. *IEEE Trans. Antennas Propag.*, Vol. AP-14, No. 3, pp. 302-307

Yu, Y. Z. & Dou, W. B. (2008a). Scalar and vectorial properties of diffraction-free Bessel beams. *2008 Asia Pacific Microwave Conference*, APMC2008, Hongkong & Macau, China.

Yu, Y. Z. & Dou, W. B. (2008b). Vector analyses of nondiffracting Bessel beams. *Progress In Electromagnetics Research Letters*, Vol. 5, pp. 57–71

Yu, Y. Z. & Dou, W. B. (2008c). Generation of Bessel beams at mm- and submm-wave bands using binary optical elements. *2008 Global Symp. on Millimeter Waves Proc.*, Nanjing, China, pp. 115-118.

Yu, Y. Z. & Dou, W. B. (2008d). Generation of Bessel beams at mm- and sub mm-wavelengths by binary optical elements. *Int. J. Infrared Milli. Waves*, Vol. 29, No. 7, pp. 693–703

Yu, Y. Z. & Dou, W. B. (2008e). Comparison of propagation of apertured Bessel and Gaussian beams. *2008 Int. Conf. Commun. Technol. Proc.*, Hangzhou, China, pp. 217–220

Yu, Y. Z. & Dou, W. B. (2009). Generation of pseudo-Bessel beams at THz frequencies by use of binary axicons. *Opt. Express*, Vol. 17, No. 2, pp. 888-893

25

Receiver Front-End Architectures – Analysis and Evaluation

Pedro Cruz[#], Hugo Gomes[#*] & Nuno Carvalho[#]
[#]*Universidade de Aveiro (Instituto de Telecomunicações)*
Aveiro, Portugal
**Instituto Politécnico de Leiria (ESTG)*
Leiria, Portugal

1. Introduction

In today's world, the exponential growth from communications between people/companies in different places (at same time), the increasing requirement to measure and control all processes, the analysis in real time, the mandatory requirement to provide of information and entertainment data to electronic devices that must be increasingly smaller and more complex, requires a continuous and nonstop searching for new technologies with greater capacity, lower cost, reduced size and improved reliability.
The communications systems based on radio-frequency (RF) transmission are one of the greatest examples of this challenging demand. These systems, present in almost all equipment used in daily life as mobile phones, notebooks, wireless sensors, among other, require an increasing versatility and ability to storage of data, huge transmission rates of information and size reduction.
All this need for more sophisticated equipment, along with a greater number of services available in a single equipment (preferably portable), besides the drastic size reductions from the electronic components, requires a constant search for new architectures and new materials in order to maximize the features offered.
One of the most important parts from a RF system device is this receiver architecture. In receivers, the entry block has a key role in performance and reliability of the system. Any unresolved issue caused by this block, generates enormous problems in the following blocks of the receiver's architecture. For this reason, considering the constant increase of services available for the same frequency bands, associated with the growing number of users for each service, the entry receiver architecture must be capable to resolve issues such as blocking problems, peak-to-average power ratio (PAPR) problems, among others. In other hand, must be capable to offer good selectivity, sensitivity, lower energy consumption for a small price.
This chapter is organized in the following way. Firstly, a general review about the most common receiving architectures is done, emphasizing its main advantages and drawbacks. Moreover, some enhancements to these architectures are also presented and its principal benefits are explained, such as Hartley and Weaver configurations. This section ends with

some considerations about the implementation of adaptable wideband architectures and multi-standard operation. In the following section several interference problems as blocking and PAPR problems will be analyzed. Furthermore, a few techniques of PAPR reduction are overviewed to receiver application. After that, two possible application fields of these concepts are addressed, wherein two projects are shown regarding the radio-frequency identification (RFID) and software-defined radio (SDR) systems. Finally, the concluding remarks are drawn.

2. Review of Receiver Front-Ends Architectures

This chapter is intended to make a review of the main receiver's architectures known, show the main applications and study their main advantages and limitations (Besser & Gilmore, 2003), (Razavi, 1997), (Razavi, 1998).

2.1 Super-Heterodyne Receiver

The most common configuration used in RF receivers is the well known super-heterodyne architecture (Fig. 1). This configuration is based in two down-conversion stages, i.e., the RF received signal is first demodulated to an intermediate frequency (IF) and then converted to baseband signal. The received signal (Fig. 2a) is first filtered by a pre-selection filter and (after amplified by the low-noise amplifier, (LNA)) passes through another filter to reduce the image frequency effects before the first translation from RF to IF (Fig. 2b e 2c). After this stage, the signal is again filtered and demodulated to baseband (Fig. 2d), where it is converted to the digital domain where it can be processed. In this stage some architectures make an I/Q modulation in order to achieve better amplitude/phase information from the signal received.

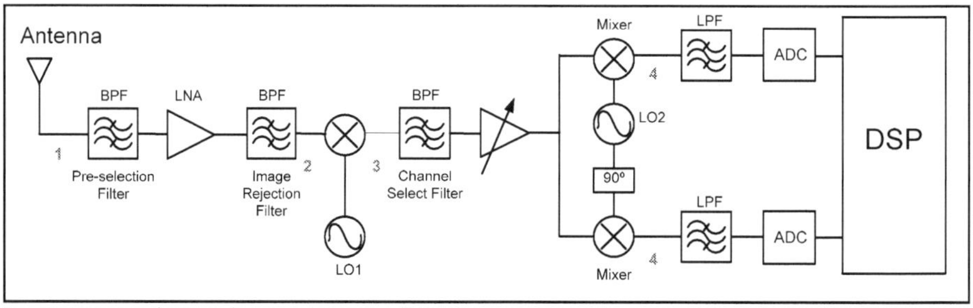

Fig. 1. A super-heterodyne receiver architecture

As referred above, this architecture is currently adopted in most radio receivers due to the availability of low-cost narrowband RF and IF components with low power consumption. Furthermore, this architecture can ensure good levels of sensitive (allows lower power signal at receiver input for which there is sufficient signal-to-noise ratio at the receiver output), selective (better ability to separate the desired band from signals received at other frequencies) and is immune to most DC problems affecting homodyne architecture.

However, super-heterodyne receivers have a number of substantial problems. The most important problem in this architecture is the cancellation from the image frequency. For a

good signal/image ratio is imperative that the image rejection filter has reduced transition band. To achieve this goal these filters must be performed by high-Q discrete components (SAW or ceramic filters), unpractical in today's IC technologies. For this reason, is not possible a full integration on-chip, resulting problems like perfect LNA 50 Ω load, noise figure and non-linear behaviour in discrete components. Another way to solve image frequency problem is the use of cancellation architectures such as Hartley or Weaver, presented in section 2.5.
Despite its greater complexity, the fact that it is designed for a specific channel (in a particular wireless standard) prevents the expansion of the receiving band.

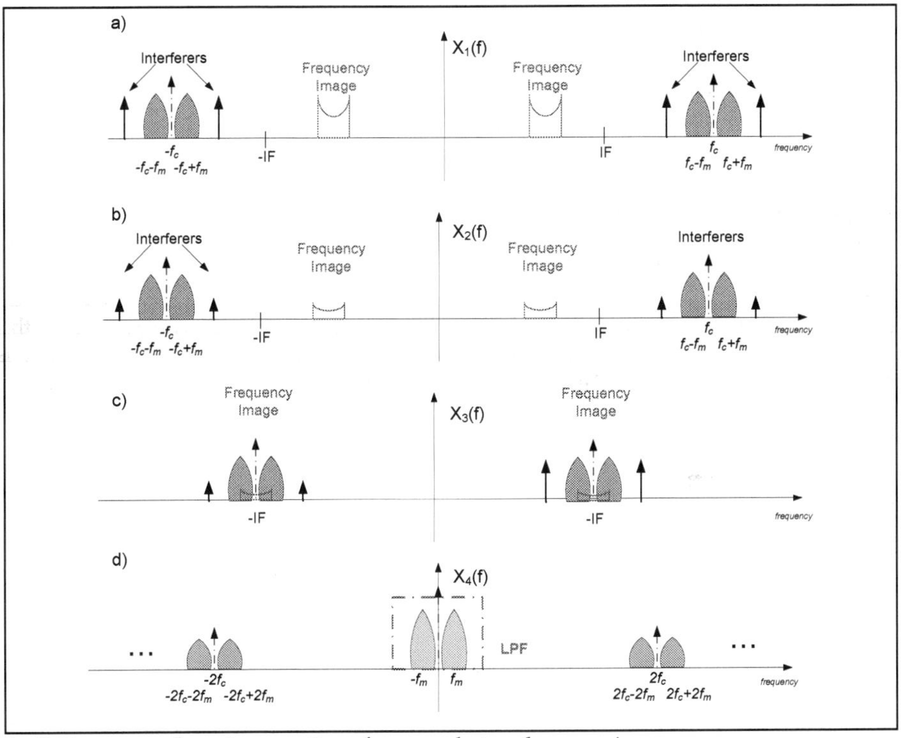

Fig. 2. Frequency domain operation of a super-heterodyne receiver

2.2 Zero-IF Receiver

Another typical receiver's architecture is the zero-IF receiver (Park et al., 2006), also known as homodyne receiver (Fig. 3). This architecture is a simplified version of the super-heterodyne, because instead of two down-conversion stages, it converts the RF signal directly to baseband. The received signal (Fig. 4a) is selected at RF by a band-pass filter, and then it is amplified by an LNA, as in the previous architecture (Fig. 4b). Finally, it is directly down converted to DC by a mixer (or two mixers with a delay of 90° between them) and converted to the digital domain using a straightforward analogue-to-digital converter

(ADC), (Fig. 4c). Compared to the last architecture this has a clear reduction in the number of analogue components and guarantees a high level of integration thanks to its simplicity.

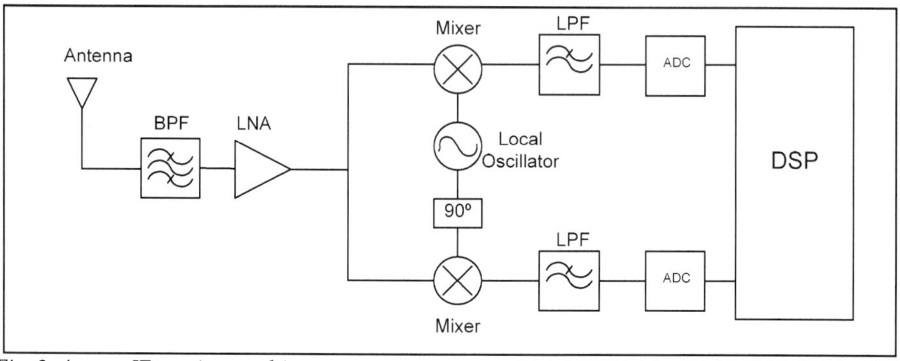

Fig. 3. A zero-IF receiver architecture

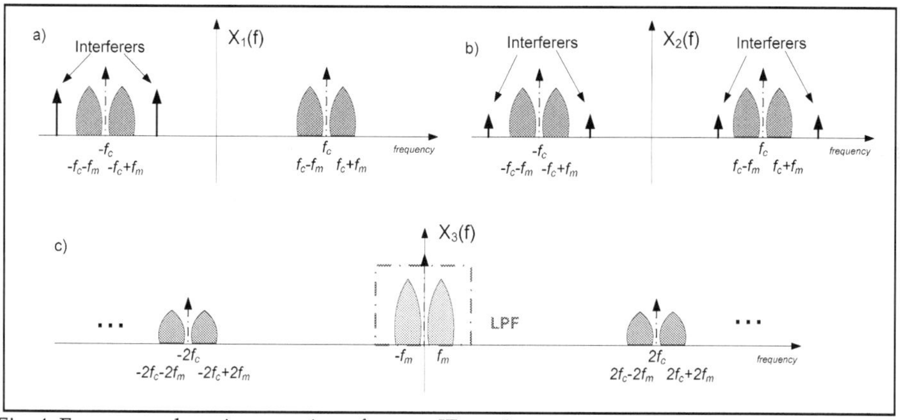

Fig. 4. Frequency domain operation of a zero-IF receiver

Despite its simplicity compared to the super-heterodyne architecture, many components of the zero-IF receiver are more complex to deploy. In addition the direct translation to DC can generate several problems that strongly conditioned the use of this architecture over the super-heterodyne. Problems as DC offset, such as local oscillator (LO) leakage due to non-ideal isolation between the port from the mixer or interferer leakage due to non-ideal isolation between the port from the mixer, I/Q mismatch due to errors in I/Q modulation, Even/Odd-order distortion caused by non-linear components generated several products in harmonic frequencies, specially second-order intermodulation products that are generated around DC and large flicker noise of the mixer can easily corrupt the output signal since it is a baseband signal. Some techniques to reduce these problems associated with the increasing integration of the constituent components of the zero-IF architecture has contributed to better performance and increased use.

2.3 Low-IF Receiver

A similar configuration to the previous one is the low-IF receiver (Adiseno et al., 2002), Fig. 5, in which the RF signal is mixed down to a nonzero low or moderate IF (few hundred kHz to several MHz) instead of going directly to DC, using quadrature RF down-conversion. This solution tries to combine the advantages from the zero-IF receiver and the super-heterodyne receiver. Like zero-IF receiver, the received signal (Fig. 6a) passes through a channel-selection filter at RF and is amplified by a LNA (Fig 6b). After this similar step, the signal is down converted to a low IF, instead of zero IF (Fig. 6c), and used an image suppression block in order to cancel the negative effects from frequency image. Finally, an ADC converts the signal to digital domain, allowing the use of digital signal processing algorithms. In some low-IF architectures the image suppression block is transferred to the digital domain.

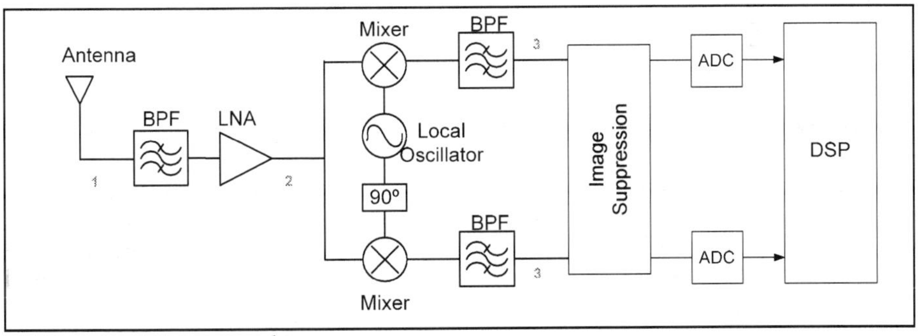

Fig. 5. A low-IF receiver architecture

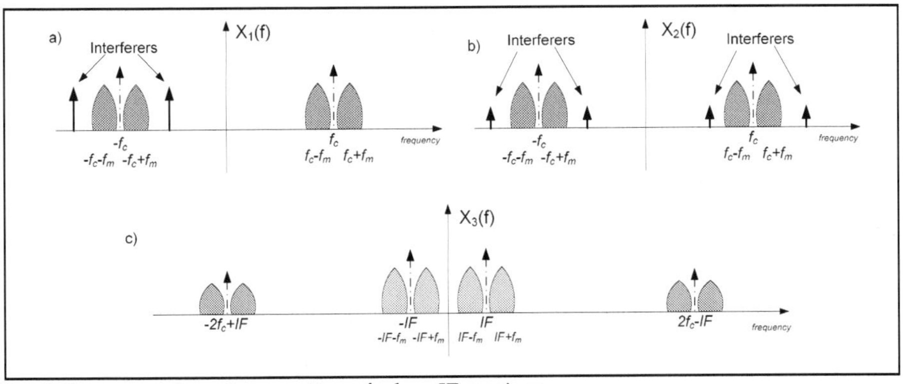

Fig. 6. Frequency domain operation of a low-IF receiver

This architecture still allows a high level of integration (advantage from zero-IF) but does not suffer from the DC problems (advantage from super-heterodyne), since the desired signal is not situated around DC. However, this architecture continued to suffer from the image frequency and I/Q mismatch problems (with greater impact than in previous architectures) and the ADC power consumption is increased since now a high conversion rate is required.

2.4 Band-Pass Sampling Receiver

An alternative to the previous configurations is the band-pass sampling receiver (Vaughan et al., 1991), (Akos et al., 1999), Fig. 7. In this architecture, the received signal is filtered by an RF band-pass filter that can be a tuneable filter or a bank of filters, and then it is amplified using a wideband LNA. The signal is then converted to the digital domain by a high sampling rate ADC and digitally processed. All the previous architectures use analogue circuitry to down-convert the incoming RF signal to I and Q baseband components, and obtaining good matching in analogue circuits is not easy. In that sense, this architecture considers firstly an analogue to digital conversion and then the I/Q demodulation in the digital domain. In this way the I/Q matching is an all digital task and obtaining sufficient matching accuracy in two digital signal paths is easily accomplished. One means to generate these digitally I/Q signals is to use the Hilbert transform (Tsui, 1995). Thus, as in the low-IF architecture, here we can take advantage of digital signal processing to alleviate some issues of the analogue front-end. Moreover, pushing the analogue-to-digital conversion closer to the antenna provides an increased flexibility.

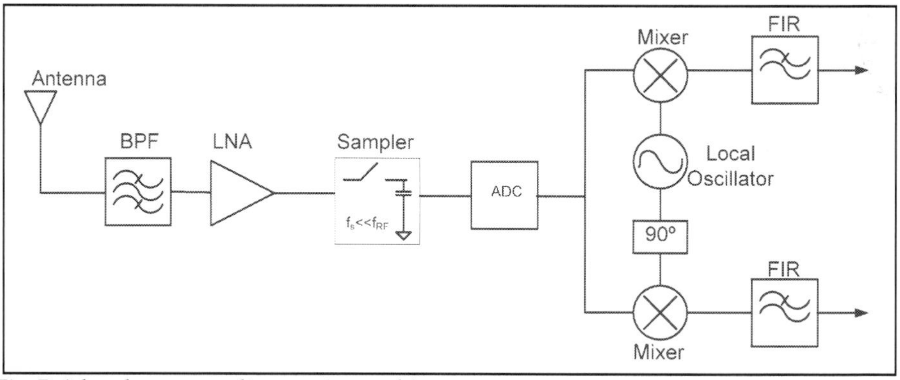

Fig. 7. A band-pass sampling receiver architecture

This configuration is based on the fact that all energy from DC to the input analogue bandwidth of the ADC will be folded back to the first Nyquist zone $[0, f_S/2]$ without any mixing down conversion needed because a sampling circuit is replacing the mixer module. In Fig. 8, is shown the frequency domain operation of the band-pass sampling receiver. Whether a correct sampling frequency is chosen, it is possible to receive more than one RF signal at same time and then make its processing in the digital domain. Nevertheless, it is mandatory to include RF band-pass filtering in order to avoid overlap of other undesirable signals.

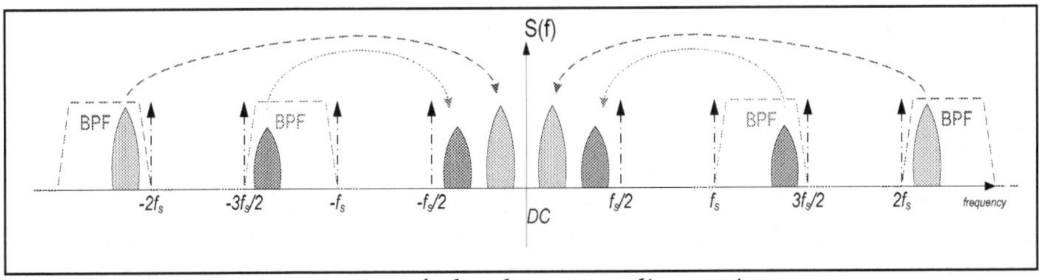

Fig. 8. Frequency domain operation of a band-pass sampling receiver

As was described in (Akos et al., 1999), it is possible to pinpoint the resulting intermediate frequencies, f_{IF}, based on the following relationship

$$if \quad fix\left(\frac{f_C}{f_S/2}\right) \quad is \quad \begin{cases} even, & f_{IF} = rem(f_C, f_S) \\ odd, & f_{IF} = f_S - rem(f_C, f_S) \end{cases} \tag{1}$$

where f_C is the carrier frequency, f_S is the sampling frequency, *fix(a)* is the truncated portion of argument *a*, and *rem(a,b)* is the remainder after division of *a* by *b*.
In this case, the RF band-pass signal filtering plays an important role because it must reduce all signal energy (essentially noise) outside the Nyquist zone of the desired frequency band that otherwise would be aliased. If not filtered, the signal energy (noise) outside the desired Nyquist zone is folded back to the first zone together with the desired signal, producing a degradation of the signal-to-noise ratio (SNR). This may be given by

$$SNR = 10.\log_{10}\left(\frac{S}{N_i + (n-1).N_0}\right) \tag{2}$$

where *S* represents the desired-signal power, N_i and N_0 are in-band and out-of-band noise, respectively, and *n* is the number of aliased Nyquist zones.
The advantage of this configuration is the sampling frequency needed and the subsequent processing rate are proportional to the information bandwidth, rather than to the carrier frequency. This reduces the number of components required. However, some critical requirements exist. For example, the analogue input bandwidth of the sample and hold circuit inside the ADC must include the RF carrier, which is a serious problem, considering the sampling rate of modern ADCs. Clock jitter can also be a vital problem when high frequencies implementations are considered.

2.5 Other Feasible Architectures

In addition to the main receiver architectures presented, there are other architectures / sub-architectures that are currently used or being developed. Some of these architectures are adaptations from older configurations that have been rearranged and use to very small circuits. As an example, the simple detector receiver (or envelope detector), used in the first AM radios, was reused for the development of very small tags (with very small consumption) on RFID systems or even as power meters. Other architectures are use as support solutions to the previous receiver architectures problems, such as Hartley and Weaver configurations.

This section thus seeks to make a brief presentation of these architectures and a general review of all the architectures presented main advantages and limitations.

The envelope detector configuration (Fig. 9) is the most simple receiver architecture used because does not need of problematic components like mixers and local oscillators and the topology is very simple and cheap. The down-conversion method used in this architecture is based in the strong nonlinear behaviour from the diode. An interesting property of nonlinear systems is the spectral regrowth capability, which means that the system has the capability of create frequency components in the output signal that do not exist at the input side. As a result of this propriety, the received signal will present at diode output several replicas from the original signal in the harmonics and baseband frequencies. With the help of a low pass filter it is possible to eliminate all undesired frequencies and achieve the down-conversion of the desired signal without the use of mixers and local oscillators.

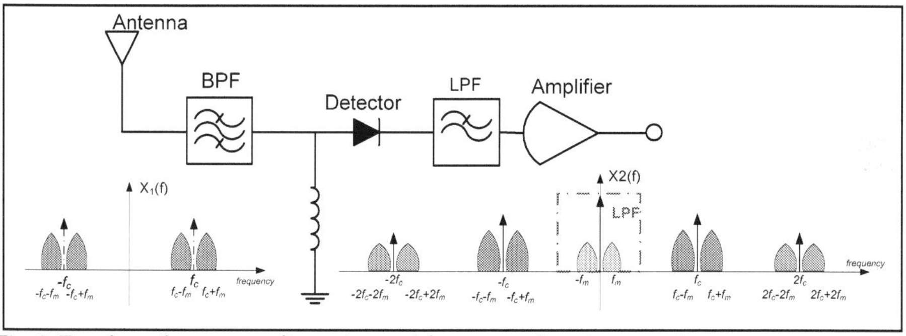

Fig. 9. Envelope detector configuration and frequency domain operation

This architecture is very useful in systems that require an extremely low energy consume such as passive RFID tags. The mains disadvantages of this receiver are the extreme intolerance to interferences, some DC problems (as zero-IF receiver) and very low sensitivity and selectivity, among others.

As was referred above, the Hartley and Weaver configurations are sub-architectures specially tailored to reduce/cancel the problems with the frequency image that affects strongly the super-heterodyne receiver. These architectures are similar but in last year's Weaver architecture have gained an advantage over the Hartley architecture, especially because of their better performance when integrated into IC.

The Hartley architecture (Fig. 10) uses an I/Q modulation in first stage and, after the low-pass filter, make a 90° time shift (Hartley shift) to invert the negative part from the spectrum. The sum of components I and Q results in the cancellation of the image frequency, and also strengthen the desired signal. The main advantages from this architecture are good image rejection ratio (IRR) and the immunity to load problems that results from the needless of high quality discrete components (needed in super-heterodyne). Although this configuration is very sensible to I/Q mismatch and the non-linear behaviour from the shift-by-90° and adder blocks can be extremely deteriorative for the output signal. The 90° shift block is also difficult to realize and its behaviour is severely affected by the variation of the discrete components used in its design.

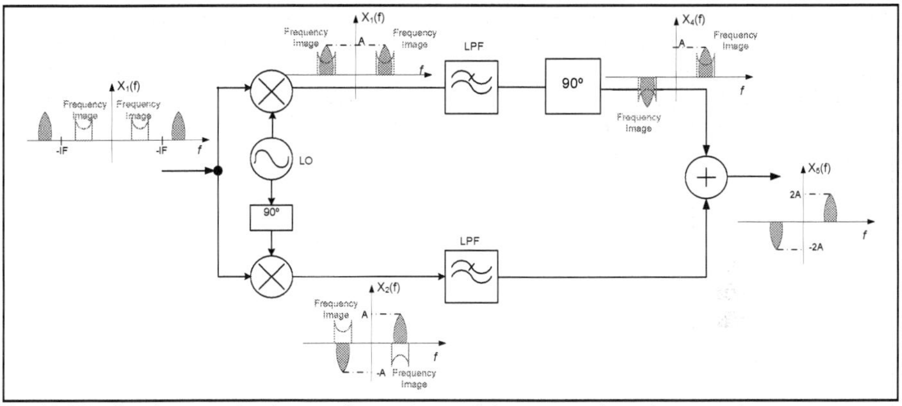

Fig. 10. Hartley image rejection configuration and frequency domain operation

The Weaver architecture is similar to Harley in first stage. Although, the shift-by-90° block is replaced by a second I/Q modulation. The two resulting signals can be subtracted and achieve the desired signal (with cancelation from the frequency image).

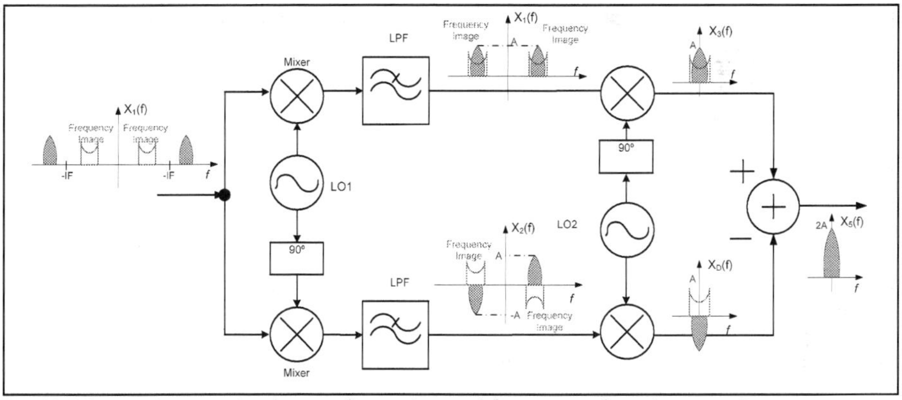

Fig. 11. Weaver image rejection configuration and frequency domain operation

The Weaver arrange has the advantage that do not depend from a difficult and no reliability shift-by-90° block and bring some flexibility to architecture because with a simple switch from the adder it is possible to achieve the desired signal or frequency image signal. With a most certain digitalization from several part from the receiver architectures, this configuration can ensure better results in technologies such as SDR. Nevertheless, the Weaver architecture has some limitations. Firstly, suffer the same I/Q mismatch problem as Hartley. Secondly, the huge number of mixers and LO amounts the energy consumption

and the cost of the topology. Finally, the temperature and process variation can significantly degrade the desired signal.
As a conclusion from the previous study of receiver architectures, in Table 1 are summarized the main advantages and major problems for each architecture.

Architecture	Advantages	Major Problems
Super-Heterodyne	- Selectivity - Sensitivity - Immune DC problems	- Image frequency - I/Q mismatch - High quality discrete components - Perfect LNA 50Ω load - Complexity - Noise figure - Nonlinear behaviour in components
Zero-IF	- Simplicity - IC integration	- Strong DC problems - I/Q mismatch - Even/Odd distortion - Flicker noise
Low-IF	- No DC problems - Simplicity - Less high quality discrete components	- I/Q mismatch - Image frequency - Requires high performance ADC
Band-Pass Sampling	- Flexibility - Signal manipulation - Low cost, circuit area - Minimize DC problems and RF problems (in digital domain)	- Susceptibility to clock aperture jitter - Noise figure degradation - Aperture distortion - Power consumption
Simple Detector	- Simplicity - Low cost	- Huge degradation with interferes - Low selectivity, sensitivity - Some DC problems
Hartley	- Good IRR - Less discrete components - Reduce load problems	- I/Q mismatch - Shift-by-90° block and adder - Variation of R and C in RC-CR network - Increased number of components
Weaver	- Similar to Hartley - Avoid RC-CR network	- Huge number of mixers - I/Q mismatch - Dependent VCO - Strong adjacent channel interferes - Increased number of components

Table 1. Comparison of previous receiver architectures

Other architectures being proposed for use in the actual and future receivers involve use of direct RF sampling techniques based on discrete-time analogue signal processing to receive the signal, such as the ones developed in (Staszewski et al., 2004), (Muhammad et al., 2005). These methods are still in a very immature stage but should be further studied due to their potential efficiency in implementing reconfigurable receivers.

2.6 Path to Future Receivers

The holy grail of future RF architectures is that they will be able to receive any type of signal despite its bandwidth and dynamic range. Even though it is considered a holy grail, the path is moving towards multi-norm, multi-standard radios that are supported in SDR, and thus on that are capable of receiving a huge range of bandwidth combined with very different power levels, and thus dynamic range approaches.

This radios will move fast to be all digital, and completely defined by software. But, in order to achieve these master goals, much research and innovation still be needed. For instance in the receiving unit, the radio should have a very wide bandwidth ADC for gather and convert all the signals from analogue to digital, and this ADC should have a strong dynamic range associated, since it should receive low power signals combined with high power ones, and considering that if the radio has to receive several different signals, they should not combine each other.

In the transmitter side, the path is moving fast to the all digital PA, where the output signal is mainly a PWM modulated waveform that will traverse a switching amplifier, allowing an almost 100% efficiency, and the focus will be put on the output filter that will convert the digital waveform to analogue one, prior to feed the antenna.

Of course the last proposals are far from being possible, but researchers are moving toward there slowly.

3. Sources of Interference

In this section, we will give a detailed explanation about the complex problem related to interference between the strong transmitted signal and the weak received signal desired. This phenomenon is known as blocking. Moreover, we will analyze the impact of the signal peak-to-average power ratio, PAPR, in the receiving architectures. Another important point to be evaluated is the complexity increase that comes from the multi-standard implementation of the addressed receiving architectures. Both of these problems will provoke a limitation in the receiver's dynamic range. Moreover, some PAPR reduction techniques that can be applied in the receiver side will be described.

3.1 Blocking Problem

The co-existence of various technologies in the wireless spectrum has always led to problems of interference between systems. The regulatory authorities (whether national or international) try to avoid those interferences with rules and restrictions to RF systems that coexist in the same frequency band or adjacent frequency bands. However, even following all these standards and specifications, the presence of interference in a particular system is unavoidable, usually leading to degradation (or loss) of its quality and functionality. The dynamic range of a receiver front-end can be defined as the ratio between the maximum tolerable signal to the minimum detectable signal. This fact advises us that a receiver must be able to detect the weak desired signal and also have sufficient dynamic range to not be blocked by an undesired strong signal.

Thus, considering any of the previous presented architectures, the first nonlinear component that appears in the receiving chain is the LNA. This component is always nonlinear, since for a certain input the output signal is no longer proportional, neither follows the superposition principle (Pedro & Carvalho, 2003).

So, if we assume that a certain system wants to receive a signal with a desired power of, for instance, -70 dBm but another system that is adjacent to this one is emitting an interfering signal with a power of about 50 dBm, we can clearly conclude that if any intermodulation products arise in the receiver chain it will completely corrupt the desired signal.
In order to better understand this problem we will consider the previous cited example and analyze that in a detailed way. Therefore, the two signals are received and the interference signal will be attenuated by the receiver input filter, for which we have attributed an out-of-band rejection of around 50 dB. Although, a cascade of filters can be used to increase the out-of band attenuation, it will degrade the noise figure and also increase the insertion loss. That way, the interfering signal will be attenuated by around 50 dB and arrive to the input of the LNA with a power of 0 dBm. In Fig. 12 we can see a possible frequency domain representation of this situation.

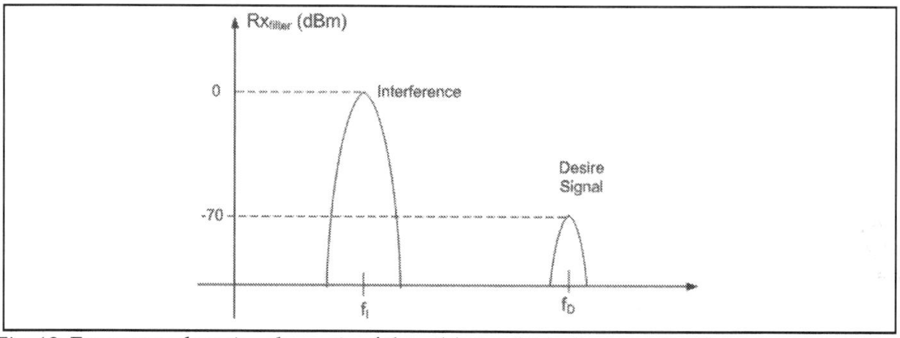

Fig. 12. Frequency domain schematic of the addressed interference situation

Thus, to study the possible intermodulation products in the example shown we will assume that the LNA is characterized by the following transfer function:

$$y_{LNA}(t) = a_1 x_{LNA}(t) + a_2 x_{LNA}{}^2(t) + a_3 x_{LNA}{}^3(t) \tag{3}$$

where x_{LNA} is the signal at the input port of the LNA, y_{LNA} is the output signal of the LNA, and a_1, a_2, and a_3 are model products coefficients.
Then, supposing that these two signals are two simple sinusoids, expression (4), we came up with a huge number of components in the outside of the LNA.

$$x_{LNA}(t) = A_1 \cos(w_1 t + \phi_1) + A_2 \cos(w_2 t + \phi_2) \tag{4}$$

where A_1 and A_2 are the amplitude components of interference and desire signal, and w_1 and w_2 are the respective frequency values.
Fig. 13 shows a complete overview of the frequency domain signal that is generated in the output of the LNA. If we look carefully to the third-order behaviour, we see that it will have a mixing product that falls exactly in the same frequency (w_2) of the desire signal. While the second-order and some of third-order intermodulation products can be eliminated by a filter that follows the LNA, there are the in-band third-order intermodulation products that

cannot be filtered out. Because we are assuming a very strong interference, A_1 is much higher than A_2, even if the LNA had very small intermodulation products (very low a_3), the interfering signal will continue to strongly affect the signal reception. As a result, development of techniques and sub-systems able to cancel, or at least mitigate, these adverse effects on radio receivers is of extreme importance.

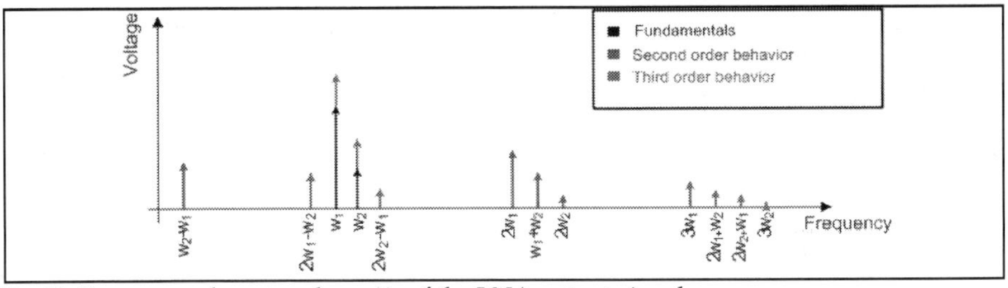

Fig. 13. Frequency domain schematic of the LNA output signal

3.2 PAPR Problem

PAPR is known since several years until now, actually the first known reference to this problem was made in (Landon, 1936), when he was studying the noise characteristics. The concept of PAPR is the relationship between the maximum value of the peak power and the average power of a given signal, and is a measure of great interest in actual wireless communications signals. It can be used for RF signal evaluation as well as for baseband signal evaluation (Bauml et al., 1996). Normally, this figure of merit is given in decibels and is calculated by using equation (5), where NT represents the number of samples considered for the PAPR evaluation.

$$PAPR_{dB} = 10.\log_{10}\left(\frac{\max\limits_{0\le t\le NT}|x(t)|^2}{\underset{0\le t\le NT}{mean}|x(t)|^2}\right) \tag{5}$$

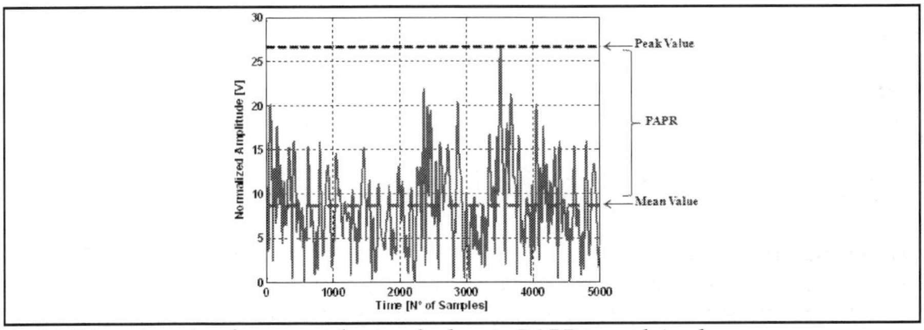

Fig. 14. Time-domain of an example signal wherein PAPR is explained

The previous figure shows an explanation of the PAPR calculation for a given example signal. There we can observe that a number of 5000 samples were used to determine the respective peak and mean values and in that way provide an estimate of the PAPR value.
The evaluation of the impact of PAPR in the communication systems components is mainly made through the analysis of complementary cumulative distribution function (CCDF) curves (Agilent App. Note, 2000), in which we define a certain percentage in the CCDF curve to pinpoint the reached PAPR value. The CCDF curves are closely related to the probability density function (PDF) of the signal because they are obtained by means of CCDF = 1 - CDF, where CDF is the cumulative distribution function that is obtained directly from the PDF statistics, as shown in equation (6):

$$cdf(x) = \int_{-\infty}^{a} pdf(x)dx \tag{6}$$

These curves provide a statistical description of the power levels in the signal and show how much time the signal spends at or above a certain power level. In the y-axis is represented the percent of time the signal power is at or above the power specified by the x-axis. An example of a CCDF curve is given in Fig. 15 using the previous presented signal. Analyzing the figure it is possible to affirm that the signal power exceeds the average by around 6 dB for 1% of the time (dashed red line) and also reach almost 8 dB for 0.01% of the time (dashed-point green line).

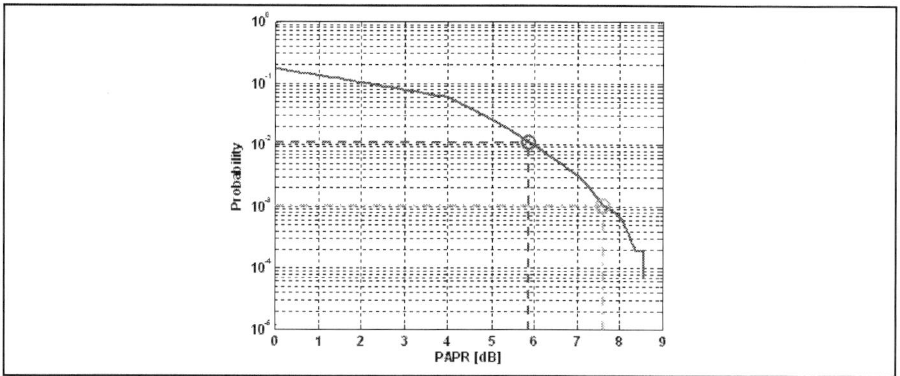

Fig. 15. CCDF curve for the previous presented example signal

Moreover, in actual wireless communications solutions, such as WiMAX (Worldwide Interoperability for Microwave Access) or 3GPP-LTE (Long Term Evolution), the receiving stages should deal simultaneously with a significant number of different signals due to the fact that they use an adaptive coded modulation algorithm (Goldsmith & Chua, 1998), which selects the best digital modulation format to apply based in a specific condition, for instance, SNR value. These digital modulation formats could vary from simple modulations as binary phase-shift keying (BPSK) to more complex formats as 64-QAM (quadrature amplitude modulation). Looking at Fig. 16, we can detect that these different modulation formats will produce signals with slightly different PAPR values. This fact will impose different restrictions to the receiving components projected.

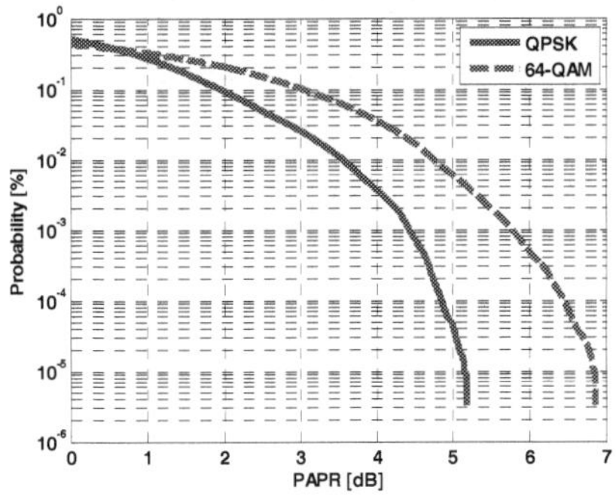

Fig. 16. CCDF curves for different signal modulation formats

Thus, it is mandatory to develop receiver configurations that can be capable to deal with the highest PAPR signal in order to not incite distortion generation. Adding to that fact, most of today standards has adopted the orthogonal frequency division multiplexing (OFDM) due to its spectral efficiency and capability to transmit high data rates over broadband radio channels. The main disadvantage of this technique is that it exhibits a high PAPR, which can be up to N times the average power (where N is the number of carriers).

So, the conjugation of both digital vector modulation formats and OFDM based schemes will lead to high values of PAPR, which limits the power that can be received or transmitted without distortion. This could lead to a necessity of using circuits with linear characteristics within a large dynamic range, otherwise the signal clipping at high levels would yield a distortion of the received signal. In fact, this PAPR problem immediately degrades the quality of the received signal because it will impose a degradation of the signal-noise ratio, SNR, in receiver ADC's accordingly to (7). If we allow the clipping of the signals peaks, then immediately the nonlinear distortion raises (Cruz et al., 2008).

$$SNR_{dB} = 6.02N + 1.76 - \alpha + 10.\log_{10}(2.OSR) \tag{7}$$

where N is the number of bits, a the PAPR and OSR the over-sample ratio.

One possible solution to this problem is to use crest factor minimization techniques (addressed in the following section), but in that case a care should be taken in order not to degrade either the nonlinear distortion, neither the error vector magnitude of the received symbols.

At the same time, a much more complex problem appears when we take into account the multi-carrier and multi-standard receiver or transmitter operation. Regarding the multi-carrier receiver operation, for instance, GSM (Global System for Mobile Communications) and W-CDMA (Wideband-Code Division Multiple Access) use a large number of channels and if we try to receive several channels at same time using a wideband receiver we will suffer from a problem created by the high PAPR of this multi-channel signal. In Fig. 17 is shown the value of PAPR of a GSM signal using one and three channels, where the complexity of the three GSM channels significantly degrades the overall signal PAPR.

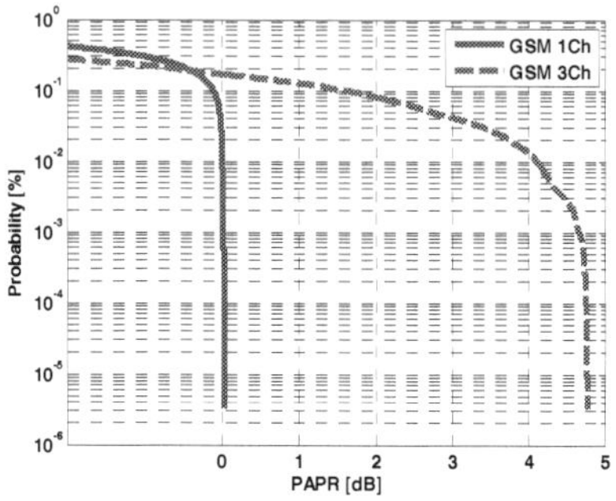

Fig. 17. CCDF curves for a 1-channel and 3-channels GSM signals

Concerning now on the multi-standard receiver operation we can also state that the PAPR of the multi-mode signals is always much higher than the single-mode configuration, as was demonstrated in (Cruz & Carvalho, 2008). Thus, in order to assess the complexity that we will be dealing with several combinations of usually used wireless standards were produced (using an arbitrary waveform generator), such as Wi-Fi and WiMAX (since they both work at near 2.4/2.5 GHz) and GSM-1800 with W-CDMA. In Table 2 are presented the relevant characteristics of the generated signals.

Signal Type	Frequency	Bandwidth	Multiplexing	Modulation
IEEE 802.11g	2.45 GHz	22 MHz	OFDM	64-QAM
IEEE 802.16e	2.502 GHz	14 MHz	OFDM	64-QAM
W-CDMA	1.9 GHz	3.84 MHz	-	п/4-QPSK
GSM-1800	1.81 GHz	200 kHz	-	GMSK

Table 2. Characteristics of the generated signals

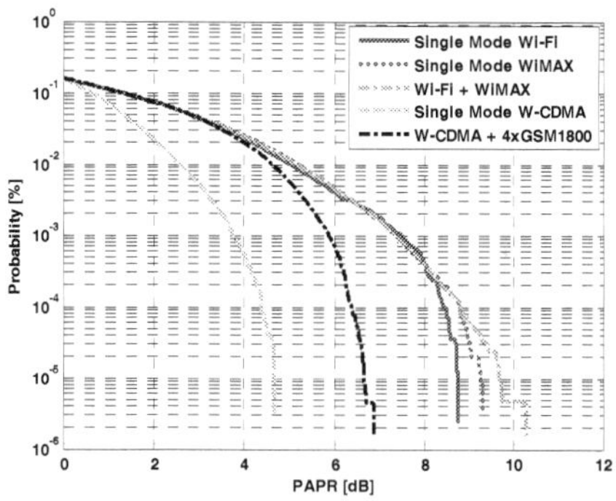

Fig. 18. CCDF curves for different multi-standard signal configurations

Fig. 18 shows the respective CCDF curves of each signal configuration (alone or in multi-mode operation) and it can be viewed that the multi-mode signals will produce more restrictions in terms of PAPR to the receiver components.

3.3 PAPR Reduction – *Required Techniques*

As was mentioned in the previous section with the constant development of the wireless world there will be need to solve or to minimize the problems that have been referred. Since several years ago the notion of PAPR reduction is being an important topic of researching in the scientific area. The increase in PAPR in multi-carrier systems becomes so complex that all the possible schemes for its minimization have become a goal for wireless system design engineers.

The approaches for that minimization spans from special coding for PAPR minimization (Han & Lee, 2005), tone reservation (Tellado & Cioffi, 1998), tone injection (Han et al., 2006), clipping followed by filtering (Vaananen et al., 2002), change of the constellation diagram (Krongold & Jones, 2003), and others based on representation of the signals to transmit (Han & Lee, 2003), like partial transmit sequences (PTS) interleaving or selected mapping (SLM).

Recently, the study has been centred not on how to minimize the PAPR exclusively, but on how to have a balance, on the minimum PAPR, transmission-rate and bit-error-rate (BER) optimization. We will focus our analysis in the techniques that can be used in the receiver for PAPR minimization instead of general techniques for PAPR reduction.

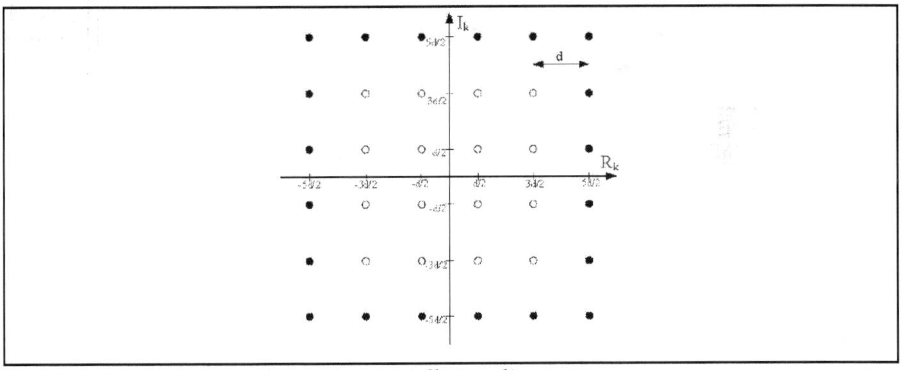

Fig. 19. Tone Injection based on extra constellation diagrams

Firstly, we have the Tone Injection technique (Han et al., 2006), in which the main idea is to create several constellation diagrams that can be dynamically chosen in order to reduce the PAPR. In fact, the addition of new constellation points in the constellation diagram is equivalent of injecting new tones. This new constellation point in the augmented constellation could be selected as to maintain the time waveform at a stable PAPR.

The idea from an OFDM point of view is nothing more than to add constants C to the OFDM symbol, which are carefully selected to reduce the PAPR and to not increase the BER. The effect of the added constant is to increase the constellation magnitude so that each of the point in the original constellation is mapped into the expanded constellation.

The values of C are $C_k = p_k D + jq_k D$, where p_k and q_k are chosen to minimize the PAPR value, and the constant D is known both at the transmitter and the receiver. Fig. 19 presents this solution for the case of a 16-QAM signal.
In Fig. 19, what we can see is that the black points could also be transmitted, but no new information is added. This means that we can transmit the same digital symbol either using the white points or the black points for the same base information bit, so the modulator have some redundancy, which is chosen in order to minimize the PAPR.
The main problem is the increase in BER. Nevertheless, the augmented capability to reduce PAPR is quite satisfactory.
Other possible available technique is the Tone Reservation (Tellado & Cioffi, 1998), where the underneath idea is to reserve, that means, to select some sub carriers in order that the overall RF signal has a reduced PAPR. In DSL communication systems this is normally done in the low SNR tones, since they will not be very important for the overall signal demodulation. So, in this case, we will add some information, C, to the unused tones to reduce the overall PAPR in the time domain scenario. The unused tones are called the reserved tones and normally do not carry data or they cannot carry data reliably due to their low SNR. It is exactly these tones that are used to send optimum vector C that was selected to reduce large peak power samples of OFDM symbols. The method is very simple to implement, and the receiver could ignore the symbols carried on the unused tones, without any complex demodulation process, neither extra tail bits.
Other simple but important technique is known as Amplitude Clipping plus Filtering (Vaananen et al., 2002), which is obviously the one that can achieve improved results and is less complex to apply. Nevertheless the clipping increases the occupied bandwidth and simultaneously degrades significantly the in-band distortion, giving rise to the increase of BER, due to its nonlinearity nature. The technique is based mainly on the following procedure: if the signal is below a certain threshold, then we let the signal as is, at the output, nevertheless if it passes that threshold then the signal should be clipped as is presented in expression (8).

$$y = \begin{cases} x & |x| \le A \\ Ae^{j\phi(x)}, & |x| \ge A \end{cases} \tag{8}$$

where $\phi(x)$ is the phase of the input signal x.
The main problem of this technique is that somehow we are distorting the signal generating nonlinear distortion both in-band and out-of-band. The in-band distortion cannot be filtered out, and some form of linearizer should be used or other form of reconstruction of the signal prior to the reception block. The out-of-band emission, usually called spectral regrowth, can be filtered out, but the filtering process will increase again the PAPR. For that reason, some algorithms are used sequentially with clipping and filtering in order to converge to a minimum value. This technique can be further associated with other schemes to improve the PAPR overall solution.
Finally, we describe a scheme that is called Companding / Expanding technique (Jiang et al., 2005), which is very similar to clipping, but the signal is not actually clipped, but rather companded or expanded accordingly to its amplitude. This technique was used since the

analogue telephone lines were the voice was companded in order to reduce its dynamic range problems encountered through the transmission over the copper lines. Most of the authors have dedicated their time to select the optimum form of the companding function in order to simultaneously reduce the PAPR and improve the BER performance. Fig. 20 presents one of these schemes implementation.

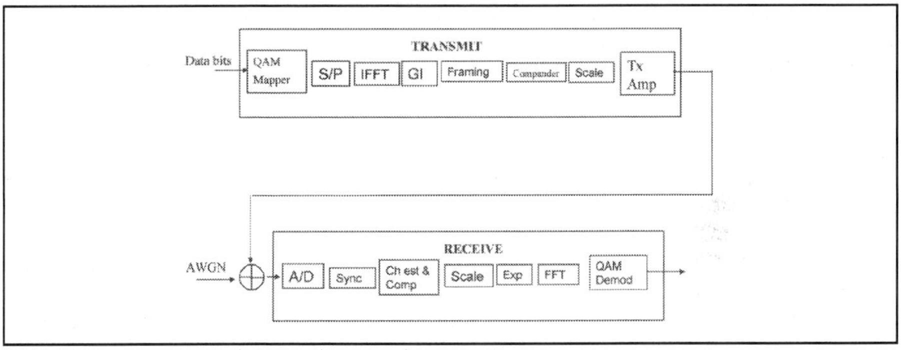

Fig. 20. Companding and Expanding implementation

One possibility for the companding function is the well-known μ-law, expression (9).

$$F(x) = \operatorname{sgn}(x)\frac{\ln(1 + u|x|)}{\ln(1 + u)}, \quad -1 \le x \le 1 \tag{9}$$

The drawbacks of this solution are similar to the clipping technique, but in this case the nonlinear distortion can be somehow post-distorted at the receiver more efficiently, since the nonlinearity is not as severe as the clipping form.

4. Example Applications

In this section, we will present possible real-world applications of several of previous described receiving architectures, in which we will describe some evaluated experiments. These include configurations that are being used in emergent fields, such as RFID and SDR systems. In these fields the multi-standard reception and the receiver PAPR minimization techniques analyzed can bring attractive improvements.

4.1 Radio Frequency Identification Applications

An RFID system is basically composed of two main blocks: the TAG and the READER (Fig. 21).

Fig. 21. RFID system

The Tag (or transponder) is a small device that serves as identifier of a person or an object in which it was implemented. When asked by the reader, returns the information contained within its small microchip. It should be noted, however, that despite this being the most common method, there are active tags that transmit information without the presence of the reader. The reader can be considered the "brain" of an RFID system. It is responsible for liaison between external systems of data processing (computer-data based) and the tags, it is also their responsibility to manage the system.

There are typically three main groups of tags: the passive, semi-passive (or semi-active) and active ones. These names derive from the needing of an internal battery for Tag's operation and transmission of signal. From these three types of Tags which will be addressed here is the semi-passive, to have a configuration very similar to the envelope detector architecture presented above. The spectral regrowth capability from the nonlinear behaviour of the diode is used in this topology, but instead of using the second harmonic product in baseband (like an envelope detector) it will use the third harmonic products (intermodulation products) that fall close to the original signal. The operational principle of the proposed approach is depicted in Fig. 22.

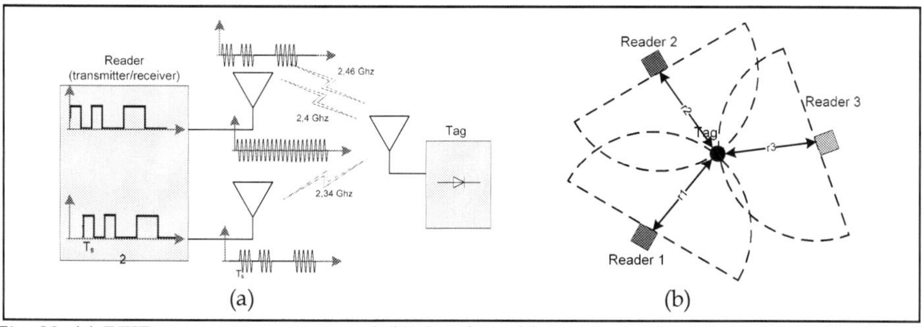

Fig. 22. (a) RFID system operation and (b) developed location method

The operational principle is as follows:

- The READER send an RF signal, at ω_2, modulated by a pseudo-random sequence and in a different frequency, ω_1, an un-modulated carrier RF signal.
- When the signal arrives to the TAG, a RF transceiver demodulates it and re-modulated in a different carrier and re-emitted to the air interface.

- The READER has a receiver tuned to this frequency, which allows to receive a replica of the transmitted signal.
- Now the two pseudo-random signals, the transmitted one, and the received one, could be compared in time, and the time of travel is calculated.
- This time delay indicates the distance between the READER and the TAG. Obviously, this distance is the ray of semi-circle with centre in the READER. For a correct location of the TAG, at least three different READERs are needed, as shown in Fig. 22(b).

This is a very simple procedure to locate the RFID. The use of an simple diode to generate a third harmonic product that can be used to re-emitted the signal back to the reader, prevents the process of demodulation and subsequent modulation of the data, do not need for local oscillators and reduce the number of a mixer, resulting a huge savings in energy consumption and cost of the components involved.

As seen, the only energy required in the Tag is the strictly necessary for the polarization of the diode. The entire RF path (reception and re-transmission) only use the energy of the signal received from the reader. In addition, this architecture enables the operation in full-duplex system, because the reader sends and receives on different frequencies allowing the simultaneous emission and reception.

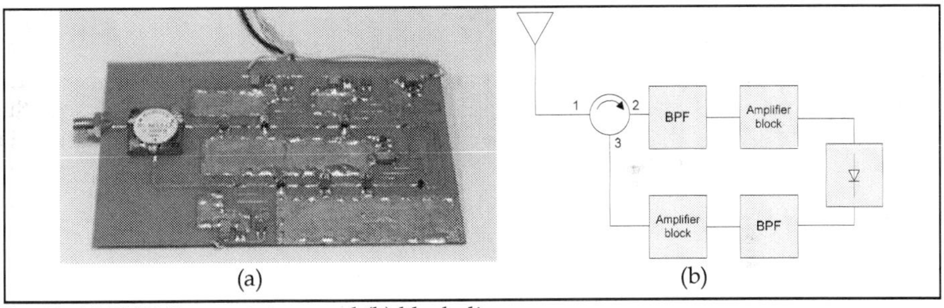

Fig. 23. (a) RFID Tag prototype and (b) block diagram

In Fig. 23 is presented the prototype of this simple envelope detector modified to this particularly case and its block diagram. The simple architecture and the small number of components could enable the full integration, creating an almost passive tag that would allow a location in real-time in full-duplex mode.

A more detailed description and some simulated and laboratory results can be found in any of these references (Gomes & Carvalho, 2007), (Gomes & Carvalho, 2008).

4.2 Software Defined Radio Applications

In order to demonstrate the application of the previous overviewed receiver architectures in SDR field, we have implemented, as an example, a band-pass sampling receiver, Fig. 7, using laboratory instruments. We used a fixed band-pass filter to select the fifth Nyquist zone to avoid aliasing of other undesired signals. This was followed by a commercially available wideband (0.5 - 1000 MHz) LNA with a 1 dB compression point of +9 dBm, an approximate gain of 24 dB, and a noise figure of nearly 6 dB. We used a commercially available 12-bit pipeline ADC that has a linear input range of approximately +11 dBm with an analogue input bandwidth of 750 MHz. Due to some limitations of the arbitrary waveform generator used for the clock signal, a clock frequency of 100 MHz was utilized. The input RF frequency was in the fifth Nyquist zone, more precisely at f_{RF} = 220 MHz. In that sense, considering the clock frequency referred and the sample and hold circuit (inside the ADC) behaviour this RF signal was folded back to the first Nyquist zone, and fell in an intermediate frequency of f_{IF} = 20 MHz, obtained with equation (1). The feature of sub-sampling operation of the ADC, depicted in Fig. 8, was discussed in (Cruz et al., 2008) wherein the authors clearly demonstrate an ADC operating in a sub-sampled configuration obtaining very similar results in all of the Nyquist zones evaluated. Furthermore, in order to obtain accurate measurement results we used the set-up proposed in (Cruz et al., 2008a) shown in Fig. 24, to completely characterize our receiver, mainly in terms of nonlinear distortion.

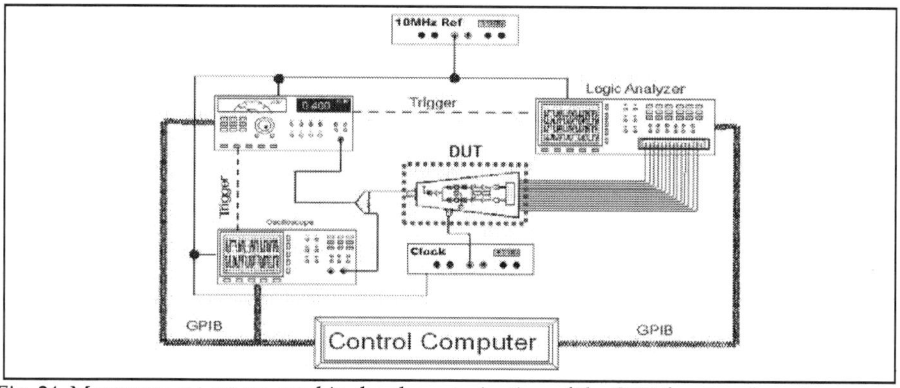

Fig. 24. Measurement set-up used in the characterization of the SDR front-end receiver

As can be seen from this set-up, the input signal was acquired by a sampling oscilloscope, while the output signal was acquired by a logic analyzer. The measured data were then post-processed using a commercial mathematical software package in the control computer. Then, we carried out measurements when several multisines having 100 tones with a total occupied bandwidth of 1 MHz were applied. We produced different amplitude/phase arrangements for the frequency components of each multisine waveform. In fact, these signals were intended to mimic different time-domain-signal statistics and thus provide different PAPR values (Remley, 2003), (Pedro & Carvalho, 2005). A WiMAX (IEEE 802.16e standard, 2005) signal was also used as the SDR front-end excitation. In this case, we used a single-user WiMAX signal in frequency division duplex (FDD) mode with a bandwidth of 3 MHz and a modulation type of 64-QAM (¾).

Fig. 25 presents the measured statistics for each excitation (multisines and WiMAX). The *Constant Phase* multisine is the one where the relative phase difference is 0° between the tones, yielding a large value of 20 dB PAPR. On the other hand, the uniform and normal multisines have uniformly and normally distributed amplitude/phase arrangements, respectively. These constructions yield around 2 dB PAPR for the uniform case and around 9 dB PAPR for the normal case. As can be observed in Fig. 25 the WiMAX signal is similar to the multisine with normal statistics.

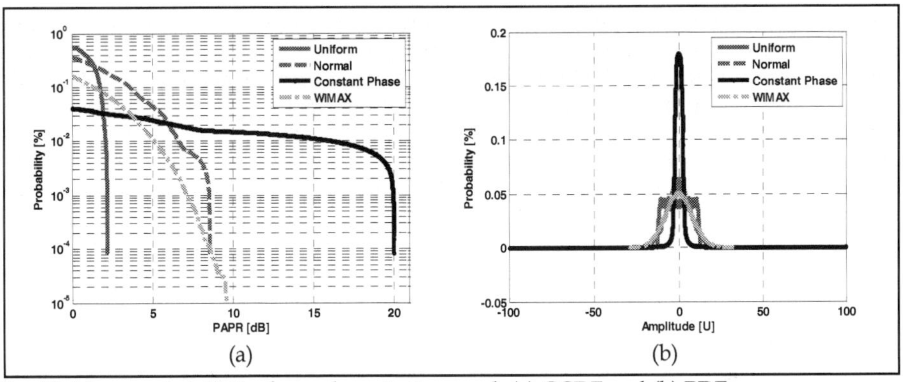

Fig. 25. Measured statistics for each excitation used, (a) CCDF and (b) PDF

Fig. 26 presents the measured results at the output of the SDR receiver using the logic analyzer, where the left graph shows the total power averaged over the excitation band of frequencies, while the right graph shows the total power in the upper adjacent channel arising from nonlinear distortion.

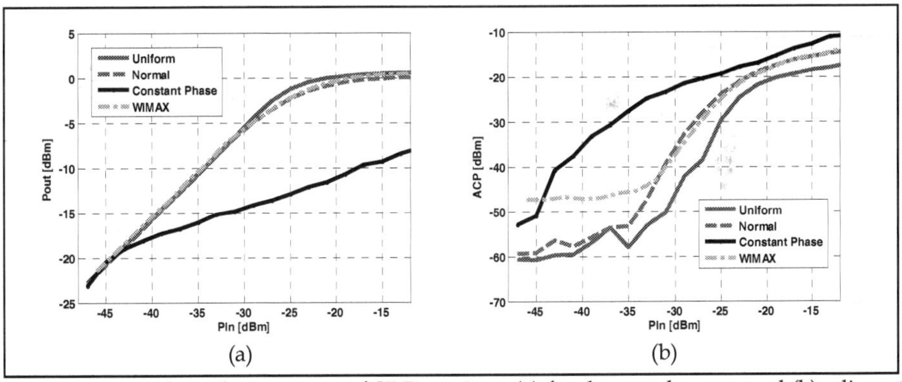

Fig. 26. Measured results at output of SDR receiver, (a) fundamental power and (b) adjacent channel power

It is clear that the signal with constant-phase statistics deviates from linearity at a much lower input power level than for the other cases since the PAPR of that signal is much higher and so clipping occurs at a relatively low input level. As well, the adjacent channel power is significantly higher for the constant phase case than for the others. As expected, the WiMAX signal performs very similarly to the multisine with normal statistics, both in the fundamental output power and in the adjacent channel power for a medium/large-signal operation (after around -30 dBm in its input). This happens because both signals have similar statistical behaviours. The higher small-signal adjacent channel power observed in the WiMAX signal compared to the multisine measurements is due to the intrinsic characteristics of this signal that is based on an OFDM technique, which results in a significantly higher out-of-channel power. The obtained results allow us to stress that the signal PAPR could completely degrade the overall performance of such type of receiver in terms of nonlinear distortion and thus being a very important parameter in the design of a receiver front-end for SDR operation. Another point that is an open problem and should be evaluated is the characterization of SDR components, which is only possible with the utilization of a mixed-mode instrument as the one implemented in (Cruz et al., 2008a).

5. Summary and Conclusions

In this chapter we have presented a review of the mostly known receiver architectures, wherein the main advantages and relevant disadvantages of each configuration were identified. We also have analyzed several possible enhancements to the receiver architectures presented, which include Hartley and Weaver configurations, as well as new receiver architectures based in discrete-time analogue circuits.

Moreover, the main interference issues that receiver front-end architectures could experience were shown and analyzed in depth. Furthermore, some PAPR reduction techniques that may be applied in these receiver front-ends were also shown. In the final section, two interesting applications of the described theme were presented.

As was said, the development of such multi-norm, multi-standard radios is one of the most important points in the actual scientific area. Also, this fact is very important to the telecommunications industry that is expecting for such a thing. Actually, this is what is being searched for in the SDR field where the motivation is to construct a wideband adaptable radio front-end, in which not only the high flexibility to adapt the front end to simultaneously operate with any modulation, channel bandwidth, or carrier frequency, but also the possible cost savings that using a system based exclusively on digital technology could yield. It is expected that this chapter becomes a good start for RF engineers that wants to learn something about receivers and its impairments.

6. Selected Bibliography

Adiseno; Ismail, M. & Olsson, H. (2002). A Wideband RF Front-End for Multiband Multistandard High-Linearity Low-IF Wireless Receivers, *IEEE Journal of Solid-State Circuits*, Vol. 37, No. 9, September 2002, pp. 1162-1168, ISSN: 0018-9200

Agilent Application Note (2000). Characterizing Digitally Modulated Signals with CCDF Curves, No. 5968-6875E, Agilent Technologies, Inc., Santa Clara, USA

Akos, D.; Stockmaster, M.; Tsui, J. & Caschera, J. (1999). Direct Bandpass Sampling of Multiple Distinct RF Signals, *IEEE Transactions on Communications*, Vol. 47, No. 7, July 1999, pp. 983-988

Bauml, R.; Fischer, R. & Huber, J. (1996). Reducing the peak-to-average power ratio of multicarrier modulation by selected mapping, *Electronic Letters*, 1996, Vol. 32, pp. 2056-2057

Besser, L. & Gilmore, R. (2003). *Practical RF Circuit Design for Modern Wireless Systems*, Artech House, ISBN 1-58053-521-6, Norwood, USA

Cruz, P.; Carvalho, N.B. & Remley, K.A. (2008), Evaluation of Nonlinear Distortion in ADCs Using Multisines, *IEEE MTT-S International Microwave Symposium Digest*, pp. 1433-1436, ISBN: 978-1-4244-1780-3, Atlanta, USA, June 2008

Cruz, P.; Carvalho, N.B.; Remley, K.A. & Gard, K.G. (2008). Mixed Analog-Digital Instrumentation for Software Defined Radio Characterization, *IEEE MTT-S International Microwave Symposium Digest*, pp. 253-256, ISBN: 978-1-4244-1780-3, Atlanta, USA, June 2008

Cruz, P. & Carvalho, N.B. (2008). PAPR Evaluation in Multi-Mode SDR Transceivers, *38th European Microwave Conference*, pp. 1354-1357, ISBN: 978-2-87487-006-4, Amsterdam, Netherlands, October 2008

Goldsmith, A. & Chua, S. (1998). Adaptive Coded Modulation for Fading Channels, *IEEE Transactions on Communications*, Vol.46, No. 5, May 1998, pp. 595-602, ISSN: 0090-6778

Gomes, H.; Carvalho, N.B. (2007). The use of Intermodulation Distortion for the Design of Passive RFID, *37th European Microwave Conference*, pp. 1656-1659, ISBN: 978-2-87487-001-9, Munich, Germany, October 2007

Gomes, H.; Carvalho, N.B. (2009). RFID for Location Proposes Based on the Intermodulation Distortion, *Sensors & Transducers journal*, Vol. 106, No. 7, pp. 85-96, July 2009, ISSN 1726-5479

Han, S.H. & Lee, J.H. (2003). Reduction of PAPR of an OFDM Signal by Partial Transmit Sequence Technique with Reduced Complexity, *IEEE Global Telecommunications Conference*, pp. 1326-1329, ISBN: 0-7803-7974-8, San Franscisco, USA, December 2003

Han, S.H. & Lee, J.H. (2005). An Overview of Peak-to-Average Power Ratio Reduction Techniques for Multicarrier Transmission, *IEEE Wireless Communications*, Vol. 12, No. 2, pp. 56-65, April 2005

Han, S.H.; Cioffi, J.M. & Lee, J.H. (2006). Tone Injection with Hexagonal Constellation for Peak-to-Average Power Ratio Reduction in OFDM, *IEEE Communications Letters*, Vol. 10, No. 9, pp. 646-648, September 2006, ISSN: 1089-7798

IEEE 802.16e standard (2005). Local and Metropolitan Networks - Part 16: Air Interface for Fixed and Mobile Broadband Wireless Access Systems, 2005

Jiang, T.; Yang, Y. & Song, Y. (2005). Exponential Companding Technique for PAPR Reduction in OFDM Systems, *IEEE Transactions on Broadcasting*, Vol. 51, No. 2, pp. 244-248, June 2005, ISSN: 0018-9316

Krongold, B.S. & Jones, D.L. (2003). PAR Reduction in OFDM via Active Constellation Extension, *IEEE Transactions on Broadcasting*, Vol. 49, No. 3, pp. 258-268, September 2003, ISSN: 0018-9316

Landon, V.D. (1936). A Study of the Characteristics of Noise, *Proceedings of the IRE*, Vol. 24, No. 11, pp. 1514-1521, November 1936, ISSN: 0096-8390

Muhammad, K.; Ho, Y.C.; Mayhugh, T.; Hung, C.M.; Jung, T.; Elahi, I.; Lin, C.; Deng, I.; Fernando, C.; Wallberg, J.; Vemulapalli, S.; Larson, S.; Murphy, T.; Leipold, D.; Cruise, P.; Jaehnig, J.; Lee, M.C.; Staszewski, R.B.; Staszewski, R.; Maggio, K. (2005). A Discrete Time Quad-Band GSM/GPRS Receiver in a 90nm Digital CMOS Process, *Proceedings of IEEE 2005 Custom Integrated Circuits Conference*, pp. 809-812, ISBN 0-7803-9023-7, San Jose, USA, September 2005

Park, J.; Lee, C.; Kim, B. & Laskar, J. (2006). Design and Analysis of Low Flicker-Noise CMOS Mixers for Direct-Conversion Receivers, *IEEE Transactions on Microwave Theory and Techniques*, Vol. 54, No. 12, December 2006, pp. 4372-4380, ISSN: 0018-9480

Pedro, J.C. & Carvalho, N.B. (2003). *Intermodulation Distortion in Microwave and Wireless Circuits*, Artech House, ISBN 1-58053-356-6, Norwood, USA

Pedro, J.C. & Carvalho, N.B. (2005). Designing Multisine Excitations for Nonlinear Model Testing, *IEEE Transactions Microwave Theory and Techniques*, Vol. 53, No. 1, pp. 45-54, January 2005, ISSN: 0018-9480

Razavi, B. (1997). Design Considerations for Direct–Conversion Receivers. *IEEE Transactions on Circuits and Systems - II: Analog and Digital Signal Processing*, Vol. 44, No. 6, June 1997, pp. 428-435, ISSN 1057-7130

Razavi, B. (1998). Architectures and Circuits for RF CMOS Receivers, *Proceedings of IEEE 1998 Custom Integrated Circuits Conference*, pp. 393-400, ISBN 0-7803-4292-5, Santa Clara, USA, May 1998

Remley, K.A. (2003). Multisine Excitation for ACPR Measurements, *IEEE MTT-S International Microwave Symposium Digest*, pp. 2141-2144, ISBN: 0-7803-7695-1, Philadelphia, USA, June 2003

Staszewski, R.B.; Muhammad, K.; Leipold, D.; Chih-Ming Hung; Yo-Chuol Ho; Wallberg, J.L.; Fernando, C.; Maggio, K.; Staszewski, R.; Jung, T.; Jinseok Koh; John, S.; Irene Yuanying Deng; Sarda, V.; Moreira-Tamayo, O.; Mayega, V.; Katz, R.; Friedman, O.; Eliezer, O.E.; de-Obaldia, E.; Balsara, P.T. (2004). All-Digital TX Frequency Synthesizer and Discrete-Time Receiver for Bluetooth Radio in 130-nm CMOS, *IEEE Journal of Solid-State Circuits*, Vol. 39, No. 12, December 2004, pp. 2278-2291, ISSN: 0018-9200

Tellado, J. & Cioffi, J.M. (1998). Peak Power Reduction for Multicarrier Transmission, *IEEE Global Telecommunications Conference*, Sydney, Australia, Nov. 1998.

Tsui, J. (1995). *Digital Techniques for Wideband Receivers*, Artech House, ISBN 0-89006-808-9, Norwood, USA

Vaananen, O.; Vankka, J. & Halonen, K. (2002). Reducing the Peak-to-Average Ratio of Multicarrier GSM and EDGE Signals, *IEEE International Symposium on Personal, Indoor and Mobile Radio Communications*, pp. 115-119, ISBN: 0-7803-7589-0, Lisbon, Portugal, September 2002

Vaughan, R.; Scott, N. & White, D. (1991). The Theory of Bandpass Sampling, *IEEE Transactions on Signal Processing*, Vol. 39, No. 9, September 1991, pp. 1973-1984, ISSN: 0090-6778

26

Microwave Measurement of the Wind Vector over Sea by Airborne Radars

Alexey Nekrasov
Taganrog Institute of Technology of the Southern Federal University
Russia,
Hamburg University of Technology
Germany

1. Introduction

The oceans of the Earth work in concert with the atmosphere to control and regulate the environment. Fed by the sun, the interaction of land, ocean, and atmosphere produces the phenomenon of weather and climate. Only in the past half-century meteorologists have begun to understand weather patterns well enough to produce relatively accurate, although limited, forecasts of future weather patterns. One limitation of predicting future weather is that meteorologists do not adequately know the current weather. An accurate understanding of current conditions over the ocean is required to predict future weather patterns. Until recently, detailed local oceanic weather conditions were available only from sparsely arrayed weather stations, ships along commercial shipping lanes and sparsely distributed oceans buoys (Long, et al, 1976).

The development of satellite and airborne remote sensing has improved the situation significantly. Satellite remote sensing has demonstrated its potential to provide measurements of weather conditions on a global scale as well as airborne remote sensing on a local scale. Measurements of surface wind vector and wave height are assimilated into regional and global numerical weather and wave models, thereby extending and improving our ability to predict future weather patterns and sea/ocean surface conditions on many scales.

A pilot also needs operational information about wind over sea as well as wave height to provide safety of hydroplane landing on water.

Many researchers solve the problem of remote measuring of the wind vector over sea actively (Moore & Fung, 1979), (Melnik, 1980), (Chelton & McCabe, 1985), (Feindt, et al, 1986), (Masuko, et al, 1986), (Wismann, 1989), (Hildebrand, 1994), (Carswell, et al, 1994). On the global scale, the information about sea waves and wind, in general, could be obtained from a satellite using active microwave instruments: Scatterometer, Synthetic Aperture Radar (SAR) and Radar Altimeter. However, for the local numerical weather and wave models as well as for a pilot on a hydroplane to make a landing decision, the local data about wave height, wind speed and direction are required.

Research on microwave backscatter by the sea surface has shown that the use of a scatterometer, radar designed for measuring the surface scatter characteristics, allows for an estimation of sea surface wind vector because the normalized radar cross section (NRCS) of the sea surface depends on the wind speed and direction. Based on experimental data and scattering theory, a significant number of empirical and theoretical backscatter models and algorithms for estimation of the sea wind speed and direction from satellite and airplane have been proposed (Long, et al, 1976), (Moore & Fung, 1979), (Melnik, 1980), (Chelton & McCabe, 1985), (Masuko, et al, 1986), (Wismann, 1989), (Hildebrand, 1994), (Carswell, et al, 1994), (Wentz, et al, 1984), (Young, (1993), (Romeiser, et al, 1994). The accuracy of the wind direction measurement is ±20°, and the accuracy of the wind speed measurement is ±2 m/s in the wind speed range 3-24 m/s.

SAR provides an image of the roughness distribution on the sea surface with large dynamic range, high accuracy, and high resolution. Retrieval of wind information from SAR images provides a useful complement to support traditional wind observations (Du, et al, 2002). Wind direction estimation amounts to measuring the orientation of boundary-layer rolls in the SAR image, which are often visible as image streaks. The sea surface wind direction (to within a 180° direction ambiguity) is assumed to lie essentially parallel to the roll or image-streak orientation. Wind speed estimation from SAR images is usually based on a scatterometer wind retrieval models. This approach requires a well-calibrated SAR image. The wind direction estimated from the European remote sensing satellite (ERS-1) SAR images is within a root mean square (RMS) error of ±19° of in situ observations, which in turn results in an RMS wind speed error of ±1.2 m/s (Wackerman, et al, 1996).

The radar altimeter also provides the information on the sea wind speed, which can be determined from the intensity of the backscattered return pulse, and on the sea wave height, which can be deduced from the return pulse shape. At moderate winds (3–12 m/s), the wind speed can be measured by the altimeter with an accuracy of about ±2 m/s. The typical accuracy of radar altimeter measurements of the significant wave height is of the order of ±0.5 m (or 10 %, whichever is higher) for wave heights between 1 and 20 m (Komen, et al, 1994). Unfortunately, altimeter wind measurements yield wind velocity magnitude only, and do not provide information on wind direction.

Mostly narrow-beam antennas are applied for such wind measurement. Unfortunately, a microwave narrow-beam antenna has considerable size at Ku-, X- and C-bands that hampers its placing on flying apparatus. Therefore, a better way needs to be found.

At least two ways can be proposed. The first way is to apply the airborne scatterometers with wide-beam antennas as it can lead to the reduction in the antenna size. The second way is to use the modified conventional navigation instruments of flying apparatus in a scatterometer mode, which is more preferable.

From that point of view, the promising navigation instruments are the airborne radar altimeter (ARA), the Doppler navigation system (DNS) and the airborne weather radar (AWR). So, the principles of recovering the sea surface wind speed and direction, using those navigation instruments are discussed in this chapter.

2. Principle of Near-Surface Wind Vector Estimation

Radar backscatter from the sea surface varies considerably with incidence angle (Hildebrand, 1994). Near nadir is a region of quasi-specular return with a maximum of NRCS that falls with increasing the angle of incidence. Between incident angles of about 20° and 70°, the NRCS falls smoothly in a so-called "plateau" region. For middle incident angles, microwave radar backscatter is predominantly due to the presence of capillary-gravity wavelets, which are superimposed on large gravity waves on the sea surface. Small-scale sea waves of a length approximately one half the radar wavelength are in Bragg resonance with an incident electromagnetic wave. At incidence angles greater than about 70° is the "shadow" region in which NRCS falls dramatically, due to the shadowing effect of waves closer to the radar blocking waves further away.

The wind blowing over sea modifies the surface backscatter properties. These depend on wind speed and direction. Wind speed U can be determined by a scatterometer because a stronger wind will produce a larger NRCS $\sigma^\circ(U,\theta,\alpha)$ at the middle incidence angle θ and a smaller NRCS at the small (near nadir) incidence angle. Wind direction can also be inferred because the NRCS varies as a function of the azimuth illumination angle α relative to the up-wind direction (Spencer & Graf, 1997).

To extract the wind vector from NRCS measurements, the relationship between the NRCS and near-surface wind, called the "geophysical model function", must be known. Scatterometer experiments have shown that the NRCS model function for middle incidence angles is of the widely used form (Spencer & Graf, 1997)

$$\sigma^\circ(U,\theta,\alpha) = A(U,\theta) + B(U,\theta)\cos\alpha + C(U,\theta)\cos(2\alpha)\,, \tag{1}$$

where $A(U,\theta)$, $B(U,\theta)$ and $C(U,\theta)$ are the Fourier terms that depend on sea surface wind speed and incidence angle, $A(U,\theta) = a_0(\theta)U^{\gamma_0(\theta)}$, $B(U,\theta) = a_1(\theta)U^{\gamma_1(\theta)}$, and $C(U,\theta) =$ $= a_2(\theta)U^{\gamma_2(\theta)}$; $a_0(\theta)$, $a_1(\theta)$, $a_2(\theta)$, $\gamma_0(\theta)$, $\gamma_1(\theta)$ and $\gamma_2(\theta)$ are the coefficients dependent on the incidence angle.

As we can see from (1), an NRCS azimuth curve has two maxima and two minima. The main maximum is located in the up-wind direction, the second maximum corresponds to the down-wind direction, and two minima are in cross-wind directions displaced slightly to the second maximum. With increase of the incidence angle, the difference between two maxima and the difference between maxima and minima become so significant (especially at middle incidence angles) that this feature can be used for retrieval of the wind direction over water (Ulaby, et al, 1982).

In the general case, the problem of estimating the sea surface wind navigational direction ψ_w consists in defining the main maximum of a curve of the reflected signal intensity (azimuth of the main maximum of the NRCS $\psi_{\sigma^\circ_{max}}$)

$$\psi_w = \psi_{\sigma^\circ_{max}} \pm 180^\circ\,, \tag{2}$$

and the problem of deriving the sea surface wind speed consists in determination of a reflected signal intensity value from the up-wind direction or from some or all of the azimuth directions. The azimuth NRCS curve can be obtained using the circle track flight for a scatterometer with an inclined one-beam fixed-position antenna or the rectilinear track flight for a scatterometer with a rotating antenna (Masuko, et al, 1986), (Wismann, 1989), (Carswell, et al, 1994).
Also, the wind speed over sea can be measured by a scatterometer with a nadir-looking antenna (altimeter) using, for instance, the following NRCS model function at zero incident angle $\sigma^{\circ}(U,0^{\circ})$ (Chelton & McCabe, 1985).

$$\sigma^{\circ}(U,0^{\circ})[dB] = 10(G_1 + G_2 \log_{10} U_{19.5}), \tag{3}$$

where G_1 and G_2 are the parameters, $G_1 = 1.502$, $G_2 = -0.468$; $U_{19.5}$ is the wind speed at 19.5 m above the sea surface. A comparison of altimeter wind speed algorithms together with (3) is represented in (Schöne & Eickschen, 2000).
Thus, the scatterometer having an antenna with inclined beams provides the information on both the wind speed over sea and the wind direction, and the scatterometer with a nadir-looking antenna allows estimating only the sea surface wind speed and provides no information on the wind direction.

3. Wind Vector Measurement Using an Airborne Radar Altimeter

3.1 Airborne Radar Altimeter

The basic function of the ARA is to provide terrain clearance or altitude with respect to the ground level directly beneath the aircraft. The ARA may also provide vertical rate of climb or descent and selectable low altitude warning (Kayton & Fried, 1997).
Altimeters perform the basic function of any range measuring radar. A modulated signal is transmitted toward the ground. The modulation provides a time reference to which the reflected return signal can be reflected, thereby providing radar-range or time delay and therefore altitude. The ground represents an extended target, as opposed to a point target, resulting in the delay path extending from a point directly beneath the aircraft out to the edge of antenna beam. Furthermore, the beam width of a dedicated radar altimeter antenna must be wide enough to accommodate normal roll-and-pitch angles of the aircraft, resulting in a significant variation in return delay.
The ARA is constructed as FM-CW or pulsed radar. The frequency band of 4.2 to 4.4 GHz is assigned to the ARA. The frequency band is high enough to result in reasonably small sized antennas to produce a 40° to 50° beam but is sufficiently low so that rain attenuation and backscatter from rain have no significant range limiting effects. Typical installations include a pair of small microstrip antennas for transmit and receive functions (Kayton & Fried, 1997).

3.2 Beam Sharpening

As the ARA has a widebeam antenna and wind measurements are performed with the antennas having comparatively narrow beams (beamwidth of $4^{\circ} - 10^{\circ}$), to apply the ARA for wind vector estimation the beam sharpening technologies should be used.

Lately, to sharpen the effective antenna beams of real-aperture radars avoiding the size enlargement of their antennas, Doppler discrimination along with range discrimination have been employed. An example of application of such a simultaneous range Doppler discrimination technique is the conically scanning pencil-beam scatterometer performing wind retrieval (Spencer, et al, 2000a). When simultaneous range Doppler processing is used, the resolution cell is delineated by the iso-Doppler and iso-range lines projected on the surface, where the spacing between the lines is the achievable Doppler or range resolution respectively. As the beam scans, the azimuth resolution is the best at the side-looking locations and is the coarsest at the forward and afterward locations. A conceptual description of such a scatterometer has been described in (Spencer, et al, 2000b).

Another example of employing the simultaneous range Doppler discrimination technique is the delay Doppler radar altimeter developed at the Applied Physics Laboratory of the Johns Hopkins University (Raney, 1998). The delay Doppler altimeter uses coherent processing over a block of received returns to estimate the Doppler frequency modulation imposed on the signals by the forward motion of the altimeter. Doppler analysis of the data allows estimating their along-track positions relative to the position of the altimeter. It follows that the along-track dimension of the signal data and the cross-track (range or time delay) dimensions are separable. In contrast to the response of a conventional altimeter having only one independent variable (time delay), the delay Doppler altimeter response has two independent variables: along-track position (functionally related to Doppler frequency) and cross-track position (functionally related to time delay). After delay Doppler processing, these two variables describe an orthonormal data grid. With this data space in mind, delay Doppler processing may be interpreted as an operation that flattens the radiating field in along-track direction. Unfortunately, a cross-track ambiguity takes place under measurements, as there are two possible sources of reflections (one from the left side and another from the right side), which have a given time delay at any given Doppler frequency (Raney, 1998).

Recently, the sensitivity of signals from the Global Positioning System (GPS) to propagation effects was found to be useful for measurements of surface roughness characteristics from which wave height, wind speed, and direction could be determined. The Delay Mapping Receiver (DMR) was designed, and a number of airborne experiments were completed. The DMR includes two low-gain (wide-beam) L-band antennas: a zenith mounted right-hand circular polarized antenna, and a nadir mounted left-hand circular polarized (LHCP) antenna. It is assumed that a downward-looking LHCP antenna intercepts only the scattered signal and is insensitive to the direct signal. By combining code-range and Doppler measurements, the receiver distinguished particular patches of the ocean surface illuminated by GPS signal that, in fact, is the delay Doppler spatial selection. The estimated wind speed using surface-reflected GPS data collected at a variety of wind speed conditions showed an overall agreement better than 2 m/s with data obtained from nearby buoy data and independent wind speed measurements derived from satellite observations. Wind direction agreement with QuikSCAT measurements appeared to be at the 30° degree level (Komjathy, et al, 2001), (Komjathy, et al, 2000).

3.3 Wind Vector Estimation Using an Airborne Radar Altimeter with the Antenna Forming the Circle Footprint

As the radar altimeter and the scatterometer are required on board of an amphibious airplane, their measurements should be integrated in a single instrument. One of the ways of such integration is to use a short-pulse wide-beam nadir-looking radar, like an airborne Wind-Wave Radar (Hammond, et al, 1977), but with additional Doppler filters. Here, only a short-pulse scatterometer mode of estimating the wind vector by such an airborne altimeter is considered (Nekrassov, 2003).

Let a flying apparatus equipped with a scatterometer (altimeter) having a nadir-looking wide-beam antenna make a horizontal rectilinear flight with the speed V at some altitude H above the mean sea surface, the antenna have the same beamwidth θ_a in both the vertical and horizontal planes, forming a glistening zone on the sea surface, and then transmit a short pulse of duration τ at some time $t=0$ (Fig. 1). If the surface is (quasi-) flat, the first signal return, from the nadir point, occurs at time $t_0 = 2H/c$, where c is the speed of light. The trailing edge of the pulse undergoes the same interactions as the leading edge but delayed in time by τ. The last energy is received from nadir at time $t_0 + \tau$, and the angle for the pulse-limited footprint is $\theta_p = \sqrt{c\tau/H}$. For larger values of time, an annulus is illuminated. The angular incident resolution $\Delta\theta$ is the poorest at nadir, and it improves rapidly with the time from the nadir point.

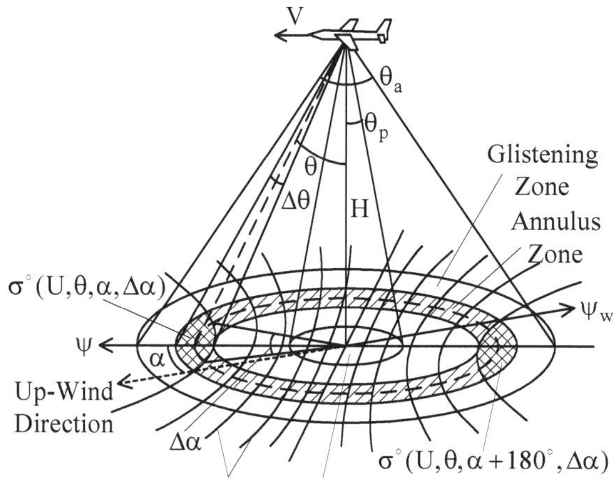

Fig. 1. Two-cell geometry of wind vector measurement by the ARA with the antenna forming a circle footprint

Let the NRCS model function for middle incident angles (annulus zone) be of the form (1) and the NRCS model function for the pulse-limited footprint be of the form (3). Then, the following algorithm to estimate the wind vector over the sea surface can be proposed.

The wind speed can be obtained by means of nadir measurement, for instance, from (3) and converted to a height of measurement of 10 m ($U_{10} = U$), which is mostly used today; for a neutral stability wind profile using the following expression (Jackson, et al, 1992)

$$U_{10} = 0.93U_{19.5} = 0.93 \cdot 10^{[\log_{10}\sigma^\circ(U,0^\circ) - G_1]/G_2}, \qquad (4)$$

or using (1), the average azimuthally integrated NRCS obtained from the annulus zone $\sigma^{\circ}_{an}(U,\theta)$ can be represented in the following form (Nekrassov, 2002)

$$\sigma^{\circ}_{an}(U,\theta) = \frac{1}{2\pi}\int_{0}^{2\pi}\sigma^{\circ}(U,\theta,\alpha)d\alpha = A(U,\theta) \tag{5}$$

and then the wind speed can be found from the following equation

$$U = \left(\frac{A(U,\theta)}{a_0(\theta)}\right)^{1/\gamma_0(\theta)} = \left(\frac{\sigma^{\circ}_{an}(U,\theta)}{a_0(\theta)}\right)^{1/\gamma_0(\theta)}. \tag{6}$$

This method of wind speed estimation allows averaging the power reflected from whole annulus area. However, NRSC values from essentially different athimuthal directions are required to derive a wind direction.

It is necessary to note that the dependence of measured NRCS value on the angular size of a pulse-limited footprint should be taken into account, if the narrow-beam NRCS model function is used. Therefore, the nadir NRCS data obtained by an altimeter having a nadir-looking wide-beam antenna should be corrected in case of a pulse-limited footprint angular size is over approximately $5^{\circ} - 6^{\circ}$ (Nekrassov, 2001).

Now assume that narrow enough Doppler zones could be provided by means of Doppler filtering (Fig. 1). Then, the intersection of an annulus with a Doppler zone would form a spatial cell that discriminates the signal scattered back from the appropriate area of the annulus in the azimuthal direction. Employing Doppler filtering, which provides the azimuthal selection under the measurements with the azimuth resolution (azimuth angular size of a cell) $\Delta\alpha$ in the directions of 0° and 180° relative to the flying apparatus' course as represented by Fig. 1, the wind direction can be derived. To provide the required azimuth angular sizes of the cells, the frequency limits of the fore-Doppler filter $F_{D1.f}$ and $F_{D2.f}$ and of the aft-Doppler filter $F_{D1.a}$ and $F_{D2.a}$ (relative to the zero-Doppler frequency shift) should be as follows

$$F_{D1.f} = -F_{D1.a} = \frac{2V}{\lambda}\sin\left(\theta - \frac{\Delta\theta}{2}\right),$$
$$F_{D2.f} = -F_{D2.a} = \frac{2V}{\lambda}\sin\left(\theta + \frac{\Delta\theta}{2}\right), \tag{7}$$

where λ is the radar wavelength.

At low speed of flight the Doppler effect is not so considerable as at higher speed of flight, and so such locations of the selected cells allows to use the maximum Doppler shifts available. Unfortunately, the coarsest azimuth resolution

$$\Delta\alpha = 2\arccos\left(\frac{\sin(\theta - \Delta\theta)}{\sin\theta}\right) \tag{8}$$

takes place in that case, and the NRCS model function $\sigma^{\circ}(U,\theta,\alpha,\Delta\alpha)$, which considers the azimuth angular size of a cell, should be used.

$$\sigma^{\circ}(U,\theta,\alpha,\Delta\alpha)=\frac{1}{\Delta\alpha}\int_{\alpha-0.5\Delta\alpha}^{\alpha+0.5\Delta\alpha}\sigma^{\circ}(U,\theta,\alpha')d\alpha'$$
$$=A(U,\theta)+k_1(\Delta\alpha)B(U,\theta)\cos\alpha+k_2(\Delta\alpha)C(U,\theta)\cos(2\alpha), \quad (9)$$

where $k_1(\Delta\alpha)$ and $k_2(\Delta\alpha)$ are the coefficients dependent on the azimuth angular size of a cell

$$k_1(\Delta\alpha)=\frac{2\sin(0.5\Delta\alpha)}{\Delta\alpha},$$
$$k_2(\Delta\alpha)=\frac{\sin\Delta\alpha}{\Delta\alpha}. \quad (10)$$

Let $\sigma^{\circ}(U,\theta,\alpha,\Delta\alpha)$, and $\sigma^{\circ}(U,\theta,\alpha+180^{\circ},\Delta\alpha)$ be the NRCS obtained with the fore-Doppler and aft-Doppler filters from the cells corresponding to the maximum value of the Doppler shift (Fig. 1). Then, the speed of wind can be found from (6), and two possible wind directions $\psi_{w.1,2}$ can be found as the following

$$\psi_{w.1,2}=\psi-\alpha_{1,2}\pm180^{\circ}, \quad (11)$$

where $\alpha_{1,2}$ are two possible up-wind directions

$$\alpha_{1,2}=\pm\arccos\left(\frac{\sigma^{\circ}(U,\theta,\alpha,\Delta\alpha)-\sigma^{\circ}(U,\theta,\alpha+180^{\circ},\Delta\alpha)}{2k_1(\Delta\alpha)B(U,\theta)}\right). \quad (12)$$

Unfortunately, an ambiguity of the wind direction takes place in the measurement. Nevertheless, this ambiguity can be eliminated by recurring measurement after 45° change of the flying apparatus' course. The nearest wind directions of pairs of wind directions measured before and after course changing will give the true wind direction.

3.4 Wind Vector Estimation Using an Airborne Radar Altimeter with the Antenna Forming the Ellipse Footprint

To eliminate need of measurements with two different courses of flight under estimation of the sea surface wind speed and direction by the ARA, a modified beam shape forming the ellipse footprints should be used (Nekrasov, 2008a).

Let the antenna beam is wide enough, then two annulus zones at incidence angles θ_1 and θ_2 could be formed as shown by Fig. 2. They will have angular incidence widths $\Delta\theta_1$ and $\Delta\theta_2$ respectively. Now, let the altimeter antenna form an ellipse footprint so that the longer axis of the footprint is rotated by 45° from the horizontal projection of the longitudinal axis of a flying apparatus as shown in Fig. 3.

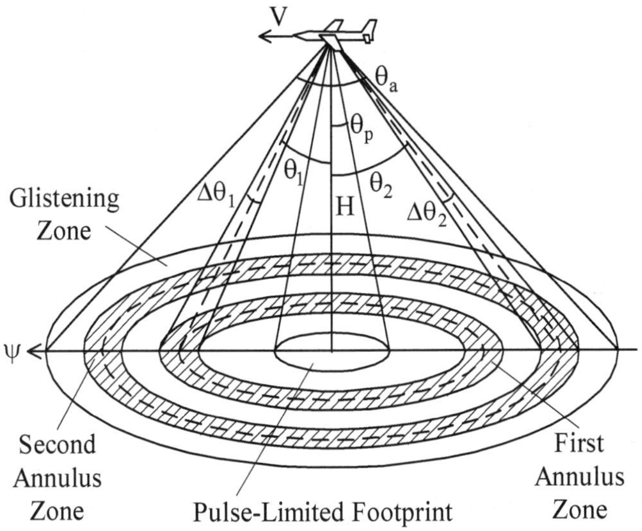

Fig. 2. Forming the annulus zones

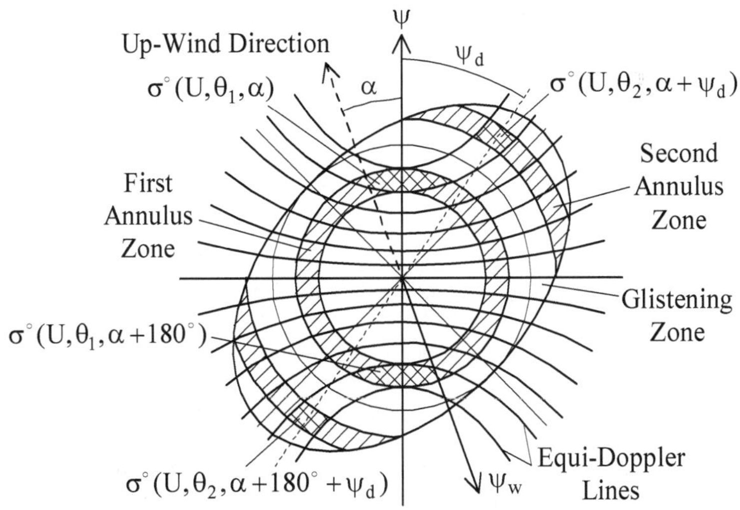

Fig. 3. Geometry of wind vector measurement by ARA having the antenna with the different beamwidth in the vertical and horizontal planes, forming the ellipse footprint, when the longer axis of the footprint is rotated by 45° from the horizontal projection of the longitudinal axis of a flying apparatus

Then, two annulus zones at incidence angles θ_1 and θ_2 ($\theta_1 < \theta_{a.h} < \theta_2 < \theta_{a.v}$) could be formed, and the range Doppler selection can facilitate the identification of cells with the NRCS $\sigma^\circ(U,\theta_1,\alpha)$ and $\sigma^\circ(U,\theta_1,\alpha+180^\circ)$ corresponding the azimuth directions α and $\alpha+180^\circ$ from the first annulus, and the identification of cells with the NRCS $\sigma^\circ(U,\theta_2,\alpha+\psi_d)$ and $\sigma^\circ(U,\theta_2,\alpha+180^\circ+\psi_d)$ corresponding the azimuth directions $\alpha+\psi_d$ and $\alpha+180^\circ+\psi_d$ from the second annulus.

To provide the required azimuth angular sizes of cells of the first annulus $\Delta\alpha_1$ and the second annulus $\Delta\alpha_2$, as shown in Fig. 4, the frequency limits of the fore-Doppler filter $F_{D1.f}$ and $F_{D2.f}$, and of the aft-Doppler filter $F_{D1.a}$ and $F_{D2.a}$ (relative to the zero-Doppler frequency shift) should be as follows

$$F_{D1.f} = -F_{D1.a} = \frac{2V}{\lambda}\sin\left(\theta_1 - \frac{\Delta\theta_1}{2}\right),$$
$$F_{D2.f} = -F_{D2.a} = \frac{2V}{\lambda}\sin\left(\theta_1 + \frac{\Delta\theta_1}{2}\right); \tag{13}$$

$$F_{D1.f} = -F_{D1.a} = \frac{2V}{\lambda}\sin\theta_2\cos\left(\psi_d + \frac{\Delta\alpha_2}{2}\right),$$
$$F_{D2.f} = -F_{D2.a} = \frac{2V}{\lambda}\sin\theta_2\cos\left(\psi_d - \frac{\Delta\alpha_2}{2}\right). \tag{14}$$

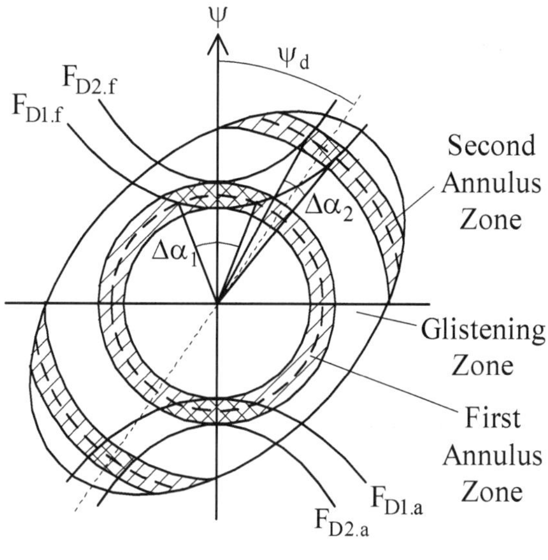

Fig. 4. Forming the selected cells and their angular sizes in horizontal plane

The azimuth angular size of cells of the first annulus is

$$\Delta\alpha_1 = 2\arccos\left(\frac{\sin(\theta_1 - \Delta\theta_1)}{\sin\theta_1}\right). \tag{15}$$

From (13) and (14), we obtain the azimuth location of cells of the second annulus ψ_d, $180^\circ + \psi_d$, and their angular size in the horizontal plane

$$\psi_d = 0.5\left[\arccos\left(\frac{\sin(\theta_1 - 0.5\Delta\theta_1)}{\sin\theta_2}\right) + \arccos\left(\frac{\sin(\theta_1 + 0.5\Delta\theta_1)}{\sin\theta_2}\right)\right],$$

$$\Delta\alpha_2 = \arccos\left(\frac{\sin(\theta_1 - 0.5\Delta\theta_1)}{\sin\theta_2}\right) - \arccos\left(\frac{\sin(\theta_1 + 0.5\Delta\theta_1)}{\sin\theta_2}\right). \tag{16}$$

The speed of wind can be found from (6). Two possible up-wind directions $\alpha_{1an.1,2}$ can be found from the NRCS values obtained from cells of the first annulus, and another two possible up-wind directions $\alpha_{2an.1,2}$ can be found from the NRCS values obtained from cells of the second annulus (Nekrasov, 2007)

$$\alpha_{1an.1,2} = \pm\arccos\left(\frac{\sigma^\circ(U,\theta_1,\alpha) - \sigma^\circ(U,\theta_1,\alpha+180^\circ)}{2k_1(\Delta\alpha_1)B(U,\theta_1)}\right),$$

$$\alpha_{2an.1,2} = \pm\arccos\left(\frac{\sigma^\circ(U,\theta_2,\alpha+\psi_d) - \sigma^\circ(U,\theta_2,\alpha+180^\circ+\psi_d)}{2k_1(\Delta\alpha_2)B(U,\theta_2)}\right). \tag{17}$$

The nearest up-wind directions of pairs of the up-wind directions obtained (one from $\alpha_{1an.1,2}$ and one from $\alpha_{2an.1,2}$) will give the true up-wind direction α, and then, the navigational direction of wind can be found

$$\psi_w = \psi - \alpha \pm 180^\circ. \tag{18}$$

3.5 Conclusion to Wind Vector Estimation Using an Airborne Radar Altimeter

The study has shown that the wind vector over sea can be measured by means of an ARA employed as a nadir-looking wide-beam short-pulse scatterometer in conjunction with Doppler filtering. Such a measuring instrument should be equipped with two additional Doppler filters (a fore-Doppler filter and an aft-Doppler filter) to provide the spatial selection under the wind measurements.

For the two-cell geometry of wind vector estimation, when the spatially selected cells are located in the directions of 0° and 180° relative to the flying apparatus' course, an ambiguity of the wind direction appears in the measurement. Nevertheless, to find the true wind direction, a recurring measurement after 45° change of the flying apparatus' course is required. The nearest wind directions of pairs of wind directions obtained before and after

course changing will give the true wind direction. To avoid such inconvenience under estimation of the sea surface wind speed and direction by the ARA, a modified beam shape forming the ellipse footprints should be used.
Such an altimeter should operate at a Ku-band (or at least at a C-band) using a horizontal transmit and receive polarization. A lower radar wavelength provides Doppler selection at a lower speed of flight, and at the Ku-band, the upwind/downwind and upwind/crosswind differences in the NRCS values at middle incidence angles (for wind speed of 3 to 24 m/s) are greater than at the lower bands. Horizontal transmit and receive polarizations provide greater upwind/downwind differences in the NRCS values at middle incidence angles than the vertical polarizations. Incidence angle of the second annulus zone should tend to 45°, and the incidence angle of the first annulus zone should be no less than 20°. The antenna should have different beamwidths in the vertical and horizontal planes ($\theta_{a.v} > \theta_{a.h}$) and form the ellipse footprint so that the longer axis of the footprint is rotated by approximately 45° from the horizontal projection of the longitudinal axis of a flying apparatus. It is desirable that the antenna is installed so that the longer axis of the ellipse footprint coincides with the azimuth locations of cells of the second annulus in operating regime.

4. Doppler Navigation System Application for Estimation of the Wind Speed and Direction

4.1 Doppler Navigation System

DNS is the self-contained radar system that utilizes the Doppler effect (Doppler radar) for measuring the ground speed and drift angle of flying apparatus and accomplishes its dead-reckoning navigation (Sosnovskiy & Khaymovich, 1987). The internationally authorized frequency band of 13.25 to 13.4 GHz has been allocated for airborne Doppler navigation radar. A center frequency of 13.325 GHz of the band corresponds to a wave length of 2.25 cm. This frequency represents a good compromise between too low a frequency, resulting in low-velocity sensitivity and large aircraft antenna size and beam widths, and too high a frequency, resulting in excessive absorption and backscattering effects of the atmosphere and precipitation. (Earlier Doppler radars operated in two somewhat lower frequency bands, i.e., centered at 8.8 and 9.8 GHz, respectively, but now these bands are no longer used for stand-alone Doppler radars.) (Kayton & Fried, 1997).
Measurement of the wind vector and drift angle of flying apparatus is based on change of a Doppler frequency of the signal reflected from the underlying surface, depending on a spatial position of an antenna beam. Usually, an antenna of the DNS has three beams (λ-configuration; beams 1, 2, and 3) or four beams (x-configuration; beams 1, 2, 3, and 4) located in space as represented in Fig. 5. An effective antenna beamwidth is of 3° to 10° (Kolchinskiy, et al, 1975). Power reasons (DNS should operates over water as well as over land) and sensitivity of the DNS to velocity influence a choice of a mounting angle of a beam axis in the vertical plane θ_0.
Fig. 6 shows curves of the NRCS versus incidence angle for radar system operating in the frequency band (Ke-band) currently assigned to Doppler navigation radar (Kayton & Fried, 1997). It is seen from the curves that for most types of terrain the NRCS decreases slowly with increase of the beam incidence angle. However, for water surfaces, the NRCS falls radically as the incidence angle increases and assumes different values for different conditions of sea state or water roughness. For the typical Doppler-radar incidence angles of

15° to 30° (Kolchinskiy, et al, 1975), the NRCS is considerably smaller for most sea states than for land and decreases markedly for the smoother sea state. Therefore, a conservative Doppler-radar design is based on an NRCS for the smoothest sea state over which the aircraft is expected to navigate. (Very smooth sea states are relatively rare).
There are two basic antenna system concepts used for drift angle measurement. These are the fixed-antenna system, which is used in most modern systems, and the track-stabilized (roll-and-pitch-stabilized) antenna system. For physically roll-and-pitch-stabilized antenna systems, the value of an incidence angle remains essentially constant and equal to the chosen design value. For fixed-antenna system, a conservative design is based on the NRCS and range for the largest incidence angle that would be expected for the largest combination of pitch and roll angles of the aircraft (Kayton & Fried, 1997).

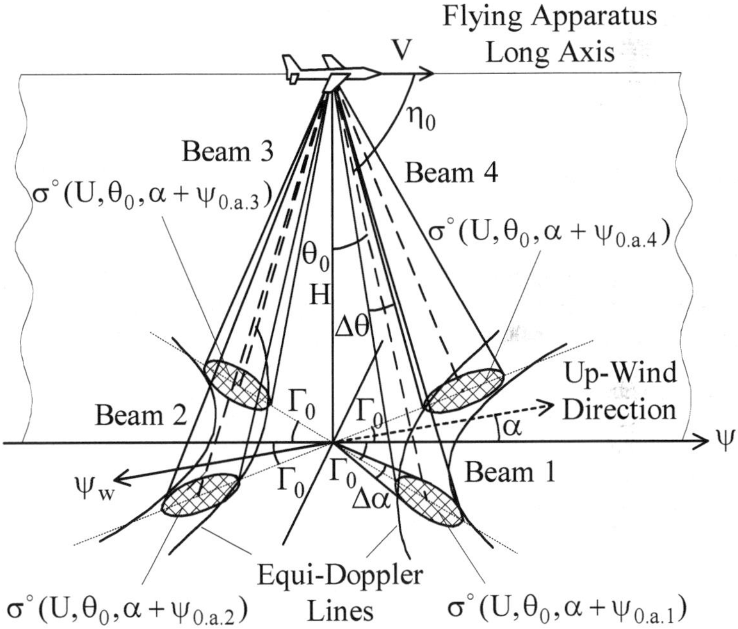

Fig. 5. Typical spatial location of the DNS beams: λ-configuration (beams 1, 2, and 3) and x-configuration (beams 1, 2, 3, and 4)

The choice of a mounting angle of a beam axis in the inclined plane η_0 (nominal angle between antenna longitudinal axis and central beam direction) represents a compromise between high sensitivity to velocity and over-water accuracy, which increases with smaller mounting angles of a beam axis in the inclined plane, and high signal return over water, which increases for larger mounting angles of a beam axis in the inclined plane. Most equipments use a mounting angle of a beam axis in the inclined plane of somewhere between 65° and 80° (Kayton & Fried, 1997). The choice of a mounting angle of a beam axis in the horizontal plane Γ_0 depends on the desired sensitivity to drift, which tends to increase

with increasing that mounting angle. For the typical Doppler-radar, mounting angles of a beam axis in the horizontal plane are of 15° to 45° (Kolchinskiy, et al, 1975).
The relationship among those mounting angles is (Kayton & Fried, 1997)

$$\cos\eta_0 = \cos\Gamma_0 \cos\theta_0 . \tag{19}$$

The mounting angle of a beam axis in the horizontal plane should satisfy the following condition $\Gamma_0 > \beta_{dr.max}$, where $\beta_{dr.max}$ is the maximum possible drift angle (Sosnovskiy & Khaymovich, 1987). The mounting angle of a beam axis in the inclined plane is defined by requirements to the width of a Doppler spectrum of the reflected signal Δf_D, which depends on the effective antenna beamwidth in the inclined plane $\theta_{a.incl}$; $\theta_{a.incl} \approx 5^\circ$ for DNS. The relative width of a Doppler spectrum $\Delta f_D / F_D$ is given by (Davydov, et al, (1977)

$$\frac{\Delta f_D}{F_D} = \frac{\theta_{a.incl}}{\sqrt{2}} \tan\eta_0 , \tag{20}$$

where F_D is the Doppler frequency, $F_D = \frac{2V_g}{\lambda}\cos\eta_0$, V_g is the aircraft velocity relative to the ground. To perform high accuracy measurements with the DNS, the following condition should be provided (Davydov, et al, (1977)

$$\frac{\Delta f_D}{F_D} \le 0.1 \div 0.2 . \tag{21}$$

Thus, from (20) and (21), the mounting angle of a beam axis in the inclined plane should satisfy the following condition

$$\eta_0 \le \arctan\left[(0.1 \div 0.2)\frac{\sqrt{2}}{\theta_{a.incl}}\right]. \tag{22}$$

From (22), assuming that the effective antenna beamwidth in the inclined plane is typical and equal to 5°, the condition of choice the mounting angle of a beam axis in the inclined plane is

$$\eta_0 \le 58.3^\circ \div 72.2^\circ . \tag{23}$$

Then, using (19), the areas of admissible mounting angles of beam axes could be obtained. Lower limits corresponding to the maximum admissible mounting angles of beam axis in the inclined plane and area of typical mounting angles of beam axes in the vertical and horizontal planes are represented in Fig. 7 (Nekrasov, 2008b). Trace 1 and trace 2 are the lower limits corresponding to the maximum admissible mounting angles of beam axis in the inclined plane of 58.3° (lower limit of high accuracy of measurement at $\Delta f_D / F_D = 0.1$) and 72.9° (lower limit of sufficient high accuracy of measurement at $\Delta f_D / F_D = 0.2$), respectively. A dash line displays the area of typical mounting angles of beam axes in the vertical and horizontal planes.

Fig. 7 demonstrates that for typical mounting angles of beam axes, sufficient high accuracy of measurement by the DNS is provided for the most part of the area of typical mounting angles in the vertical and horizontal planes. The measurement accuracy rises with increase of the beam incidence angle in the vertical plane.

The DNS multi-beam antenna allows selecting a power backscattered by the underlying surface from different directions, namely from directions corresponding to the appropriate beam relative to the aircraft course ψ, e.g. $\psi_{0.a.1}$, $\psi_{0.a.2}$, $\psi_{0.a.3}$, and $\psi_{0.a.4}$, as shown in Fig. 5. Each beam provides angular resolutions in the azimuthal and vertical planes, $\Delta\alpha$ and $\Delta\theta$ respectively. As three or four NRCS values obtained from considerably different azimuth directions are quite enough to measure the wind vector over water by intensity of reflected signal (Nekrassov, 1997), an airborne DNS can be used as a multi-beam (three- or four-beam) scatterometer for recovering the near-surface wind speed and direction. For this purpose, an airborne DNS having the following mounting angles of antenna beam axes $\theta_0 = 30^\circ$ and $\Gamma_0 = 30^\circ \div 45^\circ$, or $\theta_0 > 30^\circ$ ($\theta_0 \to 45^\circ$) and $\Gamma_0 = 30^\circ \div 45^\circ$, could be used. The second case requires a heightened transmitted power in comparison with the first case. Nevertheless, it allows a better usage the anisotropic properties of the water surface scattering at middle incidence angles to measure the near-surface wind vector, and also to increase an accuracy of measurement of typical DNS parameters.

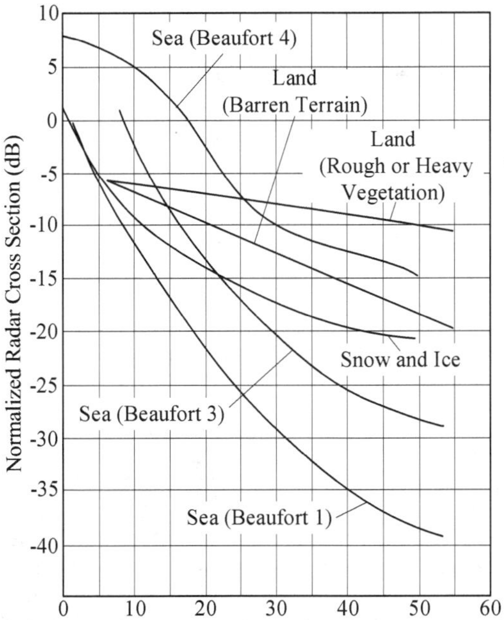

Fig. 6. Backscattering cross section per unit surface area (NRCS) versus incidence angle for different terrains at Ke-band (Kayton & Fried, 1997)

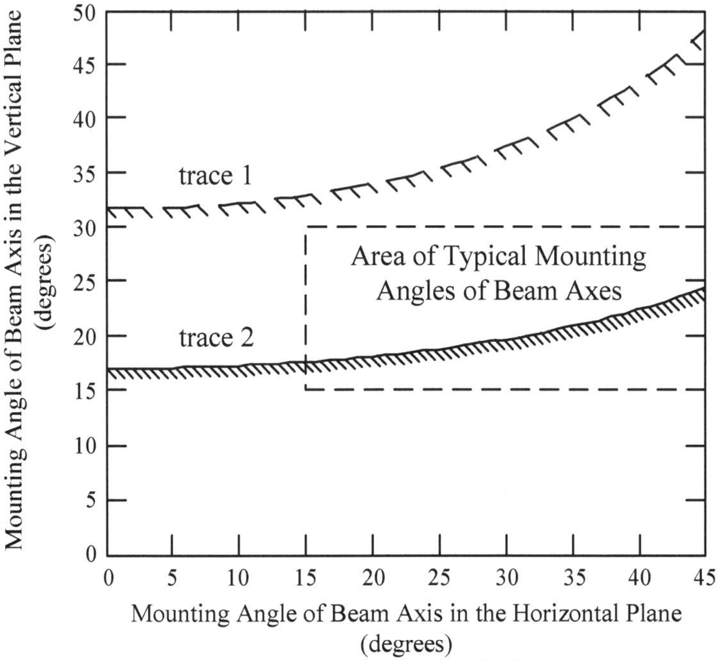

Fig. 7. Lower limits corresponding to the maximum admissible mounting angles of beam axis in the inclined plane and area of typical mounting angles of beam axes in the vertical and horizontal planes: trace 1 is the lower limit corresponding to the maximum admissible mounting angle of beam axis in the inclined plane of 58.3° (lower limit of high accuracy of measurement at $\Delta f_D / F_D = 0.1$); trace 2 is the lower limit, which corresponds to the maximum admissible mounting angle of beam axis in the inclined plane of 72.9° (lower limit of sufficient high accuracy of measurement at $\Delta f_D / F_D = 0.2$); dash line is the contour of the area of typical mounting angles of beam axes in the vertical and horizontal planes.

4.2 Wind Vector Estimation Using a Doppler Navigation System with the Fixed-Antenna System

Let a flying apparatus equipped with a DNS make a horizontal rectilinear flight with the speed V at some altitude H above the mean sea surface, the DNS use a multi-beam fixed-antenna system (physically non-stabilized to the local horizontal), and so, the values of the incidence angles of beams and beam locations in azimuthal plane are not equal to the chosen design values. Then, the actual azimuth direction of beam N $\psi_{\theta\gamma.a.N}$ relative to the aircraft course (aircraft ground track) and the actual incidence angle of beam N $\theta_{\theta\gamma.a.N}$ are

$$\psi_{\theta\gamma.a.N} = \begin{cases} \arctan\left(\dfrac{\tan(\arctan(\tan\theta_0 \sin\psi_{0.a.N}) + \gamma_{fa})}{\tan(\arctan(\tan\theta_0 \cos\psi_{0.a.N}) + \theta_{fa})}\right) \\ \quad \text{for} \quad \tan(\arctan(\tan\theta_0 \cos\psi_{0.a.N}) + \theta_{fa}) \geq 0, \\ \pi + \arctan\left(\dfrac{\tan(\arctan(\tan\theta_0 \sin\psi_{0.a.N}) + \gamma_{fa})}{\tan(\arctan(\tan\theta_0 \cos\psi_{0.a.N}) + \theta_{fa})}\right) \\ \quad \text{for} \quad \tan(\arctan(\tan\theta_0 \cos\psi_{0.a.N}) + \theta_{fa}) < 0, \end{cases} \tag{24}$$

$$\theta_{\theta\gamma.a.N} = \arctan\left(\sqrt{\tan^2(\arctan(tg\theta_0 \sin\psi_{0.a.N}) + \gamma_{fa}) + \rightarrow}\right.$$

$$\left.\overline{\rightarrow \tan^2(\arctan(\tan\theta_0 \cos\psi_{0.a.N}) + \theta_{fa})}\right), \tag{25}$$

where $\psi_{0.a.N}$ is the azimuthal mounting angle of the beam axis N relative to the aircraft course ψ, $\psi_{0.a.1} = \Gamma_0$, $\psi_{0.a.2} = 180^\circ - \Gamma_0$, $\psi_{0.a.3} = 180^\circ + \Gamma_0$, $\psi_{0.a.4} = 360^\circ - \Gamma_0$, γ_{fa} is the roll angle of flying apparatus (right roll is positive), θ_{fa} is the pitch angle of flying apparatus (pull-up is positive).

For example, five-degree roll and pitch combinations at a mounting angle of a beam axis in the vertical plane of 30° and an arbitrary mounting angle of a beam axis in the azimuthal plane may lead to a beam axis shift up to 6.4° in the vertical plane and up to 14.4° in the azimuthal plane. The same roll and pitch combinations at a mounting angle of a beam axis in the vertical plane of 45° lead to a lesser beam axis shift up to 5.5° in the vertical plane and up to 10.6° in the azimuthal plane. So, that possible difference of the actual beam axis angles from the mounting angles should be taken into account under a measuring algorithm development, and it is desirable that the mounting angle of a beam axis in the vertical plane tends to 45°.

Let the sea surface wind blow in direction ψ_w, and the angle between the up-wind direction and the aircraft course is α. Let the NRCS model function for middle incident angles be of the form (1). In case of the selected cell is narrow enough in the vertical plane, that is true for the DNS, the NRCS model function for middle incidence angles (1) can be used without any correction for wind measurement while the azimuth angular size of a cell is up to $15^\circ - 20^\circ$ (Nekrassov, 1999). If the DNS would be a multi-beam roll-and-pitch-stabilized antenna system, the NRCS values obtained with beams 1, 2, 3 and 4 would be as fallows $\sigma^\circ(U,\theta_0,\alpha+\psi_{0.a.1})$, $\sigma^\circ(U,\theta_0,\alpha+\psi_{0.a.2})$, $\sigma^\circ(U,\theta_0,\alpha+\psi_{0.a.3})$ and $\sigma^\circ(U,\theta_0,\alpha+\psi_{0.a.4})$ respectively. As the DNS has a fixed-antenna system, the NRCS values obtained with beams 1, 2, 3 and 4 are $\sigma^\circ(U,\theta_{\theta\gamma.a.1},\alpha+\psi_{\theta\gamma.a.1})$, $\sigma^\circ(U,\theta_{\theta\gamma.a.2},\alpha+\psi_{\theta\gamma.a.2})$, $\sigma^\circ(U,\theta_{\theta\gamma.a.3},\alpha+\psi_{\theta\gamma.a.3})$ and $\sigma^\circ(U,\theta_{\theta\gamma.a.4},\alpha+\psi_{\theta\gamma.a.4})$ respectively. Then, the following algorithm to estimate the wind vector over the sea surface can be proposed.

The wind speed and up-wind direction are found by solving a system of equation for a three-beam DNS

$$\begin{cases}\sigma^{\circ}(U,\theta_{\theta\gamma.a.1},\alpha+\psi_{\theta\gamma.a.1})=A(U,\theta_{\theta\gamma.a.1})+\\ \qquad B(U,\theta_{\theta\gamma.a.1})\cos(\alpha+\psi_{\theta\gamma.a.1})+\\ \qquad C(U,\theta_{\theta\gamma.a.1})\cos(2(\alpha+\psi_{\theta\gamma.a.1})),\\ \sigma^{\circ}(U,\theta_{\theta\gamma.a.2},\alpha+\psi_{\theta\gamma.a.2})=A(U,\theta_{\theta\gamma.a.2})+\\ \qquad B(U,\theta_{\theta\gamma.a.2})\cos(\alpha+\psi_{\theta\gamma.a.2})+\\ \qquad C(U,\theta_{\theta\gamma.a.2})\cos(2(\alpha+\psi_{\theta\gamma.a.2})),\\ \sigma^{\circ}(U,\theta_{\theta\gamma.a.3},\alpha+\psi_{\theta\gamma.a.3})=A(U,\theta_{\theta\gamma.a.3})+\\ \qquad B(U,\theta_{\theta\gamma.a.3})\cos(\alpha+\psi_{\theta\gamma.a.3})+\\ \qquad C(U,\theta_{\theta\gamma.a.3})\cos(2(\alpha+\psi_{\theta\gamma.a.3})).\end{cases} \tag{26}$$

and for a four-beam DNS (Nekrasov, 2008b)

$$\begin{cases}\sigma^{\circ}(U,\theta_{\theta\gamma.a.1},\alpha+\psi_{\theta\gamma.a.1})=A(U,\theta_{\theta\gamma.a.1})+\\ \qquad B(U,\theta_{\theta\gamma.a.1})\cos(\alpha+\psi_{\theta\gamma.a.1})+\\ \qquad C(U,\theta_{\theta\gamma.a.1})\cos(2(\alpha+\psi_{\theta\gamma.a.1})),\\ \sigma^{\circ}(U,\theta_{\theta\gamma.a.2},\alpha+\psi_{\theta\gamma.a.2})=A(U,\theta_{\theta\gamma.a.2})+\\ \qquad B(U,\theta_{\theta\gamma.a.2})\cos(\alpha+\psi_{\theta\gamma.a.2})+\\ \qquad C(U,\theta_{\theta\gamma.a.2})\cos(2(\alpha+\psi_{\theta\gamma.a.2})),\\ \sigma^{\circ}(U,\theta_{\theta\gamma.a.3},\alpha+\psi_{\theta\gamma.a.3})=A(U,\theta_{\theta\gamma.a.3})+\\ \qquad B(U,\theta_{\theta\gamma.a.3})\cos(\alpha+\psi_{\theta\gamma.a.3})+\\ \qquad C(U,\theta_{\theta\gamma.a.3})\cos(2(\alpha+\psi_{\theta\gamma.a.3})),\\ \sigma^{\circ}(U,\theta_{\theta\gamma.a.4},\alpha+\psi_{\theta\gamma.a.4})=A(U,\theta_{\theta\gamma.a.4})+\\ \qquad B(U,\theta_{\theta\gamma.a.4})\cos(\alpha+\psi_{\theta\gamma.a.4})+\\ \qquad C(U,\theta_{\theta\gamma.a.4})\cos(2(\alpha+\psi_{\theta\gamma.a.4})).\end{cases} \tag{27}$$

Then, the wind direction can be found as follows

$$\psi_w=\alpha\pm180^{\circ}. \tag{28}$$

4.3 Wind Vector Estimation Using a Doppler Navigation System with the Roll-And-Pitch-Stabilized Antenna System

Let a flying apparatus equipped with a DNS make a horizontal rectilinear flight with the speed V at some altitude H above the mean sea surface, the DNS use a roll-and-pitch-stabilized antenna (physically stabilized to the local horizontal) and so the value of an incidence angle θ remains essentially constant and equal to the chosen design value θ_0. Let the aircraft velocity vector V be located along the intersection of the local horizontal plane and the local vertical plane through the longitudinal axis of aircraft (condition of no-drift angle and no-climb angle), that means that the aircraft flight is horizontal and the aircraft course ψ is the same as the aircraft's ground track. The directions of the DNS beams 1, 2, 3, 4 relative to the aircraft course are $\psi_{0.a.1}$, $\psi_{0.a.2}$, $\psi_{0.a.3}$, and $\psi_{0.a.4}$, respectively (Fig. 5.). Let the sea surface wind blow in direction ψ_w, and the angle between the up-wind direction and the aircraft course is α. Let the NRCS model function for middle incident angles be of the form (1). Then, the NRCS values obtained with beams 1, 2, 3, 4 are $\sigma^\circ(U,\theta_0,\alpha+\psi_{0.a.1})$, $\sigma^\circ(U,\theta_0,\alpha+\psi_{0.a.2})$, $\sigma^\circ(U,\theta_0,\alpha+\psi_{0.a.3})$ and $\sigma^\circ(U,\theta_0,\alpha+\psi_{0.a.4})$ respectively, and the wind speed and up-wind direction are found by solving a system of equation (26) for a three-beam DNS, or (27) for a four-beam DNS. Then, the wind direction can be found from (28).

Let a mounting angle of a beam axis in the horizontal plane is equal to 45°. As the DNS antenna is roll-and-pitch-stabilized, the directions of the DNS beams remains constant, and the NRCS values obtained with beams 1, 2, 3, 4 are $\sigma^\circ(U,\theta_0,\alpha+45^\circ)$, $\sigma^\circ(U,\theta_0,\alpha+135^\circ)$, $\sigma^\circ(U,\theta,\alpha+225^\circ)$, $\sigma^\circ(U,\theta_0,\alpha+315^\circ)$. Then, the following algorithm to estimate the wind vector over the sea surface can be proposed.

The wind speed can be found from the following equation (Nekrasov, 2005)

$$U=\left(\frac{A(U,\theta_0)}{a_0(\theta_0)}\right)^{1/\gamma_0(\theta_0)}=$$

$$\left(\frac{\sigma^\circ(U,\theta_0,\alpha+45^\circ)+\sigma^\circ(U,\theta_0,\alpha+135^\circ)+}{4a_0(\theta_0)}\right.\rightarrow$$

$$\left.\frac{\sigma^\circ(U,\theta_0,\alpha+225^\circ)+\sigma^\circ(U,\theta_0,\alpha+315^\circ)}{}\right)^{1/\gamma_0(\theta_0)}. \qquad (29)$$

To find the wind direction, at first, the space of possible solutions could be divided into four quadrants. The quadrant, which contains the solution, should be found, and then, the unique wind direction can be obtained as follows

$$\begin{gathered}
\textit{if } \sigma^\circ(U,\theta_0,\alpha+45^\circ)>\sigma^\circ(U,\theta_0,\alpha+225^\circ)\\
\textit{and } \sigma^\circ(U,\theta_0,\alpha+135^\circ)\geq\sigma^\circ(U,\theta_0,\alpha+315^\circ) \Rightarrow \psi_w=\psi+45^\circ+\alpha_q\pm180^\circ,\\
\textit{if } \sigma^\circ(U,\theta_0,\alpha+45^\circ)\leq\sigma^\circ(U,\theta_0,\alpha+225^\circ)\\
\textit{and } \sigma^\circ(U,\theta_0,\alpha+135^\circ)>\sigma^\circ(U,\theta_0,\alpha+315^\circ) \Rightarrow \psi_w=\psi+225^\circ-\alpha_q\pm180^\circ,\\
\textit{if } \sigma^\circ(U,\theta_0,\alpha+45^\circ)<\sigma^\circ(U,\theta_0,\alpha+225^\circ)\\
\textit{and } \sigma^\circ(U,\theta_0,\alpha+135^\circ)\leq\sigma^\circ(U,\theta_0,\alpha+315^\circ) \Rightarrow \psi_w=\psi+225^\circ+\alpha_q\pm180^\circ,\\
\textit{if } \sigma^\circ(U,\theta_0,\alpha+45^\circ)\geq\sigma^\circ(U,\theta_0,\alpha+225^\circ)\\
\textit{and } \sigma^\circ(U,\theta_0,\alpha+135^\circ)<\sigma^\circ(U,\theta_0,\alpha+315^\circ) \Rightarrow \psi_w=\psi+45^\circ-\alpha_q\pm180^\circ,
\end{gathered} \qquad (30)$$

where α_q is the angle of the wind in the quadrant,

$$\alpha_q = \begin{cases} 0^\circ, & A_1 \geq 1 \\ 0.5 \arccos A_1, & -1 < A_1 < 1, \\ 90^\circ, & A_1 \leq 1 \end{cases} \quad (31)$$

$$A_1 = \frac{\sigma^\circ(U,\theta_0,\alpha+45^\circ) + \sigma^\circ(U,\theta_0,\alpha+225^\circ) - 2A(U,\theta_0)}{2C(U,\theta_0)}. \quad (32)$$

4.4 Conclusion to Wind Vector Estimation Using a Doppler Navigation System

The analysis of the DNS, the backscatter model function and the geometry of wind vector measurements have shown that the wind vector over sea can be measured by an airborne DNS that has three- or four-beam fixed or roll-and-pitch-stabilized antenna system and employed as a multi-beam scatterometer.

For this purpose, an airborne DNS having the following mounting angles of antenna beam axes $\theta_0 = 30^\circ$ and $\Gamma_0 = 30^\circ \div 45^\circ$, or $\theta_0 > 30^\circ$ ($\theta_0 \to 45^\circ$) and $\Gamma_0 = 30^\circ \div 45^\circ$, could be used. The second case requires a heightened transmitted power in comparison with the first case. Nevertheless, it allows a better usage the anisotropic properties of the water surface scattering at middle incidence angles to measure the near-surface wind vector, to increase an accuracy of measurement of typical DNS parameters, and also to decrease a beam axis deviation due to roll and pitch influence.

5. Measuring the Wind Vector by an Airborne Weather Radar

5.1 Airborne Weather Radar

AWR is radar equipment mounted on an aircraft for purposes of weather observation and avoidance, aircraft position finding relative to landmarks, and drift angle measuring (Sosnovsky, et al, 1990). The AWR is necessary equipment of any civil airplane. It must be obligatory installed on all civil airliners. All military transport aircrafts are usually equipped by weather radars too. Due to the specificity of airborne application, designers of avionics systems always try to use the most efficient progressive methods and reliable engineering solutions that provide flight safety and flight regularity at harsh environment (Yanovsky, 2005).

The development of the AWR is mainly associated with growing functionalities on detection of different dangerous weather phenomena. The radar observations involved in a weather mode are magnitude detection of reflections from clouds and precipitation and Doppler measurements of the motion of particles within a weather formation. Magnitude detection allows determination of particle type (rain, snow, hail, etc.) and precipitation rate. Doppler measurements can be made to yield estimates of turbulence intensity and wind speed. Reliable determination of the presence and severity of the phenomenon known as wind shear is an important area of study too (Kayton & Fried, 1997).

Nevertheless, the second important assignment of the AWR is providing a pilot with navigation information using earth surface mapping. In this case a possibility to extract some navigation information that allows determining aircraft position with respect to a geographic map is very important for air navigation. Landmark's coordinates relative to the airplane that are measured by the AWR give a possibility to set flight computer for exacter and more efficient fulfilment of en-route flight, cargo delivery, and cargo throw down to the given point. These improve tactical possibilities of transport aircraft, airplanes of search-and-rescue service, and local airways (Yanovsky, 2005).

Other specific function of the AWR is interaction with ground-based responder beacons. New functions of the airborne weather radar are detection and visualization of runways at approach landing as well as visualization of taxiways and obstacles on the taxiway at taxiing.

Certainly, not all of the mentioned functions are implemented in a particular airborne radar system. Nevertheless, the airborne weather radar always is a multifunctional system that provides earth surface surveillance and weather observation. Usually, weather radar should at least enable to detect clouds and precipitation, select zones of meteorological danger, and show radar image of surface in the map mode.

AWRs or multimode radars with a weather mode are usually nose mounted. Most AWRs operate in either X- or C-band (Kayton & Fried, 1997). The λ^{-4} dependence of weather formations on carrier wavelength λ favours X-band radar for their detecting. At the same time, the X-band provides the performance of the long-range weather mode better than Ku-band. The AWR antenna, in the ground-mapping mode, has a large cosecant-squared elevation beam where horizontal dimension is narrow (2° to 6°) while the other is relatively broad (10° to 30°), and it sweeps in an azimuth sector (up to ±100°) (Kayton & Fried, 1997), (Sosnovskiy & Khaymovich, 1987). The scan plane is horizontal because of the antenna is stabilized (roll-and-pitch-stabilized). Those features allow supposing that the AWR in the ground-mapping mode can be also used as a scatterometer for the wind speed and direction retrieval over water.

5.2 Wind Vector Estimation Using an Airborne Weather Radar Having a Narrow Scanning Sector

Let a flying apparatus equipped with an AWR make a horizontal rectilinear flight with the speed V at some altitude H above the mean sea surface, an AWR operate in the ground-mapping mode as a scatterometer, the radar antenna have different beamwidth in the vertical $\theta_{a.v}$ and horizontal $\theta_{a.h}$ planes ($\theta_{a.v} > \theta_{a.h}$) as shown in Fig. 8, and scan periodically through an azimuth in a narrow sector (narrower than ±90° but no narrower than ±45°) as shown in Fig. 9. Also let a delay selection be used to provide a necessary resolution in the vertical plane.

Then, the beam scanning allows selecting a power backscattered by the underlying surface for given incidence angle θ from various directions in an azimuth sector, e.g. from directions $\alpha - 45^\circ$, α and $\alpha + 45^\circ$ relative to the up-wind direction as represented in Fig. 9. Angular (narrow horizontal beamwidth) selection in the horizontal plane along with the delay selection provide angular resolutions in the azimuthal and vertical planes, $\Delta\alpha$ and $\Delta\theta$ respectively. As three or four NRCS values obtained from considerably different azimuth directions are quite enough to measure the wind vector over water by intensity of reflected signal (Nekrassov, 1997), AWR can be used as a scanning scatterometer for recovering the near-surface wind speed and direction.

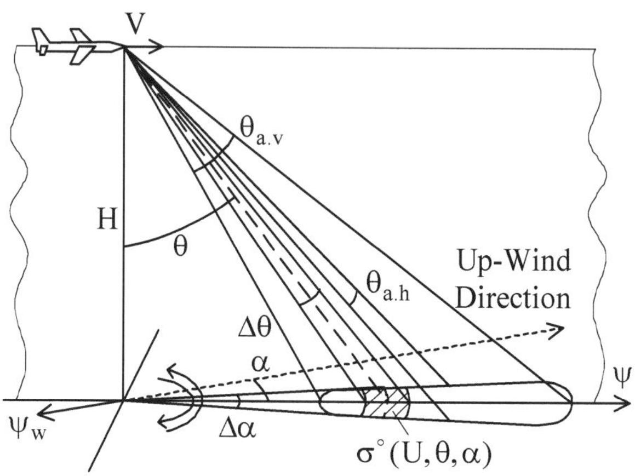

Fig. 8. Airborne weather radar beam and selected cell geometry

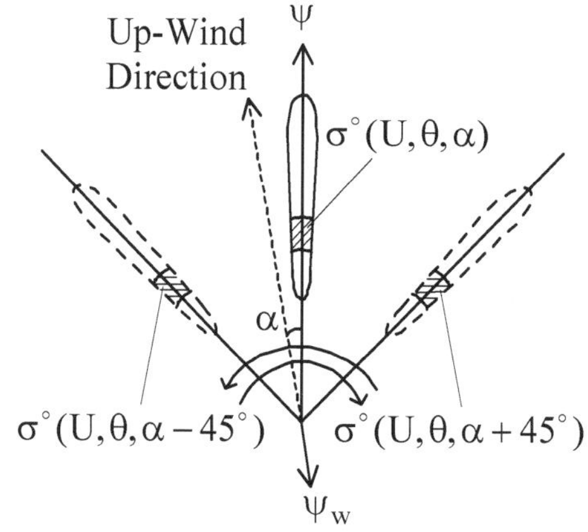

Fig. 9. Scanning beam footprints in a narrow sector and selected cells

Let the sea surface wind blow in direction ψ_w, and the angle between the up-wind direction and the aircraft course ψ is α. Let the NRCS model function for middle incident angles be of the form (1). In case of the selected cell is narrow enough in the vertical plane, the NRCS model function for middle incidence angles (1) can be used without any correction for wind measurement while the azimuth angular size of a cell is up to $15^\circ - 20^\circ$ (Nekrassov, 1999).

Let the NRCS values be obtained only from directions $\alpha - 45^\circ$, α and $\alpha + 45^\circ$. They are $\sigma^\circ(U,\theta,\alpha - 45^\circ)$, $\sigma^\circ(U,\theta,\alpha)$, and $\sigma^\circ(U,\theta,\alpha + 45^\circ)$ respectively. Then, the following algorithm to estimate the wind vector over the sea surface can be proposed.

The wind speed and up-wind direction are found by solving the following system of equations (Nekrasov & Labun 2008)

$$\begin{cases} \sigma^\circ(U,\theta,\alpha-45^\circ) = A(U,\theta) + B(U,\theta)\cos(\alpha-45^\circ) + \\ \qquad C(U,\theta)\cos(2(\alpha-45^\circ)), \\ \sigma^\circ(U,\theta,\alpha) = A(U,\theta) + B(U,\theta)\cos\alpha + \\ \qquad C(U,\theta)\cos(2\alpha)), \\ \sigma^\circ(U,\theta,\alpha+45^\circ) = A(U,\theta) + B(U,\theta)\cos(\alpha+45^\circ) + \\ \qquad C(U,\theta)\cos(2(\alpha+45^\circ)). \end{cases} \tag{33}$$

The system of equations (33) could be solved approximately using searching procedure within the ranges of discrete values of possible solutions. Then, the wind direction can be found from (28).

It is necessary to note that the AWR in the mode of the wind vector measurement should use the horizontal transmit and receive polarization, and provide the incidence angle of the selected sells $\theta \to 45^\circ$ that is explained by better usage of the anisotropic properties of the water surface scattering at middle incidence angles and also by power reasons. For water surfaces, the NRCS falls radically as the incidence angle increases and assumes different values for different conditions of sea state or water roughness while, for most other types of terrain, the NRCS decreases slowly with increase of the beam incidence angle (Kayton & Fried, 1997). Or at least, the incidence angle of the selected sells should be in the range of validity for the NRCS model function (1), and should be out of the "shadow" region.

5.2 Wind Vector Estimation Using an Airborne Weather Radar Having a Wide Scanning Sector

Let a flying apparatus equipped with an AWR make a horizontal rectilinear flight with the speed V at some altitude H above the mean sea surface, the AWR operate in the ground-mapping mode as a scatterometer, the radar antenna have different beamwidth in the vertical $\theta_{a.v}$ and horizontal $\theta_{a.h}$ planes ($\theta_{a.v} > \theta_{a.h}$) as shown in Fig. 8, and scan periodically through an azimuth in a sector of ±90° or wider as shown in Fig. 10. Also let a delay selection be used to provide a necessary resolution in the vertical plane.

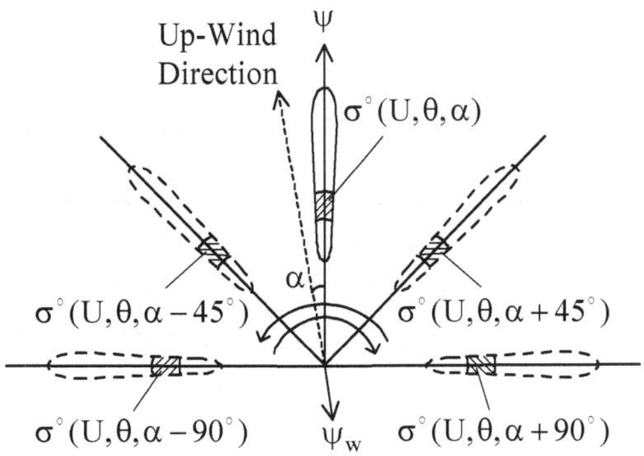

Fig. 10. Scanning beam footprints in a wide sector and selected cells

Then, the beam scanning allows selecting a power backscattered by the underlying surface for given incidence angle θ from various directions in an azimuth sector, e.g. from directions $\alpha-90^\circ$, $\alpha-45^\circ$, α, $\alpha+45^\circ$, and $\alpha+90^\circ$ relative to the up-wind direction as represented in Fig. 10. Angular (narrow horizontal beamwidth) selection in the horizontal plane along with the delay selection provide angular resolutions in the azimuthal and vertical planes, $\Delta\alpha$ and $\Delta\theta$ respectively.

Let the sea surface wind blow in direction ψ_w, and the angle between the up-wind direction and the aircraft course ψ is α, and the NRCS values be obtained only from directions $\alpha-90^\circ$, $\alpha-45^\circ$, α, $\alpha+45^\circ$, and $\alpha+90^\circ$. They are $\sigma^\circ(U,\theta,\alpha-90^\circ)$ $\sigma^\circ(U,\theta,\alpha-45^\circ)$, $\sigma^\circ(U,\theta,\alpha)$, $\sigma^\circ(U,\theta,\alpha+45^\circ)$, and $\sigma^\circ(U,\theta,\alpha+90^\circ)$ respectively. Then, the following algorithm to estimate the wind vector over the sea surface can be proposed.

Using the measuring geometry, equation (1), and taking into account that the azimuth angular size of the selected sells are narrow enough, the following system of equations can be written down (Nekrasov, 2009)

$$\begin{cases} \sigma^\circ(U,\theta,\alpha-90^\circ)=A(U,\theta)+B(U,\theta)\cos(\alpha-90^\circ)+ \\ \qquad C(U,\theta)\cos(2(\alpha-90^\circ)), \\ \sigma^\circ(U,\theta,\alpha-45^\circ)=A(U,\theta)+B(U,\theta)\cos(\alpha-45^\circ)+ \\ \qquad C(U,\theta)\cos(2(\alpha-45^\circ)), \\ \sigma^\circ(U,\theta,\alpha)=A(U,\theta)+B(U,\theta)\cos\alpha+C(U,\theta)\cos(2\alpha)), \\ \sigma^\circ(U,\theta,\alpha+45^\circ)=A(U,\theta)+B(U,\theta)\cos(\alpha+45^\circ)+ \\ \qquad C(U,\theta)\cos(2(\alpha+45^\circ)), \\ \sigma^\circ(U,\theta,\alpha+90^\circ)=A(U,\theta)+B(U,\theta)\cos(\alpha+90^\circ)+ \\ \qquad C(U,\theta)\cos(2(\alpha+90^\circ)). \end{cases} \tag{34}$$

From the sum of the first and the fifth equations of (34) we have

$$\cos 2\alpha=\frac{\sigma^\circ(U,\theta,\alpha-90^\circ)+\sigma^\circ(U,\theta,\alpha+90^\circ)-2A(U,\theta)}{2C(U,\theta)}. \tag{35}$$

From the sum of the second and the fourth equations of (34) we obtain

$$\cos\alpha=\frac{\sigma^\circ(U,\theta,\alpha-45^\circ)+\sigma^\circ(U,\theta,\alpha+45^\circ)-2A(U,\theta)}{\sqrt{2}B(U,\theta)}. \tag{36}$$

Substitution of $\cos 2\alpha$ from (35) and $\cos\alpha$ from (36) into the third equation of system (34) gives the following equation

$$A(U,\theta) = \frac{1}{\sqrt{2}}\sigma^\circ(U,\theta,\alpha) - \frac{1}{2}\left(\sigma^\circ(U,\theta,\alpha-45^\circ) + \sigma^\circ(U,\theta,\alpha+45^\circ)\right) - \frac{1}{2\sqrt{2}}\left(\sigma^\circ(U,\theta,\alpha-90^\circ) + \sigma^\circ(U,\theta,\alpha+90^\circ)\right). \quad (37)$$

The wind speed over water can be calculated from (37). Then, two possible up-wind directions relative the course of the flying apparatus can be found from (36). They are

$$\alpha_{1,2} = \pm\arccos\left(\frac{\sigma^\circ(U,\theta,\alpha-45^\circ) + \sigma^\circ(U,\theta,\alpha+45^\circ) - 2A(U,\theta)}{\sqrt{2}B(U,\theta)}\right). \quad (38)$$

The unique up-wind direction α relative the course can be found by substitution of the values α_1 and α_2 into the first and the fifth equations of the system of equations (34). Finally, the wind direction ψ_w can be found from (18).

5.4 Conclusion to Wind Vector Estimation Using an Airborne Weather Radar

The analysis of the AWR, the backscatter model function and the geometry of wind vector measurements have shown that the wind vector over sea can be measured by an AWR employed in the ground-mapping mode as a scatterometer scanning periodically through an azimuth in a narrow sector (narrower than ±90° but no narrower than ±45°) and a wide sector (±90° or wider), in addition to its typical meteorological and navigation application.
The AWR in the mode of the wind vector measurement should use the horizontal transmit and receive polarization, and provide the incidence angle of the selected sells $\theta \to 45^\circ$ that is explained by better usage of the anisotropic properties of the water surface scattering at middle incidence angles and also by power reasons. Or at least, the incidence angle of the selected sells should be in the range of validity for the NRCS model function (1), and should be out of the "shadow" region.

6. Conclusion

The study has shown that the ARA, DNS and AWR can be used as measuring instruments for the wind vector over water in addition to their typical navigation applications. The ARA should be employed as a nadir-looking wide-beam short-pulse scatterometer in conjunction with Doppler filtering. The DNS having the three- or four-beam fixed or roll-and-pitch-stabilized antenna system should be used as a multi-beam scatterometer. The AWR may be employed in the ground-mapping mode as a scatterometer scanning periodically through an azimuth in the narrow or wide sector.
As no commercial equipment exists so far, the results obtained can be used for creation of a new airborne radar system for stand-alone and simultaneous measurements of the water surface backscattering signature and the wind vector over water, including applications to amphibian aircraft safe landing on the water surface, in particular under search and rescue missions or fire-fighting operations in the fire risk coastal areas that could help to save the human lives and environment.

7. Acknowledgment

I would like to express my sincere thanks to Prof. Dr.-Ing. Klaus Schünemann, Prof. Dr.-Ing. Arne Jacob and Prof. Dr.-Ing. Udo Carl (Hamburg University of Technology), Prof. Dr.-Ing. Reinhard Knöchel (Christian-Albrechts-University of Kiel), Dr. Wolfgang Rosenthal (GKSS Research Center, Geesthacht), Prof. Dr. hab. Adam Krężel (University of Gdańsk), Prof. Dr. J. Pereira Osório (University of Porto), Prof. Dr. ir. Peter Hoogeboom and Prof. Dr. ir. Leo P. Ligthart (Delft University of Technology), Prof. Dr. Maurizio Migliaccio, Prof. Dr. Paolo Corona and Prof. Dr. Renato Passaro (University of Naples „Parthenope"), Dr. Ján Labun and Assoc. Prof. RNDr. František Olejník (Technical University of Košice) for their research opportunity provided, and to the Ministry of Education and Science of Russia, the German Academic Exchange Service (DAAD), the Information Processing Center (OPI), Poland, the Institute for International Scientific and Technological Co-operation (ICCTI), Portugal, the Netherlands Organization for Scientific Research (NWO), the National Research Council of Italy (CNR), the Slovak Academic Information Agency (SAIA) for their research grants and fellowships.

8. References

Carswell, J.R.; Carson, S.C.; McIntosh, R.E.; Li, F.K.; Neumann, G.; McLaughlin, D.J.; Wilkerson, J.C.; Black, P.G. & Nghiem, S.V. (1994) Airborne scatterometers: Investigating ocean backscatter under low- and high-wind conditions, *Proceedings of the IEEE*, Vol. 82, No. 12, pp. 1835-1860.

Chelton, D.B. & McCabe, P.J. (1985) A review of satellite altimeter measurement of sea surface wind speed: With a proposed new algorithm, *Journal of Geophysical Research*, Vol. 90, No. C3, pp. 4707-4720.

Davydov, P.S.; Zhavoronkov, V.P.; Kashcheyev, G.V.; Krinitsyn, V.V.; Uvarov, V.S. & Khresin, I.N. (1977) *Radar Systems of Flying Apparatuses*, Transport, Moscow, 352 p., *in Russian*.

Du, Y.; Vachon, P.W. & Wolf, J. (2002) Wind direction estimation from SAR images of the ocean using wavelet analysis, *Canadian Journal of Remote Sensing*, Vol. 28, No. 3, pp. 498-509.

Feindt, F.; Wismann, V.; Alpers, W. & Keller, W.C. (1986) Airborne measurements of the ocean radar cross section at 5.3 GHz as a function of wind speed, *Radio Science*, Vol. 21, No. 5, pp. 845-856.

Hammond, D.L.; Mennella, R.A., & Walsh, E.J. (1977) Short pulse radar used to measure sea surface wind speed and SWH, *IEEE Transactions on Antennas and Propagation*, Vol. 25, No. 1, pp. 61-67.

Hildebrand, P.H. (1994) Estimation of sea-surface wind using backscatter cross-section measurements from airborne research weather radar, *IEEE Transactions on Geoscience and Remote Sensing*, Vol. 32, No. 1, pp. 110-117.

Jackson, F.C.; Walton, W.T.; Hines, D.E.; Walter, B.A. & Peng, C.Y. (1992) Sea surface mean square slope from K_u-band backscatter data, *Journal of Geophysical Research*, Vol. 97, No. C7, pp. 11411-11427.

Kayton M. & Fried W.R. (1997) *Avionics Navigation Systems*, John Wieley & Sons, New York, 773 p.

Kolchinskiy, V.Ye.; Mandurovskiy, I.A. & Konstantinovskiy, M.I. (1975) *Autonomous Doppler Facilities and Systems for Navigation of Flying Apparatuses,* Sovetskoye Radio, Moscow, 432 p., *in Russian.*

Komen, G.J.; Cavaleri, L.; Donelan, M.; Hasselmann, K.; Hasselmann, S. & Janssen, P.A.E.M. (1994) *Dynamics and Modelling of Ocean Waves,* Cambridge University Press, Cambridge, 532 p.

Komjathy, A.; Armatys, M.; Masters, D.; Axelrad, P.; Zavorotny, V.U. & Katzberg, S.J. (2001) Developments in using GPS for oceanographic remote sensing: Retrieval of ocean surface wind speed and wind direction, *CD Proceedings of the ION National Technical Meeting,* Long Beach, CA, USA, 9 p.

Komjathy, A.; Zavorotny, V.; Axelrad, P.; Born, G. & Garrison, J.L. (2000) GPS signal scattering from sea surface: Wind speed retrieval using experimental data and theoretical model, *Remote Sensing of Environment,* Vol. 73, No. 2, pp. 162–174.

Long, D.G.; Donelan, M.A.; Freilich, M.H.; Graber, H.C.; Masuko, H.; Pierson, W.J.; Plant, W.J.; Weissman, D. & Wentz, F. (1996) Current progress in Ku-band model functions, Brigham Young Univ., USA, Tech. Rep. MERS 96-002, 88 p.

Masuko, H.; Okamoto, K.; Shimada M. & Niwa, S. (1986) Measurement of microwave backscattering signatures of the ocean surface using X band and Ka band airborne scatterometers, *Journal of Geophysical Research,* Vol. 91, No. C11, pp. 13065–13083.

Melnik, Yu.A. (1980) *Radar Methods of the Earth Exploration,* Sovetskoye Radio, Moscow, 264 p., *in Russian.*

Moore, R.K. & Fung, A.K. (1979) Radar determination of winds at sea, *Proceedings of the IEEE,* Vol. 67, No. 11, pp. 1504–1521.

Nekrasov, A. (2005) On possibility to measure the sea surface wind vector by the Doppler navigation system of flying apparatus. *Proceedings of RADAR 2005,* Arlington, Virginia, USA, pp. 747-752.

Nekrasov, A. (2007) Measurement of the wind vector over sea by an airborne radar altimeter, which has an antenna with the ellipse beam shape, *Proceedings of APMC 2007,* Bangkok, Thailand, pp. 91-94.

Nekrasov, A. (2008a) Measurement of the wind vector over sea by an airborne radar altimeter having an antenna with the different beamwidth in the vertical and horizontal planes, *IEEE Geoscience and Remote Sensing Letters,* Vol. 5, No. 1, pp. 31-33.

Nekrasov, A. (2008b) Measuring the sea surface wind vector by the Doppler navigation system of flying apparatus that has a four-beam fixed-antenna system, *Proceedings of RADAR 2008,* Adelaide, Australia, pp. 493-498.

Nekrasov, A. (2009) Measurement of the sea surface wind vector by the airborne weather radar having a wide scanning sector, *Accepted to Radar 2009,* Bordeaux, France, 4 p.

Nekrasov, A. & Labun, J. (2008) About measurement of the sea surface wind vector by the airborne weather radar, *Acta Avionica,* Vol. X, No. 16, pp. 98-103.

Nekrassov, A. (1997) Measurement of sea surface wind speed and its navigational direction from flying apparatus, *Proceedings of Oceans'97,* Halifax, Nova Scotia, Canada, pp. 83–86.

Nekrassov, A. (1999) Sea surface wind vector measurement by airborne scatterometer having wide-beam antenna in horizontal plane, *Proceedings of IGARSS'99,* Hamburg, Germany, Vol. 2, pp. 1001–1003.

Nekrassov, A. (2001) Measurement of the sea surface wind speed and direction by an airborne microwave radar altimeter, GKSS Report No. GKSS/2001/38, Geesthacht, Germany, 17 p.

Nekrassov, A. (2002) On airborne measurement of the sea surface wind vector by a scatterometer (altimeter) with a nadir-looking wide-beam antenna, *IEEE Transactions on Geoscience and Remote Sensing*, Vol. 40, No. 10, pp. 2111-2116.

Nekrassov, A. (2003) Airborne measurement of the sea surface wind vector by a microwave radar altimeter at low speed of flight, *IEICE Transactions on Electronics*, Vol. E86-C, No. 8, pp. 1572-1579.

Raney, R.K. (1998) The delay/Doppler radar altimeter, *IEEE Transactions on Geoscience and Remote Sensing*, Vol. 36, No. 5, pp. 1578–1588.

Romeiser, R.; Schmidt, A. & Alpers, W. (1994) A three-scale composite surface model for the ocean wave-radar modulation transfer function, *Journal of Geophysical Research*, Vol. 99, No. C5, pp. 9785–9801.

Schöne, T. & Eickschen, S. (2000) Wind speed and SWH calibration for radar altimetry in the North Sea, *CD Proceedings of the ERS-Envisat Symposium*, Gothenburg, Sweden, p. 8.

Sosnovskiy, A.A. & Khaymovich, I.A. (1987) *Radio-Electronic Equipment of Flying Apparatuses*, Transport, Moscow, 256 p., *in Russian.*

Sosnovsky, A.A.; Khaymovich, I.A.; Lutin, E.A. & Maximov, I.B. (1990) *Aviation Radio Navigation: Handbook*, Transport, Moscow, 264 p., *in Russian.*

Spencer, M.W. & Graf, J.E. (1997) The NASA scatterometer (NSCAT) mission, *Backscatter*, Vol. 8, No. 4, pp. 18–24.

Spencer, W.M.; Tsai, W.-Y. & Long, D.G. (2000a) High resolution scatterometry by simultaneous range/Doppler discrimination, *Proceedings of IGARSS 2000*, Honolulu, Hawaii, USA, pp. 3166–3168.

Spencer, W.M.; Wu, C. & Long, D.G. (2000b) Improved resolution backscatter measurements with the SeaWinds pencil-beam scatterometer, *IEEE Transactions on Geoscience and Remote Sensing*, Vol. 38, No. 1, pp. 89–104.

Ulaby, F.T.; Moore, R.K. & Fung, A.K. (1982) *Microwave Remote Sensing: Active and Pasive*, Volume 2: *Radar Remote Sensing and Surface Scattering and Emission Theory*, Addison-Wesley, London, 1064 p.

Wackerman, C.C.; Rufenach, C.L.; Shuchman, R.A.; Johannessen, J.A. & Davidson, K.L. (1996) Wind vector retrieval using ERS-1 synthetic aperture radar imagery, *IEEE Transactions on Geoscience and Remote Sensing*, Vol. 34, No. 6, pp. 1343–1352.

Wentz, F.J.; Peteherych, S. & Thomas, L.A. (1984) A model function for ocean radar cross sections at 14.6 GHz, *Journal of Geophysical Research*, Vol. 89, No. C5, pp. 3689-3704.

Wismann, V. (1989) Messung der Windgeschwindigkeit über dem Meer mit einem flugzeuggetragenen 5.3 GHz Scatterometer, Dissertation zur Erlangung des Grades eines Doktors der Naturwissenschaften, Universität Bremen, Bremen, Germany, 119 S.

Yanovsky, F.J. (2005) Evolution and prospects of airborne weather radar functionality and technology, *Proceedings of ICECom 2005*, Dubrovnik, Croatia, pp. 1–4.

Young, I.R. (1993) An estimate of the Geosat altimeter wind speed algorithm at high wind speeds, *Journal of Geophysical Research*, Vol. 98, No. C11, pp. 20275–20285.

27

Passive Microwave Remote Sensing of Rain from Satellite Sensors

Sante Laviola and Vincenzo Levizzani
Institute of Atmospheric Sciences and Climate – National Research Council
Bologna, Italy

Abstract
The purpose of this chapter is to offer an accurate treatment of relevant aspects of satellite remote sensing of precipitation using passive microwave (PMW) radiometers. Microwave observations of the Earth's system differ substantially from those based on optical and infrared wavelengths. Visible (VIS) and infrared (IR) instruments essentially sense the cloud top properties by measuring reflected or emitted radiation. Microwave frequencies, on the contrary, possess greater penetrating capabilities than optical radiation and can thus be exploited to investigate cloud internal properties by retrieving the interaction of hydrometeors with the radiation field. This fact is particularly true in the remote sensing of precipitation, where the impact of the volume of raindrops on the radiative field is directly linked to the total extinction of incident radiation.
The following paragraphs will be focused on one hand on theoretical considerations at the foundations of thermal radiation processes and on the other the attention will be centered on a treatment of practical aspects of retrieving precipitation in the microwave bands.
Particular attention will be dedicated in the first part of the chapter to the radiative transfer theory in the microwaves by using the Rayleigh-Jeans formulation and a description of the absorption and scattering processes associated with the Mie theory with the approximation of extinction from poly-disperse media as proxies for natural media. This section will create a valid substrate to understand and theoretically evaluate the impact of atmospheric constituents such as precipitation types or hydrometeor phases and sizes on the natural radiation emitted from the Earth. Furthermore, the theoretical considerations of this section will be compared in the second part of the chapter with real measurements. Making use of a wide suite of microwave frequencies of a new generation of PMW sensors flying on board polar orbiting satellites the attenuation of the microwave signal due to rain clouds will be discussed possibly discerning the contribution to the total radiation of the emissivity from various surfaces falling into the satellite field of view. Finally, a new microwave high-frequency method to retrieve and classify precipitation types will be presented.

1. Introduction

Remote sensing from satellite-based sensors has established itself as the key method for the observation and monitoring of the planet Earth because of several reasons. From a practical point of view, observations from orbital platforms guarantee continuous measurements of very wide regions of the globe especially over inaccessible areas such as impervious mountains or open oceans where measurement campaigns are often difficult, expensive and even dangerous. Scientifically speaking, since a satellite platform generally "carries" a large instrumental suite covering several frequency-channels, ery diverse observations and studies can be planned at same time exploiting data from a single space mission. Finally, especially for last generation of geostationary and low-orbiting satellites, orbital sounding can be done at high temporal and spatial resolutions both day and night allowing for a continuous monitoring of a phenomenon during all its evolutional stages. Satellite rainfall retrieval acquires even more value when associated to more conventional observing systems, which were inadequate to correctly quantify precipitating phenomena both at the local and global scales. A variety of schemes using VIS and IR satellite data have been applied to the problem of precipitation estimation (Levizzani, 2003; Levizzani et al., 2007; Hsu et al., 2007) and to retrieve cloud microphysical properties (Rosenfeld, 2007; Cattani et al., 2007). Nevertheless, due to the very low penetrating capabilities of dense medium approaches based on these are restricted to observations of cloud top. For this reason, their implementation into retrieval techniques of precipitation often result in erroneous estimations of rain amount especially during light rain episodes when cloud top temperatures are comparable to those of non-rainy clouds. Microwave radiation is, in general, affected in a minor way by most clouds, particularly for wavelengths above the centimeters. From the precipitation retrieval perspective, microwaves offer the enormous advantage to guarantee a direct relationship between the radiation field and the bulk of hydrometeors "illuminated" by the upwelling radiation. On this basis, many retrieval techniques originally thought for VIS and IR wavelengths were implemented with microwave frequencies when they became available at the end of the 1970s. Skillful combinations of observations from high spatial and temporal resolution VIS-IR geostationary sensors with rainfall-measuring capabilities of microwave radiometers are employed in the so called "blended techniques". Such methods, based on the calibration of IR brightness temperatures with microwave rain signatures (e.g., Turk & Mehta, 2007) or on the advection of microwave precipitation by using thermal IR signal (Joyce et al., 2004), are normally used to reconstruct precipitation tracks over regions not regularly covered by polar orbiting satellites. Therefore, it is quite intuitive to adopt improvement strategies microwave precipitation retrievals for their input to statistical methods as well as to numerical models. The last fifteen years have witnessed to the great progress in microwave retrieval methods as a direct consequence of the improvement of microwave orbital instruments as regards their ground resolution, which went from some hundred to few tens of kilometers, and also from the point of view of the number of available channels aboard the spacecraft. Of equal importance is the ever increasing number of flying satellites that have considerably reduced the global satellite passing time lags down to 3 hours.

In the following we present the topical aspects of microwave remote sensing of precipitation by firstly discussing the theoretical concepts and secondly applying them to some case studies. Since the aim of this work is mainly devoted to the practical application of microwave remote sensing, the argued theory will be followed by examples that describe roughly but clearly can be observed and measured from satellite in a certain situations where either absorption or scattering extinguish the incoming radiation in a medium. The second part of the chapter contains a brief description of a new technique to retrieve rainfall based on microwave high frequencies of the five-channel passive sensor Advanced Microwave Sounding Unit-B (AMSU-B). Particularly, this method exploits rain signature within the water vapor absorption band at 183.31 GHz to infer precipitation rates and a series of thresholds to classify rain types and categorize non-rainy pixels into snow cover, cloud liquid water and small cloud droplets/water vapor classes. Finally, the performances of the method will be presented in three case studies where the model was tested in very different situations as to the genesis of the cloud system and consequently as to rainfall type and regime.

2. Fundamental principles of microwave remote sensing from satellite sensors

Passive microwave remote sensing, which has largely developed over the last few years thanks to the improvement in the spatial resolution and in the variety of channel-frequencies on board of last generation sensors, can be summarized using a few keywords: quasi-transparency, surface emissivity, absorption-scattering. This oversimplification of the problem is not due to the fact that microwave sounding is simpler than that based on other approaches such as that based on IR or VIS wavelengths, but it refers to some aspects of the observed scene that are more unambiguously identified with microwave frequencies than with other shortwave methods. This is due to the existing direct link between the measured observable and the perturbation induced to the radiation field along the wave propagation path. This signal displacement at the nominal frequency value is directly correlated to the bulk of the observed parameter. This fact will be more elucidated in section 2.2 where an overview of the radiative transfer equations in the approximation of microwaves (MRTE) will be discussed. The final part of the section will be centered on radiation extinction by absorption and scattering both from the theoretical and the observational point of view.
Nevertheless, the radiative transfer theory will be introduced by some fundamental concepts on the emissivity in section 2.1. Microwave observation is greatly susceptible to surface features: the variation of moisture content or vegetation coverage and the differences in salinity levels or the presence of sea ice drastically contaminate microwave remote sensing. Section 2.1 will propose an overview on emission properties of surfaces and an interesting example will be illustrated where a snow covered terrain shows the same scattering signal of convective clouds, thus falsely identifying an area over land as intensely precipitating.

2.1 An overview on the emissive properties of surfaces

Microwave measurements from space are extremely sensitive to a wide range of surface and atmospheric properties. Surface type, different terrain coverage and local variation of atmospheric parameters, can deeply affect space observations often introducing errors in the retrieval methods because of the incorrect evaluation of local conditions. The main goal of this section is to highlight the major "problems" that affect satellite observations with PMW sensors and in particular some examples will be shown of ambiguous situations where the same signal could be equally associated to more than one different contributor.
Unlike the thermal IR spectral region where the black body approximation often well describes the real behavior of emitters, in the microwaves an emitting surface must be considered as a grey body so that its emissivity value is usually lower than unit. An accurate

theoretical treatment of the physics of dielectric materials will be omitted since the aim of this paper is to offer a practical observational guide from satellite-based microwave sensors. We will limit ourselves to describe the effect of superficial emissivity variations by considering the observed surfaces as "cold" and "warm". These two categorizations are by no means enough, because several intrinsic and superficial features contribute to determine the emissivity value ε and consequently to deviate the behavior of a real body from the Planck's law.

The observed variability in microwave radiances for homogeneous land surfaces is normally caused by variations in skin temperature and surface emissivity, while the variability for open seawater is attributed to the atmospheric constituents such as columnar water vapor, temperature profiles and presence of cloud liquid water. These just very general considerations really contain the justification about the use of terms "cold" and "warm". The land surface emissivity being higher ($\varepsilon \approx$ 0.80-1.00) than ocean's ($\varepsilon \approx$ 0.40-0.60) appears as a "warm" object. Nevertheless, unlike for the ocean, land emission variability is strictly linked to the strong temporal and spatial variations of soil features as roughness, vegetation cover and moisture content. It is thus very complex to model surface properties in the microwave from arid surfaces to dense vegetation or snow and consequently it is difficult to discern between the surface and atmospheric contributors to the upwelling radiation. The impact of the different surface type on the temperature and humidity retrievals has been quantified by English (1999); in these studies microwave emission errors for different continental surfaces is evaluated by using a mathematical technique to potentially extend the low-altitude sounding information over solid surfaces. Other authors have developed computational scheme to improve the mathematical description of surface emissivity for several land types: bare soil (Shi et al., 2002), vegetation canopy (Ferrazoli at al., 2000) and snow-covered terrain (Fung, 1994).

Over open ocean the substantially stable and uniform "cold" background emphasizes more the extinction of upwelling radiation by atmospheric constituents and the contribution of various elements to the total radiation depression are reasonably well separated. Sea surface emissivity is largely determined by dielectric properties of seawater through the Fresnel equation and, especially for a drier atmosphere, the surface has a larger effect on the measured radiance. Many authors have developed models to predict the dielectric constant of seawater in order to improve the retrieval method of atmospheric parameters. Klein and Swift (1977), for example, proposed an improved model for the dielectric constant developed on the basis of measurements at L-band and S-band. Their equations provide an adequate description of the dielectric constant with an accuracy within 0.3 K but model performances largely decrease at higher microwave frequencies. Other studies based on radiometric airborne observations of the ocean-roughened surface (Guillou et al., 1996) have extended and validated existing sea emissivity models at higher frequencies 89 and 157 GHz. Likewise, laboratory experiments with an aqueous NaCl solution and synthetic seawater modeling (Ellison et al., 1998) have demonstrated that the assessment of sea surface emissivity for the interpretation of radar and radiometer data necessarily requires accurate permittivity measurements (better than 5%) of natural seawater in the frequency range 40-100 GHz.

In the last fifteen years with the increasing number of satellite platforms hosting increasingly higher spatial resolution new generation microwave sensors, the use of orbital instrument data became more widespread. A multisensor satellite approach, based on the

Defense Meteorological Satellite Program (DMSP) instrumental suite, is followed by Greenwald & Jones (1999) to compare satellite observations of ocean surface emissivity at 150 GHz together with selected permittivity models to evaluate the accuracy of retrieval schemes and the impact of atmospheric parameters on final retrievals. Stephen & Long (2005) and Banghua et al. (2008) have modeled microwave emissions of the Sahara desert by using data from the Tropical Rainfall Measuring Mission (TRMM) Microwave Imager (TMI) and the emissivity values over snowy soil with data of the Advanced Microwave Sounding Unit (AMSU) on board the National Oceanic and Atmospheric Administration (NOAA) satellites, respectively.

An example of the radiometric response of the NOAA/AMSU-B frequency range 89-190 GHz to the emission for several surface and atmospheric contributors is reported in Fig. 1. An analysis of the images in the window frequencies at 89 GHz (top-left) and 150 GHz (top-middle) evidences the striking contrast between land and open water numerically denounced by a brightness temperature discrepancy over 50 K at 89 GHz. Similarly, the coldest background widely enhances the presence of cloud liquid water at 89 GHz close to Spanish, Italian and Northern Europe coastlines. This characteristic is attenuated at 150 GHz, whose weighting function "peaks" around 1 km above surface, and thus warmer atmospheric layers partially mask cloud liquid signatures. An interesting aspect of Fig. 1 is related to the land emissivity changes. Observing the image at 150 GHz a brightening structure is extensively distributed in the middle of the image. In the same location but at 89 GHz this region is related to the Alps and Apennines whereas at 190 GHz (top-right) it almost disappears except over higher mountain tops. The similarity between satellite images and daily snow cover map unmistakably suggests that snowy terrain is the main responsible of significant reduction of the Earth's emissivity. Because of the underlying freezing surface, low-layers water vapor, which generally absorbs radiation at 150 GHz smoothing the effects of surface emissivity, is more or less totally condensed over snow cover pack forming a sort of "dry-zone" in the first layers above ground. This assertion is also corroborated by mixing ratio measurements retrieved by three sample radiosonde stations (red dots in Fig. 1). As a consequence of these drier conditions the weighting function lowers close to the surface largely enhancing the effects of scattering by ice particles of fallen snow. The final result is that the brightness temperature of the upwelling radiation reaching the satellite drastically decreases from 40 K to 70 K over the Alps and Apennines. In addition, it must be said that the signal extinction of snow cover at 150 GHz is quite similar to that of scattering by ice on cloud top with an enormous errors during rain pixel classification. A different behavior is shown in the 89 GHz channel, where the upward radiance varies from 20 K to 70-80 K over mountain with increasing surface roughness. Finally, the 190 GHz channel sounding the absorption of water vapor around 2 km in general is less affected by surface emissivity variations. Nevertheless, when local dry condition establish this frequency senses closer to the surface and it can sense more surface effects. This condition is frequently observed over polar regions where dry profiles constrain opaque frequencies around 183.31 GHz to sound atmospheric layers near the frozen surface.

Our experiments, take us to develop a series of thresholds based on a combination of the above frequencies with the scope to improve snow cover pixel detection and reduce false rain signals into the retrieval method presented in section 4.3. An example of our snow cover product, obtained by using frequencies thresholds proposed in the central part of Fig. 1, is shown on the same figure (bottom-right). The application of a snow cover filter, which also distinguishes between wet and dry snow, has significantly reduced the number of misclassifications and gave us the possibility to apply the method also at higher latitudes with a substantial improvement of the algorithmic performances.

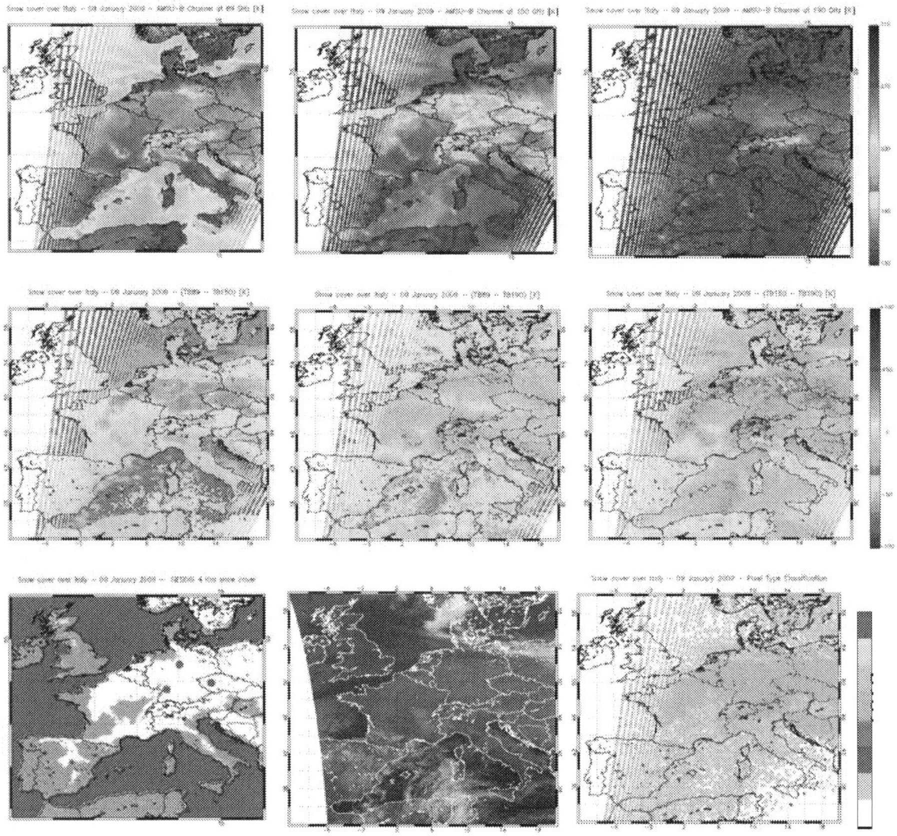

Fig. 1. NOAA-16 AMSU-B soundings on 09 January 2009, 0520 UTC, at 89 GHz (top-left), 150 GHz (top-middle) and 190 GHz (top-right), and corresponding MSG-SEVIRI image at 10.8 µm (bottom-middle). The snow cover pack is more clearly enhanced at 150 GHz with respect to other frequencies. Nevertheless, the combination of these frequencies can be used to detect snow. The snow mantle (bottom-left) is better highlighted with the threshold (BT_{89} - BT_{150}) (middle-left) but since the same values are quite similar to rainy ones the simultaneous application of tests based on (BT_{89} - BT_{190}) (middle-center) and (BT_{150} - BT_{190}) (middle-right) can be skillfully used to discern rainy from snow pixels. An example of snow cover map applied to the 183-WSL retrieval scheme is shown on bottom-right where green, Chartreuse green and lime-green are flags for snowfall, dry snow cover and wet snow cover, respectively; red and yellow dots refer to convective and stratiform precipitation; blue and cyan represent cloud liquid water and cloud droplets and finally white is the label for no-data.

2.2 The Radiative Transfer Equation

The radiative transfer equation is a mathematical description of the spatial-angular distribution of monochromatic radiation intensity I_ν which, at a certain instant t and at the frequency band ν, propagates into a medium across cross section A, in the observation direction Ω along the path s. The intensity of radiation varies while this passes through the medium. In particular, the energy of the incoming beam will decrease due to the absorption by the medium substance and to the deviation of a fraction of the radiation from the original trajectory due to the scattering in all directions. At the same time, the thermal radiation emission by the volume of material will enhance the energy balancing the net energy flux losses by the extinction processes. A brief phenomenological discussion on the radiation interaction properties with the material medium will be presented hereafter; the reader interested to a rigorous analysis should refer to more specialized books (e.g., Chandrasekhar, 1960). This general treatment of the properties of the energy interactions with matter, obtained by referring to the radiative transfer formulation discussed in Sharkov (2003), will allow us to readily focus on the practical scopes of this chapter by discussing the approximations of microwave radiative transfer and quantifying the extinction of the Earth's emission by natural disperse media such us clouds and rain observed from satellite in terms of brightness temperatures. Finally, the above theoretical and phenomenological concepts will be ideally combined in a method for the estimation of ground rainfall intensities through exploiting absorption and scattering mechanisms by hydrometeors.

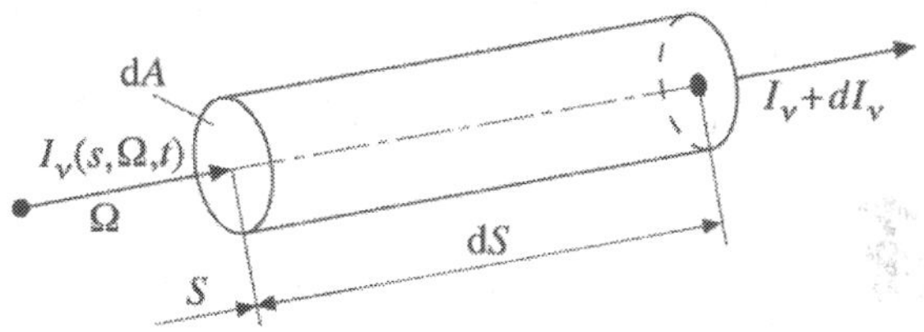

Fig. 2. Representation of the simple cylindrical geometry used to describe the total energy transformation from the initial intensity I_ν to the final $I_\nu + d_\nu$.

If we consider an elementary volume $dAdS$ in the form of a cylinder with the main axis coincident with the radiation path s (Fig. 2), the variation of flux intensity when the incoming radiation passes through the elementary path ds is represented by the quantity:

$$dI_\nu(s,\Omega)dAd\Omega d\nu dt \tag{2.2.1}$$

where dA, $d\Omega$, $d\nu$ and dt correspond to elementary crossed surface, solid angle of energy propagation direction, frequency band in the vicinity of ν and unit of time, respectively.
Let us indicate with W the increase of the radiation I_ν passing through the above considered volume. The quantity

$$W_\nu dAdsd\Omega d\nu dt \tag{2.2.2}$$

represents the enhanced energy of an incident beam into the elementary cylindrical volume *dAds* with respect to the direction Ω and relative to the time interval *dt* and frequency band *dv*. From the combination of the (2.2.1) and (2.2.2), the quantity W_ν is derived in terms of the incoming intensity energy variation to unit path

$$\frac{dI_\nu(s,\Omega)}{ds} = W_\nu \qquad (2.2.3)$$

By considering an absorbing, emitting and scattering medium, the quantity W_ν can be written in the explicit formulation of interaction mechanisms as follows:

$$W_\nu = W_E - W_A + W_{IS} - W_{AS} \qquad (2.2.4)$$

This relationship represents the balance equation between the increment (positive terms) and decrement (negative terms) of the energy during the interaction whit material substance. In particular, the first term to right-hand side represents the increasing of radiation energy per unit time, volume, solid angle and frequency due to the emission of radiation, if the local thermodynamic equilibrium (LTE) is guaranteed and then the Kirckoff's law domain is established; it will be related to the Planck function and spectral absorption by following the relationship

$$W_E = \gamma_\nu(r) I_{\nu B}[\nu, T(r)] = \gamma_\nu(r)\left[\frac{2h\nu^3 n^2}{c_0^2}\frac{1}{[\exp(h\nu/kT)-1]}\right] \qquad (2.2.5)$$

where $\gamma_\nu(r)$ characterizes the spectral absorption coefficient of the substance per unit of the radiation propagation path length, while the term in square brackets describes the Planck function in terms of frequency for a transparent substance with a refractive index *n* and temperature *T*. A strong approximation to linearly represent the Planck distribution is usually assumed in remote sensing practical applications at longer wavelengths (i.e., smaller frequencies) as in the radio-frequency regime. Derived by Rayleigh and Jeans, this reformulation of the Planck's formula can be achieved in the case where $h\nu/kT \ll 1$. After expanding in a Taylor series the exponential term of the black-body equation, the Rayleigh-Jeans radiation law can be obtained rewriting the (2.2.4) as

$$I_\nu(\nu,T) = \frac{2\nu^3 h}{c_0^2}\frac{1}{\left[\left(1+\frac{h\nu}{kT}+\ldots\right)-1\right]} \approx \frac{2\nu^2 n}{c_0^2}kT \qquad (2.2.6)$$

This new formulation of Planck's law allows to directly calculate the radiative transfer in terms of brightness temperature (T_{BB}) linking the fist term on the left-hand side to the properties of medium and its physical temperature on the right-hand side.

The second term of the (2.2.4) corresponds to the energy losses caused by the radiation absorption by a medium that, for a volume element in LTE and in the unit time, solid angle and frequency, can be written as

$$W_A = \gamma_\nu(s) I_\nu[s,\Omega] \tag{2.2.7}$$

The third and fourth terms describe the balance of radiation energy diffused in all direction by the scattering mechanisms. Specifically, the quantity W_{IS} takes into account the radiation scattered by the medium in the direction of the observer (positive) that, for an isotropic medium and purely coherent scattering, can be expressed as

$$W_{IS} = \frac{1}{4\pi}\sigma_\nu(s)\iint_{4\pi} I_\nu(s,\Omega')p_\nu(\Omega')d\Omega' \tag{2.2.8}$$

while the quantity W_{AS} is related to radiation losses for the reason that the energy beams are deflected along the main direction Ω. In terms of the unit of time, volume, solid angle and frequency, it can describe by the following equation

$$W_{AS} = \sigma_\nu(s) I_\nu[s,\Omega] \tag{2.2.9}$$

where the quantities $\sigma_\nu(s)$ and $p_\nu(\Omega')$ represent the spectral scattering coefficient and spectral phase function normalized to unit, respectively. By substituting the explicit relationships into the compact formulation (2.2.4), we have

$$\begin{aligned}&\frac{dI_\nu(s,\Omega)}{ds} + [\gamma_\nu(s)+\sigma_\nu(s)]I_\nu(s,\Omega) = \\ &= \gamma_\nu(s) I_{\nu B}[T(s)] + \frac{1}{4\pi}\sigma_\nu(s)\iint_{\Omega'=4\pi} I_\nu(s,\Omega')p_\nu(\Omega')d\Omega'\end{aligned} \tag{2.2.10}$$

that, in more compact form, could be written as

$$\frac{1}{\beta_\nu(s)}\frac{dI_\nu(s,\Omega)}{ds} + I_\nu(s,\Omega) = S_\nu(s) \tag{2.2.11}$$

where

$$S_\nu(s) = (1-\omega_\nu)I_{\nu B}[T(s)] + \frac{1}{4\pi}\sigma_\nu(s)\iint_{\Omega'=4\pi} I_\nu(s,\Omega')p_\nu(\Omega')d\Omega' \tag{2.2.12}$$

$$\beta_\nu(s) = \gamma_\nu(s) + \sigma_\nu(s) \tag{2.2.13}$$

$$\omega_\nu = \frac{\sigma_\nu(s)}{\gamma_\nu(s) + \sigma_\nu(s)} \qquad (2.2.14)$$

In these relations $S_\nu(s)$ is called the source function, $\beta_\nu(s)$ is the spectral extinction coefficient and $\omega_\nu(s)$ is the spectral albedo.

In the following part two useful examples will be proposed to better elucidate the theoretical concepts expressed above. In particular, the complete equation (2.2.10) will be specialized for particular cases of a purely scattering medium and of a solely absorbing and emitting medium and each of them will be described with the help of real satellite images Nevertheless, it is essential to clarify that the cases to which the satellite images refer in the example are quite close to the ideal cases of the theoretical description of the atmospheric extinction processes and simply represent a rough fitting of the theory. Many aspects predicted by the theory are neglected on purpose to simplify the treatment and concentrate the interests to the core of the problem.

If we observe a hypothetical real purely scattering medium, namely where the thermal radiation does neither absorb nor emit such is the case of the frozen top of cold rainy clouds, equation (2.2.14) will be banally simplified as $\omega_\nu(s)= 1$. With this simplification, the term related to Planck' s emission in equation (2.2.12) completely disappears and the total extinction coefficient (2.2.13) becomes $\beta_\nu(s)= \sigma_\nu(s)$. Equation (2.2.14) can be rewritten as:

$$\frac{1}{\sigma_\nu(s)}\frac{dI_\nu(s,\Omega)}{ds} + I_\nu(s,\Omega) = \frac{1}{4\pi}\sigma_\nu(s)\iint_{\Omega'=4\pi} T_\nu(s,\Omega')p_\nu(\Omega')d\Omega' \qquad (2.2.11\text{-a})$$

This is an integro-differential equation and its analytical solution does not exist. Several methods often based on approximated formulation of the (2.2.11) could be found in more specialized books.

On the other hand, absorbing and emitting media differ from purely scattering ones because they absorb external radiation and re-emit it in the same direction without scattering extinction by the substance constituents. Small cloud droplets, water vapor and precipitating clouds with few ice crystals on top or totally deprived of them (warm rain) can be virtually considered as an absorbing/emitting medium. For such media, where $\omega_\nu(s)= 0$, the equation (2.2.11) assumes the form:

$$\frac{1}{\beta_\nu(s)}\frac{dI_\nu(s,\Omega)}{ds} + I_\nu(s,\Omega) = I_{\nu B}[T(s)] \qquad (2.2.11\text{-b1})$$

That, solved in terms of radiation energy intensity, becomes

$$I_\nu(s,\Omega) = I_0\exp(-\beta s) + \int_0^s I_{\nu B}[T(s)]\exp(-\beta s')ds' \qquad (2.2.11\text{-b2})$$

where the first term represents the amount of absorption of external radiation by the medium described by the boundary intensity radiation I_0 and an exponential decreasing law of incoming radiation into the medium; the integral term expresses the radiation variation emitted from the surface at the temperature T along the path length s.

In order to show the effect of scattering and absorption on a real satellite measurement it can be useful to consider the images in Fig. 3 and 4. Specifically, those images refer to the soundings at high frequencies of the AMSU-B PMW sensor. Considering that AMSU-B channels are ranged in the scattering domain (generally scattering effects increase for $\nu > 60$ GHz whereas for $\nu < 60$ GHz the radiation extinction is conventionally attributed to the absorption processes), many sensitivity studies (Bennartz & Bauer, 2003) have demonstrated that in absence of strong scatterers liquid cloud droplets largely absorb radiation at 89 GHz while ice hydrometeors on the top of clouds act as scatterers more at 150 GHz and 190 GHz than at the other frequencies. Furthermore, our experiments demonstrate that when a light precipitation episode is sensed, usually associated to stratiform rain with raindrop sizes comparable to non-rainy cloud droplets, the sensitivity to the absorption at 89 GHz is more marked than the scattering signal at 150 GHz. Therefore, making use of these basic considerations we report an example of intense scattering by large ice crystals during the evolutional stages of a Mesoscale Convective System (MCS) over the Mediterranean Sea and an example of absorption by light stratiform rain and cloud liquid water over the North-Eastern England Sea.

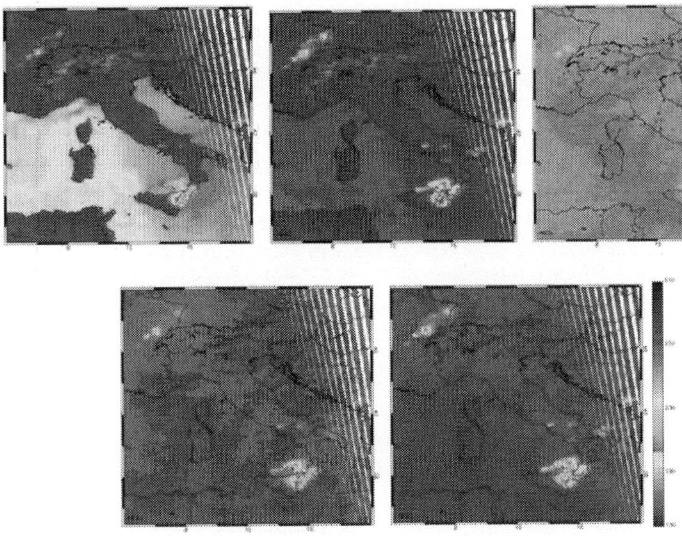

Fig. 3. Mesoscale Convective System over Southern Italy on 22 October 2005 as observed by the NOAA-15/AMSU-B at 89 , 150 , 184, 186 and 190 GHz anticlockwise from top left panel. Neglecting a small radiation absorption by surrounding droplets and water vapor molecules more evident at the opaque frequencies, the convective region can be considered as a purely scattering medium. At 150 GHz the brightness temperature depression was registered above 100 K with respect to its nominal value.

In the case of deep convection it is worth noting how the ice particle bulk depresses upwelling radiances, expressed as brightness temperatures in unit of Kelvin, at all frequencies from surface (89 GHz and 150 GHz) to the top of the troposphere (at 184 GHz the weighting function "peaks" at about 8 km lowering down to 2 km at 190 GHz) denouncing a system vertically well developed.
Besides, it is interesting observe that the signal depression is enhanced at 150 GHz (comparable with measurement at 190 GHz) where the signal extinction is quantifiable over 100 K with respect to the channel's nominal value. In the practical use of satellite remote sensing, the properties of this frequency combined to those of other channels such as the 89 GHz and 190 GHz are often exploited to discern ice signature in the clouds and possibly correlate probability information related to the conversion of melting ice into rainfall at the ground (Bennartz et al., 2002; Laviola & Levizzani, 2008).

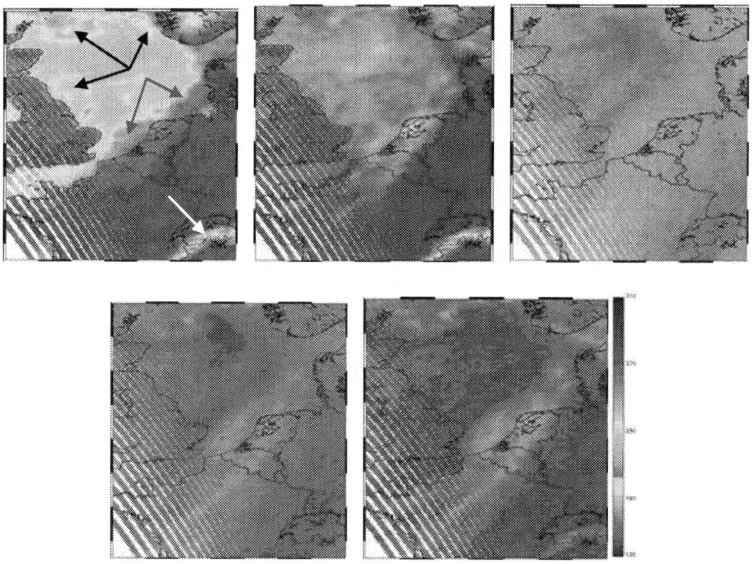

Fig. 4. Quasi-pure stratiform system over Belgium and cloud liquid water over North-Eastern England Sea on 17 January 2007 as observed by the NOAA-15/AMSU-B at 89 , 150, 184, 186 and 190 GHz anticlockwise from top left panel. The strong contrast at 89 GHz allows to observe water clouds over open sea (black arrows) whereas the change in emissivity highlights snow cover over Alps both at 89 and 150 GHz (also at 190 GHz). At higher opaque frequencies (186 and 184 GHz) the absorption of middle and high layer water vapor can only detected.

Referring to Fig. 4, the observed satellite radiance attenuation is mainly due to the absorption and emission of small cloud particles and hydrometeors. Nevertheless, a more realistic description of the situation would have to take into account that the variation of upwelling radiation is certainly due to the combination of absorption and scattering by a mixture of liquid and ice hydrometeors and disperse liquid particles. By the same token, in

the previous cases the absorption due to water vapor and small cloud droplets, which typically surround precipitating clouds as a halo, was not considered because it is small enough with respect to scattering of radiation by ice crystals on cloud top.
Referring once more to the case of Fig. 4, warm cloud spots at 89 GHz due to the absorption of water clouds over open water (black arrows) and stratiform rainy clouds (blue arrow) over the coastline also sounded at 150 and 190 GHz can be clearly distinguished. It is interesting to compare extinction intensities at 89 and 150 GHz both from the absorption and scattering point of view by using in that order open sea liquid cloud and snow cover over the Alps (white arrow) as terms of comparison. At 89 GHz over the sea the strong contrast between cold sea surface (≈ 200 K) and warmer liquid clouds (≈ 250) bring to a net difference of about 50 K whereas over land the difference due to the scattering of snowy terrain is quantified in about 60-70 K. At 150 GHz the discrepancy between the cold sea surface and liquid water clouds can be evaluated in about 10 K while the change in surface emissivity over land induces a satellite brightness temperature depression up to 70 K increasingly describing the strong sensitivity of that frequency to the scattering.

3. Impact of precipitation on microwave measurements

In the approximation of disperse media theory, natural systems such as dust, fog, clouds, rain particles are considered as heterogeneous polydisperse media consisting of mixtures of substances and/or different thermodynamic phases. Assuming a particle size density function $n(r)$described by the M-P function (Marshall & Palmer, 1948) and a drop terminal velocity $V(r)$ depending on particle radius r, rainfall rates will be proportional to the fourth or third moment of the drop density function. From the radiative point of view, when incident radiation interacts with precipitating hydrometeors all particles present in any elementary volume are totally irradiated and consequently the incoming radiation is extinguished both by absorption and scattering processes at the same time.
Passive microwave rainfall estimations are carried out by exploiting either absorbed or scattered signals from raindrops or a combination of the two as is the case of the 183-WSL method. In the hypothesis of warm rain rainfall is estimated through the emission associated with absorption by liquid hydrometeors through Kirchoff' s law. In this case, raindrops absorption and emission provide a direct physical relationship between rainfall and the measured microwave radiances. With increasing precipitation intensities, scattering by large drops becomes dominant with respect to absorption and the observed radiation appears drastically depressed for a downward-viewing observer.
A more realistic situation is represented by rainclouds formed by a mixture of liquid, frozen and eventually supercooled hydrometeors. Since scattering is primarily caused by ice hydrometeors aloft the emitted signal by liquid drops is substantially blocked by intense scattering and its contribution to the total extinction significantly decreases with the increase of the frozen bulk. Measured radiances are therefore indirectly related to the rain mass and consequently the estimations become less correlated with falling rain below cloud base.
This situation is often observed during the development of intense convections (see Fig. 3) typically associated with heavy rain events. The case of liquid rain drops discussed before can be roughly associated with the stratiform systems (see Fig. 4) whose light precipitation is linked more to the absorption of water droplets than to the scattering of small crystals which form on cloud top.

This theoretical argument associated with the treatments of previous paragraphs is useful to understand the behavior of the 183-WSL with respect to condensing water vapor. When the newly nucleated droplets surround a developing rainy region, they can act as embryos for the development of small rain drops. Depending on the updraft strength such droplets can be dragged inside the cloud core thus contributing to the cloud's precipitation formation mechanisms or can freely evolve into light rain constrained to the border of the main cloud body. The small size of the buoyant drops in these bordering areas determines the signal extinction, particularly at 183.31 GHz, to be characterized by absorption rather than by scattering.

4. High frequency method to retrieve rainrates

A new rainfall estimation method, named 183-WSL (Laviola & Levizzani, 2008, 2009a), is now described based on the high frequency water vapor absorption bands at 183.31 GHz of AMSU-B sensors on board NOAA and EUMETSAT Polar System (EPS) satellites, which is conceived to retrieve rainrates over land and sea. AMSU-B is the second module of the AMSU passive MW across-track scanner operating into the frequency range from 90 up to 190 GHz with a spatial resolution of 16 km at nadir view (Saunders et al., 1995; Hewison & Saunders 1996).

An emission approach at 183.31 GHz is adopted to infer surface precipitation because one of our major targets is the estimation of warm rain. The 183-WSL retrieval scheme (Laviola & Levizzani, 2009b), based on a suite of brightness temperature (BT) thresholds, distinguishes and classifies convective and stratiform precipitation while filtering out condensed water vapor and snow cover on mountain top, which particularly affects more opaque superficial channels (i.e., 190 GHz).

4.1 The sensitivity at 183.31 GHz to surface emissivity and rainy cloud altitudes

The 183.31 GHz bands are mainly dedicated to the sounding of the atmospheric water vapor amount (Kakar, 1983; Wang et al., 1989). However, several studies have demonstrated the effects of clouds on these frequencies and their possible application into rainfall retrieval schemes. Note that the use of PMW information is necessary to detect rainy systems or correct and integrate IR measurements, for instance in the blended techniques. However, their use is limited because of the variability of surface emissivity (ε). Grody et al. (2000) proposed a few algorithms based on different land type studies to evaluate surface emissivity using AMSU data.

Here we choose radiative transfer results with different values of surface emissivity, which refers to land if ε is typically > 0.6 and to water in the other cases, to quantify the effect of surface on AMSU-B channels.

Fig. 5 shows the simulated brightness temperatures for all AMSU-B frequencies as a function of surface emissivity in clear sky conditions. The results are obtained by a adding/doubling radiative transfer model (Evans et al., 1995a,b) running with mid-latitude profiles and coupled to Rosenkranz (2001) approach for the computation of the absorption at selected frequencies.

As expected, the signal around 89 and 150 GHz has strong surface contributions showing a deep depression near low emissivity values and converging about to the same brightness temperature when $\varepsilon=1$ (dry-land). Therefore, the decreasing surface emissivity from dry-land values to water bodies' enhances the influence of atmospheric moisture on these channels. Another significant aspects of Fig. 6 is that, since their weighting functions are peaked beyond 2 km altitude, the three moisture channels are little or not at all affected by different surface emissivities thus suggesting their application both over land and over the sea.

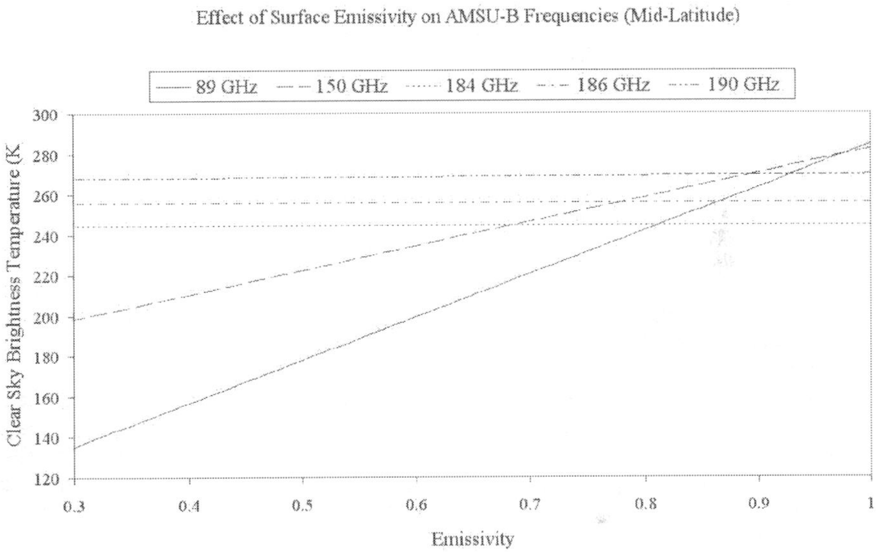

Fig. 5. Emissivity effects on the AMSU-B channels. The two window channels are strongly dependent on the surface emissivity showing an increasing value up to 280 K from the simulated sea surface (ε = 0.50) to dry land (ε = 1.00). At moisture frequencies, where the weighting functions are higher than window ones, the surface emissivity effect is low.

Other sensitivity studies not reported here have emphasized that, when moving towards higher latitudes where the atmosphere is less optically thick, the contribution of surface emissivity affects more and more the measurement particularly at 190 GHz where also thinner ice clouds can modify the signal.

Precipitating cloud altitude is another important variable affecting the AMSU-B brightness temperatures. We studied the behavior of AMSU-B moisture channels as a function of the position of a rainy cloud in the troposphere. All simulations have been carried out using radiosonde temperature and humidity profiles screening out the possible cloud formations along balloon trajectory with the threshold suggested by Karstens et al. (1994). The cloud structure is built adopting a Marshall-Palmer's water drop size distribution (Marshall & Palmer, 1948) and the Mie theory to solve the scattering equations. In agreement with the weighting function distribution, which peak between 2 and 8 km, our results show that only rainy clouds positioned above 2 km altitude interact with three AMSU-B opaque frequencies at 183.31 GHz and the interaction will be always more intense as the cloud becomes thicker.

4.2 Physical basis of the 183-WSL algorithm

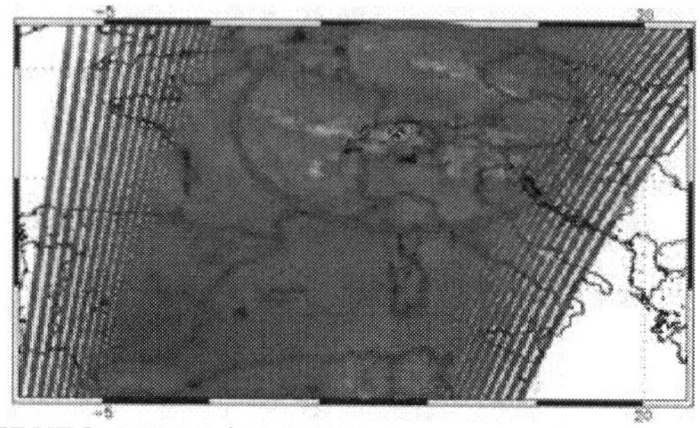

Fig. 6. 11 June 2007, 0957 UTC. NOAA/AMSU-B 183.31 7 GHz brightness temperatures of a stratiform system over France. The blue circle contains the detected low rainrate clouds.

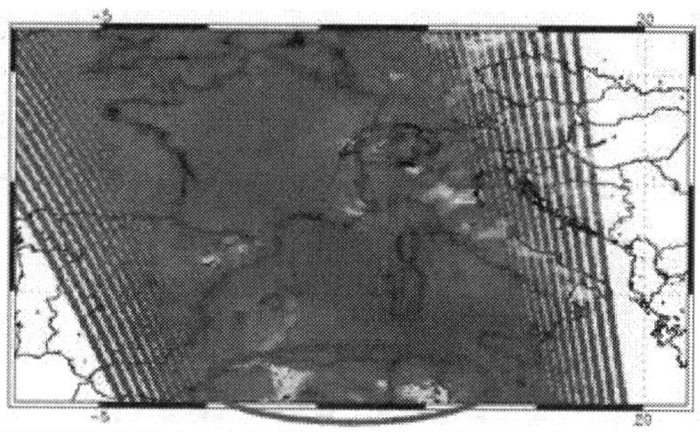

Fig. 7. 12 June 2007, 1457 UTC. NOAA/AMSU-B 183.31 7 GHz brightness temperatures of a deep convective system over the coast of North Africa. The blue circle contains the two convective cores.

A substantial number of precipitating systems forming in the lower atmospheric layers at mid latitudes are formed for the large portion by water drops grown through the collision and coalescence mechanism because cloud temperature does not reach values low enough for the droplets to start freezing. This implies that the vertical rain profile is a few km thick and that falling rain will presumably be light and persistent.

On the other hand, strong updrafts typical of the warm season are capable of transporting water drops up to the tropopause level giving rise to deep convective columns, which convey a large amount of cloud water to the ground through heavy showers. These two kinds of precipitation systems induce very different BT responses in the MW spectral range as observed in Fig. 6 and 7 where the soundings of a stratiform and of a convective system at 183.31 ± 7 GHz are shown, respectively. In the first, low rain clouds absorb the Earth radiation showing a moderate cold area corresponding to BTs in the 240-250 K range. The second situation refers to a deep convective system over Africa consisting of two cores, which strongly depress the BT reading of the instrument (≈ 200 K) because of the scattering of large ice crystal located at cloud top.

The 183-WSL algorithm is based on a linear combination of the AMSU-B opaque channels and it detects rainrates (in mm h^{-1}) over land and sea by sounding cloud features from 1-2 km up to the top of the troposphere according to the channels weighting functions. Note that, however, since our studies have shown that when a light-rain stratiform system forms large amounts of the surrounding water vapor absorbs the 183 GHz radiation inducing false rain signals, a suite of thresholds is introduced to reduce these spurious effects (see Table I). In addition, tests carried out during the winter season highlight that the scattering signal at 183.31 ± 7 GHz relative to the snow cover on mountain top (the Alps in this case) is comparable to the ice scattering signature at the top of convective cloud.

Classification	Land (K)	Sea (K)
Water vapor/Snow cover	$\Delta T < 3$	$\Delta T < 0$
Stratiform rain	$3 < \Delta T < 10$	$0 < \Delta T < 10$
Convective rain	$\Delta T > 10$	$\Delta T > 10$

Table 1. Classification thresholds based on the window channel differences $\Delta T = (T_{89} - T_{150})$

4.3 The 183-WSL algorithm: retrieval design and performances

The 183-WSL work design is schematically described by four steps. The first step is dedicated to ingesting and processing the satellite data stream. All relevant information, namely BTs, surface type (land/sea/mixed), satellite local zenith angles, topography, are separated from the overall data stream and arranged for input into the 183-WSL processing chain. The second and third steps are constituted by the modules 183-WSLW and 183-WSLS/C, respectively, which operate to discriminate rainy from non rainy pixels and classify precipitation type as a stratiform or convective on the basis of threshold values calculated for land/mixed and sea surfaces. This step is currently been improved adding a new module to classify cloud liquid water by estimating the amount of water in terms of the Liquid Water Path (183-LWP) and with a snow cover mask able to recognize snowy soils and categorize those pixels as wet or dry snow. These improvements (not included in the 183-WSL version used for the examples of this chapter) were needed to reduce the number of false rain signals especially during winter season when also snowy terrain deeply scatter the upwelling radiation similarly to ice hydrometeor signatures. Finally, the last step of the 183-WSL algorithm computes the final rainrate product in unit of mm h^{-1}.

The proposed case studies exemplify different situations in which rainfall was retrieved and classified by means of the two 183-WSL convective/stratiform modules (183-WSLC/S). In the first case the values of the scattering index (SI) introduced by Bennartz et al. (2002) to build four rain intensity classes were used for comparison. As expected, the agreement between the SI and the 183-WSLC (convective) is higher than the one between the SI and the 183-WSLS (stratiform). The reason refers to the nature of the SI that retrieves only the probabilities of surface rainrates due to melting of scattering ice crystals. Therefore, the scattergrams related to the stratiform portion and to water vapor should be intended as light-rain low-SI values. At the same time the water vapor distribution threshold based on the BT differences between 89 and 150 GHz should be < 0 K (sea) and < 3 K (land).

The other two cases show a comparison between the 183-WSL and retrievals of the Goddard Profiling (GPROF) algorithm (Kummerow et al., 1996, 2001). A good agreement is found particularly in the case of intense rainfall. When ice crystals form during deep convective rain development the increase of scattered radiation is better observed by both algorithms with respect to the case of extinction caused by light rain, for which the 183-WSL algorithm shows more sensitivity than GPROF.

4.3.1 Saharan dust causing red rain over Bulgaria

On 23 and 24 May 2008 an intense dust plume from Sahara overflying Greece and the Black Sea interacted with an Atlantic Front generating persistent "red" rain over Bulgaria (Fig. 8, white arrows). The strongly scattering but non-precipitating hydrometeors (water vapor around dust over the Mediterranean Sea) are filtered out by the computational scheme (8-f). The incoming Atlantic front generates deep convection over Italy with rainrate estimations around 10 mm h^{-1}. Note that the classification thresholds correctly flag as precipitating those pixels where BTs are greater than 3 K (8-a) as compared with the MODIS-COT (Moderate Resolution Imaging Spectroradiometer-Cloud Optical Thickness) product shown in 8-b. Rain classification in Fig. 9 (left) shows that the 183-WSL low rainrates (< 5 mm h^{-1}) are associated with scattering signatures (Bennartz et al., 2002) < 30 K whereas rainrates classified as heavy (> 5 mm h^{-1}) are correlated with the highest SI values. On the middle and right of Fig. 9, rain distribution with longitudes and rain types on the basis of classification thresholds are respectively proposed.

4.3.2 Severe storm over Italy: 183-WSL *vs* GPROF/AMSR-E

During the severe storm of June 2007 we have tested the 183-WSL performances both in the convective portion and in moderate rain conditions that characterized the various sectors of the storm, with light rain being not very frequent. The 183-WSL underestimates rainfall with respect to GPROF/AMSR-E. From the analysis of the discrepancy graphs (vertical bars) an increasing displacement is noted with increasing rain intensities. This is possibly due to the different nature of the algorithms. In the case of moderate rain (Fig. 10-a) the precipitating areas are quite similar over the southern Mediterranean Sea; over the northern sector GPROF drastically underestimates and this is true for the other cases. Note that the convective system coming from SE (Fig. 10-b) is well described but 183-WSL precipitation presents a more continuous pattern from the convective core to the borders. In case of lighter rain (Fig. 10-c) the 183-WSL captures more rainy areas than GPROF/AMSR-E, especially over the Alps.

4.3.3 Hurricane Dean: 183-WSL *vs* GPROF/TMI

Cyclone Dean was a classic seasonal tropical system forming over the Cape Verde islands, passing close to Jamaica and pouring rain on the coast of the Yucatan as a category 5 hurricane. Figure 11 shows the cyclone development stages retrieved by the 183-WSL algorithm (top) and TRMM 2A12 product from the best coincident overpasses (bottom). By comparing the TMI and the 183-WSL products a reasonable agreement can be observed although a more comprehensive study needs to be carried out.

The scattergrams at the bottom of Fig. 11 depict a generally good correlation between the two retrieval techniques. Nevertheless, some other studies of ours describe a slight overestimation of the 183-WSL but this is probably due to the large water vapor absorption characteristic of this kind of extreme event.

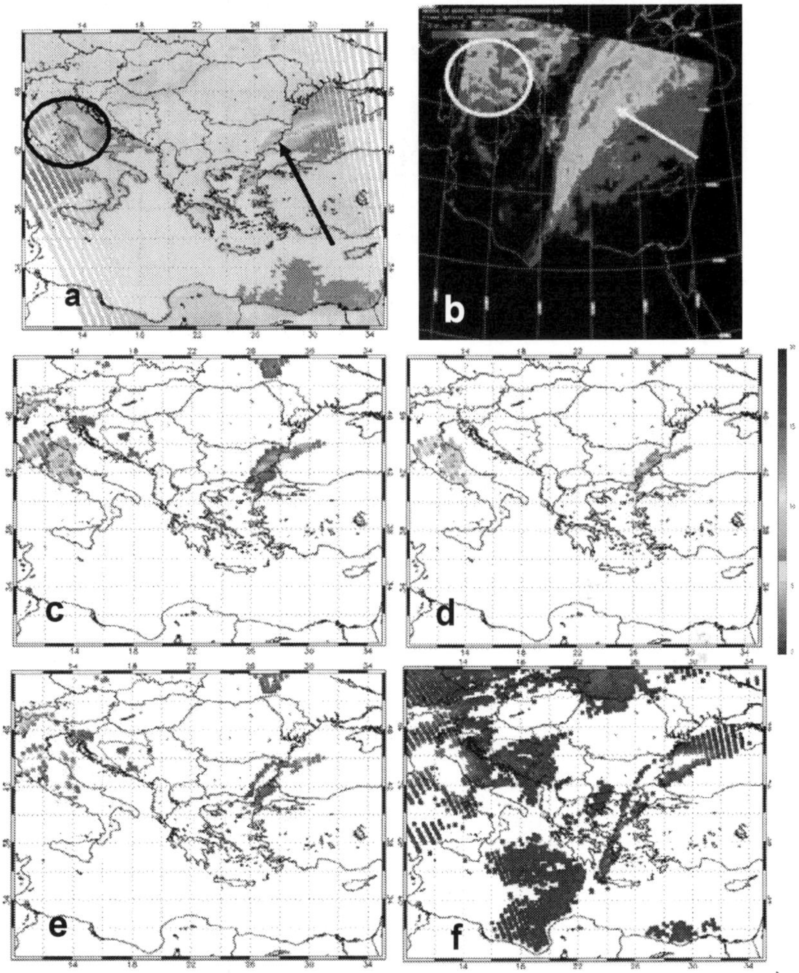

Fig. 8. "Red rain" over Bulgaria on 23 March 2008 0920 UTC. The classification thresholds (a) correctly discriminates between rainy areas, generally characterized by high values of Cloud Optical Thickness (MODIS-COT in b), from non-rainy. From (c) to (f): 183-WSL rainrate estimations, 183-WSLC convective rain, 183-WSLS stratiform rain and 183-WSLW condensed water vapor removed from the computations.

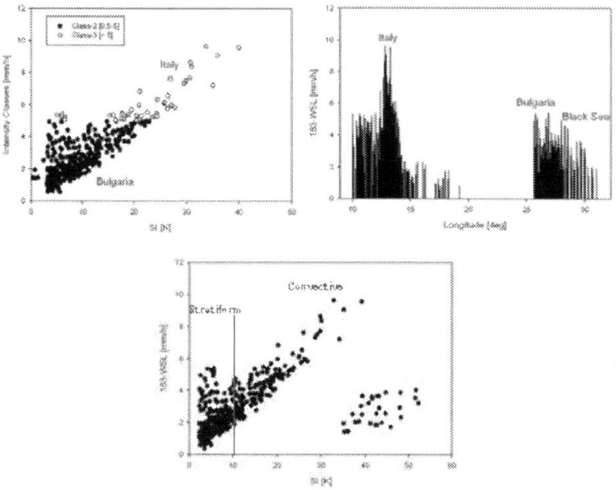

Fig. 9. Scattergrams of the case study in Fig. 8. In figure left, a comparison between classified 183-WSL rain intensities in class 1 [0-5 mm h-1] and class 2 [> 5 mm h-1] and the scattering index (SI) values is shown. Note that rainfall intensities belonging to class are associated to lower SI values (SI< 30 K) whereas rainrates > 5 mm h-1 correspond to SI values around 50 K. Figures middle and right describe rainfall distribution with latitude and rain types, respectively.

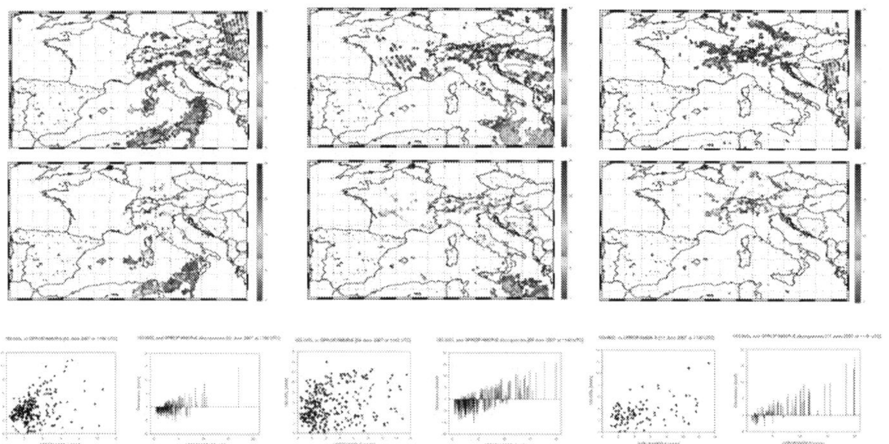

Fig. 10. Severe storm over Italy on 2 June 1156 UTC (left), 4 June 1143 UTC (middle) and 11 June 1149 UTC (right). The 183-WSL rainrates (top) are compared with those from GPROF/AMSR-E (bottom). Vertical bars describe an increasing displacement of the 183-WSL estimations with increasing rain intensities. The large dispersion of the scattergrams can be justified observing that GPROF drastically underestimates rain intensities where light rainfall is detected.

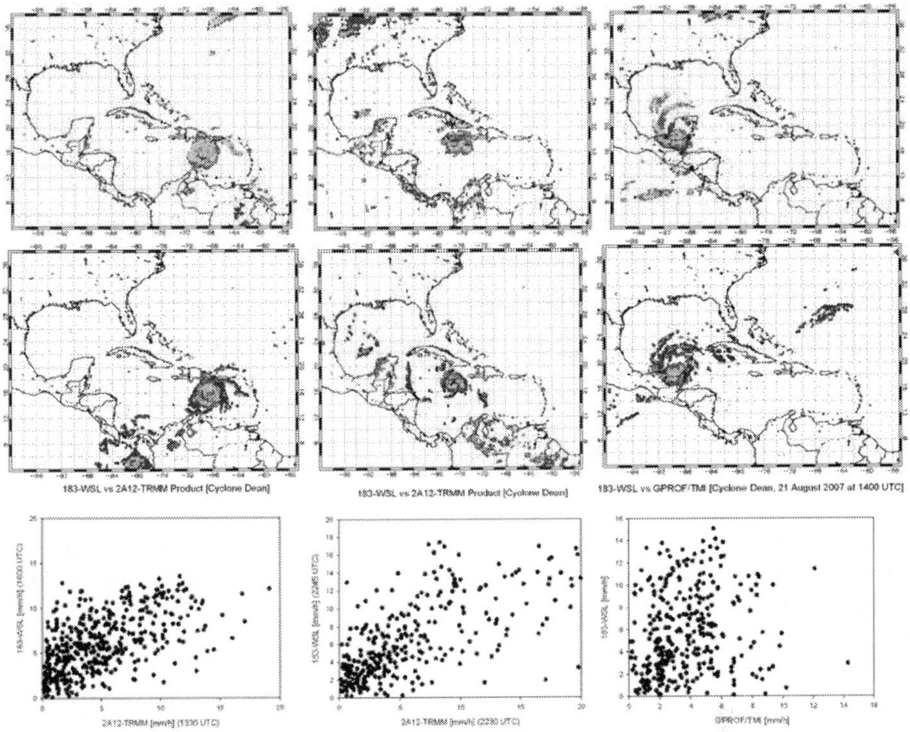

Fig. 11. Cyclone Dean, 18-21 August 2008. The 183-WSL retrievals (top) and corresponding TRMM 2A12 product images on 18 at 1300 UTC, 19 at 2200 UTC and 21 at 1400 UTC, respectively. Diagrams clearly show the increasing of correlation between the 183-WSL and TRMM 2A12 to increase of rain rates.

5. Summary and conclusion

The most important aspects of passive microwave remote sensing has been explored both from theoretical and for operational point of view. The chapter does not rigorously treat the physical principles of PMW remote sensing, but uses theory as a reference point to correctly interpret and describe satellite observations. For this reason, the first sections were contributed to focus the attention on two fundamental themes that must be taken into account when using microwave radiometers: surface emissivity and radiation extinction processes. Compared to optical and IR wavelengths where surface contribution varies between 0.80 ÷ 1.00, microwaves are very susceptible to changes in surface conditions. Over ocean, the substantially stable emissive surface ensures that microwave soundings of atmospheric parameters are quite consistent within a strategy of rainrate retrieval. Over land areas, the passive microwave observations yield to significantly less quantitative measures of rainfall because the effects of surface emission variability can drastically affect measurements and consequently the retrieved products. Those surface effects are more marked in the case of absorption by liquid raindrops where a low-emissivity background (i.e. cold) is required in order to make observations of the emitted radiation associated through Kirchhoff's law to the absorption regime. Liquid hydrometeors are the dominant contributors to the absorption and emission, providing a direct physical relationship between rainfall and the observed microwave radiances. Examples of the quasi-pure absorption process can be found in the Tropics where the dissolving of a deep convection system induces the development of stratified warm precipitation where collision-coalescence formation mechanisms accrete raindrops.

In the case of scattering, rainfall amount is indirectly estimated via the scattering of radiation by liquid and ice particles. Because scattering is mainly due to frozen hydrometeors located on the top of clouds, the ice pack aloft largely blocks emission by liquid raindrops below. Consequently, the upwelling radiation is not directly correlated to the bulk of rain and rainfall intensity is deduced as a result of an effective measure of the radiation-raindrops interaction but as a probability function of scattered radiation-rainfall. Retrieval methods based on scattering approach, however, allow for the observation of precipitation over any background with the limits of being very robust during rain events characterized by large amount of ice crystals on cloud top (see the MCS in Fig. 3) and almost "blind" during rain episodes where less or no-ice particles are formed (i.e., light stratiform rain or warm rain).

To practically discuss the differences of retrieval techniques based on absorption or scattering approaches we have proposed in the second part of the chapter the results of the algorithm 183-WSL, which basically works via absorption mechanisms but behaves increasingly similarly to scattering algorithms with the increment of ice aggregates in the cloud. The scattergrams in Fig. 9 quantitatively demonstrate that for rainrates belonging to the light-moderate intensity class the distribution is quite disperse indicating low correlation between the 183-WSL retrievals and co-located scattering index (SI) values. With increasing rainfall intensities the agreement 183-WSL-SI improves (see empty dots). Another example is shown in Fig. 10 where low rainfall intensity values are observed and numerically quantified by vertical bar diagrams.

These case studies on one hand highlight the differences between two retrieval methods based on absorption and substantially pure scattering processes and from the other they open up the road to future planned studies. Our numerous investigations actually reveal the robustness of the 183-WSL algorithm in many different situations where precipitating events are often characterized by light rains or snow covered terrain, two extreme circumstances for passive microwave observations. The strength of such results can be extended to a more general discussion on the use of high frequency microwaves to better delineate low rainrate regions and to inspect frozen soils. In addition, our studies also demonstrate that the suite of frequencies between 90 and 190 GHz, are suitable to study rainfalls mainly resulting from non ice-phase process with size spectral range lower than millimeter where these frequencies offer useful information to identify and possibly measure "warm rain" processes.

6. References

Banghua, Y.; Weng, F., & Meng, H. (2008). Retrieval of snow surface microwave emissivity from the advanced microwave sounding unit, *J. Geophys. Res.*, 113, D19206, doi:10.1329/2007JD009559, pp. 1-23.

Bennartz, R. & Bauer P. (2003). Sensitivity of microwave radiances at 85–183 GHz to precipitating ice particles, *Radio Sci.*, 38(4), 8075, doi:10.1029/2002RS002626.

Bennartz, R., Thoss A., Dybbroe, A., & Michelson D. (2002). Precipitation analysis using the Advanced Microwave Sounding Unit in support of nowcasting applications, *Meteorol. Appl.*, 9, pp. 177-189.

Cattani, E., Melani, S., Levizzani, V., & Costa, M. J. (2007). The retrieval of cloud top properties using VIS-IR channels. *Measuring precipitation from space - EURAINSAT and the future*, Advances in Global Change Research, 28, Springer, Dordrecht, pp. 79-96.

Chandrasekhar, S. (1960). Radiative transfer, *Dover Publications Inc.*, New York, pp. 393.

Ellison, W.; Balana, A.; Delbos, G., Lamkaouchi, K., Eymard, L., Guillou, C., & Prigent, C. (1998). New permittivity measurements of seawater, *Radio Sci.*, 33, no. 3, pp. 639-648.

English, S. J. (1999). Estimation of temperature and humidity profile information from microwave radiances over different surface types, *J. Appl. Meteor.*, 38, pp. 1526-1541.

Evans, F., & Stephens G. F. (1995a). Microwave radiative transfer through clouds composed of realistically shaped ice crystal. Part I: single scattering properties, *J. Atmos. Sci.*, 52, pp. 2041-2057.

Evans, F., & Stephens G. F. (1995b). Microwave radiative transfer through clouds composed of realistically shaped ice crystal. Part II: single scattering properties, *J. Atmos. Sci.*, 52, pp. 2058-2072.

Ferrazoli, P., Wigneron, J. P., Guerriero, L., & Chanzy, A. (2000). Multifrequency emission of wheat: Modeling and application, *IEEE Trans. Geosci. Remote Sens.*, 38, pp. 2598-2607.

Fung, A. K. (1994). Microwave scattering and emission models and their applications, *Artech House.*, pp. 573.

Greenwald, T. J., & Jones, A. S. (1999). Evaluation of seawater permittivity models at 150 GHz using satellite observations, *IEEE Trans. Geosci. Remote Sens.*, 37, no. 5, pp. 2159-2164.

Grody, N. C., Weng, F., & Ferraro R. R. (2000). Application of AMSU for obtaining hydrological parameters, *In: Microwave Radiometry and Remote Sensing of the Earth's Surface and Atmosphere*, P. Pampaloni and S. Paloscia, Eds.,USP Int. Science Publishers, Utrecht, pp. 339-352.

Guillou, C., English, S. J., Prigent, C., & Jones, D. C. (1996). Passive microwave airborne measurements of the sea surface response at 89 and 157 GHz, *J. Geophys. Res.*, 101, no. C2, pp. 3775-3788.

Hewison, T. J., & Saunders, R. W. (1996). Measurements of the AMSU-B antenna pattern. *IEEE Trans. Geosci. Remote Sensing*, 34, pp. 405-412.

Hsu, K-L., Hong, Y., & Sorooshian, S. (2007). Rainfall estimation using a cloud patch classification map. *Measuring precipitation from space - EURAINSAT and the future*, Advances in Global Change Research, 28, Springer, Dordrecht, pp. 329-342.

Joyce, R. J., Janowiak, J. E., Arkin, P. A., & Xie, P.(2004). CMORPH A Method that produces global precipitation estimates from passive microwave and infrared data at high spatial and temporal resolution. *J. Hydrometeor.*, 5, pp. 487-503.

Kakar, R. K. (1983). Retrieval of clear sky moisture profiles using the 183 GHz water vapor line, *J. Climate Appl. Meteor.*, 22, pp. 1282-1289.

Karstens, U., Simmer, C., & Ruprecht, E. (1994). Remote sensing of cloud liquid water, *Meteor. Atmos. Phys.*, 54, pp. 157-171.

Klein, L. A., & Swift, C. T. (1977). An improved model for the dielectric constant of sea water at microwave frequencies, *IEEE Trans. Antennas Propag., AP25(1)*,pp. 104-111.

Kummerow, C. D., Olson, W. S., & Giglio, L. (1996). A simplified scheme for obtaining precipitation and vertical hydrometer profiles from passive microwave sensors, *IEEE Trans. Geosci. Remote Sens.*, 34, pp. 1213–1232.

Kummerow, C. D., Hong, Y., Olson, W. S. Yang, S., Adler, R. F., McCollum, J., Ferraro, R., Petty, G., Shin, D. B., & Wilheit, T. T. (2001). The evolution of the Goddard Profiling Algorithm (GPROF) for rainfall estimation from passive microwave sensors, *J. Appl. Meteor.*, 40, pp. 1801-1820.

Levizzani, V. (2003). Clouds and rainfall by visible-infrared radiometry. *Remote Sensing of Atmosphere and Ocean from Space: Models, Instruments and Techniques*, Advances in Global Change Research, 13, Springer, pp. 127-143.

Levizzani,V., Bauer, P., & Turk, F.J. (2007). Measuring precipitation from space - EURAINSAT and the future, *Springer*, Dordrecht, pp. 722.

Laviola, S., & Levizzani, V. (2008). Rain retrieval using 183 GHz absorption lines. *IEEE Proc. MicroRad 2008, 10th Specialist Meeting on Microwave Radiometry and Remote Sensing of the Environment*, Firenze, 11-14 Mar., doi: 10.1109/MICRAD.2008.4579505.

Laviola, S., & Levizzani, V. (2009a). Observing precipitation with AMSU-B opaque channels: the 183-WSL algorithm. *Proc. 4th Workshop Int. Precipitation Working Group*, Beijing, China, 13-17 October. [available at http://www.isac.cnr.it/~ipwg/meetings/beijing/4th-IPWG-Proceedings-web-March-2009.pdf].

Laviola, S., & Levizzani, V. (2009b). Using water vapor lines to retrieve rainrates: the 183-WSL algorithm. *J. Appl. Meteor. Climat.*, submitted.

Marshall, J. S., &.Palmer, W. McK. (1948). The distribution of raindrops with size, *J. Meteor.*, 5, pp. 165-166.

Rosenfeld, D. (2007). Cloud top microphysics as a tool for precipitation measurements. *Measuring precipitation from space - EURAINSAT and the future*, Advances in Global Change Research, 28, Springer, Dordrecht, pp. 61-77.Rosenkranz, P. W. (2001). Retrieval of temperature of moisture profiles from AMSU- and AMSU-B measurements, *IEEE Trans. Geosci. Remote Sens.*, 39, pp. 2429-2435.

Shi, J., Chen, K. S., Li, Q., Jackson, T. J., & O'Neil, P. E. (2002). A parameterized surface reflectivity model and estimation of bare-surface soil moisture with L-band radiometer, *IEEE Trans. Geosci. Remote Sens.*, 40, pp. 2674-2686.

Saunders, R. W., Hewison T. J., Stephen, S. J., & Atkinson, N. C. (1995). The radiometric characterization of AMSU-B, *IEEE Trans. Microwave Theory and Techniques*, 43, pp. 760-771.

Sharkov, A. E. (2003). Passive microwave remote sensing of the Earth. Physical foundations, *Springer and Praxis Publishing Publication*, pp. 557.

Stephen, H., & Long, D. G. (2005). Modeling microwave emissions of erg surface in the Sahara desert, *IEEE Trans. Geosci. Remote Sens.*, vol. 43, no. 12, pp. 2822-2830.

Turk, F. J., & Mehta, A. V. (2007). Toward improvements in short-time scale satellite-derived precipitation estimates using blended satellite techniques. *Measuring precipitation from space - EURAINSAT and the future*, Advances in Global Change Research, 28, Springer, Dordrecht, pp. 281-96.

Wang, J. R., Wilheit, T.T., & Chang, L. A. (1989). Retrieval of total precipitable water using radiometric measurements near 92 and 183 GHz, *J. Appl. Meteor.*, 28, pp. 146-154.

Use of GTEM-cell and Wire Patch Cell in calculating thermal and non-thermal biological effects of electromagnetic fields

Marija Salovarda Lozo and Kresimir Malaric
Faculty of Electrical Engineering and Computing, Zagreb
Croatia

1. Introduction

The increasing use of technology in our everyday environment has led to the ubiquitous presence of electromagnetic fields. These leeds to many new researches in past ten years which deal with biological effects of high frequency fields, especialy GSM frequency band. For that purposse developed are appropiate exposure systems.

In 1996, the World Health Organization (WHO) established the guidelines for quality Electromagnetic field experiments, where the importance of well-defined and characterised exposure conditions was emphasised.

Various system designs have been used for the exposure. Most of them belong to one of the following basic design ideas: Transverzal Electromagnetic (TEM) cell, RF chambers, radial transmission lines (RTL), waveguides or Wire Patch Cells (WPC).

In this chapter focus would be on G-TEM cell and Wire Patch Cell, their construction, characteristics, usage, advantages and disadvantages.

Performed were several models and tests with different power output levels in order to see field distribution inside system. As most of biological effects caused by RF electromagnetic radiation are thermal, temperature distribution and increase were also noted in order to separate temperature rise because of exposure system itself and temperature rise due to electromagnetic field levels.

2. Biological effect of Electromagnetic fields

Exposure to electromagnetic fields is not a new phenomenon. However, due to the increasing use of technology in our everyday environment has led to the ubiquitous presence of electromagnetic fields. Such fields arise wherever there is a voltage or a current. All types of radio broadcasting and TV transmitters produce electromagnetic fields, and they also arise in industry, business and the home, where they affect us even if our sense organs perceive nothing. Everyone is exposed to a complex mix of weak electric and magnetic fields, both at home and at work.

Generally we can divide biological effects of electromagnetic radiation on non-thermal end thermal effects. Non-thermal biological effects are related to low frequency electromagnetic field and are mostly manifested as induced currents in human body. Thermal biological effects occur due to hifg frequency electromagnetic fields. Radio frequency radiation interacts with matter by causing molecules to oscillate with the electric field. This interaction is most effective for molecules that are polar (have their own internal electric field) such as water. The water molecule loses this rotational energy via friction with other molecules and causes an increase in temperature. This effect is the basis for microwave cooking (Michelson et al, 1987).

Temperature increase during exposure to high frequency electromagnetic waves depends on: the specific area of the body exposed and the efficiency of heat elimination; intensity of field strength; duration of exposure; specific frequency or wavelength; and thickness of skin and subcutaneous tissue. Each frequency in the electromagnetic spectrum is absorbed by living tissue at a different rate, called the specific absorption rate or SAR , which has units of watts per kilogram (W/kg). Related to human thermoregulation, organs with the least blood flow are most endangered. The eyes are particularly vulnerable to RF energy in the microwave range, and prolonged exposure to microwaves can lead to cataracts.

The levels of radiofrequency fields to which people are normally exposed are very much lower than those needed to produce significant heating. The heating effect of radiowaves forms the underlying basis for current guidelines. Scientists are also investigating the possibility that effects below the threshold level for body heating occur as a result of long-term exposure.

The IEEE and many national governments have established safety limits for exposure to various frequencies of electromagnetic energy based on SAR.

$$SAR = \frac{\sigma E^2}{\rho} \tag{1}$$

Where E is RMS value of the electric field strength in the V/m, σ is the conductivity of body tissue in S/m and ρ is density of body tissue in Kg/m^3

In the area of biological effects and medical applications of non-ionizing radiation approximately 25,000 articles have been published over the past 30 years. However, some gaps in knowledge about biological effects exist and need further research (WHO).

2.1 Regulation and standards

When the problem of electromagnetic pollution became evident, government and many non-governmental organizations came to solution to this problem with limitation of field strength. Over the years this limitations are getting stricter as measurements results showed growing number of new radiation sources. Individual countries have different approaches to the limits stipulated in the various regulations, standards, norms and recommendations.

In Table 1. Are exposure EMF limits by international organizations:

ICNIRP (International Commission on Non-ionizing Radiation Protection), **IEEE** (Institute of electrical and electronics engineering), **CENELEC** (European Committee for Electrotechnical Standardization).

Frequency		FCC occup.	ICNIRP occup.	DIV V EXPO-Range 2
50/60 Hz	Power delivery	----	500 μT (*B*-Field)	1.251 μT (*B*-Field)
27 MHz	CB radio, diathermy	68.2 V/m	61.0 V/m	61.4 V/m
100 MHz	FM radio	61.4 V/m	61.4 V/m	61.4 V/m
433 MHz	Industrial applications	73.4 V/m	62.4 V/m	62.4 V/m
900 MHz	Cell, pager	106 V/m	90.0 V/m	90.0 V/m
2.45 GHz	Microwave, industry	137 V/m	137 V/m	137 V/m
6 GHz	Digital radio	137 V/m	137 V/m	137 V/m
20 GHz	Satellite transmission	137 V/m	137 V/m	13 7 V/m

Table 1. Exposure limits in comparison

3. Exposure systems

Progress in understanding biological effects of non-ionizing electromagnetic radiation is closely related to measurement capability.
To be able to measure any effect of electromagnetic radiation it is necessary to have appropriate exposure system. During last ten years more various exposure systems where developed for research of biological effects and this systems are from constantly adopting accoring needs. Most of them belong to one of the following basic design ideas: Transverzal Electromagnetic (TEM) cell, RF chambers, radial transmission lines (RTL), waveguides or Wire Patch Cells (WPC).
Well defined exposure conditions are essential to optain reproducible results ant they are an precondition for the repeatability of studies.
All enviromental requirements for the specific experiment must be strcly complied to it.
Field distribution should be homogenious, i.e., the deviation from homogenety should be as small as possible.
In qualitative comparison of five differend exposure sistems (Table 2.) Schoenberg et al. (2001) came to conclusion that best results were achived by wire patch cell and RF chambers for the exposure of cells in homogenous suspension in 60mm Petri dishes and T-75 flasks. But none of discussed approaches enabled exposure with reasonable homogenity.
The decision which wxposure system is most appropriate for certan in vitro study depends on numerical and technical requirements. Therfore Schoenberg et al, recommend close cooperation of biological and engineering experts in designing any in vitro exposure setup.

	TEM Cell	RF Chamber	RTL	Waveguide	Wire Patch cell
Frequency range	< 1 GHz	Up to several GHz	< 3 GHz	0.7-2 GHz	0.7-2 GHz
Number of dishes	2	>20	20	4-10	8(1)
Preferred Polarization	E(2)	k(3)	E	E(2)	E
Efficiency	Low	Medium	Medium	High	High
Power requirements	High	High	Medium	Low	Low
Inhomogeneity	Low	High	Medium	Medium	Medium
Complexity	Low	High	Medium	Low	Low
Size	Small	Large	Medium	Medium	Small
System cost	High	High	Moderate	Moderate	Moderate
Environmental control	Incubator	Self-built	Self-built	Incubator	Incubator
Electromagnetic shielding	Self-contained	Self-built	Self-built	Self-contained	Self-build
Exposure control	Power meter	Power meter	Power meter	Power meter	Field probe(4)

(1) 3 5mm Petri dishes placed into 60 mm Petri dishes with same medium height.
(2) k and H polarizations lead to poor homogeneity and are not further considered
(3) E polarization leads to excessive power requirements and is not further considered
(4) Field probe must be used for control in resonating structures
(Schoenborn et al, 2001)

Table 2. Qualitative comparison of the performance of five designs for the exposure of plated cells in 60 mm Petri dishes.

3.1 Transverse electromagnetic (TEM) Cell

A transverse electromagnetic (TEM) cells ware the most often used in bioelectromagnetic experiments, EMC investigations and other studies as a source of standard electromagnetic field. It was developed by Litovitz et al. [1993] and utilized by Penafiel et al. [1997] for the exposure of cells in T25 flasks at 835 MHz-

TEM cell is an exposure system that combines characteristics as : generation of uniform field or a homogeniuos plane wave, usable cell volume for placing samples, shielding from electromagnetic influence both from outside and inside, it operates over a wide frequency band, generates a planewave, field strength can be computed from the radiofrequency (RF) power travelling the cell.

TEM sells can be made in various sizes to siut the experimental requirements. The width of the cell must be less then one-half the wavelenght at the highest operating frequency to prevent higher-order-mode or multimoding effects. They are in use for upper frequency limits of 50, 150, 300, 500, and 1000MHz. TEM cells used for frequencies in Giga-hertz area are called Giga-hertz transverse electromagnetic GTEM cells.

Field strenghts which can be achived in TEM cells go from very low to extremely high values depending on the input power.
At higher frequencies the cells are compact, portable, simple to build, and can be used for broadband operation up to higher frequency limit. The major disadvantage of the TEM cell is the restriction on specimen size imposed by the cell dimensions, which is inversely proportional to the upper frequency limit (Donaldson et al, 1978)

Fig. 1. Example of TEM cell.

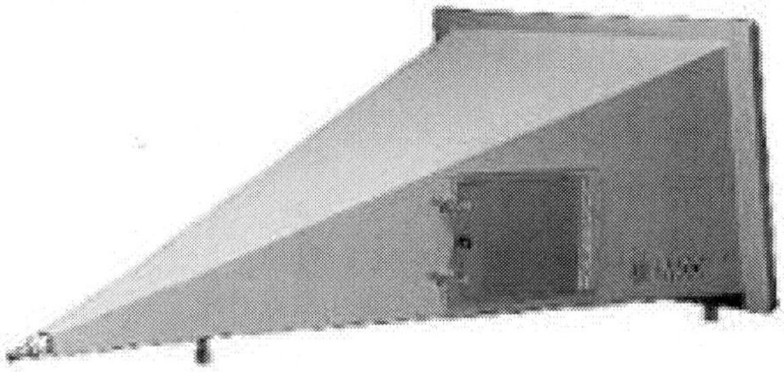

Fig. 2. Example of GTEM-cell.

3.2 Wire Patch Cell

A wire patch cell has been designed for exposing cell cultures during in vitro experiments studying possible effects of mobile radio telephone. It is based on the wire patch antenna which works at 900 MHz with a highly homogeneous field inside the antenna cavity.
The designed cell structure is symmetric and provides a rather homogeneous field distribution in a large area around its centre. this small open device is easy to construct and fits into an incubator. AS the wire patch cell is a radiating element with the same radiating pattern as a dipole, and thus some absorbing materials are necessary around the system when used for in vitro experiments. Secondly, because of its narrow bandwidth, it is difficult to maintain its working frequency. (Leval et al, 2000)
The 900 MHz WPC consist of two square plates (the ground plane and roof) of the same size (15 x 15 cm2), spaced 2,9 cm apart using four metalic grounding contacts located at each corner of the cell. A coaxial cable is located at centre of the cell, with the ewternal conductor

connected to the centre of the ground plane and the inner one passing trough the plate and then connecting to the roof.
Dimensions of radiating structure are related to the signal wavelenght and inversely proportional to the frequency (Ardoino et al, 2004).

Fig. 3. Example of Wire Patch Cell containing four Petri dishes.

4. GTEM cell and WPC modelling

Before any constructing begins it is recommended to perform computer modelling of desired exposure system to investigate all needed requirements which are demanded from such systems, for example homogeneous field distribution, frequency range, etc .
As most of biological effects caused by high frequency electromagnetic radiation are thermal, temperature distribution and increase should be also taken into account in order to separate temperature rise because of exposure system itself and temperature rise due to electromagnetic field levels.
If temperature increase due tu exposure systems construction it self is noted it is necessary to consider implementation of some cooling system before performing electromagnetic exposure test. Othervise test results can be missinterpreted.
With QuickField software we have modelled GTEM cell and WPC. Our main goual was to determine possible temeperature increase in these two exposure systems.
QickField is a finite element analysis software package. Main application include computer simulations of electromagnetic fields for scientific and industrial purposes.
Modeled GTEM cell is 40cm high and 60cm wide. Antenna source is placed at ¾ of height. Modeled 1800MHz WPC has dimensions 20x20 cm^2 (model is 2 dimensional), with Petri dishes (radius 3.5cm) placed 5 cm from septum (centre). Dimensions of 1800MHz WPC were suggested by Ardoino et al., 2004 (Salovarda et al, 2007,2008).

4.1 Finite element method

The Finite element method is used to solve complex, nonlinear problems in electromagnetism.[8][9] The first step in finite-element analysis is to divide the analyzed configuration into small homogeneous elements. The model contains information about the device geometry, material constants, excitations and boundary constraints. In each finite element, a linear variation of the field quantity is assumed. The corners of the elements are called *nodes*. The goal is to determine the field quantities at the nodes.

The Finite-element analysis technique solves the unknown field quantities by minimizing energy functional. The energy functional is an expression describing all the energy associated with the configuration being analyzed. For 3-dimensional, time-harmonic problems this functional may be represented as

$$F = \int_{v} \frac{\mu|H|^2}{2} + \frac{\varepsilon|E|^2}{2} - \frac{J \cdot E}{2j\omega} dv \tag{2}$$

The first two terms in the integrand represent the energy stored in the magnetic and electric fields and the third term is the energy dissipated (or supplied) by the conduction currents. By expressing H in terms of E and by setting the derivative of this functional with respect to E equal to zero, an equation of the form $f(J,E) = 0$ is obtained. A kth order approximation of the function f is then applied at each of the N nodes and the boundary conditions are enforced, resulting in the system of equations.

$$\begin{bmatrix} J_1 \\ J_2 \\ \cdot \\ \cdot \\ J_n \end{bmatrix} = \begin{bmatrix} y_{11} & y_{12} & \cdot & \cdot \\ y_{21} & y_{22} & \cdot & \cdot \\ \cdot & \cdot & \cdot & \cdot \\ \cdot & \cdot & \cdot & \cdot \\ \cdot & \cdot & \cdot & y_{nn} \end{bmatrix} \begin{bmatrix} E_1 \\ E_2 \\ \cdot \\ \cdot \\ E_n \end{bmatrix} \tag{3}$$

The values of J on the left-hand side of this equation are the source terms. They represent the known excitations. The elements of the Y-matrix are functions of the analyzed problem geometry and the boundary constraints. Since each element only interacts with elements in its own neighborhood, the Y-matrix is sparse.

The terms of the vector on the right side represent the unknown electric field at each node. These values are obtained by solving the system of linear equations. In order to have a unique solution; it is necessary to constrain the field strength at all boundary nodes. (Malaric et al. 2005)

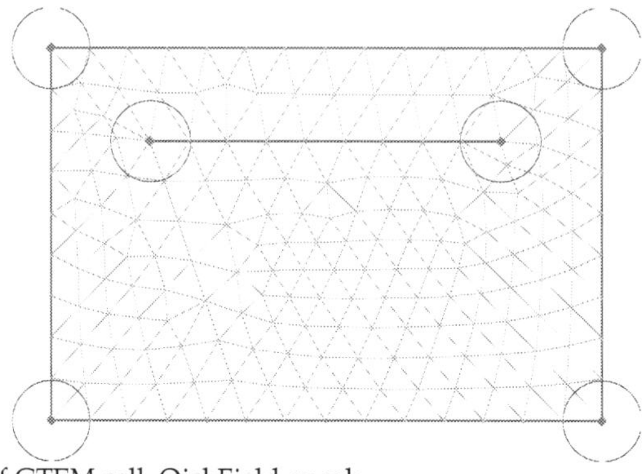

Fig. 4. FEM model of GTEM cell, QickField mesh

Applied mesh is shown in Figure 4 and the field values at the nodes have been calculated. In each finite element, a linear variation of the field quantity is assumed. The corners of the elements are called *nodes*. The goal is to find the electric field values at these nodes. Other parameters, such as the magnetic field, induced currents, and power loss can be obtained from the electric field values.

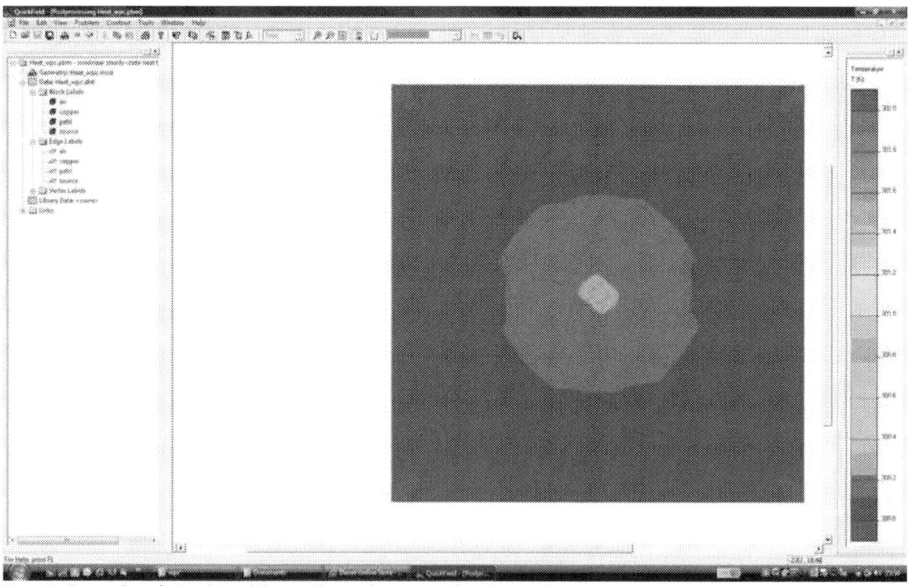

Fig. 5. Sample of QickField window showing final result.

4.2 Temperature distribution inside GTEM cell

The characteristics of GTEM-cell are: 50 Ω input impedance, inner conductor at 3/4 height, inner height to width ratio equals 2/3 and angle septum/bottom plate is 15°, with angle septum/top plate being 5°. The septum as well as coating is made of copper. The N type connector is placed at the end of the tapered section. If necessary, the connector can be replaced. The septum is supported by dielectric material. At the other end, there are the pyramidal absorbers 0.25 m long which are used for electromagnetic wave termination and two parallel 100 Ω, distributed resistive load for current termination. Figure 2 shows cross-session of the used GTEM-cell.

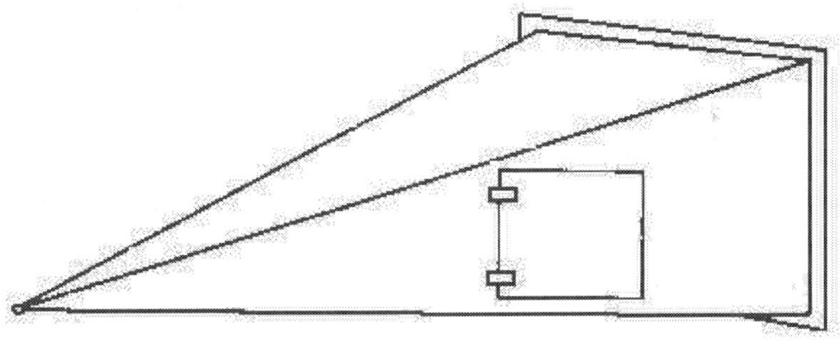

Fig. 5. GTEM-cell designed at Faculty of Electrical Engineering and Computing (FER)

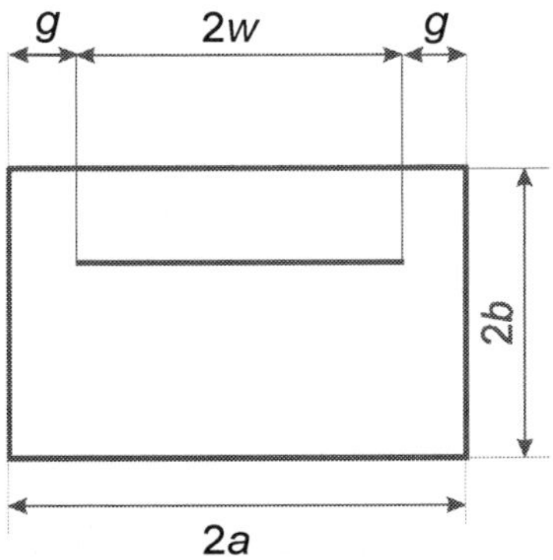

Fig. 6. Base of GTEM-cell designed at FER

By changing source values we wanted to se what is critical field value where temperatures are increased more than 1° C (1K) in the middle 15 cm below septum. Surrounding temperature in all cases was 300 K. Fig.3 and 5 show the result of numerical modeling analysis of E-strength and fig.4 and 6. results of temperature distribution in cell due to E field. Warmer colors on figures are higher values of E field or temperatures.

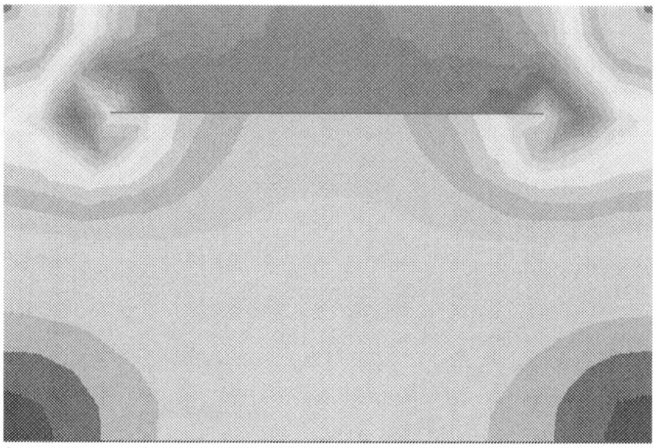

Fig. 7. E field distribution in of simulated GTEM. E=3.28V/m in the middle 15cm below septum.

Fig. 8. Temperature distribution due to E field below the septum from fig.7. Temperature in the middle 15cm below septum is 300.003 K (E=3.28V/m)

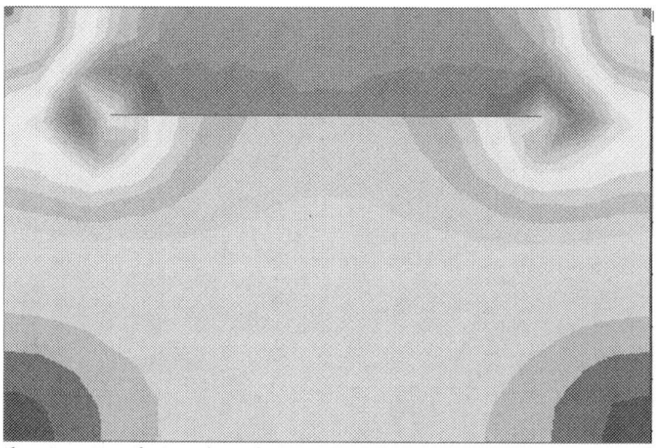

Fig. 9. E field distribution in of simulated GTEM. E=23V/m in the middle 15cm below septum.

Fig. 10. Temperature distribution due to E field below the septum from fig.9. Temperature in the middle 15cm below septum is 300.144 K (E=23V/m)

Fig. 11. Temperature in the middle 15cm below septum is 301.161 K (E=65.63V/m).

Source Voltage (V)	**E strenght (V/m)**	**Power density (mW/cm^2)**	**ΔT ($T - T^0$) K, T_0=300K**
1	3.28	0.00285376	0.003
7	23	0.14032202	0.144
20	65.63	1.14150507	1.161

Table 3. E field in the middle 15 cm below the septum

In case when E field 15 cm below septum is 23 V/m (fig. 5,6,7), temperature on same place is 300.144 K . Similar situation was measured in reference [10], when they observed influence of EMF on Lemna Minor growth in GTEM cell. Plants were exposed to field strength of 23 V/m for 4 hours. The temperature of plants which were placed 15 cm below septum at the beginning and at the end of measurements varied no more than ± 0.1° C. This measured temperature corresponds to our model in same environment (Tkalec et al, 2005).

4.3 Temperature distribution inside WPC

Performed were several tests with different power output level in order to see increase of temperature on places where petri dishes (samples) are. Starting temperature (T_0) in all cases was chosen to be 300 K.

Warmer colors in figures below indicate higher values of E-field as well as of temperature.

Fig. 12. Vertical E field distribution in side WPC for source 10V.

Fig. 13. Horizontal E field distribution in side WPC for source 10V.

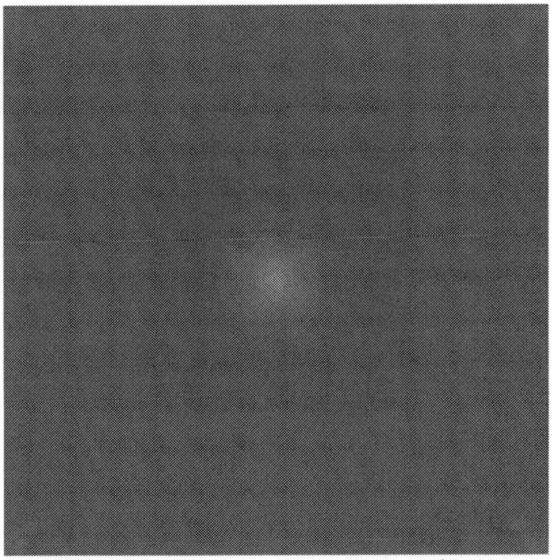

Fig.14. Temperature distribution for 2V source. Temperature value in centre of WPC is 300.226K (E-field = 26.84V/m). Temperature value 6.5 cm from septum (centre Petri dish) = 300.006K (E-field = 4.4V/m).

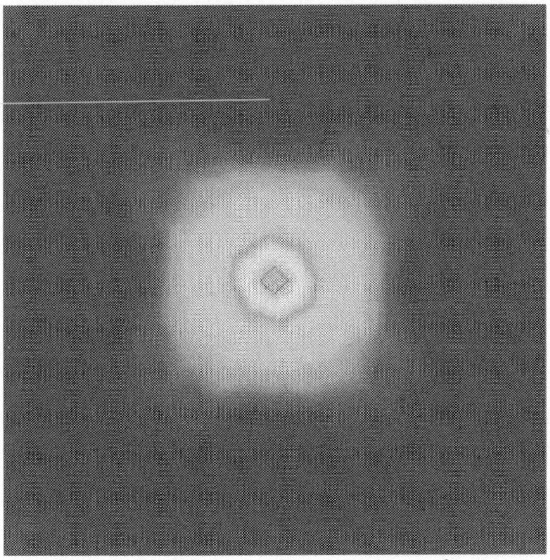

Fig. 15. Temperature distribution for 5V source. Temperature value in centre of WPC is 301.140K (E-field = 67.11V/m). Temperature value 6.5 cm from septum (centre Petri dish) = 300.049K (E-field = 10.9V/m).

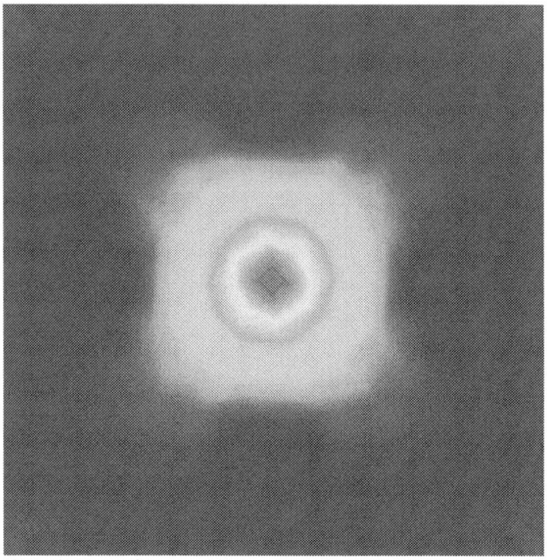

Fig. 17. Temperature distribution for 10V source. Temperature value in centre of WPC = 305.54K. Temperature value 6.5 cm from septum (centre Petri dish) = 300.195K.

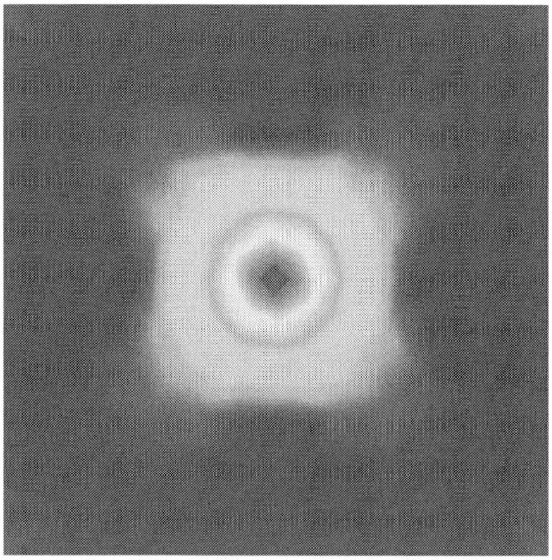

Fig. 18. Temperature distribution for 30V source. Temperature value in centre of WPC = 334.82K. Temperature value 6.5 cm from septum (centre of Petri dish)= 301.52K.

Source Voltage (V)	E strenght in the middle of petri dish, 6.5 cm from septum (V/m)	Power density in the middle of petri dish, 6.5 cm from septum , (mW/cm^2)	ΔT (T - T^0) K, T_0=300K
1	2.2	0.0013	0.002
2	4.4	0.0051	0.006
5	10.9	0.0319	0.049
10	21.9	0.1278	0.195
20	43.9	0.5114	0.738
30	65.8	1.1507	1.517

Table 4. E-field and temperature increase in 6.5 cm from septum (centre of Petri dish).

4. Conclusion

By comparing WPC simulation results with simulation result performed on G-TEM cell (40cm high and 60cm wide, antenna source placed at ¾ of height) similarities are shown.
The low E field levels (2V/m - 11V/m) should not cause any significant change of the temperature, but temperature rise for higher E-field levels should not be ignored (65.8 V/m = ΔT 1.517K).
We see significant temperature increase close to septum (e.g. 65.8 V/m = ΔT 34.82K). Therefore it is recomended to implement cooling system in order to avoide possible biological effects from heating rather then electromagnetic field itself. One exaple of cooling could be two water spiral-plate jackets placed on external faces of WPC. (Ardoino et al., 2004.) Next our step is to build WPC according gained knowlage. Then we will perform measurements of temperature rise depending on field strenghts both in GTEM cell andWPC.

Fig. 19. Exaple of WPC with cooling system as two water spiral-plate jackets placed on external faces of WPC.

5. References

Ardoino, L., Lopresto, V., Mancini, S., Pinto, R., Lovisolo, A. (2004). 1800 MHz In Vitro exposure device for experimental studies on the effects of mobile communication systems, *Radiation Protection Dosimetry* Vol. 112, No. 3, pp. 419-428 3.2

Leval, L., Leveque, P.H., Jecko, B., (2000). A new in vitro device for the mobile frequency 900 MHz, *Bioelectromagnetics,* No. 22, 255-263.

Malaric, K., Bartolic, J., Malaric, R. (2005). Immunity measurements of TV an FM/AM receiver in G-TEM-cell, *Measurement,* No. 38, pp. 219-229, Elsevier Ltd

Michelson, S.M.; Lin J.C. (1987). *Biological Effects and Health Implications of Radiofrequency Radiation,* Plenum Press, ISBN 0-306-41580-1, New York. 2. 3.

Salovarda, M., Malaric K., Malaric, R. (2007). Temeprature distribution inside GTEM-cell for biomedical experiments, *Tehnicki glasnik, Casopis Veleucilista u Varazdinu,*Vol.1, No.1-2; 16-19, ISSN 1846-6168

Salovarda, M., Malaric, K. (2008). Temperature Distribution Inside Wire Patch Cell, *Proceedings of the 14th Conference on Microwave Techniques COMITE 2008,* pp. 125-128, ISBN 978-1-4244-2137-4, ,Prag Czechoslovakia Section IEEE, April 2008.

Schoenborn, F., Poković, K., Burkhardat, M., Kuster, N. (2001). Basis for Optimization of In Vitro Exposure Apertures for Health Hazard Evaluations of Mobile, *Communications,* No.22, 547 - 559.

Tkalec, M., Malaric, K., Pavlek-Kozina, B. (2005). Influence of 400, 900, and 1900 MHz, Electromagnetic Fields on Lemma minor Growth and Peroxidise Activity. *Bioelectromagnetics,* No.26, 185-193.

BIOELECTRIC EFFECTS OF LOW-FREQUENCY MODULATED MICROWAVE FIELDS ON NERVOUS SYSTEM CELLS

María J. Azanza, A. del Moral and R. N. Pérez-Bruzón
University of Zaragoza
Spain

1. Introduction. Cell biology fundamentals of the bioelectric behaviour of neuron plasma membrane underlining the biophysics of the neuron bioelectric impulse.

1.1 Introduction.

In this chapter we present a comprehensive study of the action of applied magnetic fields (MF) (static, alternating of extremely low frequency (ELF) (0.1-217 Hz) and ELF MF modulating microwave (MW) carriers of 9.6 (100 and 800 Hz) and 13.6 GHz (2-100 Hz)). The target is the neuron membrane, and the underlying mechanism explaining the MF effect is the Ca^{2+} ion liberation from its membrane cytosolic stores. The chapter is organized as follows. After studying the bioelectric action potential across membrane, enlarging Hodgkin-Huxley master equations with the Ca^{2+} diffusion current and solving them considering the membrane as a Kirchoff knot, we describe the experimental set-ups. Next we describe and discuss the main effects discovered: bioelectric frequency inhibition and excitation, neuron synchronization under ELF MF, frequency resonance effect ("frequency window" effect), ELF modulated MW demodulation explanation, and also the effect of such a MW on human astrocytoma cells proliferation. A general model based on membrane phospholipids superdiamagnetism (SD) and Ca^{2+} liberation under Coulomb explosion (CE) is described and successfully corroborated by our experimental results. In particular our SD+CE model explains well the so called *frequency window effect*, where only a narrow band-width is effective in the AC MF interaction with the biological systems.

1.2 Molecular structure of plasma membrane.

Neurons cell biology and molecular foundations of bioelectric and synaptic-mediated processes are well defined in terms of: membrane surface structure and membrane molecular components.

According to the lipid globular-protein fluid mosaic model (Singer and Nicolson, 1972) cell membranes main components -lipids and proteins- are organized in a phospholipid (PP) bilayer, being proteins either embedded in one or two of the hemilayers, or located entirely exposed at the external or internal cell surfaces, being attached to the bilayer by a covalent linkage. The PP bilayer is asymmetric, due in part to the electrical charge lipid molecule distribution, i.e. negatively charged phosphatidylserine (PS) and glycolipids (GL) that are almost exclusively located in the inner and outer halves of the membrane respectively (Rothman and Lenard, 1977) (Fig.1). PS molecules, with their additional functional groups and net negative charge -e in the polar head, interact with cytosolic Ca^{2+} (Papahadjopoulos and Bangham, 1966). Precise information about the exact content of PS in neuron membrane

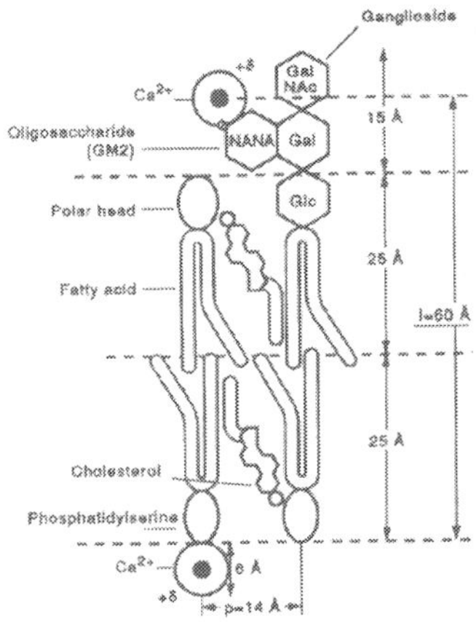

Fig. 1. Model of neuron membrane, showing the PS and a NANA glycolipid with the ganglioside head GM2, both heads negatively charged with –e (modified from Alberts et al., 1989). Bound to the heads are water solvated Ca^{2+}, the overall heads having an effective positive charge +δ. Interposed between the lipids are cholesterol molecules. Also shown are the PP long molecules. The crucial lengths intervening in the model of MF interaction with plasma membrane (see § 3), $l \cong 60$ Å and $p \cong 14$ Å, are shown (del Moral and Azanza, 1992).

is scarce but it is known that they amount ≅ 14% of the half cytosolic lipid bilayer molecules (Alberts et al., 1989). Gangliosides, defined as sialic acid-containing glycosphingolipids, account for an estimated 5-10% of the total lipids in the neuronal plasma membrane, or about 10-20 % in the external half of the bilayer, where they are located. GL can carry one or more -e charges (polyanionic heads) depending upon the number of N-acetylneuraminic acid (sialic acid, NANA) residues, which represent in turn binding sites for Ca^{2+}. Interposed between the lipids are cholesterol molecules with dielectric constant $\varepsilon_r = 2.21$.

1.3 Neuron bioelectric impulse biophysics.

Neurons plasma membranes are organized to generate and propagate *bioelectric* signals, i.e. waves of voltage, the *nerve impulses,* in response to a stimulus. In a resting neuron, at both sides of the plasma membrane, there exits an electric potential difference, *resting potential,* being the inner membrane surface (i) negatively charged (≅ - 60 mV) with respect to the external one (o). Transmembrane potential is produced by a different concentration of ions at both sides of the membrane, as given by well known Nernst equation, $E = (RT/zF)\ln(c_o/c_i)$, where c_i and c_0 are the relevant ion concentrations.

Considering transmembrane diffusible ions e.m.f., there exists in the cytosol a high concentration of K^+ ions while in the extracellular fluid there are higher concentrations (R is the gas constant, F, de Faraday and z, the ion valence) of Na^+, Ca^{2+} and Mg^{2+} ions, being Cl^- ions concentrations variable depending on the kind of neuron. As a stimulus is applied to the membrane, a stimulation threshold arising, arising a stimulation threshold, transmembrane voltage is reversed becoming the inner surface positive, *depolarization*. In few tens of ms the original inner negative potential is recovered (Fig.2). Such a voltage modifications have the shape of an impulse were we can distinguish three main phases: depolarization (D), repolarization (R) and hyperpolarization (H), each one conveyed by a kind of ion. Membrane stimulation induces the opening of either Na^+- or Ca^{2+}-channels. Ions diffuse inside de membrane raising their equilibrium potential according to Nernst equation. As the membrane is depolarizing K^+-channels are opened, allowing the reversion of transmembrane potential. The sorting of K^+ ions charges the membrane negatively, *repolarization*, getting the negative value for K^+ equilibrium potential, *hyperpolarization*. Finally the $3Na^+/2K^+$-exchange pump, plus sorting of Cl^- ions, recovers the initial resting transmembrane potential. Through the non-polar (dielectric) lipid bilayer, ions diffuse across protein-channels. Three kinds of ionic channels are operating in membranes, leakage channels (L), voltage- and ligand-activated channels.

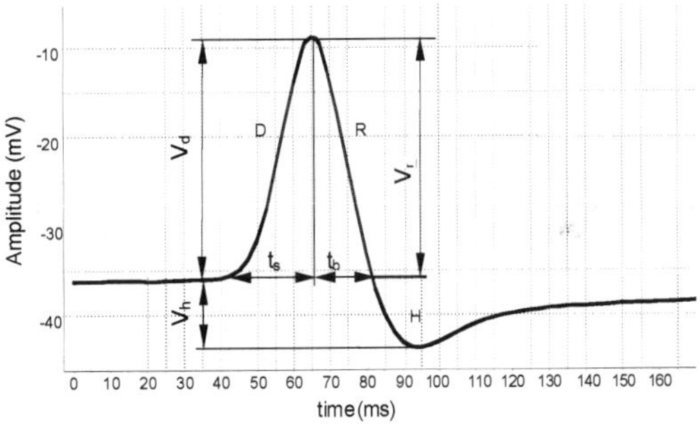

Fig. 2. D: depolarization; R: repolarization; H: hyperpolarization. V_d depolarization voltage; V_h hiperpolarization voltage; V_r repolarization voltage; t_s stimulation time; t_h hyperpolarization time. (Pérez-Bruzón, 2006).

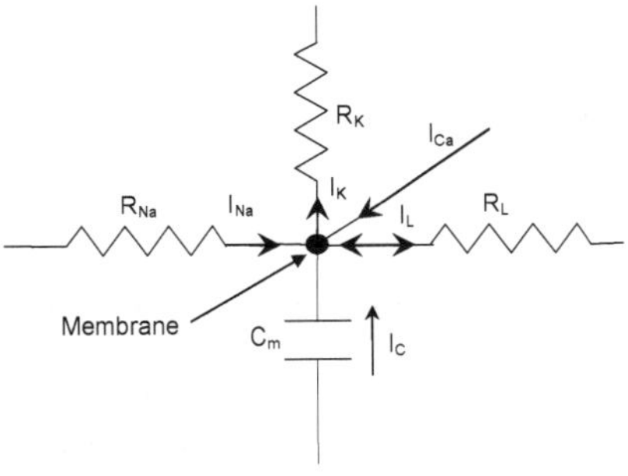

Fig. 3. Plasma membrane as a knot, where ionic currents concur.

The D and H main trams of bioelectric impulse in neurons (Fig. 2) are controlled by the well known Hodgkin & Huxley (HH) equation (Hodgkin and Huxley, 1952). This equation has been usually referred to an equivalent electric circuit of in parallel conductances, membrane capacitance, C_m and DC generators, the latter being mainly the Nernst equilibrium Na^+ and K^+ e.m.f., E_{Na} and E_K, due to the ions electrochemical gradients (other e.m.f. are due to Ca^{2+} and Cl^- ions). Consideration of this network by meshes does not allow its easy solution, and we will consider the membrane as a Kirchoff knot where the currents concur (Fig. 3). Therefore HH equation in the presence of an applied magnetic field, B_{eff}, takes the knot law of charge conservation (no charge accumulation in membrane),

$$C_m(dV/dt) + g_{Na}m(t)^3h(t)(V - E_{Na}) + g_K n(t)^4(V - E_K) + g_L(V - V_L) - I_{Ca}(B_{eff}, t) = 0\ , \qquad (1)$$

where V is the transmembrane voltage, g_i (i =Na, K, L) the channels conductances. m and n are the HH channel excitatory and h inhibitory functions, of microscopic origin not yet well known, although the phenomenological needed powers four, point out to four independent processes, acting for the opening (m, n) and closing (h) of corresponding channels. Leakage channels and voltage/ligand operated ones are probably responsible for the setting of the threshold voltage, V_s but current through them is weak and here neglected. Finally, for our purpose, HH currents have been supplemented by the Ca^{2+} current produced by MF (*HH magnetic* (HHM) equation) as we shall discuss below. We will solve eq.[1] in the *relaxation time, τ, approximation* for the HH functions, where e.g. we assume that

$$dn/dt = -n(t)/\tau_K \qquad (2)$$

where n(t) is assumed to be proportional to the number of K^+-channels which remain closed at time t. Integration of [2] taking t = 0 at the beginning of R process plus H process, yields $n(t) = n_0\exp(-t/\tau_K)$. Similarly taking t = 0 at the beginning of D process we obtain that function $m(t) = m_0\exp(-t/\tau_{Na})$. Otherwise the inhibition function at D process follows the equation $dh/dt = +h(t)/\tau_{inh}$, of integral $h(t) = h_0\exp(+t/\tau_{inh})$. We will now obtain the membrane voltage V (t) dependence, partitioning the impulse in the mentioned regimes.

Repolarization and hyperpolarization: these two processes follow one after other and it is well known that in the R+H process only K^+-channels are open and therefore [1] becomes, $C_m(dV/dt) + g_K n(t)^4(V - E_K) - I_{Ca}(B_{eff}, t) = 0$,which integration after substitution of n(t) yields

$$V_K(t) = E_K + (E_{Na} - E_K)\exp\left[-\left(g_K n_0^4 \tau_K/4C_m\right)\left(1 - e^{-4t/\tau_K}\right) + \int_0^t dt' I_{Ca}(B_{eff}, t')/(V_K(t') - E_K)\right], \qquad (3)$$

which is an integral equation in V_K (t) with kernel $I_{Ca}(B_{eff}, t)$. We will show below (from [8] and [10]) that $I_{Ca}(B_{eff}, t) = -(N(0)f(B_{eff} = 0)q_{Ca}/\tau_{Ca})\exp(-\alpha B_{eff}^2)\exp(-t/\tau_{Ca})$, where N(0) is the initial Ca^{2+} ion number in a burst and τ_{Ca} the Ca^{2+} relaxation time (diffusion time in the cytoplasm) (t origin in [3] is taken at $V(t) = E_{Na}$, origin of R). For comparison with experimental results in single neurons, it is useful to work in frequency domain,ω so that we will obtain the *frequency spectrum* of the spontaneous impulse $V_K(t)$. Fourier transform (FT) of [3] exp[...] function is unknown, but for $t < \tau_K$ first exponential can be series expanded, so obtaining

$$V_K(t) \approx E_K + (E_{Na} - E_K)\left[1 - \left(g_K n_0^4 \tau_K / 4C_m\right)\left(1 - e^{-4t/\tau_K}\right) + \int_0^t dt' I_{Ca}\left(B_{eff}, t'\right) / \left(V_K\left(t'\right) - E_K\right)\right]. \quad (4)$$

The ω spectrum of [4] *spontaneous* $V_K(t)$ ($I_{Ca} = 0$) is obtained by Fourier transforming $V_K(t)$ around a central frequency ω_0, characteristic of the impulse, yielding (except for a Dirac $\delta(\omega - \omega_0^*)$ artefact introduced by the exponential series cut-off)

$$V_K(\omega) = A \Big/ \left[\left(\omega - \omega_0\right)^2 + (\Delta\omega/2)^2\right], \quad (5)$$

where $A \equiv g_K n_0^4 \tau_K / 4C_m$ and $\Delta\omega/2 = 2\pi/\tau_K$ the HMHW, which provides τ_K. Therefore the impulse spectrum is the familiar *Lorentzian* function, taking its maximum value at $\omega = \omega_0$. Eqs. [4] and [5] can be easily extended to the real situation of having different types of K^+-channels (up to seven in *Helix aspersa* neurons (Azanza et al., 2008)), but this extension is not very suitable for comparison with the impulse because of the too large number of parameters involved.

Depolarization: this process follows after the refractory time and threshold voltage establishment, and since involved Na^+ channels are operated by voltage, inclusion of Ca^{2+} current sums only a term to $V_{Na}(t)$. But also retarded in time K^+ channels are opened, although being in small number during D tram their current can be neglected. The HHM relevant equation is then $C_m(dV/dt) + g_{Na} m(t)^3 h(t)(V - E_{Na}) - I_{Ca}(B_{eff}, t) = 0$, which in presence of MF yields another integral equation. Integration followed by the first exponential expansion as before yields

$$V_{Na}(t) \approx E_{Na}\left[1 - \left(g_{Na} m_0^3 h_0 \tau_{eff} / 3C_m\right) \exp(-t/\tau_{eff}) + \int_0^t dt' I_{Ca}\left(B_{eff}, t'\right) / \left(V_{Na}\left(t'\right) - E_{Na}\right)\right], \quad (6)$$

where the relaxation time is given by $\tau_{eff}^{-1} = \tau_{Na}^{-1} - \tau_{inh}^{-1}/3$, since the inhibition and activation are independent processes. As before the ω-spectrum of spontaneous $V_{Na}(t)$ is Lorentzian of $\Delta\omega/2 = 2\pi/\tau_{eff}$, and $A \equiv g_{Na} m_0^3 h_0 \tau_{eff} / 3C_m$. Extension to different kinds of Na^+-channels is not worthwhile because of above mentioned reason. Ca^{2+} and Cl^- channels operated by voltage as well would be treated in the same way to Na^+ ones, but as mentioned before their associated currents can be safely neglected.

2. Biophysical experiments.

2.1 Experiments made on single unit neurons from *Helix aspersa* (mollusc) brain ganglia by applying static (SMF) and alternating (ELF) magnetic fields .

Since experiments under low intensity SMF and alternating AC-ELF MF ones are intimately related in their interpretation with the ones carried out under modulated MW fields it is important to present them, in order to fully understand the neuron behaviour under the latter. We will briefly describe the experimental set-ups for the three kinds of experiments, as follows.

2.1.1 Experimental set-up for exposure to SMF.

Brain ganglia (about 6 mm^3 of volume) (Fig. 4) were placed in the centre of an electromagnet polar pieces (Fig. 5). Nervous ganglia were immersed in molluscs Ringer solution.

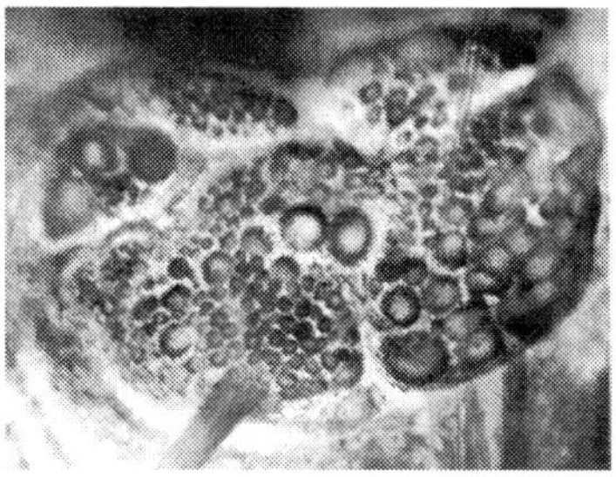

Fig. 4. Microelectrode inside neuron F1. (from Kerkut et al., 1975).

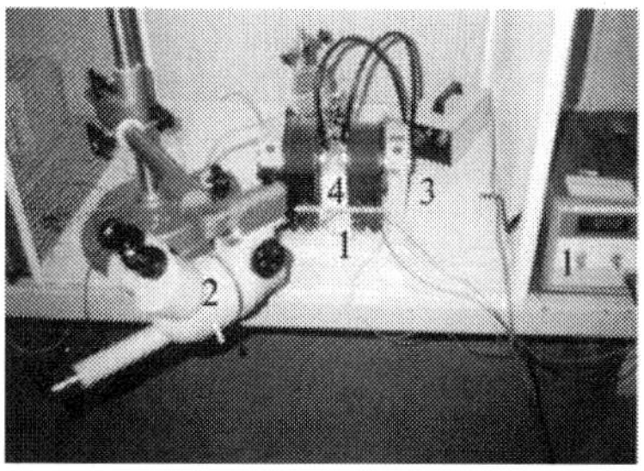

Fig. 5. SMF application. 1: Power supply and electromagnet polar pieces. 2: Microscope. 3: Cold light. 4. Brain ganglia in the centre of polar pieces. 5. Microelectrode.

Intracellular electrophysiological activity from single neurons was recorded in real time with glass microelectrodes (tip diameter < 0.5 μm, tip resistance 2-20 MΩ), filled with a conducting 1M potassium acetate solution (pH 6.8) (Fig.5). Intensities of applied SMF were in the range of 1.0 mT up to 0.7 T (Azanza, 1988; 1989; 1990; Azanza and del Moral, 1994 1995; 1996). Applied MF -either static or alternating- were perpendicular to local geomagnetic field (GMF) lines. Set-up was disposed inside a Faraday cage.

2.1.2 Experimental set-up for exposure to ELF-MF.

Brain ganglia samples, were disposed between a pair of Helmholtz coils as above described for exposure to SMF (Fig. 6). Applied ELF-MF were of: frequencies between 0.1 and 217 Hz and AC amplitude between 0.2 μT up to 15 mT. Experiments at AC, μT amplitude, were performed inside a Mumetal chamber (Fig. 7). The screening was of 100 times, relative to the values of local geomagnetic field (GMF). The AC amplitude inside the Mumetal chamber, was of 0.1μT with respect to the ambient AC field of 0.2 μT (Azanza and del Moral, 1998; Azanza et al., 2001, 2002; Calvo et al., 1999a, b; Pérez-Bruzón, 2006).

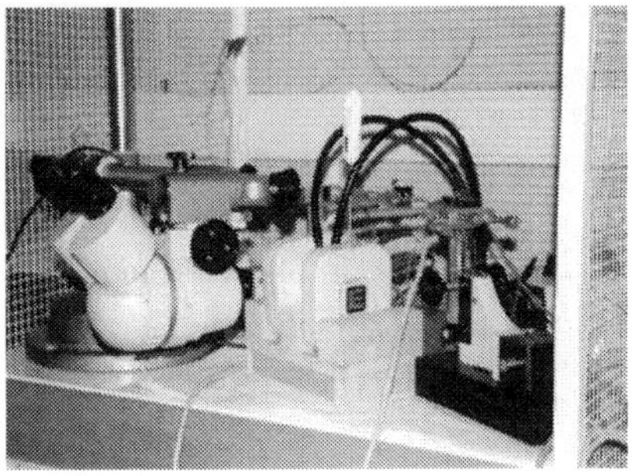

Fig. 6. Experimental set-up for application of ELF-MF. The neuron sample is placed between a pair of Helmholtz coils.

Fig. 7. Exposure to ELF-MF of 0.2 μT, 2μT and 12 μT were performed inside a Mumetal screening chamber (4). (3) Cold light. (2) Helmholtz coils. (1) Brain sample.

2.2 Experiments made on single unit neurons from *Helix aspersa* (mollusc) brain ganglia by applying 13.6 GHz microwaves, modulated by ELF-EMF.

2.2.1 Experimental set-up and dosimetry

Helix aspersa brain ganglia were maintained as described above for SMF and ELF-MF experiments. For exposure to EMF of 13.6 GHz the ganglion bath was placed within a resonant, open, toroidal cavity (Fig. 8). The resonant cavity (Figs. 8 and 9) is made of a 1 mm thickness dielectric ring of FR4, cooper metallized on both surfaces, which are in turn aluminium short-circuited in their external edge for forming the cavity. The MW field was generated by a home made Gunn diode oscillator, which modulates in amplitude the high frequency voltage by an ELF frequency signal voltage between 2-100 Hz. The MW-MF is homogeneous within an area of about 4 mm^2 around the cavity centre, where the ganglion is accurately positioned. The MW EF ($E_0 \cong 3.5$ V/m) is polarized along Oz axis (Figs. 8 and 9) and is homogeneous within the cavity height.

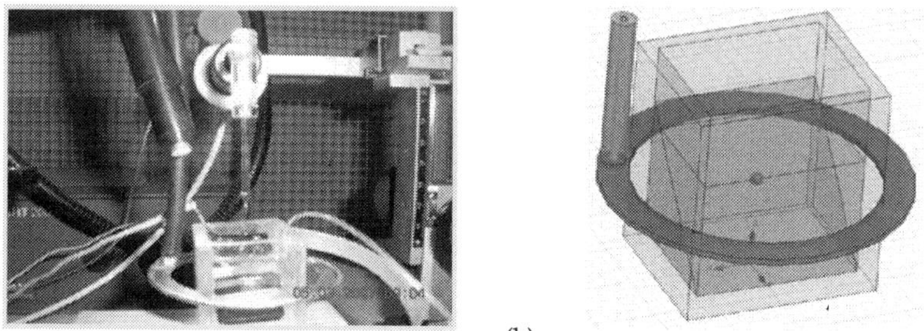

(a) (b)

Fig. 8. (a) Set-up for MW-MF exposure and schematic diagram of the set-up. Nervous ganglion was accurately positioned at the cavity centre, where the field is quite homogeneous. b) Idealized set-up for dosimetry calculations.

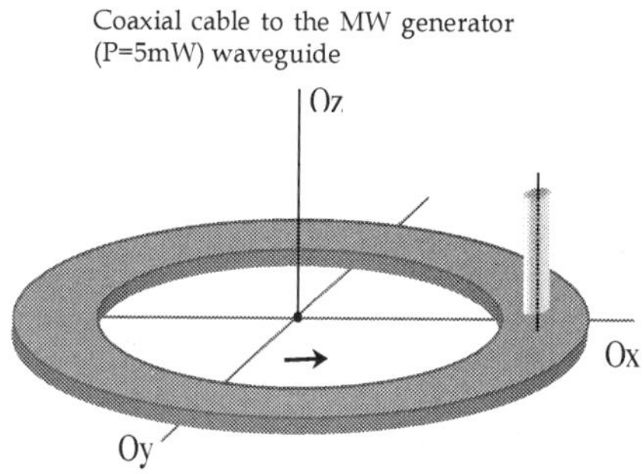

Fig. 9.- Toroidal cavity (mode TEM): external radius r_e = 2.5 cm; internal r_i = 2cm; h=1 mm. Magnetic field **H** is polarized in the cavity plane (the one of the biological sample), along the coaxial cable to the MW generator (P=5mW) waveguide. On Ox axis containing the feeding port. Electric field, **E** is normal to plane.

The MW signal was extracted using a rectangular waveguide, working in a dominant TE_{10} mode, followed by a coaxial cable (50 Ω), so that the mode becomes TEM, the cable being connected to the cavity by BNC gold plated connector. Modulation depth was fixed at 90%. MW frequency of 13.6 GHz was measured using a MW spectrum analyzer E4407B (Agilent) and the generator output power of 5 mW was measured using a power meter 4231A (Boonton). Typical Poynting vector at the cavity center was S ≅ 0.35 W/m^2. Typical peak value of H_o ≅ 0.10 Am^{-1} (= 1.25 mOe) at the *Helix* brain ganglia position (cavity centre) (note this intensity is close to the lowest one applied in our ELF alone experiments). The bioelectric impulses were Fourier spectrum analysed using computer standard methods (Chart v 4.1.2 program for Windows, ADInstruments). It is also worthwhile to mention that the applied MF in the electrophysiological experiments was of the same order of magnitude that the applied to astrocytes in our experiments of irradiation performed within the GTEM anechoic chamber (Fig. 21B).

The values of SAR (Fig. 10) and measured temperature increase of sample, between 0.0258 and 0.0261 °C show that the experiments have been carried out under non thermal conditions. Therefore measurable *thermal effects are not expected.* Dosimetry calculations

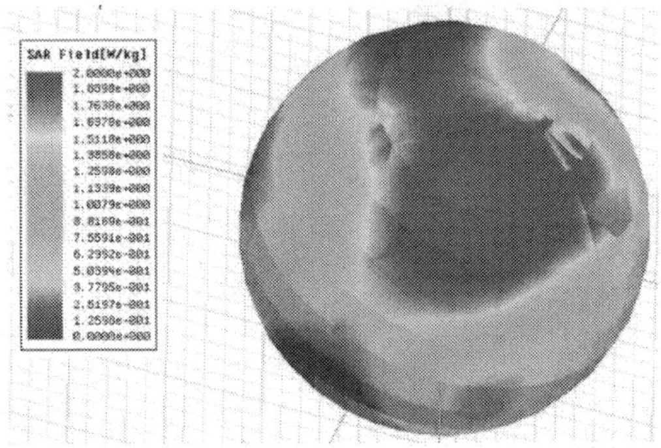

Fig. 10. SAR in the surface of the sphere (nervous ganglion). The mean value of SAR in the volume occupied by the sphere is 2.02×10^{-3} W/Kg.

were made by using the method of finite elements in frequency domain implemented in commercial package ANSOFT HFSS.

Although the open cavity radiates some of the injected electromagnetic power (5 mW) to the exterior it has been shown that it keeps an EMF distribution similar to the closed cavity one (field distribution is only perturbed within the metallic ring). As the chamber with Ringer solution and nervous ganglia is introduced inside the toroid some field attenuation is expected due to the conductivity of the saline solution. Also dominant polarizations in the centre of the applicator are perturbed with respect to the empty applicator. Calculated temperature variations, ΔT, in the Ringer bath solution under applied MW are similar to the values measured with respect to a control Ringer solution not illuminated with MW. The measurements were made with a calibrated $R_0 = 100\ \Omega$ (0°C) Pt-resistor thermometer (0.01°C precision, resistance temperature coefficient α =0.03.850 x10^{-2} Ω/°C between 0-100°C) and a multimeter (0.0001 ohms resolution) using the four point technique for temperature dependent resistance measurement. Temperature was obtained from linear interpolation, $t=(R_t - R_{0)})/\ \alpha R_{0.}$

2.3 Experimental Results

2.3.1 Main experimental observations by application of SMF and ELF (0.1-50 Hz) magnetic fields.

We have observed that the behaviour of an individual neuron, against an applied MF, either static or alternating, is not random but fixed for a mapped neuron: stimulation, decrease of the activity and eventual inhibition and slow response or no response. Magnetic fields, either SMF or ELF-MF, induce effects which reproduce normal, spontaneous, activities of neurons. Applied MF seem to work as switchers, they switch on/switch off the spontaneous activities. Responses of excitation/inhibition are shortened under applied MF.

Under applied SMF, about the 70% of neurons show a Ca^{2+}-dependent modification of the *spikes-frequency* (see spike in Fig. 2), with non appreciable modification of spike shape. For the 50 % of neurons the frequency decreases and eventually are inhibited. For the 20 % of neurons the frequency increases, being stimulated. For the remainder 30 % of neurons very slow or no responses are observed. After long time exposures, spikes-amplitude decrease through a mechanism dependent on the progressive inactivation of the $3Na^{+}$-$2K^{+}$-ATP-ase pump (Azanza and del Moral, 1996). We have observed higher neurons sensitivity under applied ELF-MF. For the 56% of neurons they are inhibited. About the 26% of neurons are stimulated and about 18 % of neurons show slow or no responses. Spikes frequency responses are more frequent than spikes amplitude responses. Also neurons show a much higher sensitivity to frequency variations than to amplitude variations of applied MF (Pérez-Bruzón, 2006).

Searching for the origin of stimulation/inhibition induced effects on neurons, we were able to experimentally show that MF somehow induces the liberation of Ca^{2+} ions in the cytosol. Depending on *neuron type* the increased free cytosolic calcium concentration ($[Ca^{2+}]_i$) produces: i) the increase of neuron membrane conductance for K^{+} ions (g_K) through Ca^{2+}-dependent-K^{+}-channels in turn promotes the sorting out of K^{+} ions to the extracellular fluid, hence the hyperpolarization and so *inhibition* of neuron activity; ii) the increase of positive charge directly induces the Ca^{2+}-dependent-membrane depolarization, promoting in turn the neuron *stimulation.* We have shown *mimic effects* between the induced ones by MF and the induced by increased $[Ca^{2+}]_i$, after a set of key experiments: i) by promoting the entrance of Ca^{2+} ions into the cytoplasm increasing by seven times the Ringer Ca^{2+} concentration (Azanza and del Moral, 1988, 1994); ii) by promoting the liberation of Ca^{2+} ions from the endoplasmic reticulum into the citosol with caffeine -agonist of ryanodine receptors- (Azanza, 1989, 1990; Azanza and del Moral, 1994) and iii) by promoting the entrance of Ca^{2+} into the citosol through NMDA-receptors activated by glutamate (Calvo, 2003; Azanza et al. 2009). The most important conclusion is that *inhibition and stimulation are Ca^{2+}-dependent processes,* neuron-specific and are the result of membrane molecular structure expressed in terms of: kind, localization and relative density of ionic channels in plasma neuron, as we have shown by the characterization of *Helix* channels by immunocytochemistry (Azanza et al. 2008).

Main observations under exposure to ELF-MF were as follows:

2.3.1.1 - Synchronization of at least pairs of neurons activity defined as a *coincidence in spikes frequency and firing rhythm* in time (Azanza et al., 2002, 2009). One of the most striking behaviour was oscillatory and recruitment activities observed after some time under exposure to sinusoidal ELF-MF. These characteristics of neuron activity are the expression of a kind of *synchronizing* activity of neurons relatively far away one each other but integrated in a small network (Fig. 11). Connexin proteins which make *gap* contacts between neuron-neuron and neuron-glia cells are the main responsible for synchronization in mammals brain. In our studies by simultaneously recording the bioelectric activity from pairs of neurons we have observed that synchronization occurs in the 27 % of pairs of neurons studied (Azanza et al., 2002). We have studied the expression of connexin 26 by immunocytochemistry methods and shown that it is expressed in only the 4% of neurons in all the *Helix* suboesophagic ganglia (Azanza et al., 2007). These results plus the comparison of synchronization recordings with the ones mediated by neurotransmitters in synapsis are a strong support in favour of the participation of MF in the synchronization process (Azanza et al., 2009). The synchronization encompasses clusters of e.g. 7 and 13 neurons, surrounding a central one. The calculated neuron number in the cluster using the model of § 3.2 agrees remarkably with the experimentally inferred number (Azanza et al. 2002).

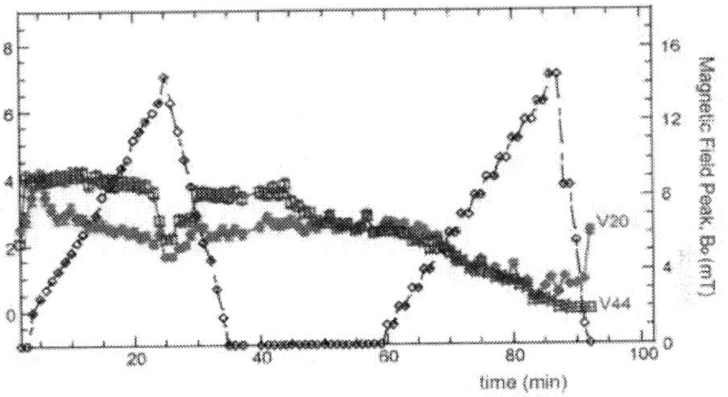

Fig. 11. Experiments were made by simultaneously recording the activity from neuron pair V44 (□, blue) and V20 (♦, red). Under exposure to 50 Hz, 0.5-15 mT EMF (◊), frequency increases reaching the same value in 2 min. As the applied ELF-MF amplitude increases the frequency of V44 did no change, but frequency for V20 goes down to its initial value. For ≈15 mT the frequency decreases sharply in parallel for both neurons, reaching a minimum value. As ELF-MF amplitude decreases the frequency for both neurons increases in parallel reaching the initial, spontaneous, value. At min. 35 the MF is switched off, no changes in the firing frequencies are observed. After 6 min (min. 41), the frequencies for both neurons start approaching, reaching same value at min 50. Synchronizing activity remained for about 32 min., disappearing when the applied field goes down. (Calvo et al., 2002; Calvo 2003).

2.3.1.2 - Frequency window effect: the neuron firing frequency, f , decreases with the applied MF frequency, f_M, as a *Lorentzian*, centred at about the spontaneous, f_0 , one (Figs. 12 and 13) (Pérez-Bruzón et al., 2004; Pérez-Bruzón, 2006; Azanza et al., 2007b).

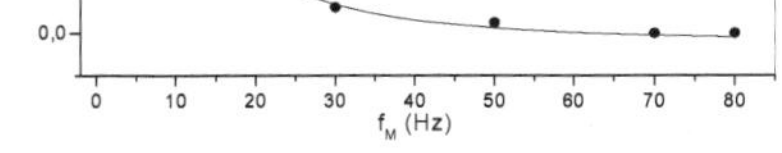

Fig. 12. Neuron F1. Lorentzian (line) fits the variation of neuron f (expressed in spikes/s) vs. field frequency, f_M. f_0=2.5 Hz, $\Delta f_{1/2}$= 9.9 Hz.

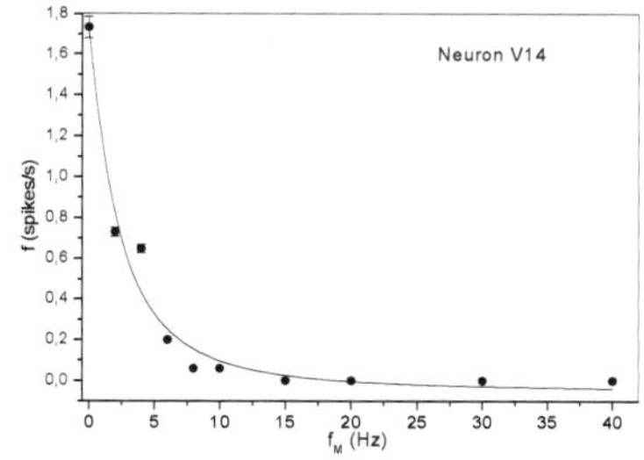

Fig. 13. Neuron V14. Lorentzian (line), fits the variation of neuron *f* (expressed in spikes/s) vs. field frequency, f_M. f_0=2.0 Hz, $\Delta f_{1/2}$= 2.7 Hz.

2.3.1.3 - Resonance effect: we have experimentally shown in molluscan brain single neurons that as the frequency of the applied MF, f_M, was coincident with the main frequency, f_0 of the Fourier spectrum of the spontaneous bioelectric activity voltage impulse, the neuron firing frequency showed a maximum, an effect so called *frequency resonance* (Fig. 14) (Pérez-Bruzón et al., 2004; Pérez-Bruzón, 2006; Azanza et al., 2007b).

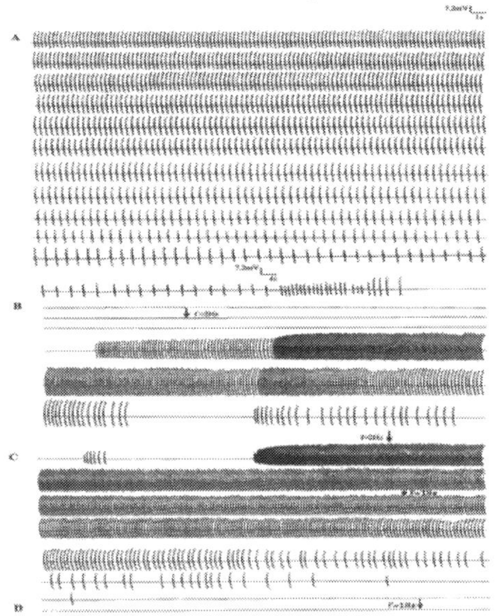

Fig. 14. Neurone V19. **A**): Spontaneous f = 2.4 spikes/s, frequency and amplitude progressively decrease, being the neuron activity completely inhibited after 6 min. of recording. **B**): ELF-MF of 1 mT-2 Hz, was applied for 10 min. With 4 min delay the neuron activity (frequency) was stimulated, spikes amplitude also increasing. C): ELF-MF of 1 mT-2 Hz was applied, the frequency and amplitude increased for a second time. As 1 mT-1 Hz was applied, the neuron frequency was progressively decreasing and neuron activity completely inhibited. Experiment duration: was of (Pérez-Bruzón, 2006).

2.3.1.4 - Demodulation effect: the purpose of our research by applying MW electromagnetic fields (EMF) amplitude modulated (90%) by ELF-EMF was to separate out the possible effect of the MW from the one induced by modulated ELF-EMF within a wider range of frequencies, i.e. 2-100 Hz.
The exposure of neurons to MW modulated by ELF-MF MF between 2 and 20 Hz and 20 Hz have shown that are the ELF-MF the responsible for the elicited responses (Figs. 15a and 16), a so called *demodulation effect*. Main observation was *no effect under the carrier*, f_c=13.6 GHz, but *"frequency resonances" at low frequencies*, e.g. f_M =16 Hz (Figs. 15a, 16), similar to the case of only ELF application, i.e. also with *Lorentzian profiles* (Fig. 17) (Azanza et al., 2007b; del Moral et al., 2008). The effect is a *"frequency resonance"* of *Lorentzian* shape, when the *MF frequency matches the characteristic frequency (-ies) of the neurone impulse Fourier spectrum* (Figs. 15b and 18b). We should stress that a *"frequency resonance"* is a maximum in the spectrum $f = f(f_M)$, where f is the bioelectric or spike frequency repetition. In neuron V14 two frequency resonances are observed at f_M = 4 and 16 Hz (Fig. 16). On Fig. 17 we can see *Lorentzian* fits to the f_M = 4 and 16 Hz resonances in neuron V14 (Fig. 16). As we will see this is an important observation upon which to base the model proposed in § 3.4 for the effect of ELF amplitude modulated MW upon neuron bioelectric activity. Note that the resonance observed in *not* an amplitude one ("spring" resonance).

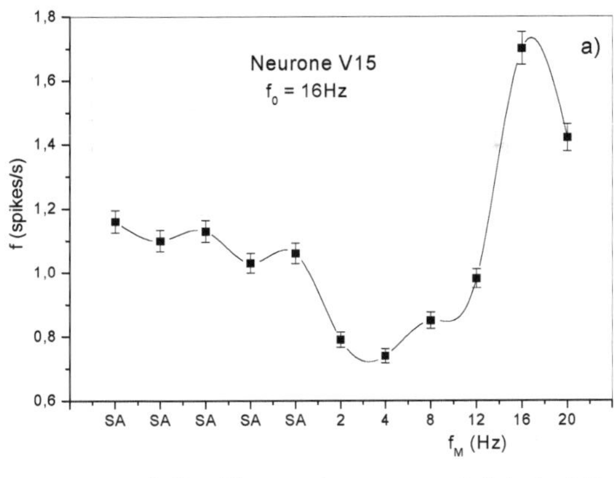

Fig. 15. a) SA, spontaneous activity. The carrier was modulated at 2, 4, 8, 12, 16, 20 Hz. Neuron V15 shows a *resonance* effect at 16 Hz. b) Spontaneous activity Fourier spectrum gives a maximum for 16.4 Hz.

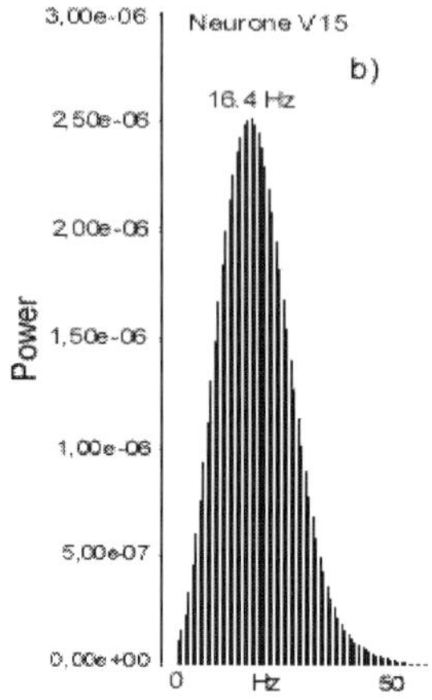

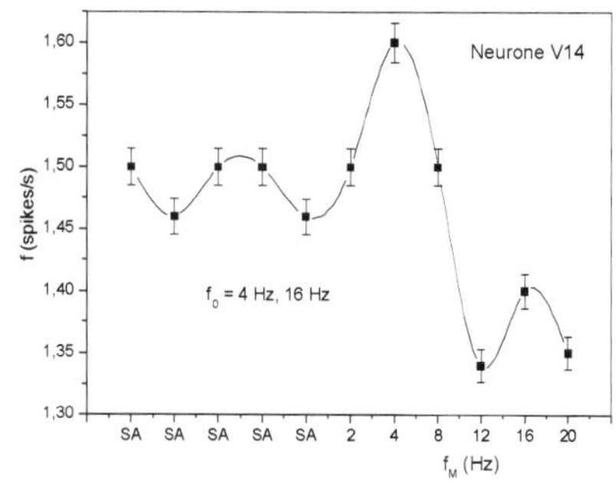

Fig. 16. Frequency *resonance* effect showing maxima at 4 and 16 Hz from neuron V14: frequency window effect.

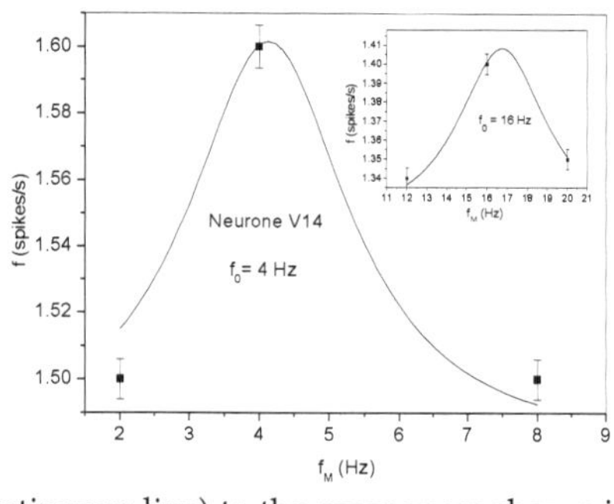

Fig. 17. Lorentzian fits (continuous line) to the resonances shown in Fig 16. The HMHWs are respectively 1.4 and 1.6 Hz.

The resonance effect seems to be neuron specific. In the experiment on neuron F32 (Fig. 18a), caffeine (3mM) does not induce any Ca^{2+}-dependent activity (Azanza, 1989). When Ringer solution is added in order to remove caffeine, we observe a small increment in neurone bioelectric activity. This increment is not relevant from the statistical point of view. MW carrier alone (C) was applied and then the carrier modulated by ELF from 2 to 100 Hz frequencies. *Resonances* at 4 Hz and 50 Hz are observed. Fourier spectrum (Fig. 18b) gives a maximum for 4.2 Hz which is coincident with a maximum neurone frequency (4 Hz) (Fig. 18a). Resonances at 12 and 50 Hz are observed for other kinds of neurons (Fig. 19).

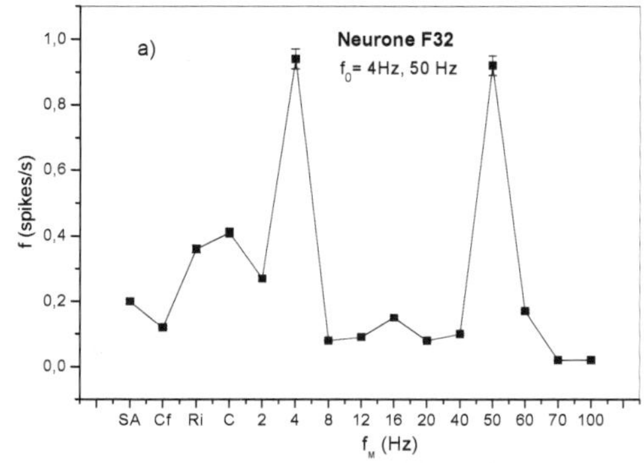

Fig. 18a. SA, spontaneous activity. Cf, caffeine. Ri, Ringer solution. C, MW carrier, induces a non significant modification of bioelectric activity. *Resonances* at 4 Hz and 50 Hz are observed.

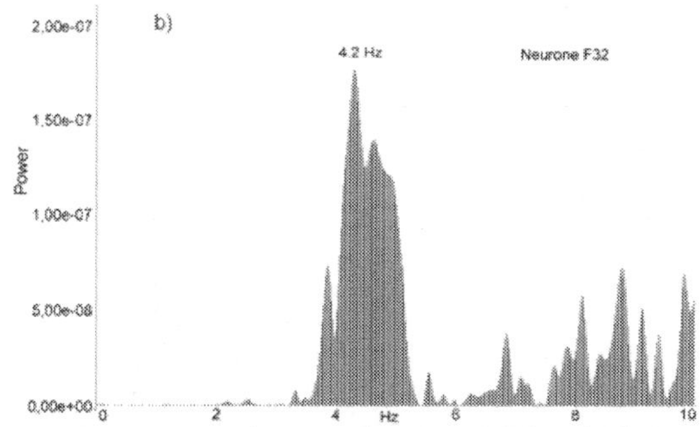

Fig. 18b. Fourier spectrum give one maximun value at 4.2 Hz which is coincident with maximum neurone frequency. A filter to avoid 50Hz noise prevents getting the correspondent maximum.

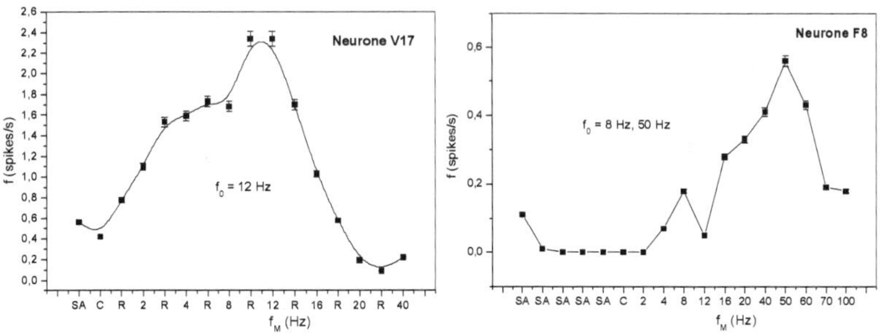

Fig. 19. Resonances at 12 Hz (neuron V17) and 8 and 50 Hz (Neuron F8).

Two conclusions can be summarized from the above experiments:

i) Neurone plasma membrane (see § 3.4) seems to behave as a physical system able to resonate as the MF frequency matches some characteristic one of the bioelectric impulse other than the spontaneous neurone frequency (Azanza et al., 2007b). An approach has been made for the interpretation of our resonance results consistent in a membrane depolarization due to the increase of cytosolic Ca^{2+} concentration. Ca^{2+} is detached from plasma neurone membrane through the superdiamagnetism (SD) and Ca^{2+} Coulomb explosion (CE) as explained in § 3.4 (del Moral and Azanza, 1992; Azanza and del Moral, 1994).
ii) Neurone bioelectric activity is highly sensitive to *low frequency* applied alternating MF modulating a MW carrier in the ten of GHz range. *Extremely low frequency modulated MW radiation* at non-thermal level of field power density (ΔT increase in bath lower than 0.01°C) modifies neurons bioelectric firing frequency, in a *resonant* way. The *resonance* appears when the ELF applied MF is close to a characteristic *frequency of the impulse train Fourier spectrum* (*not* to the firing frequency, del Moral et al., 2008).
Stimulatory effects by MW modulated by ELF-EMF have been described on human volunteers electroencephalogram recordings (EEG). 400 MHz 100 % modulated in the EEG physiological spectrum, at frequencies of 7, 14 and 21 Hz showed increased alpha (8-13 Hz), and beta (13-30 Hz) rhythms. Alpha and beta rhythms were also activated by MW modulation at 40 Hz and 70 Hz (Hinrikus et al., 2005). Similarly to observations on humans EEG we have got resonances at frequencies in the alpha and beta rhythms, values much higher than the spontaneous *Helix* neurons frequency (0.1-8.0 spikes/s). Our conclusion is that the *frequency resonant effect* must be the expression of an intrinsic biophysical property common to molluscan and human plasma *membrane* neurons which appears when the ELF applied MF is close to a *characteristic frequency of the bioelectric impulse train Fourier spectrum*. These observations could explain the effects observed on human EEG.

2.4 Experiments made on astrocytes from human astrocytoma tumour submitted to 9.6 GHz amplitude modulated by low power ELF-MF of 100 and 800 Hz.

Another kind of experiments has been performed consisting in the study of glia cell (human astrocytes) proliferation process under also ELF amplitude modulated MW EMF, that we will briefly discuss. The underlying mechanism to explain them may be also the Ca^{2+} detaching from membranes.

Brain neurons and astrocytes are cells of crucial interest for the research of potential effects of MW produced by communication systems. Astrocytes are a physical support for neurons in human brain; they feed neurons by supplying metabolites from blood; they provide a neurotransmitters and ions buffer system to the brain, and with endothelial brain vessels membrane makes the *blood brain barrier* (BBB) and are able to proliferate, being the responsible for more than 90% of human cellular brain tumours (gliomas). Any modification in any of their activities will potentially produce negative effects on brain function and human health.

The aim of our work has been to characterise the effects of short MW pulses upon the physiology of astrocytes in culture by means of cellular and biochemical studies trying to characterize any possible toxic effect by comparing the results obtained on non-exposed, standard sham-control conditions, with the ones obtained under exposure to MW.

2.4.1 Experimental set-up and dosimetry.

Experiments were performed on astrocytes from human astrocytoma (Clonetics line 1321N1). Cells were maintained in culture as an adherent monolayer in a humidified atmosphere of 5% CO_2 at 37°C in a standard incubator. After 6 days cultured, cells were transferred to the GTEM-incubator for exposure to MW inside a horn shape GTEM cell, where the TEM radiated MW is from a flat strip line along a border. MW were produced with a solid state MW generator (100 KHz-20 GHz range), provided with a versatile modulator of different wave profiles (modulation depth 90%), followed by a high power (50W maximum output) MW travelling-wave tube amplifier, followed by a directional coupler, which injects the MW signal into the GTEM chamber through a 50 ohm coaxial cable (Fig. 20). ELF modulation was kept for all irradiations at 90%. Direct and reflected from chamber powers were monitorized via a diode bridge. GTEM chamber is provided with anechoic walls to reduce unwanted reflections. The EMF-MW mode was the TEM one, same as usually in wireless telecommunication.

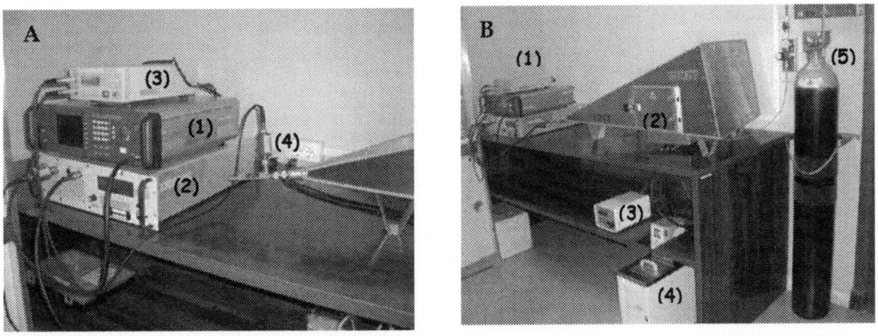

Fig. 20. **A.** (1) Microwave generator (GIGATRONICS 2520A). (2) Medium Power Wide band amplifier (T186-50). (3) RF Power meter (BOONTON 51013). (4) Directional coupler: frequency range: 4-18GHz (COU-BD418 G 50W-35). **B.** (1) Electronic equipment connected to the GTEM - Cell (2). Olympus - incubator regulators: (3) CO_2 regulator; (4) Temperature regulator; (5) CO_2 container.

Two cells culture flaks were placed into the GTEM-cell where the EMF is rather homogeneous, with their longitudinal axis in the same direction of EM wave incidence. In Fig. 21 are shown the simulation layout for the dosimetry calculation for the whole system and SAR calculation along an observation line at 50 μm height from the bottom of the flask (the estimated cell layer thickness). EF calculated values were: 0.05-0.25 and 0.10- 0.35 V/m for the two flaks respectively. The calculated MF were: 4.6-8.8 A/m, and 4.8-9.1 A/m for the two flaks. MF and EF intensities measured, are of the same order of magnitude to the ones that we are applying in our electrophysiology studies on neurons. The EF was normal to the plane of the cells monolayer and the incident MF in the horizontal plane of the monolayer. Calculated temperature increases due to MW power exposure were below $3x10^{-5}$ °C. *Therefore we can assume negligible thermal effects in the cells* (for detailed descriptions of dosimetry see Pérez-Bruzón et al., 2009).

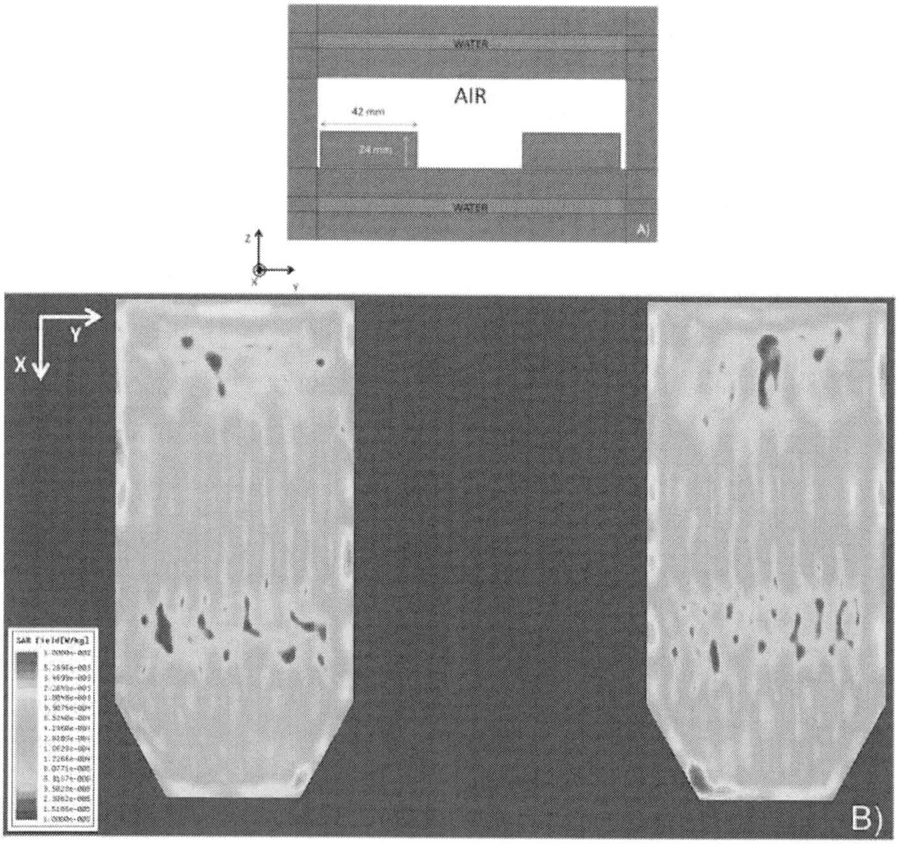

Fig. 21. **A)** Transversal section of 2 Falcon flasks inside the GTEM- incubator system. **B)** SAR simulation in a plane at 50 μm height from incubator base. The mean SAR value from left to right was $3.99x10^{-4}$ W/Kg and $4.08x10^{-4}$ W/Kg (incidence direction +X).

2.4.2 Experimental results

In our experiments we have observed an statistics significant *increased* cell proliferation rate *of about 43%* under 24h exposure to pulsed MW in two experimental conditions: 9.6 GHz carrier frequency, pulse width 100 and 120 ns, pulse repetition frequency 100 and 800 Hz, pulse repetition interval, 1.25 and 10 ms, power 0.34 and 0.60 mW, EF strength 1.25 and 1.64 V/m, MF strength 3.3×10^{-3} A/m (41.4 μOe) and 4.35×10^{-3} A/m (54.6 μOe). Searching for biomarker proteins of astrocytes cell cycle and apoptosis, we found that Hsp-70 and Bcl-2 antiapoptotic proteins were expressed in control and treated samples while an increased expression for connexin 43 proteins was found in exposed samples.

These results open the Ca^{2+} pathway for an explanation of *increased proliferation*. Signal cascade of astrocyte apoptosis may enclose modifications of intracellular calcium concentration ($[Ca^{2+}]_i$), an interesting confluence of molecular pathways to explain the effects under exposure to EMF in two kind of cells of the nervous system: of nervous: neurons and astrocytes. By considering the mechanisms that initiate cell cycle reactions, the induction of cell *stress* could be related to an increased release of hsp (heat-shock) proteins, which need not to be induced by heat production only. Hsps proteins play a critical role in the regulation of the cell cycle, proliferation and apoptosis. The synthesis of Hsp can be produced under several stress conditions: alcohol, oxidative stress, osmotic pressure change, toxic chemicals and exposure to low-frequency EMF (Pipkin et al., 1999; Leszczynski et al., 2002). Our future research will be devoted to study the quantification of Hsp-70, Bcl-2 and Cx43 proteins, in control and irradiated samples, possible increased $[Ca^{2+}]_i$ concentration process and the implication of the anti-apoptotic increase of Bcl-2 protein promoting cell survival (Takuma et al., 2004).

3. Biophysical models.

3.1 Cell membrane as target for electromagnetic field interactions.

The EMF, in the frequency range we are applying, i.e. ELF of 0.1-217 Hz, and 100 and 800 Hz modulating MW carrier and, MW of 9.6 and 13.6 GHz, of low power (SAR induced on biological samples of 4.00×10^{-1} mW/Kg and 2.02×10^{-3} W/Kg respectively), do not posses sufficient EM-energy density to cause ionization (as in the X-ray, U.V and γ ranges of the EM spectrum) or appreciable thermal effects (see dosimetry data). In fact in the range of ELF, the heating effects are eventually due to the production of eddy currents and to the water dielectric relaxation, mainly of α-type, operating at frequencies between ≈ 0 Hz and ≈ 100 Hz (see Azanza and del Moral, 2004 for details). According to the Poynting theorem the maximum average EM energy deposition (i.e. without radiation losses) in a living system is limited by the time average EM density energy, i.e.

$$e_m = \frac{1}{2}\left[\varepsilon H_{rms}^2 + \mu H_{rms}^2\right] \tag{7}$$

where E_{rms} is the EF and H_{rms} the MF r.m.s intensities respectively, and ε and μ, the dielectric constant and magnetic permeability of the biological medium, respectively. Then for an ELF field with a MF amplitude of say ≈ 0.3% Oe, $e_m \approx 2 \times 10^{-15}$ eV/m^3, and e.g. for a neurone of *Helix aspersa*, with diameter ≈ 100 μm, this means that the energy available to the entire cell is only of ≈ 3×10^{-3} eV if whole energy were deposited in the cell. This energy is really negligible either against the atomic ionization energy for the whole cell atoms (of the order of 10 eV/atom, the Van der Waals binding energies of ≈ 0.06 eV/molecule or the hydrogen bond energies of ≈ 0.16 eV/molecule), or the specific heat (1cal/mole K) for the water within the cell, to produce any noticeable thermal effects (the above energy is around 10^{-8} times smaller than that needed one to rise the temperature of the cell by 1K at room temperature).

An ELF-EMF, both through the EF and through the MF by means of the Faraday induction effect, will produce electric currents in the ionic aqueous solution surrounding the cell. However, the cell membrane is a strong dielectric ($\varepsilon_r \approx 6$) barrier for the passage of those currents to the intracellular medium, except for a small reduced fraction. Therefore if the ELF-EMFs have to produce any effects inside the living cells perhaps they must be through subtle alterations at the membrane level which in turn should convey signals across the membrane body to produce any eventual biochemical and physiological response. The unique characteristics of biological membranes make them the first candidates when searching for EMF interaction sites. Membranes from specialized cells couple the stimuli, carried out by chemical mediators as hormones and neurotransmitters, with the cytosolic machinery inducing in turn physiological responses. Those transduction complexes enclose specific receptor proteins, transductive proteins (i.e. G-proteins) and enzymes (i.e. adenylate cyclase or phospolipase C), which enormously amplify the initial weak impact promoted by the binding of the signalling molecules to their specific receptors. In this sense we can talk about the cell membranes as powerful "amplifiers" of electrochemical or biochemical events occurring in their surroundings. As a consequence of the chemical mediator- receptor interaction, effector molecules at cytosolic levels are triggered -the so called "second messengers"- inducing in turn specific metabolic changes. The best known second messenger molecules are: cyclic-adenosine-monophosphate (cAMP) (Sutherland, 1972); diacylglycerol and inositol trisphosphate (Downes, 1983; Berridge and Irvine, 1984; Nishizuka, 1984); Ca^{2+} ions, central to our model presented in § 3.4 (Nahorski, 1988; Eberhard and Holz, 1988); arachidonate (Irvine, 1982; Loeb and Gross, 1986; Axelrod et al., 1988) and free fatty acids (Ordway et al., 1991).

Therefore in considering the mechanisms of interaction between ELF-EMF and cell membranes we have to take into account the possibility of a direct effect of ELF-EMF on receptor- and transductive-proteins and enzymes and interaction mechanisms. We cannot ignore either the phospholipid bilayer, which mainly constitutes the physical support of the membrane or the glycocalyx, which is an extracellular plasma membrane component essential in the interactions between the cell and its extracellular surroundings. Same can be said about the *inner* surface molecules, which capture free positive ions, e. g. Ca^{2+}.

Non-linear-non-equilibrium processes have been considered as essential ones, at critical steps in the transmembrane signal coupling (Adey 1986, 1988 a, b), for the *"amplification"* of the weak EM energy conveyed by the ELF-EMF interacting with the plasma membrane in order to produce e.g. Ca^{2+} liberation from their membrane stores (as discussed in § 3.4) (del Moral and Azanza, 1992). The mechanism here proposed actually is not an energy amplification *one but nevertheless a cooperative* mechanism within PP clusters (see below). The assumption of having the cell membrane in a non-equilibrium or metastable state of potential energy provides for the possibility that a weak energy EMF stimulus can produce a significant perturbation of the membrane. On the other hand a physical or physicochemical non-linear "device" produces upon a "small" input signal S_i, a sort of amplification giving as output a "small" signal $S_0 = GS_i$, with G>1 only if the "characteristic" curve output vs input is non-linear (Bleaney and Bleaney, 1965). However, an alternative or concomitant way for amplification is *cooperativism* among the molecular components of the membrane surface. Through a cooperative mechanism, the release of ELF-EM energy at a point in the membrane can be enlarged by the interaction of such a point with another one receiving the released energy and feeding back to the first point part of such energy output plus the one received by itself, within a closed-cycle loop (feedback). It is worthwhile to say that studies on the electrical properties of the squid giant axon (Cole and Curtis, 1939), which led to the well-known Hodgkin-Huxley equations, already showed clearly the non-linear characteristics of excitable biological membranes (Hodgkin and Huxley, 1952).

We should also clearly distinguish between two different ways of interpreting the action of EMF on cell membranes: through the *electric* field action or through the *magnetic* one, although when this is *time varying* (EMF), most theories attach cell effects to the electric one, directly impressed by the EMF and the one induced by the accompanying time-dependent MF (Faraday induction). Nevertheless it is noteworthy that when applying EMF of ELF to cells and tissues, the devices used to apply the quasistationary fields are such that the field is an electric one, except for few cases where the EMF was applied with coils, fed by AC currents, and then it is only when we can talk about the application of a quasistationary MF. In either situation the accompanying field, magnetic in the former case and electric in the latter, is negligible.

We have experimentally shown that increased free *intracellular* Ca^{2+} ions concentration promoted under static (Azanza and del Moral, 1994) and ELF-MF (Pérez-Bruzón et al., 2004) induces modification of the bioelectric activity on neurons. MF of the order of magnitude used in our experiments ($\approx$ 1mT-0.7 T in experiments applying static MF and 0.2 μT-15 mT in experiments applying ELF-MF) are incapable of opening Ca^{2+}-membrane channels or producing by ionization the liberation of Ca^{2+} from the binding sites on the neurone membrane. Nevertheless if we consider the combined effect of *superdiamagnetism* and *Coulomb explosion* upon Ca^{2+} ions, they can be liberated on both *sides* of the membrane (del Moral and Azanza, 1992). The model is based upon the strong repulsion that the Ca^{2+} ions, attached to both sides of the membrane phospholipids (Fig.1), suffer when the *nearest neighbour* (*N.N.*) *opposite* extremes of such diamagnetic dipoles (Fig.22) come close enough to suffer Coulombic repulsion forces with a energy higher than the ionic binding energy to the membrane surface, ε_b. According to the described membrane structure, such electrostatic interaction can only occur between PS-GL pairs. These charged heads are immersed in water with a very high dielectric constant ($\varepsilon_r \cong 80$), giving to this structure low electrostatic repulsive energy, and therefore great stability at its ground state (GS), i.e. without EMF application or against thermal fluctuations.

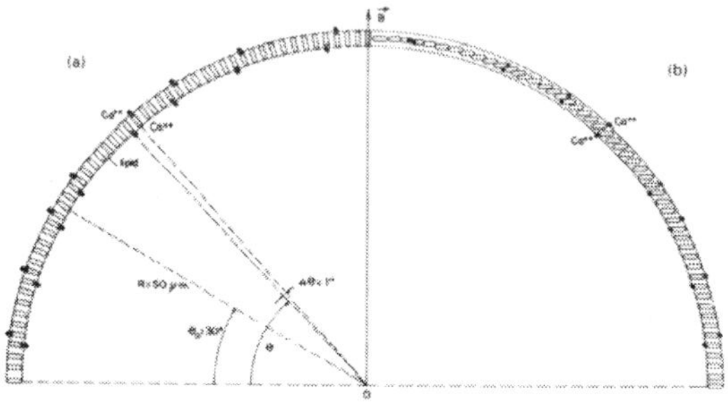

Fig. 22. (a) Schematic layout (not at scale) for the neuron membrane model, showing the lipid molecules (sticks). Some (N.N.)PP ($\cong$ 2% in our model), with attached Ca^{2+} ions are shown. The direction of the applied MF ***B***, as well as the polar angle,θ, of the radial PP molecules phospholipids (PP) molecules molecules and the angle θ_0 below which there is not possible Ca^{2+} liberation are also shown. (b) The same membrane under an applied magnetic field **B**, where diamagnetic PP have fully rotated becoming with their long axes orthogonal to ***B*** and the Ca^{2+} charged heads approach, for a field stronger than a saturation one, B_0 (Azanza and del Moral, 1994).

3.2 Superdiamagnetism (SD) and Ca^{2+}coulomb explosion (CE) model under AC magnetic fields.

This model has been widely tested in single neurons and simple neurone networks under static and ELF MF respectively, where in the latter the MF also induces a *synchronized* bioelectric activity within those networks, as mentioned before (Azanza et al. 2002; Azanza et al., 2005).

The model explains well the neuron bioelectric activity inhibition, i.e. the decrease of firing frequency under applied MF due to Ca^{2+} ions liberation from inner membrane surface. Neuron excitation under applied MF is also thought to be due to increase of Ca^{2+} in the cytosol, that increases the voltage difference across membrane and opens the voltage operated Na^{+} (and Ca^{2+} too) channels, giving rise to depolarization, due to the entrance of such ions. The model inhibition contemplates three ingredients: i) The anisotropy of the diamagnetic susceptibility tensor components, $\tilde{\chi}$ (< 0) of the long PP "rods", or difference $\Delta\chi = \chi_{\parallel} - \chi_{\perp}$, between the directions parallel (||) and perpendicular ($\perp$) to the PP axis (the same applies for protein channels). ii) The well proved cluster formation in the membrane liquid crystal of *correlated* PP long axes through their electric quadrupolar moments, $\tilde{Q}_i$ interaction, of pair (i, j) correlation function $C_Q = <\tilde{Q}_i\tilde{Q}_j> - <\tilde{Q}_i><\tilde{Q}_j>$, by

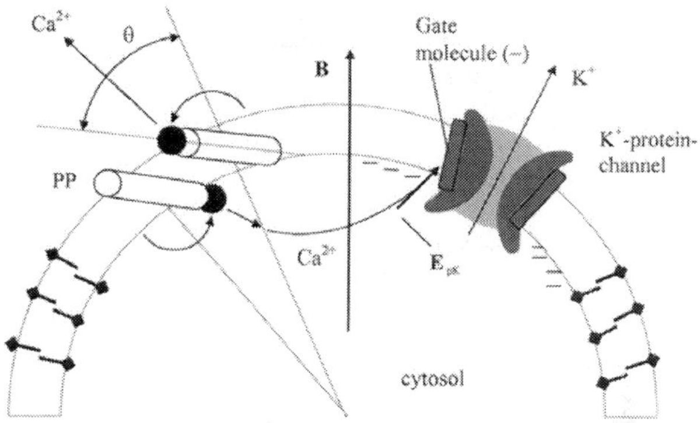

Fig. 23. Two nearest-neighbour Ca^{2+}-charged phospholipids (rods) rotate under their assumed opposite magnetic torques, approaching the Ca^{2+} ions (black circles), attached to the PP negatively charged heads (lozenges). The ions become simultaneously detached from the membrane surfaces when their weak ionic bonds to the heads are broken due to Ca^{2+}-Ca^{2+} Coulomb repulsion. Within the cytosol the Ca^{2+} ions diffuse towards the K^+-protein channels, which are opened when Ca^{2+} is captured (within the Debye shielding length, λ_D) by the "gate" molecule (calmodulin), giving rise to the K^+ outwards current (hyperpolarization) (del Moral et al., 2008).

which the PPs *cooperatively* rotate out from the MF **B** axis (PP electric dipolar moment is very weak). $\langle ... \rangle$ is the ensemble thermal average. The correlation length, ξ usually exceeds a single neurone, *via* the PPs of the interposed glia membranes between neurons, and through the gap junctions (Azanza et al. 2007a). This phenomenon is called *superdiamagnetism* (SD). iii) When the Ca^{2+} ions attached to their PP bilayer inner and outer binding sites (polar heads) happen to be nearest-neighbours (NN, in number N_{nn} per cluster face, with estimated $N_{nn} \cong 0.007 N_c$, N_c being the PP number per cluster), and the NN PPs suffer opposite (with clearly 1/2 probability) magnetic torques $\tau_m = \pm \mathbf{m} \times \mathbf{B}$, the weak ionic bindings are broken by their mutual Coulomb repulsion, of energy ε_{coul}. This produces a *simultaneous* detaching (*Coulomb explosion, CE*) of Ca^{2+} pairs (Fig.23 for a view of the mechanisms involved). Note that the Ca^{2+} water solvating and dielectric membrane negative electric images formation reduce the Ca^{2+} effective charge) (for more details see Azanza and del Moral, 1994; del Moral et al. 2008).

The main result from the SD-CE model is the *field intensity dependence* of the neurone bioelectric frequency, $f(B_{eff}, T)$. This frequency is controlled by chemistry mass action law between Ca^{2+} and membrane binder radical, R^- (sialic acid outside and phosphatidylserine inside), i.e. $[Ca^{2+}][R^-] = k(T, B_{eff})[CaR]$, where k is the kinetics constant. Thus $f \propto [R^-]$ becomes inversely proportional to the number of Ca^{2+} ions detached per cluster, $N^c_{Ca^{2+}}$ = $N_{nn} \exp(-\Delta E_c / k_B T)$, where $\Delta E_c(\theta) = -(N_c \varepsilon_m + N_{nn} \varepsilon_{coul})$ is the *dynamic* Peierls's energy barrier (i.e. changing with the PP rotation) to be overcome by the Ca^{2+} ion in order for the PP to steadily rotate. Moreover under AC MF the cell impulse H process (where the cytosol becomes more negative due the K^+ ions sorting out, Fig.2) is modified by the Ca^{2+} ions (in number of four) binding to the K^+ protein-channel (more specifically to the calmodulin "gate" molecule) and opening it due to the calmodulin electrical unfolding (Babu et al., 1985). Therefore it should be $f \propto 1/N_{Ca^{2+}}$, to first order. Summing now $N^c_{Ca^{2+}}$ from all the PP clusters in membrane(s), the final result is that the neurone bioelectric frequency varies with the RMS MF as

$$f(B_{eff}, T) = f(0,T)\exp(-\alpha B_{eff}^2), \quad (8)$$

where $\mathbf{f(0,T)}$ is the *spontaneous* frequency and the important parameter model $\alpha = N_c|\chi_\perp|V/2\mu_0 k_B T$, that encompasses the PP cluster physical properties and membrane temperature, T, and allows the experimental determination of N_c, or PP number in a membrane(s) cluster (del Moral and Azanza 1992; Azanza and del Moral 1994). As mentioned before for ELF AC MF (quasistatic) we have introduced in [8] the RMS B_{eff}, since magnetic energy density stored in the membrane is $B(t)^2/2\mu_0$, of time average $B_{eff}^2/2\mu_0$. Eq.[8] has been widely and firmly tested in single neurons of *Helix aspersa* for static (del Moral and Azanza 1992) and ELF weak MF (Azanza and del Moral 1998) and for temperature, T modification (Pérez-Bruzón, 2006). The exponential law [8] has been tested for *static* MF application and in Fig. 24 we plot the logarithm of *f* against B^2 for several mapped neurons, for B between 0 and 0.7 T and where we distinguish two linear regimes, with different slopes, due to the different N_c (at the higher fields the clusters are likely fractured and N_c decreases). If we assume N_c= 5 x 10^6 PP/cluster (as deduced from SQUID magnetization measurements in erythrocyte membranes (Azanza et al., 1993)) we obtain a value of $\chi_\perp$ rather close to the measured value, giving good selfconsistence to the model (Azanza and del Moral, 1994). Note that at a critical field all activity is abolished through a first order transition. Now for weak ELF MF the observed firing frequency, *f*, in *Helix* neurons follows a dependence $f(B_{eff}) \cong f(0)(1-\alpha B_{eff}^2)$, which precisely is the obtained one by performing a series expansion of [8] for $\alpha B_{eff}^2 << 1$ (Azanza and del Moral., 1996).
Regarding to *f* (T) dependence, observations in mapped *Helix* neuron F47 show that *f* first decreases with increasing T at fixed B_{eff}, in disagreement with eq.[8] (see Fig.25). The reason is that those neurons belong to the ≈ 26% of studied ones where *f increases* with increasing

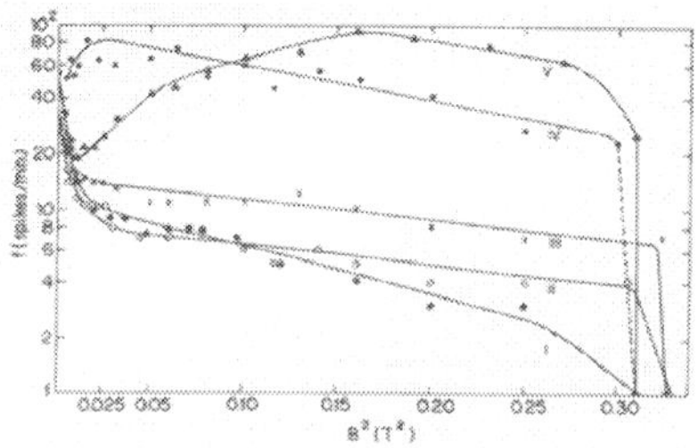

Fig. 24. The logarithm of the firing frequency for several *Helix* neurons vs. the square of the applied magnetic field, **B**. The slope of the linear portions gives $\alpha = N_c V|\chi_\perp|/2\mu_0 k_B T$, with Nc underlying a sudden change at ≈ 0.05T (Azanza and del Moral, 1994).

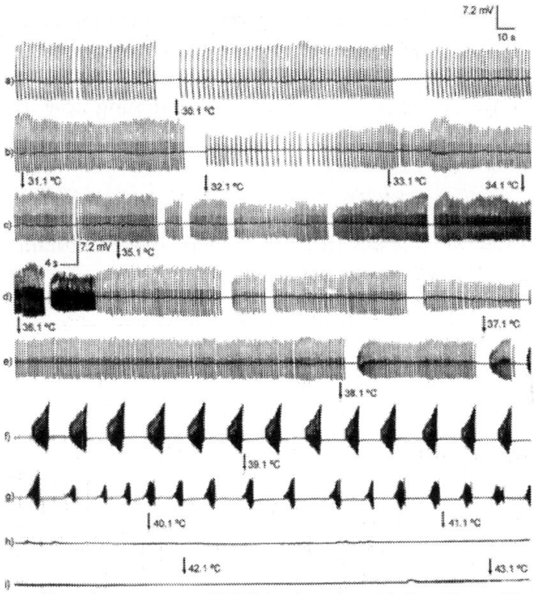

Fig. 25. Temperature effect for neurone F47 bioelectric activity. Transitions at 30°C and 37°C within the membrane liquid crystal are observed.

B_{eff} (Azanza and del Moral 1994). The responsible mechanism is that the by MF detached Ca^{2+} ions depolarize the membrane, cytosol becoming more positive, so opening Na^{+} and/or Ca^{2+} channels operated by voltage (bioelectric activity *stimulation* mechanism). Also D amplitude V_d *decreases*, the calculated variation in V_d being $\Delta V_d(B_{eff}) = -(4\pi/N_c)E_p \exp(+\alpha B_{eff}^2)$, where E_p is the pump e.m.f., due to opposite to PP protein pump rotation (since $|\chi_\parallel| < |\chi_\perp|$), where PP partially hidden the protein pump, desactiving it (Azanza and del Moral 1996) (see Fig. 26). Also observed are two transitions in the form of increasing

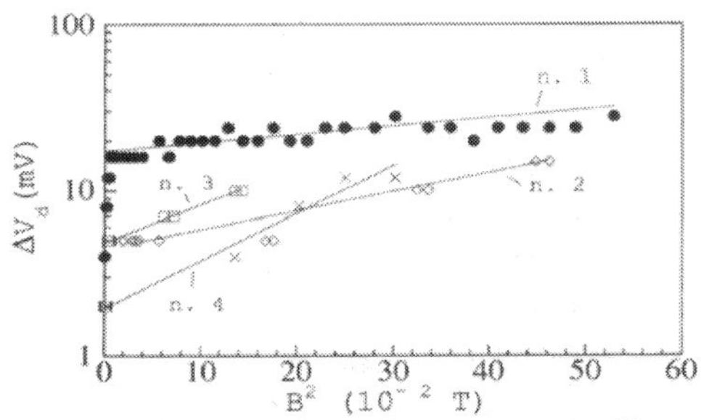

Fig. 26. Semilog plot of depolarization voltage decrease versus B^2; n corresponds to several studied neurons (Azanza and del Moral, 1996.

a ΔT small interval, at $T_{f1} \approx 30°C$ and $T_{f2} \approx 37°C$, corresponding to phase transitions within the membrane liquid crystal. After the second transition *f* decreases with T increase, now in agreement with (8), although *f* temperature behaviour requires a deeper investigation, and PP liquid crystal viscosity energy dissipation, $\varepsilon_v = -\eta_r(T)\theta(t)$ introduction in the SD -CE model (η_r is the membrane viscosity coefficient).

3.3 Bioelectric impulse shape and frequency spectrum.

Comparison with experiments in single neurons: we will now compare our HHM model of impulse shape (§ 1.2) with the electrophysiological experiments performed on *Helix aspersa* single unit neurons, a good bench for present studies as we have already noticed. Thus in Fig.27 we present the spontaneous ($B_{eff} = 0$) R+H voltage time time variation for two mapped neurons, fitted by the approximate solution eq. (4), the agreement being reasonable, but where we do not reproduced the sigmoidal variation at the ends, due to the series cut-off in eq. (3). The more "accurate" frequency used "sigmoidal" fit by $\left(1-e^{-t/\tau_K}\right)^4$ is also shown, but its basis is purely phenomenological. We took $E_K = -75\,mV$, $E_{Na} = +50\,mV$ (this e.m.f. rectified by the delayed K^+ channels), $g_K = 1.6 \times 10^{-7}\,m^{-2}\,\Omega^{-2}$ and $C_m = 4 \times 10^{-2} Fm^{-2}$, and from the fits we obtained the n_0 and τ_K values quoted in Table 1. Clearly we can not identify parameters n_0 with the number of K-protein channels (KP), with a density of ≈ 7 KP/ μm^2, which for a neurone of 100 μm diameter yields $\approx 2x10^4$ KP. In Fig.28 we show the frequency spectrum of a bioelectric impulse of neurone V19, together with the fitted theoretical one by eq. (5), using the parameter values of Table 1, the agreement being excellent, the same happening for other neurons (not shown). Under applied AC MF we have observed that the shape of the impulse becomes *unmodified,* which means that the solution of full integral eq. (3) with I_{Ca} inclusion is not needed. Integral eqs. (3) and (6) can be easily transformed into second order linear differential equations of well known solutions, not given here. In Fig. 28 is shown the Fourier frequency spectrum of R+H neuron impulse tram, fitted by a *Lorentzian* fuction as our model predicts (see § 3.4 below).

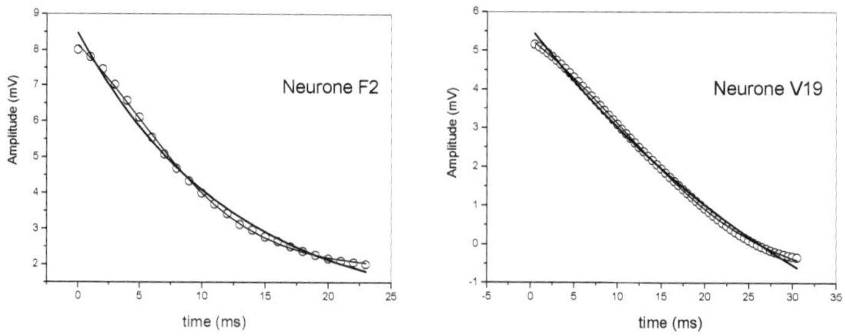

Fig. 27. Experimental (o) and model (thick line) R+H voltage time variations for neurons F2 and V19; thin line, phenomenological sigmoidal variation.

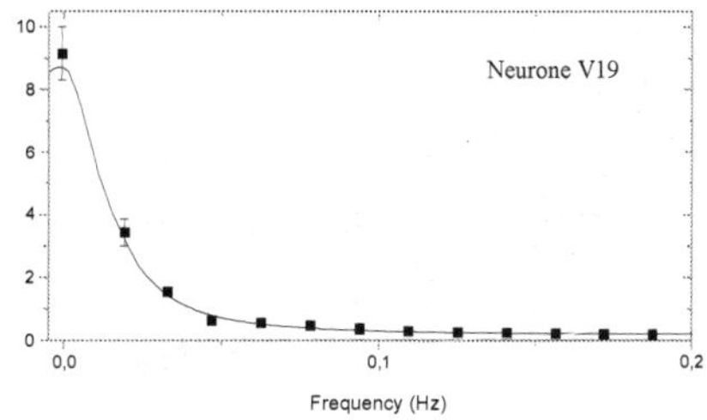

Fig. 28. Frequency Fourier spectrum of R+H impulse tram. Experiment (■) and Lorentzian, *L*(f), model fit $\mathcal{L}$(f) (full line).

Neurone	n_0	τ_K (ms)	$(m_0^3 h_0)^{1/4}$	τ_{eff} (ms)
F1	200	33.0	51	92.7
F2	188	49.4	45	149.9
V3	202	45.0	49	109.6
V14	272	12.4	58	57.0
V19	155	156.7	41	222.8

Table 1. Initial values of n, m, and h HH functions and K^+,τ_K, and Na^+,τ_{eff}, relaxation times.

Similarly in Figs. 29 we show the D voltages for the same neurons impulses, fitted by eq. (6), using the above parameter values and $g_{Na} = 1.9 \times 10^{-7} m^{-2} \Omega^{-2}$, from the fits obtaining the values of $(m_0^3 h_0)^{1/4}$ and τ_{eff} also quoted in Table 1. Values of $(m_0^3 h_0)^{1/4}$ are larger than n_0 ones, and same above consideration apply to them. Also Na^+ relaxation times, τ_{eff} are larger than τ_K, although in the impulse times $t_S < t_b$ (Fig.27) because $V_{Na}(t)$ is interrupted at the smaller (abs.value) E_{Na} than E_K for $V_K(t)$. In Fig.30 is again shown the frequency spectrum of $V_{Na}(t)$for neurone V-19, and the fit by the corresponding *Lorentzian*. D voltage is unmodified by applied AC MF and again solving of R+D equation under MF with I_{Ca} term is not needed.

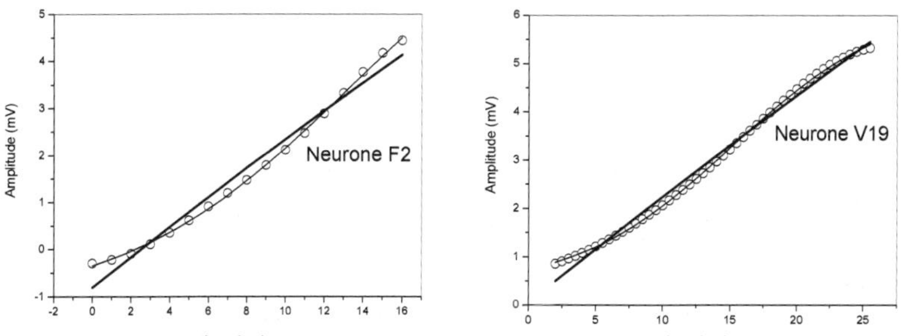

Fig. 29. Ibidem Fig. 27 for the depolarisation (D) process; lines: thick, model fit; thin, phenomenological sigmoid.

Overall our HHM model explains well the spontaneous time dependence of the bioelectric impulse and its frequency spectrum. Demonstration that AC MF produced voltage, V_{Ca} variation is negligible is a mathematical problem of HHM differential equation resolution

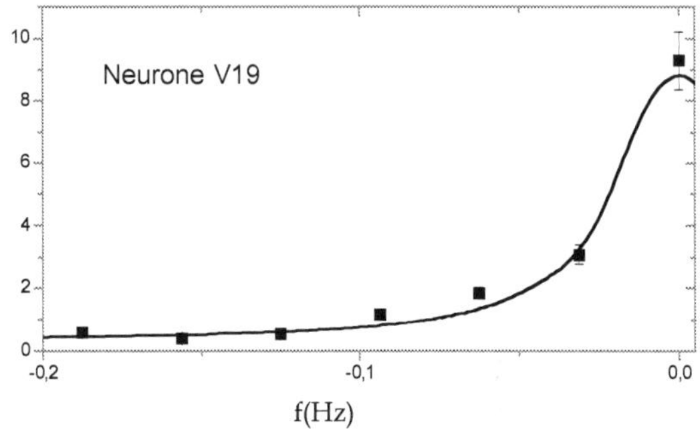

Fig. 30. Frequency spectrum (■) for impulse depolarization of neurone V-19. Line is the Lorentzian fit, $\mathcal{L}(f)$.

as mentioned before. The main predicted effect of $I_{Ca}(B_{eff}, t)$ is upon the neurone bioelectric *frequency*, as we have previously shown that it occurs experimentally. Therefore we should say that the MF (either static, AC of ELF and ELF modulated MW) effect is quite *subtle* upon neuron bioelectric activity.

3.4 Magnetic field frequency dependence of biolectric activity: frequency window effect.-

The experiments initiated in 1975 by Adey and co. (Bawin et al., 1978) about the effects of ELF MF upon neural tissue are a breakthrough in the understanding of how AC ELF MF interacts with this tissue, apart from its *negligible* Joule heating. They prepared newborn chicken brain slices and embedded them in a physiological HCO_3^- water solution doped with radioactive $^{45}Ca^{2+}$ as marker. The tissue was then irradiated with a radiofrequency (RF) field of 147 MHz, *amplitude modulated* by an ELF MF (of amplitude 25 - 30 nT) in the interval 0.5 - 35 Hz, observing an increase of $^{45}Ca^{2+}$ efflux from the tissue. The experiments demonstrated two things: i) the RF (147 MHz) electromagnetic field (EMF) does *not* produce a measurable efflux increase; ii) a calcium efflux increase was observed for the tissue irradiated with the modulated wave, but only within an interval of about 5-25 Hz, so called *frequency window* effect (FWE) (see Fig.31). We have also found a FWE in *Helix aspersa* brain, irradiated with microwaves of 9.6 GHz amplitude modulated between 2-100 Hz, as already

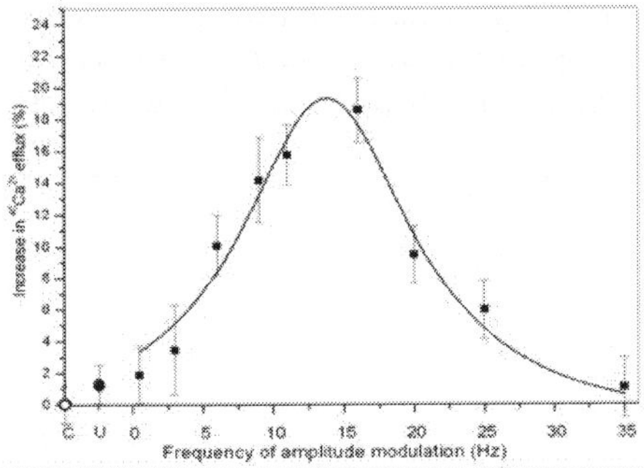

Fig. 31. The points (•) are the experimental $^{45}Ca^{2+}$ efflux increase from chicken brain under application of 147 MHz EMF carrier (intensity 0.8 mW/cm2), amplitude modulated by a MF of frequency, f_M between 0.5-35 Hz and $B_0 \cong 30$ nT (Bawin et al. 1978b). The curve is the theoretical Lorentzian, fitted according to the model eq. (9) (symbols C (O) and U (•) respectively correspond to control and unmodulated EM wave experiments.

briefly described in § 2.3.1.2. Since Ca^{2+} electrochemical gradient, E_{Ca} displaces these ions to the cell interior, the observed efflux was interpreted as Ca^{2+} liberation from the external membrane surface. FWE was afterwards found in many other kinds of cells and experimental conditions, in particular for the bioelectric *frequency*, *f* dependence with the applied ELF MF frequency, f_M in *Helix aspersa* single neurons (Pérez-Bruzón et al., 2004). Our new observation is that the calcium efflux closely follows a *Lorentzian* curve, written now in the normalized form,

$$\varphi(\omega_M) = \varphi(\omega_0)(\Delta\omega/2)^2 / \left[(\omega_M - \omega_0)^2 + (\Delta\omega/2)^2\right] , \tag{9}$$

where ω_0 is the frequency at the maximum efflux $\phi(\omega_0)$. Effectively, in Fig.31 the continuous line is the fit by (9) to the experimental calcium efflux (■ points), and where $f_0 = \omega_0/2\pi \cong 14$ Hz and $\Delta f = \Delta\omega/2\pi = 14.8$ Hz.

The quantitative explanation of such a FWE, although profusely mentioned and discussed since then (Azanza and del Moral, 1994), remained unknown. Although those workers considered that the electric field of the ELF EMF was the responsible for the effect, it seems now clear that it is the magnetic field the responsible one (Pérez-Bruzón et al., 2004). This conclusion also stems from our experiments performed upon single neurons of *Helix aspersa*, submitted to an AC MF, of amplitude $B_0 = 0.1\mu$T-1 mT in the range of 0.1- 100 Hz, as above mentioned.

The *Lorentzian* frequency, f_M dependence either of the calcium efflux to the extra-cellular fluid from chicken neurons, $\phi(\omega_M)$ or the bioelectric frequency dependence $f(f_M)$ in *Helix aspersa* neurons suggest a common origin for the time dependence of the mechanism involved in the Ca^{2+} ions detaching from their binding sites and their final sequestration or capture. This dependence merely is that the amount of Ca^{2+} ions either freed to the external or to the cytosol sides from the membrane *must* vary in the form

$$N(t) = N(0)\exp(-t/\tau_{Ca}) \ , \tag{10}$$

for an applied ELF MF starting at t = 0. This is so *because the Fourier transform of a Lorentzian function is an exponentially time decaying function* (i.e. a relaxation mechanism), with relaxation time $\tau_{Ca} = 2/\Delta f$, as we have shown in §§ 1.3 (theory) and 2.3.1.2 (experiment). This is our main point in favour of a *resonance* effect for the FWE. The time τ_{Ca} is the one required for performing the process of Ca^{2+} liberation from its membrane binders, mainly Ca^{+} diffusion within the external or cytosol fluids and final Ca^{2+} sequestration either by a protein channel (or by other cytosolic structures) or incoming to the radioactivity counter for the externally freed $^{45}Ca^{2+}$ ions. In their diffusion the Ca^{2+} ions traverse an Einstein RMS distance $\sqrt{\langle \ell^2 \rangle} \cong \sqrt{D\tau_{Ca}}$, the one for a random walk, where *D* is the diffusion coefficient of the ion in the fluid (see e.g. Nelson, 2004). Therefore we come up with $N(t)$ obeying a dynamic equation of relaxation $dN/dt = -N/\tau_{Ca}$, as we would expect for a *two states* system (freed Ca^{2+} and bonded Ca^{2+} to its sites) in stationary equilibrium, with Ca^{2+} decay between the two states. For the Ca^{2+} ions freed to the extra-cellular fluid they will end up fully thermalized and dissolved in it, increasing its concentration ($^{45}Ca^{2+}$ efflux in Adey & Bawin's experiment). For the Ca^{2+}ions liberated to cytosol, they will diffuse and finally will have a certain probability of being captured by a K^{+}-protein channel through the calmodulin (CM) attractive electric field, $\mathbf{E}_{pK}$ (see Fig.23, only acting close to CM due to the $\mathbf{E}_{pk}$ shielding beyond the Debye length, λ_D. The diffusion Ca^{2+} current density has the form

$$J_{Ca^{2+}}(t) = \left(N_{Ca^{2+}}(0) f(B_{rms}) q_{Ca^{2+}}\right)\exp\left(-t/\tau_{Ca^{2+}}\right) \tag{11}$$

where $N_{Ca^{2+}}(0)$ is the Ca^{2+} extra-concentration liberated in a supposedly Gaussian "burst", which was previously *calculated* and yields ≈ 10 times the at rest Ca^{2+} concentration.
We can quantitatively express the above considerations by Fourier transforming the observed *Lorentzian* function *L* (ω_M), which represents either the efflux $\phi(\omega_M)$ or the bioelectric frequency $\omega(\omega_M)$ dependences, around the neurone spontaneous frequency, ω_0, (or around the frequency for the maximum in the Fourier spectrum of the firing spikes wave) i.e.

$$N(t) = \int_{-\infty}^{+\infty} L(\omega_M)\, \omega\left(B_{eff} = 0\right)\exp\left(-\alpha B_{ef}^2\right)\exp\left(-i\left(\omega_M - \omega_0\right)t\right) d\omega_M = \tag{12}$$

$$\omega(B_{eff} = 0)\exp\left(-\alpha B_{eff}^2\right)\int_{-\infty}^{+\infty}\left(2(\Delta\omega/2)/(\omega_M - \omega_0)^2 + (\Delta\omega/2)^2\right)\exp\left(-i(\omega_M - \omega_0)t\right) d\omega_M = \omega(B_{eff} = 0)\exp\left(-\alpha B_{eff}^2\right)\exp\left(-t/\tau_{Ca}\right)$$

If we now call $N(0) = \omega(B_{eff} = 0)\exp(-\alpha B_{eff}^2)$ to the initially (at t = 0) detached Ca^{2+} ion number within a Gaussian burst, we end up with the Ca^{2+} relaxation (10) (note, in (12) $\omega(B_{eff} = 0)$ is dimensionless). Since τ_{Ca} is related to $\Delta\omega/2$, which is experimentally accessible from the $\boldsymbol{L}(\omega_M)$ spectra, we can determine that time from experiment.

The central frequency ω_0 in (12) is assumed to be the spontaneous average bioelectric frequency, so that we obtain a *resonance* or maximum of calcium efflux when $\omega_M = \omega_0$. This resonance mechanism should not be confused with the unspecific stochastic resonance model one. This model is based also on a two-states system, but assuming a critical noise which is modulated by the AC driving agent and reaches a threshold value (see e.g. Wellens et al., 2004). A difficulty with this model is that the *physical* mechanism involved in the weak AC agent interaction with the system is neither specified nor quantified. Therefore it becomes of questionable application to the distribution of spontaneous bioelectric frequencies, $D(\omega_0)$ (Fourier spectrum) for the biological system (membrane), also *Lorentzian* (setting $\omega_M = 0$ in (9) and (12)). In fact, and beyond the spontaneous impulse Fourier transform, such a distribution has been directly inferred from our experiments with *Helix aspersa* brain neurons, in the form of repetitive narrow bursts of higher frequency (Fig.32), localized in time when f_M is increased, and superposed to the main $f(f_M)$ *Lorentzian* decrease below f_0 (Pérez-Bruzón, 2006).

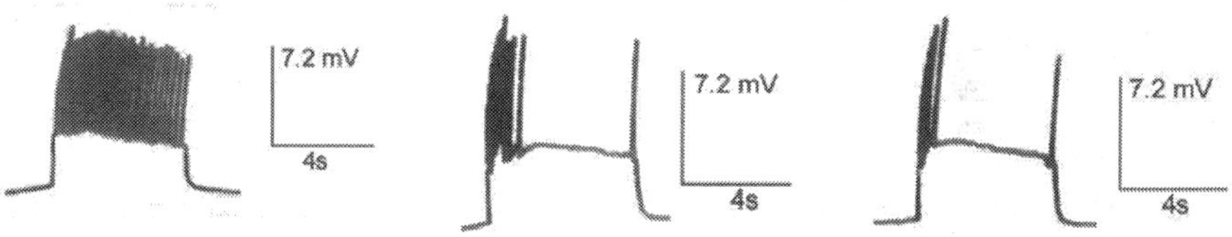

Fig. 32. Neuron D-5 spontaneous bursts of variable f, superposed to the main spike firing train.

In Fig.31 we effectively showed the excellent fit by (9) of the calcium efflux increase vs. f_M, and which means that our SD-CE and Ca^{2+} kinetics model gives a physical and *quantitative* explanation to the Adey & Bawin's FWE observation (Bawin et al., 1978). According to our assumption chicken cerebral neurons average bioelectric frequency should be $f_0 \cong 14$ impulses/s, to some extent comparable with the mollusc neurons with $f_0 \approx 0.1\text{-}8.0$ impulses/s. Similarly in Fig.33 we present the bioelectric frequency f vs. f_M variation for

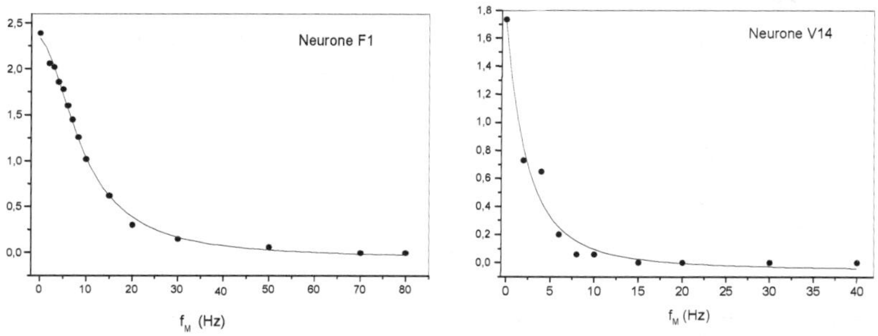

Fig. 33. Variation of bioelectric frequency, f with MF frequency, f_M. Experiment (•); lines are Lorentzian fits $L(f_M)$, with f_0=2.5, 2.0 Hz and $\Delta f/2$= 9.9, 2.7 Hz for neurons F1 and V14 respectively.

Helix aspersa brain mapped neurons F1 and V14, under AC MF of B_0=1 mT. The fit of $f(f_M)$ by $L(\omega_M)$ is also rather good. Therefore in our ELF amplitude modulated MW experiments ω_0 is the frequency corresponding to the maximum of the impulse train Fourier spectrum

(see Figs. 15b and 18b). Then the modulated MW experiments have the ability of *picking up the impulse voltage structure,* whereas AC-MF ELF experiments apparently only pick up the firing repetition frequency, f, which just decreases with increasing, f_M, although also with Lorentzian shape The reason for those different behaviours is currently unknown.
Now, as above mentioned, the bioelectric activity is commanded by the AC MF Ca^{2+} ions internally detached towards the cytosol, which join the K^+-protein channels and open them, giving rise to the bioelectric impulse of K^+ ions sorting out. Therefore this mechanism should be also operative in the chicken brain bioelectric activity, and therefore both experiments reveal the Ca^{2+} simultaneous detaching from *both surfaces* of the membrane. Bioelectric phenomena where both membrane sides are involved are known to be important in biological systems because they couple external membrane interactions with cytoplasm, our case through Ca^{2+} Coulomb explosion. Besides the determined Ca^{2+} relaxation times, τ_{Ca} are 135 ms (chicken brain) and between 93-365 ms for the studied neurons of *Helix aspersa.* An *ab-initio* calculation of the Ca^{2+} relaxation time, τ_{Ca} is difficult, if we consider the mentioned above kinetics involved. In fact a first principles calculation of the K^+ and Na^+ relaxation times in HH equations is still an open problem, relaxation times left as adjustable parameters as we assumed in § 2.3.1.2. However for τ_{Ca} if we estimate the mean diffusion length of Ca^{2+} in water, taking $D \approx 10^{-9}\,m^2s^{-1}$, the typical diffusion coefficient for small molecules in water (see e.g. Nelson, 2004), we obtain $\sqrt{\langle \ell^2 \rangle} \approx 30-60\,\mu m$, which are reasonable values for the studied neurons of average diameter d $\approx 100\,\mu m$ (Kerkut et al., 1975; Azanza et al., 2002). This agreement is a good crosschecking of our model.
A final check of the SD-CE model suitability is the obtaining of the cluster PP numbers, N_c in *Helix aspersa* neurons. This was done in § 3.2 from our applied SMF experiments (Fig. 34). This can be done also by determining the parameter α from the slopes of the above linearized $f(B_0)$ mentioned plots (§ 3.2) from time synchronized neurone pairs under f_M = 50 Hz MF field (Azanza et al., 2002; Azanza et al., 2005) (also from the SMF experiments of Fig. 24). In Fig. 34 are shown such a plots for the pair V13-V23 of the visceral ganglion. Taking $\chi_{\perp} \cong -0.56 \times 10^{-7}$, determined in erythrocyte membranes by combined SQUID

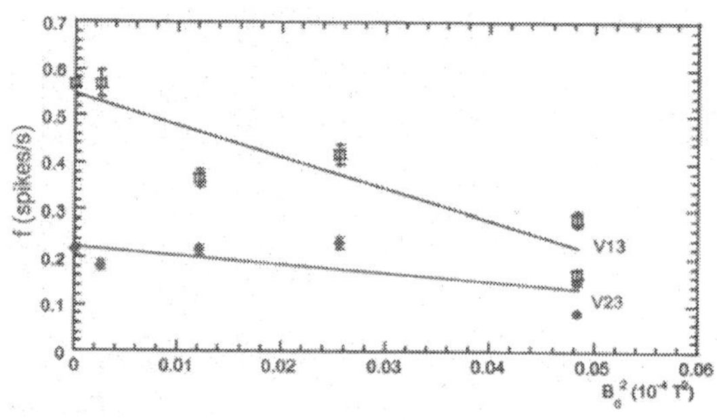

Fig. 34.- Plots of the bioelectric frequency, f vs. B^2_{eff} (f_M = 50 Hz) for *Helix aspersa* neurone pair V23-V13 (mapped by Kerkut et al., 1975), showing *synchronization* under MF. From the line slopes is determined the α parameter. Notice the f coincidence for $B_0 \cong 2$ mT.

magnetometry ($\Delta\chi$) and electrophysiological experiments as mentioned before (del Moral and Azanza 1992; Azanza et al., 1993), $V \approx 5\times 10^{-28}\,m^3$, $T \cong 293K$ and α values, obtained from the slopes, we respectively obtain $N_c \approx$ 4 and 1 $x10^{12}$ PP in a membrane cluster. Considering that the numbers of PP in such a membranes, of average diameters 120 and 103 μm respectively, are $N_{pp} \approx$ 2 and 1.5 $x10^{11}$ (Azanza et al., 2002; Azanza et al., 2005), that numbers throw about $c = N_c/N_{pp} \approx$ 42 and 16 correlated neurons respectively, firing in *synchronized* way with the probe ones, i.e. forming "giant" PP clusters of linear extension $\approx 300\,\mu m$ (the visceral ganglion has a diameter of about 600 μm and *Helix* brain has $\approx 2\times 10^4$ neurons) (Kerkut et al., 1975). Therefore neurones *form correlated small networks* under ELF weak MF, a noticeable effect. The synchronizing mechanism beyond a single neuron is mediated by the glia tissue connecting proteins, which connect in between the phospholipids membranes of two adjacent neurons via a "domino effect" of rotating PP. Observation of the existence of such connections in *Helix* brain ganglia using immunocytochemistry techniques has been actually realized (Azanza et al., 2007).

3.5 The demodulating neurone membrane.

A main issue is indeed to provide for an explanation of why the neurone only senses the low frequencies, f_M when an *amplitude modulated* magnetic field, B(t) of MW high frequency (carrier, f_c) is applied, in the form,

$$B(t) = B_c[1 + m\cos\omega_m t]\cos\omega_c t \equiv B_c[1 + G(t)]\cos\omega_c t \qquad (13)$$

were $m = B_m/B_c$ is the modulation ratio, $G(t) = m\cos\omega_m t$, $B^{(m)} = B_0^{(m)}\cos\omega_m t$ is the ELF modulating MF and B_c the MW carrier MF amplitude. Our proposed model is based again on the Ca^{2+} detaching magnetic torque, quadratic (*non-linear*) in B, and Ca^{2+} diffusion within the cytosol, mainly the latter. As it is well known any demodulation process, i.e. elimination of the carrier wave, passes through the existence of some non-linear characteristic of the demodulating device. Those processes operate in fact as a detection or demodulating "device", although rather unusual and peculiar, as follows.

The diamagnetic energy of a PP cluster under modulated field B(t) is given by

$$E_M = -\left(VN_{eff}^c/2\mu_0\right)B(t)^2\left[\chi_\perp + \Delta\chi\cos^2\theta(t)\right] \qquad (14)$$

where N_{eff}^c is the Ca^{2+} detaching effective number of PP in a cluster, as explained in § 3.2. Therefore the magnetic torque upon a PP cluster (in fact upon the applied field induced magnetic moment) becomes $\Gamma_M = -(\partial E_M/\partial\theta) = \Gamma_0 \sin 2\theta \sin 2B(t)^2$, where $\Gamma_0 \cong V\Delta\chi N_{eff}^c/2\mu_0$ and

$$B(t)^2 = (B_{rms,c}^2)[1 + 2G(\omega_m t) + G^2(\omega_m t)](1 + \cos(2\omega_c t)) \qquad (15)$$

Therefore the acting magnetic torque, Γ_M has two components: i) The carrier frequency, f_c (modulated) torque:

$$\Gamma_c = F\,(1 + 2G + G^2)\,B_{rms,c}^2\cos 2\omega_c t \qquad (16)$$

purporting the high frequencies $2\omega_c$, $2\omega_c \pm \omega_m$, $2\omega_c \pm 2\omega_m$, and where $\omega_m/\omega_c \approx 10^{-8}$ and $F \equiv \Gamma_0 \sin 2\theta$. ii) The modulating frequency, f_M field torque

$$\Gamma_m = F\, B^2_{rms,c}\left[1 + 2G(\omega_m t) + G^2(\omega_m t)\right] , \tag{17}$$

with ELF $\omega_m = 2\pi/f_m$. Now the angle $\theta(t)$ between the PP axis and MF **B** has a complicated motion equation to solve, mainly because of the complex time dependent Ca^{2+}-Ca^{2+} Coulomb repulsion torque $\Gamma_{rep}(\theta)$. The motion equation is

$$I_{PP}\ddot{\theta}(t) + \beta\dot{\theta}(t) + \kappa\theta = \Gamma_M(\theta) + \Gamma_{rep.}(\theta) \tag{18}$$

where I_{pp} is the PP inertia moment, β the PP rotation damping coefficient (mainly due to membrane viscosity) and κ the restoring parameter representing the Coulomb Ca^{2+}- Ca^{2+} repulsion, although the PP rotate as a cooperative cluster due to superdiamagnetism.
We will assume no PP acceleration, weak viscosity in the PP membrane and weak Coulomb repulsion (except at the short distance of the Coulomb explosion), and then we may assume that PP follows the applied field and then $\theta(t) \cong \omega_{field} t$ .Therefore the effective torque finally becomes of the form

$$\Gamma_M = \{\Gamma_c(t) + 4\Gamma_0 B^{(m)}_{rms} B^{(c)}_{rms} \cos\omega_m t + \Gamma_0 {B^{(m)}_{rms}}^2 \cos^2\omega_m t\}\sin 2\theta(t). \tag{19}$$

Now the main model assumption is that when period of the carrier torque, $T_c \approx$ 75ps is much smaller that the Ca^{2+} relaxation (diffusion) time, τ_{Ca} (100-400 msec) towards the K^+ channel (or other cytosolic sites), the detached Ca^{2+} ion is recaptured and it is ineffective: *so carrier torque becomes rectified.* However there can be effective detaching with the modulating field, since its period, T_m and τ_{Ca} are comparable, i.e. $\tau_{Ca} \approx 100 - 350$ms against $T_m \approx$ 1 s - 100 ms or $T_m/2$.
Therefore the *only effective* torque for an effect of ELF MWmodulated MF on a PP membrane is:

$$\Gamma^{eff}_M = (V\Delta\chi N^c_{eff}/\mu_0)\{\tfrac{2}{m} {B^{(m)}_{rms}}^2 \cos^2\omega_m t \cdot \sin\omega_m t + {B^{(m)}_{rms}}^2 \cos^3\omega_m t \cdot \sin\omega_m t\} . \tag{20}$$

A schematic picture of the two magnetic torque components in (20) is shown in Fig.35, where only in the non-hatched time intervals in each T_m period, Ca^{2+} detaching occurs because of the favourable PP Ca^{2+} charged heads positions (see skechted drawings). Therefore (20) expresses that the only MF being "effectively" acting upon the neuron (membrane) is the ELF modulating one, the magnetic torque acting upon the PP membrane plus Ca^{2+} diffusion and sequestration being both the demodulating cell membrane "device". Our current model does not include thermal effects in the demodulation process since Peierls energy barrier between the two Ca^{2+} states has been omitted on a first approximation. Summarizing we propose that the combined action of the *non-linear* in field, **B** magnetic torque upon the PPs, which liberate Ca^{2+} from the inner membrane surface plus the *slow diffusion* of Ca^{2+} towards the K^+ channels in order to open them, constitute the *demodulation mechanisms* in membranes in order to the membrane *sensing only the ELF modulating signal.*

Fig. 35. The two components of the rectified effective torque acting upon a phospholipid cluster. Ca^{2+} detaching only occurs for the $T_m=1/f_m$ period times interval shown. For the grey marked times detaching does not occur. Also showns in the Ca^{2+} diffusion current, given by [11].

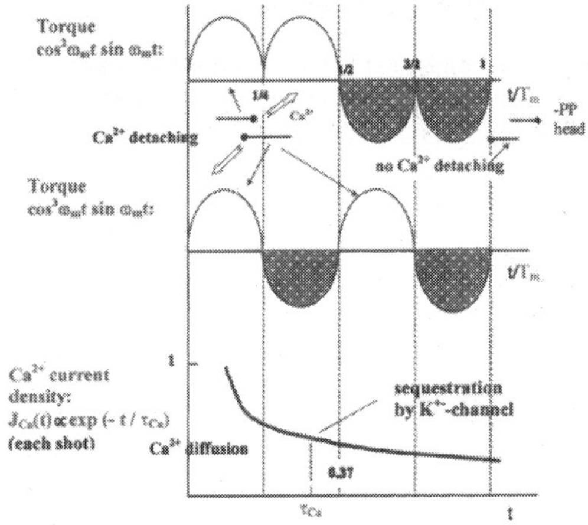

Fig. 35. The two components of the rectified effective torque acting upon a phospholipid cluster. Ca^{2+} detaching only occurs for the $T_m=1/f_m$ period time intervals shown in white. For the grey marked times detaching does not occur. Also shown is the Ca^{2+} diffusion current, given by (11).

4. Conclusions

In this chapter we have shown that the main effect of low strength MF (static, alternating of ELF and ELF MF modulated MW) upon neural tissue is the modification of the neuron bioelectric activity *frequency*, either reducing or increasing it depending on the neuron nature. The general underlying mechanism is the *Ca^{2+} ion liberation from the cytosolic membrane through a mechanism which involves the phospholipid (PP) field induced diamagnetism, the cooperative PP action superdiamagnetism, the Ca^{2+} ion Coulomb explosion from their membrane stores, either increasing the cytosol charge (bioelectric frequency excitation) or reducing the firing frequency (inhibition process) through the Ca^{2+} opening of K^+ channels.* This has been proved by comparing our electrophysiological experiments with our purposely made biophysical model, which explains in a fully quantitative way the performed experiments. A model is also presented which shows how *the neuron membrane acts as a demodulating system* for the ELF amplitude modulated applied MW and that also underlines the superdiamagnetic effect, the Ca^{2+} liberation and the *non-linear* dependence with the MF strength, B of the upon PP's exerted magnetic torque. Demodulation of ELF modulated MW in neurons becomes in this way a rather subtle mechanism, manifested in the form of the observed *frequency resonances* (no of "spring kind") of membrane with the applied ELF MF modulating the MW carrier. Neuron membranes more than acting as amplifying structures for the electromagnetic wave interaction behaves as cooperative diamagnetic structures in their interaction with either static or dynamic *low frequency* magnetic fields.

ACKNOWLEDGEMENTS
We are grateful to Professor R. Gómez and co. of University of Granada for the MW dosimetry simulations. We gratefully acknowledge financial support from the Spanish Ministry of Defence under Project ERG 101.103, from the "Fundación Humanismo y Ciencia" (Madrid) and from "Diputación General de Aragón" under Project B43.

5. References

Adey, W.R. (1986). The sequence and energetics of cell membrane transductive coupling to intracellular enzyme systems. *Bioelectrochem.Bioenergetics*, 15, 447-456.

Adey, W.R. (1988a). Electromagnetic fields, cell membrane amplification and cancer promotion. In: *Non-ionizing Electromagnetic Radiations and Ultrasound*, pp. 80-110. Ed.NCRP: Bethesda.

Adey, W.R. (1988b). Cell membranes: The electromagnetic environment and cancer promotion. *Neurochem. Res.* 13, 671-677.

Alberts, B.; Bray, D.; Lewis, J.; Raff, M.; Roberts, K. & Watson, J.D. (1989). In: *Molecular Biology of the Cell.* Garland: New York.

Axelrod, J.; Burch, R.M. & Jelsema, C.L. (1988). Receptor-mediated activation of phospholipase A2 via GTP-binding proteins: Arachidonic and its metabolites as second messengers. *Trends Neurosci.* 11, 117-123.

Azanza, M.J. & del Moral, A. (1988). Effects of static magnetic fields on isolated neurons. *J. de Phys.* (París). 12: C8-2059-2060.

Azanza, M.J. (1989). Steady magnetic fields mimic the effect of caffeine on neurons. *Brain Res.* 489: 195-198.

Azanza, M. J. (1990). Characterization of neuronal membrane K^+ and Ca^{2+} channels operated under steady magnetic fields exposure. *J.Magn.Magn.Mat.* 83: 527-529.

Azanza, M.J., Blott, B.H. del Moral, A., and Peg, M.T. (1993). Measurement of the red blood cell membrane magnetic susceptibility. *Bioelectrochem. Bionergetics.* 30: 45-53.

Azanza, M.J. & del Moral, A. (1994). Cell membrane biochemistry and neurobiological approach to biomagnetism. *Prog.Neurobiol. 44*: 517-601. Review.

Azanza, M.J. & del Moral, A. (1995). Neuron firing frequency dependence on the static magnetic field intensity. *J. Magn.Magn.Mat.* 140-144, 1464-1465.

Azanza, M.J. & del Moral, A. (1996). Isolated neuron amplitude spike decrease under static magnetic fields. *J. Magn. Magn. Mat.* 157-158: 593-594.

Azanza, M.J. & del Moral, A. (1998). ELF-magnetic field induced effects on the bioelectric activity of single neurone cells. *J. Magn. Magn. Mat.* 177-181: 1451-1452.

Azanza, M.J.; Calvo, A.C. & del Moral, A. (2001). 50 Hz Sinusoidal Magnetic fields induced effects on the bioelectric activity of single unit neurone cells. *J.Magn.Magn.Mat. 226-230*: 2101-2103.

Azanza, M.J.; Calvo, A.C. & del Moral, A. (2002). Evidence of synchronization of neurons activity of molluscan brain ganglia induced by alternating 50 Hz applied magnetic field. *Electro-Magnetobiology.* 21: 221-232.

Azanza, M.J.; Pérez-Bruzón, R.N.; Calvo, A.C. & del Moral A. (2005). Elemental neuron network dynamics under applied sinusoidal magnetic fields, Proceedings of BioEM 2005, pp. 312-314, June 19-24, Dublín, Ireland.

Azanza, M.J.; Pes, N.; Pérez-Bruzón, R.N.; Aisa, J.; Raso, M.; Junquera, C.; Lahoz, M.; Maestú, C.; Martinez-Ciriano, C.; Pérez Castejón, C.; Vera Gil, A. & del Moral, A. (2007a). Localization of connexins and glia cells of the Helix aspersa suboesophageal brain ganglia by immunocitochemistry. *Histol.Histopathol.* 22: 497-504.

Azanza, M. J.; Pérez-Bruzón, R .N. & del Moral, A. (2007b). Frequency resonance effect of under low-frequency weak magnetic fields. *J. Magn. Magn. Mat.* 310: 2865-2867.

Azanza, M.J.; Pérez Castejón, C.; Pes, N.; Pérez-Bruzón, R.N.; Aisa, J.; Junquera, C.; Maestú ,C.; Lahoz, M.; Martinez Ciriano, C.; Vera Gil, A. & del Moral, A. (2008). Characterization by immunocytochemistry of ionic channels in *Helix aspersa* suboesophageal brain ganglia. *Histol. Histopathol.* 23 397-406.

Azanza, M.J.; Calvo, A.C.; Pérez-Bruzón RN, Maestú, C. & del Moral, A. (2009). Neuron network dynamics under applied low frequency magnetic fields. Submitted.

Babu, Y.S.; Sack, J.S.; Greenough, T.J.; Bugg, C.E.; Means, A.R. & Cook, W.J. (1985). Three-dimensional structure of calmodulin. *Nature*. 315: 37-40.

Bawin, S.M.; Sheppard, A. & Adey, W.R. (1978). Possible mechanism of weak electromagnetic field coupling in brain tissue. *Bioeletrochem. Bioenergetics*. 5: 67-76.

Berridge, M.J. & Irvine, R.F. (1984). Inositol triphosphate, a novel second messenger in cellular signal transduction. *Nature* 312, 315-321.

Bleany, B.I. & Bleany, B. (1965). Electricity and Magnetism. Oxford University Press: Oxford

Calvo, A.C. & Azanza, M.J. (1999). Electrophysiologic responses of snail under applied 50 Hz alternating magnetic fields. *Electro- Magnetobiology*. 18: 305-312.

Calvo, A.C. (2006). Doctoral Thesis. University of Zaragoza, Spain, unpublished.

Cole, K.S. & Curtis, H.J. (1939). Electric impedance of the squid giant axon during activity. *J.Gen.Physiol.* 22, 649-670.

del Moral, A. & Azanza, M.J. (1992). Model for the effect of static magnetic field. *J. Magn. Magn. Mat.* 114: 240-242.

del Moral, A.; Azanza, M.J. & Pérez–Bruzón, R.N. (2006). Models of neurone dynamics: spontaneous and under ELF alternating magnetic fields. "4th International Workshop: Biological Effects of EMFs", ISBN 960-233-172-0, pp-594-603, Creta, 2006.

del Moral, A.; Pérez-Bruzón, R.N. & Azanza, M.J. (2008). Frequency windows in neurones under low frequency modulated microwaves. *Proc. of the 38th European Microwave Conference, Amsterdam, The Netherlands, 27-31 Oct.* Ed Horizon House Publications Ltd, pp 83-86.

Downes, C.P. (1983). Inositol phospholipids and neurotransmitter-receptor signalling mechanisms. *Trends Neurosci.* 6, 313-316.

Eberhard, D.A. & Holz, R.W. (1988). Intracellular Ca^{2+} activates phospholipase C. *Trends Neurosci.* 11, 517-521.

Hinrikus, H.; Bachmann, M.; Tomson, R. & Lass, J. (2005). *The Environmentalist* **25,** 187.

Hodgkin, A.L. & Huxley, A.F. (1952). A quantitative description of membrane current and its application to conduction and excitation in nerve. *J.Physiol.* 117, 500-544.

Irvine, R.F. (1982). How is the level of free arachidonic acid controlled in mammalian cells?. *Biochem.J.* 204, 3-16.

Kerkut, G.A.; Lambert, J.D.C.; Gayton, R.; Loker, J.E. & Walker, R.J. (1975). Mapping of nerve cells in the ganglia of *Helix aspersa* . *Comp. Biochem. Physiol.* 50A: 1-25.

Leszczynski, D.; Joenväärä, S.; Reivinen, J. & Kuokka, R. (2002). Non-thermal activation of the hsp27/p38MAPK stress pathway by mobile phone radiation in human endothelial cells: molecular mechanism for cancer and blood-brain barrier-related effects. *Differentiation* 70, 120-129.

Loeb, L.A. & Gross, R.W. (1986). Identification and purification of sheep platelet phospholipase A_2 isoforms. Activation by physiologic concentrations of calcium ions. *J. Biol.Chem.* 261, 10467-10470.

Nahorski, S.R. (1988). Inositol phosphates and neuronal calcium homeosthasis. *Trends Neurosci.* 11, 444-448.

Nelson, P. (2004). *Biological Physics, Energy, Information, Life,* Freeman, New York.

Nishizuka, Y. (1984). The role of protein kinase C in cell surface signal transduction and tumor promotion. *Nature* 308, 693-698.

Ordway, R.W. ; Singer, J. & Walsh, J.V. (1991). Direct regulation of ion channels by fatty acids. *Trends Neurosci.* 14, 96-100.

Papahadjopoulos, D. & Bangham, A.D. (1966). Biophysical properties of phospholipids. II Permeability of phosphatidylserine liquid crystals to univalent ions. *Biochem. Biophys.Acta* 126, 185-188.

Pérez Castejón, C.; Pérez-Bruzón, R.N.; Llorente, M.; Pes, N.; Lacasa, C.; Figols, T.; Lahoz, M.; Maestú, C.; Vera, A.; del Moral, A. & Azanza, M.J. (2009). Exposure to ELF-pulse modulated X-band microwaves increases *in vitro* human astrocytoma cells proliferation. *Histol.Histopathol. in press.*

Pérez-Bruzón, R.N.; Azanza, M.J.; Calvo, A.C. & del Moral, A. (2004). Neurone bioelectric activity under magnetic fields of variable frequency in the range of 0.1-80 Hz. *J. Magn.Magn.Mat.*272-276, 2424-2425.

Pérez-Bruzón, R.N. (2006). Doctoral Thesis. University of Zaragoza, Spain, unpublished.

Pérez-Bruzón R.N.; Pérez Castejón, C.; Llorente, M.; Pes, N.; Lacasa, C.; Figols, T.; Lahoz, M.; Maestú, C.; Vera Gil, A.; del Moral A & Azanza, M.J. (2009). Design, dosimetry analysis and validation of an incubator for the exposure of *in vitro* human astrocyte cells to X-band microwaves in a GTEM-cell. Submitted 2009.

Pipkin, J.L.; Hinson, W.G.; Young, J.F.; Rowland, K.L.; Shaddock, J.G.; Tolleson, W.H.; Duffy, P.H. & Casciano, D.A. (1999). Induction of stress proteins by electromagnetic fields in cultured HL-60 cells. *Bioelectromagnetics* 20, 347-357.

Rothman, J.E. & Lenard, J. (1977). Membrane asymmetry. *Science* 195, 743-753.

Singer, S.J. & Nicolson G.L.(1972). The fluid mosaic model of the structure of cell membranes. *Science* 175, 720-731.

Sutherland, E.W. (1972). Studies on the mechanism of hormone action. *Science,* 177, 401-408.

Takuma, K.; Baba, A. & T. Matsuda. (2004). Astrocyte apoptosis: implications for neuroprotection. *Prog. Neurobiol.* 72, 111-127.

Wellens, T.; Shatokhin, V. & Buchleitner, A. (2004). Stochastic resonance. *Rep. Prog. Phys.* 67: 4545.

Rain Attenuation on Terrestrial Wireless Links in the mm Frequency Bands

Vaclav Kvicera and Martin Grabner
Czech Metrology Institute
Czech Republic

1. Introduction

Millimetre wave terrestrial wireless communication systems are becoming an increasingly important part of the telecommunication infrastructure. They offer a higher transmission capacity thanks to a higher available bandwidth when compared to systems operating in lower frequency bands. Moreover, the frequency spectrum is heavily occupied at lower frequencies by a great number of competing services. Therefore the radio communication systems operating at frequencies higher than 30 GHz will be utilized more frequently in the future.

The performance of all radio systems in general is related to the propagation of electromagnetic waves between the transmitter and receiver. The propagation of millimetre electromagnetic waves through the medium of the lower troposphere is influenced by weather conditions. Rain attenuation belongs among the most important adverse propagation effects impairing the performance of fixed terrestrial wireless systems operating in millimetre wave frequency bands. The study of the rain attenuation phenomenon is therefore indispensable for their efficient planning and utilization.

Rain attenuation characteristics on millimetre wave radio systems are the subject of current experimental research (ITU-R, 2008). In this chapter, the availability objectives for fixed wireless links are briefly discussed, rain attenuation characteristics are defined and the geographical dependence of rain attenuation is demonstrated. Then the specific results obtained from rain attenuation measurements in several mm wave bands are presented. Frequency, polarisation and path length scaling of rain attenuation statistics in mm wave bands are analysed.

2. Availability Objectives for Fixed Wireless Links

Terrestrial fixed wireless links form a part of the global telecommunication network. In order to achieve the reasonable availability and stability of network connections, the basic performance criteria defined for these links has to be required. The availability performance of a particular fixed wireless link is defined roughly as the fraction of time in which data transmission through the link is fully operating. The availability of the link can be limited by both a device malfunction and by propagation effects. Rain attenuation is generally considered to be the most important effect impairing the millimetre wave links. When the received signal is attenuated due to the scattering and absorption of electromagnetic waves on the rain drops and the attenuation value exceeds the link power margin (also called the fade margin), the link transmission is interrupted. It follows that the availability performance of the particular mm wave link depends on the rain attenuation statistics and the power margin of the link.

2.1 Overview of ITU recommendations on the availability of fixed wireless links

The International Telecommunication Union, Radiocommunication Sector (ITU-R) recommends the quality and availability objectives applicable for fixed wireless links within a telecommunication network (Rec. ITU-R F.1703, 2008). These objectives should be met on the links used in a 27 500 km hypothetical reference path (HRP). An HRP is a model of the path between the two end terminals of the network. The HRP is divided into 2 parts - international and national portions with different quality/availability criteria. The national portion is further subdivided into long haul, short haul and access portions, again with different quality/availability objectives. The most stringent criteria are applicable for links used in the international portion of the HRP, the least stringent criteria are applicable in the access portion of the HRP. The availability performance objectives defined by ITU-R (Rec. ITU-R F.1703, 2008) are expressed by means of two parameters - the availability ratio *AR* and the mean time between outages *Mo*.

The availability ratio *AR* is the proportion of time that a path is in an available state during an observation period. The *AR* is calculated by dividing the total available time during the observation period by the duration of the observation period – usually one year. In order to distinguish available and unavailable time, the following rules are applied. A period of unavailable time begins at the outset of 10 consecutive Severely Errored Second (SES) events. These 10 seconds are considered to be part of the unavailable time. A new period of available time begins at the onset of 10 consecutive non-SES events. These 10 seconds are considered to be part of the available time. A Severely Errored Second (SES) is a one-second period which contains ≥30% Errored Blocks or at least one defect. An Errored Block (EB) is a block of bits in which one or more bits are in error. Examples of defects are loss of signal, alarm indication signal, loss of frame alignment (Rec. ITU-T G.826, 2002).

Without going further into the details described in (Rec. ITU-R F.1703, 2008), two examples will be provided of availability objectives calculated for a 30 km long link that can be formed by several radio hops connected in series. For such a link that is part of the *international* section, the *AR* should reach at least 99.985% (an unavailability of 78 min/year, i.e. 0.015%). For a link that is part the *access* section, the *AR* should equal or exceed 99.95% (an unavailability of 263 min/year, i.e. 0.05%). The further subdivision of availability objectives into the individual radio hops of the link is not specified and is under the network operator's responsibility. A conservative assumption that unavailability events on different radio hops are not correlated is often adopted. Then the unavailability time of the whole link is the sum of the unavailable times of its individual radio hops. It follows that more stringent *AR* values than those presented in the above example should be required on these hops.

The mean time between outages *Mo* and its reciprocal value – the outage intensity *OI* are parameters expressing a relatively new concept within ITU-R availability recommendations. They define how many unavailability events are tolerable on the link during the one-year

period. The idea is that a great number of short outages on the link are not desirable despite the fact that *AR* of the link meets the required objectives. Unfortunately, there is not enough knowledge at the moment that would provide an estimation of these parameters on a worldwide basis. The practical usefulness of these parameters still remains to be seen.
One should note that historically, ITU-T and ITU-R quality/availability objectives were developed under the assumption of a single protected connection between two end terminals that is established within a public switched telephone network (PSTN). This assumption is not valid in IP packet based networks however. Link operators may therefore require their own quality/availability objectives for links used in such networks. In practice, it seems that the similar or the same objectives as in ITU-R are often used because of lack of more relevant knowledge.

2.2 Fixed wireless link power budget

Once the availability objectives are specified for the designed radio path, the link parameters have to be considered in order to meet these objectives. Propagation related bit errors that occur on millimetre wave links are caused dominantly by poor detection when the received signal power falls under the receiver threshold due to the attenuation of incoming electromagnetic waves. This is why the link power budget has to be determined above all. The nominal power available at the input of the receiver can be obtained from the following formula:

$$P_r = P_t - L_t + G_t - FSL + G_r - L_r \quad (1)$$

where P_t (dBm) is the power at the output of the transmitter, P_r (dBm) is the power at the input of the receiver, G_t and G_r (dB) are transmitting and receiving antenna gains, L_t and L_r (dB) are additional losses (branching, feeder,...) in the transmitter and receiver and *FSL* (dB) is the free space loss which is dependent on the path length d (m) and wavelength λ (m):

$$FSL = 20 \log(4\pi d/\lambda). \quad (2)$$

Normally, the nominal received power P_r is much higher than the receiver threshold P_{rth} (dBm) which is an important parameter of the receiver that depends on the modulation format and on the receiver noise figure. P_{rth} is usually defined as the power at the input of the receiver that will result in a certain threshold value of bit error ratio BER, typically 10^{-6} or 10^{-3}. It follows that the fade margin F (dB):

$$F = P_r - P_{rth} \quad (3)$$

determines the maximum attenuation of the received signal that maintains a BER lower than this threshold. Given the required maximum of unavailability time percentage p (%) according to the objectives described in the previous section, rain attenuation A_p (dB) exceeded p % of time is derived from rain attenuation statistics. In order to meet the availability performance objectives, link parameters (antenna gain, link length ...) have to be adjusted so that the following condition for the fade margin is satisfied:

$$F \geq A_p. \quad (4)$$

3. Rain Attenuation

Rain that occurred along the terrestrial path causes the rain attenuation as a result of the absorption and scattering of electromagnetic waves by rain particles. The rain attenuation depends on the frequency and polarisation, the temperature, on the size distribution of raindrops, and on their fall velocity. In the mm frequency bands, i.e. for frequencies above 10 GHz, the reason of the rain attenuation is both the absorption and scattering of electromagnetic waves by rain particles. The frequency dependence of the specific rain attenuation can be obtained from (Rec. ITU-R P.838, 2008).

3.1 Rain characteristics

Rainfall intensity along the terrestrial path is inhomogeneous in space and time. The raindrops have a non-spherical shape and therefore the attenuation of the horizontally polarized waves is greater than the attenuation of the vertically polarized waves. Rainfall intensity can be measured by different types of rain gauges located at the surface of the Earth. The cumulative distributions (CDs) of the average 1-minute rain intensities are needed for the calculations of the CDs of attenuation due to rain by different methods.

3.2 Rain attenuation models

Various methods were developed for calculations of CDs of attenuation due to rain from rain intensity measurements (COST 235, 1996; ITU-R, 2008). The ITU-R recommendation (Rec. ITU-R P. 530-12, 2008)) uses an effective path length to consider the time-space variability of rain intensity along the terrestrial path. Rain attenuation exceeded at 0.01% of the time of year is calculated from the average 1-minute rain intensity exceeded at the same time percentage. The obtained value is scaled by the empirical formula to other percentages of time between 1% and 0.001%.

3.3 Monthly and yearly statistics

The obtained CDs of both the rain intensities and the rain attenuation should be the "long-term average annual", i.e. the measurements should be taken for several years (at least 3 years) to take into account the effects of both the month-to-month and year-to-year variability of rain intensities. The obtained long-term CDs of attenuation due to rain for the average year can be used for the assessment of availability performance parameters of the mm terrestrial fixed wireless link (Rec. ITU-T G.827, 2003). The obtained long-term CDs of attenuation due to rain for the average annual worst month can be used for the assessment of error performance parameters of the mm terrestrial fixed wireless link (Rec. ITU-T G.828, 2000). The "worst month" concept is described in (Rec. ITU-R P.581-2, 2008).

4. Geographical Dependence of Rain Attenuation

Rain attenuation statistics depend on local climatic conditions. Typical values of rain intensity (and to some extent also drop size distributions) vary significantly worldwide. In this section, ITU-R methods and datasets are used to give an overview of the geographical dependence of the rain attenuation of millimetre waves.

4.1 ITU-R world map of rain intensity statistics

Rain intensity R (mm/hour) is the most important quantity determining rain attenuation. ITU-R provides a global dataset of rain intensity statistics derived from local measurements as well as from global satellite observations of liquid water precipitation in the atmosphere. Figure 1 shows a world map of rain intensity exceeded 0.01% of time (about 53 minutes in a year).

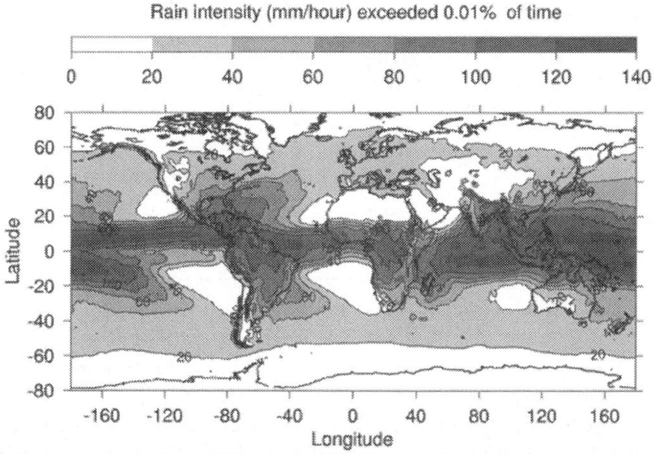

Fig. 1. Worldwide rain intensity 0.01% percentile distribution

The maximum value of the rain intensity 0.01% percentile reaches more than 130 mm/hour in tropical regions. In Europe, for example, the value is lower than 40 mm/hour over most of the continent. The observed peak values of rain intensity are even significantly higher.

4.2 ITU-R world map of rain attenuation statistics

Rain intensity increases with the concentration of raindrops within the propagation path. The higher number of scatterers - raindrops, the higher attenuation of the electromagnetic wave, thus rain attenuation generally increases with increasing rain intensity. The frequency dependence of specific attenuation for different rain intensities is depicted in Figure 2. The ITU-R model (Rec. ITU-R P.838-3, 2008) was used which is of the form:

$$\gamma = kR^{a} \tag{5}$$

where specific attenuation γ (dB/km) is related to rain intensity R (mm/hour) and frequency-dependent coefficients k and a. It is seen that in millimetre wave bands attenuation slightly increases with frequency up to about 100 GHz depending on rain intensity.

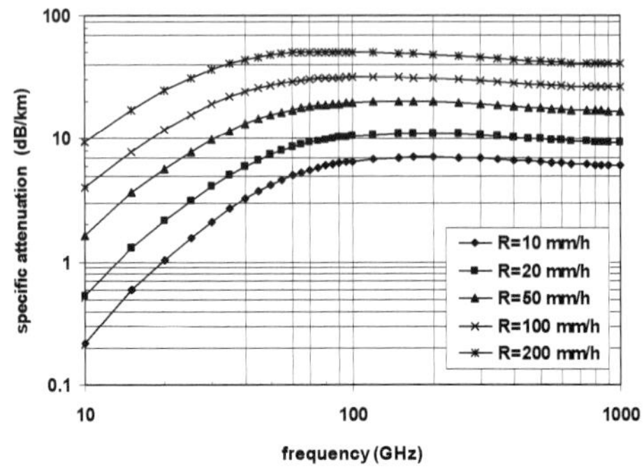

Fig. 2. Frequency dependence of specific rain attenuation for several values of rain intensity *R* (mm/hour)

ITU-R recommends the estimation procedure of rain attenuation statistics (Rec. ITU-R P.530-12, 2008) which has been validated against attenuation data obtained on the terrestrial links with operating frequencies of up to 40 GHz. The method makes certain that the estimated 0.01% percentile of specific rain attenuation $A_{0.01}$ is proportional to the value of specific attenuation γ calculated by equation (5) with $R=R_{0.01}(1)$ where $R_{0.01}$ is the 0.01% percentile of the average 1-minute rain intensity cumulative distribution observed in the planned link location. The coefficients in (5) are currently available for frequencies of up to 1000 GHz (Rec. ITU-R P.838-3, 2008) and therefore one can estimate $A_{0.01}$ in millimetre wave bands.

In Figure 3, a world map of estimated rain attenuation exceeded 0.01% of time on a 1 km long path with a frequency of 38 GHz is presented. It can be seen that in most of Europe, for example, $A_{0.01}$ exceeds the value of 5 dB/km. In Figures 4 and 5, world maps of estimated rain attenuation exceeded 0.01% of time on a 1 km long path with a frequency of 58 GHz and 93 GHz respectively are presented. At these frequencies, $A_{0.01}$ exceeds the value of 10 dB/km in most of Europe.

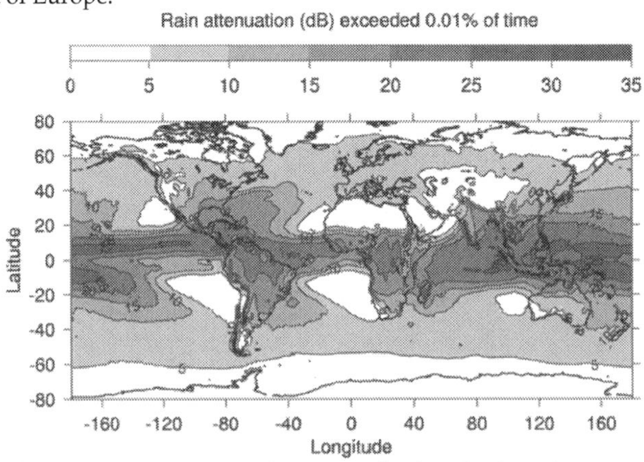

Fig. 3. Worldwide rain attenuation 0.01% percentile distribution, frequency 38 GHz, path length 1 km

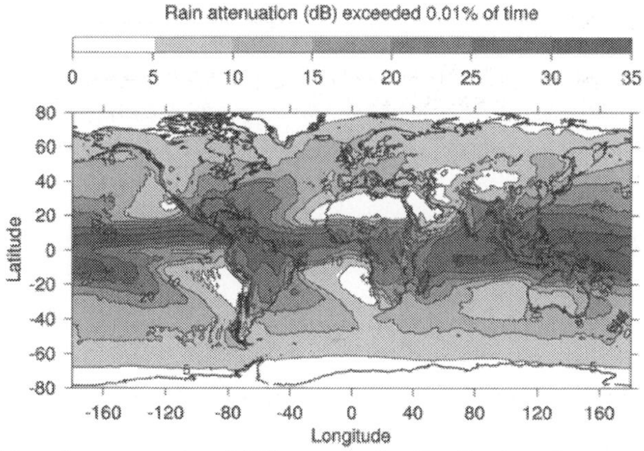

Fig. 4. Worldwide rain attenuation 0.01% percentile distribution, frequency 58 GHz, path length 1 km

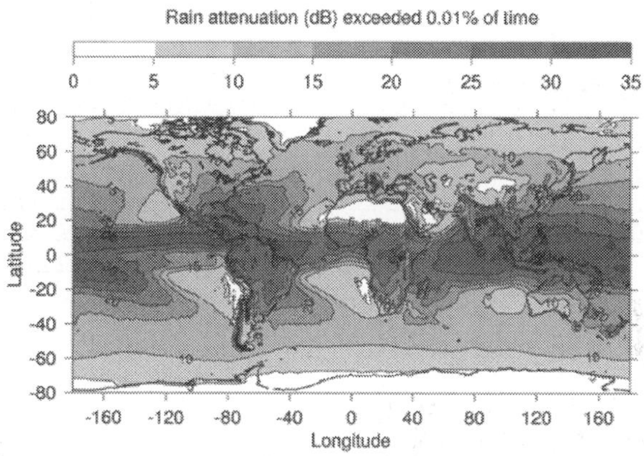

Fig. 5. Worldwide rain attenuation 0.01% percentile distribution, frequency 93 GHz, path length 1 km

4.3 Why local experimental measurements of rain attenuation statistics?

The global prediction methods of rain attenuation statistics provided by (ITU-R Rec. ITU-R P.530-12, 2008) are very useful when no sufficiently accurate local data is available especially for frequencies lower than 40 GHz. In millimetre wave bands for frequencies higher than 40 GHz, the ITU-R method still can serve well as a reliable approximation. However potential users should be aware of its inherent limited accuracy. The estimation method formulas were derived using a global fitting approach which tends to average out the errors over the world. The spatial resolution of the rain intensity dataset provided by ITU-R is 1.5 degrees in both the latitude and the longitude, which is not sufficient for the description of specific areas with extreme rain characteristics. This is a reason why it is also recommended by ITU-R to use locally-measured statistics of both rain intensities and rain attenuation whenever they are available.

5. Experimental Set-up

The used 38 GHz, 58 GHz, and 93 GHz radio systems, the meteorological measurements and the data processing are described in this section.

5.1 Terrestrial wireless systems used

Attenuation events caused by hydrometeors (rain, snow, hailstones, fog) at 38 GHz, 58 GHz and 93 GHz are measured at the Czech Metrology Institute on three parallel paths - marked as A, B, and C. On path A, a microwave system operating at 38 319.75 MHz with V polarization is used. The path length is about 9.3 km, the transmitted power is 16 dBm, and the recording margin is about 34 dB. Two microwave systems working at 58 GHz and 93 GHz are used on the parallel paths B and C with the same path length of 853m. A microwave system operating on frequency 57 650 MHz with V polarization transmitting power of 5 dBm is used on path B. The recording margin is about 24 dB thanks to the special parabolic off-set antennas used. The other microwave system is operating on path C at 93 370 MHz with V polarization. The transmitted power is 17 dBm; the recording margin is about 38 dB.

5.2 Meteorological measurements

The meteorological conditions are identified both by means of colour video-camera images of the space between the transmitter and the receiver sites and of the data obtained from a weather observation system located near the receiver site. The system is equipped with VAISALA sensors for measuring the temperature, humidity and air pressure, the wind velocity and direction, and a tipping-bucket rain gauge for the measurement of rainfall intensities. The VAISALA PWD11 device is used for the measurement of visibility. The observed meteorological conditions are continuously recorded.

The rain intensities are measured by the dynamically calibrated heated tipping-bucket rain gauge with a collector area of 500 cm^2, and the amount of rain per tip was 0.1 mm. The time of the tips was recorded with an uncertainty of 1 second. The rain gauge is situated near the receivers of the radio systems used.

5.3 Data processing

The records of received signal levels obtained on the aforementioned paths were processed statistically over a one year period from May 2007 to April 2008. The records of attenuation events were compared with the concurrent meteorological situations to identify the reason of the attenuation events. Strictly concurrent rain attenuation events occurred on three paths and only rain events were processed. The CDs of rain attenuation at 38 GHz on the 9.3 km path, 58 GHz and 93 GHz on the 853m path were obtained.

Rain intensities were processed over the same one-year period. The CD of average 1-minute rain intensities was obtained.

6. Experimental Results

The obtained monthly and annual statistics of both rain intensities and rain attenuation and the assessed availability performances of experimental links are presented in this section.

6.1 Monthly and annual statistics of rain intensities

The obtained CDs of the average 1-minute rain intensities *R(1)* for the individual months and the whole year period are given in Fig. 6. The obtained average 1-minute rain intensity for 0.01% of the time of year $R_{0.01}(1)$ is 49.5 mm/h. This rain intensity should be used for the calculation of CDs of attenuation due to rain only according to the relevant ITU-R Recommendations (Rec. ITU-R P.530-12, 2008; Rec. ITU-R P.838-3, 2008).

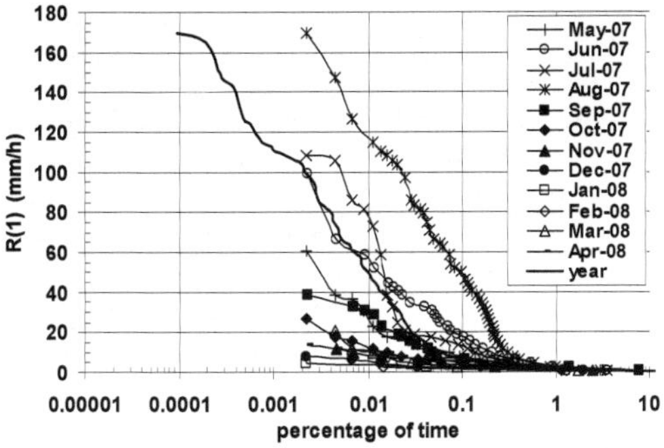

Fig. 6. Monthly and yearly CDs of rain intensities

This shows a great month-to-month variability of the rain intensity distribution. The highest measured average 1-minute rain intensity was about 170 mm/h which occurred in August 2007 which also forms the CD for the worst month in the region from 170 mm/h to 3 mm/h. The CD for the worst month for the rain intensities smaller than 3 mm/h forms the pertinent part of the CD for September.

6.2 Monthly and annual statistics of rain attenuation

The obtained monthly and yearly CDs of attenuation due to rain only at 38 319.75 MHz with V polarization on a path length of about 9.3 km are given in Fig. 7.

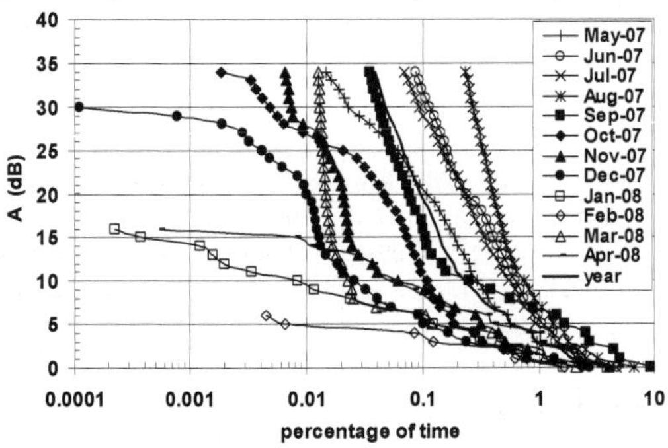

Fig. 7. Monthly and yearly CDs of attenuation due to rain at 38 GHz

A large month-to-month variability of the CDs of attenuation due to rain only caused by the large month-to-month variability of rain intensities can be observed. The CD of attenuation

due to rain only at the 38 GHz path for the worst month over the one-year period is formed for the attenuation between 34 dB and 6 dB by the pertinent part of the CD for August 2007 and for attenuation smaller than 6 dB by the pertinent part of the CD for September 2007.
The obtained monthly and yearly CDs of attenuation due to rain only at 57 650 MHz with V polarization on a path length of 853 m are given in Fig. 8.

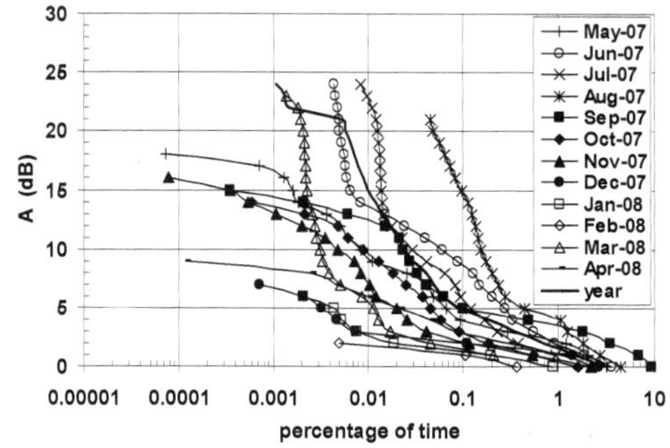

Fig. 8. Monthly and yearly CDs of attenuation due to rain at 58 GHz

The large month-to-month variability of the CDs of attenuation due to rain only caused by the large month-to-month variability of rain intensities can be observed once again. The CDs of attenuation due to rain only at the 58 GHz path for the worst month over the one-year period is formed for the attenuation between 24 dB and 21 dB by the pertinent part of the CD for July 2007, for the attenuation between 21 dB and 3 dB by the pertinent part of the CD for August 2007 and for the attenuation smaller than 3 dB by the pertinent part of the CD for September 2007.
The obtained monthly and yearly CDs of attenuation due to rain only at 93 370 MHz with V polarization on a path length of 853 m are given in Fig. 9.

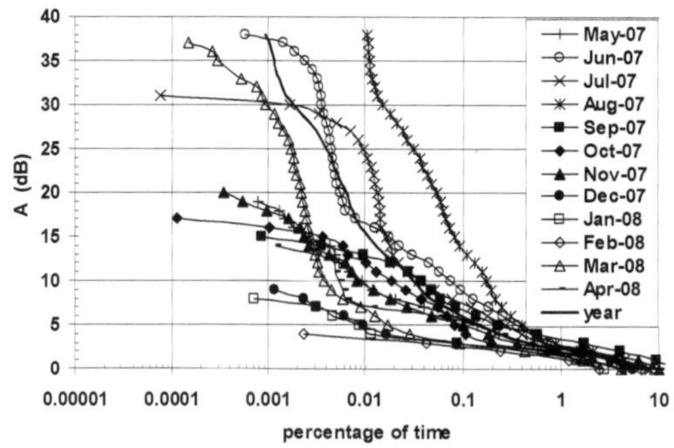

Fig. 9. Monthly and yearly CDs of attenuation due to rain at 93 GHz

The large month-to-month variability of CDs of attenuation due to rain only caused by the large month-to-month variability of rain intensities can also be observed. The CD of attenuation due to rain only at the 93 GHz path for the worst month over the one-year period is formed as for the attenuation between 38 dB and 4 dB by the pertinent part of the

CD for August 2007 and for the attenuation smaller than 4 dB by the pertinent part of the CD for September 2007.
The obtained CDs of attenuation due to rain only and the CDs of attenuation due to rain only calculated in accordance with the ITU-R Recommendation (ITU-R P.530-12, 2008) for the used frequencies and the used path lengths are shown in Fig. 10. The average 1-minute rain intensity for 0.01% of the time of year $R_{0.01}(1)$ = 49.5 mm/h obtained from Fig. 6 was used for the calculation.

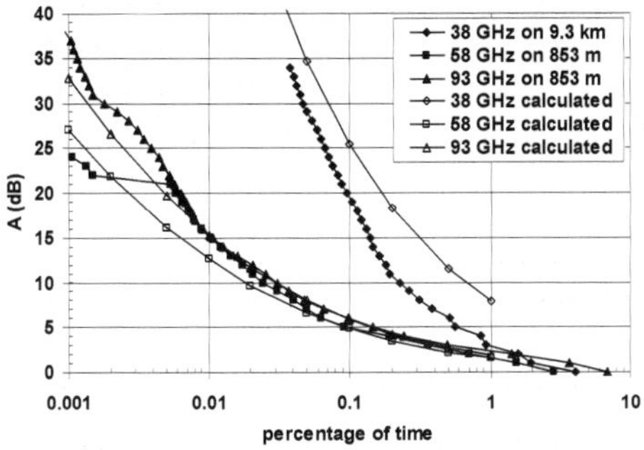

Fig. 10. Measured and calculated yearly CDs of attenuation due to rain only

The values of the measured attenuation due to rain only at 38 GHz are smaller than the calculated ones up to about 7 dB. These differences can be caused by the year-to-year variability of the rain attenuation distributions due to the year-to-year variability of the rain intensity distribution. The measured CDs due to rain only at 58 GHz and 93 GHz are very close to each other. The measured CD of attenuation due to rain only at 58 GHz is slightly over the calculated one in the region of 0.008% - 1% of the time of year. The measured CD of attenuation due to rain only at 93 GHz corresponds very well with the calculated one in the same region. For the percentages of the time of year smaller than 0.01% both CDs at 58 GHz and 93 GHz are very close to each other and are above the calculated CDs (up to about 3 dB for the CD at 93 GHz). The further inaccuracy can be caused by the fact that the ITU-R Recommendation (Rec. ITU-R P.530-12, 2008) is only considered to be valid for frequencies up to 40 GHz and path lengths up to 60 km while the lower path length limit is not mentioned.

6.3 Availability performances of experimental links

Availability performances of the three experimental links can be assessed from Fig. 10. The obtained availability performances for the three chosen fade margins of 10 dB, 15 dB, and 20 dB are given in Table 1.

Fade margin	38 GHz link	58 GHz link	93 GHz link
10 dB	99.7720%	99.9757%	99.9692%
15 dB	99.8610%	99.9897%	99.9900%
20 dB	99.9041%	99.9942%	99.9936%

Table 1. Availability performances

It can be seen that the availability performances of both experimental links at 58 GHz and 93 GHz are fully comparable to each other up to fade margins of 20 dB. Due to the fact that the 58 GHz system has a fade margin of about 24 dB only, it is not possible to compare the availability performances of both links for the fade margin greater than 20 dB. The lower availability performance of the 38 GHz link follows from the greater path length in comparison with the 58 GHz and 93 GHz links.

7. Scaling

The obtained CDs of attenuation due to rain only obtained on terrestrial paths with the different path lengths and at other frequencies and polarisations in the same climate conditions can be scaled to the required path lengths, frequencies and polarisations.

7.1 ITU-R scaling of rain attenuation

The frequency scaling and the polarisation scaling of long-term statistics of rain attenuation only are described in (Rec. ITU-R P.530-12, 2008), the path length scaling is not mentioned there.

7.2 Frequency and path length scaling of rain attenuation

A transformation method based on the ITU-R Recommendation (Rec. ITU-R P.530-12, 2008) can be applied to compare the results obtained. The simplified method was successfully used in (Tikk & Bito, 2003). The CD of rain attenuation obtained on the chosen reference path can be transformed to the other two paths for the frequencies used. The used frequency and path length scaling of 1-year statistics of rain attenuation is based on the following equation (Kvicera et al, 2009):

$$A_r = \frac{k_r\, L_r\, 0.12\, p^{\,(0.546\,+0.043\,\log_{10} p)}}{1+L_r / d_0} \left[\frac{A\,(1+L/d_0)}{k\, L\, 0.12\, p^{\,(0.546\,+0.043\,\log_{10} p)}} \right]^{\frac{a_r}{a}} \tag{6}$$

where A_r is the attenuation on the reference path, A is the attenuation on the certain path, k_r and a_r are coefficients dependent on frequency (Rec. ITU-R P.838-3, 2008) for the reference path, k and a are the same coefficients for the certain path, L_r is the reference path length, L is the path length of the certain path, d_0 is used for the calculation of the path reduction factor with $R_{0.01\%}$ = 49.5 mm/h, p is the percentage of time. Then the transformed CDs of rain attenuation on the reference path at 38 GHz, 58 GHz, and 93 GHz can be mutually compared and moreover they can be also compared with the calculated CD of rain attenuation in accordance with ITU-R Recommendation (Rec. ITU-R P.530-12, 2008).

7.2.1 Path A as the reference path

Let path A (9.3 km, 38 319.75 MHz) be considered as the reference path. The CDs of attenuation due to rain only obtained on paths B (853 m, 57 650 MHz) and C (853 m, 93 370 MHz) are scaled to reference path A in accordance with equation (6). For path A, the CD of attenuation due to rain only was calculated in accordance with the ITU-R Recommendation (Rec. ITU-R P.530-12, 2008). The average 1-minute rain intensity for 0.01% of time of year $R_{0.01}(1)$ = 49.5 mm/h obtained from Fig. 6 was used for the calculation.
The results obtained are given in Fig. 11.

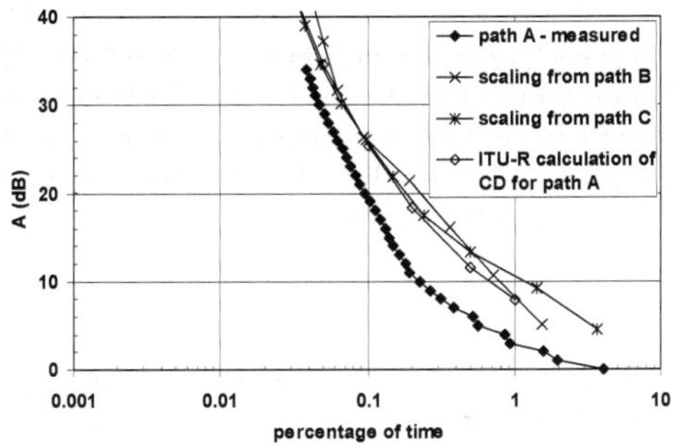

Fig. 11. Measured, scaled and calculated CDs for reference path A

It can be observed from Fig. 11 that both the scaled distributions and calculated distributions are very tight to each other and are slightly over the measured distribution (up to about 10 dB). From the point of the percentages of time, the differences between the measured distribution and the scaled and calculated distributions are not significant for attenuation values greater than 10 dB (the ratio between the percentages of time for the measured distribution and the scaled ones is smaller than factor 2).

7.2.2 Path B as the reference path

Let path B (853 m, 57 650 MHz) be chosen as the reference path. The CDs of attenuation due to rain only obtained on paths A (9.3 km, 38 319.75 MHz) and C (853 m, 93 370 MHz) are scaled to reference path B in accordance with the equation (6). For path B, the CD of attenuation due to rain only was calculated in accordance with ITU-R Recommendation (Rec. ITU-R P.530-12, 2008). The average 1-minute rain intensity for 0.01% of time of year $R_{0.01}(1)$ = 49.5 mm/h obtained from Fig. 6 was used for the calculation.

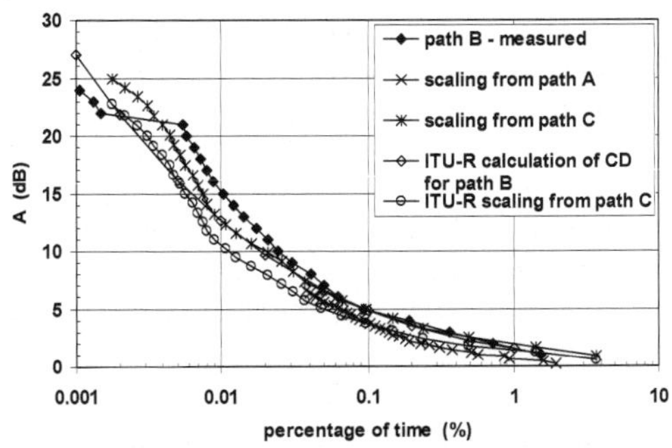

Fig. 12. Measured, scaled and calculated CDs for reference path B

In addition, the ITU-R scaling method (Rec. ITU-R P.530-12, 2008) was used for the conversion of the CD of attenuation due to rain only measured on path C to path B. The results obtained are given in Fig. 12.

The very good agreement between the scaled distributions and the measured one can be seen. The differences are smaller than 3 dB and the ratio between the percentages of time for the measured distribution and the scaled ones is smaller than 2. Both the scaled distributions from paths A and C are slightly under the measured distribution (up to about 2 dB and the ratio between the percentages of time for the measured distribution and the scaled ones is smaller than a factor of 2). The scaled distribution from path C agrees excellently with the measured distribution in the region of 0.05% - 4% of the time of year. For the percentages of time smaller than 0.05%, the scaled distribution from path C is slightly under the measured distribution and the scaled attenuation values are less than 3 dB under the measured ones. The ratio between the percentages of time for the measured and the scaled distribution is smaller than a factor of 2.

The CD due to rain only calculated in accordance with Recommendation ITU-R (Rec. ITU-R P.530-12, 2008) agrees very well with the measured distribution in the region of 0.05% - 1% of the time of year. For the percentages of time smaller than 0.05%, the calculated distribution is slightly under the measured distribution - up to about 5 dB for 0.005% of the time of year.

The CD calculated in accordance with the ITU-R scaling method (Rec. ITU-R P.530-12, 2008) lies under the measured distribution and the differences are about 5 dB for the percentages of time of year smaller than 0.01%. For the percentages of time of year greater than 0.01%, the differences are smaller.

7.2.3 Path C as the reference path

Let path C (853 m, 93 370 MHz) be considered to be the reference path. The CD of attenuation due to rain only obtained on paths A (9.3 km, 38 319.75 MHz) and B (853 m, 57 650 MHz) are scaled to reference path C in accordance with the equation (6). For path C, the CD of attenuation due to rain only was calculated in accordance with Recommendation ITU-R (Rec. ITU-R P.530-12, 2008). The average 1-minute rain intensity for 0.01% of the time of year $R_{0.01}(1)$ = 49.5 mm/h obtained from Fig. 6 was used for the calculation.

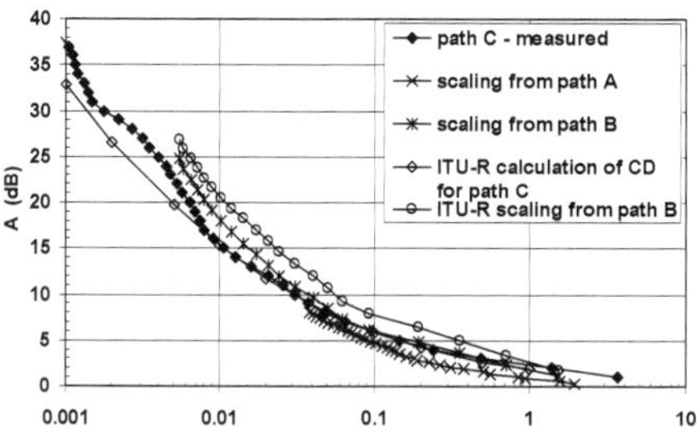

Fig. 13. Measured, scaled and calculated CDs for reference path C

The ITU-R scaling method (Rec. ITU-R P.530-12, 2008) was also used for the scaling of the CD of attenuation due to rain only measured on path C to path B. The results obtained are given in Fig. 13.

Similar results can be seen as for reference path B. There is very good agreement between the scaled distributions and the measured one. The differences are smaller than 3 dB and the ratio between the percentages of time for the measured and the scaled distribution is smaller than about a factor of 2. The scaled distribution from path A is slightly under the measured distribution and the scaled attenuation values are less than 2 dB under the measured ones. The ratio between the percentages of time for the measured and the scaled distribution is smaller than a factor of 2. For the percentages of time greater than 0.05%, the scaled attenuation values for path B agree excellently with the measured values. For the percentages of time smaller than 0.05%, the scaled distribution is slightly above the measured distribution. The scaled attenuation values are less than 3 dB above the measured ones and the ratio between the percentages of time for the measured and the scaled distribution is smaller than a factor of 2.
The CD due to rain only calculated in accordance with Recommendation ITU-R (Rec. ITU-R P.530-12, 2008) agrees excellently with the measured CD in the region of 0.01% - 1% of the time of year. For the percentages of time smaller than 0.01%, the calculated distribution is slightly under the measured distribution – up to about 3 dB for 0.003% of time of year.
The CD calculated in accordance with the ITU-R scaling method (Rec. ITU-R P.530-12, 2008) lies above the measured distribution and the differences are up to about 5 dB for the percentages of the time of year smaller than 0.01%. The differences are smaller for the percentages of the time of year greater than 0.01%.

7.2.4 Summary

Very good agreement is observed between the scaled and the calculated distributions for all three reference paths A, B, and C. Very good agreement of the scaled distributions, the calculated distributions and the measured distributions is seen for the reference paths B and C. The measured CD of attenuation due to rain only lies slightly under the scaled distributions and the calculated ones (up to about 10 dB) for the reference path A only. Nevertheless, the difference among the measured distribution and the scaled and calculated distributions is not significant for attenuation values greater than 10 dB from the point of the percentage of time due to the fact that the ratio between the percentages of time for the measured distribution and the scaled ones is smaller than a factor of 2. Scaled distributions from path A (i.e. for the reference path B and C) only lie slightly under the measured ones. Therefore it can be assumed that the method used can be used for the frequency and path length scaling for both the frequencies from 38 GHz to 93 GHz and the path lengths from 0.85 km to 9.3 km with the accuracy sufficient for the assessment of propagation conditions.

7.2.5 Scaling to other frequencies and path lengths

The described method of the frequency and path length scaling was successfully used for the assessment of CDs of attenuation due to rain only on the path between a High Altitude Platform (HAP) and an Earth base station operated on the 48 GHz band (Kvicera et al, 2009).

8. Conclusion

Terrestrial fixed wireless links form an important part of the global telecommunication network. Their availability performance and error performance are significantly influenced by weather conditions, especially by heavy rain events. An overview of the ITU recommendations related to the availability performance and error performance objectives is given. The examples of both the link power budget and the system fade margin calculations which are needed to fulfil the required availability objectives are given. Characteristics of rain that are the most important impairment factor are described and rain attenuation models are mentioned. Monthly and yearly statistics of rain attenuation which are needed for the availability performance assessment are introduced. The ITU-R world map of both rain intensity statistics and rain attenuation statistics illustrating and confirming the geographical dependence of rain attenuation are given. The meaning and the necessity of the local experimental measurement of concurrent rain intensities and rain attenuation are both explained. Both the experimental set-up of the radio systems operating at 38 GHz, 58 GHz, and 93 GHz and the concurrent meteorological measurements in Prague, the Czech Republic as well as the data-processing procedures are described in detail. The obtained experimental results, i.e. monthly and annual statistics of both rain intensities and rain attenuation are given. The availability performances based on experimental results are assessed. Due to the fact that experimental results can only be obtained at several frequencies, both the ITU-R frequency scaling method and polarisation scaling method as well as the novel path length scaling method are demonstrated.

9. References

ITU-R (2008). Radiowave propagation information for designing terrestrial point-to-points links, L. A. R. da Silva Mello & T. Tjelta, (Ed.), pp. 8-10, ITU, ISBN 92-61-12771-1, Geneva

Rec. ITU-R F.1703 (2008). Availability objectives for real digital fixed wireless links used in 27500 km hypothetical reference paths and connections, ITU, Geneva

Rec. ITU-T G.826 (2002). End-to-end error performance parameters and objectives for international, constant bit-rate digital paths and connections, ITU, Geneva

Rec. ITU-R P.838-3 (2008). Specific attenuation model for rain for use in prediction methods, ITU-R Recommendations and Reports, ITU, Geneva

COST 235 (1996). Radiowave propagation effects on next generation fixed-services terrestrial telecommunications systems, M. P. M. Hall, (Ed.), pp. 388-396, European Commission, ISBN 92-827-8023-6, Luxembourg

Rec. ITU-R P.530-12 (2008). Propagation data and prediction methods required for the design of terrestrial line-of-sight systems, ITU-R Recommendations and Reports, ITU, Geneva

Rec. ITU-T G.827 (2003). Availability performance parameters and objectives for end-to-end international constant bit-rate digital paths, ITU, Geneva

Rec. ITU-T G.828 (2000). Error performance parameters and objectives for international, constant bit rate synchronous digital path, ITU, Geneva

Rec. ITU-R P.581-2 (2008). The concept of "worst month", ITU-R Recommendations and Reports, ITU, Geneva

Tikk A. & Bito J. (2003). Site diversity gain model based on angular correlation of rain attenuation, Proceedings of Microcoll, pp. 93-96, ISBN 963 212 166 X, Budapest, Hungary, September 2003, Hungarian Academy of Sciences, Budapest

Kvicera V.; Grabner M. & Fiser O. (2009). Frequency and path length scaling of rain attenuation from 38 GHz, 58 GHz and 93 GHz data obtained on terrestrial paths, Proceedings of European Conference on Antennas and Propagation (EuCAP), [CD-ROM], ISBN 978-3-8007-3152-7, Berlin, Germany, March 2009, VDE VERLAG GMBH, Berlin